MIMO SYSTEMS, THEORY AND APPLICATIONS

Edited by **H. Khaleghi Bizaki**

MIMO Systems, Theory and Applications
Edited by H. Khaleghi Bizaki

Published by InTech
Janeza Trdine 9, 51000 Rijeka, Croatia

Publishing Process Manager Ivana Lorkovic

Technical Editor Teodora Smiljanic

Cover Designer InTech Design Team

Additional hard copies can be obtained from orders@intechopen.com

MIMO Systems, Theory and Applications, Edited by H. Khaleghi Bizaki
p. cm.
ISBN 978-953-307-245-6

For EU product safety concerns:
IN TECH d.o.o., Prolaz Marije Krucifikse Kozulić 3, 51000 Rijeka, Croatia,
info@intechopen.com or visit our website at intechopen.com.

Meet the editor

H. Khaleghi Bizaki received the B.S. and M.S. degrees in Electrical Engineering from Malek Ashtar University of Technology (MUT), Tehran, Iran, in 1998 and 2001, respectively. He received his PhD degree in Electrical Engineering (Communication system) about "precoding and Blind/Semi-Blind Estimation in MIMO Fading Channels" from Iran University of Science & Technology (IUST), Tehran, Iran, in 2008. Dr. Bizaki started his work as a researcher and assistant professor in the faculty of Electrical Engineering at MUT in 1998 and 2008, respectively. Now, he is author or co-author of more than 20 publications. His research interests include information theory, coding theory, Wireless Communication, MIMO systems, Space Time Processing, and other topics on Communication System and Signal Processing.

Contents

Preface

In recent years, it was realized that the Multiple Input Multiple Output (MIMO) communication systems seems to be inevitable in accelerated evolution of high data rates applications. The MIMO systems, have received considerable attention of researchers and commercial companies due to their potential to dramatically increase the spectral efficiency and simultaneously sending individual information to the corresponding users in wireless systems. Today, the main question is *how to include multiple antennas at transmitter and receiver side and, what are the appropriate methods of detection and signal processing strategies for specific applications?*

This book, intends to provide highlights of the current research topics in the field of MIMO system, to offer a snapshot of the recent advances in this area. This work is mainly destined to cover an overview of the major issues faced today by researchers in the MIMO related areas. Also, it is accessible to anyone with a scientific background desiring to have an up-to-date overview of this domain.

The book is written by specialists working in universities and research centers all over the world to cover the fundamental principles and main advanced topics on high data rates wireless communications systems over MIMO channels. Various aspects of theses systems are deeply discussed by emphasis of their recent applications in five part and twenty-two chapters. Moreover, the book has the advantage of providing a collection of applications that are completely independent and self-contained; thus, the interested reader can choose any chapter and skip to another without losing continuity. Each chapter provides a comprehensive survey of the subject area and terminates with a rich list of references to provide an in-depth coverage of the application at hand.

The five parts of the book is managed as follows:

Part 1 Introduction, Detection and Channel Estimation Strategies
The first part contains four chapters that investigate an introduction to MIMO systems models together with discussion about diversity, beam forming and, space time coding. The geometrical decoding in MIMO channels by name of lattice decoding, and other type of decoding such as: LS, LMMSE, ML, MAP and joint LSML are considered, too. Finally, the mathematical semi-deterministic MIMO channel model based on electromagnetic scattering and reflecting is developed and discussed in details.

Part 2 Information Theory Aspects
Part2 focus on information theory aspects of MIMO systems, including diversity-gain of MIMO systems, which highlight the trade-off between capacity and bandwidth

efficiency and rate adaptive source encoding, where the rate is adapted to follow the slow variations of the MIMO channel. Then, the capacity of MIMO system is investigated in the presence of both co-channel interference and spatial correlation. Finally, theoretical analysis for both ergodic and outage capacities of downlink transmission together with capacity analysis of uplink cellular MIMO systems by considering the co-channel interference as well as the effect of transmit power control are presented.

Part 3 Pre-processing and Post-processing in MIMO Systems
The non-linear precoder by name of Tomlinson-Harashima Precoder and beamfoming are the main core of this part. At first, the capacity of MIMO- THP in perfect and non-perfect CSI is obtained. In continue the conventional THP design is developed for imperfect, correlated and channel estimator error as robust, improved and joint optimization, respectively. Joint THP transceiver design for the multi-user MIMO downlink system under both perfect and imperfect CSI is developed, too. Iterative optimization algorithm to determination of transmit and receive beam forming weights for eigen-beam SDM in multi-user MIMO systems is discussed under constraint of both total transmit power and the maximum transmit power. The recent advance in beam forming based on finite-rate feedback from a communication-theoretic prespective is addressed as ideal and non-ideal factors of feedback link. Finally, the problem of symbol detection in Multi-Device STBC-MIMO systems is addressed. So, two evolutionary optimization methods by names, Biogeography-Based Optimization and Estimation of Distribution Algorithm are proposed to solve the problem of detection in a MD-STBC-MIMO system.

Part 4 Application and Case Studies
This part contain some advanced application of MIMO systems and some main notes in their implementations, which started by MIMO-OFDM technique that investigate the effects of phase noise in centralized and distributed narrowband MIMO systems, and discuss the feasibility of phase and frequency synchronization problem. In continue a novel threshold list subset detector that extends the List subset detector for an iterative turbo-MIMO system is considered. Then, a narrowband interference suppression technique is discussed in MIMO systems. This part terminates with fundamentals for pragmatic MIMO performance evaluation which consider some important notes in implementation from antenna and propagation perspectives.

Part 5 Implementation and Experimental Evaluation
This part starts with some practical methods for capacity measurement and hardware implementation of MIMO system with QPSK modulation. The implementation of the MIMO system together with sphere decoding and space time coding is discussed, especially with emphasis on wireless sensor network perspective. The above hardware implementation and practical measurement of MIMO system emphasis their potential of dramatically increase of spectral efficiency and their bottlenecks where should be considered in practice.

Finally, the editor would like to thank all the authors for their excellent contributions in the different areas of MIMO systems and hopes that this book will be of valuable help to the readers.

H. Khaleghi Bizaki
Iran

Part 1

Introduction, Detection and Channel Estimation Strategies

1

Advanced MIMO Techniques: Polarization Diversity and Antenna Selection

Kosai RAOOF[1], Maha BEN ZID[2,4], Nuttapol PRAYONGPUN[3] and Ammar BOUALLEGUE[4]

[1,2]UJF-Grenoble I, Gipsa Lab - UMR 5216 CNRS
[3]College of Industrial Technology
[4]National Engineering School of Tunis (ENIT), 6'Com Lab
[1,2]France
[3]Thailand
[4]Tunisia

1. Introduction

This chapter is attempted to provide a survey of the advanced concepts and related issues involved in Multiple Input Multiple Output (MIMO) systems. MIMO system technology has been considered as a really significant foundation on which to build the next and future generations of wireless networks. The chapter addresses advanced MIMO techniques such as polarization diversity and antenna selection. We gradually provide an overview of the MIMO features from basic to more advanced topics. The first sections of this chapter start by introducing the key aspects of the MIMO theory. The MIMO system model is first presented in a generic way. Then, we proceed to describe diversity schemes used in MIMO systems. MIMO technology could exploit several diversity techniques beyond the spatial diversity. These techniques essentially cover frequency diversity, time diversity and polarization diversity. We further provide the reader with a geometrically based models for MIMO systems. The virtue of this channel modeling is to adopt realistic methods for modeling the spatio-temporal channel statistics from a physical wave-propagation viewpoint. Two classes for MIMO channel modeling will be described. These models involve the Geometry-based Stochastic Channel Models (GSCM) and the Stochastic channel models. Besides the listed MIMO channel models already described, we derive and discuss capacity formulas for transmission over MIMO systems. The achieved MIMO capacities highlight the potential of spatial diversity for improving the spectral efficiency of MIMO channels. When Channel State Information (CSI) is available at both ends of the transmission link, the MIMO system capacity is optimally derived by using adaptive power allocation based on water-filling technique. The chapter continues by examining the combining techniques for multiple antenna systems. Combining techniques are motivated for MIMO systems since they enable the signal to noise ratio (SNR) maximization at the combiner output. The fundamental combing techniques are the Maximal Ratio Combining (MRC), the Selection Combining (SC) and the Equal Gain Combining(EGC). Once the combining techniques are analyzed, the reader is introduced to the beamforming processing as an optimal strategy for combining. The use of multiple antennas significantly improves the channel spectral efficiency. Nevertheless, this induces higher system complexity of the communication system and the communication system performance is effected due to correlation between antennas that need to be deployed at the same terminal. As such, the antenna selection algorithm for MIMO systems is presented. To elaborate on this point, we introduce Space time coding techniques for MIMO systems and we evaluate by simulation the performance of the communication system. Next, we emphasis on multi polarization techniques for MIMO systems. As a background, we presume that the reader has a thorough understanding of antenna theory. We recall the basic antenna theory and concepts that are used throughout the rest of the chapter. We rigorously introduce the 3D channel model over the Non Line of Sight (NLOS) propagation channel for MIMO system with polarized antennas. We treat the depolarization phenomena and we study its effect on MIMO system capacity. The last section of the chapter provides a scenario for collaborative sensor nodes performing distributed MIMO system model which is devoted to sensor node localization in Wireless Sensor Networks. The localization algorithm is based on beamforming processing and was tested by simulation. Our chapter provides the reader by simulation examples for almost all the topics that have been treated for MIMO system development and key issues affecting achieved performance.

2. MIMO literature and mathematical model

This section gives an overview of the MIMO literature. MIMO technology has been a subject of research since the last decade of the twentieth century. In 1984, Jack Winters at Bell Laboratories wrote a patent on wireless communications using multiple antennas. Jack Winters in (Winters, 1987) presented a study of the fundamental limits on the data rate of multiple antenna systems in a Rayleigh fading environment. The concept of MIMO was introduced for two basic communication systems which are a communication system between multiple mobiles and a base station with multiple antennas and another one between two mobiles with multiple antennas. In 1993, Arogyaswami Paulraj and Thomas Kailath proposed the concept of spatial multiplexing using MIMO. They filed a patent on spatial multiplexing emphasized applications to wireless broadcast. Several articles which focused on MIMO concept were published in the period from 1986 to 1995. We mainly cite the article of Emre Teletar titled "Capacity of multi antenna gaussian channels" (Telatar, 1995). This was followed by the work of Greg Raleigh and Gerard Joseph Foschini in 1996 (Foshini, 1996) which invented new approaches involving space time coding techniques. These approaches were proved to increase the spectral efficiency of MIMO systems (Raleigh & John, 1998). In 1999, Thomas L. Marzetta and Bertrand M. Hochwald published an article (Marzetta & Hochwald, 1999) which provides a rigorous study on the MIMO Rayleigh fading link taking into consideration information theory aspects. Afterwards, MIMO communication techniques have been developed and brought completely on new perspectives wireless channels. The first commercial MIMO system was developed in 2001 by Iospan Wireless Inc. Since 2006, several companies such as Broadcom and Intel have concerned a novel communication technique based on the MIMO technology for improving the performance of wireless Local Area Network(LAN) systems. The new standard of wireless LAN systems is named IEEE 802.11n. MIMO technology has attracted more attention in wireless communications. In fact, it was used to boost the link capacity and to enhance the reliability of the communication link. MIMO scheme is the major candidate technology in various standard proposals for the fourth generation of wireless communication systems. Enhanced techniques for MIMO communications led to advanced technologies for achieving successful radio transmission. It promises significant improvements in spectral efficiency and network coverage. We mainly cite multiple access MIMO systems, Ad-hoc MIMO , cooperative MIMO (Wang et al., 2010) and cooperative MIMO in sensor networks (Shuguang et al., 2004) . Note that cooperative MIMO systems use multiple distributed transmitting devices to improve Quality of Service (QoS) at one/multiple receivers. This was shown to bring saves in energy and to improve the link reliability in Wireless Sensor Network (WSN) where multiple sensor nodes can be cooperatively functioned. In the following, we introduce the mathematical model for MIMO systems. We briefly describe the flat fading MIMO channel and the continuous time delay MIMO channel model.

Flat fading MIMO channel

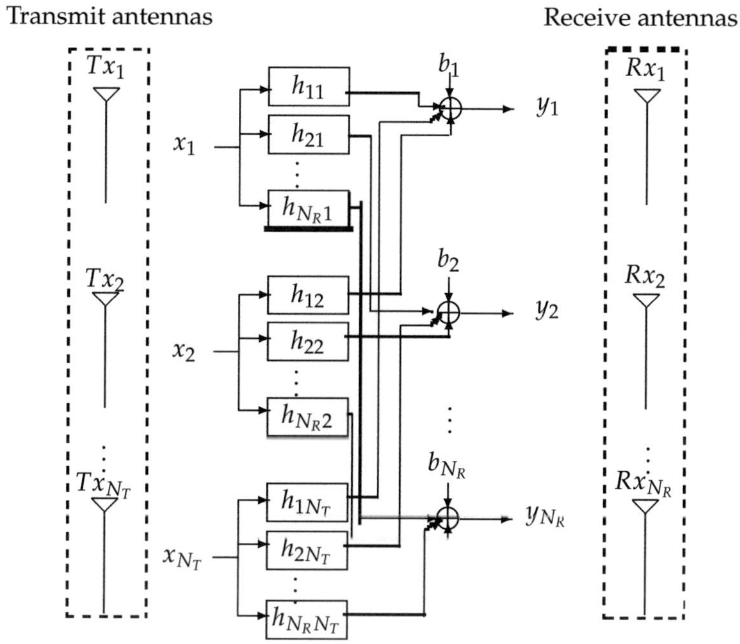

Fig. 1. Generic MIMO system model

Generic MIMO system with N_T transmit antennas and N_R receive antennas is depicted in Fig. 1. Such model is typically used for cases where the frequency domain channel transfer function remains approximately constant over the bandwidth of the transmitted waveform and is referred to as the flat fading scenario. The input output relationship for this MIMO system is defined as :

$$\mathbf{y} = \mathbf{H}\mathbf{x} + \mathbf{b} \tag{1}$$

where :

- $\mathbf{H}$ is the $(N_R \times N_T)$ complex channel matrix described as :

  $$\mathbf{H} = [\mathbf{h}_1, \ldots, \mathbf{h}_{N_T}]$$

  $\mathbf{h}_p = [h_{1p}, \ldots, h_{N_Rp}]^T$; $p = 1, \ldots, N_T$ is the complex channel vector which links the transmit antenna Tx_p to the N_R receive antennas $Rx_1, \ldots, Rx_{N_R}$.

- $\mathbf{x} = [x_1, \ldots, x_{N_T}]^T$ is the complex vector for the transmitted signal
- $\mathbf{y} = [y_1, \ldots, y_{N_R}]^T$ is the complex vector for the received signal
- $\mathbf{b} = [b_1, \ldots, b_{N_R}]^T$ is the complex vector for the additive noise signal

At the receive antenna Rx_q, the received signal is expressed as :

$$y_q = \sum_{p=1}^{N_T} h_{qp} x_p + b_q \quad ; q = 1, \ldots, N_R \tag{2}$$

In the literature, other cases of simplified MIMO systems are also explained :

- Single Input Multiple Output (SIMO) is a simplified form of MIMO systems where the transmitter system has a single antenna.
- Multiple Input Single Output (MISO) is a form of MIMO systems where the receiver system has a single antenna.
- When neither the receiver nor the transmitter has multiple antennas, the radio system is called Single Input Single Output (SISO) system.

The listed multiple antenna models are represented in Fig. 2.

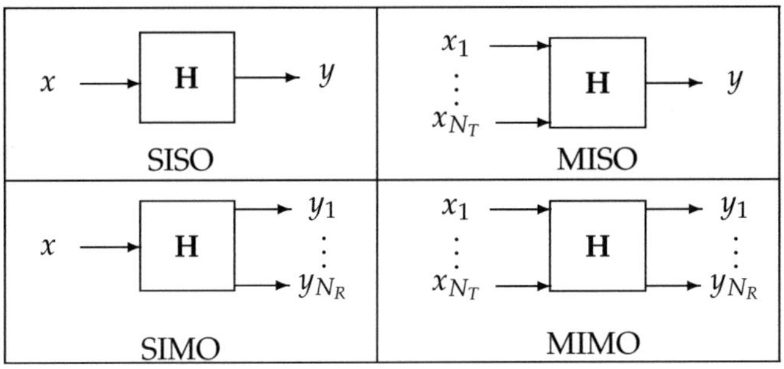

Fig. 2. Multiple antenna system

Continuous time delay MIMO channel model

The continuous time delay MIMO channel model describes the dynamic behavior of the MIMO channel. The spatio-temporel signal output $\mathbf{y}(t)$ is expressed in terms of the spatio-temporel signal input $\mathbf{x}(t)$, the $(N_R \times N_T)$ MIMO channel $\mathbf{H}$ associated time delay and the noise signal $\mathbf{b}(t)$ as :

$$\mathbf{y}(t) = \int_{\tau} \mathbf{H}(t, \tau) \, \mathbf{x}(t - \tau) \, d\tau + \mathbf{b}(t) \tag{3}$$

τ is the time delay.

3. Diversity schemes

This section is intended to present methods for improving the reliability of communication system by using different types of diversity.

3.1 Spatial diversity

The use of multiple antennas in MIMO systems improves the performance of communication systems. Signal will not suffer the same level of attenuation as it propagates along different paths. The use of multiple antennas is called spatial diversity. Joint transmit and receive diversity are carried out in MIMO systems. Nevertheless, the spatial diversity scheme can be efficiently exploited when the antenna array configuration at receive and transmit sides is properly performed to the propagation environment characteristic. This could be achieved if multiple branches which are combined are ideally uncorrelated in order to reduce probability for deep fades in fading channels.

Diversity gain

The spatial diversity systems are known for their reliability through the use of multiple receive and transmit antenna arrays. The system reliability is represented by the diversity gain. Diversity gain measures the increase of the error rate against the SNR and could be expressed as the slope of the error rate as a function of SNR when SNR tends to infinity. A tractable definition of the diversity gain is (Jafarkhani, 2005):

$$d = -\lim_{\mathrm{SNR}\to\infty} \frac{\log(P_e(\mathrm{SNR}))}{\log(\mathrm{SNR})} \tag{4}$$

$P_e(\mathrm{SNR})$ denotes the error rate measured at a fixed SNR value. A MIMO system with N_T transmit antennas and N_R receive antennas can achieve a maximum diversity gain of $N_T \times N_R$.

Multiplexing gain

Thanks to the use of multiple antennas, MIMO systems perform spatial multiplexing. Independent and separately data signals called streams are transmitted from each transmit antenna. The data streams arrived at the receiver are demultiplexed and the maximum number of independent transmission channels or degrees of freedom are $\min(N_R, N_T)$ (Zheng & Tse, 2003). Such technique leads to an increase in the system spectral efficiency without any need neither for additional bandwidth nor for additional power allocation. The spatial multiplexing order is expressed as :

$$r = \lim_{\mathrm{SNR}\to\infty} \frac{\mathrm{R}(\mathrm{SNR})}{\log(\mathrm{SNR})} \tag{5}$$

$\mathrm{R}(\mathrm{SNR})$ denotes the capacity for a given SNR value.

Diversity-Multiplexing trade-off

We should note that there is a compromise between maximizing the diversity gain so that to increase the link reliability against fading and maximizing the multiplexing gain in order to achieve the best spectral efficiency. This trade-off is expressed as :

$$d(r) = (N_T - r)(N_R - r) \quad ; r = 0, \ldots, \min(N_R, N_T) \tag{6}$$

This implies that if r pairs of antennas (Each pair consists of one transmit antenna and one receive antenna) are exploited for spatial multiplexing, it remains $(N_T - r)$ transmit antennas and $(N_R - r)$ receive antennas to be exploited for diversity gain. Nevertheless, coding techniques could be used as a solution for inherent diversity-multiplexing trade-off (Freitas et al., 2005).

3.2 Temporal diversity and Space Time processing for MIMO systems

If channel varies in time, repeated signal versions can benefit from temporal diversity if they are sent at different time intervals that is higher than the time coherence of the channel.

Space Time processing for MIMO systems

MIMO system can still achieve both spatial diversity and temporal diversity by exploiting Space Time (ST) coding (Fig. 3). Let us review the flat fading model. The complex channel matrix $\mathbf{H}(N_R \times N_T)$ is expressed as :

$$\mathbf{H} = \begin{pmatrix} h_{11} & h_{12} & \dots & h_{1N_T} \\ h_{21} & h_{22} & \dots & h_{2N_T} \\ \vdots & \vdots & \ddots & \vdots \\ h_{N_R1} & h_{N_R2} & \dots & h_{N_RN_T} \end{pmatrix}$$

Given a block time of length L, at time t, the transmitted signal is expressed as :

$$\mathbf{x}^{(t)} = [x_1^{(t)} \dots, x_{N_T}^{(t)}]^T \quad ; t = 1, \dots, \mathrm{L} \tag{7}$$

The input array signal $\mathbf{X}(N_T \times L)$ is given by :

$$\mathbf{X} = \begin{pmatrix} x_1^{(1)} & x_1^{(2)} & \dots & x_1^{(L)} \\ x_2^{(1)} & x_2^{(2)} & \dots & x_2^{(L)} \\ \vdots & \vdots & \ddots & \vdots \\ x_{N_T}^{(1)} & x_{N_T}^{(2)} & \dots & x_{N_T}^{(L)} \end{pmatrix}$$

The received signal matrix $\mathbf{Y}(N_R \times L)$ is expressed as :

$$\mathbf{Y} = \begin{pmatrix} y_1^{(1)} & y_1^{(2)} & \dots & y_1^{(\mathrm{L})} \\ y_2^{(1)} & y_2^{(2)} & \dots & y_2^{(\mathrm{L})} \\ \vdots & \vdots & \ddots & \vdots \\ y_{N_R}^{(1)} & y_{N_R}^{(2)} & \dots & y_{N_R}^{(\mathrm{L})} \end{pmatrix}$$

The noise signal matrix $\mathbf{B}(N_R \times L)$ is :

$$\mathbf{B} = \begin{pmatrix} b_1^{(1)} & b_1^{(2)} & \dots & b_1^{(\mathrm{L})} \\ b_2^{(1)} & b_2^{(2)} & \dots & b_2^{(\mathrm{L})} \\ \vdots & \vdots & \ddots & \vdots \\ b_{N_R}^{(1)} & b_{N_R}^{(2)} & \dots & b_{N_R}^{(\mathrm{L})} \end{pmatrix}$$

The input output relationship of such system is given by :

$$\mathbf{Y} = \mathbf{H} \cdot \mathbf{X} + \mathbf{B} \tag{8}$$

Thereafter, the received signal at time t at the receiving antenna Rx_q is expressed as :

$$y_q^{(t)} = \sum_{p=1}^{N_T} h_{qp} x_p^{(t)} + b_q^{(t)}; \quad t = 1, \dots, L; \quad q = 1, \dots, N_R \tag{9}$$

Thus, ST coding is a class of a linear processing design. The transmitted matrix **X** is referred as the ST code. ST codes are designed in order to achieve both maximum coding gain and diversity gain. Thereafter, two main criteria should be satisfied when ST coding is performed. These criteria are referred as the *Rank criterion* and the *Determinant criterion* (Tarokh et al., 1998). Recently many types of ST coding structures were invented. Nevertheless, ST codes

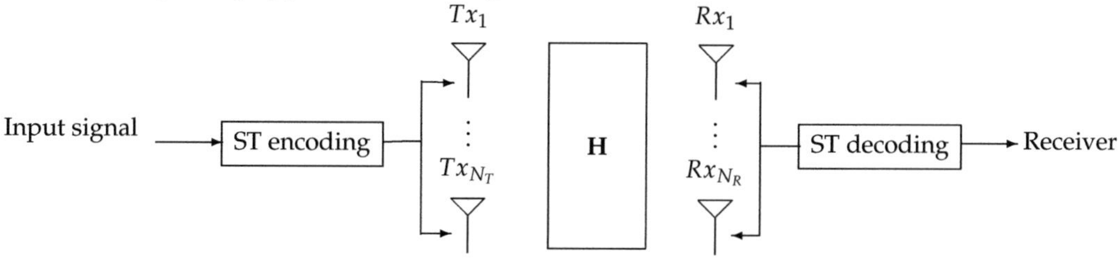

Fig. 3. ST coding for MIMO systems

may be split into two main types which are listed in the following :

1. Space Time Trellis Code (STTC) was invented by Vahid Tarokh in 1998. This coding scheme transmits multiple redundant copies of trellis code which are distributed in time and space. Vahid Tarokh gave a detailed description for trellis code construction in (Tarokh et al., 1998)
2. Space Time Block Code (STBC) aims to provide a diversity gain by transmitting block codes distributed over the transmit antennas. Development of STBC is based on complex orthogonal design. STBC for communication over Rayleigh fading channels was introduced in (Tarokh et al., 1999). The most famous orthogonal STBC (OSTBC) design is the Alamouti scheme which was invented in 1998 for a MIMO system when two-branch transmit diversity with one receive antenna and two-branch transmit diversity with two receive antennas are considered (Alamouti, 1998). At the receiver, the transmitted signal can be easily recovered due to the orthogonality of ST code. Thus, OSTBCs have received much attention from the coding community as compared to STTCs owing to their simple design and low complexity receivers (Ghrayeb, 2006). Performance analysis in terms of BER for various OSTBCs was derived in (Tran & Sesay, 2003). It was shown that we can obtain gain from OSTBCs if appropriate number of transmit antennas are deployed.

For more detailed lecture about the listed ST codes, the reader could refer to (Tarokh et al., 1998) (Tarokh et al., 1999) and (Vucetic, B. & Yuan, J., 2003). The simplest Alamouti scheme was presented in (Alamouti, 1998). For the case of two receive antennas, if we consider two symbols x_1 and x_2, then :

- At time slot 1, x_1 and x_2 are transmitted simultaneously from Tx_1 and Tx_2.
- At time slot 2, $-x_2^*$ and x_1^* are transmitted simultaneously from Tx_1 and Tx_2.

The input output relationship involving two receive antennas is expressed as :

$$\begin{pmatrix} y_1^{(1)} \\ y_2^{(1)} \\ (y_1^{(2)})^* \\ (y_2^{(2)})^* \end{pmatrix} = \begin{pmatrix} h_{11} & h_{12} \\ h_{21} & h_{22} \\ h_{12}^* & -h_{11}^* \\ h_{22}^* & -h_{21}^* \end{pmatrix} \cdot \begin{pmatrix} x_1 \\ x_2 \end{pmatrix} + \begin{pmatrix} b_1^{(1)} \\ b_2^{(1)} \\ (b_1^{(2)})^* \\ (b2^{(2)})^* \end{pmatrix} \qquad (10)$$

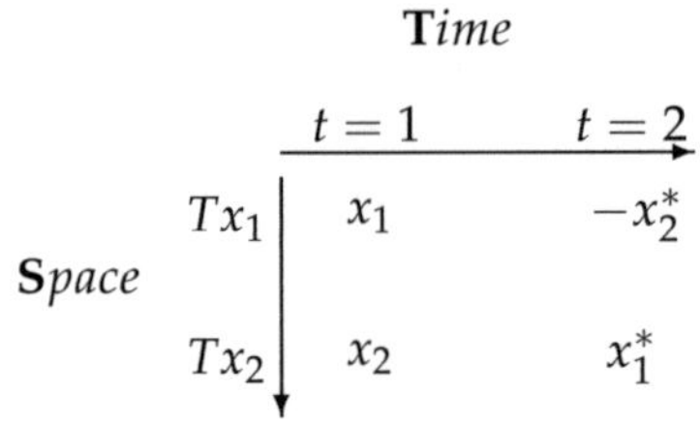

Fig. 4. Alamouti code

Let us denote :

$$\mathbf{H}\text{equ} = \begin{pmatrix} h_{11} & h_{12} \\ h_{21} & h_{22} \\ h_{12}^* & -h_{11}^* \\ h_{22}^* & -h_{21}^* \end{pmatrix}$$

The estimated transmitted signal at the receiver is given by :

$$\begin{pmatrix} \hat{x}_1 \\ \hat{x}_2^* \end{pmatrix} = \mathbf{H}^+ \begin{pmatrix} y_1^{(1)} \\ y_2^{(1)} \\ (y_1^{(2)})^* \\ (y_2^{(2)})^* \end{pmatrix} \tag{11}$$

where $\mathbf{H}^+$ is the pseudo inverse matrix of $\mathbf{H}$equ. Simulation results of the Alamouti scheme for both MIMO systems MIMO(1×2) and MIMO(2×2) in case of fading Rayleigh channel with Additive White Gaussian Noise(AWGN) are depicted in Fig. 5. Transmitted data signals are Binary Phase Shift Keying (BPSK) modulated.

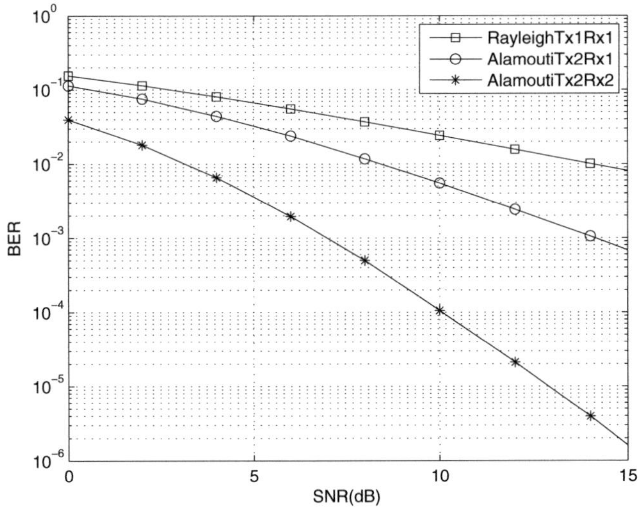

Fig. 5. BER curves for BPSK modulated signal with Alamouti coding

3.3 Frequency diversity

Frequency diversity relies on the fact that signals are transmitted on different frequencies so that different multipath structures in the propagation media are exploited. Transmitting signals on different frequencies are referred as multicarrier transmission. One special case of multicarrier transmission is Orthogonal Frequency Division Multiplexing (OFDM). OFDM has got a great interest by the researchers and it was shown that using this type of modulation gives a significant performance increase in wireless communications. OFDM modulation technique was adopted by IEEE802.11a and IEEE 802.11b wireless LAN standards. When using the OFDM technique, a single data stream is transmitted over a number of lower rate carriers. This can be considered as a form of frequency multiplexing that could be efficient for wide band communication. The signal frequency band is divided into several frequency subchannels in order to get narrow band channels. Orthogonality between different modulated carriers is imposed in order to avoid overlapping subchannels. Therefore, signals are received without adjacent carrier interference. Both transmitter and receiver are implemented using respectively the Inverse Fast Fourier Transform (IFFT) and Fast Fourier Transform (FFT) techniques. The OFDM transmission scheme introduces guard bands /cyclic prefix between the different carriers. This lowers the spectrum efficiency but it eliminates the Inter Symbol Interference (ISI). It should be noted that when using OFDM technique, channel equalization becomes simpler. Simulation of the OFDM system was performed in order to measure the performance of such technique and compare it to the single carrier system. The simulated OFDM system is given by Fig. 6.

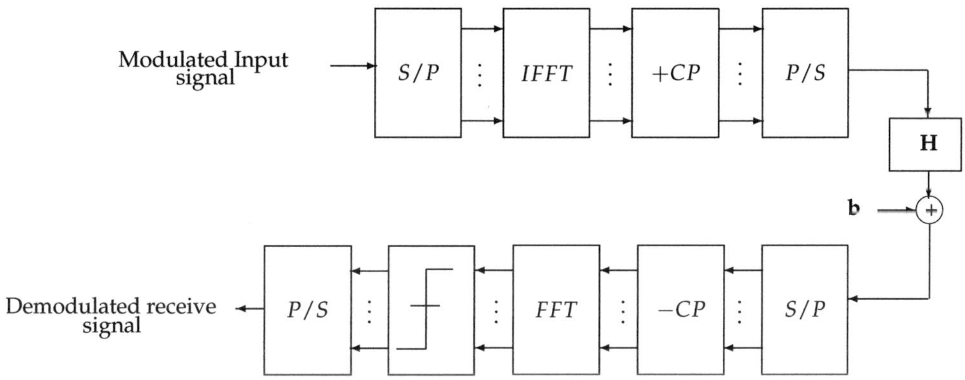

Fig. 6. General structure of OFDM system

- P/S: Parallel to serial conversion
- S/P: Serial to parallel conversion
- $+CP$: Adding the cyclic prefix
- $-CP$: Removing the cyclic prefix
- $\mathbf{H}$: Channel matrix
- $\mathbf{b}$: Additive noise

For simulation, we consider a BPSK modulation scheme, a FFT length equal to 52 and a FFT size of 64. The simulation results are given by Fig. 7. OFDM can be used in conjunction with a MIMO system. MIMO-OFDM (See Fig. 8) is interesting for high data rate systems. A design and simulation of MIMO-OFDM was introduced in (Yu et al., 2004).

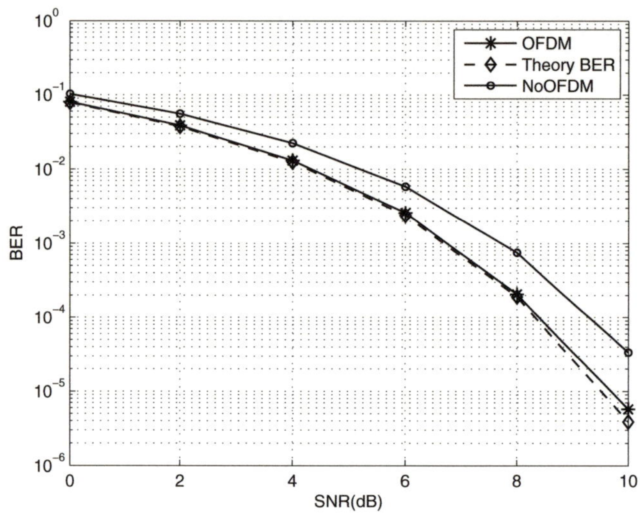

Fig. 7. BER curve for BPSK modulated signal using OFDM in AWGN channel

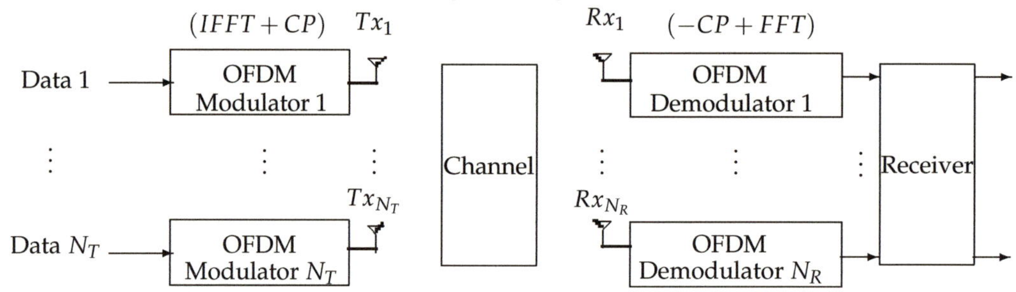

Fig. 8. MIMO-OFDM system

3.4 Pattern diversity

Pattern diversity consists of the use of several colocated antennas with different radiation patterns. This type of diversity can provide a higher gain versus a single omnidirectional antenna if antennas are enough spaced and adequately polarized.

3.5 Polarization diversity

Polarization diversity is a diversity technique where different polarizations are used. Horizontal and vertical polarizations could be used so that to provide diversity. At the MIMO receiver for example, the antennas take advantage of the multipath propagation characteristics to receive separate uncorrelated signals.

4. MIMO channel modeling

4.1 Geometry-based Stochastic Channel Models (GSCM)

Geometry-based Stochastic Channel Models(GSCM) have an immediate relation to physical reality. Such models are based on geometrical considerations, mainly scatterer locations and channel impulse response behavior. We distinguish the Double Bounce Geometry-based

Stochastic Channel Models (DB-GSCM) and the Single Bounce Geometry-based Stochastic Channel Models (SB-GSCM).

4.1.1 Double Bounce Geometry-based Stochastic Channel Models (DB-GSCM)

Geometry based stochastic channel models represent the channel in a propagation-based stochastic way in which the geometry is represented by statistical means. The GSCM is based on the concept of clusters of scatterers around the transmitter and the receiver. The scatterer locations are defined according to a random fashion that follows a particular probability distribution. Scatterers represent discrete channel paths and can involve statistical characterizations of several propagation parameters such as delay spread, angular spread, spatial correlation and cross polarization discrimination. These parameters will be detailed in the following sections. Fig. 9 shows a random geometrical two circle model in which the geometry of the scatterers follows a circular distribution. Each propagation path is able to have two times of reflection by scatterers, one at transmitter side and another one at receiver side. Local scatterers around the transmit antennas and receive antennas are respectively situated in a circle of radius R_T and a circle of radius R_R. The distance between the receive antennas and transmit antennas D is assumed to be longer than the radii R_T and R_R as depicted on Fig. 9.

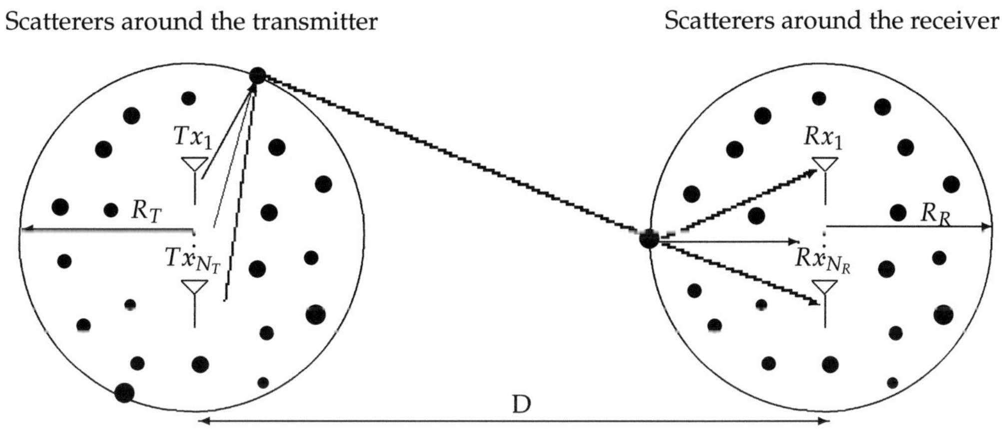

Fig. 9. Double Bounce Scattering Mechanism

4.1.2 Single Bounce Geometry-based Stochastic Channel Models (SB-GSCM)

When a single bounce of scatters is placed around the transmit antennas or the receive antennas, this is referred as SB-GSCM. SB-GSCM models are originally considered in systems where the base station is elevated and there is no local scattering obstruct while the mobile station (at the receive side) is surrounded by scatterers (Raoof & Zhou, 2009).

4.2 Stochastic channel models

Stochastic channel models can be split into three categories :

1. Correlation based models
2. Stochastic models of scatterers
3. Based propagation models

In the following, we briefly review the listed channel models.

4.2.1 Correlation models

The independent and identically distributed (i.i.d) model

The listed models are normally calculated analytically. Hence, the channel matrix $\mathbf{H}$ can be explicitly expressed. The simplest model is the i.i.d model. This supposes that multipath channels in presence of scatterers are independent and uniformly distributed in all directions. The MIMO channel coefficients are statically independent with equal variance.

The Kronecker MIMO channel model

Kronecker model assumes that spatial transmit correlation and spatial receive correlation are separable. Therefore, the full channel correlation matrix can be modeled by the Kronecker product of the transmit and receive correlation matrix. Full channel correlation matrices is expressed as :

$$\mathbf{H}\text{corr} = \mathbf{R}_{Tx} \otimes \mathbf{R}_{Rx} \tag{12}$$

where :

- $\mathbf{R}_{Tx}$= $\mathrm{E}[\mathbf{H}_{\text{Kron}}^H \mathbf{H}_{\text{Kron}}]$ is the transmit correlation matrix.
- $\mathbf{R}_{Rx}$= $\mathrm{E}[\mathbf{H}_{\text{Kron}} \mathbf{H}_{\text{Kron}}^H]$ is the receive correlation matrix.
- $\otimes$ denotes the Kronecker product. The Kronecker product for matrices $\mathbf{A}$ and $\mathbf{C}$ is defined as :

$$\mathbf{A} \otimes \mathbf{C} = \begin{pmatrix} \mathbf{A}_{11}\mathbf{C} & \mathbf{A}_{12}\mathbf{C} & \dots \\ \mathbf{A}_{21}\mathbf{C} & \mathbf{A}_{22}\mathbf{C} & \dots \\ \dots & \dots & \dots \end{pmatrix}$$

The channel matrix according to the Kronecker model is expressed as (Biglieri et al., 2007):

$$\mathbf{H}_{\text{Kron}} = \mathbf{R}_{Rx}^{1/2} \mathbf{H}_w (\mathbf{R}_{Tx}^{1/2}) \tag{13}$$

$\mathbf{H}_w$ is an i.i.d. Rayleigh fading channel. Note that there are others MIMO channel models based on the Kronecker such as the Keyhole model and Weichselberger model. A review of these models is presented in (Raoof & Zhou, 2009).

4.2.2 Stochastic scatterer model

This section gives a generic description of stochastic models of scatterers. Multipath channels are grouped into clusters according to statistical considerations. Besides, parameters for the channel impulse response are determined in a random manner without referring to the geometry of a physical medium. We mainly focus on the Saleh & Valenzuela (SVA) model (Saleh & Valenzuela, 1987). For finite numbers of clusters and multipath components, the impulse response of the SVA channel model is expressed as :

$$h(t) = \sum_{l=1}^{Lc} \sum_{k=1}^{K_m} \alpha_{k,l} exp(j\Psi_{k,l}) \delta(t - T_l - \tau_{k,l}) \tag{14}$$

where :

- L_c: Number of clusters which is Poisson distributed
- K_m: Number of multipath propagation components which are grouped into a cluster
- $\alpha_{k,l}$: Tap weight of the k-th path component of the l-th cluster
- $\Psi_{k,l}$: The phase of the k-th path component of the l-th cluster

- T_l: Delay of the l-th cluster
- $\tau_{k,l}$: Delay of the k-th multipath component relative to the l-th cluster arrival time T_l
- $\delta(.)$ is the Dirac delta function

We assume that $\tau_{k,l}$, $k = 1,\ldots,K_m$; $l = 1,\ldots,L_c$ are computed relatively to the first propagation component. Therefore, $\tau_{1,l} = 0$; $l = 1,\ldots,L_c$. Both cluster delay and multipath component delay are given by Poisson processes. The described model was also extended to the spatial domain by including direction of departure and direction of arrival. The normalized directional channel impulse response can be written as :

$$\begin{aligned} h(t,\phi_{\mathrm{T}},\phi_{\mathrm{R}}) = \frac{1}{\sqrt{L_c K_m}} \sum_{l=1}^{L_c} \sum_{k=1}^{K_m} \alpha_{kl} exp(j\Psi_{k,l}) \delta\left(t - T_l - \tau_{k,l}\right) \\ \times \delta\left(\phi_{\mathrm{T}} - \Phi_{\mathrm{T},l} - \phi_{\mathrm{T},k,l}\right) \delta\left(\phi_{\mathrm{R}} - \Phi_{\mathrm{R},l} - \phi_{\mathrm{R},k,l}\right) \end{aligned} \tag{15}$$

here :

- T_l: Initial arrival time
- $\Phi_{\mathrm{T},l}$: Mean departure angle of the l-th cluster
- $\Phi_{\mathrm{R},l}$: Mean arrival angle of the l-th cluster
- $\tau_{k,l}$: Initial arrival time with respect to the l-th cluster
- $\phi_{\mathrm{T},k,l}$: Departure angle with respect to the initial time and mean angle of the l-th cluster
- $\phi_{\mathrm{R},k,l}$: Arrival angle with respect to the initial time and mean angle of the l-th cluster

The parameters L_c and K_m are important for channel modeling design. They respectively depend on two other parameters which are the cluster decay factor and the ray decay factor. Fig. 10 shows the simulation results for the SVA model channel where four clusters of multipath components are obtained.

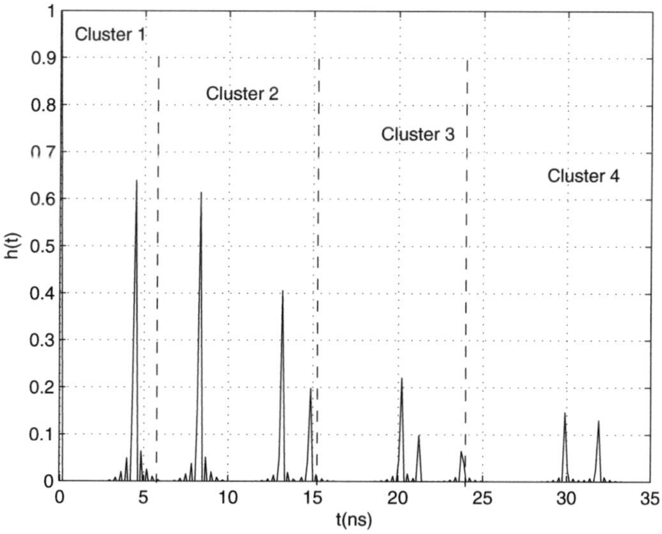

Fig. 10. Saleh & Valenzuela channel impulse response model for a SISO link

4.2.3 Geometrical propagation model

Several existing analytic models are not coinciding with real situation. In fact, they rarely consider the effect of topology structure on radio channel propagation. Geometrical-based propagation model shows a more general model in which propagation considerations are involved. For seek of brevity, we consider the example of the finite scatterer model. Let us consider the Uniform Linear Antenna (ULA) antennas at both the transmit and the receive sides. Assume that the element spacing between two close antennas at transmit and receive sides are respectively denoted by d_T and d_R. We consider finite number of scatterers that are located far away from the transmitter and the receiver. In the finite scatterer model, each path specifies a Direction of Departure (DOD) ϕ_{T} from the transmitting array and the Direction of Arrival (DOA) ϕ_{R} at the receiving array. According to these considerations, transmitting and receiving steering vectors are expressed as (Burr, 2003):

$$\begin{aligned} \mathbf{a}_{\mathrm{T}}(\theta_{\mathrm{T}}) &= [1, \exp\{-j2\pi\theta_{\mathrm{T}}\}, \cdots, \exp\{-j2\pi(N_{\mathrm{T}}-1)\theta_{\mathrm{T}}\}]^T \\ \mathbf{a}_{\mathrm{R}}(\theta_{\mathrm{R}}) &= [1, \exp\{-j2\pi\theta_{\mathrm{R}}\}, \cdots, \exp\{-j2\pi(N_{\mathrm{R}}-1)\theta_{\mathrm{R}}\}]^T \end{aligned} \tag{16}$$

- $\theta_{\mathrm{T}} = d_T \sin(\phi_{\mathrm{T}})/\lambda$
- $\theta_{\mathrm{R}} = d_R \sin(\phi_{\mathrm{R}})/\lambda$
- λ is the wavelength of radio propagation.

The discrete channel model with L_s scatterers is therefore expressed via the array steering and response vectors as:

$$\mathbf{H}_{\mathbf{S}} = \sum_{l=1}^{L_s} \beta_l \mathbf{a}_{\mathrm{R}}(\theta_{\mathrm{R},l})\, \mathbf{a}_{\mathrm{T}}^H(\theta_{\mathrm{T},l}) = \mathbf{A}_{\mathrm{R}}(\theta_{\mathrm{R}})\, \mathbf{H}_{\mathrm{P}} \mathbf{A}_{\mathrm{T}}^H(\theta_{\mathrm{T}}) \tag{17}$$

- β_l is the complex amplitude of the l-th path
- $\mathbf{H}_{\mathrm{P}} = \mathrm{diag}(\beta_1, \cdots, \beta_{L_s})$
- $\mathbf{A}_{\mathrm{T}}(\theta_{\mathrm{T},l}) = [\mathbf{a}_{\mathrm{T}}(\theta_{\mathrm{T},1}), \cdots, \mathbf{a}_{\mathrm{T}}(\theta_{\mathrm{T},L_s})]$
- $\mathbf{A}_{\mathrm{R}}(\theta_{\mathrm{R},l}) = [\mathbf{a}_{\mathrm{R}}(\theta_{\mathrm{R},1}), \cdots, \mathbf{a}_{\mathrm{R}}(\theta_{\mathrm{R},L_s})]$

5. Performance analysis of MIMO systems based on capacity

5.1 Some entropy terminologies (G.Proakis, 1995)

We briefly review in this paragraph some terminologies that we need for the channel capacity derivation.

Entropy: The entropy $\mathcal{H}(X)$ of a variable X measures the uncertainty about the realization of X. Let X be a random variable with a probability function $p(x) = P\{X = x\}, x = x_1, \ldots, x_n$ are possible values of X from a set of possible realizations χ. The entropy $\mathcal{H}(X)$ of the variable X is expressed as :

$$\begin{aligned} \mathcal{H}(X) &= \mathrm{E}[-\log_2(p(x))] \\ &= -\sum_{x\in\chi} p(x)\log_2(p(x)) \end{aligned} \tag{18}$$

E denotes the expected function.

Joint entropy: The joint entropy measures how much information is contained in a joint system of two random variables. Given two random variables X and Y with respective probability functions $p(x)$ and $p(y)$, the joint entropy is expressed as :

$$\mathcal{H}(X,Y) = -\sum_{x\in\mathcal{X}}\sum_{y\in Y} p(x,y)\log_2(p(x,y)) \tag{19}$$

Y denotes the set of possible values of y.

Conditional entropy : Suppose X and Y are random variables. Then, for any fixed value x of X, we get a conditional probability distribution on Y. We denote the associated random variable by $\mathcal{H}(Y|X)$. Conditional entropy is then :

$$\mathcal{H}(Y|X) = \mathcal{H}(X,Y) - H(X) \tag{20}$$

Mutual information : Mutual information is a quantity that measures the dependence between two arbitrary random variables. The mutual information between two discrete random variables X and Y is defined to be :

$$\begin{aligned}\mathcal{I}(X,Y) &= \mathcal{H}(X) + \mathcal{H}(Y) - \mathcal{H}(X,Y)\\ &= \mathcal{H}(Y) - \mathcal{H}(Y|X)\end{aligned} \tag{21}$$

5.2 Capacity definition based on information theory

Deterministic capacity

Channel capacity measures the maximum amount of information that could be transmitted through a channel and received with negligible error. Hence :

$$C = \max_{p(x)} \mathcal{I}(X,Y) \tag{22}$$

For a SISO link with input signal x, AWGN b and a constant channel gain h, the output signal y is expressed as :

$$y = hx + b \tag{23}$$

The mutual information is expressed as :

$$\begin{aligned}\mathcal{I}(x,y) &= \mathcal{H}(y) - \mathcal{H}(b)\\ &\leq \log_2(\pi e(P_T + \sigma_b^2)) - \log_2(\pi e\sigma_b^2)\\ &= \log_2\left(1 + \frac{P_T}{\sigma_b^2}\right)\end{aligned}$$

where :

- $P_T = \mathrm{E}\{|\mathrm{x}|^2\}$ is the transmit power
- $\sigma_b^2 = \mathrm{E}\{|\mathrm{b}|^2\}$ is the noise power
- $\log(e) = 1$

Finally, the normalized channel capacity is given by :

$$C_{\text{SISO}} = \log_2\left(1 + \frac{P_T}{\sigma_b^2}\right) \quad bits/s/Hz \tag{24}$$

This is referred as the famous Shannon's channel capacity.

Ergodic capacity

In information-theoretical sense, ergodic capacity refers to the maximum rate that communication can be achieved; assuming that the communication duration is long enough to exploit all channel state. In fact, the propagation channel varies in time. This causes the channel capacity varying in time. Consequently, to measure the fluctuating channel capacity in information theory, ergodic capacity is defined. Ergodic capacity refers to the maximum rate that can be achieved during a long observation communication to exploit all channel information. Ergodic capacity $\bar{C}_{\text{SISO}}$ is an expected value. For a SISO channel h, $\bar{C}_{\text{SISO}}$ is derived as :

$$\bar{C}_{\text{SISO}} = \mathrm{E}\left\{ \max_{p(x):\mathrm{E}\{|\mathrm{x}|^2\}\leqslant \mathrm{P_T}} \mathcal{I}(\mathrm{x},\mathrm{y}) \right\} \quad bits/s/Hz \tag{25}$$

which could be also expressed as :

$$\bar{C}_{\text{SISO}} = \mathrm{E}\left\{ \log_2(1 + \frac{P_T}{\sigma_b^2}|h|^2) \right\} \quad bits/s/Hz \tag{26}$$

Outage capacity

Outage capacity is an another statical parameter on which relies the channel performance. Outage capacity is defined as the probability that the capacity $C(h)$ is lower than a certain threshold C_{out}. Outage capacity is expressed as :

$$P_{out} = Pr\left(C(h) < C_{out}\right) \tag{27}$$

Outage probability is related to the Complementary Cumulative Distribution Function (CCDF):

$$\mathrm{CCDF} = 1 - P_{out} \tag{28}$$

In the following, we consider a MIMO system with N_T transmit antennas and N_R receive antennas. We assume that the channel is flat fading. The received signal at antenna q, y_q is expressed as :

$$y_q = \sum_{p=1}^{N_T} h_{qp}x_p + b_q \quad ; q = 1, \ldots, N_R \tag{29}$$

The MIMO channel capacity is derived as :

$$\mathcal{I}(\mathbf{x}, \mathbf{y}) = \mathcal{H}(\mathbf{y}) - \mathcal{H}(\mathbf{y}|\mathbf{x}) \tag{30}$$

$\mathcal{H}(\mathbf{y})$ and $\mathcal{H}(\mathbf{y}|\mathbf{x})$ are respectively the received signal entropy and the entropy of $\mathbf{y}|\mathbf{x}$. As the received signal $\mathbf{y}$ and the signal noise $\mathbf{b}$ are independents, $\mathcal{H}(\mathbf{y}|\mathbf{x}) = \mathcal{H}(\mathbf{b})$. Thereafter, the capacity is obtained by maximizing the received signal entropy. For MIMO capacity derivation, we denote :

- $\mathbf{R_x} = \mathrm{E}\left\{\mathbf{x}\mathbf{x}^H\right\}$: Covariance matrix of the transmit signal
- $\mathbf{R_b} = \mathrm{E}\left\{\mathbf{b}\mathbf{b}^H\right\}$: Covariance matrix of the noise signal
- $\mathbf{R_y} = \mathrm{E}\left\{\mathbf{y}\mathbf{y}^H\right\}$: Covariance matrix of the received signal

Thus :

$$\mathbf{R_y} = \mathbf{H}\mathbf{R_x}\mathbf{H}^H + \mathbf{R_b} \tag{31}$$

The mutual information is expressed as :

$$\begin{aligned}\mathcal{I}(\mathbf{x}, \mathbf{y}|\mathbf{H}) &= \log_2 \det\left(\pi e \mathbf{R_y}\right) - \log_2 \det\left(\pi e \mathbf{R_b}\right) \\ &= \log_2 \det\left\{\mathbf{I}_{N_R} + \mathbf{H}\mathbf{R_x}\mathbf{H}^H(\mathbf{R_b})^{-1}\right\}\end{aligned} \tag{32}$$

For circularly symmetric Gaussian random vectors, the mutual information is maximum and also expressed as :

$$C_{\text{MIMO}} = \max_{p(\mathbf{x}):\mathrm{E}\{\mathbf{x}^H\mathbf{x}\} \le \mathrm{P_T}} \mathcal{I}(\mathbf{x}, \mathbf{y}|\mathbf{H}) \qquad bits/s/Hz \tag{33}$$

When no CSI (Channel State Information) is available at the transmitter, equal power allocation is adopted. With the assumption that no correlation exits at the transmit side,

$$\mathbf{R_x} = \frac{P_T}{N_T}\mathbf{I}_{N_T}$$

P_T is the total power available at the transmit side.
The MIMO capacity is then expressed as :

$$C_{\text{MIMO}} = \log_2 \det\left(\mathbf{I}_{N_R} + \frac{\gamma}{N_T}\mathbf{H}\mathbf{H}^H\right) \qquad bits/s/Hz \tag{34}$$

γ is the SNR.

5.3 MIMO capacity based on SVD: CSI known at the receiver

When CSI is available at the receiver, SVD factorization is used and MIMO channel capacity could be easily derived. Let us first review the SVD technique. SVD is a factorization method for complex matrix which is widely used in signal processing. We take an $(N \times M)$ matrix $\mathbf{A}$, SVD theorem states:

$$\mathbf{A} = \mathbf{U}\mathbf{S}\mathbf{V}^H \tag{35}$$

- The eigenvectors of $\mathbf{A}\mathbf{A}^H$ make up the columns of $\mathbf{U}(N \times N)$ which is an unitary matrix ($\mathbf{U}\mathbf{U}^H = \mathbf{I}_N$).
- The singular values in $\mathbf{S}(N \times M)$ are square roots of eigenvalues from $\mathbf{A}\mathbf{A}^H$ or $\mathbf{A}^H\mathbf{A}$. The singular values are the diagonal entries of the $\mathbf{S}$ matrix and are arranged in descending order.
- The eigenvectors of $\mathbf{A}^H\mathbf{A}$ make up the columns of $\mathbf{V}$. $\mathbf{V}(M \times M)$ is also a unitary matrix ($\mathbf{V}\mathbf{V}^H = \mathbf{I}_M$).

Calculating the SVD of the MIMO channel matrix $\mathbf{H}$ leads to the following factorization :

$$\mathbf{H} = \mathbf{U}\mathbf{S}\mathbf{V}^H \tag{36}$$

We substitute $\mathbf{H}$ by its SVD decomposition. Hence, the received signal is expressed as :

$$\mathbf{y} = \mathbf{U}\mathbf{S}\mathbf{V}^H\mathbf{x} + \mathbf{b} \tag{37}$$

Let :

$$\begin{aligned}\mathbf{y}' &= \mathbf{U}^H\mathbf{y}\\ \mathbf{x}' &= \mathbf{V}^H\mathbf{x}\\ \mathbf{b}' &= \mathbf{U}^H\mathbf{b}\end{aligned} \tag{38}$$

As **U** and **V** are unitary matrix, variables $\mathbf{x}'$ and $\mathbf{b}'$ keep the same statistical densities as **x** and **b**. Therefore, the channel model ($\mathbf{y} = \mathbf{Hx} + \mathbf{b}$) could be also presented as :

$$\mathbf{y}' = \mathbf{S}\mathbf{x}' + \mathbf{b}' \tag{39}$$

- $\mathbf{y}' = [y_1', \ldots, y_{N_R}']^T$
- $\mathbf{x}' = [x_1', \ldots, x_{N_T}']^T$
- **S**=diag$(\sqrt{\lambda_1}, \ldots, \sqrt{\lambda_R}, 0, \ldots, 0)$ $\quad ; R = \min(N_R, N_T)$ is the rank of the channel matrix **H**.

Equation (1) can be rewritten as:

$$\mathbf{y}_i' = \begin{cases} \sqrt{\lambda_i}x_i' + b_i', & \text{i=1,...,}R;\\ b_i', & i\text{=}R\text{+1,...,}N_R.\end{cases} \tag{40}$$

According to the equation above, the MIMO channel consists of R uncorrelated subchannels. The covariance matrix of the signals $\mathbf{y}'$, $\mathbf{x}'$ and $\mathbf{b}'$ are expressed as :

$$\begin{aligned}\mathbf{R}_{\mathbf{y}'\mathbf{y}'} &= U^H\mathbf{R}_{\mathbf{yy}}\mathbf{U}\\ \mathbf{R}_{\mathbf{x}'\mathbf{x}'} &= V^H\mathbf{R}_{\mathbf{xx}}\mathbf{V}\\ \mathbf{R}_{\mathbf{b}'\mathbf{b}'} &= U^H\mathbf{R}_{\mathbf{bb}}\mathbf{U}\end{aligned} \tag{41}$$

and

$$\begin{aligned}\text{tr}(\mathbf{R}_{\mathbf{y}'\mathbf{y}'}) &= \text{tr}(\mathbf{R}_{\mathbf{yy}})\\ \text{tr}(\mathbf{R}_{\mathbf{x}'\mathbf{x}'}) &= \text{tr}(\mathbf{R}_{\mathbf{xx}})\\ \text{tr}(\mathbf{R}_{\mathbf{b}'\mathbf{b}'}) &= \text{tr}(\mathbf{R}_{\mathbf{bb}})\end{aligned} \tag{42}$$

The capacity of the MIMO channel is the summation of the R uncorrelated subchannels. Hence:

$$\begin{aligned}C_{\text{SVD}} &= \sum_{i=1}^{R}\log_2\left(1 + \frac{\gamma\cdot\lambda_i}{N_T}\right)\\ &= \log_2\prod_{i=1}^{R}\left(1 + \frac{\gamma\cdot\lambda_i}{N_T}\right) \quad ;\gamma = \frac{P_T}{\sigma_b^2} \quad bits/s/Hz\end{aligned} \tag{43}$$

One eigenvalue λ of $\mathbf{HH}^H$ is obtained according to the following equation :

$$(\lambda\mathbf{I}_R - \mathbf{Q})\mathbf{y} = 0 \quad ;\mathbf{y} \neq 0 \tag{44}$$

$\mathbf{Q}$ is the Wishart matrix :

$$\mathbf{Q} = \begin{cases} \mathbf{H}\mathbf{H}^H, & N_R < N_T; \\ \mathbf{H}^H\mathbf{H}, & N_R \geq N_T. \end{cases} \tag{45}$$

λ is an eigenvalue of the matrix channel $\mathbf{H}$. Hence :

$$\det(\lambda \mathbf{I}_R - \mathbf{Q}) = \mathbf{0}$$

The associate characteristic polynomial of the channel matrix is the polynomial defined by :

$$p(\lambda) = \det(\lambda \mathbf{I}_R - \mathbf{Q})$$

which is also expressed as :

$$p(\lambda) = \prod_{i=1}^{R} (\lambda - \lambda_i) \tag{46}$$

If we substitute λ by $(-N_T \frac{\sigma_b^2}{P_T})$ then :

$$\prod_{i=1}^{R} (1 + \frac{\gamma \cdot \lambda_i}{N_T}) = \det\left(\mathbf{I}_R + \frac{\mathbf{fl}}{\mathbf{N_T}}\mathbf{Q}\right) \quad ; \gamma = \frac{P_T}{\sigma_b^2} \tag{47}$$

If $N_R < N_T$ then equation (43) becomes :

$$C_{\text{SVD}} = \log_2 \det\left(\mathbf{I}_R + \frac{\gamma}{N_T}\mathbf{H}\mathbf{H}^H\right) \qquad bits/s/Hz \tag{48}$$

Finally,

$$C_{\text{SVD}} = R \cdot \log_2 \det\left(1 + \frac{\gamma}{N_T}\mathbf{H}\mathbf{H}^{\mathbf{H}}\right) \quad bits/s/Hz \tag{49}$$

Simulation results for the ergodic MIMO capacity when CSI is available at the receiver is depicted in the Fig. 11. For a MIMO system with two transmit antennas, ergodic capacity increases linearly with the number of antennas. Ergodic capacity depends on the SNR level. Plotted curves show that capacity grows with the SNR.

5.4 MIMO capacity based on Water-filling technique : CSI known at both transmitting and receiving sides

When CSI is available at both the transmitter and the receiver, an optimal power allocation could be exploited. This is referred as the water-filling technique. The main idea of water-filling strategy is to allocate more power to better subchannels with higher SNR so as to maximize the sum of data rates in all subchannels where in each subchannel the data rate is related to the power allocation by Shannon's Gaussian capacity formula $\frac{1}{2}\log_2(1 + \text{SNR})$.

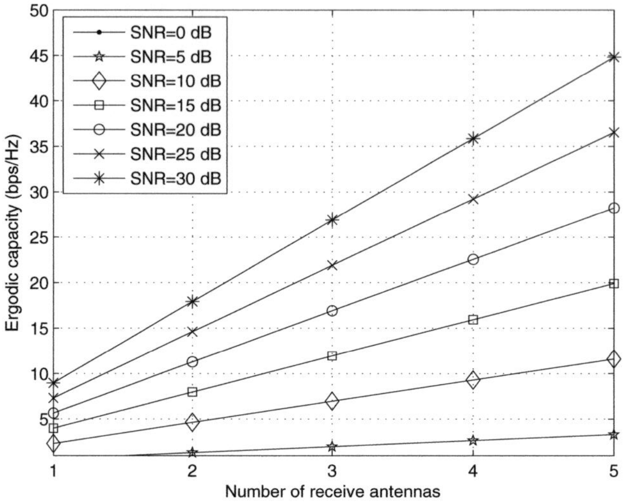

Fig. 11. Ergodic capacity derived by SVD

Water-filling strategy

The concept of the Water-filling technique (Yu & Cioffi, 2001) relies on the maximization of the MIMO channel capacity under a total transmit power constraint.

$$\sum_{p=1}^{N_T} P_p = P_T \quad ; p = 1, \ldots, N_T \tag{50}$$

The channel capacity is given by :

$$C_{\mathrm{WF}} = \sum_{p=1}^{N_T} \log_2 \left(1 + \frac{P_p}{\sigma_b^2} \lambda_p \right) \quad bits/s/Hz \tag{51}$$

Let :

$$\mathbf{Z} = \sum_{p=1}^{N_T} \log_2 \left(1 + \frac{P_p}{\sigma_b^2} \lambda_p \right) + \mathbf{L} \left(P_T - \sum_{p=1}^{N_T} P_p \right) \tag{52}$$

- $\mathbf{L}$ is the Lagrangian
- λ_p is the p-th singular value of the channel
- σ_b^2 is the noise signal variance

Allocated powers $P_p \quad ; p = 1, \ldots, N_T$ are obtained by solving :

$$\frac{\partial \mathbf{Z}}{\partial P_p} = 0 \quad ; p = 1, \ldots, N_T \tag{53}$$

Hence,

$$P_p = \mu - \frac{\sigma_b^2}{\lambda_p} \tag{54}$$

μ is a constant scalar that satisfies the total power constraint (See equation (50)).

Capacity calculation based on Water-filling

Let R be the rank of $\mathbf{H}$. The allocated power for the subchannel p is expressed as :

$$P_p = \left(\mu - \frac{\sigma_b^2}{\lambda_p}\right)^+ \quad ; p = 1, \ldots, R \tag{55}$$

where :

$$a^+ = \max(a, 0)$$

The received power at the subchannel p is then :

$$P_{rp} = \left(\lambda_p \mu - \sigma_b^2\right)^+ \tag{56}$$

Hence, the channel capacity is expressed as :

$$C_{\text{WF}} = \sum_p \log_2\left(1 + \frac{P_{rp}}{\sigma_b^2}\right) \quad bits/s/Hz \tag{57}$$

Finally, the channel capacity is :

$$C_{\text{WF}} = \sum_{p=1}^{R} \log_2\left[\left(\frac{\lambda_p \mu}{\sigma_b^2}\right)^+\right] \quad bits/s/Hz \tag{58}$$

Some numerical results for Water-Filling technique

We present some numerical results in order to simulate the performance of the water-filling technique. Let us consider a correlated MIMO(4×4) channel that follows the Kronecker model. We simulate the ergodic capacity according to both cases:

1. Equal power allocation for the transmit antennas
2. Optimal power allocation with Water-filling (WF) algorithm

Simulation results are depicted in Fig. 12. The MIMO capacity is improved by optimal power allocation strategy but stills affected by channel correlation. Simulation of the CCDF is shown in Fig. 13. The CCDF is improved by exploiting the WF technique for optimal power allocation. We present simulation results for the CCDF for two SNR values: SNR $=$ 6dB and SNR $=$ 10dB.

Water-filling technique: discussion

Water-filling provides an optimal power allocation and is an attractive strategy for capacity improvement. Nevertheless, capacity gain appears significant when more transmit antennas then receive antennas are deployed, i.e $N_R \leq N_T$. Moreover, this gain is considerable for low SNRs and is specially interesting in the case of correlated channels. Fig. 14 shows that the capacity gain is negligible for low SNRs and is almost null for high SNR values. CCDFs for the MIMO(4×2) are depicted in Fig.15 and Fig. 16 respectively for high SNR value that is equal to 18 dB and low SNR of 2 dB. Simulation results confirm that WF technique brings more performances for high noise strength and correlated MIMO channel. Finally, at high SNR value, the WF gain in ergodic capacity for MIMO$(\mathrm{N_R} \times \mathrm{N_T})$ is expressed as (Prayongpun, 2009):

$$\bar{C}_{\text{WF}} - \bar{C}_{\text{MIMO}} = \begin{cases} 0 & \text{if} \quad N_T \leq N_R \\ R \log_2\left(\frac{N_T}{R}\right) & \text{if} \quad N_T > N_R \end{cases} \tag{59}$$

where $R = \min(N_R, N_T)$.

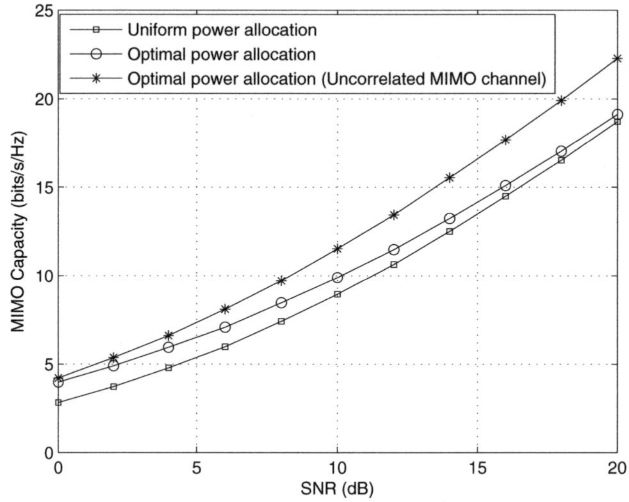

Fig. 12. MIMO(4 × 4): Capacity improvement with WF strategy-Channel correlation impact on system capacity

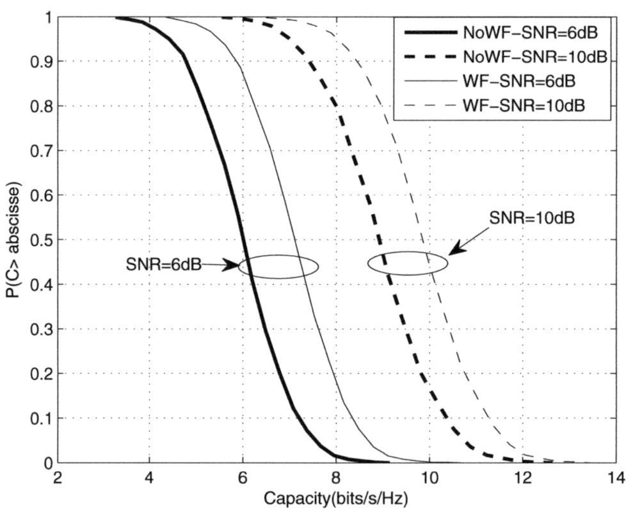

Fig. 13. CCDF for MIMO(4 × 4) with various SNR values

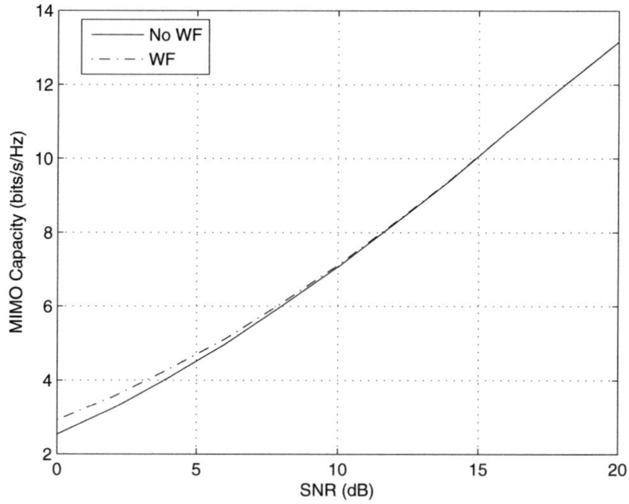

Fig. 14. Ergodic capacity for MIMO(4×2)-Kronecker channel model

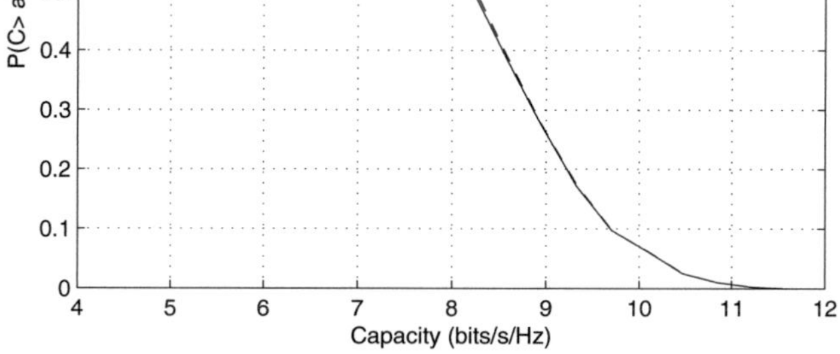

Fig. 15. CCDF for MIMO(4×2)-Kronecker channel model (SNR=18dB)

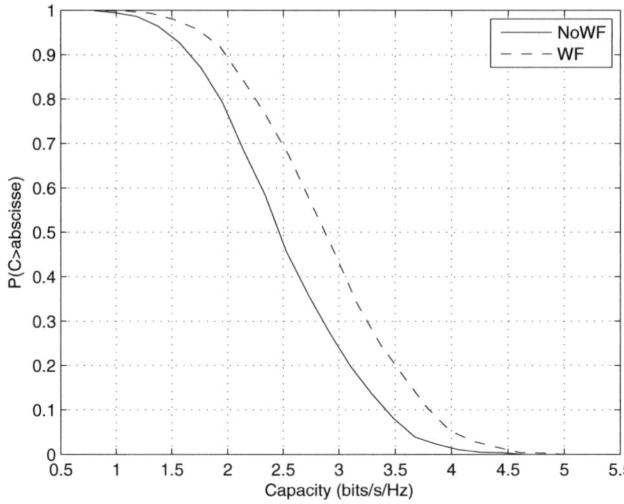

Fig. 16. CCDF for MIMO(4×2)-Kronecker channel model (SNR=2dB)

6. Combining techniques for MIMO systems

MIMO system can use several techniques at the receiver so that to combine the multiple incoming signals for more robust reception. Combining techniques are listed below :

1. Maximal Ratio Combining (MRC): Incoming signals are combined proportional to the SNR of that path signal. The MRC coefficients correspond to the relative amplitudes of the pulse replicas received by each antenna such that more emphasis is placed on stronger multipath components and less on weaker ones.
2. Equal Gain Combining (EGC) simply adds the path signals after they have been cophased (Sanayei & Nosratinia, 2004).
3. Selection Combining (SC) selects the highest strength of incoming signals from one of the receiving antennas.

Combining techniques can be carried so that to satisfy one or more targets :

1. Maximizing the diversity gain
2. Maximizing the multiplexing gain
3. Achieving a compromise between diversity gain and multiplexing gain
4. Achieving best performances in terms of Bit Error Rate (BER)
5. Maximizing the Frobenius norm of the MIMO channel and therefore the MIMO channel capacity

Let us recall the SIMO system model with N_R receive antennas. The received signal at the q-th receive antenna is expressed as :

$$y_q = h_q x + b_q \quad ; q = 1, \ldots, N_R \tag{60}$$

h_q is the q-th complex channel gain, b_q is an AWGN with zero mean and variance σ_b^2.
We keep for notations :

- P_T: Transmit signal power
- $\gamma = \frac{P_T}{\sigma_b^2}$ is the SNR

We assume channel normalization and a perfect channel estimation. We will be more interested in the combining module. Our aim is to derive the combining coefficients $g_q \quad ; q = 1, \ldots, N_R$. The output signal at the combining module can be expressed as:

$$y = x \sum_{q=1}^{N_R} g_q h_q + \sum_{q=1}^{N_R} g_q b_q \tag{61}$$

Combining technique in MIMO system is depicted in Fig.17. Combining coefficients relative to the listed techniques are given by:

Combining technique	**Combining coefficient**
MRC	$g_q = h_q^*$
EGC	$g_q = \frac{h_q^*}{\lvert h_q \rvert}$
SC	$g_q = \begin{cases} 1, & \lvert h_q \rvert \geqslant \lvert h_k \rvert, \forall k \neq q; \\ 0, & \text{otherwise.} \end{cases}$

Table 1. Combining coefficients

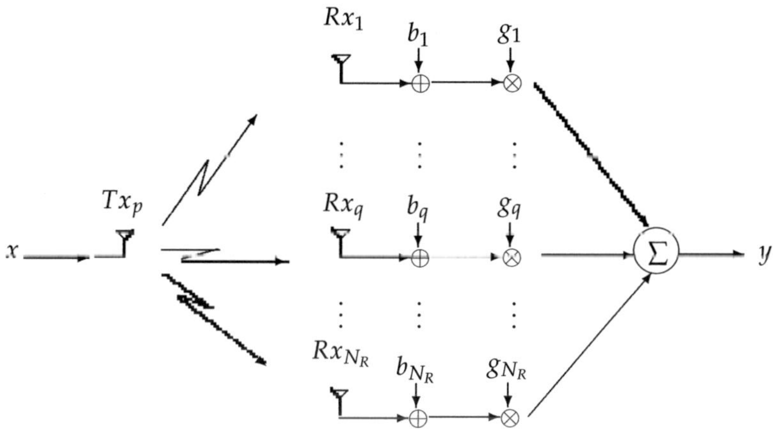

Fig. 17. SIMO system with combining technique

6.1 Maximal Ratio Combining (MRC)

The equivalent SNR of MRC has been calculated as :

$$\gamma_y = \gamma \cdot \frac{\left(\sum_{q=1}^{N_R} \lvert h_q \rvert^2 \right)^2}{\sum_{q=1}^{N_R} \lvert h_q \rvert^2} = \gamma \cdot \sum_{q=1}^{N_R} \lvert h_q \rvert^2 = \sum_{q=1}^{N_R} \gamma_q \tag{62}$$

Thus, the instantaneous SNR γ_y is expressed as the sum of the instantaneous SNR at different receive antennas. For normalized channel matrix, the SNR is then:

$$\gamma_y = N_R \cdot \gamma \tag{63}$$

The system capacity with MRC is :

$$C_{\text{MRC}} = \log_2 \left(1 + \gamma \cdot \sum_{q=1}^{N_R} |h_q|^2 \right) \qquad bits/s/Hz \tag{64}$$

6.2 Equal Gain Combining (EGC)

The instantaneous SNR is expressed as :

$$\gamma_y = \frac{\gamma}{N_R} \cdot \left(\sum_{q=1}^{N_R} |h_q| \right)^2 \tag{65}$$

Resulting capacity has been calculated as :

$$C_{\text{EGC}} = \log_2 \left(1 + \frac{\gamma}{N_R} \cdot \sum_{q=1}^{N_R} |h_q|^2 \right) \qquad bits/s/Hz \tag{66}$$

6.3 Selection Combining(SC)

The receiver scans the antennas, finds the antenna with the highest instantaneous SNR and selects it. We denote the highest received instantaneous SNR as :

$$\gamma_y = \max \left(\gamma_1, \ldots, \gamma_{N_R} \right) \tag{67}$$

The SNR at the output of the combiner for an uncorrelated channel is:

$$\gamma_y = \gamma \cdot \sum_{q=1}^{N_R} \frac{1}{q} \tag{68}$$

SC capacity is expressed as:

$$\begin{aligned} C_{\text{SC}} &= \log_2 \left(1 + \gamma \cdot \max_q |h_q|^2 \right) \\ &= \max_q \left\{ \log_2 \left(1 + \gamma \cdot |h_q|^2 \right) \right\} \quad ; 1 \leq q \leq N_R \quad bits/s/Hz \end{aligned} \tag{69}$$

The ergodic capacity curves for all three combining strategies are shown in Fig. 18. MRC yields best performances in terms of channel capacity. However, MRC is the optimal combining technique, MRC is seldom implemented in a multipath fading channel since the complexity of the receiver is directly resolvable paths (Zhou & Okamoto, 2004). In general, EGC performs worse than does MRC. Obviously, lower capacity is obtained with SC since only one Radio Frequency (RF) channel is selected at the receiver. A study of combining techniques in terms of BER was presented in (Zhou & Okamoto, 2004). MRC steel achieves the best BER performances.

7. Beamforming processing in MIMO systems

Beamforming is the process of trying to steer the digital baseband signals to one particular direction by weighting these signals differently. This is named "digital beamforming" and we call it beamforming for the sake of brevity, (Jafarkhani, 2005). The desired signal is then obtained by summing the weighted baseband signals.

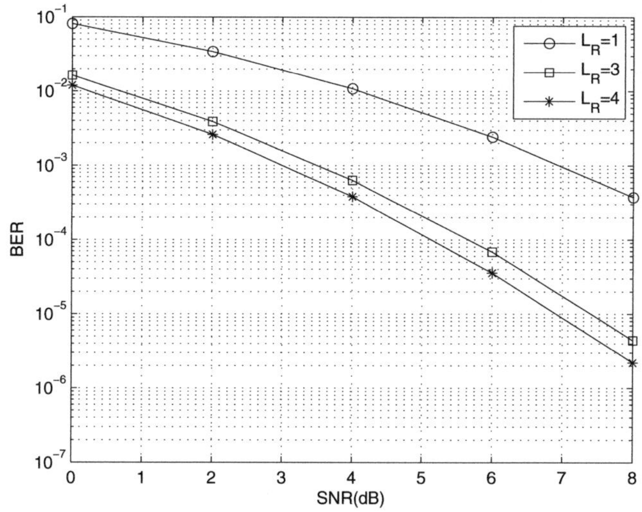

Fig. 18. Capacity for MIMO(4×1) using various combining techniques-Rayleigh fading channel

7.1 Beamforming based on SVD decomposition

In this section, we provide an overview of MIMO systems that use beamforming at both the ends of the communication link. We consider a MIMO system with N_T transmit and N_R receive dimensions. From a mathematical point of view, joint Transmit-Receive beamforming is based on the minimization (or maximization) of some cost function such as SNR maximization. This method includes determining the transmit beamforming coefficients and the receive beamforming coefficients so that to steer relatively all transmit energy and receive energy in the directions of interest. Joint Transmit Receive beamforming is illustrated in Fig. 19.

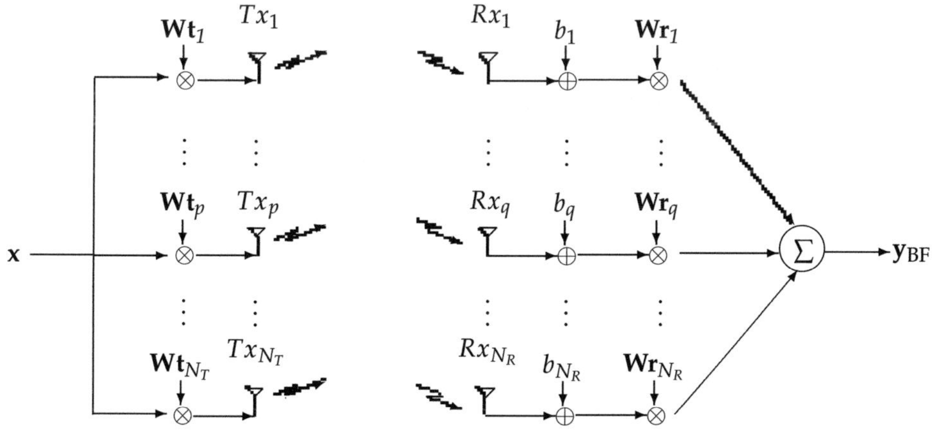

Fig. 19. Joint Transmit-Receive beamforming

- **x**: The transmit signal
- $\mathbf{Wt} = [\mathbf{Wt}_1, \ldots, \mathbf{Wt}_{N_T}]^T$: The $(N_T \times 1)$ Transmit beamforming vector

- **H**: The $(N_R \times N_T)$ channel matrix
- $\mathbf{Wr} = [\mathbf{Wr}_1,\ldots,\mathbf{Wr}_{N_R}]^T$: The $(N_R \times 1)$ Receive beamforming vector
- $\mathbf{b} = [b_1,\ldots,b_{N_R}]^T$: The $(N_R \times 1)$ Additive noise vector with variance σ_b^2
- $\mathbf{y}_{\mathrm{BF}}$: The output signal

Joint Transmit-Receive beamforming can be described by equation (70).

$$\mathbf{y}_{\mathrm{BF}} = \mathbf{Wr}^H \mathbf{H} \mathbf{W}_\mathbf{t} \cdot \mathbf{x} + \mathbf{Wr}^H \cdot \mathbf{b} \tag{70}$$

Eigen-beamforming could be performed by using eigenvectors to find the linear beamformer that optimizes the system performances. Thus, we exploit the SVD factorization for channel matrix **H** ($\mathbf{H} = \mathbf{U}\mathbf{S}\mathbf{V}^H$). Assigning **U** and **V** respectively to **Wr** and **Wt** is optimal for maximizing the SNR given by :

$$\mathrm{SNR}_{\mathrm{BF}} = \frac{\|\mathbf{Wr}^H \mathbf{H} \mathbf{Wt}\|^2 \mathrm{E}(\mathbf{x}\mathbf{x}^H)}{\sigma_b^2 \|\mathbf{Wr}\|^2}$$

When SVD factorization is applied to MIMO channel matrix, equation (70) becomes :

$$\mathbf{y}_{\mathrm{BF}} = \mathbf{S} \cdot \mathbf{x} + \mathbf{U}^H \cdot \mathbf{b} \tag{71}$$

Note that Beamforming (Ibnkahla, 2009) is considered as a form of linear combining techniques which are intended to maximize the spectral efficiency. The received SNR for communication system with beamforming is expressed as :

$$\gamma_{\mathrm{BF}} = \gamma_r \cdot \lambda_{max}(\mathbf{H})$$

λ_{max} is the maximum eigenvalue associated to matrix **S** and γ_r is the mean received SNR. Thereafter, the capacity for MIMO system with beamforming is expressed as :

$$C_{\mathrm{BF}} = \log_2 \{1 + \gamma_r \cdot \lambda_{max}(\mathbf{H})\} \quad bits/s/Hz \tag{72}$$

Simulation results for MIMO capacity where beamforming technique is performed are shown in Fig. 20. The MIMO channel capacity with beamforming is improved thanks to the spatial diversity.
Note that beamforming technique is shown to improve the performance of the communication link in terms of BER. Fig. 21 shows the plotted curves of BER as a function of SNR relative to three cases :

- System performing beamforming
- Transmission without applying beamforming
- Transmission with simply Zero Forcing (ZF) equalization

The MIMO (3×3) channel is randomly generated and input signal is BPSK modulated. We adopt the correlated MIMO channel with a spreading angle of 90° and an antenna spacing of $\frac{\lambda}{2}$. Fig. 21 shows that associated SVD beamforming technique brings the best performances in terms of BER.

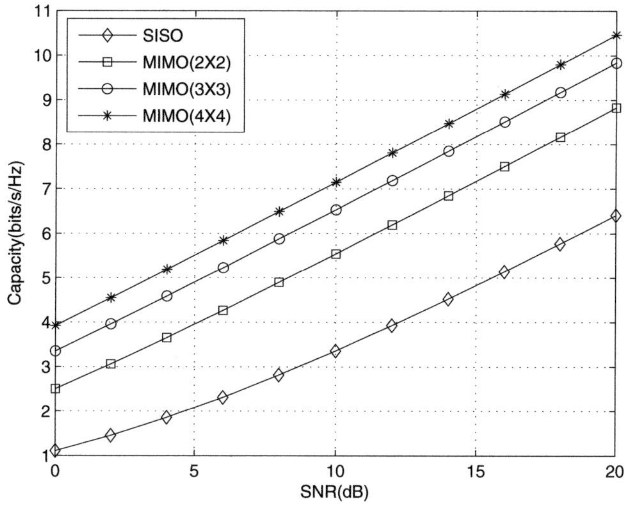

Fig. 20. Capacity of MIMO system with beamforming technique

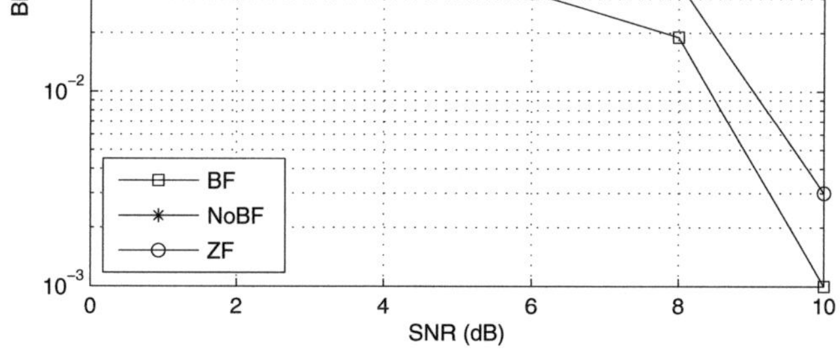

Fig. 21. SVD based beamforming technique

7.2 SINR maximization beamforming

Interference often occurs in wireless propagation environment. When several terminals are densely deployed in the coverage area, Signal to Interference Noise Ratio (SINR) grows up and efficient techniques are required to be implemented. Beamforming is an efficient strategy that could be exploited so that to mitigate interference. Maximizing the SINR criteria could be also considered so that to obtain optimal beamforming weights.

SINR maximization based beamforming in Multi user system
Model description

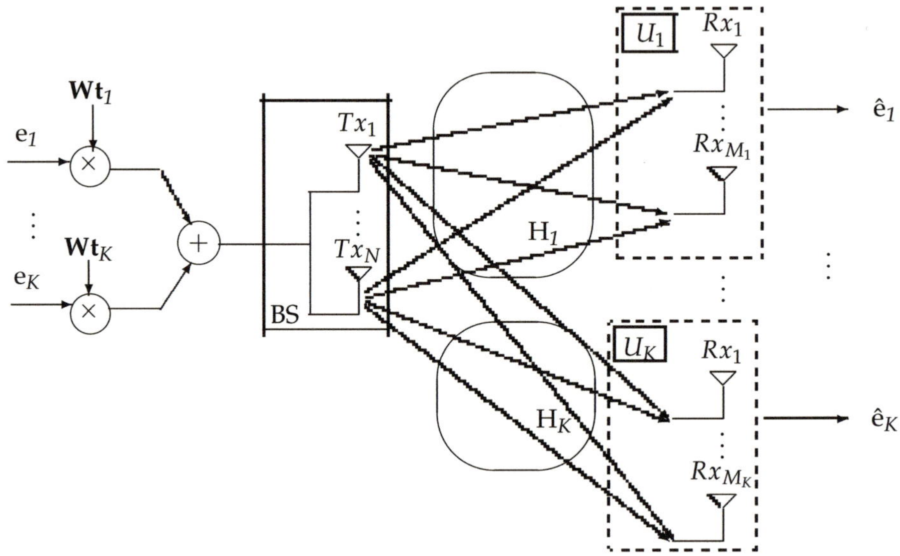

Fig. 22. Multi user system with beamforming

We denote :

- K: Number of users.
- $\mathbf{E} = [e_1, \ldots, e_N]^T$: The transmit signal vector
- $\mathbf{Wt}$=$[\mathbf{Wt}_1, \ldots, \mathbf{Wt}_K]^T$: Weight vector for beamforming
- $M_1, \ldots, M_K$ number of antennas respectively for users $U_1, \ldots, U_K$
- $\mathbf{x}$: The transmit vector signal of size $(N \times 1)$

Transmit signal is expressed as :

$$\mathbf{x} = \sum_{k=1}^{K} \mathbf{Wt}_k \cdot e_k \tag{73}$$

We assume that transmit signals and beamforming weights are normalized. The received signal (Of size $(M_i \times 1)$) by user U_i is :

$$y_i = H_i \sum_{k=1}^{K} \mathbf{Wt}_k \cdot e_k + b_i \tag{74}$$

b_i is the additive noise with variance σ_i^2. The channel matrix $H_i(M_i \times N)$ between user U_i with M_i antennas and the N antennas at the Base Station (BS) is assumed to be normalized. User $U_i \quad ; i = 1, \ldots, K$ receives the signal :

$$y_i = H_i \mathbf{Wt}_i \cdot e_i + \sum_{k=1, k \neq i}^{K} H_i \mathbf{Wt}_k \cdot e_k + b_i \tag{75}$$

At the receiver, the estimated signal for user i is:

$$\hat{e}_i = \frac{\mathbf{Wt}_i^H H_i^H y_i}{\|H_i \mathbf{Wt}_i\|} \tag{76}$$

The SINR is the ratio of the received strength of the desired signal to the received strength of undesired signals (Noise + Interference). Associated SINR to user i is expressed as :

$$SINR_i = \frac{\| H_i \mathbf{Wt}_i \|^2}{\left(\sum_{k=1, k \neq i}^{K} \| H_i \mathbf{Wt}_k \|^2 \right) + \sigma_i^2} \tag{77}$$

SINR could also be written as :

$$SINR_i = \frac{\| H_i \mathbf{Wt}_i \|^2}{\left(\frac{\sum_{k=1, k \neq i}^{K} \|\mathbf{Wt}_i^H H_i^H H_i \mathbf{Wt}_k\|^2}{\|H_i \mathbf{Wt}_i\|^2} \right) + \sigma_i^2} \tag{78}$$

Optimal beamformer weights are obtained by maximizing the Signal Leakage Ratio (SLR) metric expressed as :

$$SLR = \frac{\| H_i \mathbf{Wt}_i \|^2}{\| \tilde{H}_i \mathbf{Wt}_i \|^2} \tag{79}$$

where :

$$\tilde{H}_i = [H_1^H, \ldots, H_{i-1}^H, H_{i+1}^H, \ldots, H_K^H]^H \tag{80}$$

The optimal weights $\mathbf{Wt}_i \quad ; i = 1, \ldots, K$ are derived (Tarighat et al., 2005) as the maximum eigenvector of:

$$((\tilde{H}_i{}^H \tilde{H}_i)^{-1} (H_i^H H_i))$$

Simulation results are shown in Fig. 23. These results show that the method is optimal for determining the beamforming weights. Note that better performances in terms of BER are achieved if more transmit antennas are used.

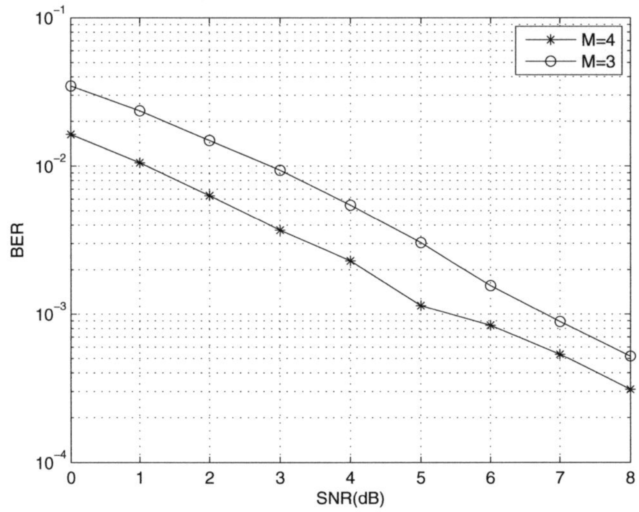

Fig. 23. Multi user BF ($K = 3, M = 3/M = 4$)

8. Processing techniques for MIMO systems: Antenna selection

MIMO system gives high performances in terms of system capacity and reliability of radio communication. Combining techniques such as MRC results in more robust system. Nevertheless, the deployment of multiple antennas would require the implementation of multiple RF chains (Dong et al., 2008). This would be costly in terms of size, power and hardware. For example, when several antennas are deployed, multiple RF chains with separate modulator and demodulator have to be implemented. To overcome these limitations, antenna selection techniques can be applied.

8.1 Antenna selection

Antenna selection technique (Ben ZID et al., 2011) is depicted in Fig. 24. We consider a MIMO system with N_T transmit antennas and N_R receive antennas. The idea of antenna selection is to select L_T antennas among the N_T transmit antennas and / or L_R antennas among the N_R receive antennas. We distinguish different forms of antenna selection:

1. Transmit antenna selection
2. Receive antenna selection
3. Hybrid antenna selection: that is when antenna selection is carried among both transmit antennas and receive antennas.

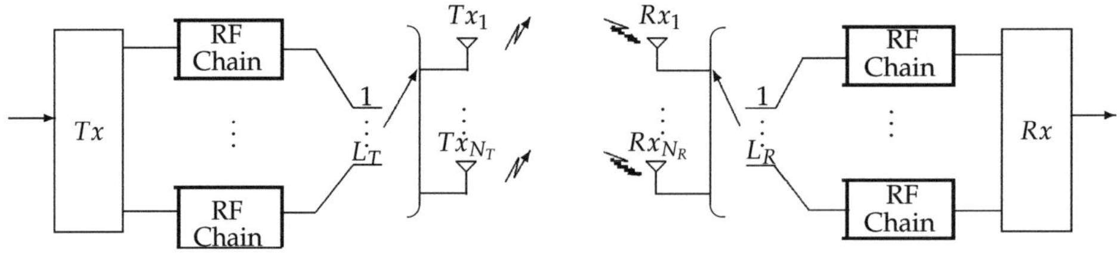

Fig. 24. Antenna selection in MIMO system

Antenna selection algorithms do not only aim to reduce the system complexity but also to achieve high spectral efficiency. When L_T antennas are selected at the transmitter and L_R antennas are selected at the receiver, the associated channel will be denoted $\mathbf{H}_S$. The capacity of such system is expressed as :

$$\begin{aligned} C_{\text{Sel}} &= \log_2 \left[\det \left(I_{L_T} + \frac{\gamma}{L_R} \mathbf{H}_S^H \mathbf{H}_S \right) \right] \\ &= \log_2 \left[\det \left(I_{L_R} + \frac{\gamma}{L_T} \mathbf{H}_S \mathbf{H}_S^H \right) \right] \end{aligned} \tag{81}$$

γ denotes the SNR. The antenna selection algorithm is intended to find the optimal subset of the transmit antennas and /or the optimal subset of the receive antennas that satisfy capacity system maximization. Nevertheless, it is obvious that the joint antenna selection at the transmitter and the receiver brings more complexity when the number of antennas increases.

Numerical results

Ergodic capacity of MIMO system with antenna selection at the transmitter and the receiver is shown in Fig. 25. For simulation purposes, we generate a Rayleigh MIMO channel with AWGN. Here, SVD factorization is applied. Plotted curves depict the ergodic capacity for the MIMO(4×4). This evidently leads to the highest system capacity. When 3 transmit antennas are selected among 4 transmit antennas and 3 receive antennas are selected among 4 receive antennas, the maximum ergodic capacity that could be achieved is plotted in function of SNR. Simulation results are also presented in the case when two antennas are selected at both the transmitter and the receiver. According to the plotted curves in Fig. 25, it is obvious that one of the important limitations of the antenna selection strategy is the important losses in capacity at high SNR regime.

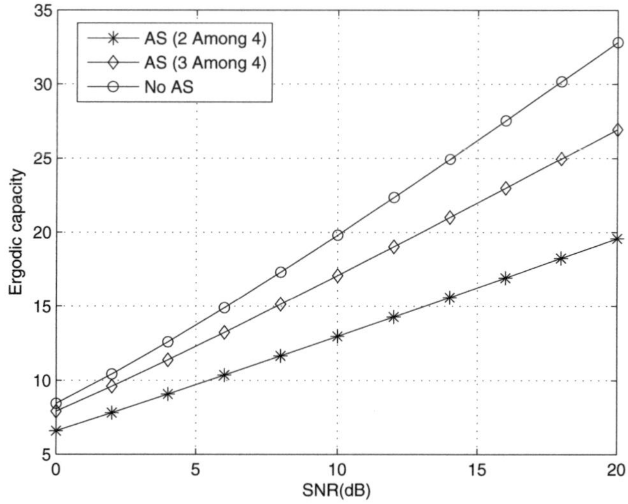

Fig. 25. Antenna selection in MIMO (4×2): Impact on ergodic capacity

8.2 Antenna selection involving ST coding

We present in this paragraph, the simulation results in terms of average BER when joint Alamouti scheme and antenna selection at the receiver are applied. The MIMO (4×2) system with a Rayleigh channel and AWGN was created. Emitted symbols are QAM (Quadrature Amplitude Modulation) modulated. The simulation model is given by the Fig. 26 ($b_1, \ldots, b_{N_R}$ denote the additive noise signals). Plotted curves concern subsets of receive antennas where $L_R = 1$ and $L_R = 3$. Simulation results show that even with only one selected antenna at the

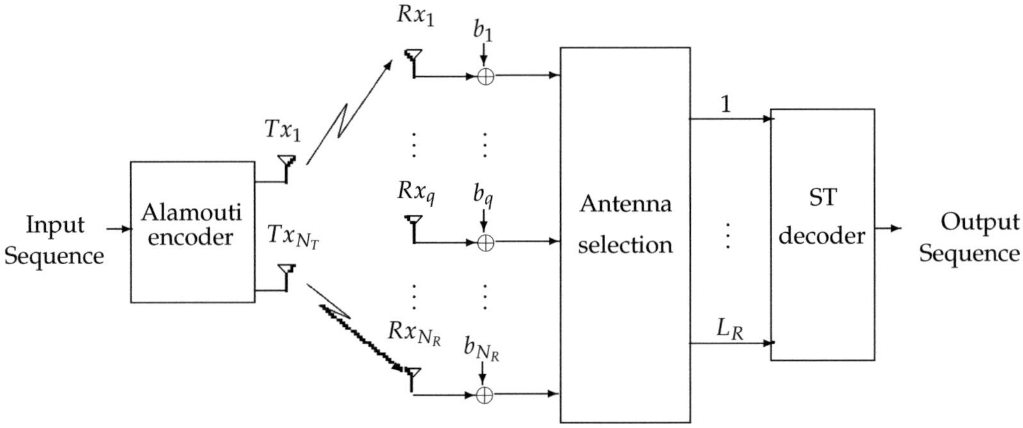

Fig. 26. MIMO system with antenna selection and Alamouti coding

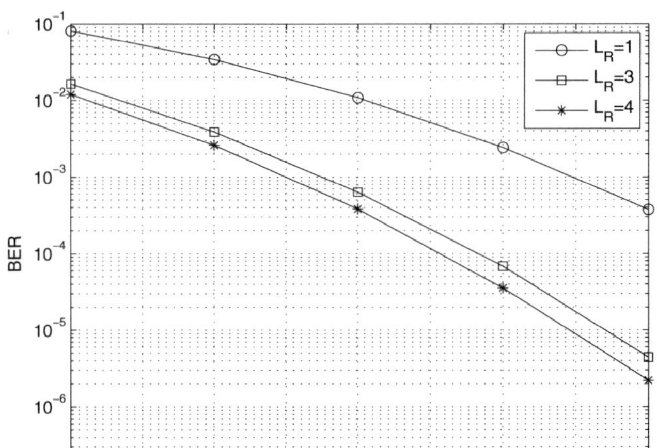

Fig. 27. Joint Alamouti coding and antenna selection in MIMO (4×2)

receiver, performances in terms of BER still satisfactory. Nevertheless, when more antennas are selected, better BER values are achieved thanks to receive diversity.

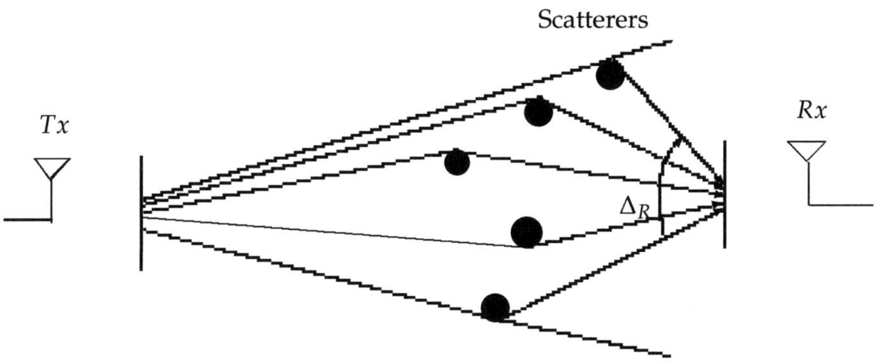

Fig. 28. Angle spread

8.3 Antenna selection in correlated MIMO channel: Angular dispersion and channel correlation

Angle spread refers to the spread of DOA of the multipath components at the transmit antenna array. When scatterers are also distributed around the receive antennas, the scattering effect leads also to an angle spreading relative to the DOA. In Fig. 28, the angle spread is denoted Δ_R. We present a SISO model rich of local scatterers. For seek of simplicity, we consider a MIMO $(N_R \times N_T)$ system with LOS channel and uniform antenna arrays at both the transmitter and the receiver.

We denote :

- $\mathbf{H}$: MIMO channel matrix
- $d_{q,p}$: distance between antenna q and antenna p
- $\rho_{q,q'}$: correlation coefficient
- λ: wavelength
- $\mathbf{R} = \mathrm{E}[\mathbf{H}^H\mathbf{H}]$ Correlation matrix
- α: Angle of arrival
- $p(\alpha)$: Probability density function of the DOA
- $\Delta_R(= 2\pi)$: Angle spread at the receiving side

When LOS propagation is assumed, the channel coefficients can be expressed as :

$$h_{qp} = e^{-j2\pi\frac{d_{q,p}}{\lambda}} \quad ; q = 1, \ldots, N_R, p = 1, \ldots, N_T \tag{82}$$

The correlation coefficient at the receiving side between two receive antennas of indexes q and q' is expressed :

$$\rho_{q,q'} = E[exp(-j2\pi\frac{d_{q,q'}sin\alpha}{\lambda})] \tag{83}$$

Formula for correlation coefficients is expressed as :

$$\rho_{q,q'} = \int_{\frac{-\Delta_R}{2}}^{\frac{\Delta_R}{2}} exp(-\frac{j2\pi d_{q,q'}sin(\alpha)}{\lambda})p(\alpha)d\alpha \quad ; q \neq q' \tag{84}$$

Evidently :

$$\rho_{q,q'} = 1 \quad ; q = q' \tag{85}$$

Following a uniform distribution, correlation coefficients can be expressed as :

$$\rho_{q,q'} = J_0(2\pi \frac{dq,q'}{\lambda}) \quad q = 1,\ldots,N_R; \quad q' = 1,\ldots,N_R \tag{86}$$

$J_0(.)$ is the zeroth order Bessel function. When antennas are uncorrelated, $\rho_{q,q'} = 0$ if $q \neq q'$. Which induces :

$$J_0(2\pi \frac{d_{q,q'}}{\lambda}) \simeq 0 \Rightarrow 2\pi \frac{d_{q,q'}}{\lambda} \simeq \pi$$

Thus, in order to mitigate correlation between antennas, antenna spacing between two antennas should be at least equal to $\frac{\lambda}{2}$. Nevertheless, antenna correlation still depends on angular dispersion. Fig. 29 presents the plotted curves of the spatial correlation as a function of the antenna spacing divided by the wavelength for various values of angle spread. According to simulation results depicted in Fig. 29, we conclude that spatial correlation between two antennas depends on antenna spacing and is reduced by higher angle spread.

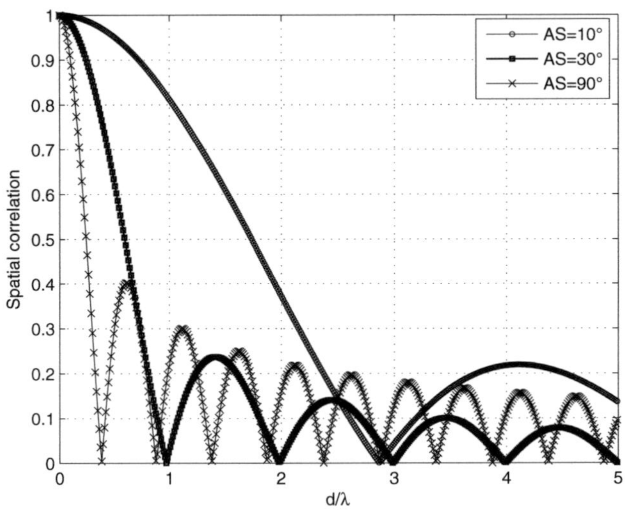

Fig. 29. Impact of angle spread on spatial correlation

Better performances in terms of BER are achieved for $AS = 30°$. This is due to the fact that for a given antenna spacing, system correlation is higher for lower angular spread. The impact of angle spread on system performances is depicted in Fig. 30.

9. Multi polarization techniques

9.1 Basic antenna theory and concepts

We present in this paragraph, some basic concepts related to antenna. A rigorous analysis of the antenna theory and the related concepts is available in (Constantine, 2005). Antenna is a transducer for radiating or receiving radio waves. It ideally radiates all the power delivered to it from the transmitter in a desired direction. The far electric field of the electromagnetic wave is written in spherical coordinates as :

$$E = E_\theta(\theta,\phi)\hat{\theta} + E_\phi(\theta,\phi)\hat{\phi} \tag{87}$$

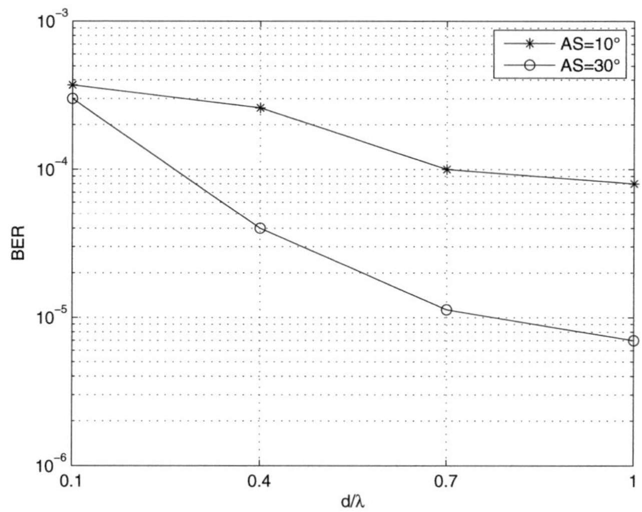

Fig. 30. Impact of angle spread on system performances

E_θ and E_ϕ are the electric field components. θ and ϕ denote respectively the elevation angle and the azimuthal angle. We distinguish two categories of antennas :

1. Omnidirectional antenna is an antenna system which radiates power uniformly
2. Dipole antenna radiates power in a particular direction.

Electric dipole could be oriented along the x axis, y axis or the z axis. Table 2 gives the expressions of the electric field components relative to each antenna orientation.

	$E_\theta(\theta,\phi)$	$E_\phi(\theta,\phi)$
x	$-\cos(\theta)\cos(\phi)$	$\sin(\phi)$
y	$-\cos(\theta)\sin(\phi)$	$-\cos(\phi)$
z	$\sin(\theta)$	0

Table 2. Radiation pattern for electric dipole

Radiation intensity

Antenna gain is defined as the ratio of the intensity radiated by the antenna divided by the intensity radiated by an isotropic antenna. Normalized radiation intensity (or Antenna gain) is :

$$\mathbf{G}(\theta,\phi) = \begin{bmatrix} G_\theta(\theta,\phi) \\ G_\phi(\theta,\phi) \end{bmatrix} = \begin{bmatrix} \dfrac{E_\theta(\theta,\phi)}{\sqrt{\dfrac{1}{4\pi}\left(\int\limits_0^{2\pi}\int\limits_0^{\pi} |E_\theta(\theta,\phi)|^2 d\Omega + \int\limits_0^{2\pi}\int\limits_0^{\pi} |E_\phi(\theta,\phi)|^2 d\Omega\right)}} \\ \dfrac{E_\phi(\theta,\phi)}{\sqrt{\dfrac{1}{4\pi}\left(\int\limits_0^{2\pi}\int\limits_0^{\pi} |E_\theta(\theta,\phi)|^2 d\Omega + \int\limits_0^{2\pi}\int\limits_0^{\pi} |E_\phi(\theta,\phi)|^2 d\Omega\right)}} \end{bmatrix} \tag{88}$$

- Ω is the beam solid angle through which all the power of the antenna would flow if its radiation intensity is constant for all angles within Ω.
- $G_\theta(\theta,\phi)$ and $G_\phi(\theta,\phi)$ are respectively the elevation antenna gain and the azimuthal antenna gain.

9.2 3D Geometric wide band channel model

The 3D Geometric wide band channel model is presented in Fig.31.

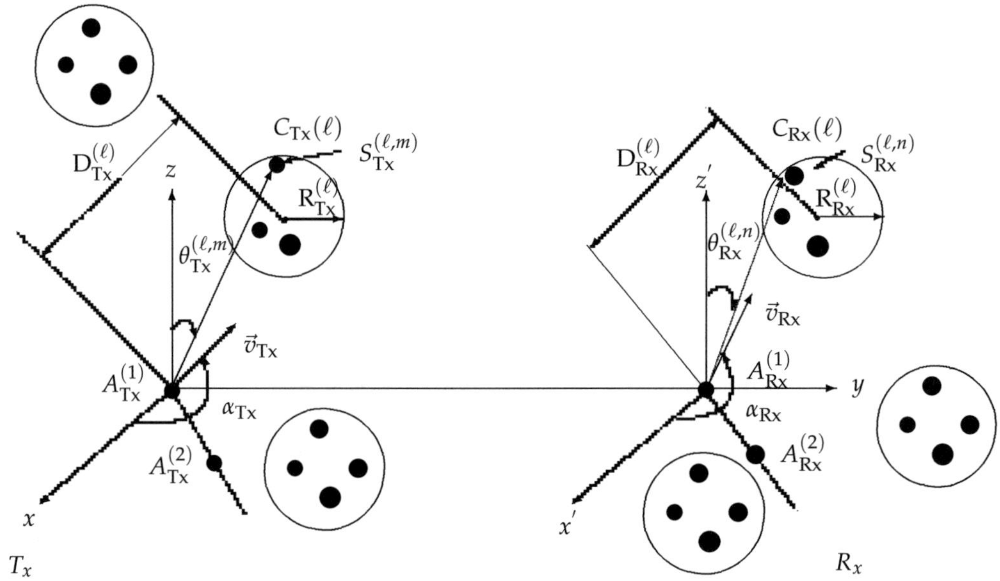

Fig. 31. 3D Geometric model for MIMO channel ($N_R = 2$, $N_T = 2$)

Two transmit antennas ($A_{\mathrm{Tx}}^{(1)}$,$A_{\mathrm{Tx}}^{(2)}$) and two receive antennas ($A_{\mathrm{Rx}}^{(1)}$,$A_{\mathrm{Rx}}^{(2)}$) are presented. Wide band MIMO channel involves several local clusters of scatterers which are distributed around the transmitter and the receiver. The cluster index is denoted ℓ, $\ell = 1,\ldots,L$. Cluster around the transmitter $C_{\mathrm{Tx}}(\ell)$ is assumed to be associated with a set of $M^{(\ell)}$ scatterers ($S_{\mathrm{Tx}}^{(\ell,m)}$; $m = 1,\ldots,M^{(\ell)}$). Cluster around the receiver $C_{\mathrm{Rx}}(\ell)$ is assumed to be associated with a set of $N^{(\ell)}$ scatterers ($S_{\mathrm{Rx}}^{(\ell,n)}$; $n = 1,\ldots,N^{(\ell)}$).
We take for notations:

- $\mathrm{R}_{\mathrm{Tx}}^{(\ell)}$: Transmit cluster radius of index ℓ
- $\mathrm{D}_{\mathrm{Tx}}^{(\ell)}$: Distance between the reference transmit antenna and the transmit cluster center
- $d_{1,\ell,m}$: Distance between antenna $A_{\mathrm{Tx}}^{(1)}$ and a scatterer $S_{\mathrm{Tx}}^{(\ell,m)}$
- $d_{2,\ell,m}$: Distance between antenna $A_{\mathrm{Tx}}^{(2)}$ and a scatterer $S_{\mathrm{Tx}}^{(\ell,m)}$
- d_{Tx}: Transmit antennas spacing
- $\mathrm{D}_{\mathrm{Tx}\leftrightarrow\mathrm{Rx}}$: Distance between the transmitter and the receiver
- $\mathrm{R}_{\mathrm{Rx}}^{(\ell)}$: Cluster radius at the receiver of index ℓ

- $D_{Rx}^{(\ell)}$: Distance between the reference antenna $A_{Rx}^{(1)}$ and the center of the cluster ℓ at the receiver.
- $d_{1,\ell,n}$: Distance between antenna $A_{Rx}^{(1)}$ and the scatterer $S_{Rx}^{(\ell,n)}$
- $d_{2,\ell,n}$: Distance between antenna $A_{Rx}^{(2)}$ and the scatterer $S_{Rx}^{(\ell,n)}$

The Non-Line of Sight (NLOS) channel coefficients in 3D Geometric model are given by (Prayongpun, 2009):

$$h_{qp}^{NLos}(t,f) = \lim_{M,N\to\infty} \sum_{\ell=1}^{L} \sqrt{\frac{PDP(\ell)}{M^{(\ell)}N^{(\ell)}}} \sum_{m=1}^{M^{(\ell)}} \sum_{n=1}^{N^{(\ell)}} G_p(\theta_{Tx}^{(\ell,m)}, \phi_{Tx}^{(\ell,m)}; \beta_{Tx}, \gamma_{Tx}) a_{\ell m}^{(p)} b_{\ell n}^{(q)}$$
$$G_q(\theta_{Rx}^{(\ell,n)}, \phi_{Rx}^{(\ell,n)}; \beta_{Rx}, \gamma_{Rx}) \exp\left\{j\left(2\pi(f_{Tx}^{(\ell,m)} + f_{Rx}^{(\ell,n)})t + \varphi_{\ell mn} + \varphi_0^{(\ell)}\right)\right\} \quad (89)$$

- $q \in \{1, \ldots, N_R\}$
- $p \in \{1, \ldots, N_T\}$
- $PDP(\ell)$ is the power delay profile which gives the intensity of a signal received through a multipath channel connecting a pair of clusters.
- $G_p(\theta_{Tx}^{(\ell,m)}, \phi_{Tx}^{(\ell,m)}; \beta_{Tx}, \gamma_{Tx})$ is the gain of antenna p with associated oriented direction $(\beta_{Tx}, \gamma_{Tx})$ and a wave propagation direction $(\theta_{Tx}^{(\ell,m)}, \phi_{Tx}^{(\ell,m)})$.
- $G_q(\theta_{Rx}^{(\ell,n)}, \phi_{Rx}^{(\ell,n)}; \beta_{Rx}, \gamma_{Rx})$ is the gain of antenna q with associated oriented direction $(\beta_{Rx}, \gamma_{Rx})$ and a wave propagation direction $(\theta_{Rx}^{(\ell,n)}, \phi_{Rx}^{(\ell,n)})$.
- $a_{\ell m}^{(p)} = \exp\{j2\pi(p-1)(d_{Tx}/\lambda)[\cos(\theta_{Tx}^{(\ell,m)})\cos(\beta_{Tx}) + \sin(\theta_{Tx}^{(\ell,m)})\sin(\beta_{Tx})\cos(\phi_{Tx}^{(\ell,m)} - \gamma_{Tx})]\}$
- $b_{\ell n}^{(q)} = \exp\{j2\pi(q-1)(d_{Rx}/\lambda)[\cos(\theta_{Rx}^{(\ell,n)})\cos(\beta_{Rx}) + \sin(\theta_{Rx}^{(\ell,n)})\sin(\beta_{Rx})\cos(\phi_{Rx}^{(\ell,n)} - \gamma_{Rx})]\}$
- $f_{Tx}^{(\ell,m)} = (|\vec{v}_{Tx}|/\lambda)\sin(\theta_{Tx}^{(\ell,m)})\cos(\phi_{Tx}^{(\ell,m)} - \alpha_{Tx})$
- $f_{Rx}^{(\ell,n)} = (|\vec{v}_{Rx}|/\lambda)\sin(\theta_{Rx}^{(\ell,n)})\cos(\phi_{Rx}^{(\ell,n)} - \alpha_{Rx})$
- $\varphi_0^{(\ell)} = -2\pi(D_{Tx}^{(\ell)} + D_{Tx\leftrightarrow Rx} + D_{Rx}^{(\ell)})/\lambda$
- $\varphi_{\ell mn} \sim U[-\pi, \pi]$

We assume that the transmitter and the receiver have motions above the plan (x, y) with relative directions α_{Tx} and α_{Rx}. In 3D Geometric model, the distances are expressed as :

$$d_{1,\ell,m} \approx D_{Tx}^{(\ell)} \quad (90)$$
$$d_{2,\ell,m} \approx D_{Tx}^{(\ell)} - d_{Tx}\cos(\theta_{Tx}^{(\ell,m)})\cos(\beta_{Tx}) - d_{Tx}\sin(\theta_{Tx}^{(\ell,m)})\sin(\beta_{Tx})\cos(\phi_{Tx}^{(\ell,m)} - \gamma_{Tx}) \quad (91)$$
$$d_{1,\ell,n} \approx D_{Rx}^{(\ell)} \quad (92)$$
$$d_{2,\ell,n} \approx D_{Rx}^{(\ell)} - d_{Rx}\cos(\theta_{Rx}^{(\ell,n)})\cos(\beta_{Rx}) - d_{Rx}\sin(\theta_{Rx}^{(\ell,n)})\sin(\beta_{Rx})\cos(\phi_{Rx}^{(\ell,n)} - \gamma_{Rx}) \quad (93)$$
$$d_{\ell,m,n} \approx D_{Tx\leftrightarrow Rx} + D_{Rx}^{(\ell)}\sin(\theta_{Rx}^{(\ell,n)})\sin(\phi_{Rx}^{(\ell,n)}) - D_{Tx}^{(\ell)}\sin(\theta_{Tx}^{(\ell,m)})\sin(\phi_{Tx}^{(\ell,m)}) \approx D_{Tx\leftrightarrow Rx} \quad (94)$$

When there are no scatterers around the transmitter or the receiver, the channel model is referred as Line of Sight (LOS).

9.3 Depolarization phenomena

The polarization of an antenna is defined as the polarization of the wave radiated by the antenna. At a given position, antenna polarization describes the orientation of the electric field. Suppose an electromagnetic wave radiated by a transmit antenna has an incident electric field E_i with two components $E_{i,\theta}$ and $E_{i,\phi}$. In each scatterer's environment, the electric field components are reflected (Fig.32). Reflected elevation and azimuth components of the incident electric field E_i are denoted $E_{r,\theta}$ and $E_{r,\phi}$.

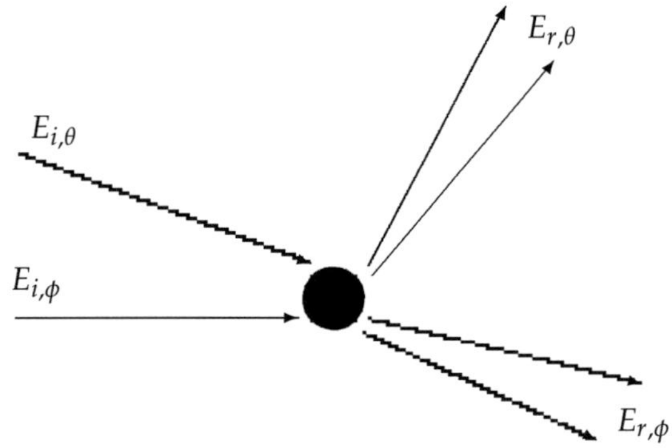

Fig. 32. Antenna depolarization

The radiation pattern is expressed as a function of the azimuth and elevation amplitudes of polarization vectors in both the elevation and the azimuth directions ($\hat{\theta}$ and $\hat{\phi}$).

$$E_i = E_{i,\theta}(\theta, \phi)\hat{\theta} + E_{i,\phi}(\theta, \phi)\hat{\phi} \tag{95}$$

The reflected electric field is expressed as :

$$E_r = \mathrm{S}E_i \tag{96}$$

where :

- E_r is the matrix notation for the reflected electric field expressed as:

$$E_r = \begin{bmatrix} E_{r,\theta} \\ E_{r,\phi} \end{bmatrix} \tag{97}$$

- S is the scattering matrix given by:

$$\mathrm{S} = \begin{bmatrix} v_{\theta\theta} & v_{\theta\phi} \\ v_{\phi\theta} & v_{\phi\phi} \end{bmatrix} \tag{98}$$

  - $v_{\theta\theta}$ and $v_{\phi\phi}$ respectively measure the co-polarization gains relative to the elevation transmission and the azimuth transmission.
  - $v_{\theta\phi}$ measures the depolarization gain of the elevation polarization relative to the azimuth polarization.
  - $v_{\phi\theta}$ measures the depolarization gain of the azimuth polarization relative to the elevation polarization.

- The matrix notation for the incident electric field is :

$$E_i = \begin{bmatrix} E_{i,\theta} \\ E_{i,\phi} \end{bmatrix} \tag{99}$$

The depolarization effect is characterized by the Cross Polarization Discrimination (XPD)(Raoof & Zhou, 2009)(Raoof & Prayongpun, 2007) which is defined as the power ratio of the co-polarization and cross-polarization components of the mean incident wave.
The polarization discrimination for the elevation transmission is :

$$XPD_\theta = \frac{\mathrm{E}|v_{\theta\theta}|^2}{\mathrm{E}|v_{\phi\theta}|^2} \tag{100}$$

The polarization discrimination for the azimuth transmission is :

$$XPD_\phi = \frac{\mathrm{E}|v_{\phi\phi}|^2}{\mathrm{E}|v_{\theta\phi}|^2} \tag{101}$$

In the following, we denote :

$$\chi_\theta = \frac{1}{XPD_\theta} \quad \text{and} \quad \chi_\phi = \frac{1}{XPD_\phi}$$

GSCM channel model involving antenna depolarization
The wide band GSCM channel modeling with antenna depolarization is given by :

$$h_{qp}^{NLOS}(t,f) = \lim_{M,N\to\infty} \sum_{\ell=1}^{L} \sqrt{\frac{PDP(\ell)}{M^{(\ell)}N^{(\ell)}}} \sum_{m=1}^{M^{(\ell)}} \sum_{n=1}^{N^{(\ell)}} a_m^{(p)} b_n^{(q)} \exp\left\{ j \left(2\pi (f_{\mathrm{Tx}}^{(\ell,m)} + f_{\mathrm{Rx}}^{(\ell,n)}) t + \varphi_{\ell mn} + \varphi_0^{\ell} \right) \right\}$$
$$\begin{bmatrix} G_{q,\theta}(\theta_{\mathrm{Rx}}^{(\ell,n)}, \phi_{\mathrm{Rx}}^{(\ell,n)}; \beta_{\mathrm{Rx}}, \gamma_{\mathrm{Rx}}) \\ G_{q,\phi}(\theta_{\mathrm{Rx}}^{(\ell,n)}, \phi_{\mathrm{Rx}}^{(\ell,n)}; \beta_{\mathrm{Rx}}, \gamma_{\mathrm{Rx}}) \end{bmatrix}^T \mathrm{S}_{\mathrm{Tx,Rx}}^{(\ell,m,n)} \begin{bmatrix} G_{p,\theta}(\theta_{\mathrm{Tx}}^{(\ell,m)}, \phi_{\mathrm{Tx}}^{(\ell,m)}; \beta_{\mathrm{Tx}}, \gamma_{\mathrm{Tx}}) \\ G_{p,\phi}(\theta_{\mathrm{Tx}}^{(\ell,m)}, \phi_{\mathrm{Tx}}^{(\ell,m)}; \beta_{\mathrm{Tx}}, \gamma_{\mathrm{Tx}}) \end{bmatrix} \tag{102}$$

The scattering matrix with the depolarization mechanism is expressed as :

$$\mathrm{S}_{\mathrm{Tx,Rx}}^{(\ell,m,n)} = \begin{bmatrix} \sqrt{\frac{1}{1+\chi_\theta^{(\ell,m,n)}}} \exp\left(j\varepsilon_{\theta\theta}^{(\ell,m,n)}\right) & \sqrt{\frac{\chi_\phi^{(\ell,m,n)}}{1+\chi_\phi^{(\ell,m,n)}}} \exp\left(j\varepsilon_{\theta\phi}^{(\ell,m,n)}\right) \\ \sqrt{\frac{\chi_\theta^{(\ell,m,n)}}{1+\chi_\theta^{(\ell,m,n)}}} \exp\left(j\varepsilon_{\phi\theta}^{(\ell,m,n)}\right) & \sqrt{\frac{1}{1+\chi_\phi^{(\ell,m,n)}}} \exp\left(j\varepsilon_{\phi\phi}^{(\ell,m,n)}\right) \end{bmatrix} \tag{103}$$

where $\varepsilon_{\theta\theta}^{(\ell,m,n)}$, $\varepsilon_{\theta\phi}^{(\ell,m,n)}$, $\varepsilon_{\phi\theta}^{(\ell,m,n)}$ and $\varepsilon_{\phi\phi}^{(\ell,m,n)}$ denote the phase offsets.

9.4 Impact of XPD on system capacity
We examine the impact of the XPD parameter on MIMO system capacity. For the seek of simplicity, we consider the (2×2) MIMO channel generated according to the Kronecker channel modeling. The dual polarized MIMO system is adopted (Fig. 33). Assuming that the CSI is known at the receiver side, the MIMO system capacity can be derived by exploiting the SVD technique. Fig.34 shows plotted curves of the CCDF for various inverse XPDs and the curve of the MIMO channel capacity as a function of the XPD. Simulation results show that

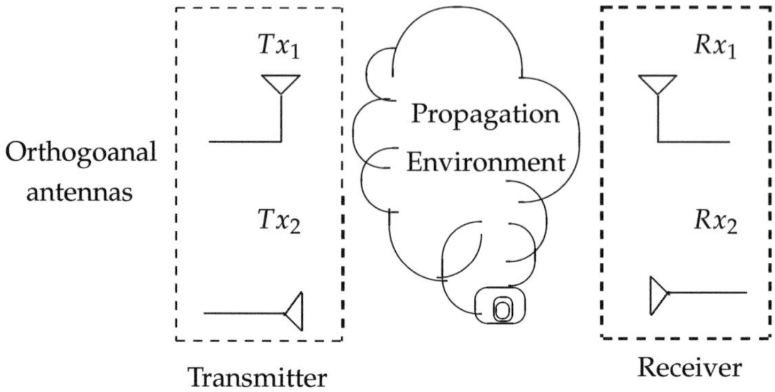

Fig. 33. Dual polarized antennas

XPD affects the performances of the MIMO system in terms of the CCDF and the capacity. The MIMO system capacity is shown to be seriously reduced for high level of the polarization discrimination. Thus, mismatch in polarization results in losses in the MIMO channel capacity.

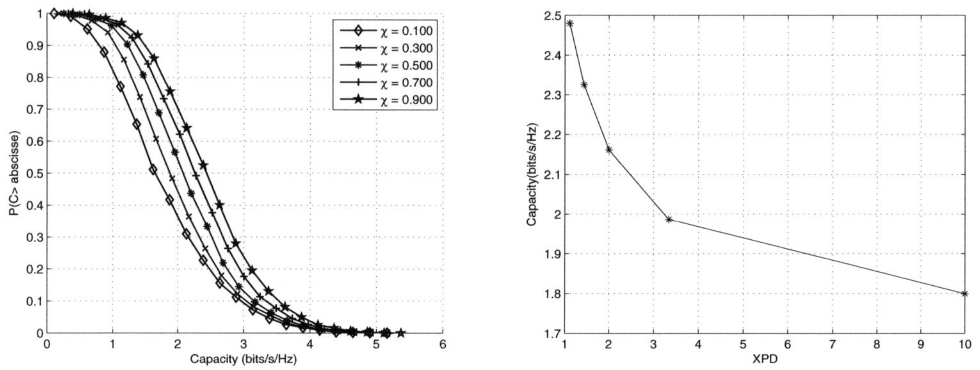

(a) (b)

Fig. 34. Impact of XPD:(a) CCDF for dual polarized MIMO system, (b) Dual polarized MIMO system capacity

9.5 GSCM channel correlation

Correlation coefficients relative to the NLOS channels $h_{qp}^{NLOS}(t,\tau)$ and $h_{\tilde{q}\tilde{p}}^{NLOS}(t,\tau)$ are expressed as :

$$R_{qp,\tilde{q}\tilde{p}}(\xi,\tau-\tau') = E[h_{qp}(t,\tau)h_{\tilde{q}\tilde{p}}^{*}(t-\xi,\tau')] \tag{104}$$

For zero phase offsets and one cluster of scatterers ($L = 1$), if $\chi_\theta = \chi_\phi = \chi$, then equation (102) reduces to :

$$R_{qp,\tilde{q}\tilde{p}}(\xi) = \lim_{M,N\to\infty} \frac{1}{MN} \sum_{m=1}^{M} \sum_{n=1}^{N} a_m^{(p)} b_n^{(q)} a_m^{(\tilde{p})^*} b_n^{(\tilde{q})^*} \exp\left\{ j\left(2\pi(f_{\mathrm{Tx}}^{(m)} + f_{\mathrm{Rx}}^{(n)})\right)\xi \right\} \tag{105}$$

$$\begin{bmatrix} G_{q,\theta}\cdot G_{\tilde{q},\theta} \\ G_{q,\phi}\cdot G_{\tilde{q},\phi} \end{bmatrix}^T \begin{bmatrix} 1 & \chi \\ \chi & 1 \end{bmatrix} \begin{bmatrix} G_{p,\theta}\cdot G_{\tilde{p},\theta} \\ G_{p,\phi}\cdot G_{\tilde{p},\phi} \end{bmatrix} \tag{106}$$

The continuous spatial correlation coefficients are given by (Prayongpun, 2009):

$$R_{qp,\tilde{q}\tilde{p}}(\xi) = \iiiint \exp\left\{j\frac{2\pi(\tilde{p}-p)d_{\text{Tx}}}{\lambda}\sin\theta_{\text{Tx}}\cos\phi_{\text{Tx}} - j2\pi f_{\text{Tx}}\xi\sin\theta_{\text{Tx}}\cos(\phi_{\text{Tx}}-\gamma_{\text{Tx}})\right\}$$
$$\exp\left\{j\frac{2\pi(\tilde{q}-q)d_{\text{Rx}}}{\lambda}\sin\theta_{\text{Rx}}\cos\phi_{\text{Rx}} - j2\pi f_{\text{Rx}}\xi\sin\theta_{\text{Rx}}\cos(\phi_{\text{Rx}}-\gamma_{\text{Rx}})\right\}$$
$$\begin{bmatrix} G_{q,\theta}\cdot G_{\tilde{q},\theta} \\ G_{q,\phi}\cdot G_{\tilde{q},\phi}\end{bmatrix}^T \begin{bmatrix} 1 & \chi \\ \chi & 1\end{bmatrix} \begin{bmatrix} G_{p,\theta}\cdot G_{\tilde{p},\theta} \\ G_{p,\phi}\cdot G_{\tilde{p},\phi}\end{bmatrix} p(\gamma_{\text{Tx}})\, p(\gamma_{\text{Rx}})\, p(\Omega_{\text{Tx}})\, p(\Omega_{\text{Rx}})\, d\gamma_{\text{Tx}} d\gamma_{\text{Rx}} d\Omega_{\text{Tx}} d\Omega_{\text{Rx}} \tag{107}$$

- Ω involves the elevation angle and the azimuth angle and $d\Omega = \sin\theta d\phi d\theta$.
- $p(\Omega_{\text{Tx}}) (= p(\theta_{\text{Tx}}, \phi_{\text{Tx}}))$ and $p(\Omega_{\text{Rx}}) (= p(\theta_{\text{Rx}}, \phi_{\text{Rx}}))$ respectively denote the probability distribution of the scatterers around the transmitter and the probability distribution of scatterers around the receiver.

The spatial correlation coefficients $R_{qp,\tilde{q}\tilde{p}}(\xi)$ is obtained as the product of three terms:

1. Transmit spatial correlation: $R_{\cdot p,\cdot\tilde{p}}(\xi)$
2. Receive spatial correlation: $R_{q\cdot,\tilde{q}\cdot}(\xi)$
3. Polarization correlation: $\mho$

where :

$$R_{\cdot p,\cdot\tilde{p}}(\xi) = \iint \begin{bmatrix} G_{p,\theta}\cdot G_{\tilde{p},\theta} \\ G_{p,\phi}\cdot G_{\tilde{p},\phi}\end{bmatrix} \exp\{j2\pi(\tilde{p}-p)d_{\text{Tx}}\sin\theta_{\text{Tx}}\cos\phi_{\text{Tx}}/\lambda\}\cdot$$
$$\exp\{-j2\pi f_{\text{Tx}}\xi\sin\theta_{\text{Tx}}\cos(\phi_{\text{Tx}}-\gamma_{\text{Tx}})\}\, p(\gamma_{\text{Tx}})\, p(\Omega_{\text{Tx}})\, d\gamma_{\text{Tx}} d\Omega_{\text{Tx}} \tag{108}$$

$$R_{q\cdot,\tilde{q}\cdot}(\xi) = \iint \begin{bmatrix} G_{q,\theta}\cdot G_{\tilde{q},\theta} \\ G_{q,\phi}\cdot G_{\tilde{q},\phi}\end{bmatrix} \exp\{j2\pi(\tilde{q}-q)d_{\text{Rx}}\sin\theta_{\text{Rx}}\cos\phi_{\text{Rx}}/\lambda\}\cdot$$
$$\exp\{-j2\pi f_{\text{Rx}}\xi\sin\theta_{\text{Rx}}\cos(\phi_{\text{Rx}}-\gamma_{\text{Rx}})\}\, p(\gamma_{\text{Rx}})\, p(\Omega_{\text{Rx}})\, d\gamma_{\text{Rx}} d\Omega_{\text{Rx}} \tag{109}$$

$$\mho = \begin{bmatrix} 1 & \chi \\ \chi & 1\end{bmatrix} \tag{110}$$

In literature, azimuth angles may follow uniform, gaussian, von Mises or Laplacien distributions. For the elevation angle, three distribution functions can be used which are uniform , cosinus or gaussian distributions. Depending on the propagation environment, we should note that both von Mises distribution for the azimuth angle and uniform distribution for the elevation angle perform better. Thus, azimuth distribution function is expressed as :

$$p(\phi_{\text{Tx}}) = \frac{\exp\{k_{\text{Tx}}\cos(\phi_{\text{Tx}}-\bar{\phi}_{\text{Tx}})\}}{2\pi I_0(k_{\text{Tx}})} \tag{111}$$

The elevation distribution function is :

$$p(\theta_{\text{Tx}}) = \frac{1}{2\Delta\theta_{\text{Tx}}} \tag{112}$$

where :

- $\phi_{\text{Tx}} \in [-\pi, \pi]$

- $\bar{\phi}_{\text{Tx}}$ is the mean azimuth angle
- $I_0(\cdot)$ is the modified Bessel function of zeroth order
- k_{Tx} is a measure of the azimuth angle dispersion. $1/k_{\text{Tx}}$ is analogous to the variance in the uniform distribution.
- $\Delta\theta_{\text{Tx}}$ is the standard deviation of the Angle spread
- $\theta_{\text{Tx}} \in [\bar{\theta}_{\text{Tx}} - \Delta\theta_{\text{Tx}}, \bar{\theta}_{\text{Tx}} + \Delta\theta_{\text{Tx}}]$ where $\bar{\theta}_{\text{Tx}}$ is the mean elevation angle

In the following, we present simulation results of the spatial correlation as a function of the antenna spacing and variable values of the term $f_{\text{Tx}}\xi$. We adopt the MIMO system with two transmit antennas and two receive antennas and we present the plotted curves for the spatial correlation coefficient relative to the transmit antennas Tx_1 and Tx_2. Plotted curves for the spatial correlation are presented in two cases :

1. Single polarized antennas (Fig. 35(a))
2. Dual polarized antennas (Fig. 35(b))

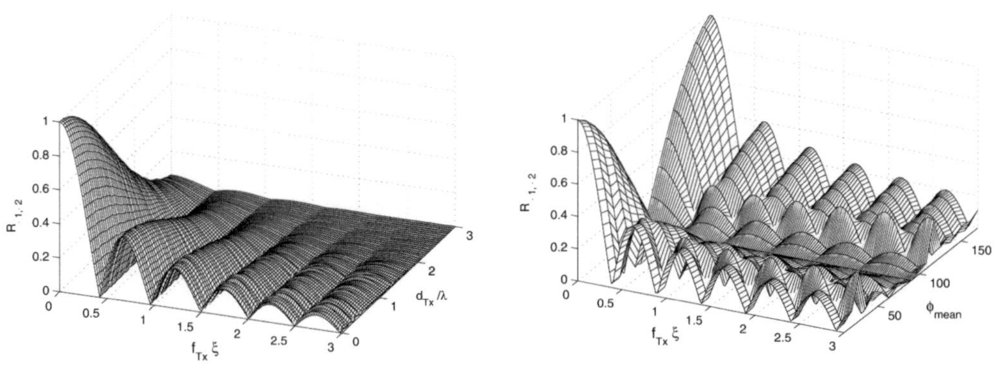

(a) (b)

Fig. 35. (a) Spatial correlation for z dipole antennas ($k_{\text{Tx}} = 50, \bar{\phi}_{\text{Tx}} = 90°, \bar{\theta}_{\text{Tx}} = 90°, \Delta\theta_{\text{Tx}} = 90°$), (b) Spatial correlation for dual polarized antennas (z dipole antenna and x dipole antenna)($k_{\text{Tx}} = 50, \bar{\phi}_{\text{Tx}} = \phi_{mean}, \bar{\theta}_{\text{Tx}} = 90°, \Delta\theta_{\text{Tx}} = 90°$)

According to Fig. 35(a) and Fig. 35(b), it is obvious that single polarized antennas system shows less correlation. When dual polarized antennas are deployed, spatial correlation which is expressed as a function of antenna spacing can be reduced. The impact of dual polarization depends on the mean azimuth angle $\bar{\phi}_{\text{Tx}} = \phi_{mean}$ and spatial correlation is shown to be negligible if ϕ_{mean} approaches 90°. Thus, spatial correlation can be nulled by deploying orthogonal antennas. Finally, we present the impact of dual polarized antennas on MIMO system capacity. The capacity gain brought by the dual polarized system relative to the single polarized system is depicted in Fig. 36. The capacity gain that could be achieved thanks to the use of dual polarized antennas is affected by the mismatch in antenna polarization. Nevertheless, for low XPDs, the dual polarization scheme considerably improves the channel capacity when comparing to the single polarized antennas system. Depolarization phenomena in MIMO system and impact of correlation on MIMO system capacity were rigorously discussed in references (Raoof & Prayongpun, 2007) and (Raoof & Prayongpun, 2005).

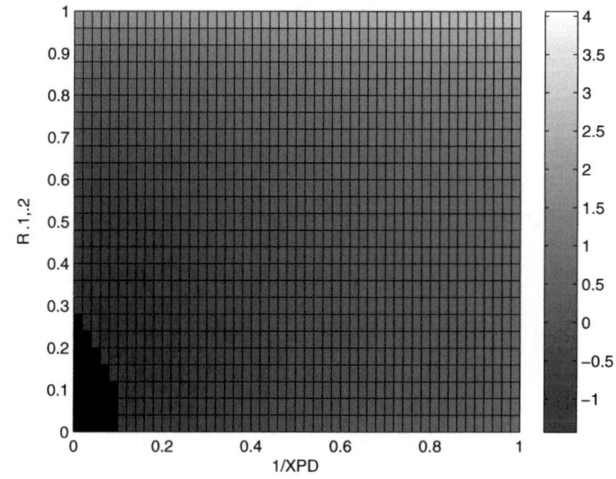

Fig. 36. The capacity gain of (2×2) dual polarized antennas

10. RSS based localization in WSN with collaborative sensors

10.1 Scenario description

We propose the scenario depicted in Fig. 37 where two clusters of wireless sensor nodes are presented. Cluster **1** consists of a source sensor S_{Tx} and a set of K sensors which are randomly distributed in the sensing area.

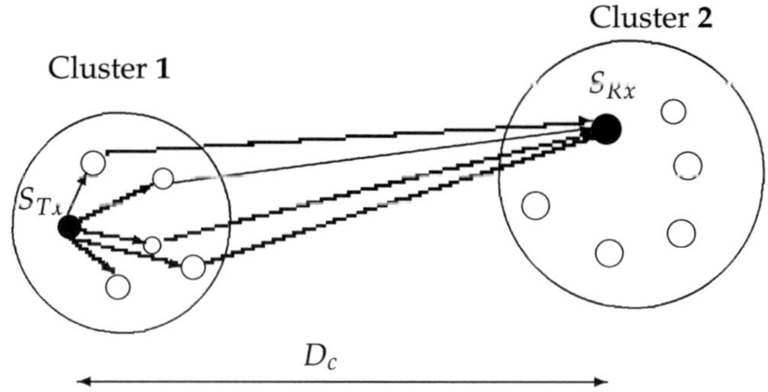

Fig. 37. Scenario description

We assume that sensor nodes are equipped with omnidirectional antennas. The source sensor S_{Tx} sends redundant BPSK modulated data signal to the sensor nodes which are located at Cluster **1**. The main problem with the presented scenario is the localization of the sensor S_{Rx}. We adopt the 2D geometrical model. Sensors are distributed in the (x y) plan and source node is arbitrary placed at the origin of the system coordinate. The geometric spherical coordinates are defined by the triplet (r, θ, ϕ). Here, $\theta = \frac{\pi}{2}$. The simulated layout for 10 sensors is depicted in Fig.38. We propose a beamforming based localization algorithm involving Received Signal Strength (RSS) measurements. We take into assumption that the radius from the origin to S_{Rx} is known. Thus, the detection of the target node relies on determining the steering vector that corresponds to the location of S_{Rx}.

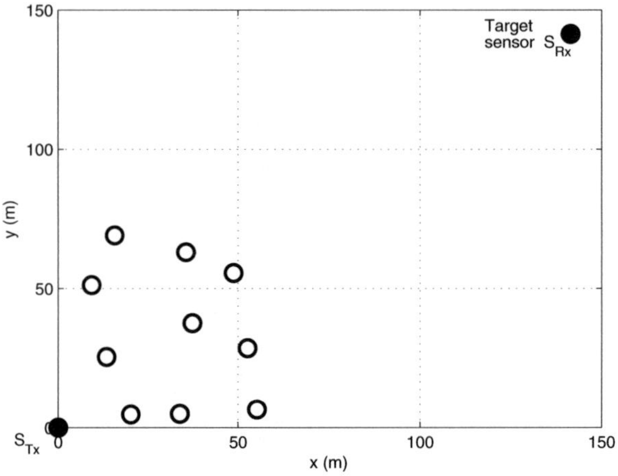

Fig. 38. Network layout (K=10)

10.2 System modeling

The wireless link between sensor S_k, $k = 1, \dots, K$ and the target sensor is modeled by LOS propagation. Each propagation link is characterized by :

1. A Rayleigh distributed attenuation denoted λ_k.
2. A delay τ_k relative to the reference sensor S_{Tx} :

$$\tau_k = d_k \cdot \frac{cos(\alpha_k)}{c} \tag{113}$$

where :

- d_k: Distance between sensor k and the source node
- c : Speed of the light
- α_k: Direction of arrival relative to the target sensor S_{Rx}

3. A dephasing ψ_k:

$$\psi_k = 2\pi f_c \cdot \tau_k \tag{114}$$

f_c denotes the frequency carrier.
Each channel gain is therefore expressed as:

$$\mathbf{h}_k = \lambda_k e^{j\psi_k} \quad ; k = 1, \dots, K \tag{115}$$

A Hadamard Direct Sequence Code Division Multiple Access (DS-CDMA) code is designed for redundant transmitted BPSK data spreading. Walsh Hadamard codes are perfectly orthogonal and employed to avoid interference among users in the propagation link. These codes are exploited for sensor identification and help to mitigate noise effect. The codes for users could be expressed as the columns (or the rows) of the Walsh-Hamadard matrix $\mathbf{C}$. The simplest Hadamard matrix codes are :

$$\mathbf{C}_1 = \mathbf{1}$$

and

$$\mathbf{C}_2 = \begin{bmatrix} 1 & 1 \\ 1 & -1 \end{bmatrix}$$

This can be extended to a generic matrix notation using a recursive technique. In fact, if **C** is a Hadamard matrix of order l (the spreading code length), then becomes the $2l$ order Hadamard matrix.

$$\begin{bmatrix} \mathbf{C} & \mathbf{C} \\ \mathbf{C} & -\mathbf{C} \end{bmatrix}$$

and

$$\mathbf{C}_{2^l} = \mathbf{C}_2 \otimes \mathbf{C}_{2^{l-1}} \tag{116}$$

where $\otimes$ is the Kronecker product and $\mathbf{C}_{2^{l-1}}, 2 \leq l$ is expressed as :

$$\mathbf{C}_{2^l} = \begin{bmatrix} \mathbf{C}_{2^{l-1}} & \mathbf{C}_{2^{l-1}} \\ \mathbf{C}_{2^{l-1}} & -\mathbf{C}_{2^{l-1}} \end{bmatrix} \tag{117}$$

Collaborative sensor nodes are presented in Fig. 39.

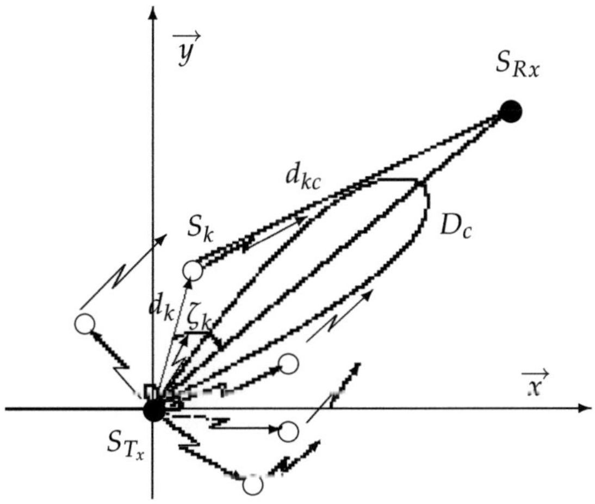

Fig. 39. Collaborative sensor nodes

- d_{kc}: Distance between sensor k and the target node
- D_c: Distance between source node and the target sensor
- ζ_k: Direction of transmitted signal vector relative to sensor k

Given a sensor S_k with cartesian coordinates (x_k, y_k), the associate spherical coordinates are :

- $r_k = \sqrt{x_k^2 + y_k^2}$
- $\phi_k = tan^{-1}(\frac{y_k}{x_k})$

Let $\beta = \frac{2\pi}{\lambda}$, the steering vector is therefore :

$$\mathbf{Vs} = [exp(-j\beta R_1), \ldots, exp(-j\beta R_K)] \tag{118}$$

Distance R_k is the Euclidean distance between the target sensor with spherical coordinates $(D_c, \frac{\pi}{2}, \phi_c)$ and sensor S_k

$$R_k = |D_c - d_k|$$

The Euclidean distance between points, A and B with respective coordinates $A(x_A, y_A)$ $B(x_B, y_B)$ is :

$$d_{euc} = \sqrt{(x_A - x_B)^2 + (y_A - y_B)^2}$$

Euclidean distance R_k can be approximately expressed as (Ahmed & Vorobyov, 2009):

$$R_k = \sqrt{D_c^2 + d_k^2 - 2d_k D_c cos(\phi_c - \phi_k)} \simeq D_c - d_k cos(\phi_c - \phi_k) = D_c - d_k cos(\zeta_k) \quad (119)$$

RSS based target sensor localization

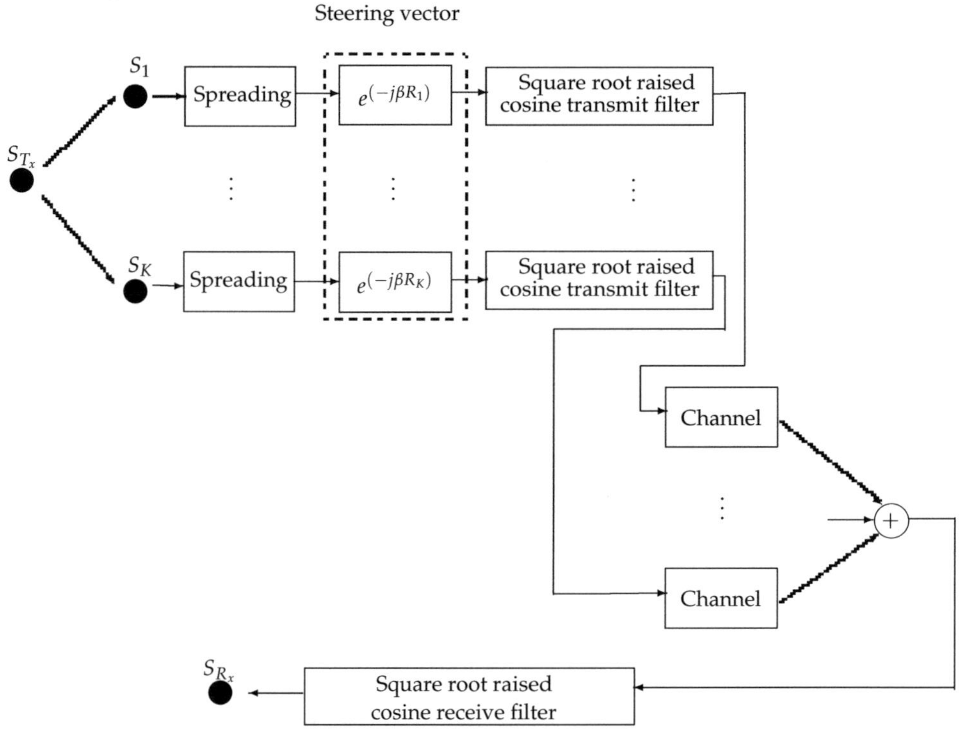

Fig. 40. The transmit location estimation process

The target sensor is said to be detected if the azimuthal coordinate ϕ_c is accurately estimated. Antennas of sensors $S_1, \ldots, S_K$ are steered in variable directions relative to a possible location of the target sensor. In this way, the steering vector becomes a function of the angle ϕ_c. The location estimation process starts by finding the location of the sensors within Cluster **1**. The set of K sensor nodes collaboratively form a common message in the direction of the target sensor node S_{Rx} (See Fig. 39). The target sensor node feedbacks the CDMA code to sensor nodes within Cluster **1**. The received data is then despreaded and broadcasted to sensor node S_{Tx}. The location of the target sensor node is estimated following the RSS measurements performed for all possible locations with respect to a scanning step. The estimated angle of the target sensor corresponds to the maximum measured RSS (measurement of the power present in a received radio signal) value. The received SNR is then evaluated and the accuracy of the proposed algorithm is evaluated in terms of the variation of the statistical parameters as a function of the SNR. The transmit localization process is depicted in Fig. 40. We apply a square root cosine matched filtering. The impulse response of the square root cosine filter is shown in Fig. 41.

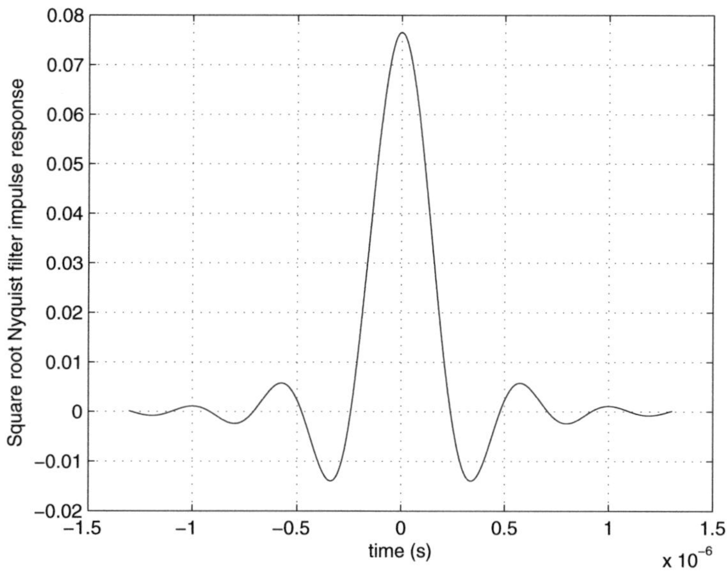

Fig. 41. Square root Nyquist filter

10.3 Simulation results

We present in this section, the simulation results in terms of statistical parameters which are obtained following the Monte Carlo method. The main simulation parameters are :

Simulation parameter	Value
Spreading factor	16
D_c	200 m
Number of angle scanning	50
Oversampling factor (Nyquist filter)	8

Mean square error

The mean square error (MSE) of an estimator is used to quantify the difference between an estimator and the true value of the quantity being estimated.

$$MSE(\phi_c) = E[(\phi_c - \bar{\phi_c})^2] \tag{120}$$

$\bar{\phi_c}$ denotes the sample average of ϕ_c.

Statistical error measurement

Statistical errors are also computed for different number of sensors. This helps to measure the deviation of the estimated location target node from its mean value. Plotted curves for statistical estimator parameters including error estimator and MSE are shown in Fig.43 and Fig.42. These are presented as a function of the received SNR. Simulation results are presented for variable numbers of sensors. The obtained results show that the target sensor location is better estimated if more collaborative sensors are used. When 16 sensors are introduced, the error estimation is almost negligible even for low SNR value. We should note that for far target sensor node, we obtain higher error estimation. Fig. 44 shows the plotted curves of the estimator error as a function of SNR for $D = 200$m and $D = 400$m. 8 collaborative sensors are deployed in the network.

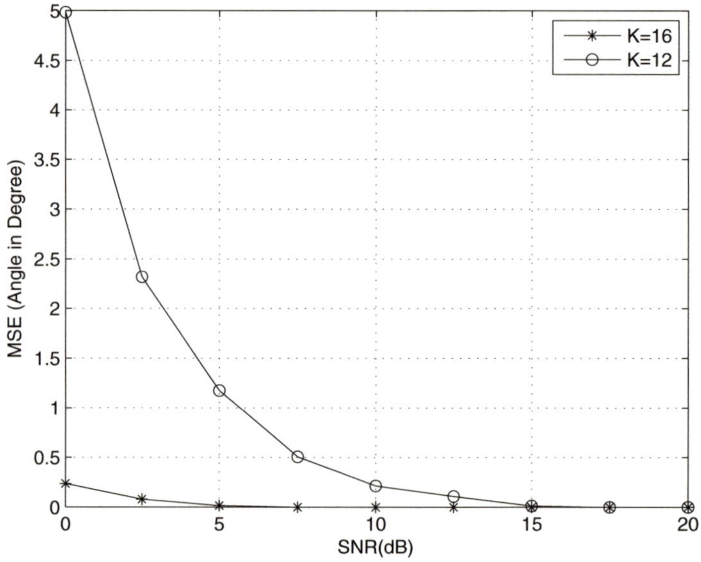

Fig. 42. The mean square error for variable number of sensors

11. Conclusion

Throughout this chapter, we showed that MIMO technology brings significant advances in spectral efficiency by employing several antennas at both ends of the communication system. The geometric channel model was rigorously described and correlated MIMO systems were also considered. Correlation occurs in rich scattering environment mainly when antennas are spaced close of each others. In this chapter, we were interested by MIMO processing techniques such as diversity combining, antenna selection and beamforming. Performances of MIMO systems involving combining techniques were evaluated. We showed that selection schemes can reduce the system hardware complexity at the mobile while keeping sufficient performance in MIMO systems. Beamforming processing was also presented. Some beamforming scenarios were studied and simulated in this chapter. Then, polarization diversity for MIMO systems was investigated and depolarization phenomena was studied and quantified by the cross polarization discrimination XPD. When orthogonal antennas are used, the spatial correlation effect can be reduced or eliminated due to low radiation pattern interference. Thus, colocated multi polarized antennas can be deployed. Capacity gain for MIMO systems with dual polarized antennas compared to single polarized antennas was evaluated. Numerical results show that for high XPD values, single polarized antennas perform better. We have finally presented a communication model for WSN involving two

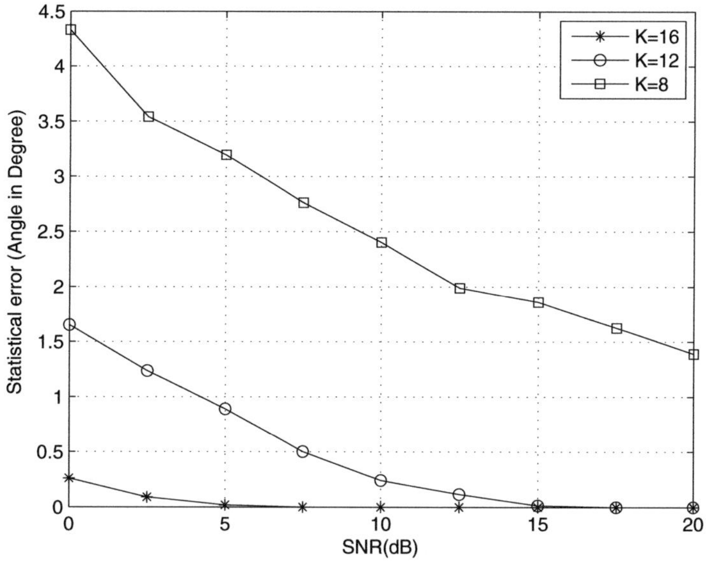

Fig. 43. Statistical error as a function of sensors number K

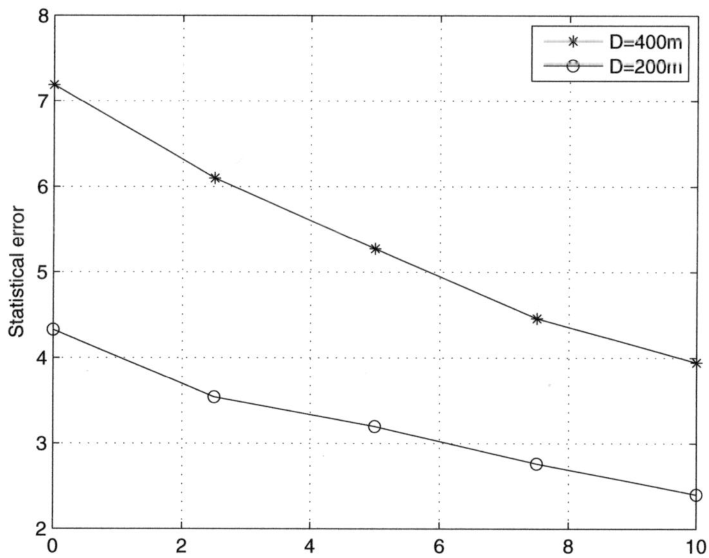

Fig. 44. Impact of distance between the source node and the target node

clusters of sensors at both the transmit side and the receive side. The set of sensors located at the transmit cluster communicating with sensor nodes located at another cluster perform a distributed MIMO system. RSS based localization exploiting beamforming technique was numerically developed. The mean square error and the estimator variance associated with the target sensor location were evaluated as a function of SNR for various parameters of simulation. The obtained results show that accurate target sensor localization based on beamforming processing and RSS measurement could be achieved if the communication model is properly configured. The performed localization technique could be also carried out under more severe channel environment. Thus, convenient antenna polarization strategy for sensors has to be determined in order to fulfill efficient target sensor localization.

12. References

Ahmed, M.F.A. & Vorobyov, S.A. (2009). Collaborative beamforming for wireless sensor networks with gaussian distributed sensor nodes, *IEEE Transactions on Wireless Communications* Vol. 8(No. 2): 638–643.

Alamouti, S.M. (1998). A simple transmit diversity technique for wireless communications, *IEEE Journal on Selected Areas in Communications* Vol. 16(No. 8): 1451–1458.

Ben Zid, M., Raoof, K. & Bouallègue, A. (2011). Sensor nodes selection in Wireless Sensor Networks over a rich scattering environment, *Accepted for presentation at The IEEE 2011 International Conference on Communications, Computing and Control Applications (CCCA'11) and for publication in its proceedings* , Tunisia.

Biglieri, E., Calderbank, R., Constantinides, A., Goldsmith, A., Paulraj, A. & Poor, H. (2007). *MIMO Wireless Communications*, Cambridge University Press, **ISBN: 978–0–521–87328–4.**

Burr, A.G. (2003). Capacity bounds and estimates for the finite scatterers MIMO wireless channel, *IEEE Journal on Selected Areas in Communications* Vol. 21(No. 5): 812–818.

Che-Wei Chang, Kothari, A., Jafri, A., Koutsonikolas, D., Peroulis, D. & Hu, Y.C. (2008). Radiating sensor selection for distributed beamforming in wireless sensor networks, *Proceedings of* Military Communications Conference (MILCOM), San Diego, pp. 1–7.

Constantine, A.B. (2005). *Antenna Theory: Analysis and Design*, John Wiley & Sons, Ltd, **ISBN: 9780471592686**, New Jersey.

Dong, Q.T., Prayongpun, N. & Raoof, K. (2008). Antenna selection for MIMO systems in correlated channels with diversity technique, *Proceedings of* 4th IEEE International Conference on Wireless Communication, Networking and Mobile Computing (WiCOM), Dalian, pp. 1–4.

Foshini, G.F. (1996). Layered space-time architecture for wireless communication in a fading environment when using multi-element antennas, *Bell Labs Technical Journal* pp. 41–59.

Freitas, W.C., Cavalcanti, F.R.P. & Lopes, R.R. (2005). Hybrid transceiver schemes for spatial multiplexing and diversity in MIMO systems, *Journal of Communication and Information Systems* Vol. 20(No. 3): 63–76.

Ghrayeb, A. (2006). A survey on antenna selection for MIMO communication systems, *Information and Communication Technologie (ICTTA)* pp. 2104–2109.

Ibnkahla, M. (2009). *Adaptive signal processing in wireless communications*, CRC Press, **ISBN: 978 1 1200 1602 1.**

Jafarkhani, H. (2005). *Space-Time Coding Theory and Practice*, Cambridge University Press, **ISBN: 978–0–511–11562–2.**

Marzetta, T.L. & Hochwald, B.M. (1999). Capacity of a mobile multiple-antenna communication link in Rayleigh flat fading, *IEEE Transactions on Information Theory* Vol. 45(No. 1): 139–157.

Prayongpun, N. (2009). Modélisation et étude de la capacité du canal pour un système multi-antennes avancé exploitant la diversité de polarisation, *Thesis dissertation-UJF-Grenoble I- Kosai RAOOF (Director)* .
URL: *http://tel.archives-ouvertes.fr/tel-00396666/en/*

Proakis, J.G. (1995). *Digital Communications*, MacGraw-Hill, Inc., **ISBN: 0–07–051726–6**.

Raleigh, G.G. & John, J.M. (1998). Spatio-temporal coding for wireless communication, *IEEE Transactions on Communications* Vol. 46(No. 3): 357–366.

Raoof, K. & Prayongpun, N. (2005). Channel capacity performance for MIMO polarized diversity systems, *Proceedings of* 2005 International Conference on Wireless Communications, Networking and Mobile Computing , pp. 1–4.

Raoof, K. & Prayongpun, N. (2007). Impact of depolarization effects on MIMO polarized wireless configuration, *Proceedings of* International Conference on Wireless Communications, Networking and Mobile Computing (WiCom), Shanghai, pp. 1–4.

Raoof, K. & Zhou, H. (2009). *Advanced MIMO systems*, Scientific Research Publishing, **ISBN:978–1–935068–02–0**.

Saleh, A. & Valenzuela, R. (1987). A statistical model for indoor multipath propagation, *IEEE Journal on Selected Areas in Communications* Vol. 5(No. 2): 128–137.

Sanayei, S. & Nosratinia, A. (2004). Antenna selection in MIMO systems, *IEEE Communications Magazine* Vol. 42(No. 10): 68–73.

Shuguang, C., Goldsmith, A.J. & Bahai, A. (2004). Energy-efficiency of MIMO and cooperative MIMO techniques in sensor networks, *IEEE Journal On Selected Areas In Communications* Vol. 22(No. 6): 1089–1098.

Tarighat, A., Sadek, M. & Sayed, A.H. (2005). A multi user beamforming scheme for downlink MIMO channels based on maximizing signal-to-leakage ratios, *Proceedings of* IEEE International Conference on Acoustics, Speech, and Signal Processing., pp. iii/1129–- iii/1132.

Tarokh, V., Jafarkhani, H. & Calderbank, A.R. (1999). Space-time block codes from orthogonal designs, *IEEE Transactions on Information Theory* Vol. 45(No.5): 1456–1467.

Tarokh, V., Seshadri, N. & Calderbank, A.R. (1998). Space-time codes for high data rate wireless communication: Performance criterion and code construction, *IEEE Transactions on Information Theory* Vol. 44(No. 2): 744–765.

Telatar, I. E. (1995). Capacity of multi-antenna gaussian channels, *European Transactions on Telecommunications* Vol. 10(No. 6): 585–595.

Tran, T.A. & Sesay, A.B. (2003). Orthogonal space-time block codes: performance analysis and comparisons, *Proceedings of* Information, Communications and Signal Processing and the Fourth Pacific Rim Conference on Multimedia, pp. 1497–1501.

Vucetic, B. & Yuan, J. (2003). *Space-Time Coding*, John Wiley & Sons, Ltd, **ISBN: 0–470–84757–3**.

Wang, C., Hong, X., Ge, X., Cheng, X., Zhang, G. & Thompson, J. (2010). Cooperative MIMO channel models: A survey, *IEEE Communications Magazine* Vol. 48(No. 2): 80–87.

Winters, J. (1987). On the capacity of radio communication systems with diversity in a Rayleigh fading environment, *IEEE Journal on Selected Areas in Communications* Vol. 5(No. 5): 871–878.

Yu, W., Cioffi, J. (2001). On constant power water-filling, *Proceedings of* IEEE International Conference on Communications (ICC), Helsinki, pp. 1665–1669.

Yu, C., Ding, Z. & Chiueh, T. (2004). Design and simulation of a MIMO OFDM baseband transceiver for high throughput wireless LAN, *Proceedings of* The 2004 IEEE Asia-Pacific Conference on Circuits and Systems, Taiwan, pp. 205 – 208.

Zheng, L. & Tse, D.N.C. (2003). Diversity and multiplexing: a fundamental tradeoff in multiple-antenna channels, *IEEE Transactions on Information Theory* Vol. 49(No. 5): 1073–1096.

Zhou, H. & Okamoto, K. (2004). Comparison of code combining and MRC diversity reception in mobile communications, *Proceedings of* 2004 IEEE Wireless Communications and Networking Conference (WCNC), pp. 943–947.

2

Geometrical Detection Algorithm for MIMO Systems

Z. Y. Shao, S. W. Cheung and T. I. Yuk
*Department of Electrical and Electronic Engineering, The University of Hong Kong
Hong Kong SAR,
China*

1. Introduction

The channel capacity and error-rate performance of MIMO systems could be improved by increasing the number of transmit antennas and receive antennas and the size of constellation used for modulation (Foschini and Gans, 1998). A main bottleneck that restricts the practical application of MIMO system is the unsatisfactory performance of the decoding algorithms, due to either high computational complexity required or poor symbol error-rate (SER) performance. Maximum-likelihood (ML) decoding which employs an exhaustive search strategy under the minimum Euclidean-distance principle can exploit all the available diversity and provide the optimum SER performance. However, its complexity increases exponentially with the number of antennas and the size of constellation used. Thus, for many cases, it is impractical to implement. Several sub-optimum decoding algorithms such as equalization-based zero-forcing (ZF) and minimum-mean-square-error (MMSE) detectors and nulling-and-cancelling detectors (NC) have been proposed for MIMO systems (Paulraj, Nabar and Gore, 2003). Although their computation complexities are dramatically less, these decoding algorithms have severe degradations in SER performances. Sphere decoding (SD) (Viterbo and Boutros, 1999) is another search-based algorithm. Unlike the exhaustive search engaged in ML decoding, SD confines the searching zone to be inside some hyper sphere constructed in the space spanned by the received lattice points. It can offer optimum SER performance with reasonable complexity. Several searching strategies such as Fincke-Pohst (Fincke and Pohst, 1985) and Schnorr-Euchner (Schnorr and Euchner, 1994) have been developed to further improve the searching efficiency in SD.
Since the minimum Euclidean-distance principle could result in an optimum SER performance, the purpose of this chapter is to introduce another perspective of reconsidering this principle from the transmit lattice space. In the space spanned by the transmit lattice points, the Euclidean distance in ML decoding is found to be related to a series of concentric hyper ellipsoids. Searching the lattice point with the minimum Euclidean distance from the received signal vector is equivalent to searching the lattice point that is passed through by the smallest hyper ellipsoid. Decoding algorithms following this perspective are often called geometrical detection (Artes, Seethaler and Hlawatsch, 2003). In this Chapter, the geometrical analysis of signal decoding for MIMO channels is presented. Then, the ellipsoid searching decoding algorithm (Shao, Cheung and Yuk, 2009) is described. It is an add-on detection algorithm to standard suboptimal detection schemes

and has better error-rate performance and higher diversity gains than the standard suboptimal detection schemes.

2. Geometrical analysis of signal decoding for MIMO channels

Consider an uncoded MIMO system with M_T transmit antennas and M_R receive antennas over a fading channel. The received signal matrix is given by:

$$\hat{\mathbf{r}} = \hat{\mathbf{H}}\hat{\mathbf{x}} + \hat{\mathbf{n}} \tag{1}$$

where $\hat{\mathbf{r}} \in \mathbb{C}^{M_R}$ is the M_R-dimensional received signal vector, $\hat{\mathbf{x}} \in \mathbb{C}^{M_T}$ is the M_T-dimensional transmitted signal vector, $\hat{\mathbf{H}} \in \mathbb{C}^{M_R \times M_T}$ is the channel matrix and is assumed to be known at the receiver, and $\hat{\mathbf{n}} \in \mathbb{C}^{M_R}$ is an independently and identically distributed (i.i.d.) zero-mean Gaussian noise vector with elements having a fixed variance. Equation (1) represents a complex transmission, and it can be transformed into a real matrix equation:

$$\begin{bmatrix} \mathbf{Re}(\mathbf{r}) \\ \mathbf{Im}(\mathbf{r}) \end{bmatrix} = \begin{bmatrix} \mathbf{Re}(\mathbf{H}) & \mathbf{-Im}(\mathbf{H}) \\ \mathbf{Im}(\mathbf{H}) & \mathbf{Re}(\mathbf{H}) \end{bmatrix} \begin{bmatrix} \mathbf{Re}(\mathbf{x}) \\ \mathbf{Im}(\mathbf{x}) \end{bmatrix} + \begin{bmatrix} \mathbf{Re}(\mathbf{n}) \\ \mathbf{Im}(\mathbf{n}) \end{bmatrix} \tag{2}$$
$$\mathbf{r} = \mathbf{H}\mathbf{x} + \mathbf{n}$$

where $\mathbf{r} \in \mathbb{R}^{N_R}$, $\mathbf{x} \in \mathbb{R}^{N_T}$, $\mathbf{H} \in \mathbb{R}^{N_R \times N_T}$ and $\mathbf{n} \in \mathbb{R}^{N_R}$.
In ML decoding, the optimal solution is given by:

$$\mathbf{x}_{ML} = \underset{\mathbf{s} \in \Omega}{\arg\min} \left\| \mathbf{r} - \mathbf{Hs} \right\|^2 \tag{3}$$

where Ω is the set of all the possible transmitted signal vectors, and the term $\left\| \mathbf{r} - \mathbf{Hs} \right\|^2$ is known as the Euclidean distance between the received vector and the transmitted vector distorted by the channel matrix.
The Euclidean distance in ML decoding can be rewritten as:

$$f(\mathbf{s}) = \left\| \mathbf{r} - \mathbf{Hs} \right\|^2 = (\mathbf{s} - \mathbf{x}_c)^{\mathbf{H}} \mathbf{M}^{-1} (\mathbf{s} - \mathbf{x}_c) \tag{4}$$

where $\mathbf{M} = (\mathbf{H}^{\mathbf{T}}\mathbf{H})^{-1}$ and $\mathbf{x}_c$ is the result of zero-forcing (ZF) equalization (Wolniansky, Foschini, Golden and Valenzuela, 1998) and can be written as:

$$\mathbf{x}_c = (\mathbf{H}^T\mathbf{H})^{-1} \mathbf{H}^T \mathbf{r} = \mathbf{x} + (\mathbf{H}^T\mathbf{H})^{-1} \mathbf{H}^T \mathbf{n} = \mathbf{x} + \tilde{\mathbf{n}} \tag{5}$$

where $(\mathbf{H}^T\mathbf{H})^{-1} \mathbf{H}^T \mathbf{n} = \tilde{\mathbf{n}}$. Substituting (4) into (3) yields:

$$\mathbf{x}_{ML} = \underset{\mathbf{s} \in \Omega}{\arg\min} \left\{ (\mathbf{s} - \mathbf{x}_c)^T \mathbf{H}^T \mathbf{H} (\mathbf{s} - \mathbf{x}_c) \right\} \tag{6}$$

It can be seen from (5) and (6) that, in the absence of noise or equivalently the transformed Gaussian noise term $\tilde{\mathbf{n}} = (\mathbf{H}^T\mathbf{H})^{-1} \mathbf{H}^T \mathbf{n}$, both ZF detector and ML decoding will result in the same correct solution. The reason why ML decoding can offer much better SER performance

than ZF lies on the fact that the transformed Gaussian noise has been minimized by the exhaustive search used in ML decoding, but the ZF results are directly distorted by the transformed Gaussian noise $\tilde{\mathbf{n}}$.
Using eigenvalue decomposition, the matrix **M** in (4) can be decomposed as:

$$\mathbf{M} = \left(\mathbf{H}^{\mathrm{T}}\mathbf{H}\right)^{-1} = \mathbf{V}\Lambda\mathbf{V}^{\mathrm{T}} \tag{7}$$

where $\Lambda = diag\left(\lambda_1, \lambda_2, ..., \lambda_{M_T}\right) \in \mathbb{R}^{N_T \times N_T}$ are the eigenvalues of **M** arranged in descending order, and $\mathbf{V} = \left[\mathbf{V}_1, \mathbf{V}_2, \cdots, \mathbf{V}_{M_T}\right] \in \mathbb{R}^{N_T \times N_T}$ is the corresponding eigenvector matrix (Samuel and Fitz, 2007).
In ML decoding, the Euclidean distance function $f(\mathbf{s})$ given in (4) geometrically represents an elliptic paraboloid (Horn and Johnson, 1985) in an $N_T + 1$ dimensional space with its axis perpendicular to an N_T dimensional subspace spanned by the symbol vectors in Ω. $\mathbf{x}_c$ is the global minimum point of the elliptic paraboloid and is located on the subspace spanned by the symbol vectors in Ω as shown in Fig. 1. It can be seen from (4) that the function $f(\mathbf{s})$ reaches its minimum value at $\mathbf{x}_c$,i.e., $f(\mathbf{s})_{\min} = f(\mathbf{x}_c) = 0$. The horizontal cross-section of the elliptic paraboloid (4) is an N_T dimensional hyper ellipsoid given by:

$$f(\mathbf{s}) = a^2 \tag{8}$$

where a^2 represents the height of the cross section above the N_T dimensional space as shown in Fig. 1. The length and direction of i-th semiaxis of the hyper ellipsoid are given as $a\sqrt{\lambda_i}$ and $\mathbf{V}_i$, respectively. With different values of a^2, a series of concentric hyper

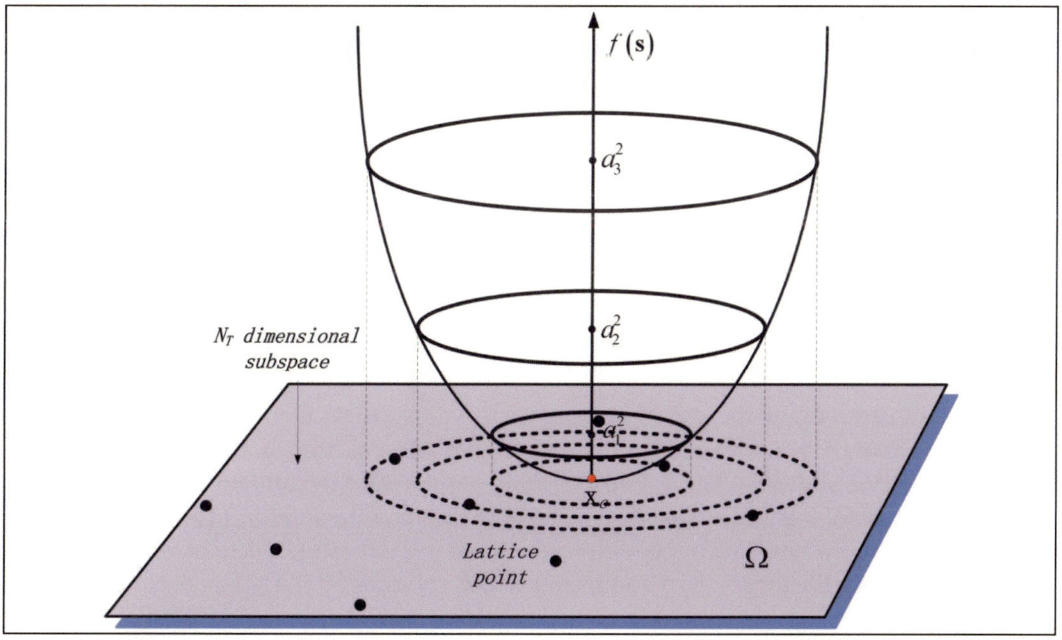

Fig. 1. Elliptic paraboloid with axis perpendicular to a subspace spanned by lattice points.

ellipsoids are obtained and can be projected onto the subspace spanned by the vectors as shown by the dash lines in Fig. 1. Thus, searching the lattice point with minimum Euclidean distance is equivalent to searching the lattice point that is passed through by the smallest hyper ellipsoid.

3. Ellipsoid-searching decoding algorithm

From section **2**, we know that $f(\mathbf{s}) = a^2$ represents a hyper ellipsoid centered at point $\mathbf{x}_c$ with the length and direction of its *i*-th semiaxis given as $a\sqrt{\lambda_i}$ and $\mathbf{v}_i$, respectively. By choosing different values of a, a group of similar hyper ellipsoids can be obtained. Thus, the solution of ML decoding must be located on a hyper ellipsoid which has the minimum surface area among these similar hyper ellipsoids.

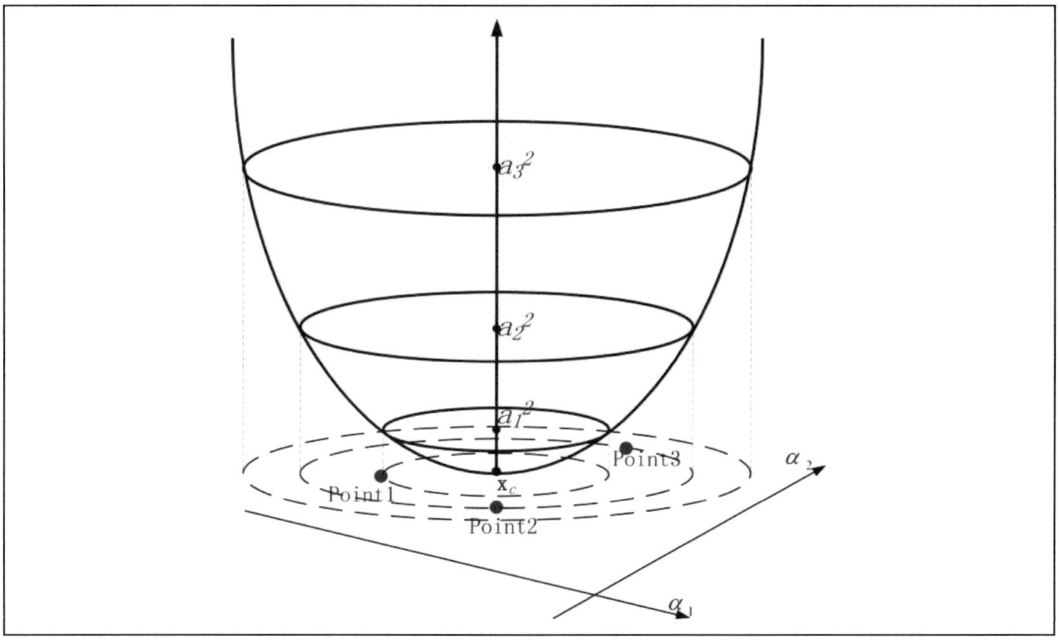

Fig. 2. Elliptic paraboloid in 3-dimensional space.

Fig. 2 shows a two dimensional lattice point space ($\alpha_1 - \alpha_2$ plane) with three lattice points Point 1, Point 2, and Point 3 as shown in the figure. With different a^2, a group of similar hyper ellipsoids can be obtained, and their projection onto the $\alpha_1 - \alpha_2$ plane are ellipses which are all centered at the point $\mathbf{x}_c$. For each lattice point, there exists an ellipse that passes through it. The corresponding ellipse of the ML solution is the one that has the minimum area. As shown in Fig. 2, Point 1 is taken to be the ML solution while Point 2 and Point 3 are not, since it is the inner-most ellipse and thus has the minimum area.

However, finding the smallest hyper ellipsoid containing the solution signal vector is not an easy task. If we use the largest hyper ellipsoid which contains all the signal vectors, then the complexity will be the same as ML decoding. Here we propose an ellipsoid-searching decoding algorithm (ESA) that uses a small hyper ellipsoid containing the solution symbol vector to start the search and then identify all the symbol vectors inside. The ESA consists of the following 3 steps:

3.1 Start with zero-forcing points

It is well known that zero-forcing (ZF) decoding is one form of linear equalization algorithm. Although it cannot offer very high performance like ML decoding, its solution however usually lies in the neighborhood of the transmit signal point. Thus we can consider choosing the hyper ellipsoid that goes through the ZF solution to start the search. First, the ZF equalized $\mathbf{x}_{zf}$ is solved. Then its corresponding a_{zf}^2 is computed. The starting hyper ellipsoid is obtained as:

$$f\left(\mathbf{x}_{zf}\right) = a_{zf}^2 \tag{9}$$

3.2 Determine a circumscribed hyper rectangle

After determining the hyper ellipsoid, the next key task is to identify whether there are any lattice points located inside this hyper ellipsoid. The axes of the N_T-dimensional rectangular coordinate system for the lattice point space are denoted as α_i- axes. Since the directions of the hyper ellipsoid's semiaxes are not in parallel with the axes of the coordinate system of the lattice point space, it is rather complicated to directly use the surface equation (9) of the hyper ellipsoid. Here we propose to use a circumscribed hyper rectangle as follows.

We set up a new N_T-dimensional rectangular coordinate system with α_i'- axes ($i = 1,2,3,\ldots,N_T$) which coincide with the i-th semiaxis of the hyper ellipsoid and has the origin coincides with the global minimum point $\mathbf{x_c}$. We use the superscript prime to denote the variables in the new coordinate system. The coordinates of the 2^{N_T} apexes of the circumscribed hyper rectangle in this new coordinate system are given by:

$$k_p' = \left[x_{p1}', x_{p2}', \ldots x_{pN_T}'\right] \tag{10}$$

where $p = 1,2,3,\ldots 2^{N_T}$, $x_{pj}' = \pm a_{zf}\sqrt{\lambda_j}$, and a_{zf} is related to the hyper ellipsoid given by (9).

It can be easily shown that, by using coordinate transformation, the coordinates of the 2^{N_T} apexes in the original lattice point space are:

$$\mathbf{k}_p^T = \mathbf{V} \cdot \left(\mathbf{k}_p'\right)^T + \mathbf{x}_c \tag{11}$$

where $\mathbf{V}$ is the eigenvector matrix in (7), and it serves as the transformation matrix:

$$\mathbf{V}_T = \left[\mathbf{V}_1, \mathbf{V}_2, \cdots, \mathbf{V}_{N_T}\right] = \begin{bmatrix} v_{11} & v_{21} & v_{31} & \cdots & v_{N_T 1} \\ v_{12} & v_{22} & v_{32} & \cdots & v_{N_T 2} \\ v_{13} & v_{23} & v_{33} & \cdots & v_{N_T 3} \\ \vdots & \vdots & \vdots & \ddots & \vdots \\ v_{1N_T} & v_{2N_T} & v_{3N_T} & \cdots & v_{N_T N_T} \end{bmatrix} \tag{12}$$

Thus the value of the i-th component of $\mathbf{k}_p$ can be obtained as:

$$x_{pi} = \sum_{q=1}^{N_T} \left(v_{qi} x'_{pq} \right) + x_{ci} \tag{13}$$

where x_{ci} is the i-th component of $\mathbf{x_c}$. Since $x'_{pq} = a_{zf}\sqrt{\lambda_q}$, the maximum and minimum boundaries in the α'_i - axes for each component in $\mathbf{k}_p$ can be expressed as:

$$x_{i\_\max} = x_{ci} + \sum_{q=1}^{N_T} \left| v_{qi} a_{zf} \sqrt{\lambda_q} \right| \tag{14.1}$$

$$x_{i\_\min} = x_{ci} - \sum_{q=1}^{N_T} \left| v_{qi} a_{zf} \sqrt{\lambda_q} \right| \tag{14.2}$$

Since the circumscribed hyper rectangle encloses the hyper ellipsoid, so any lattice point $\mathbf{s} = \begin{bmatrix} s_1 & s_2 & \dots & s_{N_T} \end{bmatrix}$ inside the hyper ellipsoid satisfies:

$$x_{i\_\min} < s_i < x_{i\_\max} \qquad i = 1,2,3,\dots,N_T \tag{15}$$

It should be noted that this is not a sufficient condition for identifying the lattice points lying inside the hyper ellipsoid.

From (15), we can obtain the possible value set $\xi_i = \{\varepsilon_{i1}, \varepsilon_{i2}, \varepsilon_{i3}, \cdots\}$ for the i-th element of the lattice points located inside the hyper ellipsoid. So the search set becomes a larger hyper rectangle that encloses the circumscribed hyper rectangle. For PAM and QAM, the elements of ξ_j are the odd numbers between $x_{j\_\max}$ and $x_{j\_\min}$, and it can be easily shown that the number of elements is:

$$Num_i = \left\lfloor \sum_{q=1}^{N_T} \left| v_{qi} a_{zf} \sqrt{\lambda_q} \right| \right\rfloor \tag{16}$$

3.3 Narrow the search set into ellipsoid

As mentioned before, the search set becomes a larger hyper rectangle and the number of lattice points inside is $\prod_{i=1, i\neq l}^{N_T} Num_i$. If there is any Num_i equals zero, then it means that there is no lattice point located inside the hyper ellipsoid. The searching process will terminate and the zero forcing point chosen before is considered as the solution.

Otherwise, assuming the possible value set ξ_ω has the largest number of elements among all the possible value sets, we form the combinations from the other $N_T - 1$ possible value sets, and then substitute each of these combinations into (9), to determine the lattice point elements of the possible value set ξ_ω that are located inside the hyper ellipsoid. In doing so, the number of combinations that need to be considered is smaller and hence lesser computation complexity. Denoting the k-th combination by:

$$\mathbf{Com}^k = \left[\varepsilon_{1,k}, \varepsilon_{2,k}, \cdots \varepsilon_{\omega-1,k}, \varepsilon_{\omega+1,k} \cdots, \varepsilon_{N_T,k}\right] \tag{17}$$

$$k = 1, 2, ..., \prod_{j=1, j\neq\omega}^{N_T} Num_j$$

where $\varepsilon_{j,k}$ represents an arbitrary element of the set ξ_j.

Geometrically, the $\mathbf{Com}^k$ is a line pierced through the hyper ellipsoid. The intersection of the line and the hyper ellipsoid consists of two points, known as $E_{\max,k}$ and $E_{\min,k}$ along the ω-th axis. Hence, the corresponding possible value set $\zeta_{\omega,k} = \{\varsigma_{\omega,1,k}, \varsigma_{\omega,2,k}, ...\}$ for the ω-th element of the lattice points are the odd numbers between $E_{\max,k}$ and $E_{\min,k}$. Thus, any lattice point that is located inside the hyper ellipsoid can be expressed as:

$$\mathbf{x}_{d,k} = \left[\varepsilon_{1,k}, \varepsilon_{2,k}, \cdots \varepsilon_{\omega-1,k}, \varsigma_{\omega,d,k}, \varepsilon_{\omega+1,k} \cdots, \varepsilon_{N_T,k}\right]^T \tag{18}$$

$$d = 1, 2, ..., n_k$$

where n_k is the number of the elements of $\zeta_{\omega,k}$ for $\mathbf{Com}^k$.

Finally, we calculate the corresponding a^2 of each lattice point $\mathbf{x}_{d,k}$ by (8). The point with the minimum a^2 is the solution.

3.4 Examples

a. 2-D lattice space

For a 2×2 8-PAM MIMO system, the lattice set is a 2-dimensional space as shown in Fig. 3, where it is assumed that the ellipse and its circumscribed rectangle have been determined using our proposed method as described previously. The semiaxes of the ellipse are in parallel with vectors $\mathbf{v}_1$ and $\mathbf{v}_2$ with lengths $a_{zf}\sqrt{\lambda_1}$ and $a_{zf}\sqrt{\lambda_2}$, respectively. The global minimum point $\mathbf{x}_c$ is marked by a triangle on the figure. The coordinates of the four apexes, A, B, C and D, in the new coordinate system are given by $A = \left(-a_{zf}\sqrt{\lambda_1}, -a_{zf}\sqrt{\lambda_2}\right)$, $B = \left(-a_{zf}\sqrt{\lambda_1}, +a_{zf}\sqrt{\lambda_2}\right)$, $C = \left(a_{zf}\sqrt{\lambda_1}, -a_{zf}\sqrt{\lambda_2}\right)$, and $D = \left(a_{zf}\sqrt{\lambda_1}, +a_{zf}\sqrt{\lambda_2}\right)$, respectively. Substituting these vectors into (13) yields the corresponding coordinates in the lattice point space. From (14), the $\mathbf{x}_1$ coordinates of points A and D are chosen as $x_{1\_\min}$ and $x_{1\_\max}$, respectively, and the $\mathbf{x}_2$ coordinates of points B and C are chosen as $x_{2\_\min}$ and $x_{2\_\max}$, respectively. Using (15), we can obtain a possible set of values along each axis, i.e., two values {1, 3} along the $\mathbf{x}_1$-axis and one value {1} along the $\mathbf{x}_2$-axis. Since the number of values along the $\mathbf{x}_1$-axis is larger than that along the $\mathbf{x}_2$-axis, we substitute $\varepsilon_{2,1} = 1$ into the hyper ellipsoid equation (9). As shown in Fig. 3, the possible value along the $\mathbf{x}_1$-axis is $\varsigma_{1,1,1} = 3$, so the point $\mathbf{x}_{1,1} = [3 \quad 1]^T$ is obtained. Since it is the only point located inside the ellipse, it would be the final solution.

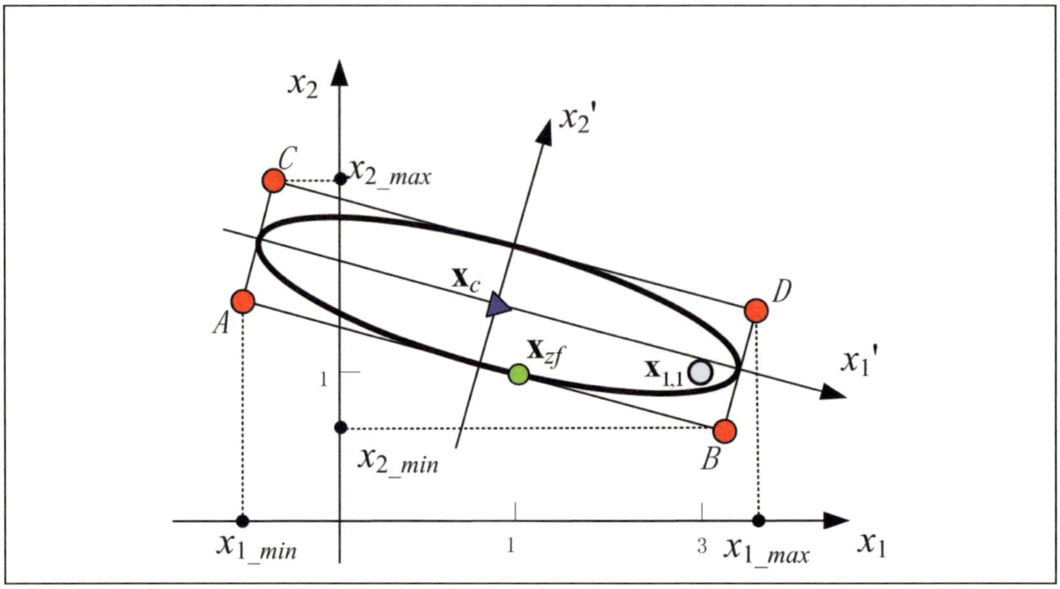

Fig. 3. 2-D lattice space example

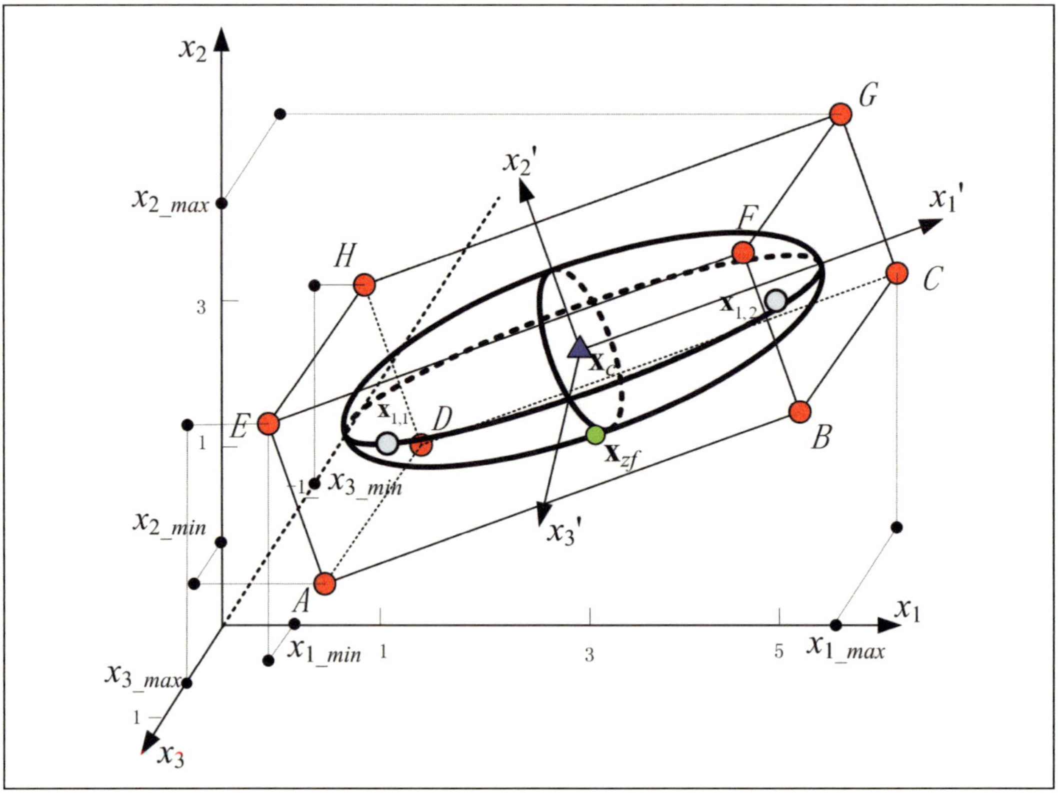

Fig. 4. 3-D lattice space example

b. 3-D lattice space

Here, we continue to consider the case of 3-dimensional lattice space, namely 3×3 8-PAM. Fig. 4 shows a 3-dimensional ellipsoid with its circumscribed rectangle which has been set up by the method introduced in section 3.2. $\mathbf{x}_c$ is the center of the ellipsoid, whose semiaxes are aligned along vectors $\mathbf{v}_1$, $\mathbf{v}_2$, $\mathbf{v}_3$, with their lengths being $a_{zf}\sqrt{\lambda_1}$, $a_{zf}\sqrt{\lambda_2}$ and $a_{zf}\sqrt{\lambda_3}$, respectively. By substituting the coordinates of the eight points A to H to (13) and (14), the boundary points x_{1_min} and x_{1_max}, x_{2_min} and x_{2_max}, x_{3_min} and x_{3_max}, which are all marked as dots, are obtained. The possible set of values along $\mathbf{x}_1$-axis is {1, 3, 5}, and the possible set of values along the $\mathbf{x}_2$-axis is {1, 3}. Along $\mathbf{x}_3$-axis, the possible set of value is {-1}. Since the number of possible values along the $\mathbf{x}_1$-axis is the largest compared to those along the other axes, we substitute $\mathbf{Com}^1 = \left[\varepsilon_{2,1}, \varepsilon_{3,1}\right] = \left[1, -1\right]$ and $\mathbf{Com}^2 = \left[\varepsilon_{2,2}, \varepsilon_{3,2}\right] = \left[3, -1\right]$ into (9) to determine $E_{\max,k}$ and $E_{\min,k}$ along the $\mathbf{x}_1$-axis. As shown in Fig. 4, the possible value set $\zeta_{1,1}$ along the $\mathbf{x}_1$-axis is {1} for $\mathbf{Com}^1$ and $\zeta_{1,2}$ is {5} for $\mathbf{Com}^2$, so the point $\mathbf{x}_{1,1} = \begin{bmatrix}1 & 1 & -1\end{bmatrix}^T$ and the point $\mathbf{x}_{1,2} = \begin{bmatrix}5 & 3 & -1\end{bmatrix}^T$ are obtained. By calculating their corresponding a^2, it can be concluded that the point $\mathbf{x}_{1,2}$ that has a smaller a^2 is taken as the final solution.

3.5 Results and conclusion

The ESA algorithm for MIMO systems has been briefly introduced. It contains three main steps: Firstly, determine the hyper ellipsoid. Secondly, find out the probable value sets for each component of the lattice point that is located in the hyper ellipsoid. Finally, search for the ML solution. In the first step, either ZF detector or MMSE detector can be selected for determining the hyper ellipsoid. In the second step, we firstly determine a loose boundary for each component of the lattice points that may be located in the hyper ellipsoid. Then, by further shrink the value set of the N_T-th component, all the redundant points can be discarded and the lattice points inside the hyper ellipsoid are exactly detected.
Since the ESA algorithm uses the same criteria (3) of ML to make decision, it can thus achieve the same performance as ML decoding. However, the ML decoding searches the entire lattice space for solution while the ESA algorithm only searches a smaller subset, thus ESA is more computation efficient. Simulation results of various algorithms on the error rate performance are shown in Fig. 5 and Fig. 6 for comparison. In the simulations, we used 4-QAM, 16-QAM , 64-QAM in Rayleigh flat fading Channels with i.i.d. complex zero-mean Guassian noise. Fig. 5 illustrates the SER performance of ESA compared with ML decoding, ZF detector and MMSE detector using 4-QAM. Fig. 6 shows the SER performance of ESA compared with ML decoding ZF detector and MMSE detector using 16-QAM and 64-QAM. The performances of ESA can achieve the same performance as the ML decoding and are much better than the sub-optimum detectors.

4. Conclusion

In this chapter, the geometrical analysis of signal decoding for MIMO channels is presented. The ellipsoid searching decoding algorithm using geometrical approach is introduced. It is an add-on to standard suboptimal detection schemes and has better SER performance and higher diversity gains compared to the standard suboptimal detection schemes. It is able to provide the same optimum SER performance as in the ML decoding but with less complexity as only a subset of the lattice points are examined.

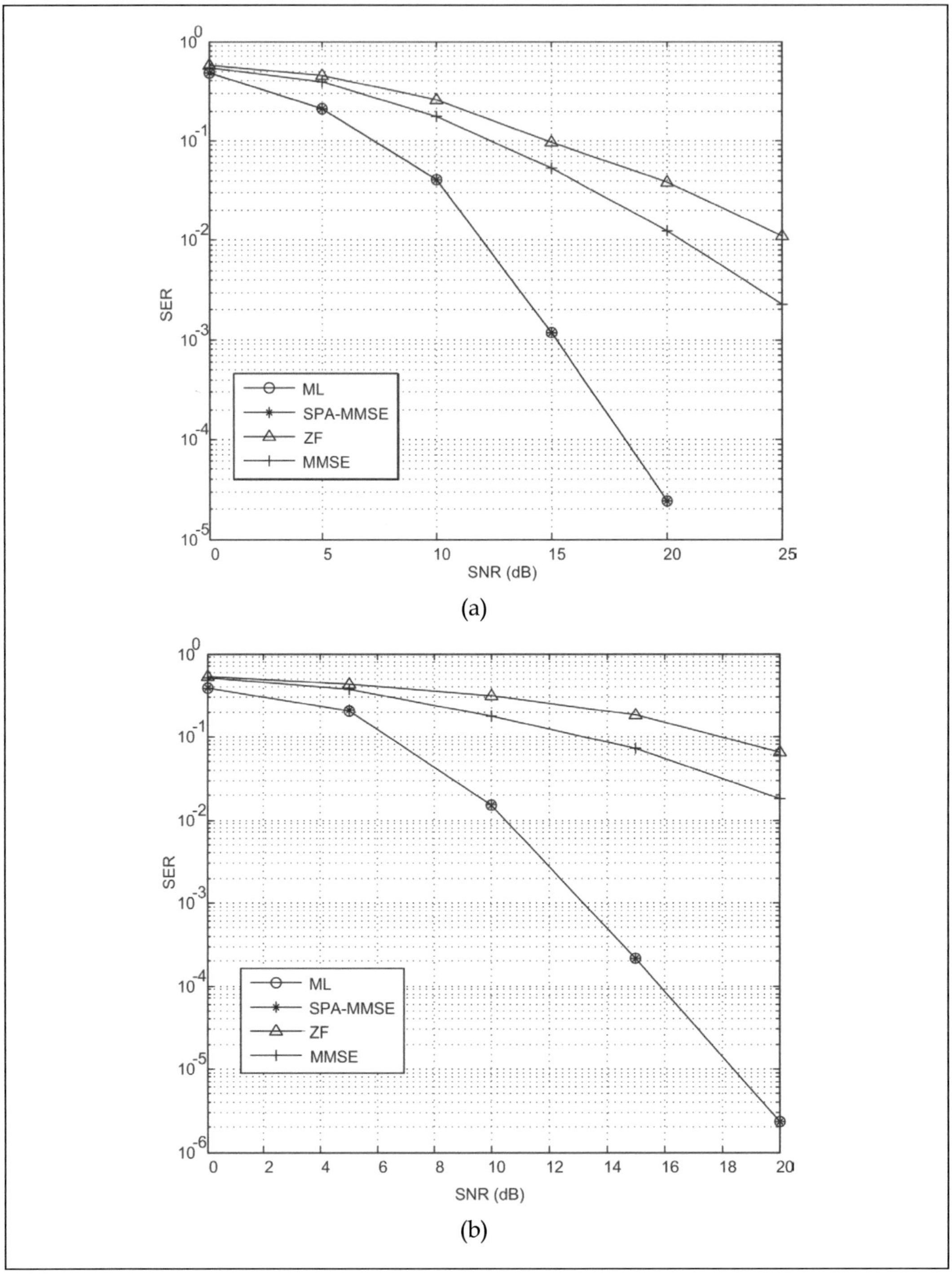

Fig. 5. Comparison of SER performance of ESA, ML decoding, ZF and MMSE using 4-QAM. (a) 4×4 MIMO systems. (b) 6×6 MIMO systems.

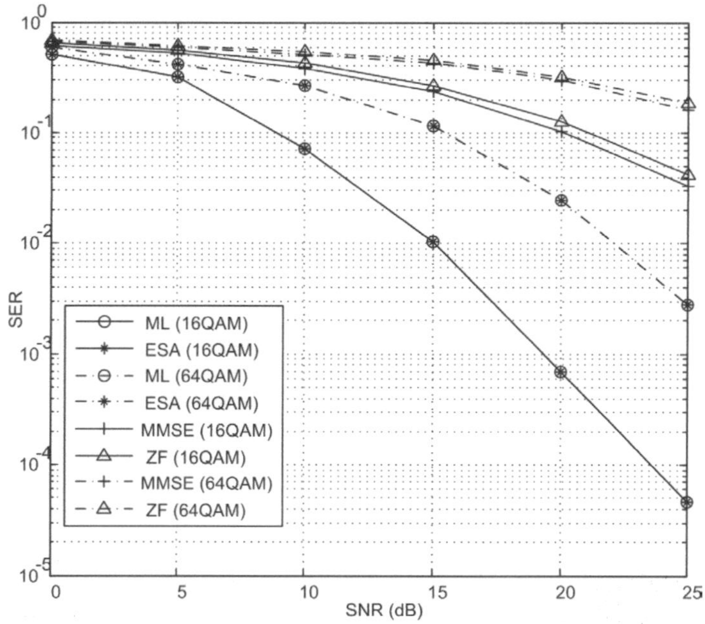

Fig. 6. Comparison of SER performance of ESA, ML decoding, ZF and MMSE using 16-QAM and 64-QAM in 4×4 MIMO system.

5. References

Fincke, U. & Pohst, M. (1985). Improved methods for Calculating vectors of short length in a lattice, including a complexity analysis, *Math. Comput.*, Vol. 44, (1985) pp.463-471, ISSN: 0025-5718

Horn R. A. and Johnson C. R. (1985). Matrix Analysis, *Cambridge University Press,* (1985) ISBN: 0-521-30586-1.

Schnorr, C.P. & Euchner, M. (1994). Lattice basis reduction: improved practical algorithms and solving subset sum problems, *Math. Program.*, Vol. 66, No. 2, (1994) pp.181-191, ISSN: 0025-5610

Foschini, G. J. & Gans, M. J. (1998). On limits of wireless communications in a fading environment when using multiple antennas, *Wireless Personal Commun.*, Vol. 6, (Mar. 1998) pp. 311-335, ISSN: 0929-6212

Wolniansky P., Foschini G. J., Golden G. & Valenzuela R. (1998). V-BLAST: an architecture for realizing very high data rates over the rich-scattering wireless channel, *International Symposium on Signals, Systems and Electronics ISSSE98*, pp. 295–300.

Viterbo, E. & Boutros, J. (1999). A Universal Lattice Code Decoder for Fading Channels," *IEEE Trans. Information Theory*, Vol. 45, No. 5, (July 1999) pp. 1639–1642, ISSN: 0018-9448.

Paulraj A.; Nabar R. & Gore D., (2003). Introduction to Space-Time Wireless Communications, *Cambridge University Press*, (May 2003), ISBN:0521826152.

Artes, H.; Seethaler, D. & Hlawatsch, F. (2003). Efficient detection algorithms for mimo channels: A geometrical approach to approximate ml detection, *IEEE Trans. Signal Processing*, Vol. 51, No. 11, (Nov. 2003) pp. 2808–2820, ISSN: 1053-587X.

Seethaler, D.; Artes, H. & Hlawatsch, F. (2003). Efficient Near-ML Detection for MIMO Channels: The Sphere-Projection Algorithm, GLOBECOM, pp. 2098–2093.

Samuel M. and Fitz M. P. (2007). Geometric Decoding Of PAM and QAM Lattices, in *Proc. IEEE Global Telecommunications Conf.*, , (Nov.2007), pp. 4247–4252.

Shao, Z. Y. ; Cheung, S. W. & Yuk, T. I. (2009). A Simple and Optimum Geometric Decoding Algorithm for MIMO Systems, *4th International Symposium on Wireless Pervasive Computing 2009*, Melbourne, Australia

3

Joint LS Estimation and ML Detection for Flat Fading MIMO Channels

Shahriar Shirvani Moghaddam[1] and Hossein Saremi[2]
[1]DCSP Research Lab., Dept. of Electrical and Computer Engineering, Shahid Rajaee Teacher Training University (SRTTU),
[2]Telecommunication Infrastructure Company (TIC),
Iran

1. Introduction

In recent years, Multi-Input Multi-Output (MIMO) communications are introduced as an emerging technology to offer significant promise for high data rates and mobility required by the next generation wireless communication systems. Using multiple transmit as well as receive antennas, a MIMO system exploits spatial diversity, higher data rate, greater coverage and improved link robustness without increasing total transmission power or bandwidth (Tse & Viswanath, 2005). However, MIMO relies upon the knowledge of Channel State Information (CSI) at the receiver for data detection and decoding. It has been proved that when the channel is Rayleigh fading and perfectly known to the receiver, the performance of a MIMO system grows linearly with the number of transmit or receive antennas, whichever is less (Numan et al., 2009). Therefore, an accurate and robust channel estimation is of crucial importance for coherent demodulation in wireless MIMO systems.

Use of MIMO channels, when bandwidth is limited, has much higher spectral efficiency versus Single-Input Single-Output (SISO), Single-Input Multi-Output (SIMO), and Multi-Input Single-Output (MISO) channels. It is shown that the maximum achievable diversity gain of MIMO channels is the product of the number of transmitter and receiver antennas. Therefore, by employing MIMO channels not only the mobility of wireless communications can be increased, but also its robustness against fading that makes it efficient for the requirements of the next generation wireless services. To achieve maximum capacity and diversity gain, some optimization problems should be considered (Yatawatta et al., 2006).

The emergence of MIMO communication systems as practical high-data-rate wireless communication systems has created several technical challenges to be met. On the one hand, there is potential for enhancing system performance in terms of capacity and diversity. On the other hand, the presence of multiple transceivers at both ends has created additional cost in terms of hardware and energy consumption. For coherent detection as well as to do optimization such as water filling and beamforming, it is essential that the MIMO channel is known. However, due to the presence of multiple transceivers at both the transmitter and receiver, the channel estimation problem is more complicated and costly compared to a SISO system. Of concern, however, is the increased complexity associated with multiple transmit/receive antenna systems. First, increased hardware cost is required to implement multiple Radio Frequency (RF) chains and adaptive equalizers. Second, increased complexity and energy is required to estimate large-size MIMO channels. Energy conservation in MIMO systems has been considered in different perspectives. For instance, hardware level optimization can be used to minimize energy. On the other hand, energy consumption can be minimized at the receiver by using low-rank equalization or/and reducing the order of MIMO systems by selection of antennas both at the receiver and transmitter, without degrading the system performance (Karami & Shiva, 2006).

In order to attain the advantages of MIMO systems and quarantee the performance of communication, effective channel estimation algorithms are needed. Many channel estimation (identification) algorithms have been developed in recent years. In the literature, three classes of methods to estimate the channel response are presented. They include Training Based Channel Estimation (TBCE) schemes relying on training sequences that are known to the receiver (Xie et al., 2007: Biguesh & Gershman, 2006: Nooalizadeh et al., 2009: Nooralizadeh & Shirvani Moghaddam, 2010), Blind Channel Estimation (BCE) methods (Sabri et al., 2009: Panahi & Venkat, 2009: Chen & Petropulu, 2001), identifying channel only from the received sequences, and Semi Blind Channel Estimation (SBCE) approaches as combination of two aforementioned procedures (Cui & Tellambura, 2007: Wo et al., 2006: Chen et al., 2007: Abuthinien et al., 2007: Khalighi & Bourennane, 2008).

One of the most usual approaches to identify MIMO CSI is TBCE. This class of estimation is attractive especially when it decouples symbol detection from channel estimation and thus simplifies the receiver implementation and relaxes the required identification conditions. In this scheme, the channel is estimated based on the received data and the knowledge of training symbols during training symbol transmit. Then, the acquired knowledge of the channel is used for data detection. TBCE schemes can be optimal at high Signal to Noise Ratios (SNRs), but they are suboptimal at low SNRs. The optimal choice of training signals is usually investigated by minimizing Mean Square Error (MSE) of the linear MIMO channel estimator. It is perceived that optimal design of training sequences is connected with the channel statistical characteristics (Hassibi & Hochwald, 2003).

Many blind channel estimation techniques can be found in the literature, and a good overview is given in (Tong & Perreau, 1998). The blind channel estimation methods can be classified into Higher-Order Statistics (HOS) based techniques (Cardoso, 1989: Comon, 1994: Chi et al., 2003) and Second Order Statistics (SOS) based techniques (Chang et al., 1997). Blind algorithms typically require longer data records and entail higher complexity.

Semi-blind channel estimation schemes, as the main core of this chapter, use a few training symbols to provide the initial MIMO channel estimation and exchange the information between the channel estimator and the data detector iteratively (Fang et al., 2007). The main steps of proposed SBCE-ML method (Shirvani Moghaddam & Saremi, 2010) are as follows:

Step 1. Initial channel estimation by using the training only;

Step 2. *Given channel knowledge, perform data detection;

*Given data decisions, perform channel estimation by taking the whole burst as a virtual training;

Step 3. Repeat step 2 until a certain stopping criterion is reached.

Several solutions have been proposed to minimize the computational cost, and hence the energy spent in channel estimation of MIMO systems. In (Yatawatta et al., 2006) authors present a novel method of minimizing the overall energy consumption. Unlike existing methods, this method considers the energy spent during the channel estimation phase which includes transmission of training symbols, storage of those symbols at the receiver, and also channel estimation at the receiver. Also they developed a model that is independent of the hardware or software used for channel estimation, and use a divide-and-conquer strategy to minimize the overall energy consumption.

In (Numan et al., 2009), a better performance and reduced complexity channel estimation method is proposed for MIMO systems based on matrix factorization. This technique is applied on training based Least Squares (LS) channel estimation for performance improvement. Experimental results indicate that the proposed method not only alleviates the performance of MIMO channel estimation but also significantly reduces the complexity caused by matrix inversion. Simulation results show that the Bit Error Rate (BER) performance and complexity of the proposed method clearly outperforms the conventional LS channel estimation method.

In (Song & Blostein, 2004), authors focused on the achievable Symbol Error Rate (SER) performance of a MIMO link with interference. Prior results on estimation of vector channels and spatial interference statistics for Code Division Multiple Access (CDMA) SISO systems. Most studies of channel estimation and data detection for MIMO systems assume spatially and temporally white interference. For example, Maximum Likelihood (ML) estimation of the channel matrix using training sequences was presented assuming temporally white interference. Assuming perfect knowledge of the channel matrix at the receiver, ordered Zero-Forcing (ZF) and Minimum Mean Squared Error (MMSE) detection were studied for both spatially and temporally white interference. However, in cellular systems, the interference is, in general, both spatially and temporally colored. This paper proposes a new algorithm that jointly estimates the channel matrix and the spatial interference correlation matrix in an ML framework. It develops a multi-vector-symbol MMSE data detector that exploits interference correlation.

In (Zaki et al., 2009), a training-based channel estimation scheme for large non-orthogonal Space-Time Block Coded (STBC) MIMO systems is proposed. The proposed scheme employs a block transmission strategy where an $N_t \times N_t$ pilot matrix is sent (for training purposes) followed by several $N_t \times N_t$ square data STBC matrices, where N_t is the number of transmit antennas. At the receiver, channel estimation (using an MMSE estimator) and detection (using a low-complexity Likelihood Ascent Search (LAS) detector) will be iterated till convergence or for a fixed number of iterations. Simulation results of this research show that good BER and high capacity are achieved by the proposed scheme at low complexities.

Joint channel estimation, data detection, and tracking are the most important issues in MIMO communications. Without joint estimation and detection, inter substream interference occurs. Joint estimation and detection algorithms used in MIMO channels are developed based on MultiUser Detection (MUD) algorithms in CDMA systems. ML is the optimum detecor in these type of joint channel estimation and data detection algorithms. In (Karami & Shiva, 2006), a new approach for joint data estimation and channel tracking for MIMO channels is proposed based on the Decision-Directed Recursive Least Squares (DD-RLS) algorithm. RLS algorithm is commonly used for equalization and its application in channel estimation is a novel idea. In this paper, after defining the weighted least squares cost function it is minimized and eventually the RLS MIMO channel estimation algorithm is derived. The proposed algorithm combined with the Decision-Directed Algorithm (DDA) is then extended for the blind mode operation. From the computational complexity point of view being $O(3)$ versus the number of transmitter and receiver antennas, the proposed

algorithm is very efficient. Also, through various simulations, the MSE of the tracking of the proposed algorithm for different joint detection algorithms is compared with Kalman filtering approach which is one of the most well-known channel tracking algorithms.
The aim of (Rizogiannis et al., 2010) is to investigate receiver techniques for ML joint channel/data estimation in flat fading MIMO channels, that are both data efficient and computationally attractive. The performance of iterative LS for channel estimation combined with Sphere Decoding (SD) for data detection is examined for block fading channels, demonstrating the data efficiency provided by the semi-blind approach. The case of continuous fading channels is addressed with the aid of RLS. The observed relative robustness of the ML solution to channel variations is exploited in deriving a block QR-based RLS-SD scheme, which allows significant complexity savings with little or no performance loss. The effects on the algorithms' performance of the existence of spatially correlated fading and Line-Of-Sight (LOS) paths are also studied. For the multi-user MIMO scenario, the gains from exploiting temporal/spatial interference color are assessed. The optimal training sequence for ML channel estimation in the presence of Co-Channel Interference (CCI) is also derived and shown to result in better channel estimation/faster convergence. The reported simulation results demonstrate the effectiveness, in terms of both data efficiency and performance gain, of the investigated schemes under realistic fading conditions. High throughput at a communication systems require high quality channel estimation at the receiver in order to provide reliable data detection, such as that performed by ML techniques. The channel estimation task is especially challenging in time varying channels, such as the one soften arising in wireless communication links.
This paper (Wo et al., 2006) deals with joint data detection and channel estimation for frequency-selective MIMO systems with focus on the analysis of the channel estimator. First, it presents a scheme alternating between joint Viterbi detection and LS channel estimation and analyze its performance in terms of unbiasedness. Since in the proposed technique the channel estimator exploits both known pilot symbols (non-blind) as well as unknown information bearing symbols (blind), this channel identification scheme is referred to as semi-blind. Second, it derives the Cramer-Rao Lower Bound (CRLB) for semi-blind channel estimation of frequency selective MIMO channels, which provides a theoretical lower bound of the achievable MSE of any unbiased estimator. By simulation the MSE performance of the proposed algorithm is evaluated and compared to the CRLB. The obtained results are universal for systems with an arbitrary number of antennas and an arbitrary channel memory length. As an example, a SBCE algorithm with LS channel estimator and ML data detector will be first introduced and analyzed. It will be shown that the presented semiblind channel estimator is biased at low SNR but tends to be unbiased at high SNR. Interestingly but reasonably, the MMSE achievable by any unbiased channel estimator at high SNR will be the same as that all data symbols are a-priori known at the receiver, but only the training symbols are known at low SNR. Simulation results show that the MSE performance of the presented SBCE coincides with the CRLB at high SNRs but exceeds CRLB at low SNRs due to biasing. Of particular interest is the SNR value where a semiblind channel estimator begin to approach the CRLB, which means that a SBCE will be able to fully exploit the channel information carried by all observations for SNRs larger than this value.
Reliable coherent communication over mobile wireless channels requires accurate estimation of time-varying multipath channel parameters. Traditionally, channel estimation is achieved by sending training sequences or using pilot channels. Recently, there is a

growing interest in training or pilot-based channel estimation for Direct Sequence CDMA (DS-CDMA) systems. In (Rizanera et al., 2005), authors address the problem of mobile radio channel estimation at high channel efficiency using a small number of training symbols. A decision aided channel estimation scheme is proposed for slow fading multipath DS-CDMA channels. The approach is an extension of single-user LS channel estimation. It is demonstrated that, due to the suggested channel estimate updating algorithm, the proposed scheme improves the channel estimation accuracy significantly. An adaptive method has been considered to provide channel estimates. In this method, the received signal is correlated with the locally generated spreading code at each multipath delay for channel estimation at each symbol interval.

By using MIMO technology an increase in the system capacity and/or an improvement in the quality of service can be achieved. The key to fully utilize the MIMO capacity relies heavily on the requirement of accurate MIMO channel estimation. This chapter have a review on TBCE as well as SBCE methods and offers some comparative simulation results. Simulations are done in different cases, MIMO 2×2 with and without space-time Alamouti coding, and also MIMO 4×4 to see the effect of the number of antenna elements. In addition, performance of different estimators, LS, Linear MMSE (LMMSE), ML and Maximum A' Posteriori (MAP) are evaluated based on BER and SER with respect to perfect channel estimator. It also proposes the proper method to estimate flat fading MIMO channels that uses LS estimator and ML detector in a joint state.

2. System model

Consider a MIMO system equipped with N_T transmit antennas and N_R receive antennas. The block diagram of a typical MIMO 2×2 is shown in Fig. 1.

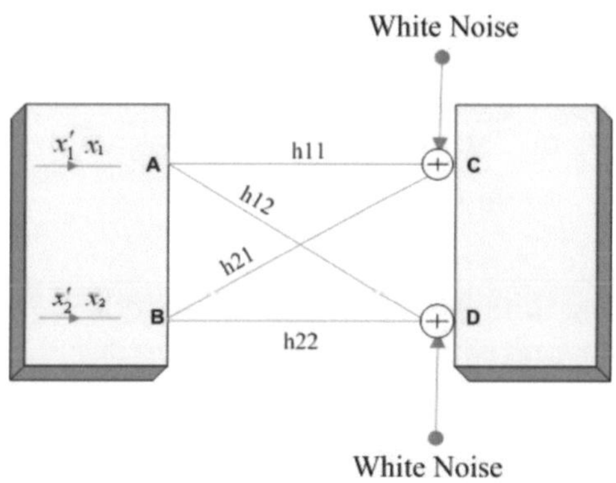

Fig. 1. General architecture of a MIMO 2×2.

where x_1, x_2 are the input (transmitted) signals of time slot 1 in locations A and B, respectively. x_1', x_2' are associated input signals of time slot 2.

It is assumed that the channel coherence bandwidth is larger than the transmitted signal bandwidth so that the channel can be considered as narrowband or flat fading. Furthermore, the channel is assumed to be stationary during the communication process of a block. Hence, by assuming the block Rayleigh fading model for flat MIMO channels, the channel response is fixed within one block and changes from one block to another one randomly. During the training period, the received signal in such a system can be written as (1)

$$\boldsymbol{Y} = H.\boldsymbol{X} + \boldsymbol{N} \tag{1}$$

where $\boldsymbol{Y}$, $\boldsymbol{X}$ and $\boldsymbol{N}$ are the complex N_R-vector of received signals on the N_R receive antennas, the possibly complex N_T-vector of transmitted signals on the N_T transmit antennas, and the complex N_R-vector of additive receiver noise, respectively. The elements of the noise matrix are independent and identically distributed (i.i.d.) complex Gaussian random variables with zero-mean and σ_n^2 variance, and the correlation matrix of $\boldsymbol{N}$ is then given by (Ma et al., 2005):

$$R = E\{\boldsymbol{N}^H.\boldsymbol{N}\} = \sigma_n^2.N_R.I_{N_P} \tag{2}$$

where $(.)^{\mathrm{H}}$ is reserved for the matrix hermitian, $E\{.\}$ is the mathematical expectation, and I_{N_P} denotes the $N_P \times N_P$ identity matrix. N_P is the number of transmitted training symbols by each transmitter antenna. The matrix H in the model (1) is the $N_R \times N_T$ matrix of complex fading coefficients. The (m,n)-th element of the matrix H denoted by $h_{m,n}$ represents the fading coefficient value between the m-th receiver antenna and the n-th transmitter antenna. Here, it is assumed that the MIMO system has equal transmit and receive antennas.
The elements of H and noise are independent of each other. In order to estimate the channel matrix, it is required that $\mathrm{N_P} \geq \mathrm{N_T}$ training symbols are transmitted by each transmitter antenna. The function of a channel estimation algorithm is to recover the channel matrix H based on the knowledge of Y and X (Shirvani Moghaddam & Saremi, 2010).
As depicted in Fig. 1, output (received) signals in locations C and D are as follow:

$$\left\{\begin{matrix} y_{n1} = h_{11}.x_1 + h_{21}.x_2 + n_1 \\ y_{n2} = h_{12}.x_1 + h_{22}.x_2 + n_2 \\ y'_{n1} = h_{11}.x'_1 + h_{21}.x'_2 + n'_1 \\ y'_{n2} = h_{12}.x'_1 + h_{22}.x'_2 + n'_2 \end{matrix}\right\} \tag{3}$$

where y_{n1}, y_{n2} are the output signals of time slot 1 in locations C and D, respectively. y'_{n1}, y'_{n2} are associated output signals of time slot 2. n_1, n_2, n'_1, n'_2 are independent Additive White Gaussian Noises (AWGN). In (Alamouti, 1998), Alamouti proposed the first space-time coding for a MIMO 2×2 system. The proposed matrix is as follow:

$$\mathrm{S}=\begin{bmatrix} s_1 & s_2 \\ -s_2^* & s_1^* \end{bmatrix} \tag{4}$$

which means that in the first time slot, s_1 and s_2 will be sent and in the second one, $-s_2^*$ and s_1^* will be transmitted. Following equations can be used to decoding process:

$$\begin{cases} \tilde{x}_1 = h_{11}^*.y_{11} + h_{12}.y_{12}^* + h_{21}^*.y_{21} + h_{22}.y_{22}^* \\ \tilde{x}_2 = h_{12}^*.y_{11} - h_{11}.y_{12}^* + h_{22}^*.y_{21} - h_{21}.y_{22}^* \end{cases} \tag{5}$$

This kind of coding is used in this research. Simulation results show its great effect on the performance of the channel estimators in both TBCE and SBCE-ML schemes.

3. Channel estimators

As illustrated in Table 1, there are many algorithms to estimate the channel response from training sequence. As shown in introduction and also (Leus & Von Der Veen, 2005: Murthy et al., 2006), LS, LMMSE, ML, and MAP are the famous and more applicable estimators. In this investigation, perfect estimator (inverse matrix) is a proper reference to compare the estimators. This reference method offers minimum BER in the case of a Rayleigh flat fading MIMO channel or AWGN.

Channel Estimator	Estimation Formula
Perfect	$H_{Perfect} = \boldsymbol{Y}.\boldsymbol{X}^{-1}$
LS	$H_{LS} = (\boldsymbol{X}^H.\boldsymbol{X})^{-1}.\boldsymbol{X}^H.\boldsymbol{Y}$
LMMSE	$H_{LMMSE} = (\sigma_n^2.C_H^{-1} + \boldsymbol{X}^H.\boldsymbol{X})^{-1}.\boldsymbol{X}^H.\boldsymbol{Y}$
ML	$H_{ML} = (\boldsymbol{X}^H.C_H.\boldsymbol{X})^{-1}.\boldsymbol{X}^H.C_H.\boldsymbol{Y}$
MAP	$H_{MAP} = (\boldsymbol{X}^H.C_n^{-1}.\boldsymbol{X} + C_H)^{-1}.\boldsymbol{X}^H.C_n^{-1}.\boldsymbol{Y}$

Table 1. Different Channel Estimators
where $(.)^{-1}$ is reserved for the matrix inverse, C_H and C_n denote channel and noise covariances, respectively.

3.1 Perfect estimator

Perfect estimator is the simplest algorithm to estimate the channel matrix. By setting the noise equal to zero in (1), the perfect approach estimates the channel matrix as

$$H_{Perfect} = \boldsymbol{Y}.\boldsymbol{X}^{-1} \tag{6}$$

Using equation (6), sub-channel responses are simply obtained by

$$\begin{cases} h_{11} = \frac{y_{n1}}{x_1} - \frac{x_2}{x_1}.\left(\frac{y'_{n1}-\frac{y_{n1}}{x_1}.x'_1}{x'_2-\frac{x_2}{x_1}.x'_1}\right) \\ h_{12} = \frac{y_{n2}}{x_1} - \frac{x_2}{x_1}.\left(\frac{y'_{n2}-\frac{y_{n2}}{x_1}.x'_1}{x'_2-\frac{x_2}{x_1}.x'_1}\right) \\ h_{21} = \frac{y'_{n1}-\frac{y_{n1}}{x_1}.x'_1}{x'_2-\frac{x_2}{x_1}.x'_1} \\ h_{22} = \frac{y'_{n2}-\frac{y_{n2}}{x_1}.x'_1}{x'_2-\frac{x_2}{x_1}.x'_1} \end{cases} \tag{7}$$

Substituting (7) back into noise-free version of (3), input signals can be expressed as

$$\begin{cases} x_{1est} = \frac{y_{n1}-h_{21}.\left(\frac{y_{n2}.h_{11}-h_{12}.y_{n1}}{h_{11}.h_{22}-h_{12}.h_{21}}\right)}{h_{11}} \\ x_{2est} = \frac{y_{n2}.h_{11}-h_{12}.y_{n1}}{h_{11}.h_{22}-h_{12}.h_{21}} \\ x'_{1est} = \frac{y'_{n1}-h_{21}.\left(\frac{y'_{n2}.h_{11}-h_{12}.y'_{n1}}{h_{11}.h_{22}-h_{12}.h_{21}}\right)}{h_{11}} \\ x'_{2est} = \frac{y'_{n2}.h_{11}-h_{12}.y'_{n1}}{h_{11}.h_{22}-h_{12}.h_{21}} \end{cases} \tag{8}$$

where x_{1est}, x_{2est} are the estimated input signals of time slot 1 in locations A and B, and x'_{1est}, x'_{2est} are associated estimated input signals of time slot 2, respectively.

3.2 LS estimator

Considering (1), LS estimator finds H_{est} so that $\boldsymbol{X}.H_{est} \approx \boldsymbol{Y}$. LS Algorithm, minimizes the Euclidian distance of $\boldsymbol{X}.H_{est} - \boldsymbol{Y}$. For this minimization we do following steps:

$$\|\boldsymbol{X}.H_{est} - \boldsymbol{Y}\|^2 = (\boldsymbol{X}.H_{est} - \boldsymbol{Y})^H.(\boldsymbol{X}.H_{est} - \boldsymbol{Y}) = (\boldsymbol{X}.H_{est})^H.(\boldsymbol{X}.H_{est}) - \boldsymbol{Y}^H.\boldsymbol{X}.H_{est} - (\boldsymbol{X}.H_{est})^H.\boldsymbol{Y} + \boldsymbol{Y}^H.\boldsymbol{Y} \quad (9)$$

By differentiating (9) with respect to H_{est} and setting the result equal to zero, it is obtained that H_{est} should satisfy the equation (10)

$$2\boldsymbol{X}^H.\boldsymbol{X}.H_{est} - 2\boldsymbol{X}^H.\boldsymbol{Y} = 0 \rightarrow \boldsymbol{X}^H.\boldsymbol{X}.H_{est} = \boldsymbol{X}^H.\boldsymbol{Y} \quad (10)$$

Finally, the LS channel estimation algorithm is based on (11)

$$H_{LS} = (\boldsymbol{X}^H.\boldsymbol{X})^{-1}.\boldsymbol{X}^H.\boldsymbol{Y} \quad (11)$$

3.3 LMMSE estimator

For linear model (1), the MMSE and LMMSE estimators are identical. So, let us minimize the estimation MSE of H. It can be expressed in the following form:

$$H_{LMMSE} = \min E\{\|H - H_{est}\|^2\} \quad (12)$$

Assuming $E(H) = 0$ and noise is AWGN, we can obtain that (12) will be minimized as

$$H_{LMMSE} = (\sigma_n^2.C_H^{-1} + \boldsymbol{X}^H.\boldsymbol{X})^{-1}.\boldsymbol{X}^H.\boldsymbol{Y} \quad (13)$$

Comparing (13) and (11), it is obvious that

$$H_{LMMSE} - H_{LS} = \sigma_n^2.C_H.\boldsymbol{X}^H.\boldsymbol{Y} \quad (14)$$

(14) shows that LMMSE needs to find an additional term compared to LS estimator. This term depends on previous data and introduces more computational complexity.

3.4 ML estimator

To identify H from (1), the ML approach maximizes (15)

$$H_{ML} = \max_H \, p(\boldsymbol{Y}|H) \quad (15)$$

where $p(\mathbf{Y}|H)$ is the conditional probability of received signal respect to channel response. It is given that the ML channel estimator (15) yields

$$H_{ML} = (\boldsymbol{X}^H.C_H.\boldsymbol{X})^{-1}.\boldsymbol{X}^H.C_H.\boldsymbol{Y} \quad (16)$$

3.5 MAP estimator

In order to estimate the channel response, in addition training bits, MAP estimator needs to find channel covariance as well as noise covariance. MAP channel estimate is in accordance with previous conditional probability $p(H|\boldsymbol{Y},\boldsymbol{X})$. MAP channel estimate can be found by solving the following equation:

$$\frac{\partial \ln(p(H|\boldsymbol{Y},\boldsymbol{X}))}{\partial H}\Big|_{H=H_{MAP}} = 0 \quad (17)$$

By using the Bay's identity (18) and solving the equation (17), MAP channel estimate can be found as (19)

$$p(H|\boldsymbol{Y},\boldsymbol{X}) = \frac{p(\boldsymbol{Y}|H,\boldsymbol{X}).p(H,\boldsymbol{X})}{p(\boldsymbol{Y}|\boldsymbol{X})} \tag{18}$$

$$H_{MAP} = (\boldsymbol{X}^H.C_n^{-1}.\boldsymbol{X} + C_H)^{-1}.\boldsymbol{X}^H.C_n^{-1}.\boldsymbol{Y} \tag{19}$$

4. Simulation results of TBCE

In order to compare the performance of LS, LMMSE, ML, and MAP estimators in TBCE for MIMO channels, three cases, MIMO 2×2 without coding, MIMO 4×4, and Alamouti coded MIMO 2×2 are simulated. Simulation results show the performance of different estimators in terms of three metrics (BER, SER, and required processing time). For the sake of simplicity and without loss of generality, we assume Rayleigh flat fading MIMO channel with AWGN, 4QAM modulation, 8 training bits for MIMO 2×2 ($N_T = N_R = 2$) and 32 bits for MIMO 4×4 ($N_T = N_R = 4$) which are generated randomly and followed by 400 data bits. It is notable that when each point in our figures is obtained by averaging over 1000 independent simulation runs, the numerical and analytical results are almost identical.
Fig. 2 shows the BER as well as SER of different estimators in the case of TBCE. As depicted, LS estimator has the better peformace (Lower BER and SER) rather than LMMSE, ML and MAP estimators and its performance is close to the perfect one.

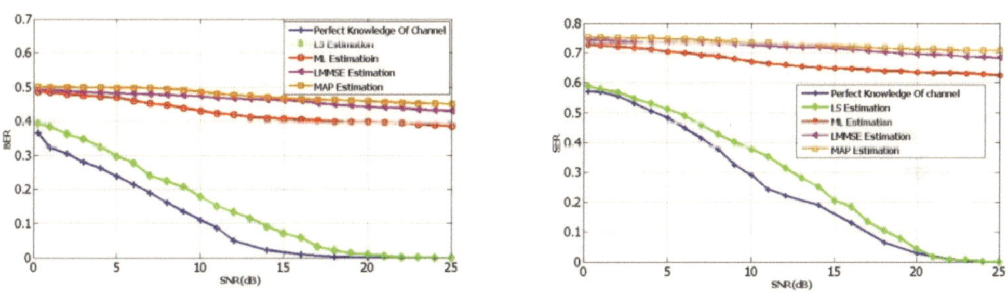

Fig. 2. Performance metrics (BER, SER) versus SNR for a MIMO 2×2 (TBCE).

As shown in Fig. 3, increasing the number of transmit antennas leads to increase the performance estimators, but it is highlighted in LS. It means, the performance of LS algorithm in a MIMO 4×4 system is improved respect to MIMO 2×2. As before, increasing the SNR is the reason for decreasing BER and SER of all estimators but it is more effective for LS one.
The BER and SER of TBCE versus SNR for various channel estimators in the case of MIMO 2×2 with Alamouti coding, are shown in Fig. 4. Comparing Fig. 4 and Fig. 2, it is observed that the BER and SER of all estimators are decreased using Alamouti coding especially at low SNRs.
Considering the processing time of TBCE equipped with prefect estimator equal to 100, Fig. 5 shows the processing time for other estimators respect to the perfect one. As expected, minimum processing time belongs to LS estimator.

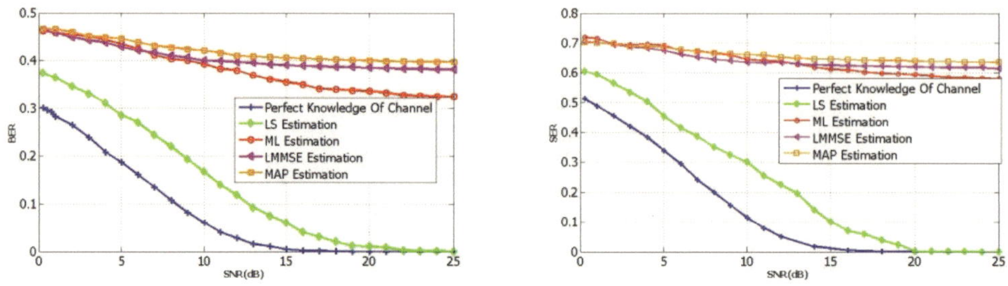

Fig. 3. Performance metrics (BER, SER) versus SNR for a MIMO 4×4 (TBCE).

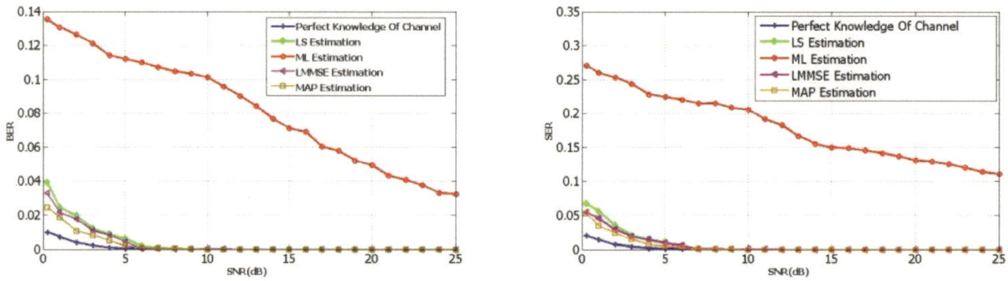

Fig. 4. Performance metrics (BER, SER) versus SNR for an Alamouti coded MIMO 2×2 (TBCE).

5. Simulation results of SBCE

For pure TBCE schemes, a long training is necessary in order to obtain a reliable MIMO channel estimate which reduces the system bandwidth efficiency considerably. SBCE-ML schemes require less computational complexity than blind methods and fewer training symbols than training-based methods, making them attractive for practical implementation. TBCE algorithms use only the training sequences to perform channel estimation, while a SBCE algorithm takes the data symbols also into account. Since the data symbols are practically unknown, before they can be used for channel estimation, the receiver has to perform detection in advance. Thus, the task of channel estimation changes into joint estimation of channel and data symbols.

By refining the channel estimate and the data decisions in a recursive manner, considerable performance gain can be achieved step by step. As depicted in Fig. 6, in an iterative structure, output of estimator is applied to detector for detecting data bits and also output of detector is applied to the estimator as virtual bits and to estimate the channel again. This iterative procedure runs until a criterion is achieved [Shirvani Moghaddam & Saremi, 2010]. For example, difference of estimation for two successive iterations is lower than a level. LS, LMMSE, ML and MAP estimators may be used in estimation part but ML detector is more attractive in semi-blind joint estimation and detection schemes. In the first step, channel response is estimated considering short training bits. Then, by using the ML detector, symbols are detected according to (20):

$$\boldsymbol{X}_{est} = \arg\min_{\widetilde{\boldsymbol{X}}\in \boldsymbol{X}} \left\{ \left\| \boldsymbol{Y} - H.\widetilde{\boldsymbol{X}} \right\|_F^2 \right\} \quad (20)$$

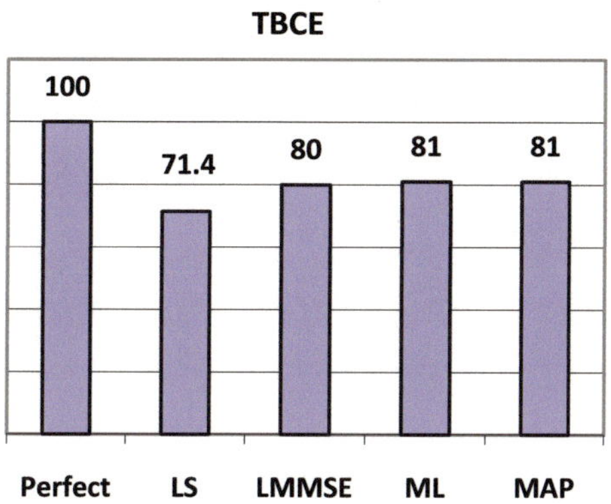

Fig. 5. Relative processing time of different estimators with respect to perfect one in a MIMO 2×2 (TBCE).

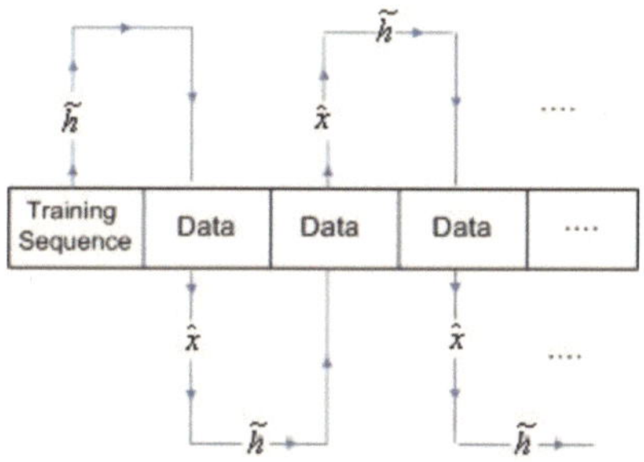

Fig. 6. Iterative structure of channel estimation and data detection in SBCE.

where H_{est} is used for detecting $\boldsymbol{X}_{est}$ and previous detected data is the virtual training sequence to next estimation. $||.||_F$ denotes the Frobenius norm. This process will be continued until (21) be satisfied.

$$\left(H_{est,i}, \boldsymbol{X}_{est,i}\right) = \left(H_{est,i-1}, \boldsymbol{X}_{est,i-1}\right) \tag{21}$$

The proposed method can be summarized as follow:

1. $i = 0: H_0$ $(i\ denotes\ the\ iteration\ index)$;

2. $i = i + 1$;

   $a. ML\ Data\ Detection$

   $b. Channel\ Estimation$

3. $Repeat\ step\ 2\ until\ \left(H_{est,i}, \boldsymbol{X}_{est,i}\right) = \left(H_{est,i-1}, \boldsymbol{X}_{est,i-1}\right)$

In the next subsections, simulation results of SBCE-ML method for a Rayleigh flat fading MIMO system in three cases, MIMO 2×2 (with and without Alamouti coding) and MIMO 4×4 are presented. For this type of channel estimation, 8 and 32 training bits are used for MIMO 2×2 and MIMO 4×4, respectively followed by 40000 data bits. simulation results of SBCE scheme are presented to find the efficient estimator with good performance (BER as well as SER) and lower processing time.

Fig. 7 illustrates the BER as well as SER of SBCE-ML using various estimators versus different SNR for a Rayleigh flat fading MIMO 2×2 channel. It is obvious that, increasing SNR is the reason for decreasing both BER and SER. As depicted, not only the performance of LS algorithm is better than other estimators but also is close to the perfect one.

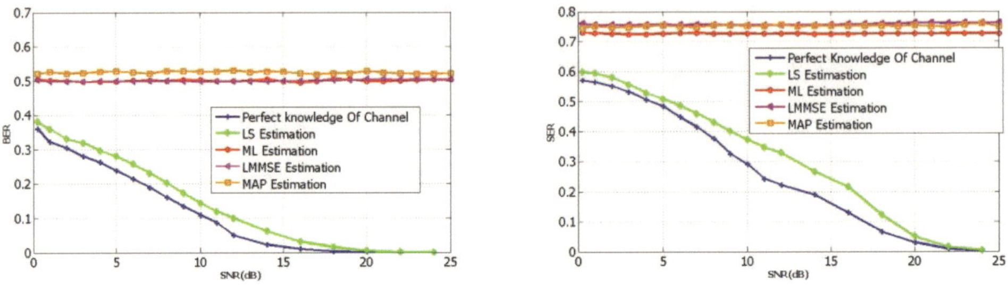

Fig. 7. Performance metrics (BER, SER) versus SNR for a MIMO 2×2 (SBCE-ML).

Increasing the number of transmit antennas leads to decreasing the performance estimators, except LS. As shown in Fig. 8, the performance of LS algorithm in a MIMO 4×4 system is improved respect to MIMO 2×2. In the other hand, a power gain or SNR improvement will be achieved. For example in SBCE-ML, transmitting power will be saved about 3 dB, if BER equals to 0.3.

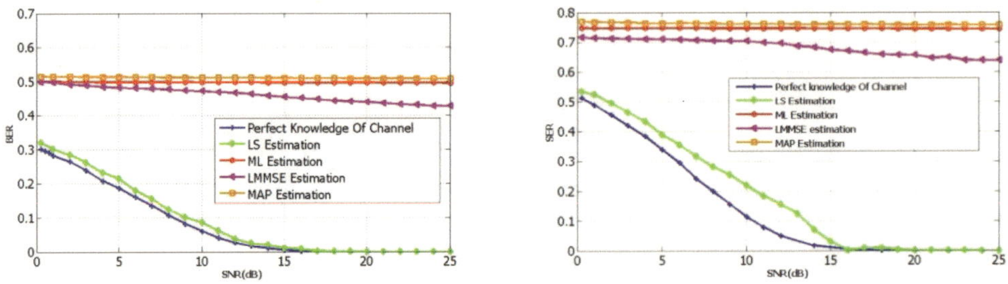

Fig. 8. Performance metrics (BER, SER) versus SNR for a MIMO 4×4 (SBCE-ML).

The BER and SER of SBCE-ML method versus SNR for various channel estimators in the case of MIMO 2×2 with Alamouti coding, are shown in Fig. 9. it is observed that the LS estimator outperforms the other estimators especially at low SNRs.

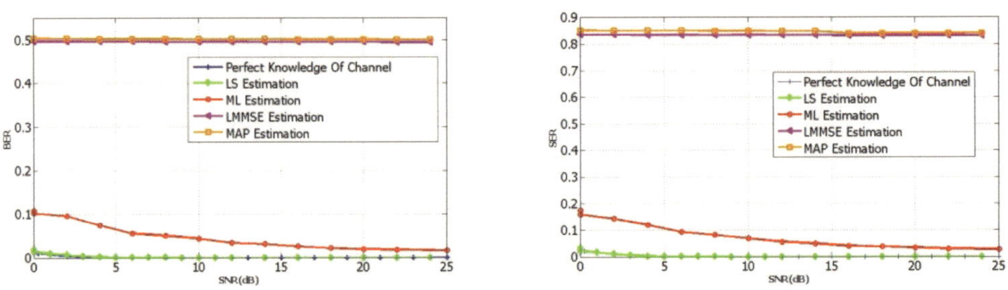

Fig. 9. Performance metrics (BER, SER) versus SNR for an Alamouti coded MIMO 2×2 (SBCE-ML).

Fig. 10 shows the processing time for different estimators (LS, LMMSE, ML, MAP) with respect to the perfect estimator in SBCE-ML scheme. In this figure, required time for perfect one is considered as 100 and other estimators' processing time is evaluated based on the perfect one. It is obvious that minimum processing time belongs to LS estimator.

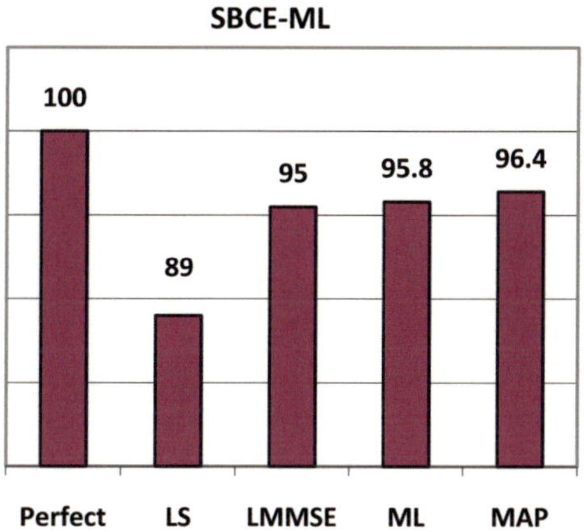

Fig. 10. Relative processing time of different estimators with respect to perfect one in a MIMO 2×2 (SBCE).

6. Comparison of LS-based TBCE and joint LS-estimation & ML-detection SBCE

Simulation results of TBCE and SBCE-ML methods show that the required processing time and both BER and SER of LS estimator compared with other estimators is much better. In this section by focusing on LS estimator, LS-based TBCE and LS-based SBCE-ML are compared in a MIMO 2 × 2 (with and without Alamouti coding) and a MIMO 4×4, for different SNRs based on BER, SER, required channel estimation processing time and relative length of training bits.

Fig. 11 indicates the BER and SER metrics of LS-based TBCE and LS-based SBCE-ML schemes for different SNRs. As shown, for both TBCE and SBCE-ML methods, increasing SNR is the reason for decreasing both BER and SER. As depicted in this figure, SBCE-ML offers a bit better performance rather than TBCE.

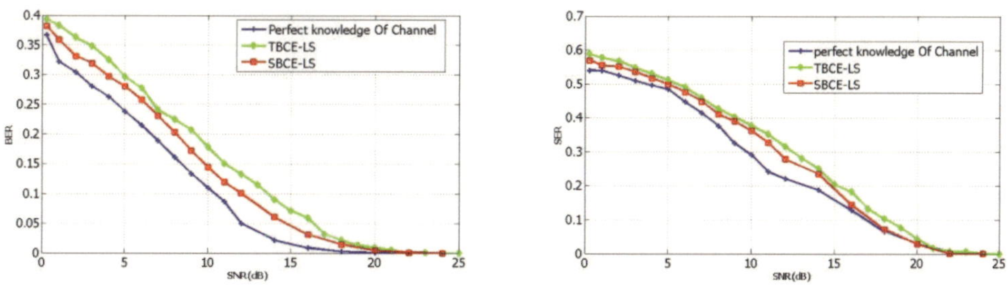

Fig. 11. Performance metrics (BER, SER) of LS-based TBCE and SBCE-ML schemes in different SNRs for a MIMO 2×2.

As shown in Fig. 12, the performance of both LS-based TBCE and SBCE-ML schemes in a MIMO 4×4 system is improved respect to MIMO 2×2. In the other hand, a power gain or SNR improvement will be achieved. For example in SBCE-ML, transmitting power will be saved about 3 dB, if BER equals to 0.3. In TBCE method, for BER equals to 0.2, transmitting power will be saved about 0.5 dB. It is worthwhile to note that the excess of transmit or/and receive antennas in MIMO systems leads to a higher capacity.

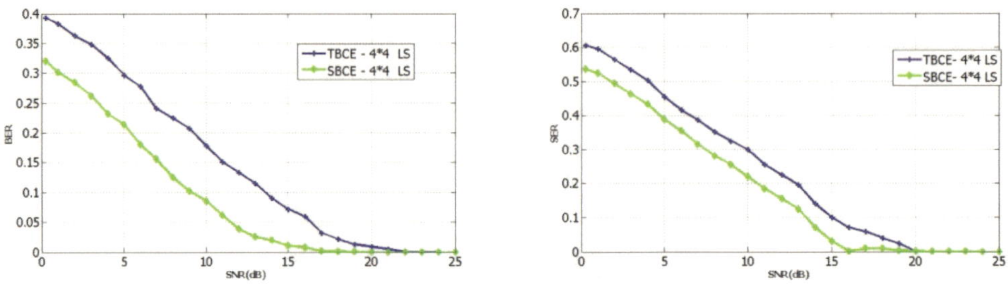

Fig. 12. Performance metrics (BER, SER) of LS-based TBCE and SBCE-ML schemes in different SNRs for a MIMO 4×4.

The BER and SER of both LS-based TBCE and SBCE-ML schemes versus SNR in the case of MIMO 2×2 with Alamouti coding, are shown in Fig. 13. As shown in this figure, when SNR equals to 0.25 dB, BER is 0.0130 for SBCE-ML and 0.0386 for TBCE. It means 3 times better performance in lowest SNRs for SBCE-ML method rather than TBCE one. At higher SNRs, the performance of LS estimator in both channel estimation schemes is analogous.

By considering the required processing time of LS-based TBCE and SBCE-ML schemes rlated to prefect estimator, Fig. 14 shows that SBCE-ML method needs 25 percent more processing time to estimate the channel than TBCE method. It is due to joint LS estimation and ML detection of SBCE method.

Fig. 15, 16 show the required training sequences in each frame of data for TBCE and SBCE-ML schemes, respectively. As depicted in Fig. 15, in TBCE method, transmitter sends 8 training bits before 400 information bits in each burst for a MIMO 2×2 system and 32 bits for a MIMO 4×4 system. Figure 16, illustrates the required number of training and information bits in SBCE-ML method for both MIMO 2×2 and MIMO 4×4. Considering the same training bits, 400 information bits in the case of TBCE method are changed to 40000 bits in SBCE-ML. As mentioned before, TBCE method needs more bits to estimate the channel because training sequences should be transmitted periodically. On the other word, SBCE-ML

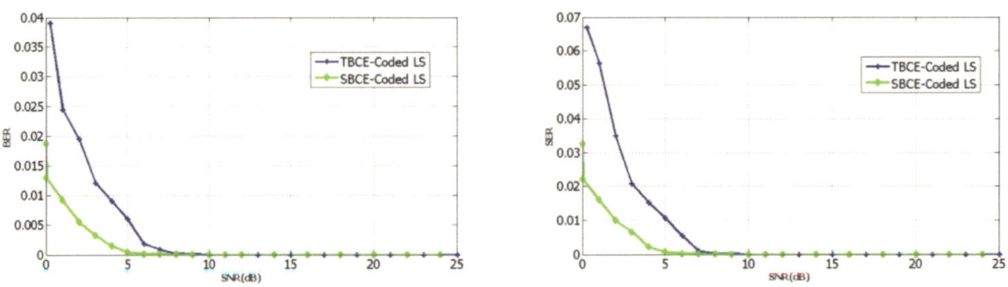

Fig. 13. Performance metrics (BER, SER) of LS-based TBCE and SBCE-ML schemes in different SNRs for an Alamouti coded MIMO 2×2.

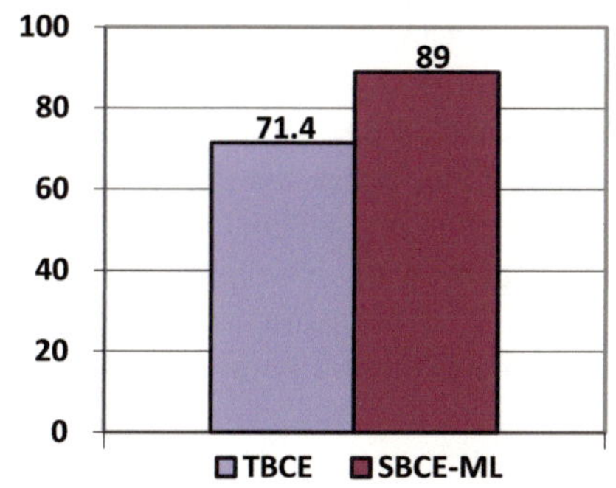

Fig. 14. Relative processing time of LS-based TBCE and SBCE-ML schemes in a MIMO 2×2.

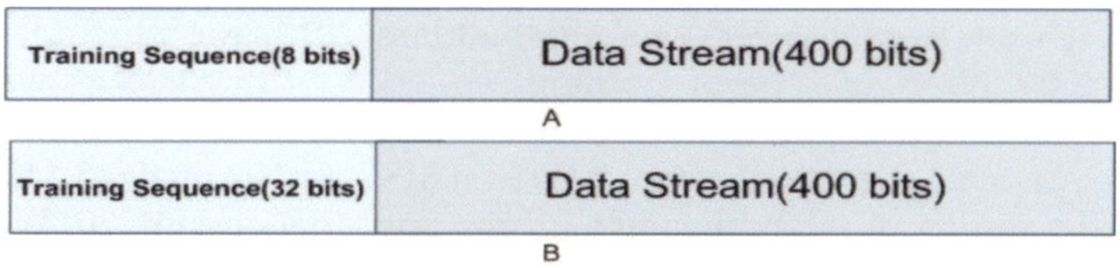

Fig. 15. The burst of LS-based TBCE. A) MIMO 2×2, B) MIMO 4×4.

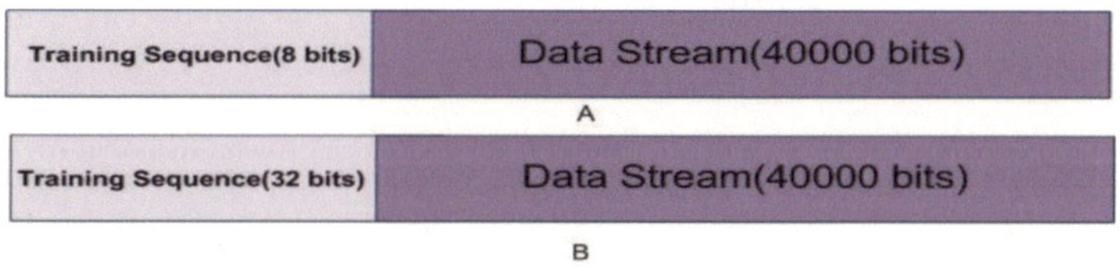

Fig. 16. The burst of LS-based SBCE-ML. A) MIMO 2×2, B) MIMO 4×4.

method needs to transmit just one training sequence. Therefore, redundancies of TBCE method are 2% and 8% for MIMO 2×2 and MIMO 4×4 systems, respectively. In the case of SBCE-ML method, redundancies are 0.02% and 0.08%, respectively. It means 100 times lower training bits for SBCE-ML respect to TBCE.

7. Conclusion

MIMO systems play a vital role in fourth generation wireless systems to provide advanced data rate. In order to attain the advantages of MIMO systems, it is necessary that the receiver and/or transmitter have access CSI. The time-varying nature of the channel typically requires the use of frequent channel retraining, which in turn increases the data overhead due to training signals, thus reducing the system's overall spectral efficiency. Hence, effective channel estimation algorithms are needed to guarantee the performance of communication.

In this chapter, training based as well as semi-blind channel estimation schemes in Rayleigh flat fading MIMO systems are investigated. After introducing LS, LMMSE, ML and MAP estimators, they are simulated in a Rayleigh flat fading MIMO channel considering AWGN. Simulation results show that LS estimator is the best choice in both TBCE and SBCE-ML schemes. This selection is due to faster processing and lower BER as well as SER of LS estimator with respect to other estimators. In addition, it is illustrated that when the number of transmitter or/and receiver antennas increases, the performance of both TBCE and SBCE-ML schemes significantly improves. Moreover, Alamouti coding has more effect on the performance of SBCE-ML rather than TBCE.

Comparing LS-based TBCE and LS-based SBCE-ML methods based on BER, SER, required training bits, and processing time, simulation results introduce most appropriate channel estimation method that uses an iterative algorithm. This new proposed method is based on LS estimator and ML detector. According to simulation results, LS-based SBCE-ML method compared to LS-based TBCE method in different SNRs offers lower BER and also SER, 25 percent higher processing time, and 100 times lower training bits.

Some new research works and simulations can be considered to extend the above mentioned results and techniques as follow:

1. Cosidering the TBCE and SBCE-ML methods for Rician flat fading MIMO channels and extending the results of (Shirvani Moghaddam & Saremi, 2010) for these channels;
2. Applying the new versions of LS algorithm, Scaled LS (SLS) and Shifted SLS (SSLS) proposed in (Nooralizadeh & Shirvani Moghaddam, 2010), for SBCE-ML scheme;
3. Considering the effect of type of training sequence, orthogonal as well as optimum (Nooralizadeh et al., 2009), in channel estimation peformance;
4. Finding the channel estimation results based on MSE (or Normalized MSE) criteria;
5. Extending the results of (Nooralizadeh & Shirvani Moghaddam, 2011) and comparing TBCE and SBCE-ML schemes in frequency selective fading MIMO channels;
6. Extending the analytical and simulation results of (Wo et al., 2006) considering the BER and SER performance metrics instead of MSE one.

8. References

Abuthinien, M.; Chen, S.; Wolfgang, A. & Hanzo, L. (2007). Joint Maximum Likelihood Channel Estimation and Data Detection for MIMO Systems, *Proceedings of IEEE International Conference on Communications (ICC'07)*, pp. 5354-5358, Glasgow, June 2007.

Biguesh, M. & Gershman, A.B. (2006). Training-Based MIMO Channel Estimation: A Study of Estimator Tradeoffs and Optimal Training Signals, *IEEE Transactions on Signal Processing*, Vol. 54, No. 3, pp. 884-893.

Cardoso, J.F. (1989). Source Separation using Higher Odered Moments, *Proceedings of IEEE International Conference on Acoustics, Speech, and Signal Processing (ICASSP'89)*, Vol. 4, pp. 2109–2112, Glasgow, May 1989.

Chang, C.-Q.; Yau, S.F.; Kwok, P.; Lam, F.K. & Chan, F.H.Y. (1997). Sequential Approach to Blind Source Separation using Second Order Statistics, *Proceedings of 1st International Conference on Information, Communications, and Signal Processing (ICICS'97)*, pp. 1608–1612, Singapore, September 1997.

Chen, B. & Petropulu, A.P. (2001). Frequency Domain Blind MIMO System Identification Based on Second and Higher Order Statistics, *IEEE Transactions on Signal Processing*, Vol. 49, No. 8, pp. 1677-1688.

Chen, S.; Yang, X.C.; Chen, L. & Hanzo, L. (2007). Blind Joint Maximum Likelihood Channel Estimation and Data Detection for SIMO Systems, *International Journal of Automation and Computing, Vol.* 4, No. 1, pp. 47-51.

Chi, C.-Y.; Chen, C.-Y.; Chen, C.-H. & Feng, C.-C. (2003). Batch Processing Algorithms for Blind Equalization using Higher-Order Statistics, *IEEE Signal Processing Magazine*, Vol. 20, No. 1, pp. 25–49.

Comon, P. (1994). Independent Component Analysis: A New Concept?, *Elsevier Signal Processing*, Vol. 36, No. 3, pp. 287–314.

Cui, T. & Tellambura, C. (2007). Semiblind Channel Estimation and Data Detection for OFDM Systems with Optimal Pilot Design, *IEEE Transactions on Communications*, Vol. 55, No. 5, pp. 1053-1062.

Fang, J.; Leyman, A.R.; Chew, Y.H. & Duan, H. (2007). Some Further Results on Blind Identification of MIMO FIR Channels via Second-Order Statistics, *Elsevier Signal Processing*, Vol. 87, No. 6, pp. 1434-1447.

Fang, J.; Leyman, A.R. & Chew, Y.H. (2005). A New Closed-Form Solution for Blind MIMO FIR Channel Estimation with Colored Sources, *Proceedings of IEEE International Conference on Acoustics, Speech, and Signal Processing (ICASSP'05)*, pp. 1049-1052, Philadelphia, March 2005.

Hassibi, B. & Hochwald, B.M. (2003). How Much Training Is Needed in Multiple-Antenna Wireless Links, *IEEE Transactions on Information Theory*, Vol. 49, No. 4, pp. 951-963.

Karami, E. & Shiva, M. (2006). Decision-Directed Recursive Least Squares MIMO Channels Tracking, *EURASIP Journal on Wireless Communications and Networking (WCN)*, Vol. 2006, Article ID 43275, pp. 1–10.

Khalighi, M.A. & Bourennane, S. (2008). Semi-Blind Single-Carrier MIMO Channel Estimation using Overlay Pilots, *IEEE Transactions on Vehicular Technology*, Vol. 57, No. 3, pp. 1951-1956.

Leus, G. & Von Der Veen, A.J. (2005). Optimal Training for ML and LMMSE Channel Estimation in MIMO Systems, *Proceedings of 13th IEEE Workshop on Statistical Signal Processing (SSP'05)*, pp. 1354-1357, France, July 2005.

Ma, X.; Yang, L. and Giannakis, G.B. (2005). Optimal Training for MIMO Frequency-Selective Fading Channels, *IEEE Transactions on Wireless Communication*, Vol. 4, No. 2, pp. 453-466.

Murthy, C.R.; Jagannatham, A.K. & Rao, B.D. (2006). Training-Based and Semiblind Channel Estimation for MIMO Systems with Maximum Ratio Transmission, *IEEE Transactions on Signal Processing*, Vol. 54, No. 7, pp. 2546-2558.

Nooralizadeh, H. & Shirvani Moghaddam, S. (2011). Appropriate Algorithms for Estimating Frequency Selective Rician Fading MIMO Systems and Channel Rice Factor: Substantial Benefits of Rician Model and Estimator Tradeoffs, *EURASIP Journal on Wireless Communications and Networking (WCN)*, 2011.

Nooralizadeh, H. & Shirvani Moghaddam, S. (2010). A Novel Shifted Type of SLS Estimator for Estimation of Rician Flat Fading MIMO Channels, *Elsevier Signal Processing*, Vol. 90, No. 6, pp. 1886-1893.

Nooralizadeh, H.; Shirvani Moghaddam, S. & Bakhshi, H.R. (2009). Optimal Training Sequences in MIMO Channel Estimation with Spatially Correlated Rician Flat Fading, *Proceedings of 2009 IEEE Symposium on Industrial Electronics and Applications (ISIEA'09)*, pp. 227-232, Malaysia, October 2009.

Numan, M.W., Islam, M.T. & Misran, N. (2009). Performance and Complexity Improvement of Training Based channel Estimation in MIMO systems, *Progress In Electromagnetics Research C (PIER-C)*, Vol. 10 (2009), pp. 1-13.

Panahi, I.M.S. & Venkat, K. (2009). Blind Identification of Multi-Channel Systems with Single Input and Unknown Orders, *Elsevier Signal Processing*, Vol. 89, No. 7, pp. 1288-1310.

Rizanera, A.; Amcaa, H.; Hacıo˘glub, K. & Ulusoya, A.H. (2005). A Decision Aided Channel Estimation Scheme, *International Journal on Electronics and Communications (AEU)*, Vol. 59 (2005), pp. 324 - 327.

Rizogiannis, C.; Kofidis, E.; Papadias, C.B. & Theodoridis, S. (2010). Semi-blind Maximum-Likelihood Joint Channel/Data Estimation for Correlated Channels in Multi User MIMO Networks, *Elsevier Signal Processing*, Vol. 90 (2010), pp. 1209–1224.

Sabri, K.; El Badaoui, M.; Guillet, F.; Adib, A. & Aboutajdine, D. (2009). A Frequency Domain-Based Approach for Blind MIMO System Identification using Second-Order Cyclic Statistics, *Elsevier Signal Processing*, Vol. 89, No. 1, pp. 77-86.

Shirvani Moghaddam, S. & Saremi, H. (2010). *A Novel Semi-Blind Channel Estimation Scheme for Rayleigh Flat Fading MIMO Channels (Joint LS Estimation and ML Detection), IETE Journal of Research, Vol. 56, No. 4, pp. 193-201.*

Shirvani Moghaddam, S. & Saremi, H. (2008). Performance Evaluation of LS Algorithm in Both Training-Based and Semi-Blind Channel Estimations for MIMO Systems, *Proceedings of 1st IFIP/IEEE Wireless Days (WD) Conference,* pp. 1-5, Dubai, November 2008.

Song, Y. & Blostein, D.B. (2004). Channel Estimation and Data Detection for MIMO Systems under Spatially and Temporally Colored Interference, EURASIP *Journal on Applied Signal Processing (ASP)*, Vol. 2004, No. 5, pp. 685–695.

Tong, L. & Perreau, S. (1998). Multichannel Blind Identification: From Subspace to Maximum Likelihood Methods, *Proceedings of IEEE*, Vol. 86, No. 10, pp. 1951–1968.

Tse, D. & Viswanath, P. (2005). *Fundamentals of Wireless Communications.* Cambridge University Press, UK.

Wo, T.; Hoeher, P.A.; Scherb, A. & Kammeyer, K.D. (2006). Performance Analysis of Maximum-Likelihood Semi-Blind Estimation of MIMO Channels, *Proceedings of 63rd IEEE Vehicular Technology Conference (VTC)*, pp. 1738-1742, Melbourne, May 2006.

Wo, T.; Scherb, A.; Hoeher, P.A. & Kammeyer, K.-D. (2006). Analysis of Semiblind Channel Estimation for FIR-MIMO Systems, *Proceedings of 4th International Symposium on Turbo Codes & Related Topics*, pp. 1-6, Germany, April 2006.

Xie, G.; Fang, X.; Yang, A. & Liu, Y. (2007). Channel Estimation with Pilot Symbol and Spatial Correlation Information, *Proceedings of IEEE International Symposium on Commun. and Information Tech. (ISCIT'07)*, pp. 1003-1006, Sydney, October 2007.

Yatawatta, S., Petropulu, A.P. & Graff, C.J. (2006). Energy-Efficient Channel Estimation in MIMO Systems, *EURASIP Journal on Wireless Communications and Networking (WCN)*, Vol. 2006, Article ID 27694, pp. 1–11.

Zaki, A.; Mohammed, S.K.; Chockalingam, A. & Sundar Rajan, B. (2009). A Training-Based Iterative Detection/Channel Estimation Scheme for Large Non-Orthogonal STBC MIMO Systems, *Proceedings of IEEE International Conference on Communications (ICC'09)*, pp. 1-5, Germany, June 2009.

4

Semi-Deterministic Single Interaction MIMO Channel Model

Arghavan Emami-Forooshani[1] and Sima Noghanian[2]
[1]University of British Columbia,
[2]University of North Dakota
[1]Canada,
[2]USA

1. Introduction

In systems which employ spatial filtering, Multiple Input Multiple Output (MIMO) systems, switched beam systems or adaptive antennas, distribution of the multipath components is important in determining the performance of the channel [Liberty & Rappaport, 1999], [Allen & Ghavami, 2005]. In this regard, intensive research efforts have been invested. Different measurement campaigns [Ranvier et al., 2007], [Chizhik et al., 2003], [Howard et al., 2002] and site specific propagation prediction methods [Seidel & Rappaport, 1994], [Anderson & Rappaport, 2004], [Gesbert et al., 2002] have been realized to characterize the wireless channel. However, to simulate these systems without using measured data or site specific propagation prediction techniques, a model must be used to generate multipath channel parameters. Therefore, a number of realistic spatial channel models are introduced and the defining equations (or geometry) are described in [Liberty & Rappaport, 1999]. However, these models are only valid for particular environments with specific assumptions. Most of these simple geometrical models such as Lee's and Geometrically-Based Single-Bounce Circular Model (GBSBC) models are only applicable to outdoor environments. In some of these models for instance, it is assumed that the transmitter (Tx) and receiver (Rx) heights are the same which is a reasonable assumption only for some outdoor applications where the Tx and Rx distance is quite large. Moreover, in these simple models scatterers' distribution is restricted into limited areas and the impact of channel (including scatterers) on changing the polarization of the electric field and also antenna pattern effect are not taken into account.
Therefore, there is a need for a general and more accurate model that is valid for both outdoor and indoor environments with different scatterers' distributions. Also a model that includes effects of changing the electric field polarization and antenna characteristics on the channel is required to make realistic conclusions about different environments.
Although ray-tracing may seem as another alternative that is more accurate in terms of scattering environment and antenna characteristics, it is site specific, i.e. it needs exact information about the study area and it is computationally intensive, needing very long runtime. If general conclusions about system configuration based on statistics of the channel are required, ray-tracing may not be a right choice as it demands to change the channel

parameters several times and evaluate and compare the results for many runs. This can be very time consuming if the runtime is too long.

In this chapter a method is introduced that can be used for channel estimation in both indoor and outdoor envrionments. The method is called Single Interaction ScaTEring Reflecting (SISTER) model [Emami, 2010]. This model is based on the method proposed in [Svantesson, 2001]. In that work, a spatio-temporal channel model for MIMO systems is proposed which is based on electromagnetic scattering and wave propagation. By studying the scattering properties of objects of simple shapes, such as spheres and cylinders, a simple function that captures the most important scattering properties is derived. A compact formulation is obtained by using a dyad notation and concepts from rough surface scattering. That model exploits the concept of positioning scattering objects and calculating the received signal including polarization properties of the channel and the antennas and 3-D wave propagation. However, it only accounts for uniformly distributed scatterers in the surrounding environment and does not include different distributions for the scatterers, reflection from the ground and antenna array factor in channel complex impulse response calculations. Moreover, it is not suitable for indoor applications since it does not take into account the reflections from the walls.

The SISTER model was developed to overcome the shortcomings of previous models mentioned above. To keep it simple, spherical shape is chosen for scatterers in order to obtain analytical expressions for scattered fields and only single interaction from each scatterer (or reflector) is considered and the interactions between scatterers (or reflectors) are neglected. Single bounce interaction has been used in some MIMO channel models such as GBSBC and Geometrical Based Single Bounce Macrocell (GBSBM) channel models [Seidel & Rappaport, 1994] and ray-tracing models [Liberty & Rappaport, 1996]. While in reality multiple interactions do exists, the level of interaction strongly depends on type propagation environment. According to [Almers et al., 2007] for picocells, propagation within a single large room is mainly determined by Line-of-Sight (LOS) propagation and single bounce reflections. However if the Tx and Rx are in different rooms, then the radio waves either propagate through the walls or they will be diffracted into the room with the Rx. The multiple-bounce can be accounted using virtual single-bounce scatterers whose position and path-loss are chosen such that they mimic multiple bounce contribution. With this approach SISTER model can be utilized for environments with significant multiple bounce propagation.

The SISTER model not only is general in terms of different fading channels and antenna configuration but also is simple and can run in a reasonable computation time. In SISTER model, scatterers are located in an enclosed area containing Tx and Rx which can have optional distance and heights. Any numbers and distributions including uniform and cluster forms can be defined for scatterers. To increase the accuracy of the model, in addition to scattering, reflections (from the ground for outdoors and from the walls for indoors) are also included in it.

2. Summarized description of the SISTER model

In SISTER model different locations, configurations, radiation patterns and polarizations can be defined for Tx and Rx antennas. Scatterers' distribution, material and size can also be defined. Simple shape of sphere is chosen for scatterers in order to obtain analytical expressions for scattered fields.

This model can be used for both indoor and outdoor applications and there is no limitation on Tx and Rx heights, separation (as long as they are in each others far field) and element spacing. In addition, both Line-of-Sight (LOS) and Non-Line-of-Sight (NLOS) cases are modeled.

Without losing the generality, it is assumed that the Mobile Station (MS) is the transmitter and the Base Station (BS) is the receiver. Therefore, the electric waves are generated at the MS and then propagate towards the scatterers (or reflectors) and finally scatter (or reflect) towards the BS. In order to use far field expressions for antennas, scatterers are located in the far field of both Tx and Rx. From antenna theory, if the distance between the antenna and the object is $r \geq \frac{2D^2}{\lambda}$ where D is the largest dimension of the corresponding antenna and λ is the wavelength, object is located in the far_field.

As mentioned earlier, to keep this model simple, only single interaction from each scatterer (or reflector) is considered and the interactions between scatterers (or reflectors) are neglected (Fig. 1).

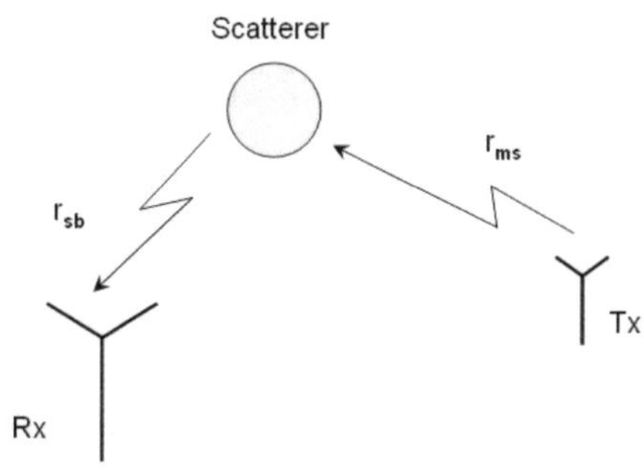

Fig. 1. Single interaction for each scatterer is considered; r_{sb} and r_{ms} are Rx and Tx distances to the scatterer, respectively.

3. Analytical calculations

In this section, different required calculations will be explained first. Then it will be shown how these calculations are used to compute the channel complex impulse response matrix (**H**-matrix) and the channel capacity. By considering an $N_T \times N_R$ -MIMO system, where N_T is the number of transmitters and N_R is the number of receivers, **H**-matrix will consist of $N_T \times N_R$ entries each of which corresponds to a different channel:

$$\mathbf{H}=\begin{bmatrix} h_{11} & h_{12} & h_{11} & h_{11} \\ h_{21} & h_{11} & h_{11} & h_{11} \\ h_{11} & h_{11} & h_{11} & h_{11} \\ h_{11} & h_{11} & h_{11} & h_{11} \end{bmatrix}_{N_T \times N_R} \quad (1)$$

where h_{ij} is the channel impulse response between i^{th} Tx and j^{th} Rx antennas.

Here, two cases of space and angle diversity are considered for the analysis. For space diversity, multiple antennas and for angle diversity, multiple simultaneous beams are assumed at both Tx and Rx.

To calculate each entry of the channel complex impulse response matrix, h_{ij}, first radiated electric field from the first Tx antenna (or beam) which is received by the first scatterer should be calculated. After that, scattered field from the scatterers which is received by the first Rx antenna (or beam) should be calculated. This procedure should be repeated for all scatterers, antennas and beams. The scattered fields then should be summed over all the scatterers at the receiver. Reflected field also should be calculated and added to the resultant field. In the LOS case, electric field for direct path between Tx and Rx should also be included in the summation.

3.1 Transmitter and receiver antenna pattern calculation

To calculate channel complex impulse response, electric field of the antenna elements used at both ends and array factor in case of using the arrays are needed. In order to take mutual coupling into account for the array case, array radiation pattern should be found by full wave analysis using one of antenna design software tools. However, for the sake of simplicity mutual coupling is not considered here. The SISTER model can be applied to different antenna patterns but for convenience, the antenna pattern which is presented here is for a half-wavelength dipole antenna.

Electric field of a *z*-directed half-wavelength dipole antenna is as follows [Balanis, 1997], [Allen & Ghavami, 2005]:

$$E_\theta = j\eta \frac{I_0 e^{-jkr}}{2\pi r} \left[\frac{\cos\left(\frac{\pi}{2}\cos\theta\right)}{\sin\theta} \right] \tag{2}$$

where E_θ, η, I_0, θ and r are electric field in $\vec{a}_\theta$ direction, intrinsic impedance of free space, current amplitude, elevation angle and radial distance of observation point. Assuming that the array axis is in *z* direction, array factor formula can be obtained by [Balanis, 1997], [Allen & Ghavami, 2005]:

$$\begin{cases} AF = \sum_{n=1}^{N} e^{j(n-1)\Psi} \\ \Psi = kd\cos\theta + \beta \end{cases} \tag{3}$$

where N, Ψ, k, d, θ and β are the number of array elements, progressive phase, wave number, elements' spacing, elevation angle of observation point and progressive phase lead current excitation, respectively.

For an array, different scan angles can be used for different MIMO elements. Recalled from antenna theory, scan angle of θ_0 can be achieved by choosing β as follows [Balanis, 1997], [Allen & Ghavami, 2005]:

$$\beta = -kd\cos\theta_0 \tag{4}$$

3.2 Scattered field calculation

Consider a sphere of radius *a* located at the origin as it is shown in Fig. 2.

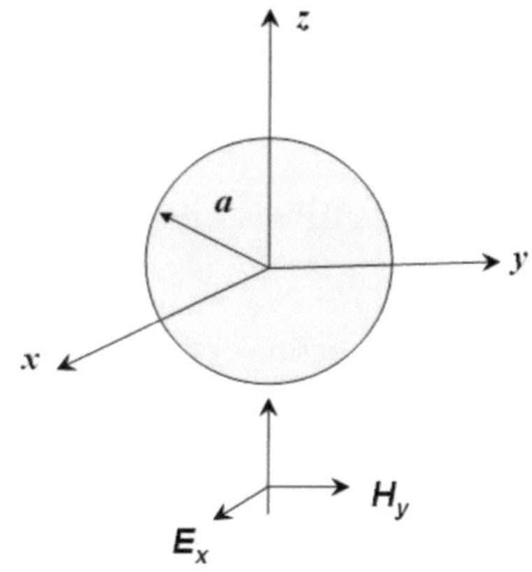

Fig. 2. A sphere of radius *a* located at the origin as a scatterer.

Assuming a uniform plane wave polarized in the *x* direction traveling along the *z*-axis is incident upon this sphere, the incident electric field is given by:

$$\begin{cases} E_i = E_0 e^{-jkr} \vec{a}_x \\ k = \omega\sqrt{\mu\varepsilon} \end{cases} \tag{5}$$

where E_0 is the incident field amplitude, ω is angular velocity, k, μ and ε are the wave number, electric permeability and permittivity of surrounding medium, respectively. Then the far-field expressions for scattered field from the spherical scatterer at a point (r, θ_i, φ_i) can be written as:

$$\vec{E}^S = E_0 \frac{e^{-jkr}}{r} (E_\theta^S \vec{a}_\theta + E_\varphi^S \vec{a}_\varphi) \tag{6}$$

where E_θ^S and E_φ^S are as follows [Svantesson, 2001]:

$$\begin{cases} \vec{E}_\theta^S = \dfrac{j\cos\varphi_i}{k} \times \sum\limits_{n=1}^{\infty} j^n \dfrac{2n+1}{n(n+1)} [a_n u_1(\theta_i) - b_n u_2(\theta_i)] \\ \vec{E}_\varphi^S = \dfrac{j\sin\varphi_i}{k} \times \sum\limits_{n=1}^{\infty} j^n \dfrac{2n+1}{n(n+1)} [a_n u_2(\theta_i) - b_n u_1(\theta_i)] \end{cases} \tag{7}$$

where $u_1(\theta_i)$ and $u_2(\theta_i)$ are:

$$\begin{cases} u_1(\theta_i) = \sin\theta_i P_n^{1'}(\cos\theta_i) \\ u_2(\theta_i) = \dfrac{P_n^1(\cos\theta_i)}{\sin\theta_i} \end{cases} \tag{8}$$

where, P_n^m is the "Associated Legendre Function" [Balanis, 1989] and assuming that the permeability of the sphere is the same as surrounding environment, a_n and b_n can be written as:

$$\begin{cases} a_n = \dfrac{-k^2 j_n(s)[s_1 j_n(s_1)]' + k_1^2 j_n(s_1)[s j_n(s)]'}{k^2 h_n^{(2)}(s)[s_1 j_n(s_1)]' - k_1^2 j_n(s_1)[s h_n^{(2)}(s)]'} \\ b_n = \dfrac{-j_n(s)[s_1 j_n(s_1)]' + j_n(s_1)[s j_n(s)]'}{h_n^{(2)}(s)[s_1 j_n(s_1)]' - j_n(s_1)[s h_n^{(2)}(s)]'} \end{cases} \tag{9}$$

where $j_n(x)$ is the "Spherical Bessel Function" of order n, $h^{(2)}{}_n(x)$ is the "Spherical Hankel Function" of the second kind of order n [Balanis, 1989] and a_n, b_n are coefficient dependent on the electrical size of the spherical scatterer and s and s_1 are defined by:

$$\begin{cases} s_1 = k_1 a \\ s = k\ a \end{cases} \tag{10}$$

where k_1, k and a are the wave number for the spherical scatterer, free space wave number and scatterer radius, respectively.
The infinite summation is approximated by taking only a limited number of terms (n_c). A rule of thumb of how many terms that should be evaluated is [Svantesson, 2001]:

$$n_c = s + 4.05 s^{1/3} + 2 \tag{11}$$

Finally in order to have large amount of scattering, electrical conductivity should be chosen high enough. Therefore, the dielectric properties of the conducting scatterers are assumed as follows:

$$\begin{cases} \varepsilon_s = \varepsilon_0 (1 - j100) \\ \mu = \mu_0 \end{cases} \tag{12}$$

where ε_0 and μ_0 are surrounding medium's (air) electrical permittivity and magnetic permeability, respectively.

3.3 Reflected field calculation

To simulate the indoor scenario, transmitter, receiver and the scatterers are located in a simple cubic room, the dimensions of which can be changed. For each single antenna at Tx and Rx in a simple cubic room, six reflecting points exist. For example for a 4×4 MIMO. for the sixteen existing channels, 96 reflection points exist. For each transmitter and receiver set, reflecting points from different walls are found in terms of the dimensions of the wall and the position of Tx and Rx.
To visualize the geometry easier, two reflecting points A_1 and A_5 corresponding to walls1 and 5 and their planes of incidence are shown in Fig. 3.
As it is shown in Fig. 4, two triangles of ABC and AB'C' are similar and hence reflecting point, A, can be obtained as follows:

$$\begin{cases} \dfrac{AB}{AB'} = \dfrac{BC}{B'C'} = \text{Known} \\ BA + AB' = BB' = \text{Known} \end{cases} \Rightarrow \quad AB \text{ and } AB' \text{ can be obtained} \tag{13}$$

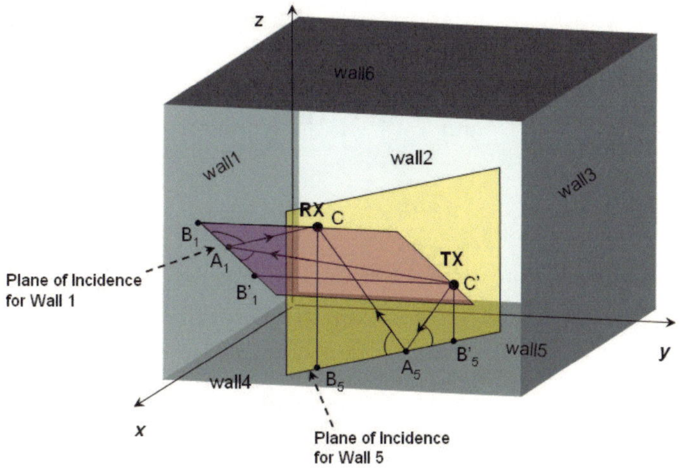

Fig. 3. 3-D geometry of two reflecting points.

where BB′ is the distance between projection points of Rx and Tx on the wall and BC and B′C′ are the distances between Rx and Tx and the wall, respectively.

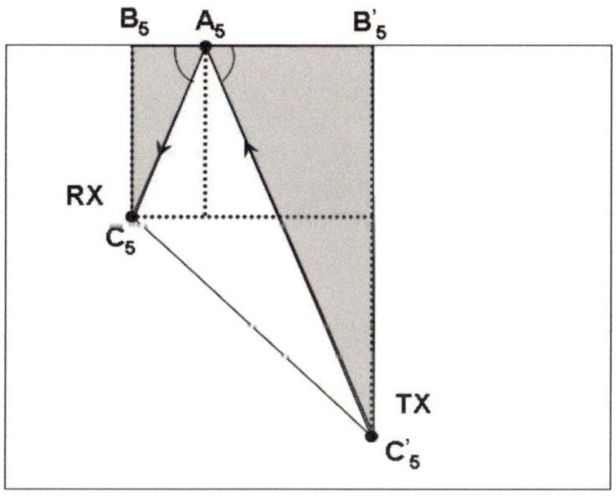

Fig. 4. 2-D Geometry of reflecting points in the plane of incidence.

As an example, A_5, reflecting point of wall5 can be found from two equations given below:

$$\begin{cases} \dfrac{A_5B_5}{A_5B'_5} = \dfrac{B_5C_5}{B'_5C'_5} = \dfrac{z_{RX} - z_{wall5}}{z_{TX} - z_{wall5}} \\ B_5A_5 + A_5B'_5 = B_5B'_5 = \sqrt{(x_{RX} - x_{TX})^2 + (y_{RX} - y_{TX})^2} \end{cases} \tag{14}$$

Other reflecting points can also be found in a similar way.

After finding all the reflecting points, the electric fields originated at Tx side and reflected from theses points and terminated at Rx side can be calculated. These fields must be added to those obtained from all scatterers and the direct path between Tx and Rx to get the total electric field. Since the reflection coefficient is different for transverse or perpendicular (Γ_{TE}) and parallel (Γ_{TM}) polarization of electric field relative to the plane of incidence, received electric field on the boundary should be decomposed into Transverse Electric (TE) and Transverse Magnetic (TM) polarizations. Plane of incidence is the plane containing both a normal to the boundary and the incident wave's propagation direction [Wentworth, 2005]. This plane is shown in Fig. 4.

To decompose electric field components for wall5, for instance, E_x and E_y, each are split into two polarizations of E_x^{TM}, E_x^{TE} and E_y^{TM}, E_y^{TE}, respectively (Fig. 5):

$$\begin{cases} E_x^{TM} = E_x \cos(\psi)\Gamma_{TM} \\ E_x^{TE} = E_x \sin(\psi)\Gamma_{TE} \end{cases} \tag{15}$$

$$\begin{cases} E_y^{TM} = E_y \sin(\psi)\Gamma_{TM} \\ E_y^{TE} = E_y \cos(\psi)\Gamma_{TE} \end{cases} \tag{16}$$

where $\psi = \arctan\frac{y_{cross}}{x_{cross}}$ and Γ_{TE} and Γ_{TM} are reflection coefficients of TE and TM polarizations, respectively and are shown in Fig. 5.

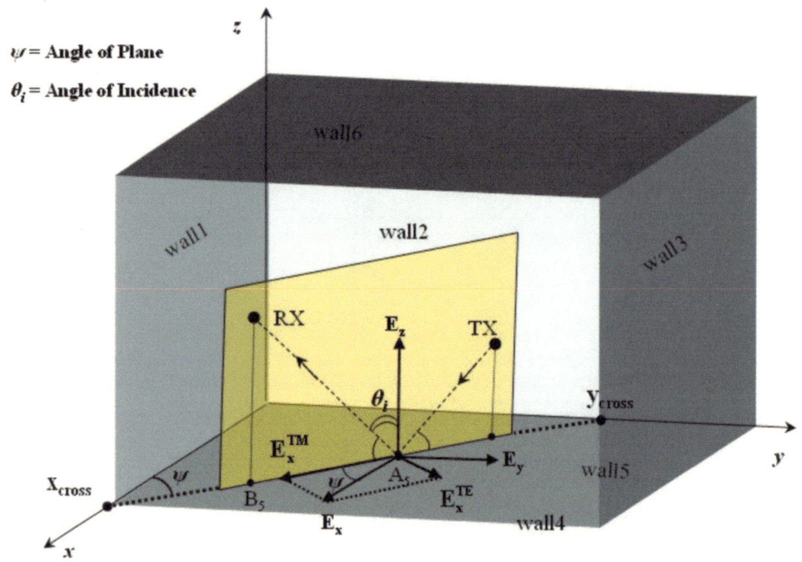

Fig. 5. 3-D view of electric field decomposition to TM and TE polarizations at the reflecting point.

Since E_z itself is the parallel component (TM), it does not need to be decomposed and hence to find its corresponding reflected field, it should be simply multiplied by Γ_{TM}.

After finding TM and TE components of reflected waves, they should be converted to previous global coordinates for further process:

$$\begin{cases} E_x' = E_x^{TM} \cos(\psi) + E_x^{TE} \sin(\psi) \\ E_y' = E_y^{TM} \sin(\psi) + E_y^{TE} \cos(\psi) \end{cases} \tag{17}$$

where $E_x^{'}$ and $E_y^{'}$ are the *x* and *y* components of the reflected electric field from wall5. The same procedure is applicable for other walls. To find Γ_{TM} and Γ_{TE}, angles of incidence and transmission are required [Wentworth, 2005]:

$$\begin{cases} \Gamma_{TE} = \dfrac{\eta_2 \cos(\theta_i) - \eta_1 \cos(\theta_t)}{\eta_2 \cos(\theta_i) + \eta_1 \cos(\theta_t)} \\ \Gamma_{TM} = \dfrac{\eta_2 \cos(\theta_t) - \eta_1 \cos(\theta_i)}{\eta_2 \cos(\theta_t) + \eta_1 \cos(\theta_i)} \end{cases} \tag{18}$$

where (η_1, η_2), (θ_i, θ_t) are the intrinsic impedances of free space and wall material and angles of incidence and transmission, respectively. Referring to Fig. 5, one can easily calculate angles of incidence and transmission for wall5 as follows:

$$\begin{cases} \theta_i = \dfrac{\pi}{2} - \arctan \dfrac{h_{Rx}}{B_5A_5} \\ \theta_t = \arcsin \dfrac{k_1 \sin(\theta_i)}{k_2} \end{cases} \tag{19}$$

where (θ_i, θ_t), h_{Rx}, (k_1, k_2) are angles of incidence and transmission, Rx height and wave number of air and wall material, respectively.

3.4 Channel capacity calculation

Assuming that the channel is unknown to the transmitter and the total transmitted power is equally allocated to all N_T antennas, the capacity of the system is given by [Foschini & Gans, 1998]:

$$C = \log_2\left((\det\left[\mathbf{I}_{N_T} + \frac{SNR}{N_T} \times \frac{\mathbf{HH}^*}{norm(HH^*)}]\right])\right) \quad bps/Hz \tag{20}$$

where $\mathbf{I}_{N_T}$ is the identity matrix, SNR is the average signal to noise ratio within the receiver aperture, N_T is the number of transmitter antennas, **H** is the $N_T \times N_R$ channel matrix and **H*** is the conjugate transpose of **H**. To calculate **H**-matrix baseband channel complex impulse response should be computed for scatterers, reflectors and direct path corresponding to each channel.

1. <u>Scatterers</u>

$$h_{scatterers} = \sum_{q=1}^{N_S} \frac{e^{-jk(|\vec{r}_{msq}| + |\vec{r}_{sqb}|)}}{(|\vec{r}_{msq}| \times |\vec{r}_{sqb}|)} [E_\theta(\vec{r}_{bs}) \cdot \vec{\ell}_{eff\theta} + E_\varphi(\vec{r}_{bs}) \cdot \vec{\ell}_{eff\varphi}] \tag{21}$$

where $N_S, \vec{r}_{msq}, \vec{r}_{sqb}, (E_\theta, E_\varphi), (\vec{\ell}_{eff\theta}, \vec{\ell}_{eff\varphi})$ are the number of scatterers, distance vector from Tx (MS) to q[th] scatterer, distance vector from Rx (BS) to q[th] scatterer, effective radiation pattern at Rx in $\vec{a}_\theta$ and $\vec{a}_\varphi$ directions (radiation patterns of Tx and Rx are included in effective radiation pattern), and effective lengths of the half-wavelength dipole in $\vec{a}_\theta$ and $\vec{a}_\varphi$ directions, respectively.

Assuming that the half-wavelength dipole antenna is connected to a matched load and current distribution is sinusoidal, two components of effective complex length of dipole can be obtained from [Collin, 1985]:

$$\begin{cases} \vec{\ell}\mathrm{eff}_{\theta} = \dfrac{\lambda}{\pi}\dfrac{E_{\theta}}{|E_0|} \\ \vec{\ell}\mathrm{eff}_{\varphi} = \dfrac{\lambda}{\pi}\dfrac{E_{\varphi}}{|E_0|} \end{cases} \tag{22}$$

where E_{θ} and $E\phi$ are the electric fields radiated by the half-wavelength dipole while it is in transmitting mode.

2. Reflectors

$$h_{\mathrm{reflectors}} = \sum_{q=1}^{N_r} \frac{e^{-jk(|\vec{r}_{mrq}|+|\vec{r}_{rqb}|)}}{(|\vec{r}_{mrq}|\times|\vec{r}_{rqb}|)}[E_{\theta}(\vec{r}_{br})\cdot\vec{\ell}_{\mathrm{eff}\theta} + E_{\varphi}(\vec{r}_{br})\cdot\vec{\ell}_{\mathrm{eff}\varphi}] \tag{23}$$

where $N_r, \vec{r}_{mrq}, \vec{r}_{rqb}, (E_{\theta}, E_{\varphi}), (\vec{\ell}_{\mathrm{eff}\theta}, \vec{\ell}_{\mathrm{eff}\varphi})$ are the number of reflectors, distance vector from Tx to q^{th} reflector (wall), distance vector from Rx to q^{th} reflector, effective radiation pattern at Rx in $\vec{a}_{\theta}$ and $\vec{a}_{\varphi}$ directions, and effective lengths of the half-wavelength dipole in $\vec{a}_{\theta}$ and $\vec{a}_{\varphi}$ directions, respectively.

3. Direct Path

To obtain direct field between Tx and Rx, the following equation is used:

$$h_{\mathrm{direct}} = \frac{e^{-jk|\vec{r}_{mb}|}}{|\vec{r}_{mb}|}[E_{\theta}(\vec{r}_{bm})\cdot\vec{\ell}_{\mathrm{eff}\theta} + E_{j}(\vec{r}_{bm})\cdot\vec{\ell}_{\mathrm{eff}\varphi}] \tag{24}$$

where $\vec{r}_{mb}, (E_{\theta}, E_{\phi}), (\vec{\ell}_{\mathrm{eff}\theta}, \vec{\ell}_{\mathrm{eff}\phi})$ are the distance vector from Tx to Rx, effective radiation pattern at Rx in $\vec{a}_{\theta}$ and $\vec{a}_{\varphi}$ directions and the effective lengths of the half-wavelength dipole in $\vec{a}_{\theta}$ and $\vec{a}_{\varphi}$ directions, respectively.

3.5 Coordinate transformations

To find the total electric field at Rx which is the last destination of the traveled wave, many coordinate transformations should be performed. Since, it is much easier to transform rectangular coordinates of local and global systems rather than spherical ones, before each transformation step, electric field in rectangular coordinate should be found.

Equation (25) is used frequently while developing the mathematical model. It is a general formula to rotate a coordinate system and convert it to the other one by knowing the angles between their axes.

$$\underbrace{\begin{bmatrix} \hat{u}_1 \\ \hat{u}_2 \\ \hat{u}_3 \end{bmatrix}}_{New\_System} = \underbrace{\begin{bmatrix} \hat{a}_1\cdot\hat{u}_1 & \hat{a}_2\cdot\hat{u}_1 & \hat{a}_3\cdot\hat{u}_1 \\ \hat{a}_1\cdot\hat{u}_2 & \hat{a}_2\cdot\hat{u}_2 & \hat{a}_3\cdot\hat{u}_2 \\ \hat{a}_1\cdot\hat{u}_3 & \hat{a}_2\cdot\hat{u}_3 & \hat{a}_3\cdot\hat{u}_3 \end{bmatrix}}_{Rotation\_Matrix} \underbrace{\begin{bmatrix} \hat{a}_1 \\ \hat{a}_2 \\ \hat{a}_3 \end{bmatrix}}_{Old\_System} \tag{25}$$

The given solution in (7) is for an x oriented field propagation along the z-axis. However, these conditions will rarely be met since the same coordinate system is used for all scatterers. By employing a local coordinate system for each object, the mentioned solution can be applied.

Different local and global coordinates are shown in Fig. 6 and defined as follows:

- Gmain (x_{Gmain}, y_{Gmain}, z_{Gmain}) is the global coordinate.
- G1 (x_{G1}, y_{G1}, z_{G1}) is a parallel coordinate system with Gmain and its origin is on the center of Tx.
- L1 (x_{L1}, y_{L1}, z_{L1}) is the local coordinate for Tx antenna and its origin is the same as that of G1 and also for this coordinate system z_{L1} is chosen along the direction of Tx dipole and x_{L1} is defined on the plane of x_{G1} and y_{G1}.
- L2 (x_{L2}, y_{L2}, z_{L2}) is the local coordinate for scatterers and its origin is on the scatterer center and for this coordinate system z_{L2} is chosen along the direction of r_{L1} and x_{L2} is chosen along the direction of $\hat{\theta}_{L1}$. r_{L1}, θ_{L1}, φ_{L1} are spherical coordinate components of each scatterer in respect to *L1* coordinate. It is worth mentioning that for each scatterer an *L2* coordinate is defined.
- *L3* (x_{L3}, y_{L3}, z_{L3}) is the local coordinate for Rx antenna the origin of which is on the center of Rx and also for this coordinate system z_{L3} is chosen along the direction of Rx dipole and x_{L3} is defined on a plane parallel to the plane of x_{Gmain} and y_{Gmain}.

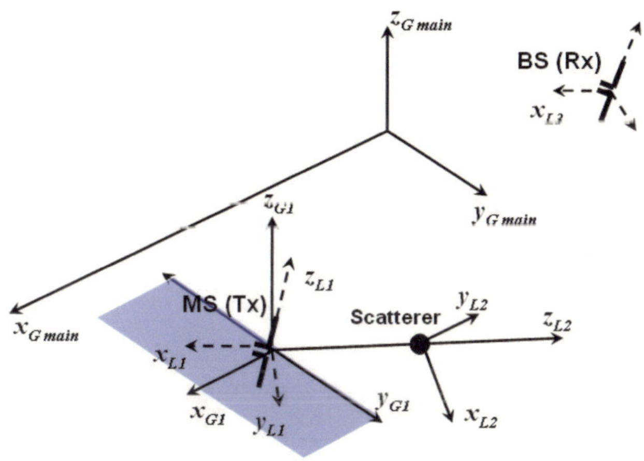

Fig. 6. Global and local coordinates and dipole antennas at both ends.

The local coordinates L1 and L3 are defined to provide the possibility of using different polarizations for Tx and Rx antennas, respectively.

Now to fulfill the condition required for using the scattering formulas, L1 coordinate system should be converted to L2 coordinate system which is the local coordinate system of each scatterer. If the scatterer is located at (r_{L1}, θ_{L1}, φ_{L1}) in respect to L1 coordinate system, to convert L1 into L2 coordinates system, one can use:

$$\begin{bmatrix} \hat{x} & \hat{y} & \hat{z} \end{bmatrix}_{L2} = \begin{bmatrix} \hat{x} & \hat{y} & \hat{z} \end{bmatrix}_{L1} \begin{bmatrix} \cos\theta_{L1}\cos\varphi_{L1} & -\sin\varphi_{L1} & \sin\theta_{L1}\cos\varphi_{L1} \\ \cos\theta_{L1}\sin\varphi_{L1} & \cos\varphi_{L1} & \sin\theta_{L1}\sin\varphi_{L1} \\ -\sin\theta_{L1} & 0 & \cos\theta_{L1} \end{bmatrix} \quad (26)$$

where θ_{L1} and φ_{L1} are scatterer's coordinates referring to L1.
If the Tx antenna type is something other than dipole or generally, is an antenna with electric field in both $\hat{\theta}$ and $\hat{\varphi}$ directions then the relation between the L1 and L2 coordinate systems is more complicated and the corresponding rotation matrix is as follows:

$$[\hat{x}\ \hat{y}\ \hat{z}]_{L2} = [\hat{x}\ \hat{y}\ \hat{z}]_{L1} \times \frac{1}{A} \times$$
$$\begin{bmatrix} E_\theta \cos\theta_{L1}\cos\varphi_{L1} - E_\varphi \sin\varphi_{L1} & -E_\varphi \cos\theta_{L1}\cos\varphi_{L1} - E_\theta \sin\varphi_{L1} & A\sin\theta_{L1}\cos\varphi_{L1} \\ E_\theta \cos\theta_{L1}\sin\varphi_{L1} + E_\varphi \cos\varphi_{L1} & -E_\varphi \cos\theta_{L1}\sin\varphi_{L1} + E_\theta \cos\varphi_{L1} & A\sin\theta_{L1}\sin\varphi_{L1} \\ -E_\theta \sin\theta_{L1} & +E_\varphi \sin\theta_{L1} & A\cos\theta_{L1} \end{bmatrix} \quad (27)$$

where E_θ, E_φ are the electric field components at each scatterer center referred to L1 and θ_{L1} and φ_{L1} are scatterer's coordinates and $A=\sqrt{E_\theta^2+E_\varphi^2}$. Equation (27) is simplified to rotation matrix in (26) if Tx antennas has electric field only in θ direction.
Finally, after all conversions of coordinate systems, the vectors which are necessary to find channel complex impulse response such as electric fields and effective lengths should be converted to the main global coordinate which is specified as G_{main} in Fig. 6.

4. Verifying the SISTER model

To verify the obtained results from developed model, "Wireless Insite" software by Remcom Inc. [Remcom Inc., 2004] is used. This software is a three-dimensional ray tracing tool for both indoor and outdoor applications which models the effects of surrounding objects on the propagation of electromagnetic waves between Tx and Rx.
In order to accomplish this verification, different steps have been taken. First, only a direct path between Tx and Rx is considered for a Single Input Single Output (SISO) system and received power is verified by both Friis equation and ray tracing tool.
It is assumed that a half wavelength dipole antenna (Gain=2.16dBi) is used at both ends, Tx-Rx distance is 2.7m, both Tx and Rx heights are 1.5m and transmitted power is 0dBm (1mW). For the mentioned system configuration, numerical results obtained from both proposed mathematical model and ray tracing are summarized in Table 1.

	$P_{received}$	$\|E_z\|$ (V/m)	Phase E_z (degree)
SISTER Model	-44.362 dBm (3.663×10^{-8} W)	0.117	76.917
Ray Tracing	-44.350 dBm (3.673×10^{-8} W)	0.117	73.496
Friis Equation	-44.337dBm (3.684×10^{-8} W)	-------	-------

Table 1. Numerical results for a SISO system.

As it can be seen the result obtained from the SISTER model matches well with a fractional error less than 0.006 with both ray tracing tool and also Friis transmission equation given in (28) [Balanis, 1997]:

$$\frac{P_r}{P_t} = \left(\frac{\lambda}{4\pi R}\right)^2 G_r G_t \tag{28}$$

where P_r, P_t, λ, R, G_r and G_t are received power, transmitted power, wavelength, Tx-Rx distance and Rx and Tx antenna gains, respectively.

In the next step (Fig. 7) one wall is added to the previous system configuration and the reflected ray is evaluated as well. For this case, summarized results can be found in Table 2 which again shows an acceptable match with those of the ray tracing. The same procedure to validate the reflected field has been done for all six walls and all have shown good match.

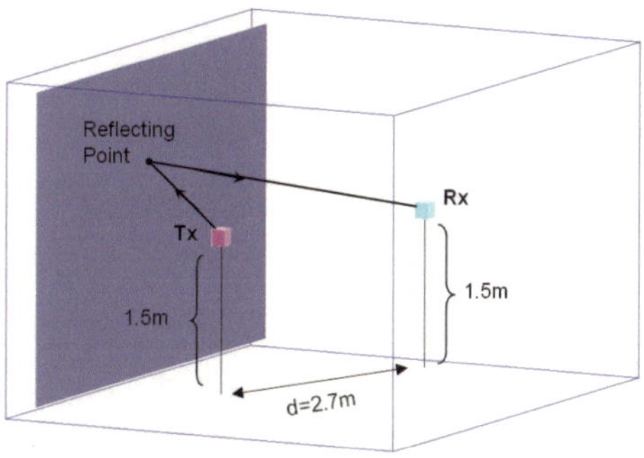

Fig. 7. Ray tracing visualization of a SISO system in an indoor environment considering reflection from one wall.

	$P_{received}$	$\|E_z\|$ (V/m)	Phase E_z (degree)
SISTER Model	-48.442 dBm (1.432×10^{-8} W)	0.073	-115.719
Ray Tracing	-48.461 dBm (1.425×10^{-8} W)	0.073	-121.210

Table 2. Numerical results for a SISO system configuration shown in Fig. 7

Channel capacity for the MIMO system configuration illustrated in Fig. 8 is compared for both proposed model and ray tracing tool. Fig. 9 shows the results for three cases; direct path only, reflected paths only, total paths.

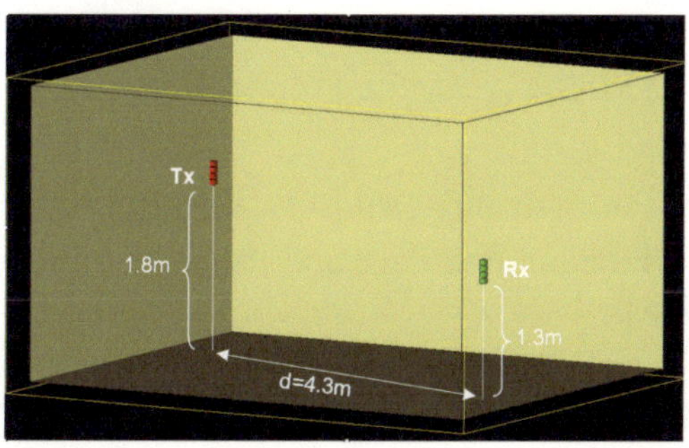

Fig. 8. Ray tracing visualization of a 4×4-MIMO system in an indoor environment considering six walls.

As the final step to verify the results, the capacity of MIMO systems with different $N_T \times N_R$ antenna numbers are evaluated in an outdoor environment for NLOS case and the results are compared with Rayleigh model for similar antenna numbers. Fig. 10 shows the capacities obtained from simulated Rayleigh channel by MATLAB and SISTER model applied to an outdoor NLOS environment with 30 scatterers for different numbers of antennas.

As these results show good agreement with both ray tracing tool and Rayleigh model is achieved.

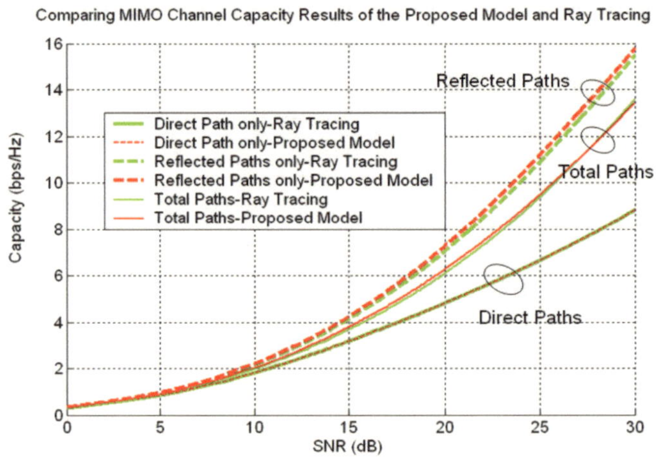

Fig. 9. Comparing MIMO channel capacity obtained from SISTER model and ray tracing tool for different rays.

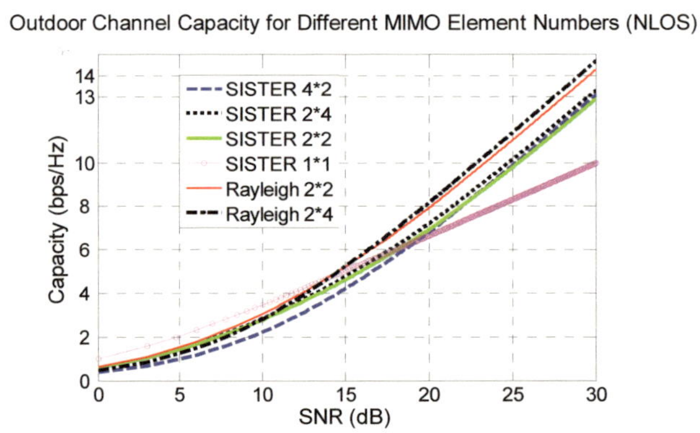

Fig. 10. Comparing channel capacity obtained from SISTER model and Rayleigh model.

The MIMO configuration is the same as Fig.8 and the room dimensions are $5 \times 4 \times 3$ m^3 and a wall exists to block the LOS path.

5. Results of applying SISTER model for different scenaris

Although the SISTER model is sufficiently general to be applied to any distributions and locations for the scatterers, here we concentrate only on picocell environments.

Moreover, "Angle Diversity" which is a new promising solution and has recently attracted considerable attention in MIMO system designs [Allen et al., 2004] is also evaluated model and compared with well-known "Space Diversity" method by applying the SISTER. In this method, instead of multiple antennas used in space diversity case, multiple simultaneous beams are assumed at both sides. The main advantage of this technique comparing is that it allocates high capacity not to all the points in space, but the desired ones. This results in minimum undesired interference. The main difficulty in such systems, however, is the beam cusps (beam overlaps) [Allen & Beach, 2004] and finding the optimal angles where the different beams should be directed towards. We have investigated the use of antenna array in angle diversity case to implement the narrow beams needed in this method. We also have addressed some problems with beam cusps which introduce correlations in MIMO channels, and suggested some solutions to overcome this problem.
Here, various results are presented which are ultimately useful to set the system design parameters and to evaluate and compare the performance of MIMO systems using space or angle diversity for both outdoor and indoor environments. Due to space limitations only some of the results are presented here and more results can be found in [E.Forooshani, 2006].

5.1 SISTER results for outdoor environments

Outdoor system specifications considered are summarized in Table 3. Tx refers to transmitter and Rx refers to receiver antennas. Without loosing the generality, it is assumed that mobile set (MS) is the transmitter and the base station (BS) is the receiver side. All simulations are done based on working frequency of 2.4GHz. For results shown in Figs 11-15, a 4×4 MIMO system is considered.
Two common scatterer distributions for outdoor environments are uniform distribution around each end and cluster distribution, as shown in Fig. 11(a) and Fig. 11(b), respectively.

	Tx (MS) height	Rx (BS) height	Relative height of Tx and Rx	Distance between Tx and Rx
Outdoor System	24λ (3m)	40λ (5m)	16λ (2m)	102λ (13m)

Table 3. Outdoor system specifications.

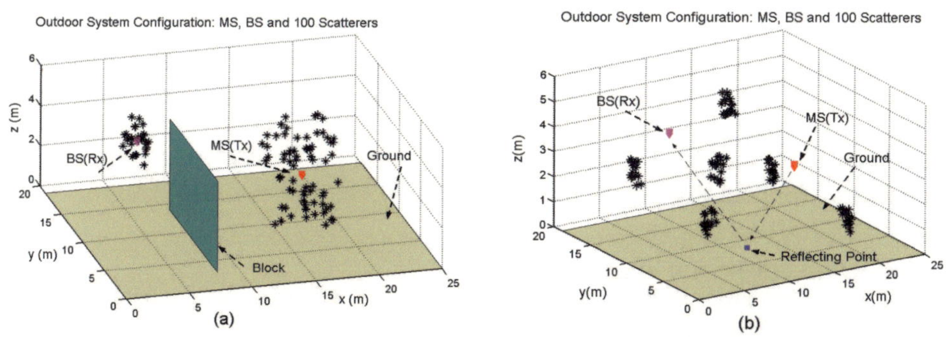

Fig. 11. Outdoor system configuration for: (a) NLOS scenario with uniformly distributed scatterers around both ends, (b) LOS scenario with cluster form scatterers in a cubic volume ($200\lambda \times 150\lambda \times 50\lambda$ or 25×18.75×6.25, m^3).

5.1.1 Impact of ground material

For outdoor environment, impact of two types of ground material, high and low conductive ones (Fig. 12) are investigated. Reflection from the high conductive ground contributes as much as the direct path and its presence can suppress the effect of direct path and hence increase the capacity comparing to the low conductive ground case. It also shows that for a ground with conductivity more than 100 S/m, capacity is mainly controlled by the reflected path from the ground and scatterers do not contribute much in the channel capacity.

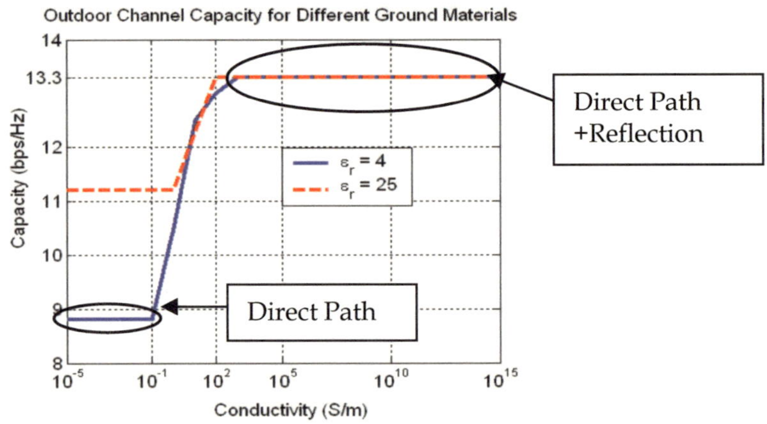

Fig. 12. Channel capacity at signal to noise ratio, SNR=30dB for different ground materials (ε_r=4, ε_r=25) considering 30 uniformly distributed scatterers, the LOS case.

5.1.2 Impact of number of scatterers

Figs. 13 and 14 show the impact of number of uniformly distributed scatterers in terms of channel capacity versus SNR. Typical number of scatterers for this study is 30. In NLOS case, it is assumed that there is no direct path but reflection from the ground exists (blocked LOS or quasi-LOS). Fig. 13 shows the LOS case. In this case reflection from the high conductive ground contributes as much as the direct path. Therefore, its presence can suppress the effect of direct path and hence increase the capacity in compare to the low conductive ground case.

For NLOS case, shown in Fig. 14, when the number of scatterer is not high (30 scatterers) reflection from the high conductive ground creates the dominant path and capacity is low. When the number of scatterers is high enough (100 scatterers), they are able to lessen the effect of reflection from the ground and in this case capacity is higher. For low conductive ground, on the other hand, the reflection from the ground is so weak that no dominant path exists and hence for both cases of 30 and 100 scatterers, channel capacity is high.

5.1.3 Comparing space and angle diversities

To compare space and angle diversity methods for a 4×4-MIMO system, a scenario consisting of four clusters of scatterers is considered. The length occupied by antenna elements is the same for both space and angle diversity methods. It is essential to keep the array length the same if we intend to have a fair comparison between the two methods in terms of system size and length. Antenna array length at both ends is 1.5λ.

For space diversity case, four antenna elements are used while in angle diversity the same four elements are used along with a Butler matrix to create four simultaneous beams with different scan angles. Assumptions made for space and angle diversity methods are summarized in Table 4.

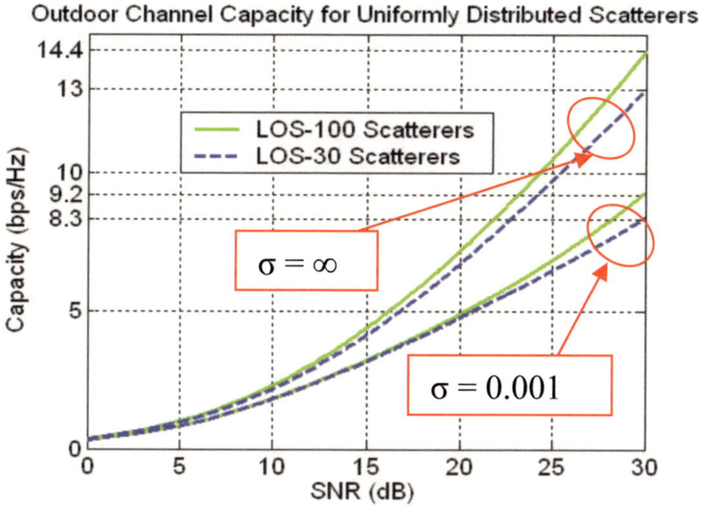

Fig. 13. Channel capacity for different number of scatterers distributed uniformly around both ends in LOS case (σ=ground's electrical conductivity, S/m).

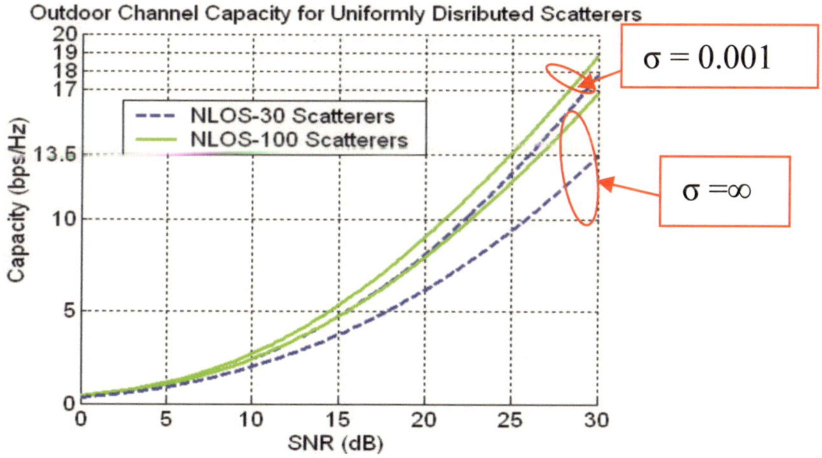

Fig. 14. Channel capacity for different numbers of scatterers distributed uniformly around both ends in NLOS case including reflection from the ground but not the direct path (σ=ground's electrical conductivity).

	Number of elements at BS	Number of elements at MS	BS element spacing (d-Rx)	MS element spacing (d-Tx)
Space Diversity	4	4	0.5λ	0.5λ
Angle Diversity	4	4	0.5λ	0.5λ

Table 4. Assumptions for space and angle diversity methods.

For space and angle diversities channel capacity is calculated based on equations (29) and (30), respectively.

$$C(SNR)=\log_2\left(\det\left[\mathbf{I}_{N_T}+\frac{SNR}{N_T}\times\frac{\mathbf{HH}^*}{\mathrm{norm}(\mathbf{HH}^*)}\right]\right) \tag{29}$$

$$C(SNR)=\log_2\left(\det\left[\mathbf{I}_{N_T}+(G_{Tx}\times G_{Rx})\frac{SNR}{N_T}\times\frac{\mathbf{HH}^*}{\mathrm{norm}(\mathbf{HH}^*)}\right]\right) \tag{30}$$

where C is the channel capacity, I_{N_T} is the Identity matrix, SNR is the signal to noise ratio, N_T is number of transmitter antennas (or beams) and **H** is the channel matrix, whose elements are calculated using the SISTER model. For space diversity h_{ij} is the path gain between antenna element i at BS and j at MS. For angle diversity each h_{ij} represents the path gain between i^{th} beam at BS and j^{th} beam at MS.

Factor (G_{Tx} × G_{Rx}) in (30) shows the array gain of angle diversity method. When an array consists of elements with the spacing of 0.5λ, then its gain is equal to the number of elements if antenna losses are ignored (G_{Tx} × G_{Rx} =4×4=16). Since it is assumed that the total power is the same for two systems, it is required to take the array gain into account while comparing capacities of two methods in terms of SNR. Note that no mutual coupling effect is assumed in this calculation.

Fig.15 shows four beams angels at MS and BS sides for angle diversity case.

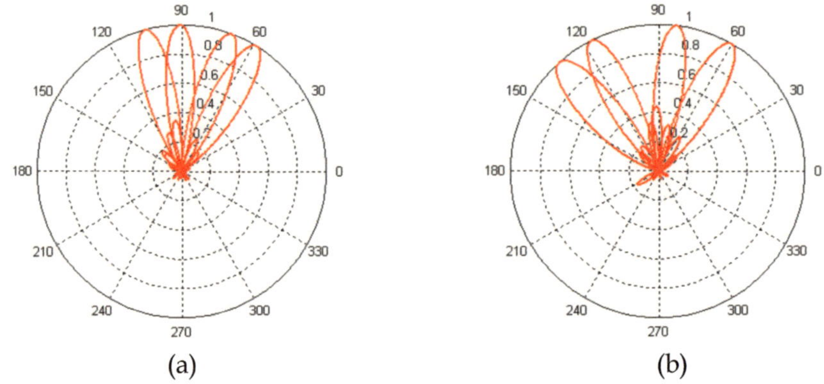

Fig. 15. Four multibeams which are pointed towards four clusters located in different θ angles (a) MS (Tx) (N-array=4, beam angles=62°, 70°, 91°, 105°), (b) BS (Rx) (N-array=4, beam angles=60°, 83°, 117°, 132°).

Table 5 and Fig. 18 (a) show singular values of normalized **H**-matrix and capacity results for both methods in LOS case, respectively. Table 6 and Fig. 16 (b) show singular values of normalized **H**-matrix and capacity results for both methods in NLOS case, respectively.
As Fig. 16 show angle diversity surpass space diversity significantly, mostly due to the array gain. Even though angle diversity often shows better channel orthogonality, improperly chosen angles caused not to achieve the maximum available capacity for the angle diversity.

	Singular Value1	Singular Value2	Singular Value3	Singular Value4
Space Div.	1.0000	0.0016	0.0004	0.0000
Angle Div.	1.0000	0.0024	0.0008	0.0000

Table 5. Singular values for 30 scatterers in 4 clusters for LOS.

	Singular Value1	Singular Value2	Singular Value3	Singular Value4
Space Div.	1.0000	0.4424	0.0062	0.0003
Angle Div.	1.0000	0.4481	0.0007	0.0000

Table 6. Singular values for 30 scatterers in 4 clusters for NLOS.

For NLOS case, the rays from Tx towards clusters behind the block are stopped which cause reduction in the number of channels. Another reason which has caused getting undesirable results for angle diversity method in both LOS and NLOS cases is the beam cusps. Considering above discussion, for the given scenario, angle diversity seems to be an appropriate alternative for space diversity which can provide similar orthogonality with less interference.

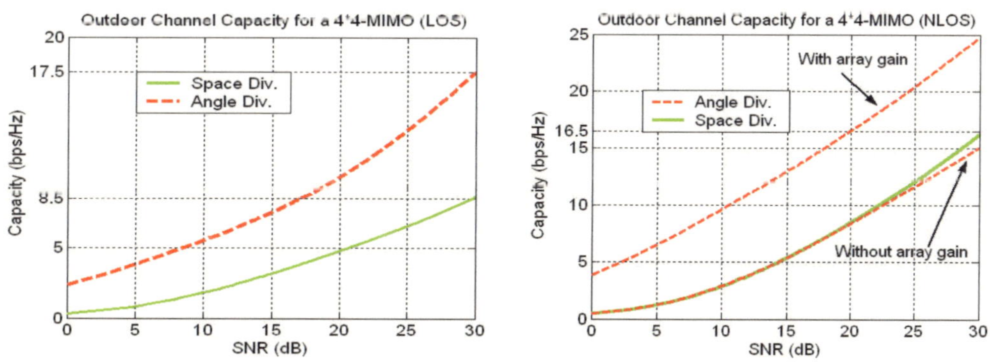

(a) (b)

Fig. 16. Channel capacity for 30 scatterers in 4 clusters for (a) LOS, (b) NLOS.

5.1.4 Impact of number of clusters

The impact of the number of clusters on the channel capacity for a NLOS scenario, similar to what was shown in Fig. 11(b) is also studied. To consider the effects of number of clusters, clusters in this configuration are located in such a way to avoid blockage by the defined obstacle in the middle of the study area. Fig. 17 shows that for a certain amount of SNR, as the number of clusters increases, at first, channel capacity increases but after a while it remains constant. This is expected as by increasing the number of clusters multipath components are increased and correlation between channels is decreased. However, after a certain point the slope of capacity increase decreases because as the space is limited the clusters are going to be closer to each other and after a while they will have overlaps. This reduces the orthogonality of the channels. These results are also in agreement with those cited in [Burr, 2003] based on "finite scatterer channel model" Also note that as the number of scatters increases and the spacing between them decreases due to the increase in mutual interactions a single interaction models such as SISTER is not accurate anymore.

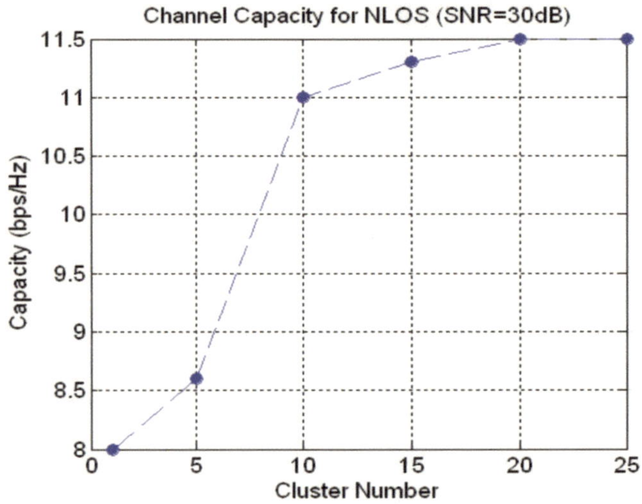

Fig. 17. Channel capacity at SNR=30 dB for different numbers of clusters which contain 10 scatterers each.

5.2 SISTER results for indoor environments

5.2.1 Office area

In order to characterize the indoor channel, the outdoor model is enhanced in such a way that it includes not only the scatterers and reflection from the ground but also reflection from the walls for a typical office area of 5×4×3 m^3. Indoor system specifications considered in this study are summarized in Table 7.

	Tx height	Rx height	Relative height of Tx and Rx	Distance between Tx and Rx	Room's dimension	Scatterers' radius	Scatterers' number
Office	10.4λ (1.3m)	14.4λ (1.8m)	4λ (0.5m)	32.24λ (4.3m)	5×4×3(m^3)	0.1m	30

Table 7. A typical office area specifications.

Two distributions of uniform and cluster form for scatterers are considered to study an office area (Fig. 18).

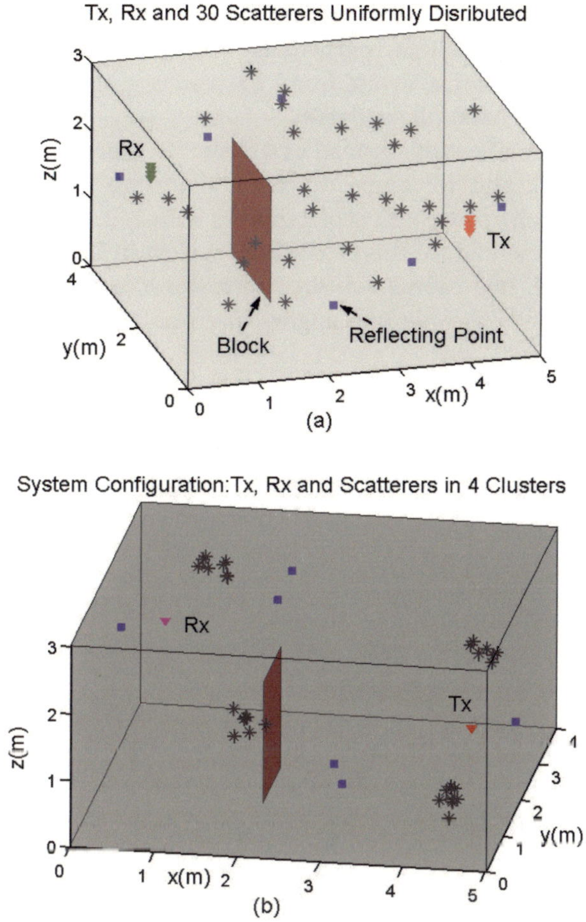

Fig. 18. An office area including Tx, Rx and 30 scatterers distributed (a) uniformly and (b) in cluster form.

5.2.2 Comparing space and angle diversities

Space and angle diversities are compared for different scenarios in [E.Forooshani, 2006] but only results for 30 uniformly distributed and cluster scatterers in indoor are presented here. Selected antenna beams in 2×2-MIMO angle diversity were (62°, 121°) for Tx and (72°, 119°) for Rx. In 4×4-MIMO systems beams were selected at (48°, 65°, 130°, 138°) for both sides. Capacities of both systems are shown in Fig. 19.

The composition of singular values is also given in Table 8. The results show that for the 4×4-MIMO system for both LOS and NLOS cases, angle diversity surpasses space diversity method in terms of channel orthogonality. Moreover, it offers array gain which leads in an increase in the capacity shown in Fig. 19(b). Based on these results, for this system, it is more convenient to apply angle diversity method since LOS and NLOS capacities are similar if the beams are selected properly while this is not true for space diversity. Furthermore, applying angle diversity helps to lessen the interference effects (compare to omnidirectional antennas, the power is directed to limited angles) in an indoor environment which is a real concern nowadays.

By try and error, it was found that, particularly for LOS case, higher capacity can be achieved by choosing angles far away from the direct path which in most cases is approximately around horizontal plane (θ=90°).

In the 2×2-MIMO for space diversity, instead of 4 elements, there are 2 elements at each end with the spacing of $3\lambda/2$ and for angle diversity; there are two arrays with λ spacing between array centers. Each array consists of 2 dipoles with $\lambda/2$ spacing.

To study angle diversity method for this 2×2-MIMO system in LOS case where 30 scatterers are uniformly distributed, two beams are directed towards the reflecting points of ceiling and the floor which actually are the two angles far from the direct path. For NLOS case,

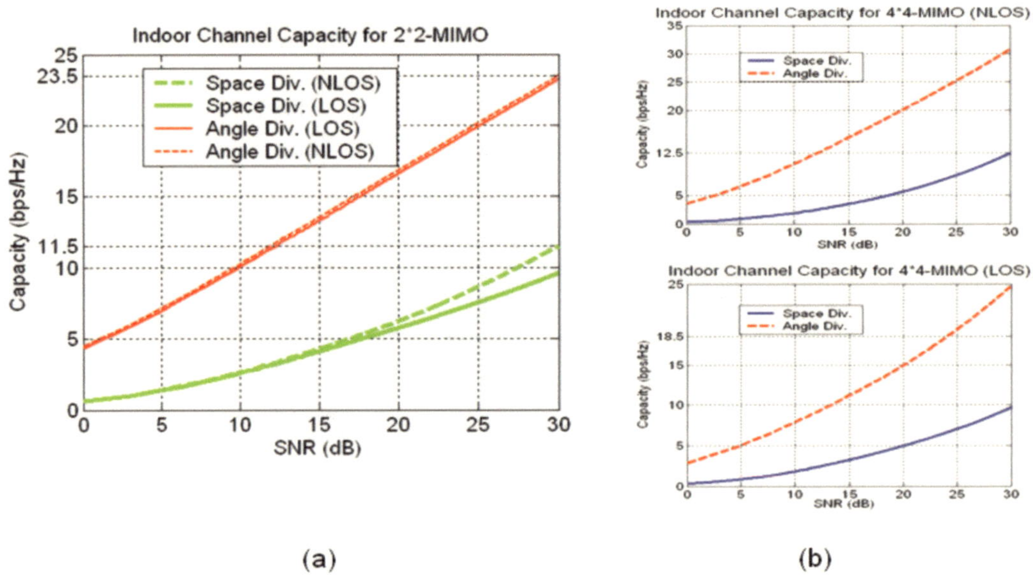

Fig. 19. Capacity for (a) 2×2-MIMO and (b) 4×4-MIMO systems.

	SV1	SV2	SV3	SV4
Space Div. (LOS) 4×4-MIMO	1.0000	0.0067	0.0008	0.0000
Angle Div. (LOS) 4×4-MIMO	1.0000	0.1120	0.0011	0.0005
Space Div. (NLOS) 4×4-MIMO	1.0000	0.0208	0.0087	0.0002
Angle Div. (NLOS) 4×4-MIMO	1.0000	0.2252	0.0658	0.0000
Space Div. (LOS) 2×2-MIMO	1.0000	0.0094	-------	-------
Angle Div. (LOS) 2×2-MIMO	1.0000	0.1529	-------	-------
Space Div. (NLOS) 2×2-MIMO	1.0000	0.0011	-------	-------
Angle Div. (NLOS) 2×2-MIMO	1.0000	0.1816	-------	-------

Table 8. Comparing singular values for the 2×2-MIMO and 4×4-MIMO systems (SV: Singular Value).

however, since no direct path exists, there is more freedom to find the desirable angles. Therefore, different angles for the NLOS case are chosen for beams that one of them is not that far from the horizontal plane.

In practical application, even though it would not be feasible to perform angle optimization every time there is a change in the Tx and Rx position, there is a possibility to develop a method for finding optimum angles. In the systems that reference signals are used even infrequently, the initial optimization based on these signals can be done and followed by updates by estimating the Angle of Arrival (AOA). The assumption in this work was that receiver has no information about the channel. This means beamforming methods that need temporal and spatial reference (training signals) is not applicable. In that case semi-blind adaptive beamforming techniques can be utilized to find the optimum angles [Allen & Ghavami, 2005]. Main concern in this work can be if the angle diversity with non-optimum angles can still outperform space diversity. Therefore, angles were chosen heuristically and no optimization was performed to find the best possible ones. The results show, for the 2×2-MIMO system similar to what was obtained for the 4×4-MIMO system, angle diversity works better for both LOS and NLOS cases. Although angle diversity for 4×4-MIMO system shows better performance, still 2×2-MIMO system gives desirable results. If one uses beamforming techniques more desirable results might be achieved.

Space and angle diversity methods are also compared for office area where scatterers are in cluster form. First beam angles were chosen based on the clusters' location and they were (61°, 77°, 103°, 121°). It can be noted that these beams are very close to each other and have some cusps. These cusps cause increase in the correlation among the channels and show decrease in channel capacity, therefore they were changed in such a way that have less cusp (43°, 73°, 108°, 136°), but they were not directed to clusters any more. This improved the capacity. The capacity results for both sets are given in Fig. 20. In general cluster location can give a good guide to find the beam angles and then by considering the cusps between beams and blockage by walls a correction should be applied to improve the capacity.

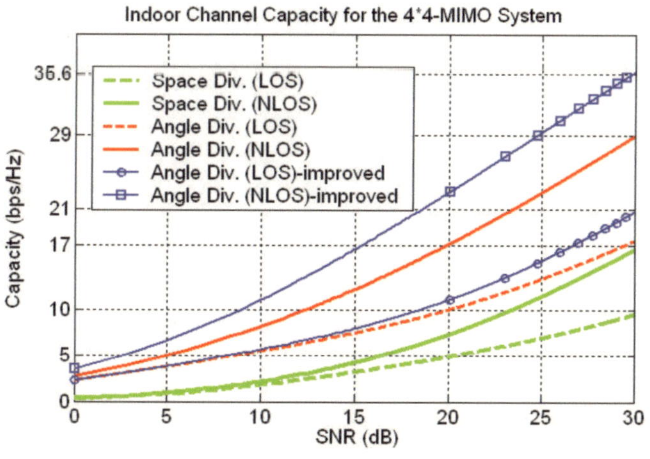

Fig. 20. Channel capacity for 30 scatterers in cluster form in the 4×4-MIMO system.

6. Conclusion

In this chapter a mathematical model to characterize wireless communication channel is developed which falls into semi-deterministic channel models. This model is based on electromagnetic scattering and reflecting and fundamental physics however it has been kept simple through appropriate assumptions.
Based on the results obtained from the SISTER model, impact of different factors on the channel capacity were studied for different scenarios which represent possible wireless MIMO systems such as Wireless Local Area Networks (WLAN) systems in real outdoor and indoor environments. Performance of space and angle diversity methods in MIMO systems are also compared and evaluated. Some of the main achievements are as follows.
The results obtained by SISTER model confirms that higher capacities are achieved for NLOS cases compare to LOS or quasi-LOS cases. However, in LOS or quasi-LOS cases where there is a single dominant path which introduces correlation among the MIMO channels, strong path's dominancy can be lessened by another strong path obtained from either a strong reflection or a resultant path of large number of scatterers and hence channel capacity will be improved. A better alternative to space diversity to improve the channel capacity (especially for LOS case) is the use of angle diversity method. This technique is a promising solution in MIMO systems whose main advantage is to allocate high capacity not to all the points but to the desired ones which results in minimum interference for undesired areas. Therefore, it can be very attractive for environments where interference is the main consideration. Probably the main advantage of angle diversity over space diversity is the similar performance of LOS and NLOS cases, while the space diversity shows a significant reduction in performance for the LOS case.
For angle diversity method in LOS case, high performance can be achieved by selecting beams such that they are not close to horizontal plane where usually a direct path exists. In fact, in LOS cases nulls of the beams should be directed towards the direct path between Tx and Rx to create decorrelated channels.
Even though angle diversity often shows better channel orthogonality, improperly chosen angles lessen the probability of obtaining the maximum achievable capacity. Therefore, choosing the right angles is very important. Improper selection can degrade the performance of a 4×4-MIMO system to that one of a 2×2-MIMO system. In general locations of clusters of scatterers can give a good guide to find the beam angles. However, after the initial selection correction has to be done to avoid beam cusps and blockage by walls. This is because the beam cusps can degrade the capacity due to increase correlation between channels. Based on this study, only in some scenarios, angle diversity shows better performance in LOS cases compare to NLOS as some scatterers which can be those with high contributions on channel orthogonality are blocked. Consequently, for most scenarios, angle diversity seems to be an appropriate alternative for space diversity which can provide similar orthogonality with less interference. Even if in some cases it shows less orthogonality still better performance than space diversity can be achieved because of higher SNR due to the array gain.

7. References

Allen, B. & Beach, M. (2004). On the analysis of switched beam antennas for the WCDMA downlink, *IEEE Trans. Veh. Technol.*, Vol. 53, No. 3, (2004), pp. 569-578.

Allen, B.; Brito, R.; Dohler, M. & Aghvami, H. (2004). Performance comparison of spatial diversity array technologies, *IEEE Trans. Consum. Electron.*, Vol. 50, No. 2, (2004), pp. 420-428.

Allen B. & Ghavami M. (2005). *Adaptive Array Systems: Fundamentals and Applications*, John Wiley & Sons, Inc., 978-0-470-86189-9, NY, USA.

Almers P.; Bonek E.; Burr A.; Czink, N.; Debbah M.; Degli-Esposti V.; Hofstetter H.; Kyosti P.; Laurenson D.; Matz G.; Molisch A. F.; Oestges C. & H. O¨ zcelik H. (2007). Survey of channel and radio propagation models for wireless MIMO systems, *EURASIP J. Wirel. Commun. Netw.*, pp. 1-19, (2007).

Anderson, C.R. & Rappaport, T.S. (2004). In-building wideband partition loss measurements at 2.5 and 60 GHz, *IEEE Trans. Wirel. Commun.* Vol. 3, No. 3, (2004), pp. 922 - 928.

Balanis, C. (1989). *Advanced Engineering Electromagnetics*, John Wiley & Sons, Inc., 0-471-621943, NY, USA.

Balanis, C. (1997). Antenna Theory Analysis and Design, John Wiley & Sons, Inc., 0-471-59268-4, NY, USA.

Burr, A. G. (2003). Capacity bounds and estimates for the finite scatterers MIMO Wireless Channel, *IEEE J. Sel. Areas Commun.*, Vol. 21, No. 5, (2003), pp. 812-818.

Chizhik, D.; Ling, J.; Wolniansky, P.W.; Valenzuela, R.A.; Costa, N. & Huber, K. (2003). Multiple-input-multiple-output measurements and modeling in Manhattan, *IEEE J. Sel. Areas Commun.*, 2003, Vol. 21, No. 3, (2003), pp. 321 - 331.

Collin, R.E. (1985). *Antennas and Radiowave Propagation*, McGraw-Hill, NY, USA.

E. Forooshani, A. (2006). *MIMO systems channel modeling and analysis*, Master of Science Thesis, University of Manitoba, Canada.

E. Forooshani, A. & Noghanian, S. (2010). Semi-deterministic channel model for MIMO systems Part-I: Model development and validation, *IET Microwave Antennas and Propag.*, Vol. 4, No. 1, (2010), pp. 17-25.

Foschini, J. & Gans, M. (1998). On the limit of wireless communications in a fading environment when using multiple antennas, *Wirel. Pers. Commun.*, Vol. 6, No. 3, (1998), pp. 311-335.

Gesbert, D.; Bolcskei, H.; Gore, D.A. & Paulraj, A.J. (2002). Outdoor MIMO wireless channels: models and performance prediction, *IEEE Trans. Commun.* , Vol. 50, No.12, (2002), pp. 1926 - 1934.

Howard, S.; Inanoglu, H.; Ketchum, J.; Wallace, M. & Walton, R. (2002). Results from MIMO channel measurements, *Proc. 13th IEEE Symp. Personal Indoor and Mobile Radio Communications*, pp. 1932 - 1936, Lisboa, Portugal, Sept. 2002.

Liberti, J.C. & Rappaport, T.S. (1996). A geometrically based model for line-of-sight multipath radio channels, *Proc. IEEE 46th, Vehicular Technology Conf.*, pp. 844 - 848, Atlanta, GA, 1996.

Liberti J.C. & Rappapaort, T.S. (1999). *Smart Antennas for Wireless Communications*, Prentice Hall, 0137192878, Upper Saddle River, NJ, USA.

Ranvier, S.; Kivinen, J. & Vainikainen, (2007). Millimeter-wave MIMO radio channel sounder, *IEEE Trans. Instrum. Meas.*, Vol. 56, No. 3, (2007), pp. 1018 - 1024.

Remcom Inc. Technical Staff (2004). *Wireless Insite*, Remcom Inc., version 2.0.5.

Seidel, S.Y. & Rappaport, T.S. (1994). Site-specific propagation prediction for wireless in-building personal communication system design, *IEEE Trans. Veh. Technol.*, Vol. 43, No.4, (1994), pp. 879 - 891.

Svantesson, T. (2001). *Antenna and Propagation from a Signal Processing Perspective*, PhD dissertation, Chalmers University of Technology, Sweden.

Wentworth, S.M. (2005). *Fundamentals of Electromagnetics with Engineering Applications*, John Wiley & Sons, 978-0-470-10575-7, 111 River Street, Hoboken, NJ, USA.

Part 2

Information Theory Aspects

5

Another Interpretation of Diversity Gain of MIMO Systems

Shuichi Ohno[1] and Kok Ann Donny Teo[2]
[1]*Hiroshima University*
[2]*DSO National Laboratories*
[1]*Japan*
[2]*Singapore*

1. Introduction

Multiple-Input Multiple-Output (or the so-called MIMO) system, which employs multiple antennas at both ends of the receiver and transmitter terminals, has been the subject of intensive research efforts in the past decade with potential application in high speed wireless communications network. This is chiefly motivated by the benefits of 1) the spatial multiplexing gain, which makes use of the degrees of freedom in communication system by transmitting independent symbol streams in parallel through spatial channels, to improve bandwidth efficiency; 2) diversity gain, which can be achieved by averaging performance over multiple path gains to combat fading, to improve channel capacity and/or bit-error rate (BER). Information theoretical analysis reveals that MIMO systems indeed offer high spectral efficiency (Foschini, 1996; Goldsmith et al., 2003; Telatar, 1999). It has been shown in (Tse and Viswanath, 2005) that the capacity of an $N_r \times N_t$ MIMO system with N_t transmit and N_r receive antennas over i.i.d. Rayleigh fading channels scales with the minimum of the number N_t of transmit antennas and the number N_r of receive antennas at the high SNR regime. With ideal capacity achieving Gaussian codes, capacity is attained by minimum mean squared error successive interference cancellation (MMSE-SIC) at the receiver (Tse and Viswanath, 2005) if the number of receive antennas is equal to or larger than the number of transmit antennas. The receive diversity achieved by endorsing multiple receive antennas have been utilized in practical communication systems. Recently, Space-Time codes have also been developed to obtain transmit antenna diversity gain (Alamouti, 1998; Caire and Shamai, 1999; Ma and Giannakis, 2003; Tarokh et al., 1999; Xin et al., 2003). Performance gains induced by different schemes of MIMO systems were comprehensively compared in (Catreux et al., 2003).

It is well-known that there is a tradeoff between multiplexing gain and diversity gain. The diversity gain is usually measured by the slope of the BER curve. Over i.i.d. Rayleigh distributed channels, the diversity order of $N_r \times N_t$ systems with linear equalization is given by $N_r - N_t + 1$ at high SNR at full multiplexing (Winters et al., 1994). This implies that given a fixed number N_t of transmit antennas, increasing the number N_r of receive antennas increases the diversity order. Conversely, given a fixed N_r, an increase in N_t (which contributes to multiplexing gain) decreases the diversity order. In (Narasimhan, 2003), by exploiting the tradeoff, an adaptive control of the number of transmit antennas and symbol constellations is proposed to improve the performance of spatial multiplexing in correlated fading channels. Moreover, theoretical analysis that shows a fundamental tradeoff between multiplexing gain and/or diversity gain including Vertical-Bell Laboratories Layered Space-Time (V-BLAST) and Space-Time Codes (STC) have been reported (Tse and Viswanath, 2005; Zheng and Tse, 2003).

Capacity or ergodic capacity, which is the capacity averaged over fading channels, are often utilized to evaluate capacity gain. On the other hand, BER or average BER, which is the BER averaged over fading channels, relate to diversity gain. These gains have been analyzed by approximate expressions for these measures at the SNR extremes, or by directly evaluating them for a particular channel probability density function (pdf), e.g., i.i.d. complex-normal distribution (Chiani et al., 2003; Marzetta and Hochwald, 1999; Smith et al., 2003). However, since the full diversity order appears only at high SNR, having higher diversity order does not necessarily mean having better performance at a particular value of SNR. Moreover, diversity gain of Rayleigh channels does not necessarily imply the existence of diversity gain for other distributed channels. In this chapter, we study universal properties of the performance of MIMO system as in (Ohno and Teo, 2007), which is independent of channel probability density functions and hold at any SNR.

We only consider the case where the performance measure is a convex or concave function of SNR. However, it is shown that important performance measures, including channel capacity and BER, are convex or concave. Thus, our results are significant. To get more insights into MIMO systems, we study capacity gain from a different point of view. A similar approach is adopted in (Ohno and Teo, 2007) to analyze the impact of antenna size of MIMO systems on BER performance with zero-forcing (ZF) equalization.

Take channel capacity for example. Let us suppose that you can install an additional receive antenna in the $N_r \times N_t$ system to construct an $(N_r + 1) \times N_t$ system. Assume that the underlying channel environment is not time-varying (i.e., static). Then, can any other gain (besides power gain) be obtained by increasing the number of receive antennas? Without the values of channel coefficients or the associated channel pdf, no one can answer this question or evaluate the possible gain correctly. Now, we look at the problem from another perspective. For simplicity, we put $N_r = 2$ and $N_t = 2$. From a 3×2 system, we can remove one receive antenna in three different ways to obtain three possible 2×2 systems. Then, we compare the performance of the original 3×2 system with the average performance of the three 2×2 systems. We show in this chapter that without the knowledge of channel coefficients and at any value of SNR, the capacity of the original 3×2 system is greater than the average capacity of the three 2×2 systems. More generally, our analysis reveals that increasing the number of receive antennas generates capacity gain even in static channels. From this, we can prove that the mean capacity with respect to channel pdf, which is mathematically equivalent to the so-called ergodic capacity for fading channels, increases as the number of receive antennas increases at any value of SNR. Our proof relies not on the channel pdf but on the concavity of the capacity function. This implies that the concavity is indispensable to obtain receive antenna diversity.

Next, we consider removing a transmit antenna from an $N_r \times N_t$ system and compare the capacity of the $N_r \times N_t$ system with the average capacity of $N_r \times (N_t - 1)$ systems. Clearly, removing one transmit antenna reduces the multiplexing gain. For comparison, we adopt the capacity per transmit antenna as a parameter. Then, we prove that reducing the number of transmit antennas improves the capacity per transmit antenna. It follows that the mean capacity per transmit antenna degrades as the number of transmit antennas increases at any value of SNR irrespective of channel pdf. This means that increasing the number of transmit antennas improves the multiplexing gain but degrades the capacity per transmit antenna. There exists a tradeoff between multiplexing gain and capacity gain regardless of channel pdf and SNR.

Although we do not evaluate how much gains there actually are, which requires the knowledge of channel coefficients or channel pdf, our results are universal in the sense that performance ordering with the number of transmit antennas and the number of receive antenna is independent of channel pdf and holds true at any value of SNR. We also study the achievable information rate of block minimum mean squared error (MMSE) equalization to obtain similar results.

2. Preliminaries and system model

We consider a MIMO transmission with N_t transmit and N_r receive antennas over flat non-frequency-selective channels. Let us define ρ/N_t as the transmit power at each transmit antenna for the $N_r \times N_t$ MIMO system. We denote the path gain from transmit antenna n ($n \in [1, N_t]$) to receive antenna m ($m \in [1, N_r]$) as h_{mn}. The path gains are assumed to be unknown to the transmitter but perfectly known to the receiver.

Let the received signal at receive antenna m be x_m. The N_r received signals are arranged in a vector as $\boldsymbol{x} = [x_1, \ldots, x_{N_r}]^T$, where $[\cdot]^T$ denotes transposition. Then, $\boldsymbol{x}$ is expressed as

$$\boldsymbol{x} = \sqrt{\frac{\rho}{N_t}} \boldsymbol{H}\boldsymbol{s} + \boldsymbol{w}, \tag{1}$$

where the $N_r \times N_t$ channel matrix $\boldsymbol{H}$, the $N_t \times 1$ combined data vector $\boldsymbol{s}$ having i.i.d. entries with unit variance, the $N_r \times 1$ vector $\boldsymbol{w}$ of zero mean circular complex additive white Gaussian noise (AWGN) entries with unit variance are respectively given by

$$\boldsymbol{H} = \begin{bmatrix} h_{11} & \ldots & h_{1N_t} \\ \vdots & \ddots & \vdots \\ h_{N_r1} & \ldots & h_{N_rN_t} \end{bmatrix}, \tag{2}$$

$$\boldsymbol{s} = \left[s_1 \ldots s_{N_t}\right]^T, \tag{3}$$

$$\boldsymbol{w} = \left[w_1 \ldots w_{N_r}\right]^T. \tag{4}$$

Let the mth row (which corresponds to the mth receive antenna) of the channel matrix $\boldsymbol{H}$ be $\boldsymbol{h}_m$ for $m \in [1, N_r]$, and the nth column (which corresponds to the nth transmit antenna) of the channel matrix $\boldsymbol{H}$ be $\tilde{\boldsymbol{h}}_n$ for $n \in [1, N_t]$ so that we can also express the channel matrix as

$$\boldsymbol{H} = \begin{bmatrix} \boldsymbol{h}_1 \\ \vdots \\ \boldsymbol{h}_{N_r} \end{bmatrix} = \left[\tilde{\boldsymbol{h}}_1 \cdots \tilde{\boldsymbol{h}}_{N_t}\right]. \tag{5}$$

The signal-to-noise ratio (SNR) at receive antenna m is found to be $\rho||\boldsymbol{h}_m||^2/N_t$, where $||\cdot||$ is the 2-norm of a vector, while the overall receive power of the symbol transmitted from antenna n, i.e., the sum of power from transmit antenna n at all receive antennas, is $\rho||\tilde{\boldsymbol{h}}_n||^2/N_t$.

With capacity achieving Gaussian codes, for a given channel $\boldsymbol{H}$, the information rate of the $N_r \times N_t$ MIMO system is expressed as (see. e.g. (Telatar, 1999; Tse and Viswanath, 2005))

$$\mathcal{C}_{N_r,N_t} = \log \left| \boldsymbol{I}_{N_r} + \frac{\rho}{N_t} \boldsymbol{H}\boldsymbol{H}^{\mathcal{H}} \right| = \log \left| \boldsymbol{I}_{N_t} + \frac{\rho}{N_t} \boldsymbol{H}^{\mathcal{H}}\boldsymbol{H} \right|, \tag{6}$$

where $(\cdot)^{\mathcal{H}}$ stands for complex conjugate transposition. Over fading channels, MIMO system offers the benefits of multiplexing gain and/or capacity/diversity gain (Larsson and Stoica, 2003; Tse and Viswanath, 2005).

For our analysis that follows, we utilize the achievable information rates of non-linear Maximum Likelihood (ML) equalization and minimum mean squared error (MMSE) equalization. MMSE equalizations at the receiver becomes available if the channel matrix has column full rank, which requires $N_r \geq N_t$.

Let us shortly review MMSE equalization for MIMO systems. If we employ block-by-block equalization, the MMSE equalizer is given by $\boldsymbol{G} = \sqrt{\frac{\rho}{N_t}} \boldsymbol{H}^{\mathcal{H}} (\frac{\rho}{N_t} \boldsymbol{H}\boldsymbol{H}^{\mathcal{H}} + \boldsymbol{I}_{N_r})^{-1}$. The equalized output is thus expressed as $\hat{\mathbf{s}} = \mathbf{G}\boldsymbol{x}$. We define the nth entry of the equalized output as $\hat{s}_n = p_n s_n + v_n$, where v_n is the effective noise contaminating the nth symbol. Then, we can show that the signal-to-interference noise ratio (SINR) of symbol n after MMSE equalization is expressed as (Kay, 1993; Tse and Viswanath, 2005)

$$\mathrm{SINR}_{N_r,N_t,n} = \frac{\rho}{N_t} \tilde{\boldsymbol{h}}_n^{\mathcal{H}} \left(\boldsymbol{I}_{N_r} + \frac{\rho}{N_t} \sum_{l=1,l\neq n}^{N_t} \tilde{\boldsymbol{h}}_l \tilde{\boldsymbol{h}}_l^{\mathcal{H}} \right)^{-1} \tilde{\boldsymbol{h}}_n. \tag{7}$$

Block-by-block MMSE equalization can be easily implemented but cannot achieve the capacity except for some special cases. Capacity is achieved by MMSE successive interference cancellation (MMSE-SIC) at the receiver. Then, SIC with optimal cancellation order is utilized in Vertical-Bell Laboratories Layered Space-Time (V-BLAST) (Foschini et al., 1999). Although cancellation order affects the BER performance, it does not change the achievable information rate (Tse and Viswanath, 2005, Chapter 8). Thus, it is convenient in what follows to only consider the simplest MMSE-SIC that does not perform the optimal ordering (i.e., arbitrary ordering) procedure. We first equalize symbols from transmit antenna 1. Then after decoding them, the contribution of the signal due to the symbol from transmit antenna 1 is reconstructed and eradicated from the received vector. The same procedure is repeated for the remaining symbols from transmit antenna 2 to transmit antenna N_t. If we denote the SINR of the equalized output at the nth step of MMSE-SIC as SINR_n^{SIC} and there is no error propagation, then the capacity in (6) can be adequately expressed as (Tse and Viswanath, 2005, Chapter 8)

$$\mathcal{C}_{N_r,N_t} = \sum_{n=1}^{N_t} \log \left(1 + \mathrm{SINR}_n^{SIC} \right). \tag{8}$$

3. Decreasing the number of receive antennas

Based on the mathematical tools in the previous section, we investigate information rates of MIMO systems when we decrease the number of receive antennas, while fixing the number of transmit antennas. As the number of receive antennas decreases/increases, the overall receive power decreases/increases, which is known as power loss/gain. Thus, it seems obvious that capacity degrades as the number of receive antennas decreases. However, the MIMO system

may have different receive power from each transmit antenna and the same conclusion is not self-evident. This begs the question: Given the "fair" condition that the overall receive power from each transmit antenna is kept constant even if the number of receive antennas decreases, does capacity decreases or increases? We study how the capacity is affected by the number of receive antennas when the overall receive power from each transmit antenna is fixed.
Let us define a sample correlation of the channel matrix $\boldsymbol{H}$ as

$$\boldsymbol{R}_{N_r,N_t} = \boldsymbol{H}^{\mathcal{H}}\boldsymbol{H} = \sum_{m=1}^{N_r} \boldsymbol{h}_m^{\mathcal{H}}\boldsymbol{h}_m. \tag{9}$$

Assuming that $N_r \geq 2$, we fix the number of transmit antennas at N_t and decrease the number N_r of receive antennas by one. When receive antenna μ is removed from the $N_r \times N_t$ system, the corresponding channel matrix is denoted as $\boldsymbol{H}^{(\mu)}$. The $(N_r - 1) \times N_t$ channel matrix $\boldsymbol{H}^{(\mu)}$ yields the $N_t \times N_t$ correlation matrix $\boldsymbol{R}^{(\mu)}_{N_r-1,N_t}$, corresponding to (9), expressed as

$$\boldsymbol{R}^{(\mu)}_{N_r-1,N_t} = \boldsymbol{H}^{(\mu)\mathcal{H}}\boldsymbol{H}^{(\mu)} = \sum_{m=1,m\neq\mu}^{N_r} \boldsymbol{h}_m^{\mathcal{H}}\boldsymbol{h}_m. \tag{10}$$

It is easy to see that the matrices $\boldsymbol{R}_{N_r,N_t}$ and $\boldsymbol{R}^{(\mu)}_{N_r-1,N_t}$ are related as

$$\sum_{\mu=1}^{N_r} \boldsymbol{R}^{(\mu)}_{N_r-1,N_t} = (N_r - 1)\boldsymbol{R}_{N_r,N_t}. \tag{11}$$

If we remove one receive antenna from the $N_r \times N_t$ system, there are N_r possible systems having $N_r - 1$ receive antennas. We compare the capacity of the $N_r \times N_t$ system with the average capacity of $(N_r - 1) \times N_t$ systems with respect to antenna selection. This average capacity is equivalent to the average capacity when we uniformly remove one receive antenna among N_r antennas, i.e., the selection of any one receive antenna has the same probability $1/N_r$.
If receive antenna μ is removed from the $N_r \times N_t$ system, then the overall receive power from transmit antenna n reduces to $\rho \sum_{m=1,m\neq\mu}^{N_r} |h_{mn}|^2/N_t$. Thus, for $(N_r - 1) \times N_t$ system, the average overall receive power from antenna n with respect to random receive antenna dropping is given by

$$\frac{1}{N_r}\sum_{\mu=1}^{N_r}\left(\sum_{m=1,m\neq\mu}^{N_r} \rho\frac{|h_{mn}|^2}{N_t}\right) = \left(\frac{N_r - 1}{N_r}\right)\rho\frac{||\tilde{\boldsymbol{h}}_n||^2}{N_t}, \tag{12}$$

which depicts a reduction in the average overall receive power from antenna n. To ensure that the average overall receive power from each transmit antenna remains constant even when the number of receive antennas is reduced by one, we increase the transmit power of the $(N_r - 1) \times N_t$ system by a factor of $\frac{N_r}{N_r-1}$, i.e., we replace ρ in (12) by $\frac{N_r}{N_r-1}\rho$. Then, for this $(N_r - 1) \times N_t$ system, the receive SNR at receive antenna m increases to $\frac{N_r}{N_r-1}\frac{\rho||\boldsymbol{h}_m||^2}{N_t}$ and hence the average overall receive power of the $(N_r - 1) \times N_t$ systems is equal to the overall receive power of the $N_r \times N_t$ system. Thus, the effects of power loss due to the reduction of the number of receive antennas disappears on the average.

The information rate of the $(N_r - 1) \times N_t$ system without receive antenna μ is expressed as

$$\mathcal{C}^{(\mu)}_{N_r-1,N_t} = \log\left|\boldsymbol{I}_{N_t} + \left(\frac{N_r}{N_r-1}\frac{\rho}{N_t}\right)\boldsymbol{R}^{(\mu)}_{N_r-1,N_t}\right|. \tag{13}$$

Then, we have from (11) that

$$\frac{\rho}{N_t}\boldsymbol{R}_{N_r,N_t} = \frac{1}{N_r}\sum_{\mu=1}^{N_r}\left(\frac{N_r}{N_r-1}\frac{\rho}{N_t}\right)\boldsymbol{R}^{(\mu)}_{N_r-1,N_t}. \tag{14}$$

At this stage, we utilize a fundamental property of log det function: its *concavity* property. Since log det is a concave function in positive definite matrices, substituting (14) into (6), we find that

$$\begin{aligned}\mathcal{C}_{N_r,N_t} &= \log\left|\boldsymbol{I}_{N_t} + \frac{1}{N_r}\sum_{\mu=1}^{N_r}\left(\frac{N_r}{N_r-1}\frac{\rho}{N_t}\right)\boldsymbol{R}^{(\mu)}_{N_r-1,N_t}\right| \\ &\geq \frac{1}{N_r}\sum_{\mu=1}^{N_r}\log\left|\boldsymbol{I}_{N_t} + \left(\frac{N_r}{N_r-1}\frac{\rho}{N_t}\right)\boldsymbol{R}^{(\mu)}_{N_r-1,N_t}\right| = \frac{1}{N_r}\sum_{\mu=1}^{N_r}\mathcal{C}^{(\mu)}_{N_r-1,N_t},\end{aligned} \tag{15}$$

where the equality holds if and only if all $\boldsymbol{R}^{(\mu)}_{N_r-1,N_t}$ for $\mu \in [1, N_r]$ can be diagonalized with the same unitary matrix.

Eq. (15) shows that for a fixed channel, the capacity of the $N_r \times N_t$ system is not smaller than the average capacity of $(N_r - 1) \times N_t$ systems taken over antenna dropping. It should be noted that the average is not taken over fading channels. For a static channel, we find another disadvantage/advantage of decreasing/increasing the number of receive antennas in addition to power loss/gain. Indeed, (15) is fundamental, from which we will see later that the mean capacity of MIMO systems is also an increasing function in the number of receive antennas at any value of SNR irrespective of channel pdf. Eq. (15) comes only from the basic property of the log det function. It is worth emphasizing that the capacity gain achieved by increasing the number of receive antennas is a direct consequence of the concavity of the log det function.

To analyze the average capacity over random channels, let us denote the channel probability density function (pdf) of channel $\boldsymbol{H}$ as $P(\boldsymbol{H})$ and similarly for $\boldsymbol{H}^{(\mu)}$ as $P(\boldsymbol{H}^{(\mu)})$. We consider the following channel characteristics:

Assumption 1.

$$P(\boldsymbol{H}^{(1)}) = P(\boldsymbol{H}^{(2)}) = \cdots = P(\boldsymbol{H}^{(N_r)}). \tag{16}$$

This implies that when any one row is removed from the $N_r \times N_t$ channel matrix, the resultant $(N_r - 1) \times N_t$ channel matrix has the same probability density function. Clearly, if the entries of $\boldsymbol{H}$ are i.i.d., then (16) holds true. However, it should be remarked that a more general class of channels which includes for example, non i.i.d. channels having correlation between channel gains, meets (16).

The mean capacity is defined as the expectation of the capacity with respect to channel pdf, i.e.,

$$E\{\mathcal{C}_{N_r,N_t}\} = \int \mathcal{C}_{N_r,N_t}P(\boldsymbol{H})d\boldsymbol{H}, \tag{17}$$

where $E\{\cdot\}$ denotes the expectation operator. This is mathematically equivalent to the so-called ergodic capacity if the channel is slowly fading and the channel statistics are ergodic. Consequently, it follows from (15) that

$$E\{\mathcal{C}_{N_r,N_t}\} > \frac{1}{N_r}\sum_{\mu=1}^{N_r}\int \mathcal{C}^{(\mu)}_{N_r-1,N_t}P(\boldsymbol{H})d\boldsymbol{H} = \frac{1}{N_r}\sum_{\mu=1}^{N_r} E\{\mathcal{C}^{(\mu)}_{N_r-1,N_t}\}, \tag{18}$$

where the equality sign is removed since the equality in (15) holds only for some special channel realizations. Under Assumption 1, $E\{\mathcal{C}^{(\mu)}_{N_r-1,N_t}\} = E\{\mathcal{C}_{N_r-1,N_t}\}$, where $E\{\mathcal{C}_{N_r-1,N_t}\}$ is the mean capacity of $(N_r-1)\times N_t$ system. Thus, we can conclude that:

Theorem 1. *Let the capacity of an $M \times N$ MIMO system be $\mathcal{C}_{M,N}$. If the MIMO channel satisfies Assumption 1, then the average capacity taken over channel pdf is an increasing function in the number of receive antennas, i.e.,*

$$E\{\mathcal{C}_{N_r,N_t}\} > E\{\mathcal{C}_{N_r-1,N_t}\}, \tag{19}$$

where $N_r \times N_t$ system and $(N_r-1)\times N_t$ system have the same receive power.

Theorem 1 clearly states the capacity gain in MIMO transmission that can be acquired by simply increasing the number of receive antennas. A special case of Theorem 1 is well-known where at high SNR, the diversity order of $N_r \times N_t$ systems over i.i.d. Rayleigh distributed channels with linear equalization is $N_r - N_t + 1$ at full multiplexing (Winters et al., 1994). Here, no approximation is made and no channel pdf is specified except for Assumption 1 to obtain Theorem 1. It is universal in the sense that (19) holds not just for a specific channel pdf but for all kinds of channel pdf meeting Assumption 1, and at all values of SNR. The capacity gain that arises from increasing the number of receive antennas always exists, since it is a result not attributed to the distinct characteristic of Rayleigh fading but attributed to the basic property of the log det function. Hypothetically, if log det were convex (which is never the case), the inequality in (15) and hence the inequality in (19) would be reverse. Thus, the concavity of the log det function is indispensable to obtain receive antenna diversity. To know how much the actual gain is, one has to evaluate the expectation using the underlying channel pdf. In some special channel pdf, e.g., complex-normal distribution, one could derive an analytical expression of the corresponding capacity gain, e.g., as in (Winters et al., 1994).

3.1 Block MMSE equalization case

Assuming that the channel matrix $\boldsymbol{H}$ is tall and has column full rank, let us analyze the achievable information rate with block MMSE equalization.

After block MMSE equalization, we have N_t parallel channels. Then, the achievable information rate, denoted as $\mathcal{C}^{B}_{N_r,N_t}$, of $N_r \times N_t$ system with block MMSE equalization can be expressed as

$$\mathcal{C}^{B}_{N_r,N_t} = \sum_{n=1}^{N_t}\log\left(1+\mathrm{SINR}_{N_r,N_t,n}\right). \tag{20}$$

If we define the (post-processing) SINR for symbol s_n after block MMSE equalization when receive antenna μ is removed as $\mathrm{SINR}^{(\mu)}_{N_r-1,N_t,n}$ for $n \in [1, N_t]$, then the achievable information rate of the $(N_r-1)\times N_t$ system is

$$\mathcal{C}^{B,(\mu)}_{N_r-1,N_t} = \sum_{n=1}^{N_t}\log\left(1+\mathrm{SINR}^{(\mu)}_{N_r-1,N_t,n}\right). \tag{21}$$

We utilize the following inequality in (Ohno and Teo, 2007):

$$\mathrm{SINR}_{N_r,N_t,n} \geq \frac{1}{N_r}\sum_{\mu=1}^{N_r}\mathrm{SINR}^{(\mu)}_{N_r-1,N_t,n}. \tag{22}$$

The R.H.S. of (22) denotes the average SINR of symbol n when one receive antenna is randomly dropped.
Since log is a concave function, we have from (22) that

$$C^B_{N_r,N_t} \geq \frac{1}{N_r}\sum_{n=1}^{N_t}\sum_{\mu=1}^{N_r}\log\left(1+\mathrm{SINR}^{(\mu)}_{N_r-1,N_t,n}\right) = \frac{1}{N_r}\sum_{\mu=1}^{N_r} C^{B,(\mu)}_{N_r-1,N_t}. \tag{23}$$

This states a deterministic yet universal characteristics of the achievable information rate of MIMO systems with block MMSE equalization. For a given channel environment, if a receive antenna is randomly dropped, the average information rate with respect to random antenna dropping degrades except for some special cases. Indeed, the average information rate depends on the number of receive antennas and *a fortiori* deteriorates as the number of receive antennas is lessened.
By using a similar derivation of Theorem 1, averaging (23) over channel pdf leads to:

Theorem 2. *Let the achievable information rate of an $M \times N$ MIMO system be $C^B_{M,N}$, when block MMSE equalization is adopted. If the MIMO channel satisfies Assumption 1, then the achievable information rate averaged over channel pdf is an increasing function in the number of receive antennas, i.e.,*

$$E\{C^B_{N_r,N_t}\} > E\{C^B_{N_r-1,N_t}\}, \tag{24}$$

where $N_r \times N_t$ system and $(N_r - 1) \times N_t$ system have the same receive power.

Theorem 2 states that capacity gain with increasing number of receive antennas exists even for block MMSE equalization. Theorem 1 as well as Theorem 2 highlight the advantage/disadvantage of MIMO system upon increasing/decreasing the number of receive antennas.

4. Decreasing the number of transmit antennas

In this section, we consider the information rate for a fixed number N_r of receive antennas when the number N_t of transmit antennas is reduced by one, assuming that $2 \leq N_t \leq N_r$. For comparison between $N_r \times N_t$ system and $N_r \times (N_t - 1)$ system, as in the previous section, we uniformly remove one transmit antenna among N_t transmit antennas, i.e., the selection of any one transmit antenna has the same probability $1/N_t$.
It is often the case that the sum of total transmit power of all transmit antennas is kept constant for different numbers of transmit antennas. But, here we fix the transmit power of each transmit antenna to be ρ/N_t. This implies that the sum of transmit power reduces from ρ to $\rho(N_t - 1)/N_t$, if one transmit antenna is removed. In this case, the overall receive power from a transmit antenna remains constant, while the average receive power at each receive antenna of $N_r \times (N_t - 1)$ system with respect to antenna dropping is $(N_t - 1)/N_t$ of the receive power at each receive antenna of the original $N_r \times N_t$ system.

For analysis, we recall block MMSE equalization and MMSE-SIC equalization. If there is no error propagation in SIC, then we have

$$\mathrm{SINR}_{N_r,N_t,n} \le \mathrm{SINR}_n^{SIC}, \tag{25}$$

where $\mathrm{SINR}_{N_r,N_t,n}$ and SINR_n^{SIC} are respectively the SINR of symbol from transmit antenna n in block MMSE equalization and in MMSE-SIC equalization. For all $n \in [1, N_t]$, the equalities in (25) hold if and only if the channel matrix $\boldsymbol{H}$ has orthogonal columns.

We can decompose the capacity of $N_r \times N_t$ system by the following manipulation:

$$\begin{aligned} \mathcal{C}_{N_r,N_t} &= \log \left| \boldsymbol{I}_{N_r} + \frac{\rho}{N_t} \sum_{n=1}^{N_t} \tilde{\boldsymbol{h}}_n \tilde{\boldsymbol{h}}_n^{\mathcal{H}} \right| \\ &= \log \left| \boldsymbol{I}_{N_r} + \frac{\rho}{N_t} \sum_{n=1, n\neq v}^{N_t} \tilde{\boldsymbol{h}}_n \tilde{\boldsymbol{h}}_n^{\mathcal{H}} \right| \left| \boldsymbol{I}_{N_r} + \left(\boldsymbol{I}_{N_r} + \frac{\rho}{N_t} \sum_{n=1, n\neq v}^{N_t} \tilde{\boldsymbol{h}}_n \tilde{\boldsymbol{h}}_n^{\mathcal{H}} \right)^{-1} \frac{\rho}{N_t} \tilde{\boldsymbol{h}}_v \tilde{\boldsymbol{h}}_v^{\mathcal{H}} \right| \\ &= \mathcal{C}_{N_r,N_t-1}^{[v]} + \log\left(1 + \mathrm{SINR}_{N_r,N_t,v}\right), \end{aligned} \tag{26}$$

where $\mathcal{C}_{N_r,N_t-1}^{[v]}$ is the capacity of the $N_r \times (N_t - 1)$ system without transmit antenna v. It follows from (26) that

$$N_t \mathcal{C}_{N_r,N_t} = \sum_{v=1}^{N_t} \left[\mathcal{C}_{N_r,N_t-1}^{[v]} + \log\left(1 + \mathrm{SINR}_{N_r,N_t,v}\right) \right]. \tag{27}$$

On the other hand, from (8) and (25), we have the relation:

$$\sum_{v=1}^{N_t} \log\left(1 + \mathrm{SINR}_{N_r,N_t,v}\right) \le \mathcal{C}_{N_r,N_t}, \tag{28}$$

where the equality holds if and only if $\mathrm{SINR}_{N_r,N_t,n} = \mathrm{SINR}_n^{SIC}$ for all $n \in [1, N_t]$, i.e., the channel matrix $\boldsymbol{H}$ is orthogonal. Combining (27) and (28) results in

$$\mathcal{C}_{N_r,N_t} \le \frac{1}{N_t - 1} \sum_{v=1}^{N_t} \mathcal{C}_{N_r,N_t-1}^{[v]}. \tag{29}$$

Capacity per transmit antenna for $N_r \times N_t$ system can be defined as $\bar{\mathcal{C}}_{N_r,N_t} = \frac{1}{N_t} \mathcal{C}_{N_r,N_t}$. Similarly for $N_r \times (N_t - 1)$ system, as $\bar{\mathcal{C}}_{N_r,N_t-1}^{[v]} = \frac{1}{N_t - 1} \mathcal{C}_{N_r,N_t-1}^{[v]}$. Then, we obtain from (29) that

$$\bar{\mathcal{C}}_{N_r,N_t} = \frac{1}{N_t} \mathcal{C}_{N_r,N_t} \le \frac{1}{N_t} \sum_{v=1}^{N_t} \frac{1}{N_t - 1} \mathcal{C}_{N_r,N_t-1}^{[v]} = \frac{1}{N_t} \sum_{v=1}^{N_t} \bar{\mathcal{C}}_{N_r,N_t-1}^{[v]}. \tag{30}$$

This means that the capacity per transmit antenna of $N_r \times N_t$ system is in general smaller than the average capacity per transmit antenna of $N_r \times (N_t - 1)$ system. The relation in Eq. (30) is satisfied for any channel (channel-independent) and for any SNR.

To get more insights, we assume that

Assumption 2.

$$P(\boldsymbol{H}^{[1]}) = P(\boldsymbol{H}^{[2]}) = \cdots = P(\boldsymbol{H}^{[N_t]}), \tag{31}$$

where $\boldsymbol{H}^{[v]}$ denotes the channel matrix when transmit antenna v is dropped from the $N_r \times N_t$ system.

Multiplying both sides of (30) by $P(\boldsymbol{H})$ and taking the average over $P(\boldsymbol{H})$, we can conclude that

Theorem 3. *Suppose an $N_r \times N_t$ system where $N_r \geq N_t \geq 2$. Let the capacity per transmit antenna of $M \times N$ MIMO system be $\bar{C}_{M,N}$. If the MIMO channel satisfies Assumption 2, then the mean capacity per transmit antenna is an decreasing function in the number of transmit antennas, i.e.,*

$$E\{\bar{C}_{N_r,N_t}\} < E\{\bar{C}_{N_r,N_t-1}\}, \tag{32}$$

where $N_r \times N_t$ system and $N_r \times (N_t - 1)$ system have the same transmit power at each antenna.

Intuitively, this result may be quite reasonable, since in the original $N_r \times N_t$ system, symbols from transmit antenna ν can be considered as an interference to symbols from transmit antenna n and the effect of symbols from transmit antenna ν is absent if transmit antenna ν is removed. For i.i.d. Rayleigh channels at high SNR, the diversity order of $N_r \times N_t$ systems is $N_r - N_t + 1$ and hence reducing N_t increases diversity order (Winters et al., 1994), while the capacity scales with $\min(N_t, N_r)$ (Tse and Viswanath, 2005). However, diversity or capacity gain at high SNR for Rayleigh channels does not imply capacity gain at all SNR for other channels. Thus, our result is not self-evident. From (32), we can find a fundamental tradeoff between bandwidth efficiency and capacity gain for any channel pdf at any value of SNR, i.e., if one increases the number of transmit antennas, then bandwidth efficiency or multiplexing gain is enhanced but the average capacity per transmit antenna is degraded.
Theorem 3 is in sharp contrast to Theorem 1. The mean capacity per transmit antenna is an *increasing* function in the number of *receive* antennas, which is easily concluded from Theorem 1, while the mean capacity per transmit antenna is a *decreasing* function in the number of *transmit* antennas.

4.1 Block MMSE equalization case

We return to block MMSE equalization case and will see that similar results for MMSE-SIC equalization also hold for block MMSE equalization.
Let us denote the (post-processing) SINR of symbol n of the $N_r \times (N_t - 1)$ system without transmit antenna ν after block MMSE equalization as $\text{SINR}^{[\nu]}_{N_r,N_t-1,n}$. It has been shown in (Ohno and Teo, 2007) that for $n \neq \nu$,

$$\text{SINR}^{[\nu]}_{N_r,N_t-1,n} \geq \text{SINR}_{N_r,N_t,n}. \tag{33}$$

Hence, removing one transmit antenna, i.e., reducing the bandwidth efficiency, improves the SINR of each symbol transmitted from the remaining antennas and hence its information rate, i.e., if we denote the achievable information rate from antenna n of $N_r \times N_t$ system and of $N_r \times (N_t - 1)$ system respectively as $C^B_{N_r,N_t,n}$ and $C^{B,[\nu]}_{N_r,N_t-1,n}$, then

$$C^{B,[\nu]}_{N_r,N_t-1,n} \geq C^B_{N_r,N_t,n}, \quad \text{for } n \neq \nu. \tag{34}$$

The achievable information rate of $N_r \times N_t$ system per transmit antenna is expressed as

$$\bar{C}^B_{N_r,N_t} = \frac{1}{N_t} \sum_{n=1}^{N_t} C^B_{N_r,N_t,n}. \tag{35}$$

Similarly, the achievable information rate of $N_r \times (N_t - 1)$ system without transmit antenna ν is

$$\bar{C}^{B,[\nu]}_{N_r,N_t-1} = \frac{1}{N_t - 1} \sum_{n=1,n\neq\nu}^{N_t} C^{B,[\nu]}_{N_r,N_t-1,n}. \tag{36}$$

Subsequently, we have

$$\begin{aligned} \bar{C}^{B}_{N_r,N_t} - \frac{1}{N_t}\sum_{\nu=1}^{N_t} \bar{C}^{B,[\nu]}_{N_r,N_t-1} &= \frac{1}{N_t}\sum_{n=1}^{N_t} C^{B}_{N_r,N_t,n} - \frac{1}{N_t}\sum_{\nu=1}^{N_t}\left(\frac{1}{N_t-1}\sum_{n=1,n\neq\nu}^{N_t} C^{B,[\nu]}_{N_r,N_t-1,n}\right) \\ &= \frac{1}{N_t}\sum_{n=1}^{N_t}\left[C^{B}_{N_r,N_t,n} - \left(\frac{1}{N_t-1}\sum_{\nu=1,\nu\neq n}^{N_t} C^{B,[\nu]}_{N_r,N_t-1,n}\right)\right]. \end{aligned} \tag{37}$$

One finds from (34) that the argument in the brackets of (37) is less than or equal to 0, which leads to

$$\bar{C}^{B}_{N_r,N_t} \le \frac{1}{N_t}\sum_{\nu=1}^{N_t} \bar{C}^{B,[\nu]}_{N_r,N_t-1}. \tag{38}$$

This shows that if one transmit antenna is randomly removed with probability $1/N_t$, for a fixed number of receive antennas, the achievable information rate per transmit antenna of the $N_r \times N_t$ system is never larger than the average achievable information rate per transmit antenna of $N_r \times (N_t - 1)$ system. Since the equality sign holds only for some special cases, on the average, reducing the number of transmit antennas improves information rate per transmit antenna. We can again find a pure tradeoff between bandwidth efficiency and capacity gain.

If we average (38) with respect to channel pdf satisfying Assumption 2, we can state that:

Theorem 4. *Suppose an $N_r \times N_t$ system with block MMSE equalization where $N_r \ge N_t \ge 2$. Let the achievable information rate per transmit antenna of $M \times N$ MIMO system be $\bar{C}^{B}_{M,N}$. If the MIMO channel satisfies Assumption 2, then the average capacity per transmit antenna is a decreasing function in the number of transmit antennas, i.e.,*

$$E\{\bar{C}^{B}_{N_r,N_t}\} < E\{\bar{C}^{B}_{N_r,N_t-1}\}, \tag{39}$$

where $N_r \times N_t$ system and $N_r \times (N_t - 1)$ system have the same transmit power per antenna.

5. Numerical simulations

To validate our theoretical findings, we perform computer simulations on the MIMO system for different antenna sizes. The results for both MMSE-SIC and block MMSE equalizations are presented. In our simulations, we always keep the average overall receive power of each symbol the same as in our theoretical analysis. We plot the information rate per transmit antenna with respect to E_b/N_0 where at each E_b/N_0, the average receive power of each symbol is kept constant regardless of the antenna configuration.

In simulations 1 and 2, we see the effect of the number of receive antennas on the information rate averaged with respect to random receive antenna dropping over a fixed channel. Fig. 1 and Fig. 2 illustrate the results for a fixed $N_t = 4$ and N_r varying from 8 to 4 for MMSE-SIC equalization (which achieves the capacity) and block MMSE equalization, respectively. As one may see from these figures, the information rate averaged with respect to random receive antenna dropping degrades with a decrease in N_r. Especially for block MMSE equalization, the degradation of the information rate is significant when N_r is decreased. These results hold true not just for this fixed channel but also for other channels we have tested, which confirm the fidelity of (15) and (23), i.e., our analysis of the ordering of the information rate based on concavity of the log det is correct.

Then, in the next two simulations, we test the impact of the number of transmit antennas on the information rate averaged with respect to random transmit antenna dropping over a fixed channel. We set $N_r = 8$ and decrease N_t from 8 to 4. The simulation results are shown in Fig. 3 and Fig. 4 for MMSE-SIC equalization and block MMSE equalization respectively. We observe that as the number of transmit antennas is reduced, the information rate averaged with respect to random transmit antenna dropping improves even when the transmit power of each transmit antenna remains the same. The information rate increase for MMSE-SIC equalization is small but is quite significant for block MMSE equalization when we decrease the number of transmit antennas. Evidently, this ordered information rate performances validate (30) and (38).

In our subsequent simulations, instead of simulating over a fixed channel, we take the average over 10^5 Rice channels of Rice factor 2 that compose of zero mean Gaussian taps with unit variance in order to verify the effect of the number of receive antennas on the information rate per transmit antenna averaged over random channels. Fig. 5 and Fig. 6 depict the results for a fixed $N_t = 4$ and N_r varying from 8 to 4 for MMSE-SIC and block MMSE equalizations respectively. We can see that the information rate averaged over random channels degrade with a decrease in N_r for both MMSE-SIC and block MMSE equalizations, demonstrating that the information rate averaged over random channels is an increasing function in the number of receive antennas. These are in good agreement with Theorem 1 and Theorem 2, which hold true for all SNR.

Lastly, to see the effect of the number of transmit antennas on the information rate per transmit antenna averaged over random channels, Fig. 7 and Fig. 8 show the results for a fixed $N_r = 8$ and N_t varying from 8 to 4 for MMSE-SIC and block MMSE equalizations respectively. The simulation results confirm Theorem 3 and Theorem 4 as the information rate averaged over random channels improves with a decrease in N_t (or equivalently, decrease in both bandwidth efficiency and multiplexing gain). In other words, the information rate averaged over random channels is a decreasing function in the number of transmit antennas. These ordered performances show unequivocally that there is an undisputed tradeoff between the information rate and bandwidth efficiency (and/or multiplexing gain).

6. Conclusions

Based on our novel point of view, we have demonstrated theoretically that under the condition of a fixed overall received power and a fixed number of transmit antennas, the information rate averaged over random receive antenna dropping and the information rate averaged over random channels degrade with a decrease in the number of receive antennas. These results are derived from the basic property of the log det function. On the other hand, for a fixed number of receive antennas, we have proven that a decrease in the number of transmit antennas translates into an amelioration in both the information rate averaged over random transmit antenna dropping as well as the information rate averaged over random channels,

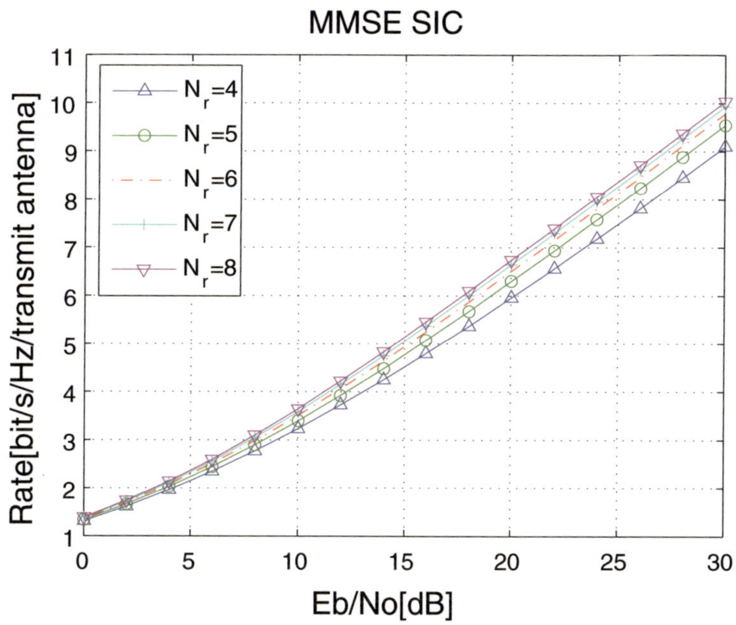

Fig. 1. Information rate per transmit antenna with respect to random receive antenna dropping for MMSE-SIC over a fixed channel for a fixed $N_t = 4$ and varying N_r.

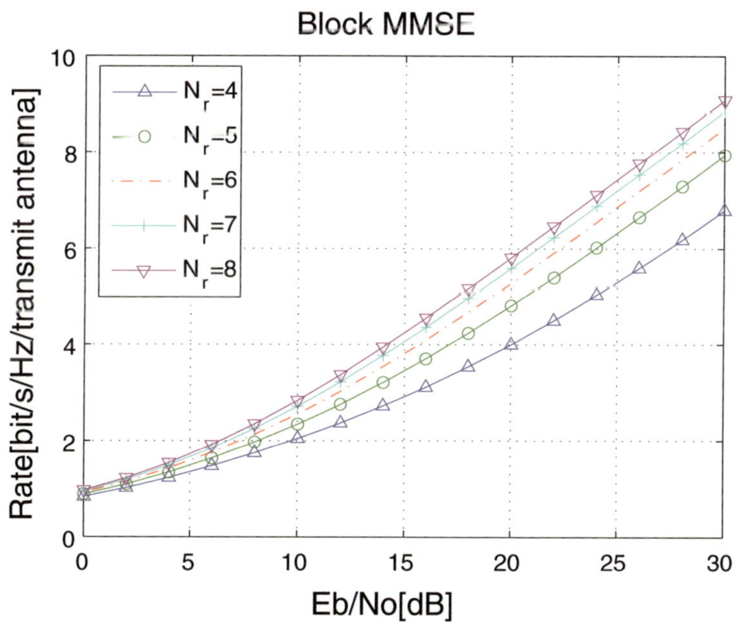

Fig. 2. Information rate per transmit antenna with respect to random receive antenna dropping for block MMSE over a fixed channel for a fixed $N_t = 4$ and varying N_r.

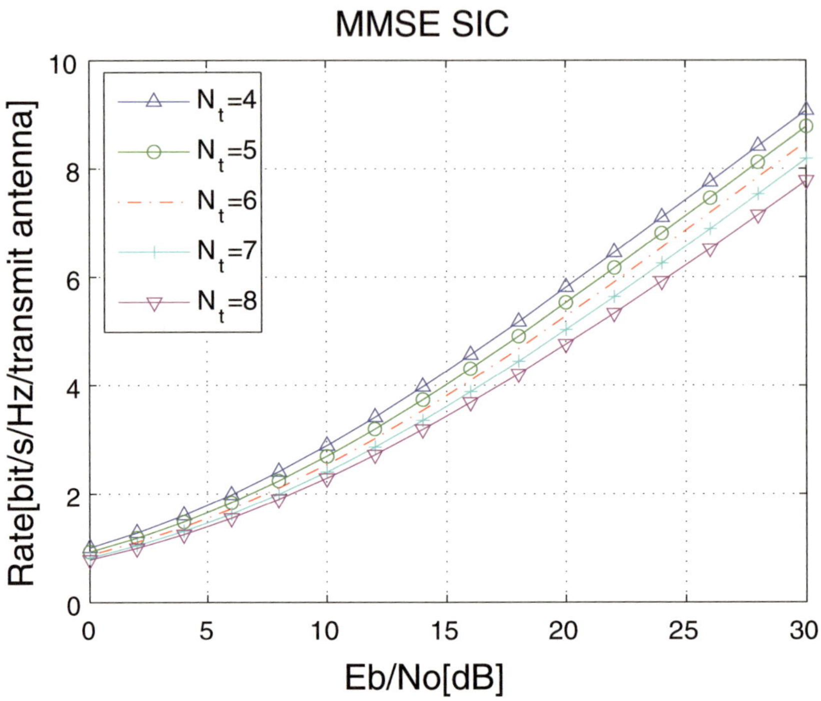

Fig. 3. Information rate per transmit antenna with respect to random transmit antenna dropping for MMSE-SIC over a fixed channel for a fixed $N_r = 8$ and varying N_t.

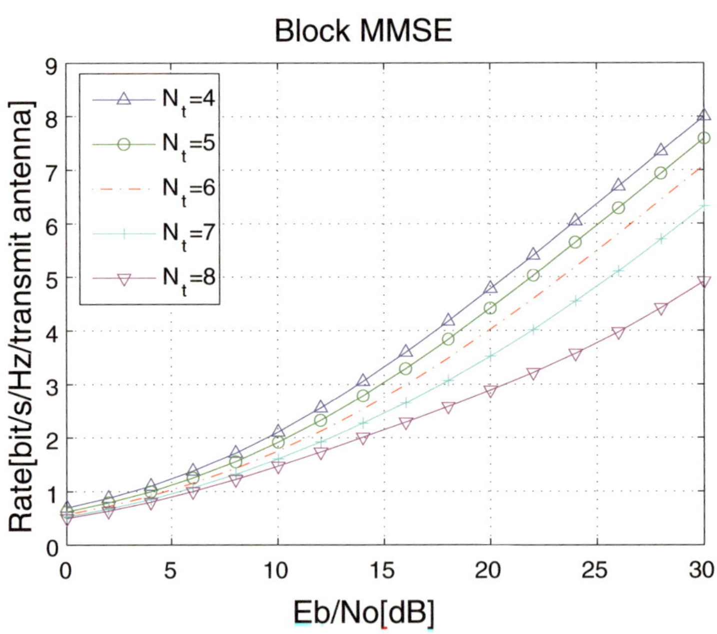

Fig. 4. Information rate per transmit antenna with respect to random transmit antenna dropping for block MMSE over a fixed channel for a fixed $N_r = 8$ and varying N_t.

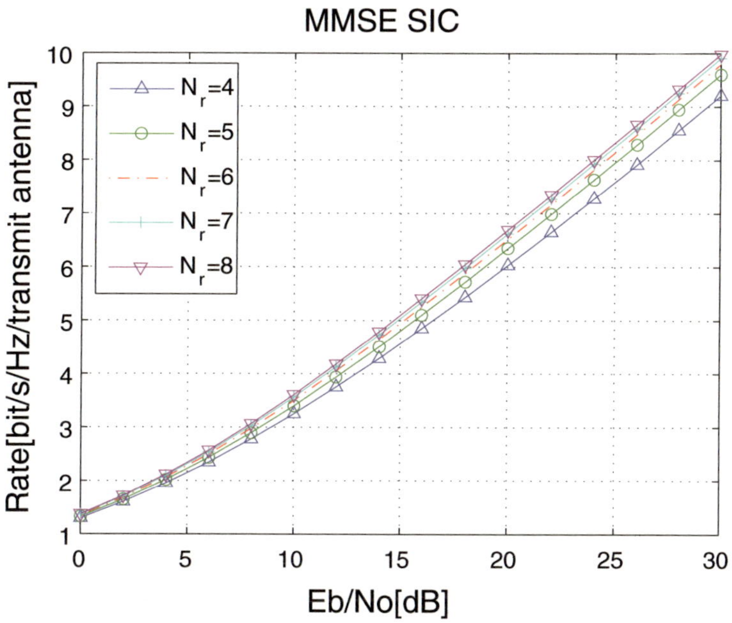

Fig. 5. Information rate per transmit antenna averaged over random channels for MMSE-SIC for a fixed $N_t = 4$ and varying N_r.

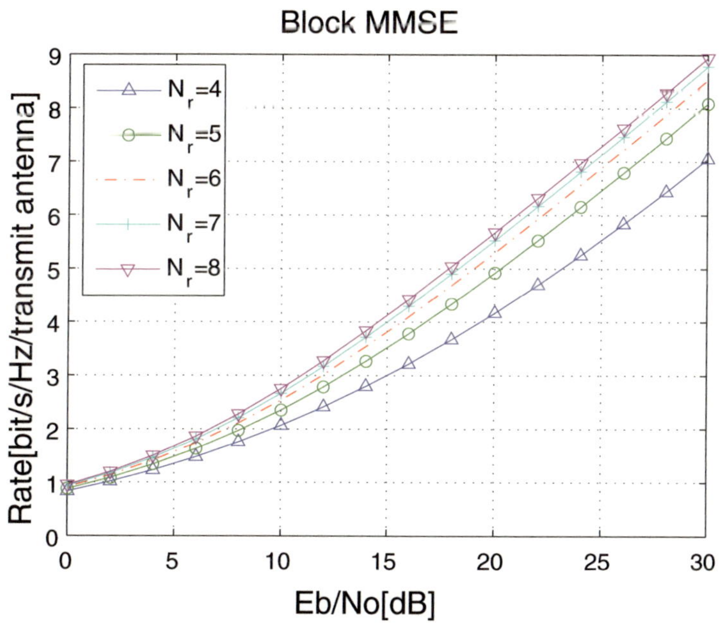

Fig. 6. Information rate per transmit antenna averaged over random channels for block MMSE for a fixed $N_t = 4$ and varying N_r.

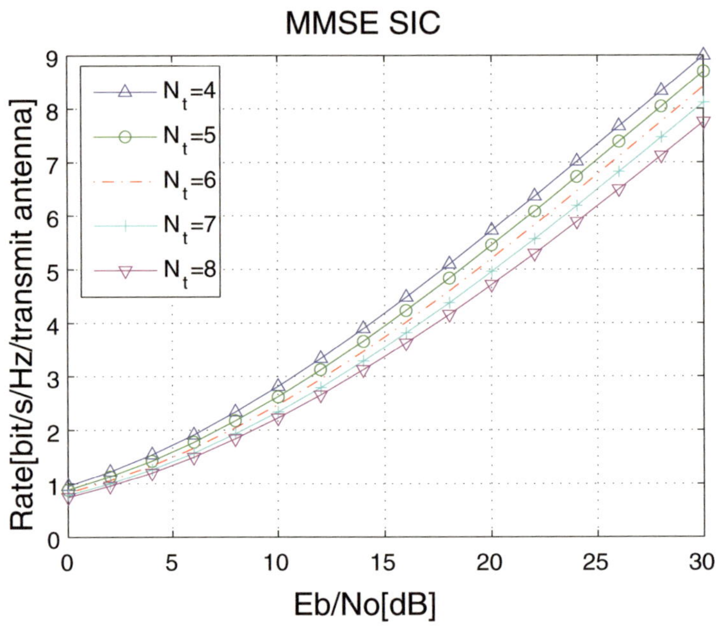

Fig. 7. Information rate per transmit antenna averaged over random channels for MMSE-SIC for a fixed $N_r = 8$ and varying N_t.

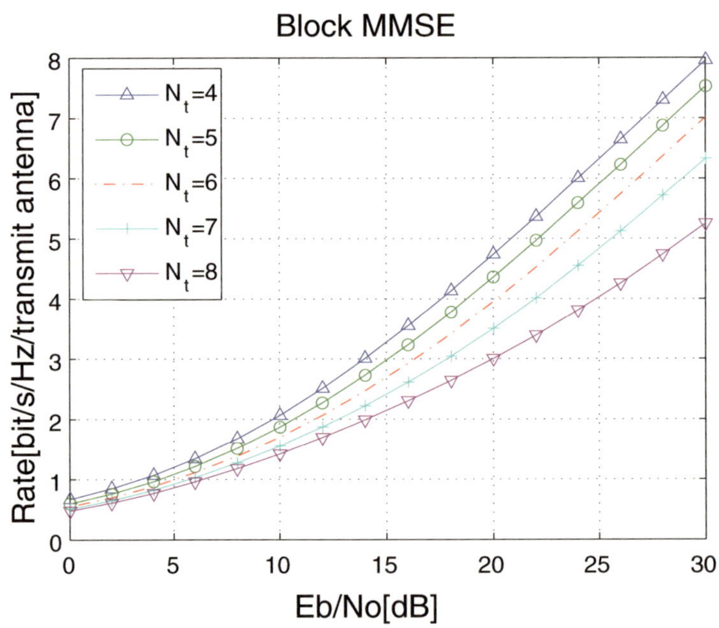

Fig. 8. Information rate per transmit antenna averaged over random channels for block MMSE for a fixed $N_r = 8$ and varying N_t.

which highlight the tradeoff between capacity and bandwidth efficiency (and multiplexing gain). All these results hold for any kind of i.i.d. channel regardless of the channel pdf and is valid at any SNR. Numerical simulations corroborated our analysis.

7. References

Alamouti, S. M. (1998). A simple transmit diversity for wireless communications, *IEEE Journal on Selected Areas in Communications* 16(8): 1451–1458.

Caire, G. and Shamai, S. (1999). On the capacity of some channels with channel state information, *IEEE Transactions on Information Theory* pp. 2007–2019.

Catreux, S., Greenstein, L. J. and Erceg, V. (2003). Some results and insights on the performance gains of MIMO systems, *IEEE Journal on Selected Areas in Communications* 21(5): 839–847.

Chiani, M., Win, M. Z. and Zanella, A. (2003). On the capacity of spatially correlated MIMO rayleigh-fading channels, *IEEE Transactions on Information Theory* 49(10): 2363–2371.

Foschini, G. J. (1996). Layered space-time architecture for wireless communication in a fading environment when using multi-element antennas, *Bell Labs Tech. J.* 1(2): 41–59.

Foschini, G. J., Golden, G., Valenzuela, R. and Wolniansky, P. (1999). Simplified processing for high spectral efficiency wireless communication employing multi-element arrays, *IEEE Journal on Selected Areas in Communications* 17(11): 1841–1852.

Goldsmith, A., Jafar, S. A., Jindal, N. and Vishwanath, S. (2003). Capacity limits of MIMO channels, *IEEE Journal on Selected Areas in Communications* 21(5): 684–702.

Kay, S. M. (1993). *Fundamentals of Statistical Signal Processing*, Vol. 1, Prentice Hall.

Larsson, E. G. and Stoica, P. (2003). *Space-time block coding for wireless communications*, Cambridge University Press.

Ma, X. and Giannakis, G. B. (2003). Full-diversity full-rate complex-field space-time coding, *IEEE Transactions on Signal Processing* 51(11): 2917–2930.

Marzetta, T. L. and Hochwald, B. M. (1999). Capacity of a mobile multiple-antenna communication link in rayleigh flat fading, *IEEE Transactions on Information Theory* 45(1): 139–157.

Narasimhan, R. (2003). Spatial multiplexing with transmit antenna and constellation selection for correlated MIMO fading channels, *IEEE Transactions on Signal Processing* 51(11): 2829–2838.

Ohno, S. and Teo, K. A. D. (2007). Universal BER performance ordering of MIMO systems over flat channels, *IEEE Transactions on Wireless Communications* 6(10): 3678–3687.

Smith, P. J., Roy, S. and Shafi, M. (2003). Capacity of MIMO systems with semicorrelated flat fading, *IEEE Transactions on Information Theory* 49(10): 2781–2788.

Tarokh, V., Jafarkhani, H. and Calderbank, A. R. (1999). Space-time block coding for wireless communications: performance results, *IEEE Trans. Communication* 17(3): 451–460.

Telatar, I. E. (1999). Capacity of multiple-antenna Gaussian channels, *Eur. Trans. Tel.* pp. 585–595.

Tse, D. and Viswanath, P. (2005). *Fundamentals of wireless communication*, Cambridge University Press.

Winters, J. H., Salz, J. and Gitlin, R. D. (1994). The impact of antenna diversity on the capacity of wireless communication systems, *IEEE Transactions on Communications* 42(234): 1740–1751.

Xin, Y., Wang, Z. and Giannakis, G. B. (2003). Space-time diversity systems based on linear constellation precoding, *IEEE Transactions on Wireless Communications* 2(2): 294–309.

Zheng, L. and Tse, D. N. C. (2003). Diversity and multiplexing: a fundamental tradeoff in multiple-antenna channels, *IEEE Transactions on Information Theory* 49(5): 1073–1096.

6

Rate-Adaptive Information Transmission over MIMO Channels

Marco Zoffoli, Jerry D. Gibson and Marco Chiani
Fellow, IEEE

1. Introduction

In the context of wireless communication, a Multiple-Input Multiple-Output (MIMO) system is a system that employs multiple antennas both at the transmitter and receiver. The first theoretical analysis of MIMO systems were developed by Winters (1987), Foschini (1996) and Telatar (1999), and since then there have been many research efforts on this subject. What mainly makes MIMO systems interesting is their potential ability to achieve an increase in system capacity or in link reliability without requiring additional transmission power or bandwidth (Goldsmith, 2005).

In this work, we focus on the utilization of MIMO systems for the lossy transmission of source information. In particular, we want to compare several different strategies for the transmission of a zero mean Gaussian source over Rayleigh-fading MIMO channels, assuming rate-adaptive source encoding. The MIMO transmission strategies are based on techniques such as Repetition coding (REP), Time Sharing (TS), the Alamouti scheme (ALM) and Spatial Multiplexing (SM) (Alamouti, 1998; Tse & Viswanath, 2006).

Depending on its characteristics, each strategy will be used either for the transmission of a Single Description (SD) or the transmission of a Multiple Description (MD) representation of this source. In SD coding, a single stream of information describing the source is transmitted over a single channel. In MD coding (Gamal & Cover, 1982), the source is represented using two different descriptions that are transmitted over two independent channels. If both descriptions are correctly received, they can be combined together at the receiver to obtain a reconstruction of the source at a certain quality. If only one of the two descriptions is correctly received, a reconstruction of the source is still possible but at a lower quality.

We consider adaptive source encoding, where the rate is adapted to follow the slow variations of the channel (due e.g. to shadowing and path loss) or the fast variations of the channel (due to fading), leading to two scenarios that we call fixed-outage and zero-outage, respectively.

In the first case, we consider Gaussian source transmission over MIMO systems when CSI is not available at the transmitter. In this scenario, since the transmitter does not have knowledge of channel state information (CSI), it does not know the instantaneous rate supported by the channel, i.e. its capacity, and hence it is not able to adapt the source coding rate to the channel conditions to ensure the decoding of the information at the receiver with an arbitrarily small probability of error. Instead, it encodes and transmits the source information using a rate chosen to achieve a selected *outage probability*. When the channel does not support the transmission of information at the chosen coding rate, data are lost and the system experiences an *outage*. We call this the *fixed-outage rate-adaptive* approach.

In the second scenario (*zero-outage rate-adaptive*), we consider the different MIMO strategies under the assumption of perfect CSI at the transmitter. In this case, the transmitter is able to follow the variations of the channel by adapting the source coding rate to the instantaneous capacity, since it is aware of the particular channel realization in every time instant. In such a situation there is no outage since the source rate is always adapted to achieve the instantaneous channel capacity (Choudhury & Gibson, 2007). This observation has a direct impact on the usefulness of the TS strategies in the zero-outage scenario. These strategies employ a time sharing approach to the transmit antennas to create independent channels from our MIMO system (Zoffoli et al., 2008a). These independent channels are then used to provide path diversity by transmitting multiple description representations of the source over them. However, path diversity is useful only if the channels are unreliable, i.e. if they suffer outages. For this reason, in the zero-outage scenario we do not consider the TS strategies.

The different strategies for both the fixed-outage and the zero-outage rate adaptation approaches are described in the following sections, where we also evaluate their performance by studying the statistics of the distortion at the receiver.

In the presence of outage, it is usually assumed either implicitly or explicitly that retransmissions will be used for data scheduled to be transmitted during an outage; indeed, choosing an operational outage rate may be associated with an acceptable retransmission rate. Although retransmissions are the natural response to outage for data sources, relying on retransmissions may or may not be appropriate for compressed voice or video for several reasons. First, it is not unusual to rely on packet loss concealment for voice and video up to some non-trivial packet loss rate. Second, it may be more desirable not to retransmit for voice and video in order to reduce latency or to maximize access point throughput. As a result, the suitable measure of performance for lossy source coding of voice and video is the average distortion of the source reproduced at the receiver. Average distortion is also the appropriate performance indicator for the zero outage rate case, since we are adapting the source coding rate to the instantaneous capacity of the channel, and it is desired to determine the reproduced quality of the source delivered to the user. Therefore, for our work here, we choose the mean squared error (MSE) fidelity criterion.

In Section II, we present the basic assumptions and set up the particular MIMO problems we are addressing. Section III contains the development of the fixed outage rate adaptive source encoding scenarios we examine, including the repetition strategy and single description source coding, the time sharing strategy and the three multiple descriptions source coding methods (no excess marginal rate, no excess joint rate, and optimized multiple descriptions source coding), the Alamouti strategy with single description source coding, and spatial multiplexing with single description source coding. Zero outage rate adaptive source encoding, wherein CSI is available at the transmitter and the source coding is adjusted to match the instantaneous capacity, is described in Section IV, including the developments and derivations of the distributions of the reconstructed source distortion for the repetition, Alamouti, and spatial multiplexing strategies. Extensive results for each of the methods and comparisons of the results are presented in Section V, while Section VI summarizes the conclusions from the work.

2. Assumptions and preliminaries

Our main goal is to discuss how MIMO techniques impact on adaptive source encoding. Although most of our results can be easily extended to cover the general $N_t \times N_r$ MIMO channel case, for the sake of simplicity we consider the frequency-flat 2×2 MIMO channel. The system is characterized by the channel matrix H, having the form

$$H = \begin{pmatrix} h_{11} & h_{12} \\ h_{21} & h_{22} \end{pmatrix}$$

Each entry h_{ij} of the channel matrix H represents the gain of the channel between the j-th transmit antenna and i-th receive antenna. These channels are assumed to be independent, random and with very slow Rayleigh fading. The h_{ij} are then i.i.d. complex Gaussian random variables with zero mean and unit variance, which remain constant over the transmission of a large number of symbols. Under these assumptions, the squared magnitude of the channel gains can be written as

$$|h_{ij}|^2 = \frac{1}{2}x_{ij}, \qquad i,j = 1,2 \tag{1}$$

where the x_{ij} are random variables distributed according to a chi-square distribution with 2 degrees of freedom (Hogg & Craig, 1970). Perfect CSI, i.e. knowledge of H, is assumed to be always available at the receiver, while the transmitter has a full or partial CSI depending on the scenario, as will be discussed later.

The total transmitted power by the transmit antennas is constrained to P_t. If both transmit antennas are transmitting simultaneously, each antenna will transmit with equal power $P_t/2$, while, if only one antenna is transmitting at a given time, it can make use of full transmit power P_t. The noise at the receiver is AWGN, with i.i.d. statistics and the same average power N at each receive antenna.

We denote with γ_{ij} the instantaneous Signal to Noise Ratio (SNR) of the signal transmitted by the j-th antenna and received by the i-th antenna. Thus,

$$\gamma_{ij} = \frac{P_t}{N}|h_{ij}|^2 = \bar{\gamma}|h_{ij}|^2, \qquad i = 1,2 \tag{2}$$

if only the j-th antenna is transmitting at a given time, and

$$\gamma_{ij} = \frac{P_t}{2N}|h_{ij}|^2 = \frac{\bar{\gamma}}{2}|h_{ij}|^2, \qquad i,j = 1,2 \tag{3}$$

if both antennas are transmitting at the same time.

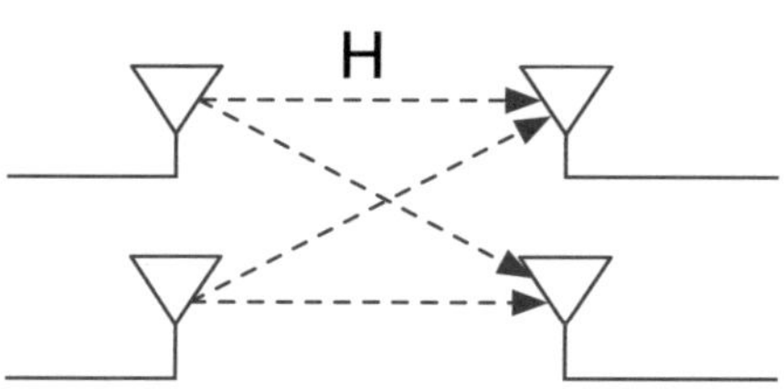

Fig. 1. 2×2 MIMO model.

The source is assumed to be a zero-mean memoryless Gaussian source with a variance normalized to unity. The system bandwidth is also assumed to be normalized to unity.

In the following, we will denote with $\bar{\gamma}$ the ratio P_t/N and with $\Gamma(z)$ and $\Gamma(a,z)$, respectively, the gamma function and the incomplete gamma function (Hogg & Craig, 1970). We will also denote with χ_k^2 the distribution of a chi-square random variable with k degrees of freedom, with $F_\chi^{(k)}(z)$ its CDF and with

$$f_\chi^{(k)}(z) = \frac{1}{\Gamma(\frac{k}{2})2^{\frac{k}{2}}} z^{\frac{k-2}{2}} e^{-\frac{z}{2}}$$

its probability density function (PDF) (Hogg & Craig, 1970).

3. Fixed-Outage rate-adaptive source encoding (FORA)

In a wireless channel, due to multipath propagation and users' mobility, the capacity is varying in time. In this section we assume that the source encoder knows only the statistical distribution of the wireless channel mutual information, and that it adapts its rate accordingly. The source encoder rate is chosen to produce a certain outage probability, determined to minimize the distortion at the received end. This could be assumed a slow-adaptive technique, since the source encoder rate will follow the variations of the channel statistics due, for instance, to shadowing and path loss changes.

3.1 Repetition

The REP strategy is based on repetition coding (Tse & Viswanath, 2006). The basic idea is to transmit the same symbol over the two transmit antennas in two consecutive time slots. In each time slot, only one of the two transmit antennas is used for transmission, while the other antenna is turned off.

Thus, in the first time slot the symbol S_1 is transmitted on the first transmit antenna and it is observed by the receiver through the two channels with gains h_{11} and h_{21}. In the second time slot, the same symbol S_1 is transmitted on the second transmit antenna and it is observed by the receiver through the two channels with gains h_{12} and h_{22}. A Maximal Ratio Combiner (MRC) (Goldsmith, 2005) is then used at the receiver to optimally combine the four signals received by the two receive antennas in the two different time slots.

The instantaneous SNR γ of the signal at the output of the MRC is given by the sum of the instantaneous SNRs γ_{ij} of its input signals (Goldsmith, 2005), that are given by Eq. (2)

$$\gamma = \sum_{i,j=1}^{2} \gamma_{ij} = \bar{\gamma} \sum_{i,j=1}^{2} |h_{ij}|^2 \tag{4}$$

In this way, a single channel is obtained from the four independent channels available in our MIMO system. This strategy is then suitable for the transmission of a SD representation of the source.

The instantaneous capacity of this single channel in [bits/channel use] is given by (Goldsmith, 2005)

$$C = \frac{1}{2}\log_2\left(1+\gamma\right) = \frac{1}{2}\log_2\left(1+\bar{\gamma}\sum_{i,j=1}^{2}|h_{ij}|^2\right) \tag{5}$$

where the factor $1/2$ arises because we are transmitting the same symbol over two consecutive time slots.

The source coding rate R_{REP} of the SD coder is chosen to be equal to the outage capacity at a given value for the outage probability P_{out}, i.e. it is chosen such that

$$Pr\left\{C < R_{REP}\right\} = P_{out}$$

Thus, with probability $1 - P_{out}$ the system is not in outage, which means that it can support the transmission at a rate R_{REP} with an arbitrarily small probability of error, since its capacity is higher than R_{REP} (Cover & Thomas, 1991). In such case, the receiver is able to reconstruct the source information with a distortion D_1 equal to (Cover & Thomas, 1991)

$$D_1 = 2^{-2R_{REP}}$$

If the system results in outage, which happens with probability P_{out}, the receiver is not able to correctly decode the transmitted information with an arbitrarily small probability of error and achieves a distortion equal to 1.
The expected distortion D at the receiver is then

$$D = (1 - P_{out})D_1 + P_{out}$$

The outage rate R_{REP}^{out}, defined as the average rate correctly received over many transmission bursts (Goldsmith, 2005), is given by

$$R_{REP}^{out} = (1 - P_{out})R_{REP}$$

3.2 Time sharing - multiple description (TS-MD)

In this strategy a TS approach is adopted to obtain two independent channels from the MIMO system. The idea behind this strategy is to transmit two different symbols over the two transmit antennas in two consecutive time slots. In each time slot, only one of the two transmit antennas is used for transmission, while the other antenna is turned off. Thus, in the first time slot the first symbol S_1 is transmitted over the first antenna and it is observed by the receiver through the two channels with gains h_{11} and h_{21}. In the second time slot, the second symbol S_2 is transmitted over the second antenna and it is observed by the receiver through the two channels with gains h_{12} and h_{22}. The receiver will then combine the two signals received in the same time slot using a MRC.
Since each received signal has a SNR given by Eq. (2), the signal at the output of the MRC in the j-th time slot has a SNR equal to (Goldsmith, 2005)

$$\gamma_j = \sum_{i=1}^{2} \gamma_{ij} = \bar{\gamma} \sum_{i=1}^{2} |h_{ij}|^2$$

In this way, two independent channels are effectively created in the two time slots, making this strategy suitable for the transmission of a MD representation of the source.
The channel at the j-th time slot has an instantaneous capacity C_j equal to (Goldsmith, 2005)

$$C_j = \frac{1}{2}\log_2\left(1 + \gamma_j\right) = \frac{1}{2}\log_2\left(1 + \bar{\gamma}\sum_{i=1}^{2}|h_{ij}|^2\right) \qquad (6)$$

where the factor $1/2$ arises because each channel is used only half of the time.

The side description rate $R_{MD}/2$, which equals the transmitted rate over each channel, is chosen to be equal to the outage capacity for a given P_{out}, i.e. is chosen such that

$$Pr\left\{C_j < \frac{R_{MD}}{2}\right\} = P_{out}$$

The expected distortion D at the receiver is then given by

$$D = (1 - P_{out})^2 D_0 + 2P_{out}(1 - P_{out})D_1 + P_{out}^2 \quad (7)$$

where D_0 and D_1 are the distortions achieved by the receiver when observing, respectively, both descriptions or only one of the two descriptions.
Depending on the type of MD coder used, D_0 and D_1 can have different expressions (Balam & Gibson, 2006) and different TS-MD strategies can be obtained. The No Excess Marginal Rate coder (MD-NMR) (Balam & Gibson, 2006) is employed in the TS-MD-NMR strategy. The side descriptions are then rate distortion optimal and the distortions have the following expressions (Balam & Gibson, 2006; Effros et al., 2004)

$$\begin{aligned} D_0 &= \frac{2^{-R_{MD}}}{2 - 2^{-R_{MD}}} \\ D_1 &= 2^{-R_{MD}} \end{aligned}$$

The No Excess Joint Rate coder (MD-NJR) (Balam & Gibson, 2006) is employed in the TS-MD-NJR strategy. Here the joint description is rate distortion optimal and the distortions have the following expressions (Balam & Gibson, 2006; Effros et al., 2004)

$$\begin{aligned} D_0 &= 2^{-2R_{MD}} \\ D_1 &= \frac{1}{2}\left(1 + 2^{-2R_{MD}}\right) \end{aligned}$$

The optimal coder (MD-OPT) (Effros et al., 2004) is employed in the TS-MD-OPT strategy. In this case, neither the side descriptions nor the joint description is rate distortion optimal, but they are chosen to minimize the expected distortion D in Eq. (7) for a given P_{out}. The distortions D_0 and D_1 are given by the following expression (Balam & Gibson, 2006; Effros et al., 2004)

$$\left(D_0, D_1\right) = \left(a, \frac{1+a}{2} - \frac{1-a}{2}\sqrt{1 - \frac{2^{-2R_{MD}}}{a}}\right)$$

with

$$a \in \left[2^{-2R_{MD}}, \frac{2^{-R_{MD}}}{2 - 2^{-R_{MD}}}\right]$$

Thus, the MD-OPT coder chooses the proper value for a to minimize the expected distortion D.
The outage rate R_{MD}^{out} is given by

$$\begin{aligned} R_{MD}^{out} &= (1 - P_{out})^2 R_{MD} + 2P_{out}(1 - P_{out})\frac{R_{MD}}{2} \\ &= (1 - P_{out})R_{MD} \end{aligned}$$

3.3 Alamouti

This strategy employs the Alamouti scheme (Alamouti, 1998; Tse & Viswanath, 2006) to obtain two independent channels from the MIMO system. Since both channels have the same gain given by $\sum_{i,j=1}^{2} |h_{ij}|^2$ (Tse & Viswanath, 2006), it is evident that it is impossible to have, for a given realization of the channel matrix H, one channel in outage and the other not in outage, i.e. both channels can only be simultaneously in outage or simultaneously not in outage.[1] This strategy is then not suitable for the transmission of a multiple description representation of the source, as also pointed out in (Effros et al., 2004). Instead, it could be used for the transmission of a single description representation, demultiplexing it into two half-rate substreams which are then transmitted over the two channels.

The signals at the output of the Alamouti decoder have the same instantaneous SNR γ, equal to the sum of the SNRs of the signals on each branch (Goldsmith, 2005). Thus, from Eq. (3) we have

$$\gamma = \sum_{i,j=1}^{2} \gamma_{ij} = \frac{\bar{\gamma}}{2} \sum_{i,j=1}^{2} |h_{ij}|^2$$

The instantaneous capacity of this system is then given by (Sandhu & Paulraj, 2000)

$$C = \log_2 \left(1 + \gamma\right) = \log_2 \left(1 + \frac{\bar{\gamma}}{2} \sum_{i,j=1}^{2} |h_{ij}|^2\right) \tag{8}$$

and the source coding rate R_{ALM} is chosen such that

$$Pr\Big\{C < R_{ALM}\Big\} = P_{out}$$

The expected distortion is then

$$D = (1 - P_{out})D_1 + P_{out}$$

where D_1 is the distortion achieved by the receiver when the system is not in outage, which is equal to (Cover & Thomas, 1991)

$$D_1 = 2^{-2R_{ALM}} \tag{9}$$

The outage rate R_{ALM}^{out} is given by

$$R_{ALM}^{out} = (1 - P_{out})R_{ALM}$$

3.4 Spatial multiplexing

In the SM strategy (Tse & Viswanath, 2006), a single symbol stream is first demultiplexed and encoded into two separate and independent substreams. Each substream is then transmitted simultaneously over each transmit antenna and, at the receiver, an optimal joint decoder is employed for retrieving the original symbol stream.

Since this strategy requires one single symbol stream, it can only be used for the transmission of a SD representation of the source.

---

[1] This is true also because the transmitted rate on each channel is the same, which is the only case of interest for us.

The instantaneous capacity achievable with this strategy is given by (Foschini & Gans, 1998)

$$C = \log_2 \det \left(I_2 + \frac{\bar{\gamma}}{2} H H^H \right) \tag{10}$$

where I_2 is the 2×2 identity matrix and H^H denotes the conjugate transpose of the channel matrix H. In a similar way as before, given the outage probability the source coding rate R_{SM} is chosen such that

$$Pr\{C < R_{SM}\} = P_{out}$$

This CDF can be computed for a general MIMO channel without resorting to MonteCarlo simulation as indicated in (Chiani, Win & Zanella, 2003). For the particular case of the 2×2 MIMO, a simple closed form expression is derived in the Appendix.
The expected distortion D at the receiver is then

$$D = (1 - P_{out})D_1 + P_{out}$$

where

$$D_1 = 2^{-2R_{SM}} \tag{11}$$

is, as usual, the distortion achieved when the system is not in outage.
The outage rate R_{SM}^{out} is given by

$$R_{SM}^{out} = (1 - P_{out})R_{SM}$$

4. Zero-outage rate-adaptive source encoding (ZORA)

In this section we assume that the source encoder knows the (instantaneous) value of the wireless channel mutual information. Thus, it encodes the source at a rate just below the mutual information, leading to the best achievable distortion at the receiver side. Note that this is a zero-outage strategy, that is expected to provide better results than the fixed-outage strategy, at the cost of increased system complexity due to the need for complete CSI at the transmitter side. Furthermore, this is a fast-adaptive technique compared to the fixed-outage, since the rate of adaptation is determined by the variations of the channel fading.

4.1 Repetition

Since transmitter side information does not increase capacity unless transmitted power is also adapted (Goldsmith, 2005), the capacity of this strategy in a given fading realization has the same expression as in (5) which can be rewritten using (1) as

$$C = \frac{1}{2} \log_2 \left(1 + \frac{\bar{\gamma}}{2} \sum_{i,j=1}^{2} x_{ij} \right) = \frac{1}{2} \log_2 \left(1 + \frac{\bar{\gamma}}{2} x_s \right)$$

where $\sum_{i,j=1}^{2} x_{ij} = x_s \sim \chi_8^2$ (Hogg & Craig, 1970).
Since the transmitter has CSI knowledge, in every time instant the source coding rate R_{REP} can be adapted to achieve the instantaneous capacity C. The distortion D_r observed at the receiver is then (Cover & Thomas, 1991)

$$D_r = 2^{-2R_{REP}} = \frac{1}{1 + \frac{\bar{\gamma}}{2} x_s}$$

which is a continuous random variable. Its expected value is

$$D_{REP} = \mathbb{E}[D_r] = \int_0^{+\infty} \frac{1}{1+\frac{\bar{\gamma}}{2}z} f_X^{(8)}(z)dz$$
$$= \frac{1}{48} \int_0^{+\infty} \frac{z^3}{2+\bar{\gamma}z} e^{-\frac{z}{2}} dz$$

which yields

$$D_{REP} = \frac{1}{6} \cdot \frac{\bar{\gamma} - \bar{\gamma}^2 + 2\bar{\gamma}^3 - e^{\frac{1}{\bar{\gamma}}} \Gamma\left(0, \frac{1}{\bar{\gamma}}\right)}{\bar{\gamma}^4}$$

The CDF $F_{REP}(d)$ of the distortion at the receiver can be derived as

$$F_{REP}(d) = Pr\left\{D_r < d\right\} = Pr\left\{x_s > \frac{2-2d}{\bar{\gamma}d}\right\}$$
$$= 1 - F_X^{(8)}\left(\frac{2-2d}{\bar{\gamma}d}\right)$$

4.2 Alamouti

This strategy (Zoffoli et al., 2008a) employs the Alamouti scheme to obtain two independent channels from the MIMO system. These two channels are then used for the transmission of a single description representation of the source, after demultiplexing it into two half-rate substreams. The capacity for the ALM strategy is given by (8) that, from (1), can be expressed as

$$C = \log_2\left(1 + \frac{\bar{\gamma}}{4} \sum_{i,j=1}^{2} x_{ij}\right) = \log_2\left(1 + \frac{\gamma}{4} x_s\right)$$

where $\sum_{i,j=1}^{2} x_{ij} - x_s \sim \chi_8^2$ (Hogg & Craig, 1970).

Using transmitter side information, the source coding rate R_{ALM} can be adjusted to follow the variations of the capacity C. Thus, the distortion at the receiver is given by (Cover & Thomas, 1991)

$$D_r = 2^{-2R_{ALM}} = \frac{1}{(1+\frac{\bar{\gamma}}{4}x_s)^2}$$

Its expected value can be evaluated as

$$D_{ALM} = \mathbb{E}[D_r] = \int_0^{+\infty} \frac{1}{(1+\frac{\bar{\gamma}}{4}z)^2} f_X^{(8)}(z)dz$$
$$= \frac{1}{6} \int_0^{+\infty} \frac{z^3}{(4+\bar{\gamma}z)^2} e^{-\frac{z}{2}} dz$$

which finally results in

$$D_{ALM} = \frac{2}{3} \cdot \frac{\bar{\gamma}\left[(\bar{\gamma}-4)\bar{\gamma}-4\right] + 4e^{\frac{2}{\bar{\gamma}}}(3\bar{\gamma}+2)\Gamma\left(0, \frac{2}{\bar{\gamma}}\right)}{\bar{\gamma}^5}$$

The CDF $F_{ALM}(d)$ of the distortion is

$$F_{ALM}(d) = Pr\Big\{D_r < d\Big\} = Pr\Big\{x_s > \frac{4-4\sqrt{d}}{\bar{\gamma}\sqrt{d}}\Big\}$$
$$= 1 - F_{\chi}^{(8)}\Big(\frac{4-4\sqrt{d}}{\bar{\gamma}\sqrt{d}}\Big)$$

4.3 Spatial multiplexing strategy

Here, a single description of the source, i.e. a single symbol stream, is first demultiplexed and encoded into two separate and independent substreams. Each substream is then transmitted simultaneously over each transmit antenna and, at the receiver, an optimal joint decoder is employed for retrieving the original symbol stream. The capacity of this strategy is given by (10).

In the Appendix it is shown that the expected distortion for ZORA over SM MIMO system is given by

$$D_{SM} = -\frac{16\Big[\bar{\gamma} - (\bar{\gamma}+2)e^{\frac{2}{\bar{\gamma}}}\Gamma\Big(0,\frac{2}{\bar{\gamma}}\Big)\Big]^2}{\bar{\gamma}^6} + \frac{8\Big[\bar{\gamma} - 2e^{\frac{2}{\bar{\gamma}}}\Gamma\Big(0,\frac{2}{\bar{\gamma}}\Big)\Big]\Big[\bar{\gamma}(\bar{\gamma}+2) - 4(\bar{\gamma}+1)e^{\frac{2}{\bar{\gamma}}}\Gamma\Big(0,\frac{2}{\bar{\gamma}}\Big)\Big]}{\bar{\gamma}^6}.$$

The CDF of the distortion observed at the receiver is obtained in the Appendix as

$$F_{SM}(d) = Pr\Big\{D_r < d\Big\} = 1 - Pr\Big\{x_1 x_2 < \frac{1}{\sqrt{d}}\Big\}$$
$$= 1 - F\Big(\frac{1}{\sqrt{d}}\Big) \qquad (12)$$

where

$$F(z) = \frac{2}{\bar{\gamma}^3}e^{\frac{4}{\bar{\gamma}}}\int_1^z e^{-\frac{2}{\bar{\gamma}}x_2}\cdot\Big\{e^{-\frac{2}{\bar{\gamma}}}\Big[\bar{\gamma}^2 - 2\bar{\gamma}\Big(x_2-1\Big) + 2\Big(x_2-1\Big)^2\Big] - e^{-\frac{2}{\bar{\gamma}}\frac{z}{x_2}}\Big[\bar{\gamma}^2 + 2\bar{\gamma}\Big(\frac{z}{x_2} - x_2\Big) + 2\Big(\frac{z}{x_2} - x_2\Big)^2\Big]\Big\}dx_2$$

5. Numerical results

5.1 Discussion for the fixed-outage strategies

We begin the discussion by comparing only the three TS-MD strategies. Then, we compare TS-MD-OPT with the remaining three strategies.

Figure 2 compares the expected distortions achievable with the TS strategies at a fixed $\bar{\gamma}$ of 10 dB. These results can be explained using the same observations we made in (Zoffoli et al., 2008b), where we considered MD strategies over two parallel and independent fading channels. For completeness, we now briefly restate here these conclusions.

As expected, TS-MD-OPT achieves the lowest distortions, since it is designed to minimize Eq. (7). At low outage probabilities, both descriptions are correctly decoded most of the time

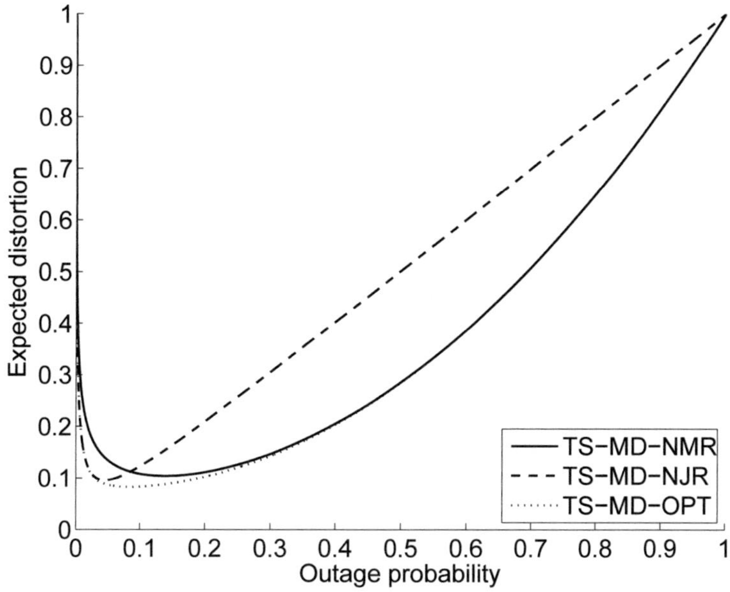

Fig. 2. FORA: expected distortion vs. outage probability for the different TS strategies with $\bar{\gamma}$ = 10 dB.

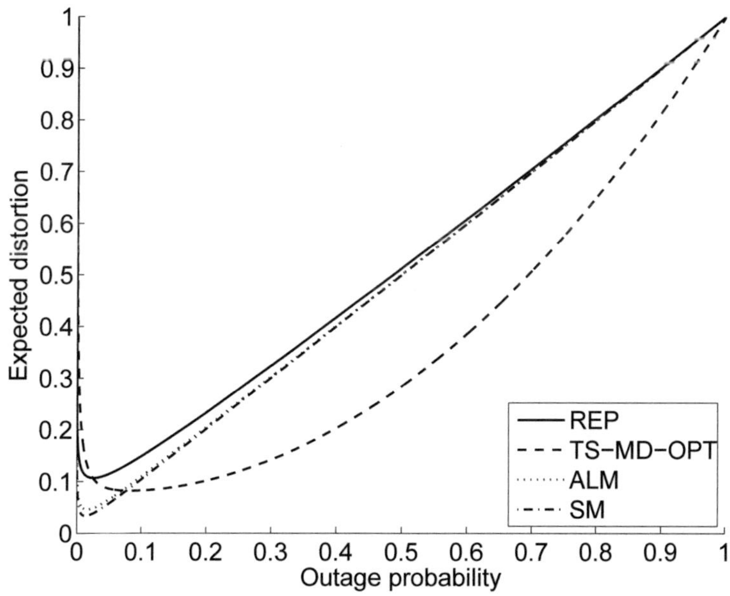

Fig. 3. FORA: expected distortion vs. outage probability for the different strategies with $\bar{\gamma}$ = 10 dB.

and optimal performance is achievable with the TS-MD-NJR strategy, since it is designed to minimize the distortion D_0. As the outage probability gets higher, the receiver becomes able to correctly decode only one description most of the time and TS-MD-NMR achieves optimal performance, since it is designed to minimize the distortion D_1.
Figure 3 compares the remaining strategies and TS-MD-OPT at a fixed $\bar{\gamma}$ of 10 dB. As can be seen, the lowest distortions are achieved with the SM strategy. However, this performance comes at the expense of complexity, mainly due to the presence of the joint decoder at the receiver. Interestingly, the ALM strategy, which can be employed for reducing this complexity, shows only a very small loss in performance with respect to SM. Looking at the source coding rates of the different strategies, reported in Fig. 4, it can be seen that ALM obtains rather high coding rates, but still significantly lower than those of SM.

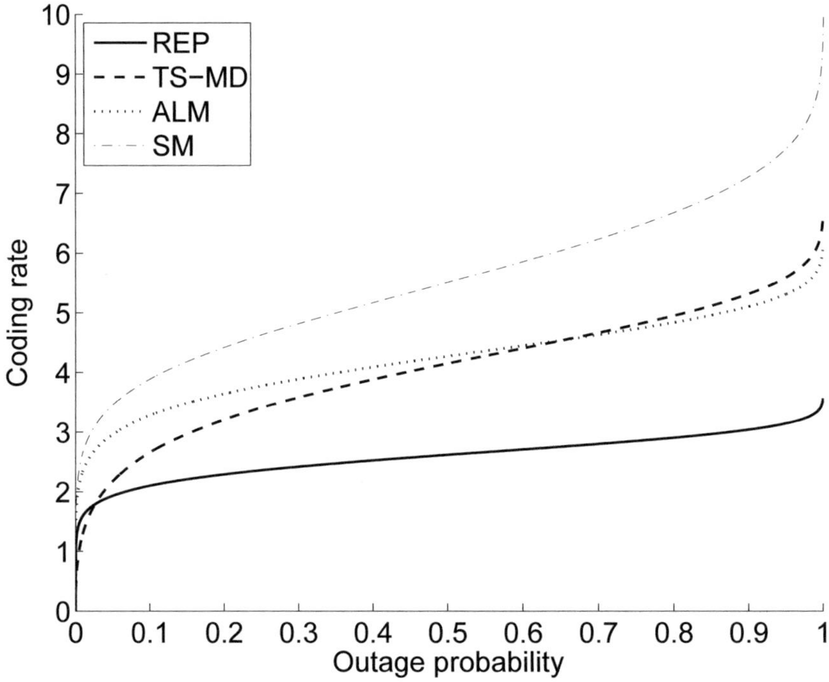

Fig. 4. FORA: source coding rates vs. outage probability for the different strategies with $\bar{\gamma}$ = 10 dB.

This observation, at a first analysis, might erroneously lead to the expectation of a more evident difference in performance between these two strategies. In fact, it must be recalled that the distortions D_1 are exponential decaying functions of the coding rate (see Eqs. (9) and (11)). So, due to this type of dependency, the distortions D_1 are very similar even though the coding rates are significantly different.
Returning to Fig. 3, as the outage probability grows, performance of SM and ALM quickly worsen and the lowest distortions become achievable with the TS-MD-OPT strategy. This happens because SM and ALM are both transmitting over a single unreliable channel, while TS-MD-OPT employs path diversity over two independent and equally unreliable channels reducing the overall system outage probability. The REP strategy has in general the worst

performance, except for very low values of outage probabilities where it performs slightly better than TS-MD-OPT.

We now consider the outage rates achievable with the various strategies, plotted in Fig. 5. In

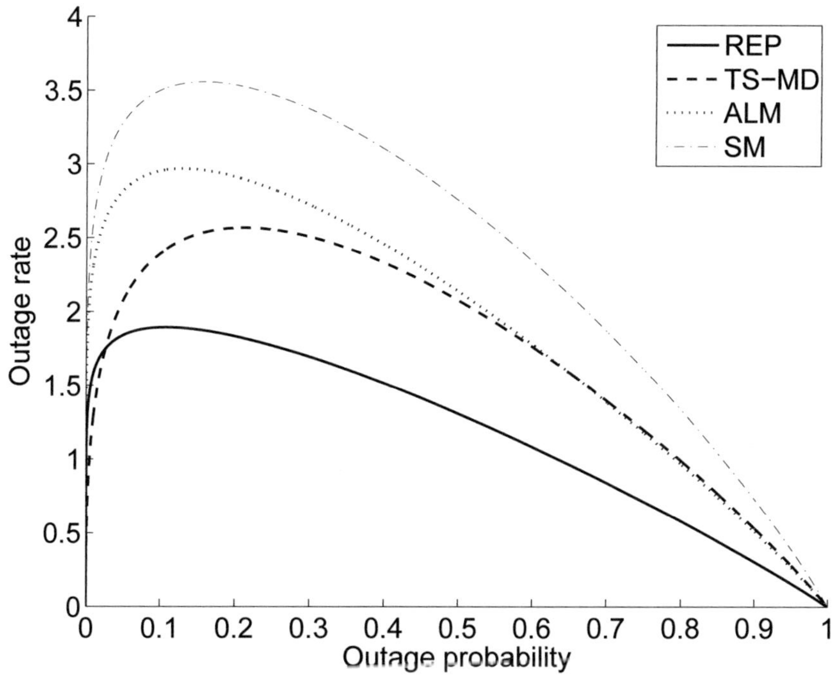

Fig. 5. FORA: outage rate vs. outage probability for the different strategies with $\bar{\gamma} = 10$ dB.

(Choudhury & Gibson, 2007) it has been shown that, when considering the lossy transmission of information over a single channel with very slow Rayleigh fading, designing the system to maximize outage rate does not lead to the minimization of the distortion at the receiver. Inspired by this observation, we want to determine if the same result applies to our MIMO case.

We denote by p_d the outage probability that minimizes expected distortion and by p_r the outage probability that maximizes outage rate. Table 1 shows the values of p_d, p_r, the corresponding distortions and the percent differences in distortion for the various strategies with $\bar{\gamma} = 10$ dB, obtained from Figs. 3 and 5. As can be seen, the outage probabilities that minimize distortion are very different from the outage probabilities that maximize outage

	p_d	p_r	$D(p_d)$	$D(p_r)$	$\Delta D\%$
ALM	0.015	0.130	0.0458	0.1372	199.41
TS-MD-OPT	0.084	0.212	0.0825	0.1058	28.24
REP	0.026	0.108	0.1069	0.1551	45.01
SM	0.013	0.157	0.0338	0.1589	369.76

Table 1. p_d, p_r and respective distortions for the various strategies with $\bar{\gamma} = 10$ dB.

rate. Thus, if the system is designed to maximize outage rate instead of minimizing distortion, suboptimal performance is achieved. In some cases this approach could result in much higher distortions, as in the SM strategy where the percent difference between the two distortions is almost 370%. TS-MD-OPT is the less sensitive strategy to this design error, but still has a distortion that is more than 28% higher than the minimum distortion.

5.2 Discussion for the zero-outage strategies

Figure 6 plots the expected distortion as a function of $\bar{\gamma}$ for the different strategies. As can be seen, the lowest distortion is achieved with the SM strategy at all values of $\bar{\gamma}$. This performance, however, comes at the expense of realization complexity, due to the presence of the joint ML decoder at the receiver (Zoffoli et al., 2008a). Similarly to the case of CSI at the receiver only of (Zoffoli et al., 2008a), if the ALM strategy is employed to reduce this complexity, only a small decrease in performance is observed, especially at high values of $\bar{\gamma}$. Significantly higher distortion is produced with the REP strategy.
Figures 7, 8 and 9 plot the CDF of the distortion for different values of $\bar{\gamma}$ respectively for the REP, ALM and SM strategies. The square markers in the plots represent the value of the expected distortion for the respective value of $\bar{\gamma}$. Interestingly, an increase in the value of $\bar{\gamma}$ not only improves the value of expected distortion for every strategy (as it appears evident also from Fig. 6), but also improves the probability $p_e(\bar{\gamma})$ of achieving that distortion. The values of $p_e(\bar{\gamma})$ for the different strategies and for different values of $\bar{\gamma}$ are reported in Table 2. It can

	$p_e(\bar{\gamma})$			
	$\bar{\gamma} = 1$ dB	$\bar{\gamma} = 3$ dB	$\bar{\gamma} = 5$ dB	$\bar{\gamma} = 10$ dB
ALM	0.630	0.658	0.684	0.731
REP	0.594	0.608	0.621	0.637
SM	0.641	0.678	0.712	0.787

Table 2. Probability of achieving expected distortion for the various strategies with different values of $\bar{\gamma}$.

be observed that, given the same increase of $\bar{\gamma}$, the increase in $p_e(\bar{\gamma})$ in the REP strategy is significantly lower than the increase in $p_e(\bar{\gamma})$ for the remaining two strategies. For example, if $\bar{\gamma}$ increases from 1 dB to 10 dB, $p_e(\bar{\gamma})$ in the REP strategy increases by about 0.04, while in ALM and SM it increases by about 0.10 and 0.14, respectively. Moreover, returning to Figs. 7, 8 and 9, an increase of $\bar{\gamma}$ also causes an increase in the slope of the CDF for all strategies, suggesting that the values of distortion become less variable as $\bar{\gamma}$ increases.
A comparison of the CDF of the distortion for the various strategies at a fixed $\bar{\gamma}$ of 5 dB is reported in Fig. 10. Both ALM and SM strategies have similar and very steep CDFs, which means that it is possible with these strategies to achieve low values of distortion with high probability. For example, with a probability of 0.9, SM achieves a distortion approximately equal to 0.05, while ALM achieves a distortion approximately equal to 0.07. The REP strategy has a much less steep CDF than the other two strategies, and indeed, with a probability of 0.9 it achieves a significantly higher distortion, approximately equal to 0.15.

5.3 Comparison between fixed-outage and zero-outage

The zero outage strategies require CSI at the transmitter, fast adaptation at the transmitter to respond to the CSI, and commensurate additional complexity compared to the fixed outage

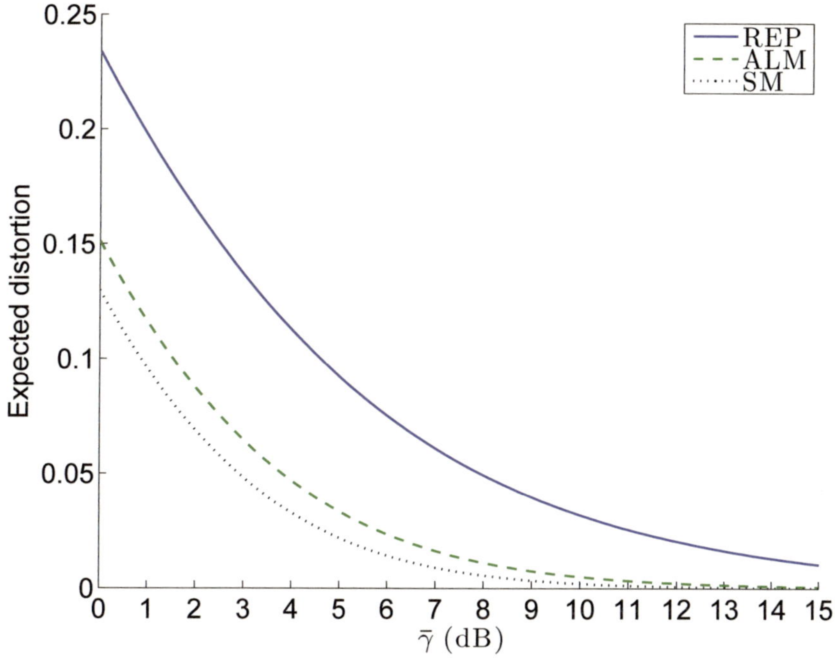

Fig. 6. ZORA: expected distortion as a function of $\bar{\gamma}$ for the different strategies.

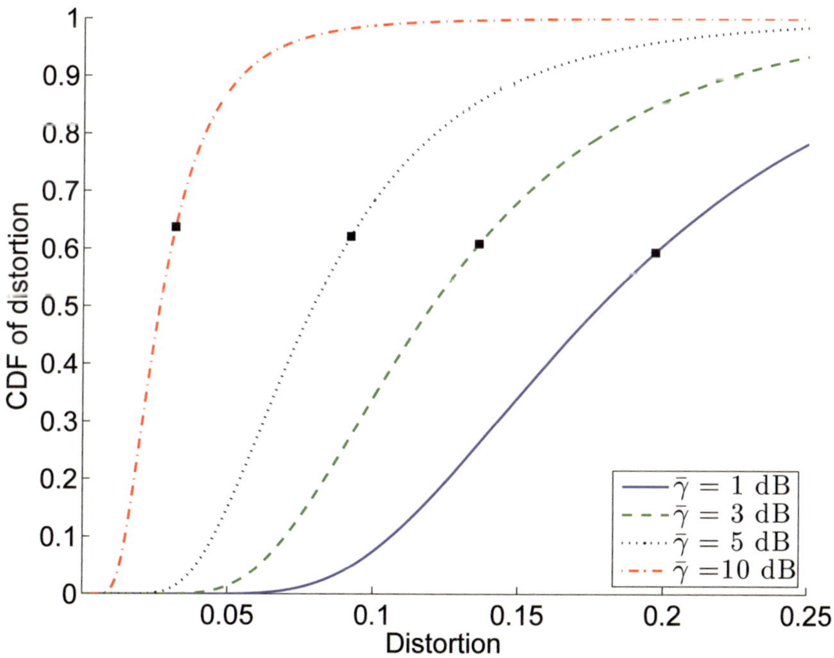

Fig. 7. ZORA: CDF of distortion for REP strategy with different values of $\bar{\gamma}$.

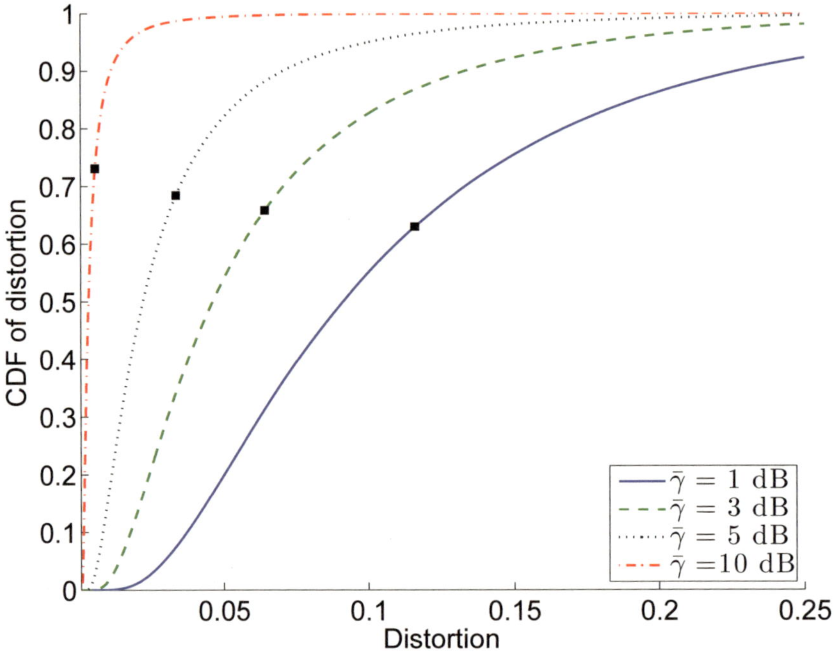

Fig. 8. ZORA: CDF of distortion for ALM strategy with different values of $\bar{\gamma}$.

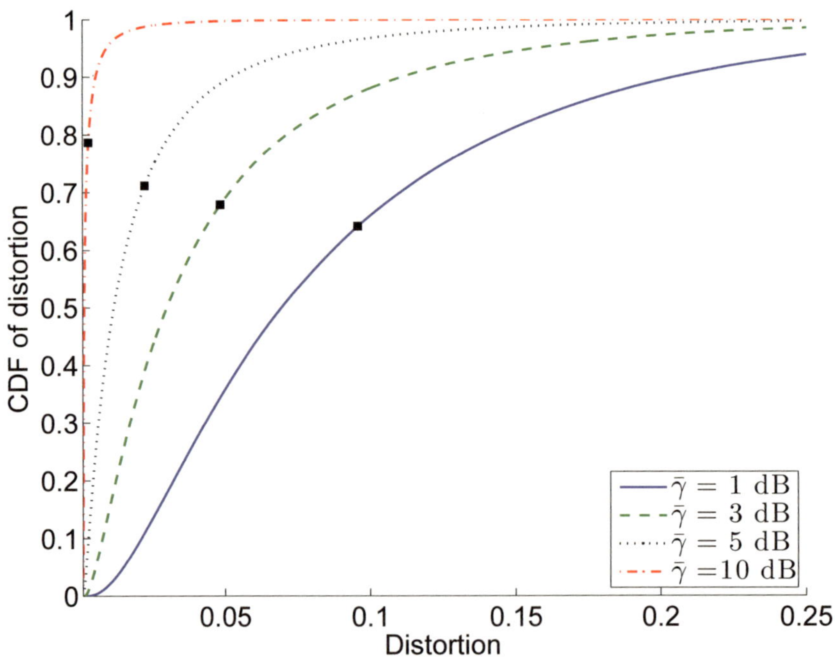

Fig. 9. ZORA: CDF of distortion for SM strategy with different values of $\bar{\gamma}$.

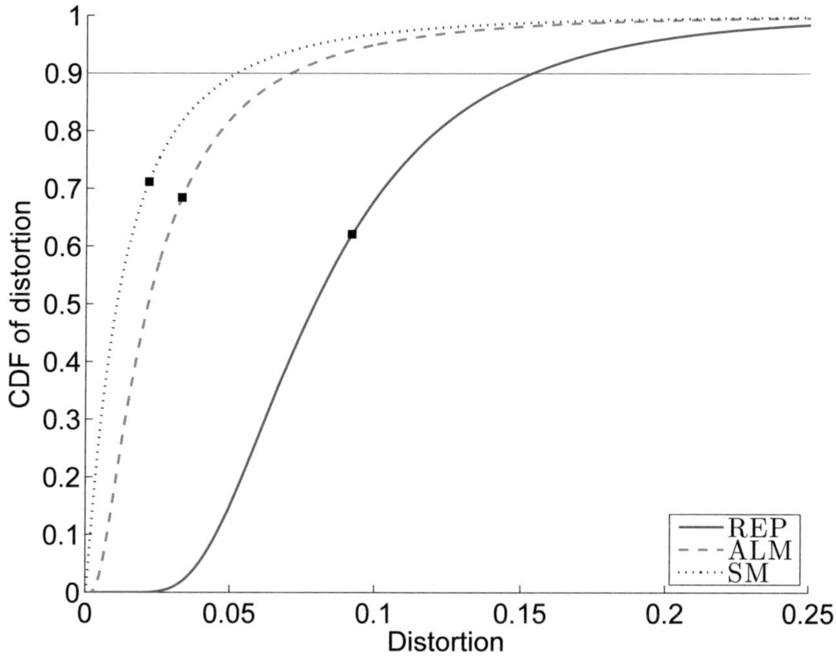

Fig. 10. ZORA: CDF of the distortion at the receiver for the different strategies at a fixed $\bar{\gamma}$ of 5 dB.

approaches. As a result, we would hope for and expect some gain in performance. Using SNR to denote the ratio P_t/N, a comparison of the FORA minimum distortion values in Fig. 3 for an SNR of 10 dB, with the minimum expected distortion achieved by the Alamouti and Spatial Multiplexing schemes at a 10 dB SNR, available in Fig. 6 for ZORA, reveals that the more complex zero outage rate strategies have more than an order of magnitude lower distortion. In fact, the Alamouti scheme under the zero outage approach achieves the same minimum average distortion as in Fig. 3 but at less than a 6 dB SNR, and the spatial multiplexing scheme under ZORA produces the minimum average distortion in Fig. 3 but at an SNR of around 4 dB. Thus, 4 to 6 dB gains in SNR are available with the more complex zero outage strategies compared to the fixed outage methods.

6. Conclusions

The results for the Fixed Outage Rate strategies used in a TS MIMO mode show that of the three MD methods, namely, the No Excess Joint Rate coder (MD-NJR), the No Excess Marginal Rate coder (MD-NMR), and the optimized MD coder (MD-OPT), the MD-OPT performs best across all outage probabilities since if follows the envelope of the other two methods. However, to optimize the MD-OPT coder, we need to know the packet loss rate to minimize the average distortion. While the TS-MD-OPT outperforms the other MD approaches studied, the TS-MD-OPT scheme performs worse for low outage probabilities than the REP, ALM, and SM MIMO schemes that use an SD coder and is only able to take advantage of the MD source coding method as the outage probability moves toward 0.1 and higher. Perhaps the most important result obtained is that the outage probability that maximizes outage rate

is much different from the outage probability that minimizes expected distortion for all of the methods, and that the performance penalty for ignoring the expected distortion in the optimization can be substantial. Thus, the usual approach of selecting the outage probability to maximize outage rate is suboptimal for lossy source coding.
For the ZORA methods, wherein CSI is available at the transmitter and the source coding rate is adapted to match the instantaneous channel capacity, it is shown that SM performs best among REP, ALM, and SM, but that ALM performs very well, particularly in light of its lower complexity. Plots of the cumulative distribution of the distortion as a function of SNR indicate that not only does an increase in SNR reduce the expected distortion, but it also increases the probability that that distortion will be achieved. Additionally, the CDFs of the distortion have a much steeper slope with increased SNR, thus implying a much narrower range of variation for the performance. The more complex and adaptive ZORA approach using transmitter CSI is also shown to achieve more than an order of magnitude reduction in average distortion over the FORA methods and to yield a reduction of 4 to 6 dB in required SNR for the same performance.

7. Appendix: Statistics of the distortion for ZORA

To determine the expression of the expected distortion at the receiver for the zero-outage strategy, we first need to introduce the characteristic function $\phi_C(z)$ of the capacity C, defined as (Chiani, Win & Zanella, 2003)

$$\phi_C(z) = \mathbb{E}\left[e^{j2\pi C z}\right] = K \det\left[U(z)\right]$$

where, for a 2×2 uncorrelated MIMO Rayleigh fading channel, $K = 1$ and $U(z)$ is a 2×2 matrix with ik-th elements given by (Chiani, Win & Zanella, 2003)

$$u_{ik}(z) = \int_0^{+\infty} x^{i+k-2} e^{-x} \left(1 + \frac{\bar{\gamma}}{2} x\right)^{j\frac{2\pi z}{\ln 2}} dx \tag{13}$$

Since the source coding rate R_{SM} is adapted, in every time instant, to the capacity C, we can now express the expected distortion D_{SM} in terms of the function $\phi_C(z)$ as (Cover & Thomas, 1991)

$$\begin{aligned} D_{SM} &= \mathbb{E}\left[2^{-2R_{SM}}\right] = \mathbb{E}\left[e^{-2C\ln 2}\right] = \phi_C\left(j\frac{\ln 2}{\pi}\right) \\ &= \det\left[U\left(j\frac{\ln 2}{\pi}\right)\right] = \hat{u}_{11}\hat{u}_{22} - \hat{u}_{12}\hat{u}_{21} \end{aligned} \tag{14}$$

where we defined $\hat{u}_{ik} = u_{ik}(j\ln 2/\pi)$. Developing the expression in (13), we get for the 2×2 MIMO

$$\hat{u}_{11} = 2 \cdot \frac{\bar{\gamma} - 2e^{\frac{2}{\bar{\gamma}}}\Gamma\left(0, \frac{2}{\bar{\gamma}}\right)}{\bar{\gamma}^2}$$

$$\hat{u}_{12} = 4 \cdot \frac{(2+\bar{\gamma})e^{\frac{2}{\bar{\gamma}}}\Gamma\left(0, \frac{2}{\bar{\gamma}}\right) - \bar{\gamma}}{\bar{\gamma}^3}$$

$$\hat{u}_{21} = \hat{u}_{12}$$

$$\hat{u}_{22} = 4 \cdot \frac{\bar{\gamma}(2+\bar{\gamma}) - 4(1+\bar{\gamma})e^{\frac{2}{\bar{\gamma}}}\Gamma\left(0, \frac{2}{\bar{\gamma}}\right)}{\bar{\gamma}^4}$$

and, after substituting these expressions into Eq. (14), we derive the final expression of D_{SM} as

$$D_{SM} = -\frac{16\left[\bar{\gamma} - (\bar{\gamma}+2)e^{\frac{2}{\bar{\gamma}}}\Gamma\left(0, \frac{2}{\bar{\gamma}}\right)\right]^2}{\bar{\gamma}^6} + \frac{8\left[\bar{\gamma} - 2e^{\frac{2}{\bar{\gamma}}}\Gamma\left(0, \frac{2}{\bar{\gamma}}\right)\right]\left[\bar{\gamma}(\bar{\gamma}+2) - 4(\bar{\gamma}+1)e^{\frac{2}{\bar{\gamma}}}\Gamma\left(0, \frac{2}{\bar{\gamma}}\right)\right]}{\bar{\gamma}^6}$$

The CDF $F_{SM}(d)$ of the distortion at the receiver is related to that of the capacity via the relationship

$$F_{SM}(d) = Pr\{D_r < d\} = Pr\left\{C > \log_2 \frac{1}{\sqrt{d}}\right\}.$$

For MIMO, the CDF of the capacity can be derived starting from the characteristic function (Chiani, Win & Zanella, 2003; Shin et al., 2006). For small MIMO systems results for the PDF and CDF can also be found in (Smith et al., 2003).

For the particular case of the 2×2 MIMO of interest here, we derive below a closed form expression for the CDF. We first need to rewrite the expression of the capacity in (10) using the singular value decomposition of H, to obtain (Telatar, 1999; Winters, 1987)

$$C = \log_2 \prod_{i=1}^{2}\left(1 + \frac{\bar{\gamma}}{2}\lambda_i\right) = \log_2(x_1 x_2)$$

where

$$x_i = 1 + \frac{\bar{\gamma}}{2}\lambda_i, \qquad i = 1,2 \tag{15}$$

and the $\lambda_1 \geq \lambda_2 \geq 0$ are the two ordered nonzero eigenvalues of the matrix HH^H, giving $x_i > 1$ and $x_1 \geqslant x_2$.

The distortion observed at the receiver is then (Cover & Thomas, 1991)

$$D_r = 2^{-2R_{SM}} = \frac{1}{(x_1 x_2)^2}$$

and $F_{SM}(d)$ results

$$F_{SM}(d) = Pr\Big\{D_r < d\Big\} = 1 - Pr\Big\{x_1 x_2 < \frac{1}{\sqrt{d}}\Big\}$$
$$= 1 - F\Big(\frac{1}{\sqrt{d}}\Big) \tag{16}$$

where $F(z) = Pr\{x_1 x_2 < z\}$.
We now define S as the set of points (x_1, x_2) such that

$$S = \Big\{(x_1, x_2) \in \mathbb{R}^2 : x_1 > 1; x_2 > 1; x_1 \geqslant x_2; x_1 x_2 < z\Big\}$$

If X is the vector $X = (x_1, x_2)$ with joint PDF $f_X(x_1, x_2)$, $F(z)$ can then be evaluated as

$$F(z) = \int_S f_X(x_1, x_2) dx_1 dx_2$$
$$= \frac{1}{2} \int_1^z \int_1^{\frac{z}{x_2}} f_X(x_1, x_2) dx_1 dx_2 \tag{17}$$

The joint PDF of X can be expressed in terms of the joint PDF $f_\Lambda(\lambda_1, \lambda_2)$ of the vector of the eigenvalues $\Lambda = (\lambda_1, \lambda_2)$ as (Papoulis & Pillai, 2001)

$$f_X(x_1, x_2) = \frac{f_\Lambda(\lambda_1, \lambda_2)}{\det(J)} = \frac{4}{\bar{\gamma}^2} f_\Lambda(\lambda_1, \lambda_2) \tag{18}$$

where J is the Jacobian matrix of X. Since from Eq. (15) we have

$$\lambda_i = \frac{2}{\bar{\gamma}}(x_i - 1), \qquad i = 1, 2$$

we can rewrite Eq. (18) as

$$f_X(x_1, x_2) = \frac{4}{\bar{\gamma}^2} f_\Lambda\left(\frac{2}{\bar{\gamma}}(x_1 - 1), \frac{2}{\bar{\gamma}}(x_2 - 1)\right) \tag{19}$$

The joint PDF of the eigenvalues of HH^H for a MIMO system with uncorrelated fading between antenna elements can be written as (Chiani, Win & Zanella, 2003; Chiani, Win, Zanella, Mallik & Winters, 2003; Telatar, 1999)

$$f_\Lambda(\lambda_1, \lambda_2) = e^{-\lambda_1} e^{-\lambda_2} (\lambda_1 - \lambda_2)^2$$

which, substituting into Eq. (19), yields

$$f_X(x_1, x_2) = \frac{16}{\bar{\gamma}^4} e^{\frac{4}{\bar{\gamma}}} e^{-\frac{2}{\bar{\gamma}}(x_1 + x_2)} (x_1 - x_2)^2$$

Substituting this last expression into (17) gives

$$F(z) = \frac{8}{\bar{\gamma}^4} e^{\frac{4}{\bar{\gamma}}} \int_1^z e^{-\frac{2}{\bar{\gamma}} x_2} \int_1^{\frac{z}{x_2}} e^{-\frac{2}{\bar{\gamma}} x_1} (x_1 - x_2)^2 dx_1 dx_2$$

which can be finally written as

$$F(z) = \frac{2}{\bar{\gamma}^3} e^{\frac{4}{\bar{\gamma}}} \int_1^z e^{-\frac{2}{\bar{\gamma}} x_2} \cdot \left\{ e^{-\frac{2}{\bar{\gamma}}} \left[\bar{\gamma}^2 - 2\bar{\gamma}\left(x_2 - 1\right) + 2\left(x_2 - 1\right)^2 \right] - e^{-\frac{2}{\bar{\gamma}} \frac{z}{x_2}} \left[\bar{\gamma}^2 + 2\bar{\gamma}\left(\frac{z}{x_2} - x_2\right) + 2\left(\frac{z}{x_2} - x_2\right)^2 \right] \right\} dx_2$$

By using this expression into (16), it is then possible to evaluate the CDF of distortion for the SM strategy.

8. References

Alamouti, S. (1998). A simple transmit diversity technique for wireless communications, *IEEE J. Sel. Areas Commun.* 16(8): 1451–1458.

Balam, J. & Gibson, J. (2006). Path diversity and multiple descriptions with rate dependent packet losses, *Proc. Information Theory and Applications Workshop*, University of California, San Diego, La Jolla, CA.

Chiani, M., Win, M. Z. & Zanella, A. (2003). On the capacity of spatially correlated MIMO Rayleigh fading channels, *IEEE Trans. on Information Theory*, 49(10): 2363–2371.

Chiani, M., Win, M. Z., Zanella, A., Mallik, R. K. & Winters, J. H. (2003). Bounds and approximations for optimum combining of signals in the presence of multiple co-channel interferers and thermal noise, *IEEE Trans. on Communications* 51(2): 296–307.

Choudhury, S. & Gibson, J. D. (2007). Information transmission over fading channels, *Global Telecommunications Conference, 2007. GLOBECOM '07. IEEE* pp. 3316–3321.

Cover, T. & Thomas, J. (1991). *Elements of Information Theory*, John Wiley & Sons.

Effros, M., Koetter, R., Goldsmith, A. & Medard, M. (2004). On source and channel codes for multiple inputs and outputs: does multiple description beat space time?, *Information Theory Workshop, 2004. IEEE* pp. 324–329.

Foschini, G. J. (1996). Layered space-time architecture for wireless communication a fading environment when using multiple antennas, *Bell Labs Tech. J.* 1(2): 41–59.

Foschini, G. J. & Gans, M. J. (1998). On limits of wireless communications in a fading environment when using multiple antennas, *Wirel. Pers. Commun.* 6(3): 311–335.

Gamal, A. & Cover, T. (1982). Achievable rates for multiple descriptions, *Information Theory, IEEE Transactions on* 28(6): 851–857.

Goldsmith, A. (2005). *Wireless communications*, Cambridge University Press.

Hogg, R. V. & Craig, A. T. (1970). *Introduction to Mathematical Statistics*, 3rd edn, NY: Macmillan.

Papoulis, A. & Pillai, S. (2001). *Probability, Random Variables and Stochastic Processes*, 4th edn, McGraw-Hill Science.

Sandhu, S. & Paulraj, A. (2000). Space-time block codes: a capacity perspective, *IEEE Commun. Lett.* 4(12): 384–386.

Shin, H., Win, M., Lee, J. H. & Chiani, M. (2006). On the capacity of doubly correlated MIMO channels, *IEEE Trans. on Wireless Commun.* 5(8): 2253–2266.

Smith, P., Garth, L. & Loyka, S. (2003). Exact capacity distributions for MIMO systems with small numbers of antennas, *IEEE Communications Letters* 7(10): 481–483.

Telatar, E. (1999). Capacity of multi-antenna Gaussian channels, *Europ. Trans. on Telecomm.* 10(6): 585–595.

Tse, D. & Viswanath, P. (2006). *Fundamentals of Wireless Communication*, Cambridge University Press.

Winters, J. H. (1987). On the capacity of radio communication systems with diversity in Rayleigh fading environment, *IEEE J. Sel. Areas Commun.* SAC-5(5): 871–878.

Zoffoli, M., Gibson, J. D. & Chiani, M. (2008a). On strategies for source information transmission over MIMO systems, *Global Telecommunications Conference, 2008. GLOBECOM '08. IEEE.*

Zoffoli, M., Gibson, J. D. & Chiani, M. (2008b). Source coding diversity and multiplexing strategies for a 2x2 MIMO system, *Information Theory and Applications Workshop*, University of California, San Diego, La Jolla, CA.

7

Analysis of MIMO Systems in the Presence of Co-channel Interference and Spatial Correlation

Dian-Wu Yue and Qian Wang
Dalian Maritime University
China

1. Introduction

Wireless communication systems employing multiple antenna elements at the transmitter and the receiver have been attracting much interest in recent years due to the significant capacity gain promised by the multiple-input multiple-output (MIMO) systems [Teletar (1999)], [Foschini & Gans (1998)]. The MIMO systems have been analyzed deeply from two different perspectives [Teletar (1999)]-[Yue & Zhang (2010)]: one concerns the evaluation of the information-theoretic (Shannon) capacity, the other concerns performance evaluation in terms of outage probability or symbol error probability of practical systems. Both of the capacity analysis and performance analysis strongly rely on random matrix theory and matrix variate distributions.

So far the capacity issues of MIMO systems have been extensively studied in the literature, yet with main focus on the scenario without interference [Teletar (1999)]-[Kiessling (2005)]. In cellular systems, however, multiple users share the same radio spectrum, which typically causes co-channel interference. It is well known that co-channel interference ultimately limits the quality of service offered to the users. There have been initial investigations for the MIMO capacity with co-channel interference in fading environments [Catreux et al. (2000)]-[Kang & Alouini (2003a)]. In particular, Song and Blostein [Song & Blostein (2002)] studied the behavior of MIMO capacity with varying number of interferers through simulations. In [Kang et al. (2007)] and [Kang & Alouini (2003a)], Kang *et al.* obtained exact closed-form expressions for the moment generating function, mean, variance of MIMO capacity. Specifically, the paper [Kang et al. (2007)] considered MIMO Rayleigh fading channels in the presence of additive noise and interferers with arbitrary average powers, but requiring that there is no spatial correlation both among transmit antenna elements and among receive antenna elements for the desired user, and there is no correlation among receive antenna elements for any interferer. Reference [Kang & Alouini (2003a)] considered MIMO Rician channels in the presence of Rayleigh co-channel interference, but requiring that there is an identical correlation structure among receive antenna elements for both the desired user and interferers, and there is no correlation among their transmit antenna elements.

In many practical situations, however, signal correlation among the antenna elements exists in realistic environments due to poor scattering conditions. A typical example of this is an uplink transmission from a mobile station (MS) to a base station (BS) , where around the

BS the number of scatterers is usually smaller than around mobile terminals, thus leading to the fact that the correlation at the BS is stronger than at the MS. Another typical example of this is a downlink transmission from a BS to a MS, where the antennas at the BS can be spaced sufficiently far to achieve uncorrelation among them. On the other hand, it is more difficult to space the antennas far apart at the mobile terminals due to physical size constraints, and consequently correlation arises among the antenna elements in such scenarios. The above factors have given us an impetus for studying the capacity of MIMO channels with interference and receive correlation [Wang & Yue (2009)]. In Section 3, we will investigate the capacity issue in the case where the MIMO channels of the desired user and co-channel interferers are all subject to Rayleigh type of fading.

A MIMO system can be configured differently. One configuration is transmit/receive diversity (TRD) which has been widely used due to its simplicity and good performance. The performance of MIMO systems with optimal TRD depends on their operational environments. Their performance in a Rayleigh fading environment without co-channel interference was investigated by Dighe *et al.* [Dighe et al. (2001)] by assuming that the MIMO channels follow independent and identical (i.i.d.) Rayleigh distribution. The resulting outage probability is expressible in the form of a determinant. This result was subsequently extended by Kang and Alouini [Kang & Alouini (2003b)] to a general case of independent, but not necessarily identically distributed, Rician fading channels. The results, again, takes the form of determinants. For the case using dual antennas at the transmitter or receiver end, they obtained [Kang & Alouini (2004a)] an explicit expression for outage probability complementing the result of Dighe [Dighe et al. (2001)]. The performance of MIMO systems with optimal TRD in the presence of co-channel interference was tackled in [Dighe et al. (2003)] and [Kang & Alouini (2004b)] under various fading environments allowing for the MIMO fading channels of the intended user and interferers to be non-i.i.d. Rician/Rayleigh, i.i.d. Rician/Rayleigh, and Rayleigh/Rayleigh. All these studies focus on MIMO systems with uncorrelated or semi-correlated antennas.

By semi-correlation, we mean that the spatial correlation exists only at one side, transmitter or receiver end, of the MIMO systems. Even for the case with semi-correlation, it is usually assumed that the intended user and interferers have the same correlation structure to simplify the mathematical analysis. In fact, the use of this assumption leads to the same mathematical treatment as the one with i.i.d. channels. The i.i.d. or uncorrelated assumption is often invalid in many practical applications. Significant correlation among the antennas exists in realistic environments due to, for example, limited spacing between antennas. Furthermore, the spatial structure (and even the fading distribution) of the interference usually differ from its counterpart for the intended user since their signals propagate over different multi-paths, suffer from different fading, and arrive at the receive antenna array with different incident angles. To handle these general fading situations, we must take different methodology [Yue & Zhang (2010)]. In Section 4, we will investigate the performance issue of MIMO systems with optimal TRD mainly over general Rayleigh/Rayleigh fading channels in a unified framework.

Throughout the paper, we use extensively relevant notations and results from multivariate statistical theory, in particular, various matrix-variate distributions. Although relevant results are available in the statistical literature [Muirhead (1982)], [Gupta & Nagar (2000)] and [Mathai et al. (1995)], they are given only for real variables. The extension of these results to their complex counterparts, as required in this paper, is straightforward. Such results, though useful for wireless communications, are not found in the open references. We therefore first summarize definitions of various special functions with complex matrix arguments and their properties in Section 2 for the use in this paper which, we hope, are also useful to researchers in the area of wireless communications.

Moreover, we will use the following notations throughout the paper. By $\mathbf{I}_n$ we denote the identity matrix of size $n \times n$ (the subscript will be omitted wherever the size of the matrix is clear from the context), $\mathbf{0}$ signifies the all-zero matrix, $\text{diag}(x_1, \ldots, x_n)$ denotes the diagonal matrix with elements $x_1, \ldots, x_n$, the determinant of the matrix $\mathbf{X}$ is denoted by $|\mathbf{X}|$ or $\det(\mathbf{X})$, $[x_{ij}]$ is a matrix with x_{ij} representing its (i,j)th element and correspondingly, $|x_{ij}|$ denotes its determinant. $\text{eig}(\mathbf{X})$ denotes the diagonal matrix of eigenvalues of $\mathbf{X}$. The symbol $\mathbf{X} > 0$ indicates that $\mathbf{X}$ is positive definite; likewise, $\mathbf{X} > \mathbf{A}$ means $\mathbf{X} - \mathbf{A} > 0$. We use notation $\text{tr}(\mathbf{X})$ to signify the trace of the square matrix $\mathbf{X}$, $\text{etr}(\mathbf{X})$ to denote $\exp(\text{tr}\mathbf{X})$, $\mathbf{X}^\dagger$ to mean the Hermitian transposition. The symbol $\otimes$ denotes the Kronecker product of two matrices, '$\sim$' means 'distributed as', $CW_p(n, \boldsymbol{\Sigma})$ is a complex Wishart distribution , $CN(\mu, \boldsymbol{\Sigma})$ is a complex vector variate Gaussian distribution, $CN_{p,q}(\mathbf{M}, \mathbf{A} \otimes \mathbf{B})$ means a complex $p \times q$ matrix variate Gaussian distribution and $\mathcal{E}_{\mathbf{X}}[\cdot]$ denotes expected value with respect to $\mathbf{X}$.

2. Definitions and properties for random functions of complex matrix arguments

2.1 Zonal polynomials

Zonal polynomials were introduced by James [James (1964)], and have become an essential tool for studying and expressing some useful special functions of matrix arguments (such as Hypergeometric functions of matrix arguments we will discuss).Using these special functions in matrix arguments, many distributions of quadratic forms can be obtained in a very compact form.

For $k > 0$, we denote $\kappa = (k_1, k_2, \ldots, k_p)$ such that $k = \sum_{j=1}^{p} k_j, k_1 \geq k_2 \geq \ldots \geq k_p \geq 0$. Then we call κ a partition of k into p parts. Partitions may be ordered lexicographically as follows. If $\kappa = (k_1, k_2, \ldots)$ and $\lambda = (l_1, l_2, \ldots)$, then we say $\kappa > \lambda$ if $k_i > l_i$ for the first index i where the partitions differ. Now let $y_1, \ldots, y_p$ be p variables. Then we say that the monomial $y_1^{k_1} \cdots y_p^{k_p}$ is of order κ and that $y_1^{k_1} \cdots y_p^{k_p}$ is of higher order than $y_1^{l_1} \cdots y_p^{l_p}$ if $\kappa > \lambda$. The degree of a monomial in p variables is the sum of degrees of the individual variables. The degree of a polynomial is the maximum degree of the monomials making up the polynomial. We denote by $\mathbf{V}_k$ the vector space of symmetric homogenous polynomials of degree k in p variables. Further let $\mathbf{V}_\kappa$ be the subspace of $\mathbf{V}_k$ defined by polynomials of order κ. Then $\mathbf{V}_k$ is the direct sum of the irreducible invariant subspaces $\mathbf{V}_\kappa$.

Defintion 1. *Let $\mathbf{V}_k$ be defined on the eigenvalues of a $p \times p$ Hermitian matrix $\mathbf{X}$. Then the polynomial $(tr\mathbf{X})^k \in \mathbf{V}_k$ has a unique decomposition into polynomials $C_\kappa(\mathbf{X}) \in \mathbf{V}_\kappa$ according to*

$$(tr\mathbf{X})^k = \sum_k C_\kappa(\mathbf{X}). \tag{1}$$

The component of $(tr\mathbf{X})^k$ in $\mathbf{V}_\kappa$, $C_\kappa(\mathbf{X})$, is called a zonal polynomial of $\mathbf{X}$.

The zonal polynomial $C_\kappa(\mathbf{X})$ is defined for all k and p, but for a partition κ of k into more than p parts, it is identically zero. The zonal polynomials have the following useful properties.

Property 1. *For a scalar a,*

$$C_\kappa(a\mathbf{X}) = a^k C_\kappa(\mathbf{X}). \tag{2}$$

Let $\kappa = (k_1, k_2, \ldots, k_p)$ be a partition of k. We will denote *the complex multivariate hypergeometric coefficient* by

$$[a]_\kappa = \prod_{i=1}^{p} (a - i + 1)_{k_i} \tag{3}$$

where $(x)_n = x(x+1)\cdots(x+n-1)$ is the Pochhammer symbol [James (1964)]. Note that $(x)_0 = 1$.

Property 2. *For p and q,*

$$\frac{C_\kappa(\mathbf{I}_p)}{C_\kappa(\mathbf{I}_q)} = \frac{[p]_\kappa}{[q]_\kappa} \tag{4}$$

Property 3.

$$C_\kappa(\mathbf{X}) = C_\kappa(\mathbf{U}\mathbf{X}\mathbf{U}^\dagger) \tag{5}$$

where $\mathbf{U} \in \mathbf{U}(p)$, *and* $\mathbf{U}(p)$ *is the group of all* $p \times p$ *complex unitary matrices.*

Property 4.

$$\int_{\mathbf{U}(p)} C_\kappa(\mathbf{U}^\dagger\mathbf{X}\mathbf{U}\mathbf{Y})[d\mathbf{U}] = C_\kappa(\mathbf{X})C_\kappa(\mathbf{Y})/C_\kappa(\mathbf{I}_p) \tag{6}$$

where $[d\mathbf{U}]$ *is the invariant measure on the unitary group* $\mathbf{U}(p)$ *normalized to make the total measure unity.*

A general formula for the coefficients of zonal polynomials has not been found. For more detail study of zonal polynomials, the reader is referred to [Muirhead (1982)].

2.2 Hypergeometric functions of matrix arguments

Many matrix variate distributions, especially central quadratic form distributions, can be written in terms in hypergeometric functions of matrix argument [James (1964)]-[Khatri (1965)]. Hypergeometric functions of matrix argument is a natural generalization of (generalized) hypergeometric functions of scalar argument, which have been used widely in the field of science and engineering.

Defintion 2. *Let* $\mathbf{X}$ *be a* $p \times p$ *Hermitian matrix. Then hypergeometric functions of one complex matrix argument is defined by*

$${}_m\tilde{F}_n^{(p)}(a_1, \ldots, a_m; b_1, \ldots, b_n; \mathbf{X}) = \sum_{k=0}^{\infty} \sum_{\kappa} \frac{[a_1]_\kappa \ldots [a_m]_\kappa C_\kappa(\mathbf{X})}{[b_1]_\kappa \ldots [b_n]_\kappa k!} \tag{7}$$

where $a_1, \ldots, a_m; b_1, \ldots, b_n$ *are arbitrary complex numbers,* $\sum_\kappa$ *denotes summation over all partition* κ.

For the conditions for convergence of the mentioned-above series, the reader is referred to [Gupta & Nagar (2000)]. From Definition 2 it follows that

$${}_0\tilde{F}_0^{(p)}(\mathbf{X}) = \operatorname{etr}(\mathbf{X}) \tag{8}$$

and

$${}_1\tilde{F}_0^{(p)}(a; \mathbf{X}) = |\, \mathbf{I} - \mathbf{X} \,|^{-a} \tag{9}$$

Defintion 3. *Let $q \leq p$. Then the hypergeometric functions of two Hermitian matrices $\mathbf{X}(p \times p)$ and $\mathbf{Y}(q \times q)$ is defined by*

$$ {}_m\tilde{F}_n^{(p,q)}(a_1, \ldots, a_m; b_1, \ldots, b_n; \mathbf{X}, \mathbf{Y}) = \sum_{k=0}^{\infty} \sum_{\kappa} \frac{[a_1]_\kappa \ldots [a_m]_\kappa C_\kappa(\mathbf{X}) C_\kappa(\mathbf{Y})}{[b_1]_\kappa \ldots [b_n]_\kappa C_\kappa(\mathbf{I}_p) k!} \tag{10} $$

The hypergeometric functions of two Hermitian matrices have the following properties.

Property 5.

$$ \int_{\mathbf{U}(p)} {}_m\tilde{F}_n^{(p,p)}(a_1, \ldots, a_m; b_1, \ldots, b_n; \mathbf{X}\mathbf{U}\mathbf{Y}\mathbf{U}^\dagger)[d\mathbf{U}] = {}_m\tilde{F}_n^{(p,p)}(a_1, \ldots, a_m; b_1, \ldots, b_n; \mathbf{X}, \mathbf{Y}). \tag{11} $$

Property 6.

$$ {}_m\tilde{F}_n^{(p,p)}(a_1, \ldots, a_m; b_1, \ldots, b_n; \mathbf{I}_p, \mathbf{X}) = {}_m\tilde{F}_n^{(p)}(a_1, \ldots, a_m; b_1, \ldots, b_n; \mathbf{X}). \tag{12} $$

For more detail study of hypergeometric functions of matrix arguments, the reader is referred to [Gupta & Nagar (2000)]. The hypergeometric functions of two Hermitian matrices can be expressed in terms of scalar hypergeometric functions [Khatri (1966)], which is the practical relevance of our some results given in follow-up parts.

2.3 Generalized Hermitian polynomials of matrix arguments

Hayakawa in 1969 gave the definition of generalized Hermite polynomial of real matrix argument $H_\kappa(\mathbf{T})$, and extended the definition to the case of two real matrix arguments: $P_\kappa(\mathbf{T}, \mathbf{A})$. Crowther in 1975 further extended it to the case of three real matrix arguments : $P_\kappa(\mathbf{T}, \mathbf{\Lambda}, \mathbf{B})$. Now we introduce the definition of generalized Hermite polynomial of complex matrix argument and its extensions. These functions of matrix arguments play an important role in the study of the distribution of some quadratic forms.

Defintion 4. *Let $\mathbf{T} : p \times q$ and $\mathbf{W} : p \times q$, be arbitrary complex matrices, then the generalized Hermite polynomial with a complex matrix argument $H_\kappa(\mathbf{T})$ which corresponds to the partition $\kappa = (k_1, k_2, \ldots, k_p)$ of k is defined as:*

$$ H_\kappa(\mathbf{T}) = \pi^{-pq} etr(\mathbf{T}\mathbf{T}^\dagger) \int_{\mathbf{W}} etr[-\mathbf{W}\mathbf{W}^\dagger - \imath(\mathbf{T}\mathbf{W}^\dagger + \mathbf{W}\mathbf{T}^\dagger)] C_\kappa(-\mathbf{W}\mathbf{W}^\dagger) d\mathbf{W} \tag{13} $$

where $\imath = \sqrt{-1}$.

It should be noted that (13) can be regarded as the Fouier transform of $\mathrm{etr}[-\mathbf{W}\mathbf{W}^\dagger] C_\kappa(-\mathbf{W}\mathbf{W}^\dagger)$. The distributions of the latent roots of a noncentral Wishart matrix and of related statistics can be expressed as series of generalized Hermite polynomials.

Defintion 5. *Let $p \leq q$, and let $\mathbf{T} : p \times q$ and $\mathbf{W} : p \times q$, be arbitrary complex matrices, and let $\mathbf{A} : q \times q$ and $\mathbf{B} : p \times p$ be Hermitian positive definite matrices; then the generalized Hermite polynomial with three complex matrix arguments $P_\kappa(\mathbf{T}, \mathbf{A}, \mathbf{B})$ which corresponds to the partition $\kappa = (k_1, k_2, \ldots, k_p)$ of k is defined as:*

$$ P_\kappa(\mathbf{T}, \mathbf{A}, \mathbf{B}) = \pi^{-pq} etr(\mathbf{T}\mathbf{T}^\dagger) \int_{\mathbf{W}} etr[-\mathbf{W}\mathbf{W}^\dagger - \imath(\mathbf{T}\mathbf{W}^\dagger + \mathbf{W}\mathbf{T}^\dagger)] C_\kappa(-\mathbf{B}\mathbf{W}\mathbf{A}\mathbf{W}^\dagger) d\mathbf{W}. \tag{14} $$

The generalized Hermite polynomial with three complex matrix arguments $P_\kappa(\mathbf{T}, \mathbf{A}, \mathbf{B})$ has several simple properties as follows.

Property 7.

$$P_\kappa(\mathbf{T}, \mathbf{I}_q, \mathbf{I}_p) = P_\kappa(\mathbf{T}, \mathbf{I}_q) = H_\kappa(\mathbf{T}). \tag{15}$$

Property 8.

$$P_\kappa(\mathbf{T}, \mathbf{A}, \mathbf{I}_p) = P_\kappa(\mathbf{T}, \mathbf{A}). \tag{16}$$

Property 9.

$$P_\kappa(\mathbf{0}, \mathbf{A}, \mathbf{B}) = (-1)^k [q]_\kappa \frac{C_\kappa(\mathbf{A}) C_\kappa(\mathbf{B})}{C_\kappa(I_q)}. \tag{17}$$

Crowther has calculated the polynomial $P_\kappa(\mathbf{T}, \mathbf{A}, \mathbf{B})$ for some special κ. With general κ, however, there is no formula available for their calculation. For more detail study of Hermitian functions of matrix argument, the reader is referred to [Gupta & Nagar (2000)] and [Mathai et al. (1995)].

3. Ergodic capacity of MIMO systems with interference and correlation

3.1 System model

We consider a wireless link equipped with t antenna elements at the transmitter and r antenna elements at the receiver. It is assumed that the system is interference-limited, and there are a total of ℓ interfering users each equipped with t_i antenna elements, $i = 1, \ldots, \ell$. The received $r \times 1$ vector at the desired user's receiver can thus be modeled as

$$\mathbf{y} = \mathbf{H}\mathbf{s} + \sum_{i=1}^{\ell} \mathbf{H}_i \mathbf{s}_i \tag{18}$$

where $\mathbf{H}$ is the $r \times t$ normalized channel complex matrix with Gaussian distribution [Gupta & Nagar (2000)]: $\mathbf{H} \sim CN_{r,t}(\mathbf{0}, \mathbf{\Sigma} \otimes \mathbf{\Psi})$, $\mathbf{\Sigma} \otimes \mathbf{\Psi}$ is the covariance matrix of random matrix $\mathbf{H}$; $\mathbf{s}$ is the $t \times 1$ transmitted data vector for the desired user with covariance matrix $\mathcal{E}(\mathbf{s}\mathbf{s}^\dagger) = \mathbf{R}_s$ and total transmitting power $\mathrm{tr}(\mathbf{R}_s) = E_s$. Similarly, for the i-th co-channel interferer, $\mathbf{H}_i$ and $\mathbf{s}_i$ are the $r \times t_i$ normalized channel matrix and the $t_i \times 1$ transmitted vector with short-term average power E_i per antenna, respectively. It is assumed that $\mathbf{H}_i \sim CN_{r,t_i}(\mathbf{0}, \mathbf{\Sigma}_i \otimes \mathbf{\Psi}_i)$ and $\mathbf{s}_i \sim CN_{t_i}(\mathbf{0}, \mathbf{R}_i)$.

Now we take a closer look at the correlation structure of $\mathbf{H}$ and $\mathbf{H}_i$ in (18). The correlations of the matrices $\mathbf{H}$ and $\mathbf{H}_i$ are specified by $\mathbf{\Sigma} \otimes \mathbf{\Psi}$ and $\mathbf{\Sigma}_i \otimes \mathbf{\Psi}_i$, respectively. Physically, $\mathbf{\Sigma}$ and $\mathbf{\Sigma}_i$ represent the $r \times r$ correlation matrices of incoming signal and interference at the receiver, respectively. Correspondingly, the transmit-antenna correlations for the desired user is characterized by the $t \times t$ correlation matrix $\mathbf{\Psi}$, whereas its counterpart for interferer i is specified by the $t_i \times t_i$ correlation matrix $\mathbf{\Psi}_i$. The structure of these correlation matrices depends on channel's fading characteristics, geometry and polarization of antenna arrays, and signal/interferers angle of arrival and spread, as described in various references [Chuah et al. (2002)].

The mentioned-above correlated MIMO channel model is one of several classical correlated MIMO channel models [Kermoal et al. (2002)]. It is very convenient for mathematical tractability, and has been used by many papers [Kiessling (2005)], [Paulraj et al. (2003)]. With it, the MIMO channel correlation is separable [Kermoal et al. (2002)], [Paulraj et al. (2003)], i.e.,

$$\mathbf{H} \simeq \mathbf{A}^{\dagger}\mathbf{H}_w\mathbf{B} \tag{19}$$

where

$$\mathbf{\Sigma} = \mathbf{A}^{\dagger}\mathbf{A} \tag{20}$$

$$\mathbf{\Psi} = \mathbf{B}^{\dagger}\mathbf{B} \tag{21}$$

and $\mathbf{H}_w \sim CN_{r,t}(\mathbf{0}, \mathbf{I}_r \otimes \mathbf{I}_t)$ is a $r \times t$ random matrix of i.i.d Gaussian elements. For simplicity, just as in [Blum et al. (2002)], [Kang et al. (2003)] and [Kang et al. (2007)], all of the interfering signals $\mathbf{s}_i$, $i = 1, \ldots, \ell$, are assumed to be not known at the desired user's receiver, and they are all modeled as complex Gaussian vectors. Hence, the interference $\sum_{i=1}^{\ell} \mathbf{H}_i\mathbf{s}_i$ conditioned on $\mathbf{H}_i, i = 1, \ldots, \ell$, is complex Gaussian with covariance matrix

$$\begin{aligned}\mathbf{R}_c &= \sum_{i=1}^{\ell} \mathbf{H}_i\mathbf{R}_i\mathbf{H}_i^{\dagger} \\ &= \mathbf{H}_I\mathbf{R}_I\mathbf{H}_I^{\dagger}\end{aligned} \tag{22}$$

where

$$\mathbf{H}_I = (\mathbf{H}_1, \cdots, \mathbf{H}_{\ell}) \tag{23}$$

and

$$\mathbf{R}_I = \mathrm{diag}(\mathbf{R}_1, \cdots, \mathbf{R}_{\ell}). \tag{24}$$

This implies that the interference is whitened by multiplying $\mathbf{y}$ by $\mathbf{R}_c{}^{1/2}$.

For analytical tractability, it is assumed firstly that $\mathbf{\Sigma}_1 = \mathbf{\Sigma}_2 = \cdots = \mathbf{\Sigma}_{\ell} = \mathbf{\Sigma}_I$. Note that our assumption is more general than in the literature [Catreux et al. (2000)]-[Kang & Alouini (2003a)] where all correlation receive matrices for interferers are identity ones, namely $\mathbf{\Sigma}_1 = \mathbf{\Sigma}_2 = \cdots = \mathbf{\Sigma}_{\ell} = \mathbf{I}$. In order to obtain easy-to-compute closed-form expressions which provide useful insight, we have to assume further that $E_1 = E_2 = \cdots = E_{\ell} = E_I$, $\mathbf{\Psi} = \mathbf{I}_t$ and $\mathbf{\Psi}_1 = \mathbf{I}_{t_1}, \mathbf{\Psi}_2 = \mathbf{I}_{t_2}, \cdots, \mathbf{\Psi}_{\ell} = \mathbf{I}_{t_{\ell}}$. These assumptions are valid when the interfering signals come from approximately same distance from the receiver and the shadowing effects are small. However, it will lead to a pessimistic estimate of system performance if the total interfering power is fixed [Ye & Blum (2005)]. Exactly, under these assumptions, what we will finally obtain is indeed a lower bound on the ergodic capacity for the general case. To make the problem mathematically tractable, these assumptions are usually adopted for the performance analysis of MIMO systems [Kang & Alouini (2003a)], [Kang & Alouini (2004b)], and [Zhang & Cui (2004)]. Moreover, we assume that perfect channel information is available to the receiver, but the transmitter has no channel information. Then the optimum $\mathbf{R}_s$ to maximize the instantaneous capacity is given by $\mathbf{R}_s = \frac{E_s}{t}\mathbf{I}_t$. For that, we can assume that $\mathbf{R}_s = \frac{E_s}{t}\mathbf{I}_t$, and $\mathbf{R}_1 = E_1\mathbf{I}_{t_1}, \mathbf{R}_2 = E_2\mathbf{I}_{t_2}, \cdots, \mathbf{R}_{\ell} = E_{\ell}\mathbf{I}_{t_{\ell}}$.

Under all these assumptions made above, we will derive some statistical expressions only with respect to the random matrix $\rho\mathbf{H}^{\dagger}(\mathbf{H}_I\mathbf{H}_I^{\dagger})^{-1}\mathbf{H}$, where $\rho = \frac{E_s}{tE_I}$. It should be noticed that from Chapter 3 of [Gupta & Nagar (2000)] for the general settings of $\{E_i\}$ and $\{\mathbf{\Sigma}_i\}$ we can approximate with high precision $\mathbf{R}_c$ by only using a single Wishart-distributed matrix (e.g., $\mathbf{H}_1\mathbf{H}_1^{\dagger}$), and thus (18) can be still used as a good approximating model for the general case.

3.2 Moment-generating function of mutual information

The instantaneous mutual information $I(\mathbf{s},\mathbf{y})$ between input vector $\mathbf{s}$ and output $\mathbf{y}$ of the MIMO link according to (18) is given by Blum et al. (2002)

$$I(\mathbf{s},\mathbf{y}) = \log_2 \mid \mathbf{I}_t + \mathbf{Q} \mid, \tag{25}$$

where

$$\mathbf{Q} = \mathbf{R}_s \mathbf{H}^{\dagger} \mathbf{R}_c^{-1} \mathbf{H}. \tag{26}$$

Now let $\tilde{\mathbf{H}} \sim CN_{r,t}(\mathbf{0}, \tilde{\mathbf{\Sigma}} \otimes \mathbf{I}_t)$ with

$$\tilde{\mathbf{\Sigma}} = \mathbf{A}^{\dagger} \mathbf{\Sigma}_I^{-1} \mathbf{A}, \tag{27}$$

here $\mathbf{A}$ is defined in (20), and let $\tilde{\mathbf{H}}_I \sim CN_{r,t_I}(\mathbf{0}, \mathbf{I}_r \otimes \mathbf{I}_{t_I})$ with

$$t_I = \sum_{i=1}^{\ell} t_i. \tag{28}$$

From the proof of Theorem 7.4.1 of [Gupta & Nagar (2000)], it can follow that $\mathbf{Q}$ can be reexpressed as

$$\mathbf{Q} = \rho \tilde{\mathbf{H}}^{\dagger} (\tilde{\mathbf{H}}_I \tilde{\mathbf{H}}_I^{\dagger})^{-1} \tilde{\mathbf{H}} \tag{29}$$

The MGF of mutual information $I(\mathbf{s},\mathbf{y})$ is defined as

$$M(\theta) = \mathcal{E}_{\mathbf{Q}}[\exp(\theta I(\mathbf{s},\mathbf{y}) / \log_2 e)] = \mathcal{E}_{\mathbf{Q}} \mid \mathbf{I}_t + \mathbf{Q} \mid^{\theta}. \tag{30}$$

Furthermore, the MGF $M(\theta)$ can be written in terms of hypergeometric functions of one matrix argument over complex field ${}_2\tilde{F}_1^{(r)}$.

Theorem 1. *Suppose that the number of receive antennas for the desired user is equal to or less than the total number of transmit antenna for the interferers, namely $r \leq t_I$. Then we have that*

$$M(\theta) = \frac{\tilde{\Gamma}_r(t+t_I)\tilde{\Gamma}_r(t_I-\theta)}{\tilde{\Gamma}_r(t_I)\tilde{\Gamma}_r(t+t_I-\theta)} \, {}_2\tilde{F}_1^{(r)}(-\theta, t; t+t_I-\theta; \mathbf{I} - \rho\tilde{\mathbf{\Sigma}}); \tag{31}$$

where $\tilde{\Gamma}_r(\cdot)$ is the complex multivariate gamma function defined by

$$\tilde{\Gamma}_r(m) = \pi^{r(r-1)/2} \prod_{i=1}^{r} \Gamma(m-i+1). \tag{32}$$

The proof of Theorem 1 is placed in 6.1.

It should be pointed out that in order to make the problem mathematically tractable, the assumption of $r \leq t_I$ is usually adopted for the performance analysis of MIMO systems [Kang & Alouini (2003a)], [Kang & Alouini (2004b)], and [Zhang & Cui (2004)]. In downlink transmission, this particularly true for small and lightweight hand-held/portable receive terminals for which the size of practical adaptive array will typically be restricted to one or two antenna elements.

A general hypergeometric function of one Hermitian matrix argument can be expressed in terms of scalar hypergeometric functions [Kiessling (2005)], which is of the practical relevance of Theorem 1.

The Vandermonde matrix with respect to a $p \times p$ diagonal matrix $\mathbf{\Lambda} = \text{diag}(\lambda_1, \lambda_2, \cdots, \lambda_p)$ with $\lambda_1 > \lambda_2 > \cdots > \lambda_p$ can be denoted by

$$V(\mathbf{\Lambda}) = (\lambda_i^{p-j}) \tag{33}$$

Lemma 1. *Let* $\mathbf{W}$ *be a* $p \times p$ *Hermitian matrix. Define* $\mathbf{\Theta} = eig(\mathbf{W}) = diag(\omega_1, \ldots, \omega_p)$ *with* $\omega_1 > \ldots > \omega_p$. *Then*

$${}_m\tilde{F}_n^{(p)}(a_1, \ldots, a_m; b_1, \ldots, b_n; \mathbf{W}) = \frac{|\mathbf{F}|}{|V(\mathbf{\Theta})|} \tag{34}$$

where $\mathbf{F} = [f_{ij}]$ *with*

$$f_{ij} = \omega_i^{p-j} \, {}_mF_n(a_1 - j + 1, \ldots, a_m - j + 1; b_1 - j + 1, \ldots, b_n - j + 1; \omega_i) \tag{35}$$

for $i, j = 1, 2, \ldots, p$.

Remark 1. *The scalar hypergeometric functions appearing in the above lemma are built-in functions in computational software packages such as Mathematica.*

Remark 2. *When some of the* ω_j*'s are equal, we obtain the results as limiting case on the right of (34) via L'Hospital's rule.*

3.3 Ergodic capacity of MIMO systems

A MGF uniquely defines a probability distribution. Once we can find the MGF of mutual information, we can determine all moments, including the practically important first moment, which is also known as ergodic capacity in MIMO literature[Catreux et al. (2000)]-[Kang et al. (2007)]:

$$C = \mathcal{E}_{\mathbf{Q}} I(\mathbf{s}, \mathbf{y}) = \mathcal{E}_{\mathbf{Q}} \log_2 | \mathbf{I}_t + \mathbf{Q} | . \tag{36}$$

The standard approach to determine the ergodic capacity is just to find first the related MGF. Specially,

$$C = \log_2 e \cdot \frac{\partial M(\theta)}{\partial \theta} |_{\theta=0}. \tag{37}$$

Furthermore, After a lengthy process (see 6.2 for details), we can obtain the following expression of the ergodic capacity of MIMO systems in terms of scalar hypergeometric functions.

Theorem 2. *Suppose that* $r \leq t_I$. *Let* $\mathbf{\Delta} = eig(\mathbf{I}_r - \rho\tilde{\mathbf{\Sigma}}) = diag(\delta_1, \delta_2, \cdots, \delta_r)$ *with* $\delta_1 > \delta_2 > \cdots > \delta_r$.
a) When $r \leq t$, *then*

$$C = \sum_{k=1}^{r} \sum_{j=0}^{t-1} \frac{\log_2 e}{t_I - k + 1 + j} + \frac{\log_2 e \cdot \sum_{h=1}^{r} | \mathbf{D}(h) |}{|V(\mathbf{\Delta})|} \tag{38}$$

where $D(h) = (d_{i,j}(h))$ *is an* $r \times r$ *matrix satisfying*

$$d_{i,j}(h) = \begin{cases} \sum_{k=0}^{j-1} \frac{(-j+1)_k (t-j+1)_k \delta_i^{r-j+k}}{(t+t_I-j+1)_k k!}, & j \neq h \\ h_{i,j} - (\sum_{b=0}^{t-j} \frac{1}{t_I+b}) \sum_{k=0}^{j-1} \frac{(-j+1)_k (t-j+1)_k \delta_i^{r-j+k}}{(t+t_I-j+1)_k k!}, & j = h \end{cases} \tag{39}$$

b) When $r > t$, then

$$C = \sum_{k=1}^{r}\sum_{j=0}^{t-1} \frac{\log_2 e}{t_I - k + 1 + j} + \frac{\log_2 e \cdot \sum_{h=1}^{t} \mid \mathbf{D}(h) \mid}{|V(\boldsymbol{\Delta})|} \tag{40}$$

where $D(h) = (d_{i,j}(h))$ is an $r \times r$ matrix satisfying

$$d_{i,j}(h) = \begin{cases} \sum_{k=0}^{j-1} \frac{(-j+1)_k (t-j+1)_k \delta_i^{r-j+k}}{(t+t_I-j+1)_k k!}, & j \neq h, j \leq t \\ \delta_i^{r-j}, & j \neq h, j > t \\ h_{i,j} - (\sum_{b=0}^{t-j} \frac{1}{t_I+b}) \sum_{k=0}^{j-1} \frac{(-j+1)_k (t-j+1)_k \delta_i^{r-j+k}}{(t+t_I-j+1)_k k!}, & j = h. \end{cases} \tag{41}$$

Here

$$h_{i,j} = \delta_i^{r-j} \frac{\Gamma(t + t_I - j + 1)}{\Gamma(t_I)\Gamma(t - j + 1)} \int_0^1 x^{t-j}(1-x)^{t_I - 1}(1 - \delta_i x)^{j-1}[ln(1 - \delta_i x) - ln(1 - x)]dx. \tag{42}$$

3.4 Numerical examples and remarks

Now we offer some numerical examples validating the analysis and showing the effect of various system parameters on the ergodic capacity of MIMO systems. For simplicity, we adopt the correlation model of exponential type (see [Loyka (2001)] and [Kiessling (2005)]) at the receiver with

$$\boldsymbol{\Sigma} = [\beta^{|i-j|}] \tag{43}$$

$$\boldsymbol{\Sigma}_I = [\beta_I^{|i-j|}] \tag{44}$$

The correlation coefficients β and β_I are for the desired user and interferers, respectively. They range from 0 to 1. Here 0 means that the correlation is the weakest, and 1 means that the correlation is the strongest. Furthermore, the SIR in dB is defined by $10 \log_{10}\left(\frac{E_s}{t_I E_I}\right)$ which characterizes the signal to interference ratio in the considered physical condition.

The ergodic capacity versus the SIR is depicted in Fig.1 where the four curves are shown for four different correlation coefficients equal to $\beta = 0.3, 0.6, 0.8, 0.9$, respectively. The considered MIMO system possesses 4 transmit antennas and 4 receive antennas with 10 interfering antennas. The correlation coefficient β_I is set at 0.4. As expected, the ergodic capacity decreases with increasing β. It can be further seen that the effect of strong correction on the capacity is significant.

Fig.2 depicts the ergodic capacity versus the SIR for four different correlations. The four curves in Fig.2 are shown for interfering correlation coefficients equal to $\beta_I = 0.3, 0.6, 0.8, 0.9$, respectively. The considered MIMO system is with 2 transmit antennas and 4 receive antennas and interfered by a user with 8 antennas. The correlation coefficient is set at $\beta = 0.5$. It can be seen from Fig.2 that the impact of correlation for interferers on the ergodic capacity increases with increased interfering correlation coefficient β_I . Therefore, the interference correlation is beneficial, especially the strong correlation.

Simulation results are included in Figs.1-2 for comparison. Each point in the simulation curves are obtained by averaging over 100,000 independent computer runs. The theoretical and simulation results are nearly identical verifying the validity of the theory. Consequently, in the following evaluations, we only consider the theoretical results.

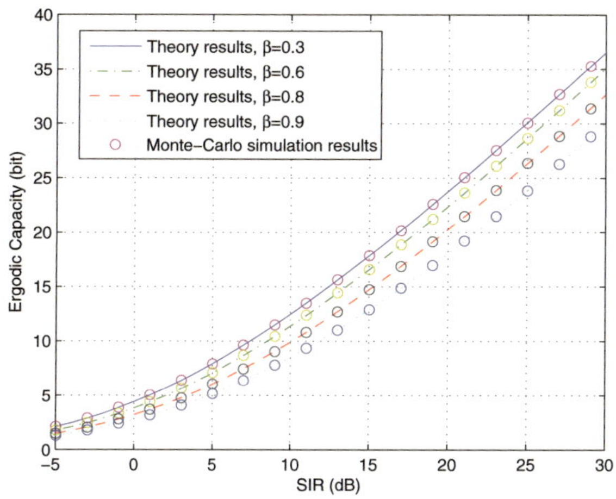

Fig. 1. Ergodic capacity versus SIR for different signal channel correlations.

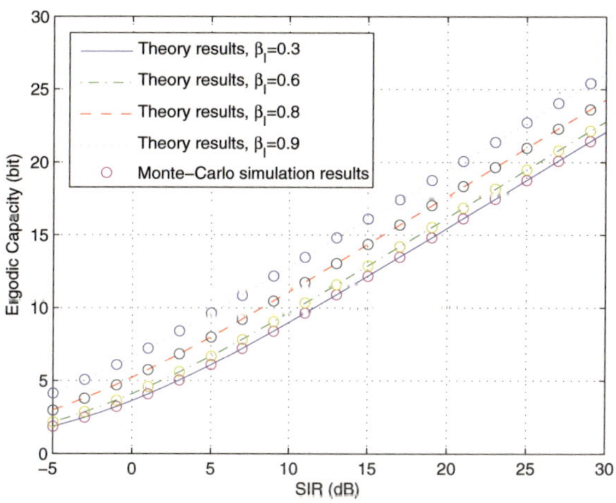

Fig. 2. Ergodic capacity versus SIR for different interfering correlations.

In Fig.3, a MIMO system with 4 transit antennas and 4 receive antennas is considered. We assume only 1 interferer is involved in this system. We observe the ergodic capacities with various interference antennas. In Fig.3, the four curves correspond to the number of total interfering transmit antennas $t_I = 4, 5, 6, 7$, respectively. It can be observed that the ergodic capacity drops as t_I increases, and the drop becomes gradually slow.

Finally, in Fig.4, we compare our analytical results (neglecting the noise component) with the Monte-Carlo simulation results with Gaussian noise involved in the corresponding physical conditions. We set the transmit power in the interest system at 30dB, and let β and β_I be qual to 0.4 and 0.8, respectively. Furthermore, we assume the system is interfered by a user with 10 antennas. We plot the curves with $t = r = 2, 3$ and 4, respectively. As shown in the figure, our

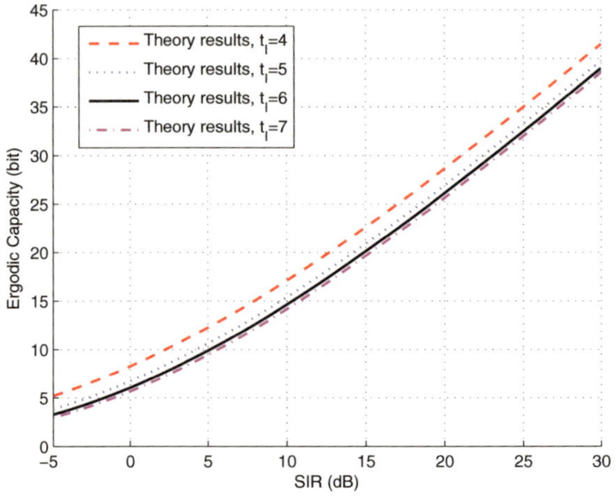

Fig. 3. Ergodic capacity versus SIR for various interfering antenna configurations.

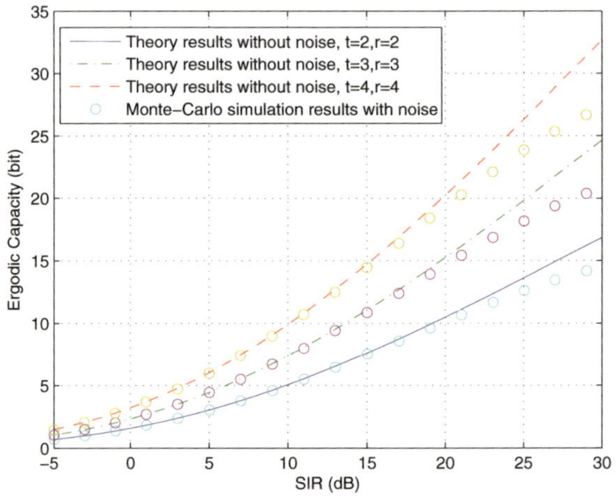

Fig. 4. Ergodic capacity versus SIR for various antenna configurations.

analytical results match the simulation results under low SIRs, however, we lose the precision gradually as SIR grows.

4. Outage performance of TRD MIMO systems with interference and correlation

4.1 System model

Suppose the intended user employs r antennas to receive signals transmitted from t antennas. The channels that link the t transmit and r receive antennas are characterized by an $r \times t$ matrix $\mathbf{H}$, which is assumed to follow the joint complex Gaussian distribution with mean matrix $\mathbf{M}$

and covariance matrix $\mathbf{\Sigma} \otimes \mathbf{\Psi}$. Symbolically, we will write

$$\mathbf{H} \sim CN_{r,t}(\mathbf{M}, \mathbf{\Sigma} \otimes \mathbf{\Psi}) \tag{45}$$

where $\mathbf{\Psi}$ and $\mathbf{\Sigma}$ define the correlation structure at the transmit and receive ends, respectively. It is assumed that the intended signal is corrupted by ℓ independent interferers, and the ith interferer transmits its signal with t_i antennas where $i = 1, \ldots, \ell$. The desired information symbol b_0 is weighted by the transmit beamformer $\mathbf{u}$ before being feeded to the t transmit antennas. The transmit beamformer is normalized to have a unit norm $\mathbf{u}^\dagger\mathbf{u} = 1$ so that the transmit energy equals $E_s = |b_0|^2$. The $r \times 1$ vector at the desired user's receiver can thus be written as

$$\mathbf{y} = b_0\mathbf{H}\mathbf{u} + \sum_{i=1}^{\ell} \mathbf{H}_i\mathbf{s}_i + \mathbf{n}, \tag{46}$$

where $\mathbf{H}_i$ is the $r \times t_i$ the channel matrix characterizing the links from the desired user's r receive antennas to the t_i transmit antennas of interferer i; and $\mathbf{s}_i$ is the symbols transmitted by interferer i, such that $\mathcal{E}[\mathbf{s}_i\mathbf{s}_i^\dagger] = E_i\mathbf{I}_{t_i}$ with E_i denoting the average symbol energy. In the way similar to defining $\mathbf{H}$, we assume

$$\mathbf{H}_i \sim CN_{r,t_i}(\mathbf{M}_i, \mathbf{\Sigma}_i \otimes \mathbf{\Psi}_i) \tag{47}$$

We assume the additive noise vector $\mathbf{n}$ to follow the $r \times 1$ complex Gaussian distribution of mean zero and covariance matrix $\mathbf{R}_n$. Conditioned on $\mathbf{H}_i, i = 1, \ldots, \ell$, the covariance matrix of interference-plus-noise component is given by

$$\mathbf{R}_c = \sum_{i=1}^{\ell} E_i\mathbf{H}_i\mathbf{H}_i^\dagger + \mathbf{R}_n. \tag{48}$$

4.2 Formulation

The TRD system transmits one symbol at a time, and employs a weighting vector $\mathbf{w}$ to combine received vector $\mathbf{y}$ to form a single decision variable. The transmit and receive weighting vectors, $\mathbf{u}$ and $\mathbf{w}$ should be chosen to maximize the output signal to interference-plus-noise ratio (SINR) at every time instant, as defined by

$$\gamma = \frac{\mathbf{w}^\dagger(\mathbf{H}\mathbf{u})(\mathbf{H}\mathbf{u})^\dagger\mathbf{w}}{\mathbf{w}^\dagger\mathcal{E}_n\left[(\sum_{i=1}^{\ell}\mathbf{H}_i\mathbf{s}_i + \mathbf{n})(\sum_{i=1}^{\ell}\mathbf{H}_i\mathbf{s}_i + \mathbf{n})^\dagger\right]\mathbf{w}} \tag{49}$$

where $\mathcal{E}_n$ denotes the expectation with respect to $\mathbf{n}$. The result of expectation equals $\mathbf{R}_c$ given in (48). Optimization of γ is the problem of Rayleigh quotient. Given the channel-state information and conditional on $\mathbf{u}$, we optimize γ with respect to $\mathbf{w}$ to obtain [Kang & Alouini (2004b)]

$$\gamma(\mathbf{u}) = \frac{\mathbf{u}^\dagger(E_s\mathbf{H}^\dagger\mathbf{R}_c^{-1}\mathbf{H})\mathbf{u}}{\mathbf{u}^\dagger\mathbf{u}} \tag{50}$$

where we have used the fact that $\mathbf{u}^\dagger\mathbf{u} = 1$ to represent the second line in the form of Rayleigh quotient. Thus, we can upper bound $\gamma(\mathbf{u})$ by

$$\gamma_{max} = \lambda_{(1)} \tag{51}$$

where $\lambda_{(1)} \geq \lambda_{(2)} \geq \cdots \lambda_{(q)}$ are the non-zero eigenvalues of the matrix product

$$\mathbf{F} = E_s \mathbf{H}^\dagger \mathbf{R}_c^{-1} \mathbf{H} \tag{52}$$

in the descending order, and $\mathbf{v}_{(1)}, \mathbf{v}_{(2)}, \cdots, \mathbf{v}_{(q)}$ are their corresponding eigenvectors. The non-ordered eigenvalues and eigenvectors will be denoted by $\lambda_1, \lambda_2, \cdots, \lambda_q$ and $\mathbf{v}_1, \mathbf{v}_2, \cdots, \mathbf{v}_q$, respectively.

The outage probability of TRD systems can be defined directly in terms of the instantaneous SINR $\gamma_{max} = \lambda_{(1)}$ or by channel capacity [Kang et al. (2003)]

$$C = \log_2(1 + \lambda_{(1)}). \tag{53}$$

Both leads to the same expression for an outage event: $\lambda_{(1)} < \Lambda$, but with the protection ratio Λ defined differently as shown by

$$\Lambda = \begin{cases} \gamma_0, & \text{outage in terms of } \gamma \\ 2^{C_0} - 1, & \text{outage in terms of } C. \end{cases} \tag{54}$$

In either case, we can write the outage probability as

$$P_{out} = \Pr\{\lambda_{(1)} < \Lambda\}. \tag{55}$$

To determine the outage performance, the central issue is to determine the probability density function (PDF) of $\lambda_{(1)}$ or equivalently, its cumulative density function (CDF).

Determination of the CDF of the principal eigenvalue of a rank-q non-negative definite matrix of the form $\mathbf{F} = E_s \mathbf{H}^\dagger \mathbf{R}_c^{-1} \mathbf{H}$ has been addressed intensively in the literature [Muirhead (1982)]. The predominant methodology, however, is to arrange the sample eigenvalues in a descending order and then to determine the PDF of the largest one. The methodology is also prevailing in the area of communications [Kang & Alouini (2004b)]. Such methodology, however, often leads to mathematically intractability except for some simple cases. In this paper, we therefore consider the non-ordered sample eigenvalues instead. The key step is to represent the outage event $\lambda_{(1)} < \Lambda$, alternatively, by virtue of non-ordered eigenvalues. To this end, we write the sample space

$$\{\mathbf{F} : \lambda_{(1)} < \Lambda)\} = \{\mathbf{F} : \cap_{i=1}^{q} (\lambda_i < \Lambda)\}. \tag{56}$$

The right-hand side is further expressible in matrix form. Hence,

$$\{\mathbf{F} : \lambda_{(1)} < \Lambda\} = \{\mathbf{F} : \mathbf{F} < \Lambda \mathbf{I}\} \tag{57}$$

where $\mathbf{F} < \Lambda\mathbf{I}$ means that $(\Lambda\mathbf{I} - \mathbf{F})$ is a positive definite matrix. The equivalence of the two expressions is obvious, in much the same way as what we do in selection combining. Let $\mathbf{V}$ denote the matrix of eigenvectors of $\mathbf{F}$. Namely, $\mathbf{V} = (\mathbf{v}_1, \cdots, \mathbf{v}_q, \cdots, \mathbf{v}_t)$. Hence we can write

$$\Lambda\mathbf{I} - \mathbf{F} = \mathbf{V}\text{diag}(\Lambda - \lambda_1, \cdots, \Lambda - \lambda_q, 0, \cdots, 0)\mathbf{V}^\dagger \tag{58}$$

The positive definiteness of $(\Lambda\mathbf{I} - \mathbf{F})$ implies that all of eigenvalues $\Lambda - \lambda_i$ are positive, and vice versa, thus showing the correctness of (57). This equivalence was previously used in Chapter 9 of [Muirhead (1982)].

We use it here to represent the outage probability yielding

$$P_{out} = \Pr\{\mathbf{F} < \Lambda\mathbf{I}\}. \tag{59}$$

The matrix representation of outage event, though simple in principle, provides a novel framework to tackle the outage issue of the optimal TRD system. The key to success along this direction is to find the joint cumulative distribution function of matrix $\mathbf{F}$.
For ease of presentation, we define variables

$$u = \max\{r, t\} \tag{60}$$

$$v = \min\{r, t\} \tag{61}$$

and the $v \times u$ complex matrix

$$\mathbf{Y} = \begin{cases} \mathbf{\Sigma}^{-1/2}\mathbf{M}\mathbf{\Psi}^{-1/2}, & r < t \\ \mathbf{\Psi}^{-1/2}\mathbf{M}^{\dagger}\mathbf{\Sigma}^{-1/2}, & t \leq r. \end{cases} \tag{62}$$

4.3 Outage performance with co-channel interference

We first proceed to operational environments with co-channel interference. For mathematical tractability, let us first simplify the interference covariance matrix given in (48). We assume that the operating environment is interference-dominated, so that the noise component is negligible. Hence, we can rewrite (48) as

$$\mathbf{R}_c = \sum_{i=1}^{\ell} E_i \mathbf{H}_i \mathbf{H}_i^{\dagger} \tag{63}$$

where $\mathbf{H}_i\mathbf{H}_i^{\dagger} \sim CW_r(t_i, \mathbf{\Sigma}_i)$. For the case with $E_1 = E_2 = \cdots = E_\ell = E_I$ and $\mathbf{\Sigma}_1 = \mathbf{\Sigma}_2 = \cdots = \mathbf{\Sigma}_\ell = \mathbf{\Sigma}_I$, it is easy to use Theorem 3.2.4 of Muirhead [Muirhead (1982)] to assert that $\mathbf{R}_c$, up to a factor of E_I, follows the Wishart distribution, as shown by

$$\mathbf{R}_c \sim CW_r(t_I, \mathbf{\Sigma}_I) \tag{64}$$

where $t_I = \sum_{i=1}^{\ell} t_i$. Clearly, this is the extension of the closure property of chi-square distribution. For the general setting of E_i's, we can accurately approximate $\mathbf{R}_c$ by using a single Wishart-distributed matrix, say $\mathbf{Q}_1$, in much the same as what we do for a sum of chi-square variables [Pearson & Hartley (1976)]. The resulting matrix $\mathbf{Q}_1$ has the following distribution

$$\mathbf{Q}_1 \sim CW_r(t_1, \mathbf{\Sigma}_1), \tag{65}$$

for which the parameters t_1 and $\mathbf{\Sigma}_1$ can be determined by equating the first two moments of $\mathbf{Q}_1$ and $\mathbf{R}_c$; for details, see Chapter 3 of [Gupta & Nagar (2000)]. From the above analysis, it follows that we can use a single a Wishart-distributed matrix, say $\mathbf{Q}_1$, to replace $\mathbf{R}_c$ to simplify the analysis. It also follows that t_1 is usually much greater than the number of antennas of the intended user. Thus, without loss of the generality, we can write the decision matrix (52) as

$$\mathbf{F} = (E_s/E_1)\mathbf{H}^{\dagger}\mathbf{Q}_1^{-1}\mathbf{H} \tag{66}$$

whereby, for a given power protection ratio Λ, the outage probability can be written as

$$\begin{aligned} P_{out}(x) &= \Pr\{\mathbf{F} < \Lambda \mathbf{I}\} \\ &= \Pr\{\mathbf{J} < x\mathbf{I}\} \end{aligned} \tag{67}$$

where $x = \Lambda E_1/E_s$ and $\mathbf{J}$ is defined in terms of random channel matrices $\mathbf{H}$ and $\mathbf{Q}_1$, as shown by

$$\mathbf{J} = \mathbf{H}^{\dagger}\mathbf{Q}_1^{-1}\mathbf{H}. \tag{68}$$

We assume the signal suffers from Rician fading so that the corresponding channel matrix $\mathbf{H} \sim CN_{r,t}(\mathbf{M}, \mathbf{\Sigma} \otimes \mathbf{\Psi})$. Suppose that the interferer employs t_1 transmit antennas such that $r \leq t_1$. We also assume that the t_1 channel-gain vectors for the interferer that link each transmit antenna to the r receive antennas are independent and identically distributed as $CN_r(\mathbf{0}, \mathbf{\Sigma}_1)$. Then, we can assert that $\mathbf{Q}_1 \sim CW_r(t_1, \mathbf{\Sigma}_1)$. Under these assumptions and by introducing the following matrix notations:

$$\mathbf{\Delta} = \begin{cases} \mathbf{\Sigma}^{-1}\mathbf{\Sigma}_1, & t \leq r \\ \mathbf{\Psi}^{-1}, & r < t \end{cases} \tag{69}$$

and

$$\mathbf{\Theta} = \begin{cases} \mathbf{\Sigma}^{-1}\mathbf{\Sigma}_1, & r < t \\ \mathbf{\Psi}^{-1}, & t \leq r \end{cases} \tag{70}$$

we can explicitly work out the outage probability defined in (67), obtaining results summarized in the following theorem. The proof of this theorem is placed in 7.1.

Theorem 3. *The outage probability of the optimal TRD system with co-channel interference is given by*

$$P_{out}(x) = d \sum_{k=0}^{\infty} \frac{x^{uv+k}}{k!} \sum_{\kappa} \frac{[t+t_1]_\kappa}{[u+v]_\kappa} P_\kappa(\mathbf{Y}, \mathbf{\Delta}, \mathbf{\Theta}) \tag{71}$$

where

$$d = \frac{\tilde{\Gamma}_v(t+t_1)\tilde{\Gamma}_v(v)}{\tilde{\Gamma}_v(t+t_1-u)\tilde{\Gamma}_v(u+v)} |\mathbf{\Delta}|^v |\mathbf{\Theta}|^u \cdot etr[-\mathbf{Y}\mathbf{Y}^{\dagger}]$$

The above generalized Hermite polynomial $P_\kappa(\cdot,\cdot,\cdot)$, though difficult in numerical calculation [Gupta & Nagar (2000)], serve as a fundamental tool in the study of the distribution of some quadratic forms. Eq.(71) is a general formula, providing a solid foundation for further study. This combination can be treated as a special Rayleigh case by setting $\mathbf{M} = \mathbf{0}$. Namely, $\mathbf{H} \sim CN_{r,t}(\mathbf{0}, \mathbf{\Sigma} \otimes \mathbf{\Psi})$. With the condition, Theorem 3 leads to the following corollary.

Corollary 1. *Let* $\mathbf{M} = \mathbf{0}$. *Then*

$$P_{out}(x) = d_1 x^{uv} {}_2\tilde{F}_1^{(u,v)}(u, t+t_1; u+v; x\mathbf{\Delta}, -\mathbf{\Theta}) \tag{72}$$

where

$$d_1 = \frac{\tilde{\Gamma}_v(t+t_1)\tilde{\Gamma}_v(v)}{\tilde{\Gamma}_v(t+t_1-u)\tilde{\Gamma}_v(u+v)} |\mathbf{\Delta}|^v |\mathbf{\Theta}|^u \tag{73}$$

The corollary is made by inserting $\mathbf{M} = \mathbf{0}$ into (71) and invoking Property 9 in Section 2 (i.e. the complex counterpart of Expression (1.8.3) in [Gupta & Nagar (2000)]).
Our concern is whether (72) can be further simplified. To this end, we note that when $r = t$, the hypergeometric function ${}_2\tilde{F}_1^{(u,v)}$ involved in (72) can be converted to scalar hypergeometric functions which are much easier to calculate by using for example, the built-in functions in Matlab, Mathematica and Maple. The simplification can be done by invoking the following lemma (see Lemma 2 in [Kiessling (2005)]).

Lemma 2. *Let* $\mathbf{A} = eig(\mathbf{X}) = diag(\lambda_1, \ldots, \lambda_p)$ *and* $\mathbf{B} = eig(\mathbf{Y}) = diag(\omega_1, \ldots, \omega_p)$ *with* $\lambda_1 > \ldots > \lambda_p$ *and* $\omega_1 > \ldots > \omega_p$*. Furthermore define*

$$\Gamma_p(p) = \prod_{i=1}^{p} \Gamma(p - i + 1), \tag{74}$$

$$\alpha_p(\mathbf{A}) = \prod_{i<j} (\lambda_i - \lambda_j) \tag{75}$$

and

$$\Psi_n^p(\mathbf{b}) = \prod_{i=1}^{p} \prod_{j=1}^{n} (b_j - i + 1)^{i-1} \tag{76}$$

for $\mathbf{b} = (b_1, b_2, \ldots, b_n)$*. Then*

$${}_m\tilde{F}_n^{(p,p)}(a_1, \ldots, a_m; b_1, \ldots, b_n; \mathbf{X}, \mathbf{Y}) = \frac{\Gamma_p(p)\Psi_n^p(\mathbf{b}) \mid \mathbf{L} \mid}{\alpha_p(\mathbf{A})\alpha_p(\mathbf{B})\Psi_m^p(\mathbf{a})} \tag{77}$$

where $\mathbf{L} = [l_{ij}]$ *with*

$$l_{ij} = {}_mF_n(a_1 - p + 1, \ldots, a_m - p + 1; b_1 - p + 1, \ldots, b_n - p + 1; \lambda_i \omega_j) \tag{78}$$

for $i, j = 1, 2, \ldots, p$.

When some of the λ_i's or ω_j's are equal, we obtain the results as limiting case on the right of (77) via L'Hospital's rule (see [Kiessling (2005)] for a detail process.)
Let us return to the general case with $r \neq t$. There is a simple method to convert this problem into the corresponding one with $r = t$. The basic skill is to obtain the exact outage probability as the result of a limiting process. The interested reader is referred to [Kiessling (2005)] for details. By the same token, we can simplify (72) to obtain an alternative expression which is much easier in numerical calculation.

Corollary 2. *Let* $\mathbf{D}_\Delta = eig(\mathbf{\Delta}) = diag(\delta_1, \ldots, \delta_u)$ *and* $\mathbf{D}_\Theta = eig(\mathbf{\Theta}) = diag(\theta_1, \ldots, \theta_v)$ *with* $\delta_1 > \ldots > \delta_u$ *and* $\theta_1 > \ldots > \theta_v$*. Then*

$$P_{out}(x) = d_2 x^{uv - u(u-1)/2} |\mathbf{Z}| \tag{79}$$

where d_2 *is defined as follows*

$$d_2 = \frac{(-1)^{u(u-1)/2}\Gamma_v(v)[\Gamma(t + t_1 - u + 1)]^v |\mathbf{\Delta}|^v |\mathbf{\Theta}|^v}{\Gamma_v(t + t_1 - u)[\Gamma(v + 1)]^v \alpha_u(\mathbf{D_\Delta})\alpha_v(\mathbf{D_\Theta})} \tag{80}$$

and the entries of matrix $\mathbf{Z} = [z_{ij}]$ *are given by*

$$z_{ij} = \begin{cases} {}_2F_1(1, t + t_1 - u + 1; v + 1; -x\theta_i\delta_j), & i \leq v; \\ (x\delta_j)^{(i-v-1)}, & i > v. \end{cases} \tag{81}$$

The expression in (71) is a general result. Its correctness can be examined by showing that the main result of [Kang & Alouini (2004b)] is one of its special cases.

Corollary 3. *Let* $\mathbf{M} = \mathbf{0}$ *and* $\mathbf{\Psi} = \mathbf{I}_t$. *Then*

$$P_{out}(x) = \prod_{i=1}^{v} \frac{|\beta(\frac{x}{1+x})| \cdot \Gamma(t+t_1-i+1)}{\Gamma(t+t_1-u-i+1)\Gamma(u-i+1)\Gamma(v-i+1)} \tag{82}$$

where $\beta(y)$ *is an* $v \times v$ *matrix function of the scalar* y *with entries*

$$[\beta(y)]_{ij} = \beta_y(u-v+i+j-1, t_1-r+1).$$

The function $\beta_y(p,q)$ is called the incomplete beta function (see [Gradshteyn & Ryzhik (1994)], Eqn.[8.391]).
This result is exactly the same as Eqn.(11) of [Kang & Alouini (2004b)]. The proof is a little complicated, yet not important to us, and thus is omitted.

4.4 Outage performance without co-channel interference

When co-channel interference is absent, we can set $E_i = 0, i = 1, \ldots, \ell$ to rewrite (48) as

$$\mathbf{R}_c = N_0 \mathbf{\Phi}_n \tag{83}$$

where $\mathbf{\Phi}_n$ has been normalized to signify the branch noise correlation matrix whereas N_0 denotes the noise variance at each branch. Now we need a difference treatment due to the replacement of the random matrix summation $\mathbf{R}_c = \sum_{i=1}^{\ell} E_i \mathbf{H}_i \mathbf{H}_i^{\dagger}$ with a constant matrix $N_0 \mathbf{\Phi}_n$ in the quadratic form $\mathbf{F}$. Nevertheless, the procedure is parallel.
Given the change in covariance matrix $\mathbf{R}_c$, we need to modify x and $\mathbf{J}$ accordingly, as shown by

$$x = \Lambda N_0 / E_s, \quad \mathbf{J} = \mathbf{H}^{\dagger} \mathbf{\Phi}_n^{-1} \mathbf{H}. \tag{84}$$

Correspondingly, matrices $\mathbf{\Delta}$ and $\mathbf{\Theta}$ are modified to

$$\mathbf{\Delta} = \begin{cases} \mathbf{\Sigma}^{-1}\mathbf{\Phi}_n, & t \leq r \\ \mathbf{\Psi}^{-1}, & r < t. \end{cases} \tag{85}$$

and

$$\mathbf{\Theta} = \begin{cases} \mathbf{\Sigma}^{-1}\mathbf{\Phi}_n, & r < t \\ \mathbf{\Psi}^{-1}, & t \leq r. \end{cases} \tag{86}$$

With these notations, we can write $P_{out} = \Pr\{\mathbf{J} < x\mathbf{I}\}$ which, after some manipulations as shown in 7.2, leads to the following result.

Theorem 4. *The outage probability of the optimal TRD system without co-channel interference is given by*

$$P_{out}(\mathbf{Q} < x\mathbf{I}) = c \sum_{k=0}^{\infty} \frac{x^{uv+k}}{k!} \sum_{\kappa} \frac{P_{\kappa}(\mathbf{Y}, \mathbf{\Delta}, \mathbf{\Theta})}{[u+v]_{\kappa}} \tag{87}$$

where

$$c = \frac{\tilde{\Gamma}_v(v)}{\tilde{\Gamma}_v(u+v)} |\mathbf{\Delta}|^v |\mathbf{\Theta}|^u \cdot etr[-\mathbf{Y}\mathbf{Y}^{\dagger}]. \tag{88}$$

An important case is Rayleigh faded signals for which $\mathbf{M} = \mathbf{0}$ and (87) can be simplified.

Corollary 4. *when* $\mathbf{M} = \mathbf{0}$, *we have that*

$$P_{out} = c_1 x^{uv} \, {}_1\tilde{F}_1^{(u,v)}(u; u+v; x\mathbf{\Delta}, -\mathbf{\Theta}) \tag{89}$$

where

$$c_1 = \frac{\tilde{\Gamma}_v(v)}{\tilde{\Gamma}_v(u+v)} |\mathbf{\Delta}|^v |\mathbf{\Theta}|^u. \tag{90}$$

This corollary's proof is similar to that of Corollary 2 and thus is omitted.
Similar to ${}_2\tilde{F}_1^{(u,v)}$, the hypergeometric function ${}_1\tilde{F}_1^{(u,v)}$ involved in (89) can be also easily calculated by representing it in terms of scalar hypergeometric functions for ease of calculation. Specifically, by using the same techniques as used by Kiessling [Kiessling (2005)], we can obtain the following corollary.

Corollary 5. *Let* $\mathbf{D}_\Delta = eig(\mathbf{\Delta}) = diag(\delta_1, \ldots, \delta_u)$ *and* $\mathbf{D}_\Theta = eig(\mathbf{\Theta}) = diag(\theta_1, \ldots, \theta_v)$ *with* $\delta_1 > \ldots > \delta_u$ *and* $\theta_1 > \ldots > \theta_v$.

$$P_{out}(x) = c_2 x^{uv - u(u-1)/2} |\mathbf{Y}| \tag{91}$$

where c_2 *is given by*

$$c_2 = \frac{(-1)^{u(u-1)/2} \Gamma_v(v) |\mathbf{\Delta}|^v |\mathbf{\Theta}|^v}{[\Gamma(v+1)]^v \alpha_u(\mathbf{D}_\Delta) \alpha_v(\mathbf{D}_\Theta)}, \tag{92}$$

and the entry of the matrix $\mathbf{Y} = [y_{ij}]$ *is given by*

$$y_{ij} = \begin{cases} {}_1F_1(1; v+1; -x\theta_i \delta_j), & i \le v; \\ (x\delta_j)^{(i-v-1)}, & i > v. \end{cases} \tag{93}$$

To examine the correctness of our results given in (89), let us consider the special case of independent noise and i.i.d. fading Rayleigh channels such that $\mathbf{\Phi}_n = \mathbf{I}$ and $\mathbf{\Psi} = \mathbf{\Sigma} = \mathbf{I}$. These conditions, when inserted into (89) and simplified, leads to (94) shown below.

Corollary 6. *Let* $\mathbf{\Phi}_n = \mathbf{I}$ *and* $\mathbf{\Psi} = \mathbf{\Sigma} = \mathbf{I}$. *Then*

$$P_{out} = \frac{|\, \mathbf{A}(x) \,|}{\prod_{k=1}^{v} \Gamma(u-k+1)\Gamma(v-k+1)} \tag{94}$$

where $\mathbf{A}(x)$ *is a* $v \times v$ *matrix function with its* (i,j)*th entries given by*

$$[\mathbf{A}(x)]_{ij} = \gamma(u - v + i + j - 1, x)$$

for $i, j = 1, 2, \ldots, v$.

This result is identical to the corresponding one in [Dighe et al. (2001)] and [Kang & Alouini (2003b)]. If we further set $v = 2$, then (94) can be rewritten as

$$P_{out} = \frac{\gamma(u-1,x)\gamma(u+1,x) - \gamma(u,x)^2}{\Gamma(u)\Gamma(u-1)}, \tag{95}$$

which is exactly the same as the known result described in [Kang & Alouini (2004a)]. Its proof is not difficult but not important and thus, is omitted.

4.5 Numerical results and remarks

The validity of Theorem 3 and Theorem 4 has been rigorously examined by showing that they include most of existing results in the literature as special cases. In this section, we examine the correctness of Corollary 1 and Corollary 4 with numerical results. For simplicity, we assume that the spatial correlation among antennas follows the exponential model with correlation between antennas p and q given by $c(p,q) = g^{|p-q|}\exp(j(p-q)\pi/12)$. Physically, $g^{|p-q|}$ denotes the correlation magnitude, and g stands for the correlation coefficient.

We assume that the receiver is equipped with r antennas for the reception of Rayleigh faded signals from t intended transmit antennas. The received signals are corrupted by Rayleigh faded interference from ℓ interferers. Thus, Corollaries 2 and 5 are applicable in theoretical evaluation. Simulation results are also included for comparison. Each point in the simulated curves is produced by averaging over at least $100,000$ independent computer runs. Throughout this section, we set $t = 4$ and $r = 2$, and assume that the correlation at the intended transmit and receive ends is characterized by g_t and g_r, respectively.

We first investigate the case with co-channel interference. For ease of illustration, assume the presence of only one co-channel interferer (i.e., $\ell = 1$) which employs t_1 antennas for transmission. Further assume that the correlation structure at the both sides of the $t_1 \times r$ interfering channel matrix is the same, characterized by g_1.

Fig.5 shows the variation of outage probability with the number of the interferer's transmit antennas. The parameter setting is: $g_t = 0.5$, $g_r = 0.9$, and $g_1 = 0.5$. The curves in the figure are for $t_1 = 2, 3, 4, 10, 14$, respectively. As expected, the outage performance becomes worse as t_1 increases, but the decrease magnitude becomes smaller and smaller. It is also observed that the simulated results coincide with their theoretical counterparts.

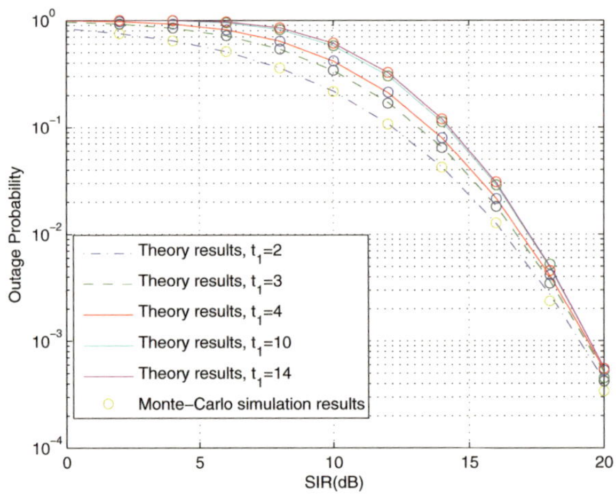

Fig. 5. Variation of outage probability with the number of interfering antennas.

The influence of the interferer's correlation coefficient on the outage probability is shown in Fig 6 where t_1 is set to 3 and the three curves are shown for $g_1 = 0.3$, 0.8 and 0.9, respectively. Other parameters are set to be $g_t = 0.5$ and $g_r = 0.95$. We observe that over the region of moderate and high SIR, the outage performance improves with increased g_1. This is is easy to understand since a higher interference correlation implies a sharper directional beam which is easier to be nullified by using interference-covariance matrix inversion involved in our quadratic form. Clearly, unlike the effect of the intended user's correlation, the spatial correlation of co-channel interference is an advantage to the outage performance of TRD systems. From these curves, we can see, again, a nearly perfect agreement between the theoretical and simulated results.

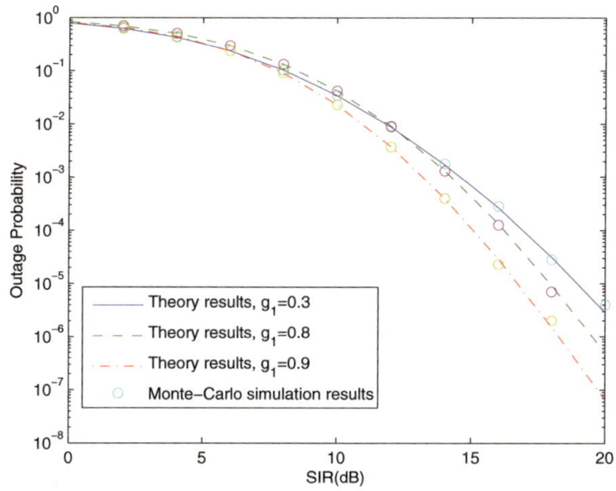

Fig. 6. Influence of interference correlation g_1 on the outage performance.

In Fig.7, the outage probability versus the number of transmit antennas under different SIRs are plotted. The parameters are set at $r = 2$, $g_t = 0.5$, $g_r = 0.9$ and $g_1 = 0.5$. The three curves in the figure are for SIR= 10dB, 15dB and 20dB, respectively. As shown in the figure, the outage performances improves almost linearly with the number of transmit antennas t increasing.

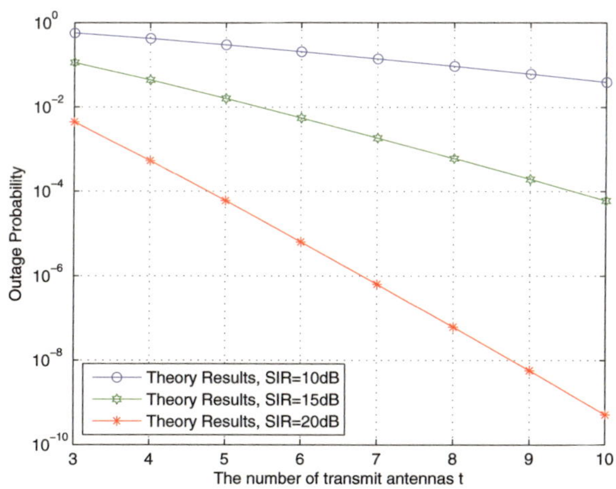

Fig. 7. Influence of signal transmit correlation on the outage probability.

Fig.8 considers the case when 2 interfering users involved. The 2 interfering channel matrixes are with the same correlation coefficient $g_1 = 0.5$, in the receive end. The equivalent t_1 and

$\mathbf{\Sigma}_1$ are determined by equating the first two moments of $\mathbf{Q}_1$ and $\mathbf{R}_c$ as we introduced in the previous section. The other parameters are set at $t = 3$, $r = 3$, $g_t = 0.5$ and $g_r = 0.9$. We observe the loss of precision as we change the interference power distribution which is denoted by a ratio $\epsilon = E_1/E_2$. It is shown in the figure that our analysis has high precision when the ratio ϵ is close to 1, however, when the ratio loses balance, say $\epsilon = 5$, the theory curve can only be considered as a lower bound of the real performance.

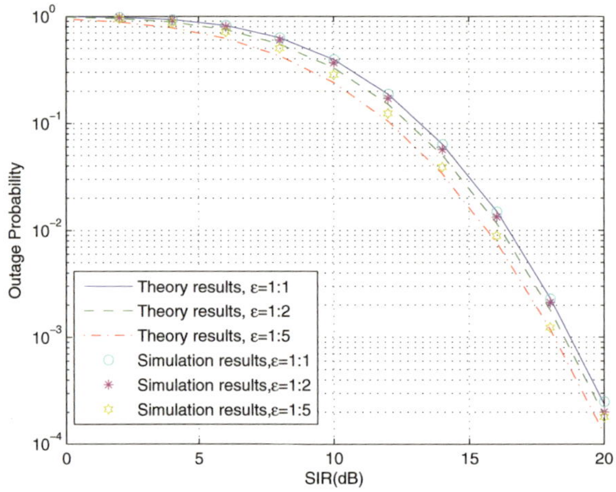

Fig. 8. Influence of the number of transmit antennas on the outage probability.

We next consider the case without co-channel interference. Fig.9 shows the outage probability as a function of SIR for different values of g_t. Here we set $g_r = 0.5$. The three curves are for $g_t = 0.1$, 0.5 and 0.9, respectively. It is clear that the outage performance drops with increased transmit correlation coefficient g_t. This is quite intuitive since high transmit correlation means the lose of more degrees of freedom in transmit diversity. A perfect agreement between simulation and theoretic results are observed again.

5. Conclusions

Wireless transmission using multiple antennas has attracted much interest due to its capability to exploit the tremendous capacity inherent in MIMO channels. However, the performance of MIMO systems is very sensitive to the presence of co-channel interference or spatial fading correlation. In this chapter, based on the theory of complex matrix variate distributions, we have investigated the performance of MIMO systems in the presence of both co-channel interference and spatial correlation. We first have derived several exact closed-form expressions of the MIMO ergodic capacity in Rayleigh fading environments, and demonstrated by experimentation the influences of co-channel interference and spatial correlation on the ergodic capacity. Then we have tackled the outage performance issue of MIMO systems with optimal transmit/receive diversity, and obtained two formulas of outage probability for general cases of Rayleigh faded signals with and without Rayleigh faded interference, respectively. Finally, we have presented numerical results to validate the theoretical analysis of outage probability. It can been found that the theoretical analysis

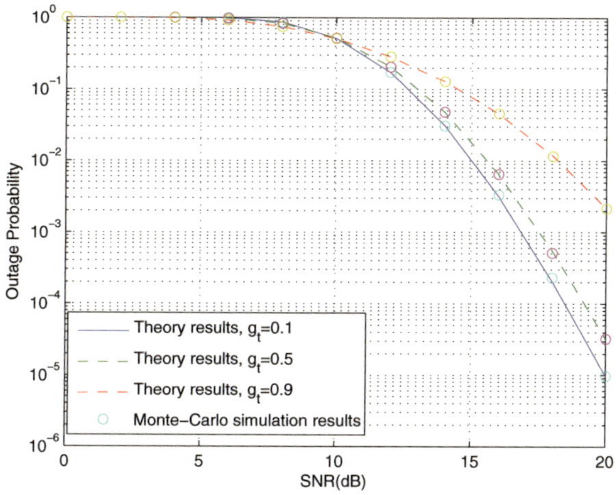

Fig. 9. Influence of the interference power distribution on the outage probability.

of MIMO systems with co-channel interference and spatial correlation depends heavily on multivariate statistics knowledge, especially the theory of matrix variate distributions.

6. Appendix: Proofs of theorem 1 and theorem 2 in section 3

6.1 Proof of theorem 1

Proof of Theorem 1 : **a)** Suppose that $t \leq r$. From Equation (61) of [Khatri (1966)], the PDF of the random matrix $\mathbf{Q}$ can be written as

$$f(\mathbf{Q}) = \frac{\tilde{\Gamma}_r(t+t_I)}{\tilde{\Gamma}_r(t_I)\tilde{\Gamma}_t(r)} |\rho \mathbf{I}_t|^{-r} |\tilde{\mathbf{\Sigma}}|^{-t} |\mathbf{Q}|^{r-t}$$
$$|\mathbf{I}_t + (q\rho)^{-1}\mathbf{Q}|^{-(t+t_I)} {}_1\tilde{F}_0^{(t,r)}(t+t_I, \mathbf{Q}(q\rho\mathbf{I}_t + \mathbf{Q})^{-1}, \mathbf{I}_r - q\tilde{\mathbf{\Sigma}}^{-1}) \tag{96}$$

where q is an arbitrary scalar constant. Let $q = \rho^{-1}$. Then we get after simplifying

$$f(\mathbf{Q}) = \frac{\tilde{\Gamma}_r(t+t_I)}{\tilde{\Gamma}_r(t_I)\tilde{\Gamma}_t(r)} |\rho\tilde{\mathbf{\Sigma}}|^{-t} |\mathbf{Q}|^{r-t}$$
$$|\mathbf{I}_t + \mathbf{Q}|^{-(t+t_I)} {}_1\tilde{F}_0^{(t,r)}(t+t_I, \mathbf{Q}(\mathbf{I}_t + \mathbf{Q})^{-1}, \mathbf{I}_r - (\rho\tilde{\mathbf{\Sigma}})^{-1}) \tag{97}$$

Make the transformation

$$\mathbf{L} = (\mathbf{I}_t + \mathbf{Q})^{-1}\mathbf{Q}, \tag{98}$$

and the Jacobian of the transformation is given by Equation (5.1.3) of [Khatri (1965)]

$$J(\mathbf{Q}; \mathbf{L}) = |\mathbf{I}_t - \mathbf{L}|^{-2t} \tag{99}$$

Thus the MGF of mutual information $I(\mathbf{s}, \mathbf{y})$ is expressed as

$$M(\theta) = \int_{\mathbf{Q}>0} |\mathbf{I} + \mathbf{Q}|^{\theta} f(\mathbf{Q}) d\mathbf{Q}$$
$$= \frac{\tilde{\Gamma}_r(t+t_I)}{\tilde{\Gamma}_r(t_I)\tilde{\Gamma}_t(r)|\rho\tilde{\mathbf{\Sigma}}|^t} \int_{0<L<\mathbf{I}_t} |\mathbf{L}|^{r-t} |\mathbf{I} - \mathbf{L}|^{t_I - r - \theta} {}_1\tilde{F}_0^{(t,r)}(t+t_I, \mathbf{L}, \mathbf{I}_r - (\rho\tilde{\mathbf{\Sigma}})^{-1}) d\mathbf{L} \tag{100}$$

Using Equation (7) of [Khatri (1966)] and Definition 2 here, we further have

$$M(\theta) = \frac{\tilde{\Gamma}_r(t+t_I)\tilde{\Gamma}_t(t+t_I-r-\theta)}{\tilde{\Gamma}_r(t_I)\tilde{\Gamma}_t(t+t_I-\theta)}|\rho\tilde{\boldsymbol{\Sigma}}|^{-t}\,{}_2\tilde{F}_1^{(t,r)}(t+t_I,r;t+t_I-\theta;\mathbf{I}_t,\mathbf{I}_r-(\rho\tilde{\boldsymbol{\Sigma}})^{-1}). \quad (101)$$

From Equation (54) of [Shin & Lee (2003)] or Property 2 in Section 2, we have

$$\frac{C_\kappa(\mathbf{I}_t)}{C_\kappa(\mathbf{I}_r)} = \frac{[t]_\kappa}{[r]_\kappa} \quad (102)$$

Therefore, we have by noting relationship between the hypergeometric function of two matrix arguments and the hypergeometric function of one matrix argument (involving Property 2 and Property 6)

$${}_2\tilde{F}_1^{(t,r)}(t+t_I,r;t+t_I-\theta;\mathbf{I}_t,\mathbf{I}_r-(\rho\tilde{\boldsymbol{\Sigma}})^{-1}) = {}_2\tilde{F}_1^{(r)}(t+t_I,t;t+t_I-\theta;\mathbf{I}_r-(\rho\tilde{\boldsymbol{\Sigma}})^{-1}) \quad (103)$$

Applying (49) of James [James (1964)] to the above expression, we further get

$$\begin{aligned} M(\theta) &= \frac{\tilde{\Gamma}_r(t+t_I)\tilde{\Gamma}_t(t+t_I-r-\theta)}{\tilde{\Gamma}_r(t_I)\tilde{\Gamma}_t(t+t_I-\theta)}|\rho\tilde{\boldsymbol{\Sigma}}|^{-t}\,{}_2\tilde{F}_1^{(r)}(t+t_I,t;t+t_I-\theta;\mathbf{I}_r-(\rho\tilde{\boldsymbol{\Sigma}})^{-1}) \\ &= \frac{\tilde{\Gamma}_r(t+t_I)\tilde{\Gamma}_t(t+t_I-r-\theta)}{\tilde{\Gamma}_r(t_I)\tilde{\Gamma}_t(t+t_I-\theta)}\,{}_2\tilde{F}_1^{(r)}(-\theta,t;t+t_I-\theta;\mathbf{I}_r-(\rho\tilde{\boldsymbol{\Sigma}})). \end{aligned} \quad (104)$$

It is obvious that

$$\frac{\tilde{\Gamma}_t(t+t_I-r-\theta)}{\tilde{\Gamma}_t(t+t_I-\theta)} = \frac{\tilde{\Gamma}_r(t_I-\theta)}{\tilde{\Gamma}_r(t+t_I-\theta)} \quad (105)$$

Thus we obtain the desired result

$$M(\theta) = \frac{\tilde{\Gamma}_r(t+t_I)\tilde{\Gamma}_r(t_I-\theta)}{\tilde{\Gamma}_r(t_I)\tilde{\Gamma}_r(t+t_I-\theta)}\,{}_2\tilde{F}_1^{(r)}(-\theta,t;t+t_I-\theta;\mathbf{I}-\rho\tilde{\boldsymbol{\Sigma}}). \quad (106)$$

b) Now we consider the case where $r \leq t$. It follows easily that

$$|\mathbf{I}+\mathbf{Q}| = |\mathbf{I}+\mathbf{F}| \quad (107)$$

where $\mathbf{F} = \tilde{\mathbf{R}}^{-1/2}\tilde{\mathbf{H}}\tilde{\mathbf{H}}^\dagger\tilde{\mathbf{R}}^{-1/2}$. In order to get an expression of $M(\theta)$, we can make use of the PDF of the random matrix $\mathbf{F}$ to replace the PDF of $\mathbf{Q}$. Based on Equation (62) of [Khatri (1965)], the PDF of the random matrix $\mathbf{F}$ is given by

$$\begin{aligned} f(\mathbf{F}) &= \frac{\tilde{\Gamma}_r(t+t_I)}{\tilde{\Gamma}_r(t_I)\tilde{\Gamma}_r(t)}|\rho\tilde{\boldsymbol{\Sigma}}|^{-t}|\mathbf{F}|^{t-r} \\ &\cdot|\mathbf{I}_r+(q\rho\tilde{\boldsymbol{\Sigma}})^{-1}\mathbf{F}|^{-(t+t_I)}\,{}_1\tilde{F}_0^{(r,t)}(t+t_I,\mathbf{F}(q\rho\tilde{\boldsymbol{\Sigma}}+\mathbf{F})^{-1},\mathbf{I}_t-q\mathbf{I}_t) \end{aligned} \quad (108)$$

where q is an arbitrary scalar constant. By taking $q \to \infty$, the PDF of $\mathbf{F}$ can be rewritten as

$$f(\mathbf{F}) = \frac{\tilde{\Gamma}_r(t+t_I)}{\tilde{\Gamma}_r(t_I)\tilde{\Gamma}_r(t)}|\rho\tilde{\boldsymbol{\Sigma}}|^{-t}|\mathbf{F}|^{t-r}\,{}_1\tilde{F}_0^{(r,t)}(t+t_I,\mathbf{F}(\rho\tilde{\boldsymbol{\Sigma}})^{-1},-\mathbf{I}_t). \quad (109)$$

From Definitions 2 and 3, we obtain with the help of Equation (90) of James [James (1964)]

$$
\begin{aligned}
f(\mathbf{F}) &= \frac{\tilde{\Gamma}_r(t+t_I)}{\tilde{\Gamma}_r(t_I)\tilde{\Gamma}_r(t)}|\rho\tilde{\boldsymbol{\Sigma}}|^{-t}|\mathbf{F}|^{t-r}\,{}_1\tilde{F}_0^{(r)}(t+t_I,(\rho\tilde{\boldsymbol{\Sigma}}\mathbf{I}_r)^{-1}\mathbf{F}) \\
&= \frac{\tilde{\Gamma}_r(t+t_I)}{\tilde{\Gamma}_r(t_I)\tilde{\Gamma}_r(t)}|\rho\tilde{\boldsymbol{\Sigma}}|^{-t}|\mathbf{F}|^{t-r}|\mathbf{I}_r+(\rho\tilde{\boldsymbol{\Sigma}}\mathbf{I}_r)^{-1}\mathbf{F}|^{-(t+t_I)}. \qquad (110)
\end{aligned}
$$

Thus the MGF of mutual information $I(\mathbf{s},\mathbf{y})$ can be expressed as

$$
\begin{aligned}
M(\theta) &= \int_{\mathbf{F}}|\mathbf{I}+\mathbf{F}|^{\theta}f(\mathbf{F})d\mathbf{F} \\
&= \frac{\tilde{\Gamma}_r(t+t_I)}{\tilde{\Gamma}_r(t_I)\tilde{\Gamma}_r(t)|\rho\tilde{\boldsymbol{\Sigma}}|^{t}}\int_{\mathbf{F}>0}|\mathbf{F}|^{t-r}|\mathbf{I}_r+\mathbf{F}|^{\theta}|\mathbf{I}_r+(\rho\tilde{\boldsymbol{\Sigma}}\mathbf{I}_r)^{-1}\mathbf{F}|^{-(t+t_I)}d\mathbf{F}. \qquad (111)
\end{aligned}
$$

Using Problem 1.18 of [Gupta & Nagar (2000)], we get the following desired result with the help of (49) of James [James (1964)]

$$
M(\theta) = \frac{\tilde{\Gamma}_r(t+t_I)\tilde{\Gamma}_r(t_I-\theta)}{\tilde{\Gamma}_r(t_I)\tilde{\Gamma}_r(t+t_I-\theta)}\,{}_2\tilde{F}_1^{(r)}(-\theta,t;t+t_I-\theta;\mathbf{I}-\rho\tilde{\boldsymbol{\Sigma}}). \qquad (112)
$$

6.2 Proof of theorem 2

Proof of Theorem 2: By Theorem 1 we get

$$
\begin{aligned}
C &= \log_2 e\cdot\frac{\partial M(\theta)}{\partial\theta}\,|_{\theta=0} \\
&= \log_2 e\cdot\frac{\partial}{\partial\theta}\{\frac{\tilde{\Gamma}_r(t+t_I)\tilde{\Gamma}_r(t_I-\theta)}{\tilde{\Gamma}_r(t_I)\tilde{\Gamma}_r(t+t_I-\theta)}\,{}_2\tilde{F}_1^{(r)}(-\theta,t;t+t_I-\theta;\mathbf{I}-\rho\tilde{\boldsymbol{\Sigma}})\} \\
&= \log_2 e\cdot\frac{\partial}{\partial\theta}\{\frac{\tilde{\Gamma}_r(t+t_I)\tilde{\Gamma}_r(t_I-\theta)}{\tilde{\Gamma}_r(t_I)\tilde{\Gamma}_r(t+t_I-\theta)}\}\,|_{\theta=0}\;{}_2\tilde{F}_1^{(r)}(0,t;t+t_I;\mathbf{I}-\rho\tilde{\boldsymbol{\Sigma}}) \\
&\quad+\log_2 e\cdot\frac{\partial}{\partial\theta}\{{}_2\tilde{F}_1^{(r)}(-\theta,t;t+t_I-\theta;\mathbf{I}-\rho\tilde{\boldsymbol{\Sigma}})\}\,|_{\theta=0} \\
&= \log_2 e(A+B) \qquad (113)
\end{aligned}
$$

In what follows, we will derive expressions of A and B in order to compute C. By (87) of James [James (1964)], we can have

$$
{}_2\tilde{F}_1^{(r)}(0,t;t+t_I;\mathbf{I}-\rho\tilde{\boldsymbol{\Sigma}}) = 1. \qquad (114)
$$

For an integer $r\leq a$, we get with the definition of gamma function

$$
\begin{aligned}
\frac{\partial}{\partial\theta}\Gamma_r(a-\theta)\,|_{\theta=0} &= \frac{\partial}{\partial\theta}\prod_{i=1}^{r}\Gamma(a-\theta-i+1)\,|_{\theta=0} \\
&= \sum_{k=1}^{r}\prod_{i=1,i\neq k}^{r}\Gamma(a-i+1)\frac{\partial}{\partial\theta}\Gamma_r(a-k-\theta+1)\,|_{\theta=0} \\
&= -\Gamma_r(a)\sum_{k=1}^{r}\psi(a-k+1) \qquad (115)
\end{aligned}
$$

Here $\psi(\cdot)$ is the digamma function defined by (8.360) of [Gradshteyn & Ryzhik (1994)]

$$\psi(x) = \frac{\Gamma'(x)}{\Gamma(x)}. \tag{116}$$

With the help of (8.365) in [Gradshteyn & Ryzhik (1994)], we can have

$$\begin{aligned} A &= \frac{\partial}{\partial\theta}\{\frac{\Gamma_r(t_I-\theta)}{\Gamma_r(t_I)}\} \mid_{\theta=0} + \frac{\partial}{\partial\theta}\{\frac{\Gamma_r(t+t_I)}{\Gamma_r(t+t_I-\theta)}\} \mid_{\theta=0} \\ &= \sum_{k=1}^{r}\psi(t+t_I-k+1) - \sum_{k=1}^{r}\psi(t_I-k+1) \\ &= \sum_{k=1}^{r}\sum_{j=0}^{t-1}\frac{1}{t_I-k+1+j} \end{aligned} \tag{117}$$

Now we consider how to compute B. From Lemma 1 it is known that

$$_2\tilde{F}_1^{(r)}(-\theta,t;t+t_I-\theta;\mathbf{I}-\rho\tilde{\mathbf{\Sigma}}) = \frac{\mid \mathbf{G} \mid}{|V(\mathbf{\Delta})|} \tag{118}$$

where $\mathbf{G} = [g_{i,j}]$ with

$$g_{i,j} = \delta_i^{r-j}\,{}_2F_1(-\theta-j+1,t-j+1;t+t_I-\theta-j+1;\delta_i) \tag{119}$$

for $i,j = 1,2,\ldots,r$. In particular, we get by (3) of James [James (1964)]

$$g_{i,j} \mid_{\theta=0} = \sum_{k=0}^{j-1}\frac{(-j+1)_k(t-j+1)_k\delta_i^{r-j+k}}{(t+t_I-j+1)_k k!}. \tag{120}$$

a) For $r \le t$, it follows with the help of (48) of James [James (1964)]

$$\begin{aligned} \frac{\partial g_{i,j}}{\partial\theta} \mid_{\theta=0} &= \delta_i^{r-j}\frac{\partial}{\partial\theta}\frac{\Gamma(t+t_I-\theta-j+1)}{\Gamma(t_I-\theta)\Gamma(t-j+1)}\int_0^1 x^{t-j}(1-x)^{t_I-\theta-1}(1-\delta_i x)^{j-1+\theta}dx \mid_{\theta=0} \\ &= \delta_i^{r-j}\frac{\Gamma(t+t_I-j+1)}{\Gamma(t_I)\Gamma(t-j+1)}\int_0^1 x^{t-j}(1-x)^{t_I-1}(1-\delta_i x)^{j-1}[ln(1-\delta_i x)-ln(1-x)]dx \\ &\quad +\delta_i^{r-j}(\psi(t_I)-\psi(t_I+t-j+1))\,{}_2F_1(-j+1,t-j+1;t+t_I-j+1;\delta_i) \\ &= \delta_i^{r-j}\frac{\Gamma(t+t_I-j+1)}{\Gamma(t_I)\Gamma(t-j+1)}\int_0^1 x^{t-j}(1-x)^{t_I-1}(1-\delta_i x)^{j-1}[ln(1-\delta_i x)-ln(1-x)]dx \\ &\quad -(\sum_{b=0}^{t-j}\frac{1}{t_I+b})\sum_{k=0}^{j-1}\frac{(-j+1)_k(t-j+1)_k\delta_i^{r-j+k}}{(t+t_I-j+1)_k k!}. \end{aligned} \tag{121}$$

Therefore, we have when $r \le t$

$$\begin{aligned} B &= \frac{\partial}{\partial\theta}\{\,_2\tilde{F}_1^{(r)}(-\theta,t;t+t_I-\theta;\mathbf{I}-\rho\tilde{\mathbf{\Sigma}})\} \mid_{\theta=0} \\ &= \frac{\sum_{h=1}^{r} \mid \mathbf{D}(h) \mid}{|V(\mathbf{\Delta})|} \end{aligned} \tag{122}$$

where $D(h) = (d_{i,j}(h))$ with

$$d_{i,j}(h) = \begin{cases} \sum_{k=0}^{j-1} \frac{(-j+1)_k(t-j+1)_k\delta_i^{r-j+k}}{(t+t_I-j+1)_k k!}, & j \neq h \\ h_{i,j} - \sum_{b=0}^{t-j} \frac{1}{t_I+b} \sum_{k=0}^{j-1} \frac{(-j+1)_k(t-j+1)_k\delta_i^{r-j+k}}{(t+t_I-j+1)_k k!}, & j = h. \end{cases} \quad (123)$$

Here $h_{i,j}$ is defined by

$$h_{i,j} = \delta_i^{r-j} \frac{\Gamma(t+t_I-j+1)}{\Gamma(t_I)\Gamma(t-j+1)} \int_0^1 x^{t-j}(1-x)^{t_I-1}(1-\delta_i x)^{j-1}[ln(1-\delta_i x) - ln(1-x)]dx \quad (124)$$

b) When $t < r$, we note that for $j > t$

$$g_{i,j} = \sum_{k=0}^{j-1-t} \frac{(-\theta-j+1)_k(t-j+1)_k\delta_i^{r-j+k}}{(t+t_I-j+1-\theta)_k k!}. \quad (125)$$

After some column operations on the determinant $|G|$, we can have for $t < r$

$$B = \frac{\sum_{h=1}^{t} \mid \mathbf{D}(h) \mid}{|V(\mathbf{\Delta})|} \quad (126)$$

where $D(h) = (d_{i,j}(h))$ with

$$d_{i,j}(h) = \begin{cases} \sum_{k=0}^{j-1} \frac{(-j+1)_k(t-j+1)_k\delta_i^{r-j+k}}{(t+t_I-j+1)_k k!}, & j \neq h, j \leq t \\ \delta_i^{r-j}, & j \neq h, j > t \\ h_{i,j} - \sum_{b=0}^{t-j} \frac{1}{t_I+b} \sum_{k=0}^{j-1} \frac{(-j+1)_k(t-j+1)_k\delta_i^{r-j+k}}{(t+t_I-j+1)_k k!}, & j = h. \end{cases} \quad (127)$$

7. Appendix: Proofs of theorem 3 and theorem 4 in section 4

7.1 Proof of theorem 3

The Distributions of quadratic forms in matrix argument have been investigated extensively by many authors. For more details, the reader is referred to [Gupta & Nagar (2000)] and [Mathai et al. (1995)]. In order to prove Theorem 3, we first extend a lemma for real data to its complex counterpart to obtain the following.

Lemma 3. *Let* $\mathbf{X} \sim CN_{m,n}(\mathbf{M}, \mathbf{\Sigma} \otimes \mathbf{\Psi})$, $\mathbf{\Sigma} > 0, \mathbf{\Psi} > 0$ *and let* $\mathbf{A}$ *be a* $n \times n$ *Hermite positive definite matrix. Then the PDF of quadratic form* $\mathbf{S} = \mathbf{X}\mathbf{A}\mathbf{X}^\dagger$ *is given by*

$$f(\mathbf{S}) = f \sum_{k=0}^{\infty} \sum_{\kappa} \frac{1}{k![n]_\kappa} \times$$
$$P_\kappa(\mathbf{\Sigma}^{-\frac{1}{2}}\mathbf{M}\mathbf{\Psi}^{-\frac{1}{2}}(\mathbf{I}_n - q\mathbf{B})^{-\frac{1}{2}}, \mathbf{B}^{-1} - q\mathbf{I}_n, \mathbf{\Sigma}^{-\frac{1}{2}}\mathbf{S}\mathbf{\Sigma}^{-\frac{1}{2}}) \quad (128)$$

where κ denotes a partition of k, $q \geq 0$, $\mathbf{B} = \Psi^{1/2}\mathbf{A}\Psi^{1/2}$, $\mathbf{I}_n - q\mathbf{B} > 0$ and

$$f = \frac{etr(-q\mathbf{\Sigma}^{-1}\mathbf{S}) \mid \mathbf{S} \mid^{n-m}}{\widetilde{\Gamma}_m(n) \mid \mathbf{\Sigma} \mid^{n} \mid \mathbf{\Psi} \mid^{m} \mid \mathbf{A} \mid^{m}} \cdot etr[-\mathbf{\Sigma}^{-1}\mathbf{M}\mathbf{\Psi}^{-1}\mathbf{M}^{\dagger}]. \tag{129}$$

Note that q is an arbitrary scalar constant. The PDF for $q > 0$ is called the Wishart type representation, and for $q = 0$ is called the power series type representation.
To prove Theorem 3, we also need two properties of the generalized Hermite polynomial with three complex matrix arguments, as described below.

Lemma 4.

$$\int_{\mathbf{S}>0} etr[-\mathbf{G}\mathbf{S}] \mid \mathbf{S} \mid^{q-p} P_{\kappa}(\mathbf{T},\mathbf{A},\mathbf{B}^{-1/2}\mathbf{S}\mathbf{B}^{-1/2})d\mathbf{S}$$
$$= \widetilde{\Gamma}_p(q,\kappa) \mid \mathbf{G} \mid^{-q} P_{\kappa}(\mathbf{T},\mathbf{A},\mathbf{B}^{-1/2}\mathbf{G}^{-1}\mathbf{B}^{-1/2}) \tag{130}$$

where

$$\widetilde{\Gamma}_p(a,\kappa) = \pi^{p(p-1)/2}\prod_{i=1}^{p}\Gamma(a+k_i-i+1). \tag{131}$$

Lemma 5.

$$\int_{0<\mathbf{S}<\mathbf{V}} \mid \mathbf{S} \mid^{q-p} P_{\kappa}(\mathbf{T},\mathbf{A},\mathbf{B}^{-1/2}\mathbf{S}\mathbf{B}^{-1/2})d\mathbf{S}$$
$$= \frac{\widetilde{\Gamma}_p(q,\kappa)\widetilde{\Gamma}_p(p)}{\widetilde{\Gamma}_p(p+q,\kappa)} \mid \mathbf{V} \mid^{q} P_{\kappa}(\mathbf{T},\mathbf{A},\mathbf{B}^{-1/2}\mathbf{V}\mathbf{B}^{-1/2}) \tag{132}$$

where $\mathbf{V}$ *is an arbitrary Hermite positive definite matrix.*

Proof of Theorem 3: We begin with the case of $t \leq r$ and determine the PDF of the quadratic form $\mathbf{J}$ in (68). Under the condition of given matrix $\mathbf{Q}_1$, by plugging $q = 0$ into (128) of Lemma 3, the conditional PDF of $\mathbf{J}$ can be expressed as

$$f(\mathbf{J})|_{Q_1} = q_0\sum_{k=0}^{\infty}\sum_{\kappa}\frac{1}{k![r]_{\kappa}} \times$$
$$P_{\kappa}(\mathbf{\Psi}^{-\frac{1}{2}}\mathbf{M}^{\dagger}\mathbf{\Sigma}^{-\frac{1}{2}},\mathbf{\Sigma}^{-\frac{1}{2}}\mathbf{Q}_1\mathbf{\Sigma}^{-\frac{1}{2}},\mathbf{\Psi}^{-\frac{1}{2}}\mathbf{J}\mathbf{\Psi}^{-\frac{1}{2}}) \tag{133}$$

where

$$q_0 = \frac{\mid \mathbf{J} \mid^{r-t}}{\widetilde{\Gamma}_t(r) \mid \mathbf{\Sigma}\mathbf{Q}_1^{-1} \mid^{t} \mid \mathbf{\Psi} \mid^{r}}\mathrm{etr}[-(\mathbf{\Sigma})^{-1}\mathbf{M}\mathbf{\Psi}^{-1}\mathbf{M}^{\dagger}]. \tag{134}$$

Then by applying Lemma 4 we carry on the expectation of $f(\mathbf{J})|_{Q_1}$ with respect to $\mathbf{Q}_1 \sim CW_r(t_1,\mathbf{\Sigma}_1)$ yielding

$$f(\mathbf{J}) = q_1\sum_{k=0}^{\infty}\sum_{\kappa}\frac{[t+t_1]_{\kappa}}{k![r]_{\kappa}} \times$$
$$P_{\kappa}(\mathbf{\Psi}^{-\frac{1}{2}}\mathbf{M}^{\dagger}\mathbf{\Sigma}^{-\frac{1}{2}},\mathbf{\Sigma}^{-\frac{1}{2}}\mathbf{\Sigma}_1\mathbf{\Sigma}^{-\frac{1}{2}},\mathbf{\Psi}^{-\frac{1}{2}}\mathbf{J}\mathbf{\Psi}^{-\frac{1}{2}}) \tag{135}$$

where

$$q_1 = \frac{|\mathbf{J}|^{r-t}\widetilde{\Gamma}_r(t+t_1)}{\widetilde{\Gamma}_t(r)\widetilde{\Gamma}_r(t_1)\,|\,\boldsymbol{\Sigma}\boldsymbol{\Sigma}_1^{-1}\,|^t|\,\boldsymbol{\Psi}\,|^r}\mathrm{etr}[-(\boldsymbol{\Sigma})^{-1}\mathbf{M}\boldsymbol{\Psi}^{-1}\mathbf{M}^{\dagger}]. \tag{136}$$

The desired outage probability is nothing but the integration of $f(\mathbf{J})$ over $\mathbf{J} < x\mathbf{I}$. The integral, however, involves matrix arguments and needs to be simplified. To this end, we invoke a property of the generalized Hermite polynomial, i.e., Lemma 5. By applying this property,setting $\boldsymbol{\Omega} = x\mathbf{I}$, and using the definitions of $\boldsymbol{\Delta}$ and $\boldsymbol{\Theta}$, we complete the proof for this case of $t \le r$.

We next consider the case of $r < t$. Let

$$\mathbf{J}_1 = \mathbf{H}_1^{\dagger}\mathbf{H}_1 \tag{137}$$

where $\mathbf{H}_1 = \{\mathbf{Q}_1^{-1/2}\mathbf{H}\}^{\dagger}$. Due to the fact

$$P_{out} = \Pr(\mathbf{J} < x\mathbf{I}_t) = \Pr(\mathbf{J}_1 < x\mathbf{I}_r), \tag{138}$$

then in this case the proof is quite similar to the proof given for the case where $t \le r$, and so is omitted.

Finally, we need the identity, $\widetilde{\Gamma}_r(t+t_1)\widetilde{\Gamma}_t(t+t_1-r) = \widetilde{\Gamma}_r(t_1)\widetilde{\Gamma}_t(t+t_1)$, to give the unified representation of (71).

7.2 Proof of theorem 4

The following property of the generalized Hermite polynomial with three complex matrix arguments is useful in the proof.

Lemma 6. *For a $p \times q$ random matrix $\mathbf{V} \sim CN(\mathbf{0}, \mathbf{I}_q \otimes \mathbf{I}_p)$,*

$$P_{\kappa}(\mathbf{T}, \mathbf{A}, \mathbf{B}) = \mathcal{E}_{\mathbf{V}}[C_{\kappa}(-\mathbf{B}(\mathbf{V} - \imath\mathbf{T})\mathbf{A}(\mathbf{V} - \imath\mathbf{T})^{\dagger})]. \tag{139}$$

where $\imath = \sqrt{-1}$.

In [Teletar (1999)], Telatar gave the following useful limiting result for a Wishart-distributed matrix sequence.

Lemma 7. *Let $\mathbf{S}_n \sim CW_r(n, \frac{1}{n}\mathbf{I}_r)$. When $n \to \infty$, then*

$$\mathbf{S}_n \to \mathbf{I}_r. \tag{140}$$

Proof of Theorem 4: Without loss of generality, we can assume from (85) and (86) that $\boldsymbol{\Phi}_n = \mathbf{I}$. Under the condition of Theorem 3, we first let $t_1 = n$ be a variable, and further let $\mathbf{Q}_1(n) \sim CW_r(n, \frac{1}{n}\mathbf{I}_n)$. Then, according to Lemma 7, we can assert that when $n \to \infty$, the TRD system with co-channel interference will reduce to the TRD without co-channel interference. Correspondingly, the outage probability of the optimal TRD system with co-channel interference (71) will reduce to the outage probability of the optimal TRD system without co-channel interference, which is just (87) in Theorem 4. Let us verify this assertion. By inserting $\boldsymbol{\Sigma}_1 = \frac{1}{n}\mathbf{I}_r$ into (71) and comparing the two expressions of (71) and (87),

we only need to prove Eqs.(141) and (142) shown below.
a) For $t \leq r$, when $n \rightarrow \infty$, then

$$\mathbf{P}_n = \frac{\widetilde{\Gamma}_r(t+n,\kappa)}{n^{rt}\widetilde{\Gamma}_r(n)} P_\kappa(\mathbf{Y}, \frac{1}{n}\mathbf{\Sigma}^{-1}, \mathbf{\Psi}^{-1}) \rightarrow P_\kappa(\mathbf{Y}, \mathbf{\Sigma}^{-1}, \mathbf{\Psi}^{-1}). \tag{141}$$

b) For $t > r$,when $n \rightarrow \infty$, then

$$\mathbf{P}_n = \frac{\widetilde{\Gamma}_r(t+n,\kappa)}{n^{rt}\widetilde{\Gamma}_r(n)} P_\kappa(\mathbf{Y}, \mathbf{\Psi}^{-1}, \frac{1}{n}\mathbf{\Sigma}^{-1}) \rightarrow P_\kappa(\mathbf{Y}, \mathbf{\Psi}^{-1}, \mathbf{\Sigma}^{-1}). \tag{142}$$

Here, we have used the fact that

$$[a]_\kappa = \frac{\widetilde{\Gamma}_m(a,\kappa)}{\widetilde{\Gamma}_m(a)}. \tag{143}$$

Based on Lemma 6 , the proof of (141) and (142) can be done by showing the validity of the following assertion. Namely, for an arbitrary $r \times r$ Hermite matrix $\mathbf{S}$ and $n \rightarrow \infty$, we have

$$\mathbf{P}'_n = \frac{\widetilde{\Gamma}_r(t+n,\kappa)}{n^{rt}\widetilde{\Gamma}_r(n)} C_\kappa(\frac{1}{n}\mathbf{\Sigma}^{-1}S) \rightarrow C_\kappa(\mathbf{\Sigma}^{-1}S). \tag{144}$$

To this end, we invoke Property 1 to simplify (144). It remains to show

$$\frac{\widetilde{\Gamma}_r(t+n,\kappa)}{n^{rt+k}\widetilde{\Gamma}_r(n)} \rightarrow 1 \tag{145}$$

whose validity can be checked by directly using the definition of $\widetilde{\Gamma}_p(a,\kappa)$ given in (131).

8. References

Telatar I. E.(1999). Capacity of multi-antenna Gaussian channels. *European Trans. on Telecomm.*, Vol. 10, No. 6, Nov.-Dec -1999, pp. 585-596, ISSN 1124-318X.

Foschini. G. J & Gans. M. J.(1998). On limits of wireless communication in a fading environment when using multiple antennas. *Wireless Personal Commun.*, Vol. 6, Number. 3, Mar- 1998, pp. 311-335, ISSN 1572-834X.

Loyka S.(2001). Channel capacity of MIMO architecture using the exponential correlation matrix. *IEEE Commun. Letters*, Vol.5, No.9, Sep- 2001, pp.369-371, ISSN 1089-7798.

Chuah C. N., et al(2002). Capacity scaling in MIMO wireless systems under correlated fading. *IEEE Trans. Inform. Theory*, Vol. 48, No. 3, Mar -2002, pp. 637-650, ISSN 0018-9448.

Chiani M., Win M. Z. & Zanella A.(2003). On the capacity of spatially correlated MIMO rayleigh-fading channels. *IEEE Trans. Inform. Theory*, Vol. 49, No. 10, Oct -2003, pp. 2363-2371, ISSN 0018-9448.

Shin H. & Lee J. H.(2003). Capacity of multiple-antenna fading channels:spatial fading correlation, double scattering, and keyhole. *IEEE Trans on Information Tehory*, Vol.49, No.10, Oct -2003, pp.2636-2647, ISSN 0018-9448.

Kiessling M.(2005). Unifying analysis of ergodic MIMO capacity in correlated rayleigh fading enviroments. *European Transactions on Telecommunications,* Vol.16, No.1, Jan -2005, pp.17-35, ISSN 1124-318X.

Catreux S., Greenstein L. J. & Dressen P. F.(2000). Simulation results for an interference-limited multiple-input multiple-output cellular system. *IEEE Commun. Letters,* Vol.4, No. 11, Nov -2000, pp. 334-336, ISSN 1089-7798.

Blum R. S., Winters J. H. & Sollenberger N. R.(2002). On the capacity of cellular systems with MIMO. *IEEE Commun. Letters,* Vol. 6, No. 6, Jun -2002, pp. 242-244, ISSN 1089-7798.

Blum R.S.(2003). Analysis of MIMO capacity with interference. *In proc. International Conference on communications, 2003. (ICC '03. IEEE),* pp.2991-2995, ISBN 0-7803-7802-4, Anchorage, USA, May 2003.

Song Y. & Blostein S. D.(2002). MIMO Channel Capacity in co-channel interference. *in Proc 21st Biennial Symposium on Communications,* pp.220-224, ISBN Kingston, Canada, May 2002.

Ye S. & Blum R. S.(2005). Some properties of the capacity of MIMO systems with co-channel interference *in Proc 2005 IEEE International Conference on Acoustics, Speech, and Signal Processing,* pp.1153-1156, ISBN 0-7803-8874-7, Philadelphia, PA, USA, March, 2005.

Kang M., Yang L. & Alouini M. -S(2003). Performance analysis of MIMO channels in presence of co-channel interference and additive Gaussian noise. *in Proc. the 35th Annual Conference on Information Sciences and Systems (CISS'2003),* Johns Hopkins University, Baltimore, MD, Mar. 2003.

Kang M., Yang L., & Alouini M.-S.(2007). Capacity of MIMO channel in the presence of co-channel interference. *Wireless Communcations and Mobile Computing,* vol.7, July -2007, pp.113-125, ISSN 1530-8677.

Kang M. & Alouini M. -S.(2003). Capacity of MIMO Rician channels with multiple correlated Rayleigh co-channel interferers. *in Proc IEEE Global Telecommun. Conf. (Globecom'2003),* pp.1119-1123, ISBN 0-7803-7974-8, San Francisco, CA, Dec. 2003.

Wang Y. & Yue D.-W.(2009). Capacity of MIMO Rayleigh fading channels in the presence of interference and receive correlation. *IEEE trans. on Vehicular Technology,* Vol.58, No.8, Oct -2009, pp.4398-4405, ISSN 0018-9545.

Dighe P. A., Mallik R. K. & Jamuar S. R.(2001). Analysis of trasmit-receive diversity in Rayleigh fading. *in Proc. IEEE Global Telecom. Conf. (Globecom' 2001),* pp. 1132-1136, ISBN 0-7803-7206-9, San Antonio, TX, Nov. 2001.

Kang M. & Alouini M.-S.(2003). Largest eigenvalue of complex Wishart matrices and performance of MIMO MRC systems. *IEEE Journal on Selected Areas in Communications: Special Issue on MIMO Systems and Applications (JSAC-MIMO),* Vol. 21, No. 3, Apr -2003, pp. 418-426, ISSN 0733-8716.

Kang M. & Alouini M.-S.(2004). A comparative study on the performance of MIMO MRC systems with and without co-channel interference. *IEEE Trans. Commun.,* Vol.52, No.8, Aug -2004, pp.1417-1425, ISSN 0090-6778.

Dighe P. A., Mallik R. K., & S. S. Jamuar S. S.(2003). Analysis of K-transmit dual-receive diversity with cochannel interferers over a Rayleigh fading channel. *Wireless Personal Commun.,* Vol. 25, No. 2, May -2003, pp. 87-100, ISSN 1572-834X.

Kang M. & Alouini M. -S.(2004). Quadratic forms in complex Gaussian matrices and performance analysis of MIMO systems with co-channel interference. *IEEE Trans. Wireless Commun.,* Vol.3, No.2, Feb -2004, pp.418-431, ISSN 1536-1276.

Yue D.- W. & Zhang Q. T.(2010). Generic approach to the performance analysis of correlated transmit/receive diversity MIMO systems with/without co-channel interference. *IEEE Trans on Information Tehory,* Vol.56, No.3, Mar -2010, pp.1147-1157, ISSN 0018-9448.

Muirhead R. J.(1982). *Aspects of Multivariate Statistical Theory,* Wiley, ISBN 0471094420, New York, 1982.

Gupta A. K. & Nagar D. K.(2000). *Matrix Variate Distributions,* Chapman& Hall/CRC, ISBN 1-58488-046-5, NewYork, 2000.

Mathai A. M., Provost S. B. & Hayakawa T.(1995). *Bininear Forms and Zonal Polynomials,* Spring-verlag, ISBN 0-387-94522-9, NewYork, 1995.

James A. T.(1964). Distributions of matrix variates and latent roots derived from normal samples. *Ann. Math. Stat.,* Vol. 35, No. 2, June -1964, pp. 475-501, ISSN 0003-4851.

Khatri C. G.(1966). On the certain distribution problems based on positive definite quadratic functions in normal vectors *Ann. Math. Statist.,* Vol. 37, No. 2, June -1966, pp.468-479, ISSN 0003-4851.

C.G.Khatri C. G.(1965). Classical statistical analysis based on a certain multivariate comlex Gaussian distribution. *Ann. Math. Statist.,* Vol.36, No. 1, Jan -1965, pp.98-114, ISSN 0003-4851.

Kermoal J. P., Schumacher L., Pedersen L. I., Mogensen P. E. & Frederiksen F.(2002). A stochastic MIMO radio channel model with experimental validation. *IEEE Journal on Selected Areas in Communications,* Vol. 20, No 6, June- 2002 pp.1211-1226, ISSN 0733-8716.

Paulraj A., Nabar R. & Gore D.(2003). *Introduction to Space-Time Wireless Communication,* Cambrige University Press, ISBN 0521826152, Cambrige, 2003.

Zhang Q. T. & Cui X. W.(2004). Outage probability for optimum combining of arbitrarily faded signals in the presence of correlated Raleigh interferers. *IEEE trans. on Vehicular Technology,* Vol.53, No.4, July -2004, pp.1043-1051, ISSN 0018-9545.

Pearson E.S. & Hartley H. O.(1976) *Biometrika Tables for Statisticians,* Biometrika Trust, vol.1, p.234, ISBN 085264700X, London, 1976.

Gradshteyn I. S. and I. M. Ryzhik I. M.(1994) *Table of Integrals, Series, and Products, 5th ed.* Academic Press, ISBN 0122947568, Orlando, FL, 1994.

8

Cellular MIMO Systems

Wei Peng
Tohoku University,
Japan

1. Introduction

Multiple-input multiple-output (MIMO) has been accepted as a promising technology for its potential to achieve low bit error rate (BER) by space time coding [1] or to achieve large capacity by multiplexing [2]. MIMO multiplexing has been widely adopted to realize high speed data communications. The capacity of MIMO systems in the point-to-point transmission without external interferences has been studied in [3, 4] to show that large capacity can be achieved in a rich scattering environment. In a cellular environment, the same frequency/frequencies can be used in neighboring cells. As a result, co-channel interference exists and the channel between the base-station (BS) and the mobile-station (MS) is changed from noise-limited channel to interference-limited channel.

Recently, the capacity of MIMO systems in the cellular environment has attracted much interest. Uplink (transmission from MS to BS) capacity with variable-rate transmissions is studied in [5]. By modeling the co-channel interference as additive white Gaussian noise (AWGN), the uplink capacity is also studied in [6, 7]. On the other hand, from the users' stand point, the downlink capacity may be more interesting. However, the results for the downlink capacity of cellular MIMO systems presented in the literature are mainly based on the simulation results. Very detailed simulation results for the downlink MIMO capacity in 3G FDD WCDMA cellular systems can be found in [8]. By assuming single-frequency-reuse (the frequency reuse factor (FRF) equals to 1), the capacity of downlink cellular MIMO systems is studied by simulations and the results are presented in [9, 10]. These results are given in terms of the number of antennas, the modulation schemes, the propagation parameters as well as the cell size. However, FRFs other than 1 are not considered. By taking various FRFs into consideration, a comparative study on the capacity of cellular MIMO systems is presented in [11]. A comprehensive comparison between the capacity of SISO, SIMO, STBC-MISO and MIMO systems in a cellular environment is made based on the simulation results.

In general, fixed FRF has been considered in cellular systems. It is reported in [12] that a flexible FRF may help to improve the capacity for cellular single antenna (SISO) systems. However, flexible design of FRF for cellular MIMO systems is rarely available in the literature.

In this chapter, the downlink capacity of cellular MIMO systems is theoretically analyzed in terms of both the ergodic and outage capacities. The theoretical results of the best and worst situation capacities suggest that the greatest capacities may be achieved by using FRF 1 or FRF 3 adaptively according to the situations. Therefore, a hybrid frequency reuse scheme is introduced to maximize the overall downlink capacity. It is shown by numerical results that the average ergodic and outage capacities can both be increased by the hybrid frequency reuse scheme when compared with the schemes using fixed FRF 1 or FRF 3. Especially, when compared with the commonly accepted FRF 1 scheme, the average outage capacity can be increased as much as 50%. Therefore, by using the hybrid frequency reuse scheme, the coverage problem of the single-frequency-reuse cellular systems can be greatly alleviated.

The rest of the chapter is organized as follows. Section 2 describes the system model of the point-to-point MIMO systems. Some useful results for the capacity of point-to-point MIMO systems are also presented. Section 3 describes the system model of the cellular MIMO systems. The currently existing frequency reuse schemes are introduced. And the ergodic and outage capacities are theoretically analyzed based on the cellular structures of different frequency reuse schemes. The hybrid frequency reuse scheme is proposed in Section 4. Numerical results are then presented in Section 5. Finally, the chapter is concluded in Section 6.

2. Point-to-point MIMO systems

A. System model

The received signal in a point-to-point MIMO system with N_t transmit and N_r receive antennas can be written as

$$\mathbf{y} = \mathbf{H}\mathbf{x} + \mathbf{n}, \tag{1}$$

where $\mathbf{H}$ is an $N_r \times N_t$ channel matrix. The elements of $\mathbf{H}$ are identical and independently distributed (i.i.d.) complex Gaussian variables with zero mean and unit variance (This means that we assume Rayleigh fading). $\mathbf{y}$ is the N_r- dimensional received signal vector. $\mathbf{x}$ is the N_t- dimensional transmitted signal vector. n is the N_r- dimensional additive white Gaussian noise (AWGN) vector with variance σ^2.

B. Capacity analysis for point-to-point MIMO systems

The capacity C of the MIMO systems from the view point of information theory is the mutual information between input signals and output signals, given by [13]

$$C = I(\mathbf{x};(\mathbf{y},\mathbf{H})) = E\{I(\mathbf{x};\mathbf{y}|\mathrm{H} = H)\}, \tag{2}$$

where $E\{\cdot\}$ represents the expectation over channel realizations and H represents the instantaneous channel matrix.

It is assumed that the receiver has perfect channel state information (CSI) but the transmitter does not. Therefore, the transmitted power is allocated equally to each transmit antenna. According to (2), the capacity for a system with N_t transmit and N_r receive antennas is generally given by [3, 4]

$$C_{MIMO} = E\left\{\log_2 \det\left(\mathbf{I}_{N_r} + \frac{P_t}{N_t\sigma^2}\mathbf{H}\mathbf{H}^*\right)\right\}, \tag{3}$$

where $P_t = E\left\{\|\mathbf{x}\|^2\right\}$ is the average total transmitted power and superscript * represents conjugate transpose. Let $\mathbf{S} = \mathbf{H}\mathbf{H}^*$ if $N_r < N_t$ and $\mathbf{S} = \mathbf{H}^*\mathbf{H}$ otherwise, $n = \max\{N_r, N_t\}$ and $m = \min\{N_r, N_t\}$. The capacity in (3) can be rewritten in respect of the eigenvalues $\lambda_1, \cdots, \lambda_m$ of the matrix S as [4]

$$C_{MIMO} = E\left\{\sum_{i=1}^{m} \log_2\left(1 + \frac{P_t}{N_t\sigma^2}\lambda_i\right)\right\}. \quad (4)$$

When the components of noise vector and transmitted signal vector are i.i.d. and $rank(\mathbf{H}) = m$, C_{MIMO} in (4) can be rewritten as [4]

$$C_{MIMO} = m\int_0^{\infty} \log_2\left(1 + \frac{P_t}{N_t\sigma^2}\lambda\right)p(\lambda)d\lambda. \quad (5)$$

Since we are assuming Rayleigh fading, the probability density function (p. d. f) of λ, $p(\lambda)$, is given by

$$p(\lambda) = \frac{1}{m}\sum_{k=0}^{m-1}\frac{k!}{(k+n-m)!}\left[L_k^{n-m}(\lambda)\right]^2 \lambda^{n-m}e^{-\lambda}, \quad (6)$$

where $L_k^{n-m}(\lambda) = \frac{1}{k!}e^{\lambda}\lambda^{m-n}\frac{d^k}{d\lambda^k}\left(e^{-\lambda}\lambda^{n-m+k}\right)$.

Increasing the number of receive antennas will increase degree of freedom and therefore improve the capacity performance. However, MSs could not employ large number of antennas due to the size limitation. Therefore, in this study, the number of the receive antennas is assumed to be equal to the number of transmit antennas. Under this assumption, the capacity in (5) becomes

$$C_{MIMO} = \int_0^{\infty} \log_2\left(1 + \frac{P_t}{N_t\sigma^2}\lambda\right)\sum_{k=0}^{m-1}\left[L_k(\lambda)\right]^2 e^{-\lambda}d\lambda, \quad (7)$$

where $L_k(\lambda) = \frac{1}{k!}e^{\lambda}\frac{d^k}{d\lambda^k}\left(e^{-\lambda}\lambda^k\right)$. Note that although the expression in (7) yields the ergodic capacity, the outage capacity can also be evaluated similarly. For example, the outage capacity $C_{10\%}$ (which represents an outage of 10%) can be obtained by

$$C_{10\%} = m\log_2\left(1 + \frac{P_t}{N_t\sigma^2}\lambda_{10\%}\right), \quad (8)$$

where $\lambda_{10\%}$ satisfies $P(\lambda < \lambda_{10\%}) = 0.1$.

3. Cellular MIMO downlink transmission

A. System model

In the cellular circumstance, there exists co-channel interference from the neighboring cells due to the frequency reuse. The received signal vector in (1) should be modified to include the path loss effect and shadowing loss effect as well as the co-channel interference as

$$\begin{aligned}\mathbf{y}_{cellular} &= d_0^{-\alpha}10^{-\xi/10}\mathbf{Hx} + \sum_{i=1}^{B} d_i^{-\alpha}10^{-\xi/10}\mathbf{H}_i\mathrm{x}_i + \mathbf{n} \\ &= d_0^{-\alpha}10^{-\xi/10}\mathbf{Hx} + \mathrm{v}\end{aligned}. \qquad (9)$$

In (9), d_0 represents the distance between the MS and the desired BS, $d_i\left(i=1,\cdots,B\right)$ represents the distance between the MS and the ith co-channel BS, α represents the path loss exponent and B is the number of considered co-channel BSs; ξ represents the shadowing loss in dB, which follows the Gaussian distribution with zero mean and standard derivation σ_ξ. It is reported in [10] that the system capacity will decrease when σ_ξ increases. However, such decrease will not exceed 50% (3dB) when σ_ξ increases from 0 dB to 8 dB. To simplify the analysis, in this study, $\xi = 0$ (no shadowing loss) is assumed; H_i represents the channel matrix between the MS and the ith co-channel BS and x_i is the transmitted signal vector from the ith co-channel BS; $\mathrm{v} = \sum_{i=1}^{I} d_i^{-\alpha}\mathrm{H}_i\mathrm{x}_i + \mathrm{n}$ represents the interference plus noise term.

The co-channel interference is usually modeled as Gaussian distributed [6, 7]. Under this assumption, v can be treated as equivalent AWGN with zero mean and variance [14]

$$\sigma_v^2 = \mathrm{var}\{\mathbf{v}\} = \mathrm{var}\left\{\sum_{i=1}^{B} d_i^{-\alpha}\mathbf{H}_i\mathrm{x}_i + \mathbf{n}\right\} = \sum_{i=1}^{B} d_i^{-\alpha}E\left\{\left\|\mathrm{x}_i\right\|^2\right\} + \sigma^2, \qquad (10)$$

where $E\left\{\left\|\mathrm{x}_i\right\|^2\right\}$ is the average total transmitted power of x_i.

B. Frequency Reuse schemes

There are two types of frequency reuse schemes where integer FRFs or non-integer FRFs are used. The non-integer FRF was recently introduced by [12]. In the following, the frequency reuse schemes with integer FRFs and non-integer FRFs will be described respectively.

a. **Frequency Reuse Scheme with Integer FRFs**

Fig. 1 shows the co-channel interference from the neighboring cells to a MS in the central cell. Integer FRFs of 1, 3, 4 and 7 are used for example. Here, the best situation and the worst situation are defined according to the received signal-to-noise- ratio (SNR). The best situation happens with the MS near the center of the cell where the desired BS locates. The worst situation happens with the MS at the boundary of the cell, where the distance between the MS and the desired BS is largest [5]. Let K represent the FRF. As shown in Fig. 1, when $K \geq 3$, the co-channel interferers in the second and above tiers are far away from the MS and therefore their interference can be ignored. However, when $K=1$, the interferers in the second tier are no more negligible. Therefore, when $K=1$, co-channel

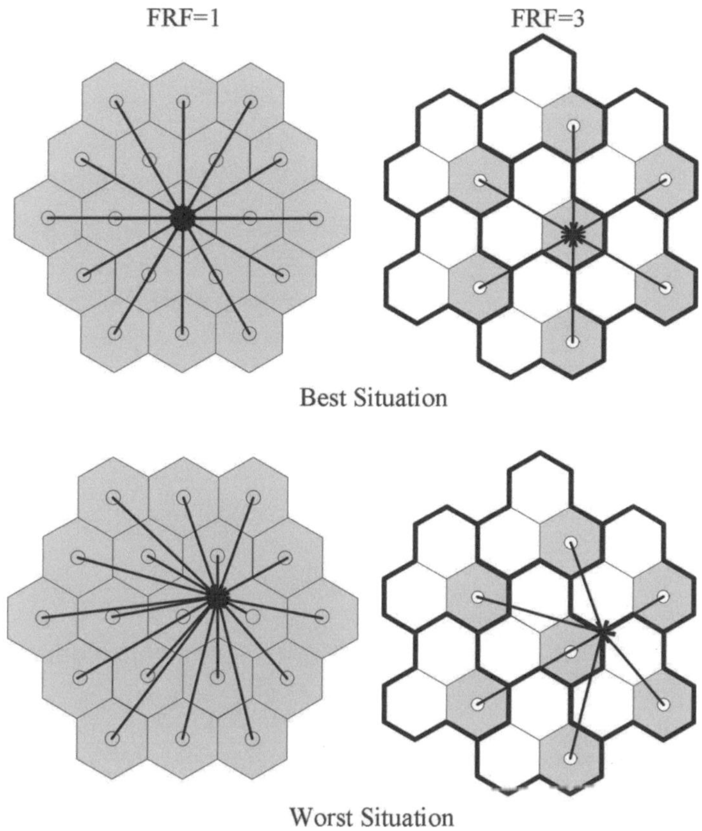

Fig. 1. Co-channel interference with integer FRFs.

interferers in both the first and second tiers will be considered. The signal-to-interference-plus-noise-ratio (SINR) at the MS is approximated by [11]

$$\Gamma_{inte,K} \approx \begin{cases} \dfrac{P_t \cdot r_0^{-\alpha} / \sigma^2 \cdot \varepsilon^{-\alpha}}{P_t \cdot r_0^{-\alpha} / \sigma^2 \cdot \left[\sum_{i=1}^{6} \left(\dfrac{d_{K,1,i}}{r_0} \right)^{-\alpha} + \sum_{i=1}^{12} \left(\dfrac{d_{K,2,i}}{r_0} \right)^{-\alpha} \right] + 1}, K = 1 \\ \dfrac{P_t \cdot r_0^{-\alpha} / \sigma^2 \cdot \varepsilon^{-\alpha}}{P_t \cdot r_0^{-\alpha} / \sigma^2 \sum_{i=1}^{6} \left(\dfrac{d_{K,1,i}}{r_0} \right)^{-\alpha} + 1}, K \geq 1 \end{cases}, \tag{12}$$

where r_0 is the cell radius, $\varepsilon = d_0 / r_0$ is the normalized distance between the MS and the desired BS; $d_{K,l,i}$ represents the distance between the MS and the ith co-channel BS in the $l^{th}(l = 1,2)$ tier when the FRF equals to K. Note that the SINR expression in (12) can be used for the MS at arbitrary positions within a cell.

b. Frequency Reuse Scheme with Non-integer FRF

In this situation, the frequencies are allocated in a more sophisticated way. Here two non-integer FRFs, 7/3 and 7/4 (following the definitions in [12]), are considered. The corresponding frequency allocation schemes are shown in Fig. 2 where symbols *f*1, …, *f*7

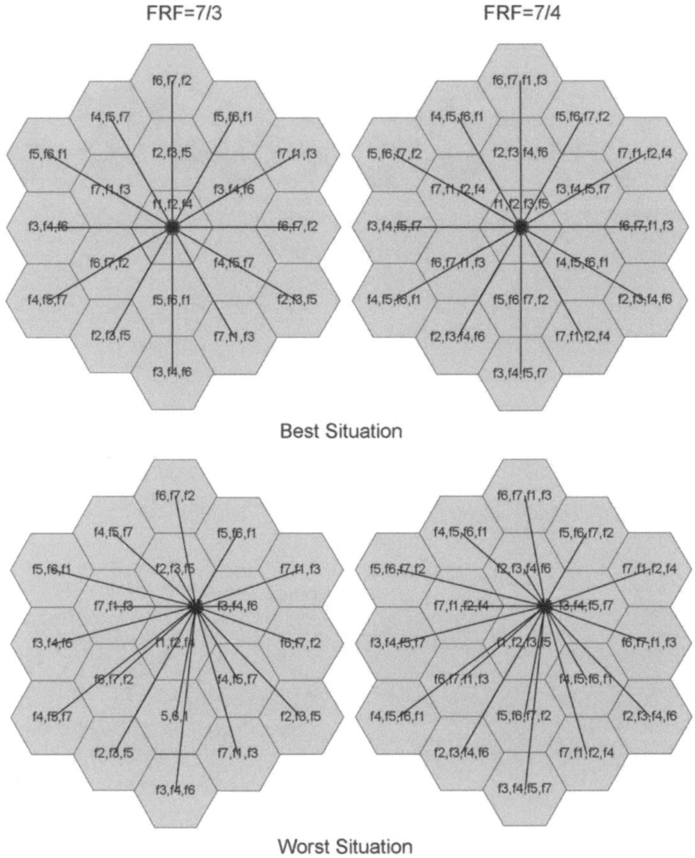

Fig. 2. Co-channel interference with non-integer FRFs.

represent the frequencies used in each cell. Take FRF=7/3 as an example. The frequency set $\{f1, f2, f4\}$ is used by the desired BS in the center. The neighboring cells use the frequency sets $\{f2, f3, f5\}$, $\{f3, f4, f6\}$, $\{f4, f5, f7\}$, $\{f5, f6, f1\}$, $\{f6, f7, f2\}$ and $\{f7, f1, f3\}$, respectively. As a result, each neighboring BS uses one frequency in common with the desired BS. Therefore, when non-integer FRF is used, the considered co-channel interferers are located similarly to the situation when FRF=1. However, the power of the interference is different. The SINR at the MS is approximated by

$$\Gamma_{non-inte,K} \approx \frac{P_t \cdot r_0^{-\alpha} / \sigma^2 \cdot \varepsilon^{-\alpha}}{\eta_K P_t \cdot r_0^{-\alpha} / \sigma^2 \cdot \left[\sum_{i=1}^{6} \left(\frac{d_{K,1,i}}{r_0} \right)^{-\alpha} + \sum_{i=1}^{12} \left(\frac{d_{K,2,i}}{r_0} \right)^{-\alpha} \right] + 1}, K \text{ is non-integer}, \quad (13)$$

where η_K is determined by the FRF K; $\eta_{7/3} = 1/3$ and $\eta_{7/4} = 1/2$.

C. Capacity analysis for cellular MIMO systems

The capacities in (7) and (8) for point-to-point MIMO systems should be modified accordingly to calculate their counterparts in cellular environment. In addition, the capacities for an MS in the cellular environment should be normalized by the total bandwidth. Therefore, the ergodic and outage capacities are given as

$$C_K \approx \frac{1}{K}\int_0^\infty \log_2\left(1+\frac{\Gamma_K}{N_t}\lambda\right)\sum_{k=0}^{N_t-1}\left[L_k(\lambda)\right]^2 e^{-\lambda}d\lambda\,, \tag{14}$$

$$C_{out\%,K} \approx \frac{1}{K}m\log_2\left(1+\frac{\Gamma_K}{N_t}\lambda_{out\%}\right). \tag{15}$$

where $\Gamma_K = \Gamma_{inte,K}$ when K is integer and $\Gamma_K = \Gamma_{non-inte,K}$ otherwise; $C_{out\%,K}$ represents the outage capacity with an outage of $out\%$ when $\mathrm{FRF}=K$ and $\lambda_{out\%}$ satisfies $P(\lambda < \lambda_{out\%}) = out\%$.

Calculating from (14) and (15), the best situation and the worst situation capacities are shown in Fig. 3 and Fig. 4 respectively. The parameters used to generate the results are as follows. The number of antennas ($N_t = N_r$) is set to be 4; the received SNR at the boundary ($P_t \cdot r_0^{-\alpha}/\sigma^2$) is set as $\{0dB, 10dB, 20dB, 30dB\}$; the path loss exponent α is set as $\{2.5, 3.5\}$; the path loss from the cell center to the cell boundary is set as -15dB (i.e., the cell radius is 3.98 km when α equals to 2.5 and 2.68 km when α equals 3.5); Finally, the FRFs are set as $\{1, 7/4, 7/3, 3, 4, 7\}$.

When considering the best situation, it can be observed from Fig. 3 that:

1. The greatest ergodic capacity is achieved by the single-frequency-reuse systems. This observation also coincides with the conclusions in [15].
2. The greatest outage capacity ($C_{10\%}$) is achieved by systems using FRF=7/4 or integer FRF=3.

When considering the worst situation, it can be observed from Fig. 4 that:

1. The greatest ergodic capacity is achieved by systems using FRF=3.
2. For low and moderate values of $P_t \cdot r_0^{-\alpha}/\sigma^2$ (0 dB or 10 dB), the greatest outage capacity ($C_{10\%}$) is achieved by systems using FRF=3.
3. For higher $P_t \cdot r_0^{-\alpha}/\sigma^2$ (20 dB or 30 dB), even greater outage capacities can be obtained by using FRFs > 3. However, the increase is insignificant.

Therefore, considering the ergodic/outage capacities under best/worst situations as a whole, the optimal capacity performance may be achieved by using FRF=1, 7/4 or 3.

4. Hybrid frequency reuse scheme for cellular MIMO systems

The widely accepted approach to design the FRF is to use a fixed FRF within the entire cell. However, from the results in Fig. 3 and Fig. 4, it is obvious that a fixed FRF cannot guarantee the greatest capacities in different circumstances. For example, $K=1$ is the optimal FRF when the MS is at the center of the cell, but it cannot support high capacity for the MSs at the cell boundary; On the other hand, $K=3$ is a good choice when the MS is at the cell boundary. However, it cannot support high capacity for the MSs near the center of the cell. Enlightened by hybrid frequency reuse scheme for cellular SISO systems in [12], we propose a hybrid frequency reuse scheme for the cellular MIMO systems. In this hybrid frequency reuse scheme, both FRF 1 and FRF 3 will be used and adaptively allocated. One possible solution to realize the hybrid frequency reuse scheme is shown in Fig. 5 where all

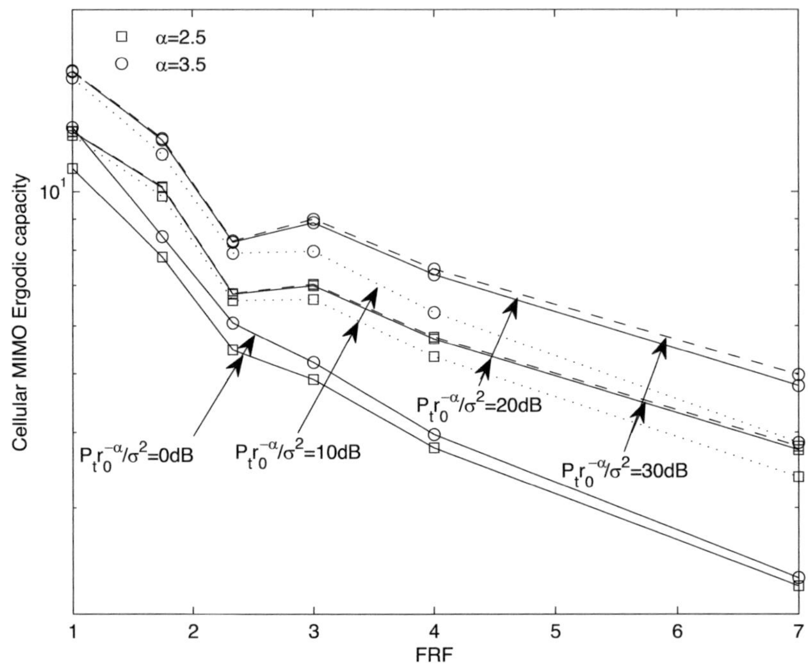

(a)

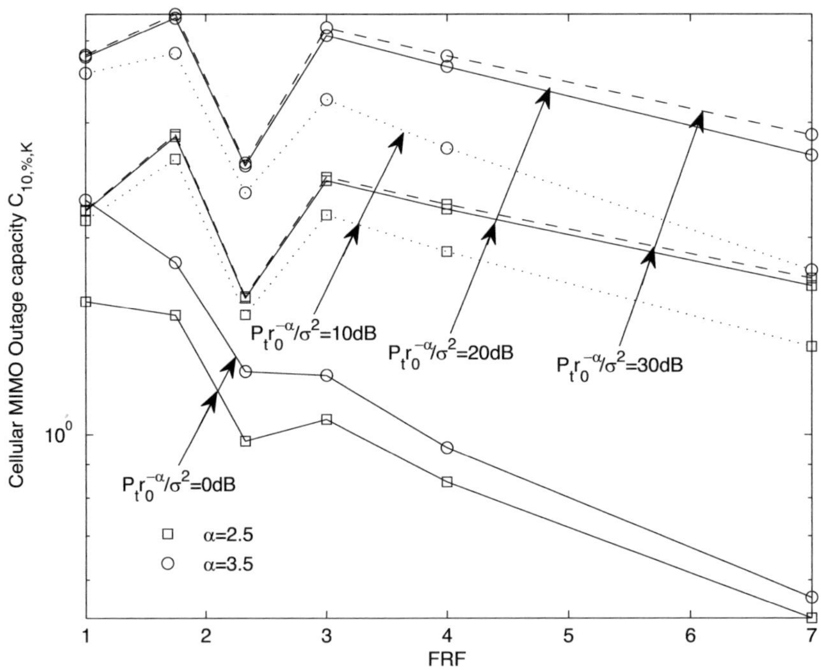

(b)

Fig. 3. Capacities of cellular MIMO systems under the best situation.

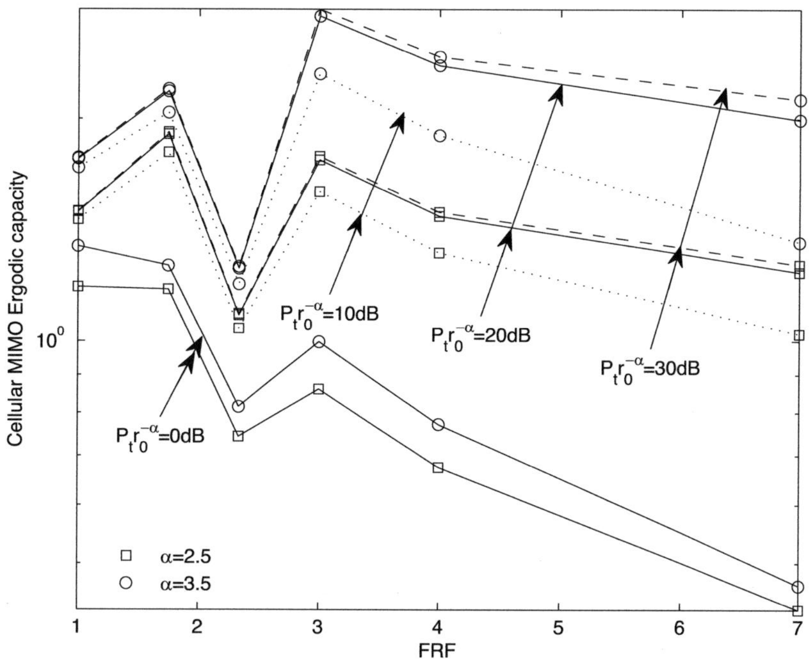

(a)

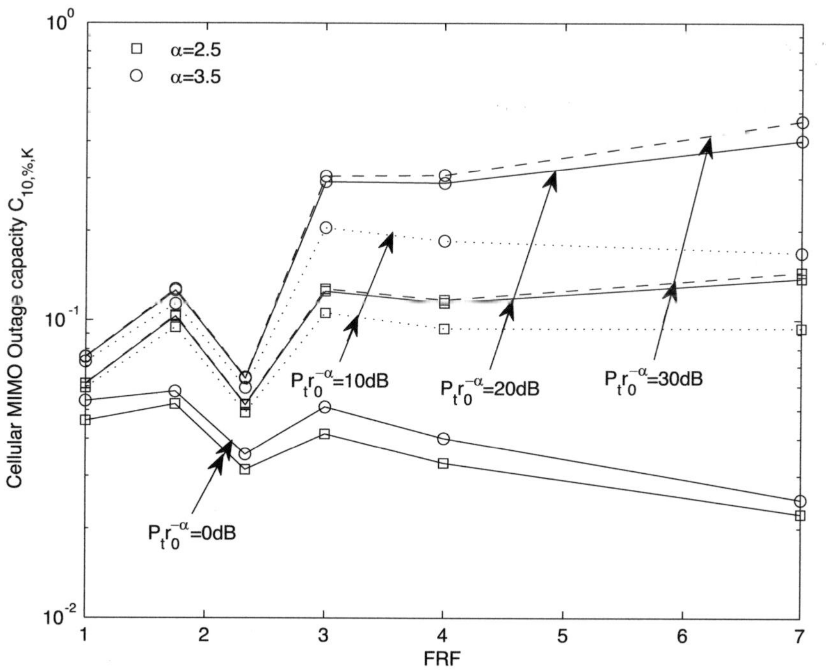

(b)

Fig. 4. Capacities of cellular MIMO systems under the worst situation.

the BSs use a frequency set of three frequencies $\{f_1, f_2, f_3\}$. For each MS, FRF 1 or FRF 3 will be used according to its position within the cell. When the MS is near the cell center, $\{f_1, f_2, f_3\}$ will all be used (FRF=1). Otherwise, when the MS is near the cell boundary, only one frequency, f_1 or f_2 or f_3, will be used (FRF=3).

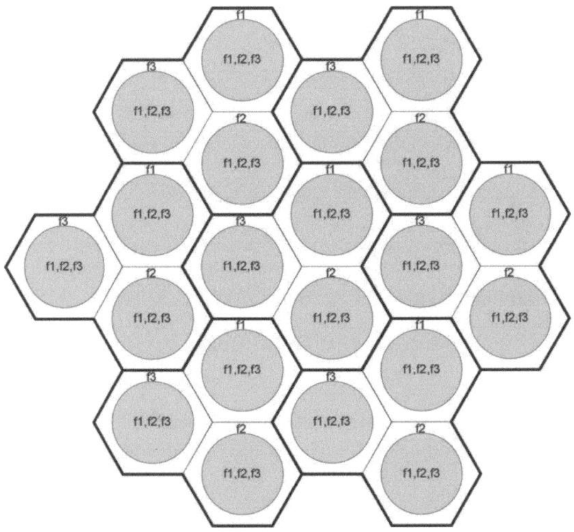

Fig. 5. Hybrid frequency reuse scheme. When the MS is near the center, three frequencies {f_1, f_2, f_3 } will be used(FRF=1); When the MS is near the boundary, only one frequency will be used (FRF=3).

For a given MS, we have determined the FRF by the following steps:

Step 1. As the pre-knowledge, the cell radius r_0, the path loss exponent α, the transmitted power P_t and the variance of the noise σ^2 should be estimated.

Step 2. For a given MS, estimate its distances to the desired MS d_0 and the co-channel MSs d_i, $i = 1, \cdots, B$.

Step 3. Calculate the SINR by (12) and then evaluate the system capacities by substituting the SINR into (14) /(15) to get the ergodic/outage capacities. The capacities for $K = 1$ and $K = 3$ will be calculated respectively to get $C_1 / C_{out,1}$ and $C_3 / C_{out,3}$.

Step 4. Select FRF so that $K = \max_K \{C_K\}$ for maximum ergodic capacity or $K = \max_K \{C_{out,K}\}$ for maximum outage capacity.

Following these steps, the FRF allocation within a cell is shown in Fig. 6 as an example where polar coordinate is used. In the figure, the hexagonal areas represent one cell, the circle areas within each cell are the areas where FRF 1 will be used. Otherwise, FRF 3 will be used. It is observed from Fig. 6 that:

1. The FRF 1 areas shrink slightly as the path loss exponent increases.
2. The FRF 1 areas shrink as the received SNR at the cell boundary ($P_t \cdot r_0^{-\alpha} / \sigma^2$) increases.

It is indicated that the co-channel interference problem cannot be alleviated by increasing the transmit power if single-frequency-reuse scheme is employed. Therefore, for the areas near the cell boundary, the proposed hybrid frequency reuse scheme will be a good solution to reduce the co-channel interference.

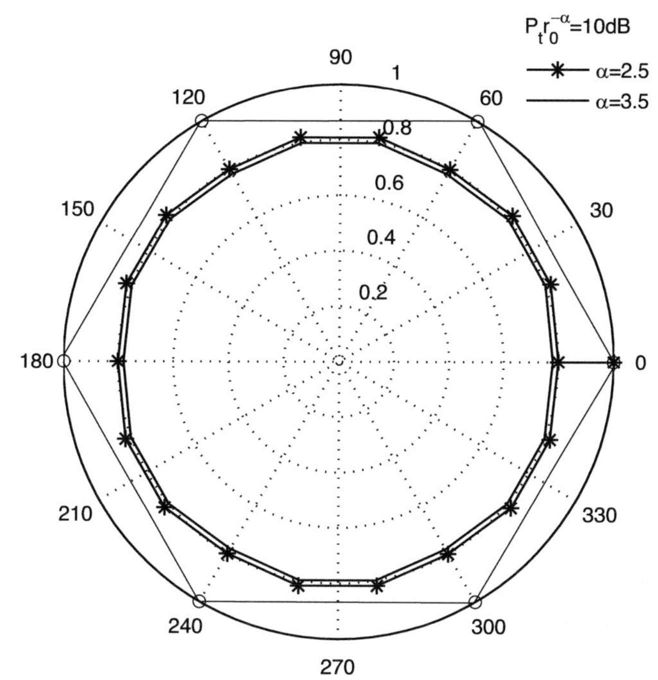

(a)

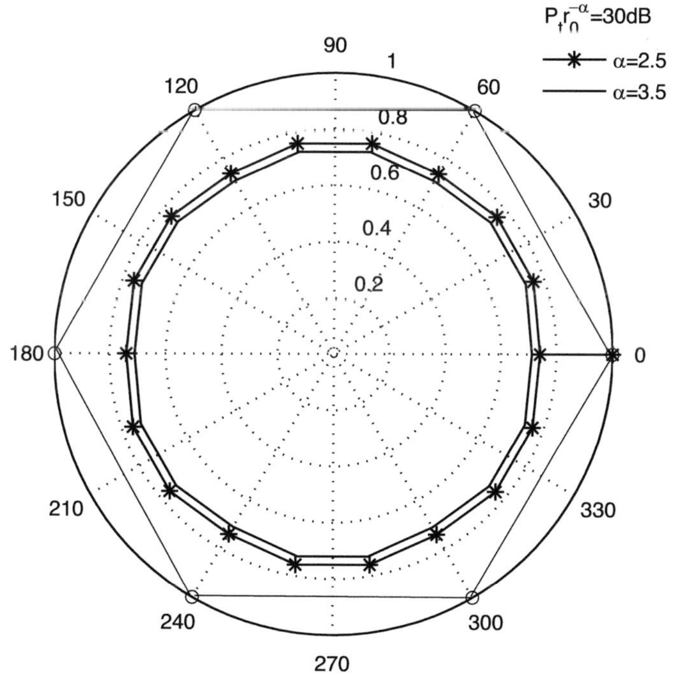

(b)

Fig. 6. FRF allocation within a cell.

5. Numerical results

It is assumed that the MS is uniformly located within a cell. The average capacities can then be calculated by averaging (14) and (15) over the entire cell. The parameters used to generate the numerical results are listed in Table I.

The capacities of cellular MIMO systems using the proposed hybrid frequency reuse scheme (hybrid FRF) are compared with those using FRF 1 and FRF 3 schemes in Fig. 7 and Fig. 8. It can be observed from that the proposed hybrid frequency reuse scheme can increase both the average ergodic and outage capacities. When compared with the FRF 1 scheme, the increase is mainly on the average outage capacities as shown in Fig. 7. The increase can be as significant as about 50% when $P_t \cdot r_0^{-\alpha}/\sigma^2$ =30 dB for the MS equipped with 8 antennas. Even for the noisy environment when $P_t \cdot r_0^{-\alpha}/\sigma^2$ =10 dB, such increase is more than 10% for the MS equipped with 2 antennas. On the other hand, when compared with the FRF 3 scheme, the increase is mainly on the average ergodic capacities as shown in Fig. 8. The increase is over 60% when $P_t \cdot r_0^{-\alpha}/\sigma^2$ =10 dB and over 47% when $P_t \cdot r_0^{-\alpha}/\sigma^2$ =30 dB.

To make it clearer, the increase of average capacities gained by the hybrid FRF over the FRF 1 scheme and FRF 3 scheme is summarized in Table II in percentage.

Number of antennas ($N_t = N_r$)	2~8
Received SNR at the cell boundary $P_t \cdot r_0^{-\alpha}/\sigma^2$	10dB, 30dB
Path loss exponent α	3.5
Path loss from the cell center to the cell boundary	-15dB
Frequency allocation schemes	FRF1, FRF 3, proposed hybrid frequency reuse scheme

Table I. Parameters

			Number of antennas						
			2	3	4	5	6	7	8
Over FRF 1	Outage	$P_t \cdot r_0^{-\alpha}/\sigma^2 = 10dB$	10.5%	12.8%	14.3%	15.3%	16.7%	17.5%	17.8%
		$P_t \cdot r_0^{-\alpha}/\sigma^2 = 30dB$	21.7%	29.8%	37.8%	43.7%	47.4%	49.5%	49.9%
	Ergodic	$P_t \cdot r_0^{-\alpha}/\sigma^2 = 10dB$	1.6%	1.6%	1.6%	1.6%	1.6%	1.6%	1.6%
		$P_t \cdot r_0^{-\alpha}/\sigma^2 = 30dB$	3.0%	3.0%	3.0%	3.0%	3.0%	3.0%	3.0%
Over FRF 3	Outage	$P_t \cdot r_0^{-\alpha}/\sigma^2 = 10dB$	13.6%	7.6%	3.8%	1.4%	0.5%	0.1%	-*
		$P_t \cdot r_0^{-\alpha}/\sigma^2 = 30dB$	2.9%	-	-	-	-	-	-
	Ergodic	$P_t \cdot r_0^{-\alpha}/\sigma^2 = 10dB$	62.9%	63.1%	63.4%	63.5%	63.6%	63.7%	63.7%
		$P_t \cdot r_0^{-\alpha}/\sigma^2 = 30dB$	47.3%	47.6%	47.8%	47.9%	48.0%	48.1%	48.1%

*represents no increase

Table II. The increase of average capacities of the proposed hybrid frequency reuse scheme

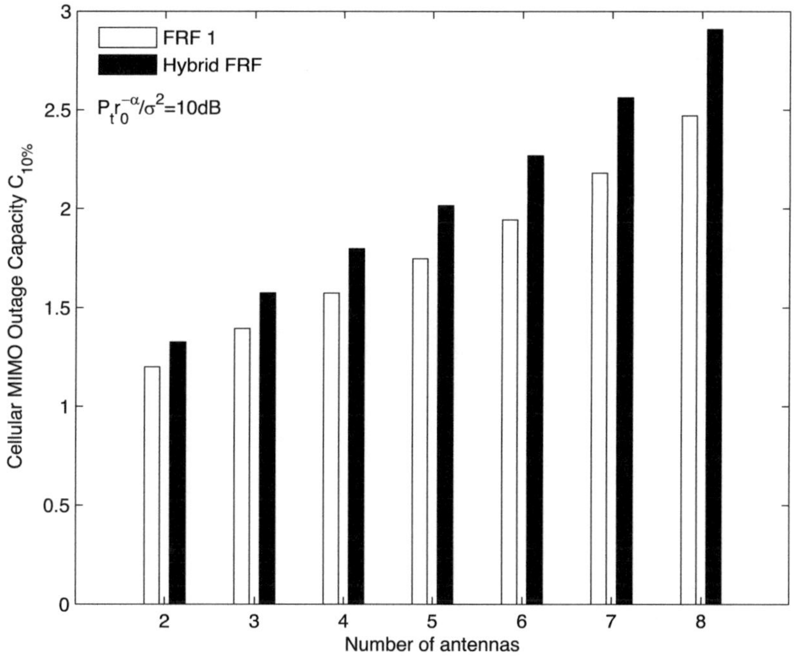

(a)

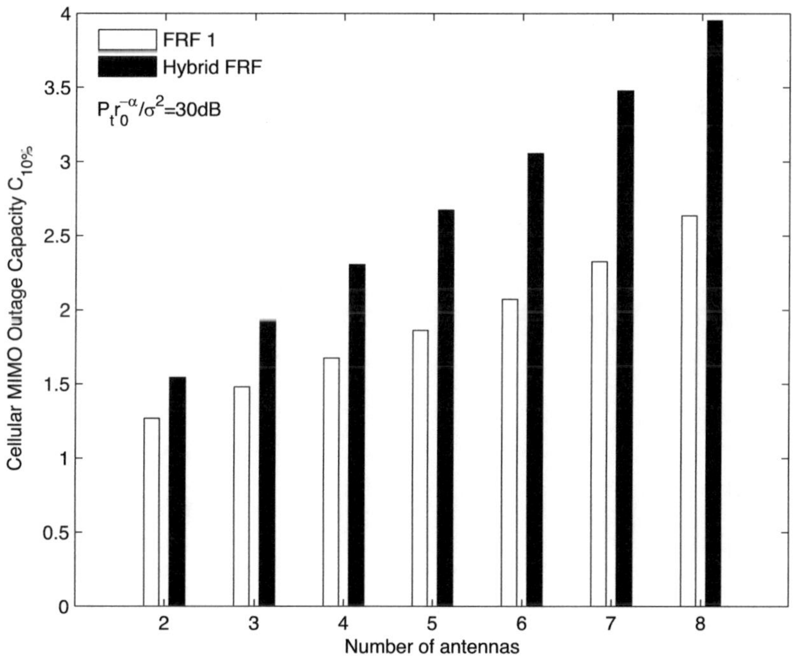

(b)

Fig. 7. Average 10% outage capacities of the cellular MIMO systems using FRF 1 scheme and the hybrid frequency reuse scheme.

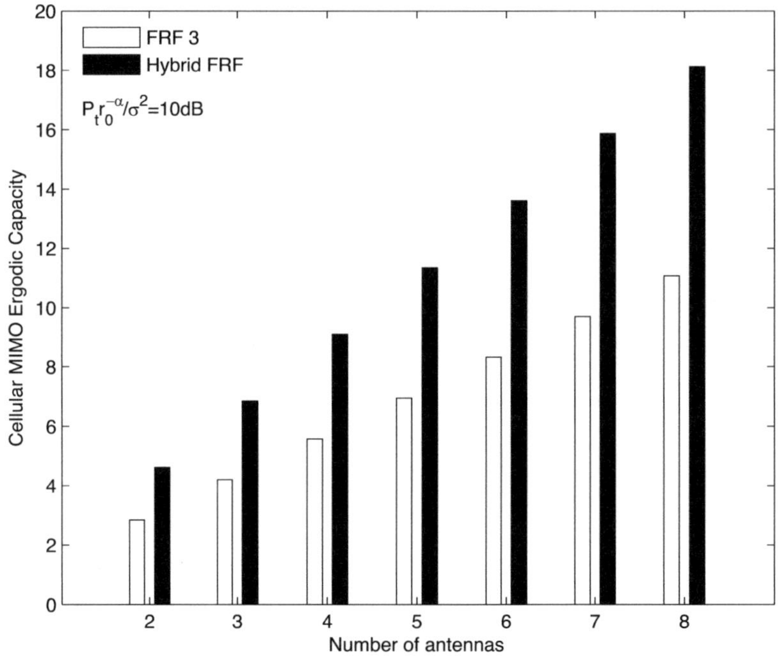

(a)

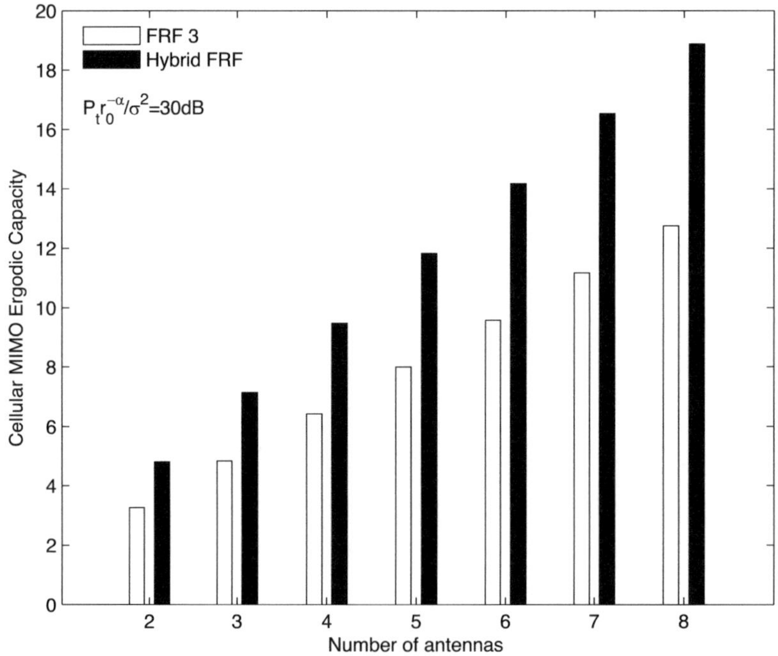

(b)

Fig. 8. Average ergodic capacities of the cellular MIMO systems using FRF 3 scheme and the hybrid frequency reuse scheme.

Remark: As we know, the coverage problem (the transmission between the BS and MS fails at the cell boundary due to the co-channel interference) has been the major problem for the commonly used single-frequency-reuse cellular systems. From the numerical results, it is seen that such problem can be greatly alleviated by using the proposed hybrid frequency reuse scheme.

6. Conclusions

In this chapter, the downlink capacity of cellular MIMO systems has been theoretically analyzed in terms of both ergodic and outage capacities. The FRF has been considered and a hybrid frequency reuse scheme has been introduced. Numerical results have shown that both the ergodic and outage capacities can be increased by the hybrid FRF scheme. Especially, when compared with the commonly used FRF 1 scheme, the outage capacity can be increased as much as 50%. Therefore, the hybrid FRF scheme can greatly alleviate the coverage problem of the single-frequency-reuse cellular systems.

7. Reference

[1] V. Tarokh, N. Sehadri and A. R. Calderband, "Space-time codes for high data rate wireless communication: Performance criterion and code constructions," IEEE Transactions on Information Theory, vol. 44, pp. 744-765, March 1998.
[2] G. J. Foschini, "Layered space-time architecture for wireless communication in a fading environment when using multielement antennas," Bell Labs Technical Journal, pp. 41-59, Autumn 1996.
[3] G. J. Foschini and M. J. Gans, "On limits of wireless communications in a fading environment when using multiple antennas," Wireless Personal Communications, vol. 6, pp. 311-335, 1998.
[4] E. Telatar, "Capacity of multi-antenna Gaussian channels," European Transactions on Telecommunications, vol. 10, pp. 585-595, November 1999.
[5] M. S. Alouini and A. J. Goldensmith, "Area spectral efficiency of cellular mobile radio systems," IEEE Transactions on Vehicular Technology, vol. 48, pp. 1047-1066, July 1999.
[6] R. S. Blum, J. H. Winters and N. R. Sollenberger, "On the capacity of cellular systems with MIMO," IEEE Communications Letters, vol. 6, pp. 242-244, June 2002.
[7] W. Matthew, B. Mark and N. Andrew, "Capacity limits of MIMO channels with co-channel interference," IEEE Vehicular Technology Conference, pp. 703-707, 2004.
[8] M. M. Matalgah, J. Qaddour, A. Sharma and K. Sheikh, "Throughput and spectral efficiency analysis in 3G FDD WCDMA cellular systems," IEEE Globecom conference, pp. 3423-3426, November 2003.
[9] S. Catreux, P. F. Driessen and L. J. Greenstein, "Simulation results for an interference-limited multiple-input multiple-output cellular sysem," IEEE Communications Letters, vol. 4, pp. 334-336, November 2000.
[10] S. Catreux, P. F. Driessen and L. J. Greenstein, "Attainable throughput of an interference-limited Multiple-Input Multiple-Output (MIMO) cellular systems ," IEEE Transactions on Communications, vol. 49, pp. 1307-1311, August 2001.
[11] K. Adachi, F. Adachi and M. Nakagawa, "On cellular MIMO spectrum efficiency," IEEE Vehicular Technology Conference, pp. 417-421, October 2007.

[12] Y. J. Choi, C. S. Kim and S. Bahk, "Flexible design of frequency reuse factor in OFCDM cellular networks," IEEE International Conference on Communications, pp. 1784-1788, May 2006.
[13] T. M. Cover and J. A. Thomas, Elements of information theory, New York: Wiley, 1991.
[14] J. G. Proakis, Digital Communications, New York: McGraw Hill, 2001.
[15] Z. Wang and R. S. Gallacher, "Frequency reuse scheme for cellular OFDM systems," IEEE Electronics Letters, vol. 38, pp. 387-388, April 2002.
[16] Wei Peng and Fumiyuki Adachi, "Hybrid Frequency Reuse Scheme for Cellular MIMO Systems," IEICE Transactions on Communications, vol. E92-B, May 2009.

Part 3

Pre-processing and Post-processing in MIMO Systems

9

MIMO-THP System with Imperfect CSI

H. Khaleghi Bizaki
Electrical and Electronic Engineering University Complex (EEEUC), Tehran, Iran

1. Introduction

In recent years, it was realized that designing wireless digital communication systems to more efficiently exploit the spatial domain of the transmission medium, allows for a significant increase of spectral efficiency. These systems, in general case, are known as Multiple Input Multiple Output (MIMO) systems and have received considerable attention of researchers and commercial companies due to their potential to dramatically increase the spectral efficiency and simultaneously sending individual information to the corresponding users in wireless systems.
In MIMO channels, the information theoretical results show that the desired throughput can be achieved by using the so called Dirty Paper Coding (DPC) method which employs at the transmitter side. However, due to the computational complexity, this method is not practically used until yet. Tomlinson Harashima Precoding (THP) is a suboptimal method which can achieve the near sum-rate of such channels with much simpler complexity as compared to the optimum DPC approach. In spite of THP's good performance, it is very sensitive to erroneous Channel State Information (CSI). When the CSI at the transmitter is imperfect, the system suffers from performance degradation.
In current chapter, the design of THP in an imperfect CSI scenario is considered for a MIMO-BC (BroadCast) system. At first, the maximum achievable rate of MIMO-THP system in an imperfect CSI is computed by means of information theory concepts. Moreover, a lower bound for capacity loss and optimum as well as suboptimum solutions for power allocation is derived. This bound can be useful in practical system design in an imperfect CSI case.
In order to increase the THP performance in an imperfect CSI, a robust optimization technique is developed for THP based on Minimum Mean Square Error (MMSE) criterion. This robust optimization has more performance than the conventional optimization method. Then, the above optimization is developed for time varying channels and based on this knowledge we design a robust precoder for fast time varying channels. The designed precoder has good performance over correlated MIMO channels in which, the volume of its feed back can be reduced significantly.
Traditionally, channel estimation and pre-equalization are optimized separately and independently. In this chapter, a new robust solution is derived for MIMO THP system, which optimizes jointly the channel estimation and THP filters. The proposed method provides significant improvement with respect to conventional optimization with less increase in complexity.

Notation: Random variables, vectors, and matrices are denoted by lower, lower bold, and upper bold italic letters, respectively. The operators E(.), diag(.), $\perp$, PDF, and CDF stand for expectation, diagonal elements of a vector, statistically independent, Probability Density Function, and Cumulative Distribution Function, respectively.

2. MIMO-BC-THP systems

2.1 Type of MIMO channels

There are three types system can be modeled as MIMO channel [1]:

a. *point-to-point MIMO channel*

This type of MIMO system is a multiple antenna scenario, where both transmitter (TX) and receiver (RX) use several antennas with seperate modulation and demodulation for each antenna. We refer this type of channel as MIMO channel (Central transmitter and receiver).

b. *multipoint-to-point MIMO Channel*

The uplink direction of any multiuser mobile communication system is an example of a MIMO system of this type. The joint receiver at the base station has to recover the individual users' signals. We will refer to this type of channel as the MIMO multiple access channel (Decentralized transmitters and central receiver).

c. *point-to-multipoint MIMO Channel*

The downlink direction of mobile multiuser communication systems is an example of what we call a MIMO broadcast channel (Central transmitter and decentralized receivers).

2.2 Precoding strategy

The main difficulty for transmission over MIMO channels is the separation or equalization of the parallel data streams, i.e., the recovery of the components of the transmitted vector $\boldsymbol{x}$ which interfere at the receiver side. The most obvious strategy for separating the data streams is linear equalization at the receiver side.

It is well-known that linear equalization suffers from noise enhancement and hence has poor power efficiency [2]. This disadvantage can be overcome by spatial decision-feedback equalization (DFE). Unfortunately, in DFE error propagation may occur. Moreover, since immediate decisions are required, the application of channel coding requires some clever interleaving which in turn introduces significant delay [2].

The above methods require CS) only at the receiver side. If CSI is (partly) also available at the transmitter, the users can be separated by means of precoding. Precoding, in general case, stands for all methods applied at the transmitter that facilitate detection at the receiver. If a linear transmitter preprocessing strategy is used, we prefer to denote it as preequalization or linear precoder. In other case we refer it as non-linear precoder.

In MIMO channels a version of DFE by name, matrix DFE is used where is a non-linear spatial equalization strategy at the receiver side. The feedback part of the DFE can be transferred to the transmitter, leading to a scheme known as THP. It is well known that neglecting a very small increase in average transmit power, the performance of DFE and THP is the same, but since THP is a transmitter technique, error propagation at the receiver is avoided [3]. Moreover, channel coding schemes can be applied in the same way as for the ideal additive white Gaussian noise (AWGN) or flat fading channel.

The analogies between temporal equalization methods (in Single Input Single Output (SISO) channels) and their direct counterparts as spatial equalization methods (in MIMO channels) are depicted in Table I [2].

		ISI channel $H(z)$ (temporal Equalization)	MIMO channel $\boldsymbol{H}$ (spatial Equalization)
linear	at Rx	Linear equalization via $1/H(z)$	Linear equalization via $\boldsymbol{H}_t^{-1}$
	at Tx	Linear pre-equalization via $1/H(z)$	Linear pre-equalization via $\boldsymbol{H}_r^{-1}$
	at Tx / Rx	OFDM/DMT, vector precoding	SVD
Non-linear	at Rx	DFE	Matrix DFE
	at Tx / Rx	THP	MIMO-THP

Table 1. Corresponding Equalization Strategies for ISI Channels and MIMO Channels.

2.3 The Principle of THP

The information theory idea behind the THP is based on Costa's "writing on dirty paper result" for interference channels [4], which can be informally summarized as follows:
"*When transmitting over a channel, any interference which is known apriori to the transmitter does not affect the channel capacity. That is, by appropriate coding, transmission at a rate equal to the capacity of the channel without this interference is possible.*"
If we extend the Costa precoding concepts for multiple antenna with Co-Antenna Interference (CAI) then THP structure can be obtained [1, 3]. Consider these subchannels in some arbitrary order. In this case, the encoding for the first subchannel has to be performed accepting full interference from the remaining channels, since at this point the interference is unknown. For the second subchannel, however, if the transmitter is able to calculate the interference from the first subchannel, "Costa precoding" of the data is possible such that the interference from the first subchannel is taken into account. Generally, in the kth subchannel considered, Costa precoding is possible such that interference from subchannels 1 to k-1 is ineffective.
We can apply this result to the MIMO channel [5]: If the precoding operation contains a Costa precoder, no interference can be observed from lower number subchannels into higher number subchannels.
Note that it is possible to transform $\boldsymbol{H}$ into a lower triangular matrix with an orthonormal operation [6]. In this way interference from lower-index subchannels into higher-index subchannels is completely eliminated, and together with Costa precoding adjusted to this modified transmission matrix, effectively only a diagonal matrix remains for the transmission. It turns out that a simple scheme for Costa precoding works analog to the feedbackpart of DFE, now moved to the transmitter side and with the nonlinear decision device replaced by a modulo-operation. This is also known as THP [7, 8], and the link between THP and Costa precoding was first explored in [9].

2.4 MIMO-THP system model

The base station with n_T transmit antenna and n_R user (in which $n_T \le n_R$) with single antenna can be considered as MIMO broadcast system. A block diagram of this MIMO system together with THP is illustrated in Fig. 1 and is briefly explained here.
The n_T dimensional input symbol vector $\boldsymbol{a}$ passes through feedback filter $\boldsymbol{B}$, which is added to the intended transmit vector to pre-eliminate the interference from previous users.

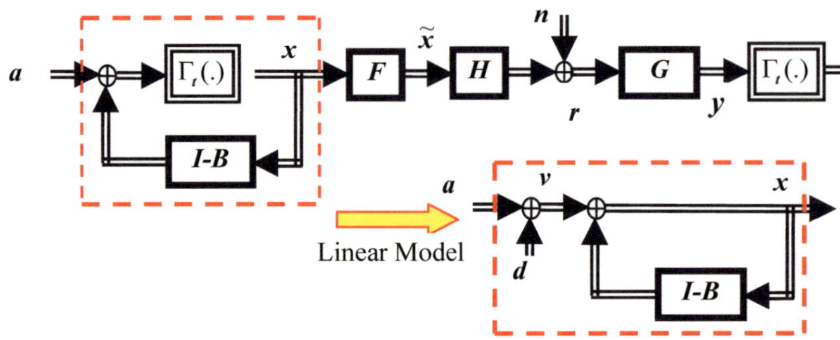

Fig. 1. THP model in a MIMO system

Then the resultant signal is fed to modulo-operator, which serve to limit the transmit power. The output signal of modulo-operator is then passed through a feed forward filter to further remove the interference from future users [10]. Finally, the precoded signal is launched in to the MIMO channel. As all interferences are taken care of at the transmitter side, the receivers at the mobile user side are left with some simple operations including power scaling (diagonal elements of matrix $\boldsymbol{G}$), reverse modulo-operation, and single user detection.
According to Fig. 1, the base band received signal can be modeled as:

$$\boldsymbol{r} = \boldsymbol{H}\widetilde{\boldsymbol{x}} + \boldsymbol{n} \tag{1}$$

where $\widetilde{\boldsymbol{x}} \in \boldsymbol{C}^{n_T \times 1}$, $\boldsymbol{r} \in \boldsymbol{C}^{n_R \times 1}$, $\boldsymbol{H} \in \boldsymbol{C}^{n_R \times n_T}$ and $\boldsymbol{n} \in \boldsymbol{C}^{n_R \times 1}$ are transmitted, received, channel and noise matrices, respectively (C denotes complex domain). The elements of the noise vector are assumed as independent complex Gaussian random variables with zero mean and variance σ^2, i.e., $\boldsymbol{n} \sim CN(0, \sigma^2 \mathbf{I}_{n_R})$. The elements of matrix $\boldsymbol{H}$ are considered as complex Gaussian random variables (i.e. flat fading case). In other words, the channel tap gain from transmit antenna i to receive antenna j is denoted by h_{ji} which is assumed to be independent zero mean complex Gaussian random variables of equal variance, that is $E[|h_{ji}|^2] = 1$.
The operation of THP is related to the employed signal constellation $\boldsymbol{A}$. Assume that in each of the parallel data streams an M-ary square constellation (M is a squared number) is employed where the coordinates of the signal points are odd integers, i.e., $\boldsymbol{A} = \{a_I + ja_Q | a_I, a_Q \in \{\pm 1, \pm 3, ..., \pm(\sqrt{M} - 1)\}\}$. Then the constellation is bound by the square region of side length $t = 2\sqrt{M}$ which is needed for modular operation [3].
Note: In the rest of the chapter, for means of simplicity, the number of transmit and receive antennas are assumed to be the same (i.e., $n_T = n_R = K$). Also, we consider the flat fading case. Whenever these assumptions are not acceptable we clarify them.
The lower triangular feedback matrix $\boldsymbol{B}$, unitary feed forward matrix $\boldsymbol{F}$ and diagonal scaling matrix $\boldsymbol{G}$ can be found by ZF or MMSE criteria as [11]. The received signal before modulo reduction can be given as:

$$\boldsymbol{y} = \boldsymbol{G}\boldsymbol{r} = \boldsymbol{G}\boldsymbol{H}\boldsymbol{F}\boldsymbol{B}^{-1}\boldsymbol{v} + \widetilde{\boldsymbol{n}} \tag{2}$$

where $\widetilde{\boldsymbol{n}} = \boldsymbol{G}\boldsymbol{n}$,and $\boldsymbol{v} = \boldsymbol{a} + \boldsymbol{d}$ is effective input data, and $\boldsymbol{d}$ is the precoding vector used to constrain the value of $\widetilde{\boldsymbol{x}}$ [13]. If ZF criterion is used, it requires $\boldsymbol{G}\boldsymbol{H}\boldsymbol{F}\boldsymbol{B}^{-1} = \boldsymbol{I}$. Thus, the processing matrices $\boldsymbol{G}$, $\boldsymbol{B}$, and $\boldsymbol{F}$ can be found by performing Cholesky factorization of $\boldsymbol{HH}^{\mathrm{H}}$ as [13]:

$$\begin{aligned} \boldsymbol{HH}^H &= \boldsymbol{RR}^H \\ \boldsymbol{G} &= \boldsymbol{diag}(r_{11}^{-1},...,r_{KK}^{-1}) \\ \boldsymbol{B} &= \boldsymbol{GR} \\ \boldsymbol{F} &= \boldsymbol{H}^{-1}\boldsymbol{R} \end{aligned} \tag{3}$$

where $\boldsymbol{R} = [r_{ij}]$ is a lower triangular matrix. The error covariance matrix can be shown as:

$$\Phi_{\tilde{n}\tilde{n}} = \mathrm{E}[(\boldsymbol{Gn})(\boldsymbol{Gn})^H] = \mathrm{diag}[\sigma_n^2 / r_{11}^2,...,\sigma_n^2 / r_{KK}^2] \tag{4}$$

i.e, the noise is white.
If MMSE criterion is used the matrix $\boldsymbol{R}$ can be found through Cholesky factorization of [5]:

$$(\boldsymbol{H}^H\boldsymbol{H} + \zeta\boldsymbol{I}) = \boldsymbol{R}^H\boldsymbol{R} \tag{5}$$

where $\zeta = \sigma_n^2 / \sigma_a^2$. The matrices $\boldsymbol{G}$, $\boldsymbol{B}$ and $\boldsymbol{F}$ can be found as:

$$\begin{aligned} \boldsymbol{G} &= \mathrm{diag}[r_{11}^{-1},...,r_{KK}^{-1}] \\ \boldsymbol{B} &= \boldsymbol{GR} \\ \boldsymbol{F} &= \boldsymbol{R}^{-H}\boldsymbol{H}^H \end{aligned} \tag{6}$$

The error covariance can be shown as:

$$\Phi_{ee} = \mathrm{E}[\boldsymbol{ee}^H] = \sigma_n^2\boldsymbol{G}^2 = \boldsymbol{diag}[\sigma_n^2 / r_{11}^2,...,\sigma_n^2 / r_{KK}^2] \tag{7}$$

i.e. error can be considered as white.
In outdated CSI case, the system model, which is considered in Fig. 1, operates in a feedback channel where the CSI is measured in downlink and fed to the transmitter in uplink channel. Time variations of channel lead to a significant outdated (partial) CSI at the transmitter. In fact there will always be a delay between the moment a channel realization is observed and the moment it is actually used by the transmitter. The effect of time variations (or delay) can be considered as: $\boldsymbol{H} = \hat{\boldsymbol{H}} + \Delta\boldsymbol{H}$, where $\boldsymbol{H}, \hat{\boldsymbol{H}}$ and $\Delta\boldsymbol{H}$ are true, estimated and channel error due to time variations [13]. We assume that the channel error has Gaussian probability density function with moments $\boldsymbol{E}[\Delta\boldsymbol{H}] = 0$ and $\boldsymbol{E}[\Delta\boldsymbol{H}.\Delta\boldsymbol{H}^H] = \boldsymbol{C}_{\Delta H}$.
According to Fig. 1, the received signal can be considered as:

$$\boldsymbol{y} = \boldsymbol{Gr} = \boldsymbol{G}(\hat{\boldsymbol{H}} + \Delta\boldsymbol{H})\boldsymbol{FB}^{-1}\boldsymbol{v} + \tilde{\boldsymbol{n}} \tag{8}$$

where $\tilde{\boldsymbol{n}} = \boldsymbol{Gn}$ and $\boldsymbol{v}$ is effective data vector [12]. If ZF criterion is used, it requires:

$$\boldsymbol{G}\hat{\boldsymbol{H}}\boldsymbol{FB}^{-1} = \boldsymbol{I} \tag{9}$$

The processing matrices $\boldsymbol{R}, \boldsymbol{G}, \boldsymbol{B}$ and $\boldsymbol{F}$ can be found by doing Cholesky factorization of $\hat{\boldsymbol{H}}\hat{\boldsymbol{H}}^H$ as [11]:

$$\begin{aligned} \hat{H}\hat{H}^H &= RR^H \\ G &= diag(r_{11}^{-1},...,r_{KK}^{-1}) \\ B &= GR \\ F &= \hat{H}^{-1}R \end{aligned} \tag{10}$$

where $e = w + \tilde{n} = G\Delta HFx + Gn$ is considered as an error vector and the term w stands for channel imperfection effect due to outdated CSI. The error covariance matrix can be obtained as:

$$\Phi_{ee} = E[ee^H] = \sigma_x^2 G(C_{\Delta H} + \zeta I)G^H \tag{11}$$

Note that, with a small channel error assumption (i.e. $C_{\Delta H} \to 0$), the error covariance matrix in an imperfect case tends to the error covariance matrix in a perfect case, i.e.

$$\Phi_{ee} = \sigma_x^2 Diag(1/r_{11}^2, 1/r_{22}^2, ..., 1/r_{KK}^2) \tag{12}$$

3. MIMO-THP capacity

The first attempt to calculation of achievable rates of THP is done by Wesel and Cioffi in [15]. The authors considered THP for discrete-time SISO consists of Inter-Symbol Interference (ISI) and AWGN. They derived an exact expression for maximum achievable information rate for ZF case and provided information bound for MMSE case. In this section, we develop the achievable rates analysis provided in [15] for MIMO-THP in flat fading channel. We obtain the maximum achievable rate and some upper and lower bounds of it for ZF and MMSE cases with perfect and imperfect CSI.

3.1 Achievable rates of point-to-point MIMO-THP

Consider a point-to-point MIMO system with THP as Fig. 2.

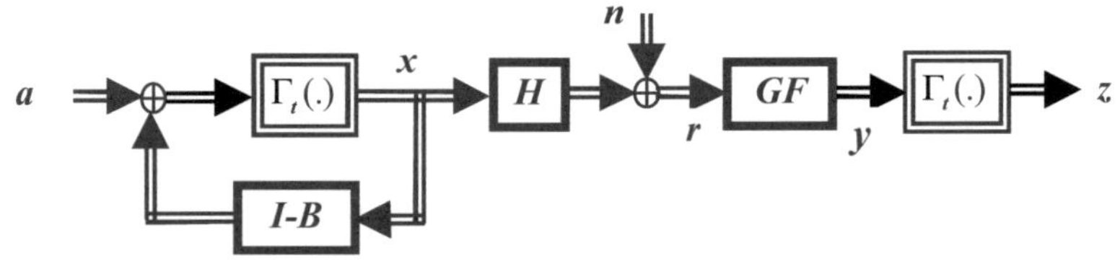

Fig. 2. THP model in a point-to-point MIMO system

The received signal vector can be expressed as:

$$z = \Gamma_t[GFHB^{-1}v + GFn] = \Gamma_t[a + w + n'] \tag{13}$$

where w is residual spatial interference after MMSE criterion on THP filters (in the ZF case $w = 0$) and $\Gamma_t(.)$ is modulo t operator so eliminate its output on interval $T = [-t_k/2, t_k/2) \times [-jt_k/2, jt_k/2)$. As $n' = GFn$ is white Gaussian noise and with the assumption that $w_i \perp w_j$ & $a_i \perp w_j$ for $\forall i \neq j$ (so that symbol $\perp$ stand for statistical independence) the received vector $\boldsymbol{r}$ can be decoupled in K parallel streams as[1] [17]:

$$z_k = \Gamma_t \left[a_k + w_k + n'_k\right] \;\; ; \;\; k = 1 \sim K \tag{14}$$

Because of the decoupling of the received information symbols in (14) and assuming independence between elements in $\boldsymbol{a}$ the mutual information between the transmitted symbols and the received signal vector can be expressed as the sum of the mutual information between elements of each vector:

$$I(\boldsymbol{a};\boldsymbol{z}) = \sum_{k=1}^{K} I(a_k; z_k) \tag{15}$$

where $I(.)$ denote mutual information. Each term in the sum is independently can be considered as:

$$\begin{aligned} I(a_k; z_k) &= h(z_k) - h(z_k \mid a_k) = h[\Gamma_t(a_k + w_k + n'_k)] - h[\Gamma_t\,((a_k + w_k + n'_k) \mid a_k)] \\ &= h[\Gamma_t(a_k + w_k + n'_k)] - h[\Gamma_t\,((w_k + n'_k) \mid a_k)] \end{aligned} \tag{16}$$

where $h(.)$ denotes differential entropy. Calculation of the above mutual information seems to be difficult and we try to find an upper and lower bound of (16) by some approximations.
Remark 1: An upper bound on the achievable rate of the channel produced by MMSE-THP of (16) can be found as [17]:

$$I_{Upper}(\boldsymbol{a};\boldsymbol{z}) = \sum_{k=1}^{K} \left(2\log_2(t) - h\left[\Gamma_t\,(n'_k)\right]\right) \tag{17}$$

Also, the upper bound can be obtained essentially by neglecting the spatial interference term w_k in (16) [17]. The lower bound depends largely on the variance of w_k [15]. A lower bound on achievable rate can be found as [17]:

$$I_{Lower}(\boldsymbol{a};\boldsymbol{z}) = \sum_{k=1}^{K} 2\log_2(t) - \log_2(2\pi\sigma^2) - \frac{\gamma^2}{\sigma^2}\log_2 e - 2\log_2(2erf(t/2\sigma)) \tag{18}$$

Thus, a truncated Gaussian [17] with variance of $\gamma^2 = \mathrm{var}(w_k + \Gamma_t(n'_k))$ produces a slightly tighter bound but requires the computation of $\mathrm{var}(\Gamma_t(n'_k))$.
Remark 2: The upper bound attained in (17) can be simplified if some approximations are allowed so that a quasi-optimal (or sub-optimal) closed form solution can be found. This approximations can be done based on the value of t/σ (See [17]).

3.2 General THP in point-to-point MIMO with perfect CSI

Whenever CSI is available at the transmitter in a communication system, since the transmitter has knowledge of the way the transmitted symbols are attenuated and distributed by the channel, it may adjust transmit rate and/or power in an optimized way.

---

[1] For MMSE case the above assumption for high value of SNR is acceptable and the above results can be true in asymptotic case, so MMSE performance for high SNR values converge to ZF [2].

For instance, in the multi-antenna scenario some of the equivalent parallel channels might have very bad transmission properties or might not be present at all. In this situation, the transmitter might want to adjust to that by either dropping some of the lower diversity order sub channels or by redistributing the data and the available transmission power to improve the average error rate. This can be done by generalization of THP concepts as GTHP by enabling different power transmission for each antenna. GTHP can be done in two main scenarios [17]:

First: If the loading is made according to capacity of system; this structure enables different transmission rate per antenna.

Second: If it is needed to ensure reliable transmission rate for each antenna, the loading should be made according to minimize error rate of system.

Here we consider two different optimization scenarios for loading strategies of THP and extend it's concept in structure that t is not constant, so the modulo interval is different for each sub channel (t_k) [17].

3.2.1 Capacity criterion

In this section, the power adaptation strategy of the second type of GTHP concept is employed. The optimal power allocation is calculated in MIMO-GTHP systems, while regarding the modulation schemes is given. If the loading is made according to capacity of system, this structure enables different transmission rate per antenna. One of the good features of this scenario is that it is scalable architecture, because it allows adding or removing transmitters without losing the precoding structure as explained in [16].

If a assumed as an i.i.d. uniform distribution on T, for such a case, x is also i.i.d. uniform on T (regardless of the choice of matrix B). Thus the transmitted power from kth antenna will be $p_k = E\{|x_k|^2\} = t_k^2/6$ (for real case $p_k = t_k^2/12$). Its corresponding rate will equal the maximum achievable mutual information in (17):

$$I(t_k) = I(a_k; z_k, t_k) = 2\log_2(t_k) - h[\Gamma_{t_k}(n'_k)] \tag{19}$$

Then the maximum achievable rate for a system with THP will be the maximum of the sum of the rates of each stream subject to a maximum total transmitted power constraint, i.e.

$$\begin{cases} C_{General} = \max\limits_{\{t_k\}} \sum\limits_{k=1}^{K} I(t_k) = \max\limits_{\{t_k\}} \sum\limits_{k=1}^{K} (2\log_2(t_k) - h[\Gamma_{t_k}(n'_k)]) \\ s.t. \quad \sum\limits_{k=1}^{K} p_k = \frac{1}{6}\sum\limits_{k=1}^{K} t_k^2 = P_T \end{cases} \tag{20}$$

In order to maximize (20) we assume that all the available streams are classified into two groups[2] (g-Gaussian and u-uniform) based on t/σ values [16]. As shown in [17], the achievable rate of streams belonging to u tends to zero; no power is assigned to these streams, i.e. $t_k = 0\ \forall k \in u$. Thus the solution of the maximization problem in (20) can be found as assigning the same power to the entire stream in g (and no power to those in u). The optimal solution can be shown to be [17]:

---

[2] Based on the value of t/σ for each stream

$$C_{General}^{Adaptive} = \sum_{k\in g} I_k\left(\sqrt{\frac{6p_T}{|g|\sigma_{n'_k}^2}}\right) \quad , \quad k \in g \tag{21}$$

where $|g|$ denotes the number of active antennas and $\sigma_{n'_k}^2$ is variance of n'_k. Then some kind of adaptive rate algorithm is necessary to achieve the maximum capacity of the GTHP.

3.2.2 Minimum SER criterion

In some application it is needed to ensure reliable transmission rate for each antenna (especially in MIMO broadcast channels). In this section we try to find the optimal sub channel power allocation in MIMO GTHP systems, while regarding the modulation schemes is given. As mentioned, for each sub channel we have:

$$z_k = a_k + n'_k \qquad k = 1,2,...,K \tag{22}$$

where we assumed that $\boldsymbol{w}_k$ tend to zero. For simplicity assume MQAM transmission in all sub channels is used. In this case the approximate average SER for a fixed channel $\boldsymbol{H}$ simply given as [17]:

$$SER \approx \sum_{k=1}^{K}(1-\frac{1}{\sqrt{M_k}})Q(\sqrt{\frac{3}{M_k-1}\frac{|r_{kk}|^2}{\sigma_{n'}^2}E_k}) \tag{23}$$

where $B_k = \frac{3}{M_k-1}\frac{|r_{kk}|^2}{\sqrt{M_k}}$ and we assumed modulation order (i.e. M_k) can be varied for each sub-channel, so that variable bit allocation is possible (that we didn't consider here). In this case we have [17]:

$$E_k = \frac{1}{B_k}W(\frac{B_k}{A_k}\Lambda) \tag{24}$$

where $A_k = \frac{M_k(M_k-1)}{(\sqrt{M_k}-1)^2|r_{kk}|^2}$ and $W(x)$ is the real valued Lambert's W-function defined as the inverse of the function $f(x)=x.e^x$; $x \geq 0$, i.e., $W(x)=a \Leftrightarrow a.e^a = x$.

Since the $W(x)$ function is real and monotonically increasing for real $x > -1/e$, the value of λ such that $\sum_{k=1}^{K}E_k(\lambda) - K = 0$ holds which can be found by using some classical methods as denoted in [17]. On the other hand, $W(x)$ is a concave and unbounded function with $W(0)=0$ and $W(x) \leq x$, the unique solution for $\boldsymbol{E} = [E_1,...,E_K]^T$ can be found by the following simple iterative procedure[14]:

i. Chose a small positive Λ which satisfy

$$\sum_{k=1}^{K}\frac{\Lambda}{A_k} \leq E_T \tag{25}$$

ii. Calculate

$$\hat{E}_T = \sum_{k=1}^{K} \frac{1}{B_k} W(\frac{B_k}{A_k} \Lambda) \tag{26}$$

iii. If $\hat{E}_T$ is not yet sufficiently close to E_T, multiply Λ by $E_T/\hat{E}_T$ and go back to step (ii).

iv. Compute $\boldsymbol{E} = [E_1,...,E_K]^T$ according to (24).

Note that since $W(x)$ for $x > -1/e$ is monotonic function, then according to relation (24) the highest power $(\max E_k)$ assign to the weakest signal so that the *SNR* value almost stay constant for all sub channels.

3.3 Achievable rate in imperfect CSI

In [17] the scheme proposed in [18] for MIMO THP system was modified by allowing variations of the transmitted power in each antenna. The authors stated the problem of finding the maximum achievable rate for this modified spatial THP scheme and found that Uniform Power Allocation (UPA) with antenna selection is a quasi-optimal transmission scheme with a perfect CSI.

In this sub-section, based on previous researches about SISO and point-to-point MIMO channels, an analytical approach to attain the maximum achievable rate bound in an imperfect CSI case is developed for broadcast channel. It will be shown that this bound depends on the variance of the residual Co-Antenna Interference (CAI) term. Moreover, it will be shown that the power allocation obtained by the UPA in [17] is sub-optimal in an imperfect CSI, too.

3.3.1 Maximum achievable rates

The received signal after modulo operation can be considered as $z = a + \Gamma_{t_k}[G\Delta HFx + \tilde{n}]$. Since x has i.i.d. distribution, $W = G\Delta HFx$ can be considered as an unknown interference with an i.i.d. distribution. Also, for such an a, z is i.i.d. uniform on T. In this case, the received information can be decoupled in K independent parallel data steams and the mutual information between k^{th} element of data vector, a_k, and the corresponding element of the received signal, z_k, is [13]:

$$I(a_k; z_k) = h(z_k) - h(z_k \mid a_k) = h(z_k) - h[\Gamma_{t_k}((w_k + \tilde{n}_k) \mid a_k)] \le \log_2(6p_k) - h\left[\Gamma_{t_k}\left(\sum_{j=1}^{K} \frac{\delta'_{kj}}{r_{kk}} x_j + \tilde{n}_k\right)\right] \tag{27}$$

where $\delta'_{kj} = [\Delta \boldsymbol{HF}]_{kj}$ and $h(.)$ denotes differential entropy. Let us define the random variable e_k as $e_k = \sum_{j=1}^{K} \frac{\delta'_{kj}}{r_{kk}} x_j + \tilde{n}_k$ where its power is $\sigma_{ek}^2 = \frac{\sigma_n^2}{r_{kk}^2} + \sum_{j=1}^{K} \frac{p_j}{r_{kk}^2} \left|\delta_{kj}\right|^2$ and $\delta_{kj} = [\Delta \boldsymbol{H}]_{kj}$. With the assumption of small error, e_k can be approximately modeled as a complex Gaussian random variable. In the case where, the above assumption is true, the mutual information expression (27) can be very well approximated as [13]:

$$I(a_k; z_k) \approx \log_2 [\chi_k]^+ \tag{28}$$

where

$$\begin{cases} \chi_k = 6r_{kk}^2 p_k / \pi e(\sigma_n^2 + \sum_{j=1}^{K} |\delta_{kj}|^2 p_j) \\ \log[x]^+ = \max[\log(x),0] \end{cases} \tag{29}$$

The achievable rates for THP in an imperfect CSI case will then be the sum of the mutual information of all K parallel steams as [13]:

$$\begin{cases} C = \max\limits_{\{p_k\}} \sum_{k=1}^{K} I(a_k;z_k) \approx \max\limits_{\{p_k\}} \sum_{k=1}^{K} \log_2[\chi_k]^+ \\ s.t. \quad \sum_{k=1}^{K} p_k = P_T \end{cases} \tag{30}$$

Observed that C (or χ_k) depends on three components: $|\delta_{kj}|^2, r_{kk}^2$ and p_k. In order to maximize χ_k, some kind of spatial ordering is necessary in order to maximize it. For this purpose, it is required to decompose $\boldsymbol{H}$ (in Cholesky factorization) so that the elements of r_{kk}^2 to be maximized (finding the ordering matrix similar to [11]).

On the other hand, it was assumed that by making small error assumption, e_k can be approximately modeled as a complex Gaussian random variable. This is equivalent to assuming $\max\limits_{j} |\delta_{kj}|^2 p_j \le \sigma_n^2 ; \forall j$. Now, we assume that the entries of error matrix are bounded as $\max\limits_{k,j} |\delta_{kj}|^2 \le \alpha_{kj} ; \forall k,j$ [13]. In addition, for the sake of simplicity and without loss of generality, we assume that $\alpha = \max\limits_{k,j} \alpha_{kj}$. Then, the power distribution that will maximize the achievable rates will be the solution of the following maximin problem:

$$\begin{cases} C = \max\limits_{p_i} \min\limits_{\delta_{ij}} \sum_{k=1}^{K} \log_2[\chi_k]^+ \\ s.t. \quad \sum_{k=1}^{K} p_k = p_T \ , \ \max\limits_{i,j} |\delta_{ij}|^2 \le \alpha \ ; \ \forall i,j \end{cases} \tag{31}$$

In order to solve the above maximin problem the worst-case is assumed, i.e. $\alpha = \max\limits_{i,j} |\delta_{ij}|^2 ; \forall i,j$. With this assumption, the minimum mutual information will be attained for each term in the summation. Then, the resulting maximization problem leads to [13]:

$$\begin{cases} C = \max\limits_{p_i} \sum_{k=1}^{K} \log_2 \left[\dfrac{6 p_k r_{kk}^2}{\pi e(\sigma_n^2 + \alpha p_T)} \right]^+ \\ s.t. \quad \sum_{k=1}^{K} p_k = p_T \end{cases} \tag{32}$$

The resulting maximization problem is a standard constrained optimization problem, and can be solved with the use of the Lagrange method in which the solution result is $p_k = const.$ It means that the p_k is independent of k, i.e the distribution of the power, in worst-case, is UPA.

Note that, if we consider different noises with different powers for each user, the distribution of power may not be the UPA.

3.3.2 Capacity loss

In the previous section, it is shown that the capacity of MIMO-THP can be obtained by the UPA. More over, it can be observed from (32) that this capacity, in worst case, depends on the channel error value (i.e. α). We define the capacity loss as difference between the capacity of MIMO-BC-THP in a perfect CSI and in an imperfect CSI, i.e. relation (32), as [19]:

$$\Delta\hat{C} = C_1 - \hat{C}_2 = \sum_{k=1}^{K}\log_2\left[\frac{6p_k r_{kk}^2}{\pi e \sigma_n^2}\right] - \sum_{k=1}^{K}\log_2\left[\frac{6p_k r_{kk}^2}{\pi e(\sigma_n^2 + \alpha\, p_T)}\right] \leq K\log_2[1+K] \quad (33)$$

The above bound for capacity loss only is valid for values of α so that the approximation of $\max_j |\delta_{kj}|^2 p_j \leq \sigma_n^2 ;\forall j$ is valid [19]. It means that this bound depends on SNR value and is acceptable for high SNR value, i.e., this capacity loss bound is asymptotic bound for worst-case in which bounds the capacity loss of MIMO-THP . It is desired to bounding the capacity loss of optimal solution of (29). Assume C_2 is the capacity of optimal solution that can be obtained by exactly analysis or by numerical simulation. In this case we can bound $\Delta C = C_1 - C_2$ as [19]:

$$\Delta C = K\log_2[1+K] - K\log_2\left[\left(1+\frac{p_T}{\sigma_n^2}\alpha\right)\right] + \varepsilon \quad (34)$$

where ε is a positive value. The lower bound can be obtained by choosing $\varepsilon = K-2$ [19]:

$$\Delta C \geq \left[K\log_2[1+K] - K\log_2\left[\left(1+\frac{p_T}{\sigma_n^2}\alpha\right)\right] + (K-2)\right]^+ \quad (35)$$

where $[x]^+ = \max[x,0]$. In simulation we refer (35) as theoretic loss.

3.4 Spatial ordering

The VBLAST-Like ordering can be used in order to improve the power loading performance of MIMO-GTHP system in Fig. 1 [1]. To do this, since the loading is based on the SNR values of the equivalent parallel sub-channels, which in turn are proportional to $|r_{kk}|^2$, the distribution of these diagonal entries is an essential parameter in power loading performance. It turns out that by introducing a permutation matrix in the decomposition of ***H***, i.e, allowing different ordering of the sub channels, the distribution of the $|r_{kk}|^2$ values can be modified as [1]:

$$\boldsymbol{P}_{opt} = \arg\min_{\boldsymbol{P}} (1/|r_{11}|^2, 1/|r_{22}|^2, \ldots, 1/|r_{KK}|^2) = \arg\min_{\boldsymbol{P}}(|\boldsymbol{G}|^2)$$

It means that in the cholesky factorization of (4), the decomposition should be made so that the square value of diagonal elements of matrix ***R*** minimized. It means that the matrix ***P*** is selected so that the column of ***H*** corresponding to minimum square value of diagonal elements of ***G*** is permuted to the left. Deleting this column from the matrix ***H***, and forming the cholesky factorization of this modified matrix, we can obtain second column of matrix **P**. Continuing this way, constantly updating **P**, the decomposition of **H** is constructed. The pseudo-code for the algorithm is given in Fig. (3).

$$
\begin{aligned}
&\textit{Initialization:} \\
&\qquad \mathbf{P} = [p_{i,j}] = 0 \\
&\textit{for } i = 1 \textit{ to } K \textit{ do:} \\
&\qquad [\mathbf{Q}\ \ \mathbf{R}] = qr(\mathbf{H}) \quad ; \quad \textit{for ZF} \\
&\qquad \mathbf{R} = chol(\mathbf{H}\mathbf{H}^H + \zeta\mathbf{I}) \quad ; \quad \textit{for MMSE} \\
&\qquad \mathbf{G} = diag(inv(\mathbf{R}))\hat{}2 \\
&\qquad l_i = \arg\min_j (g_{i,j}) \\
&\qquad P_{l_i,i} = 1 \\
&\qquad \mathbf{H}_{:,l_i} = zeros(size(\mathbf{H}_{:,l_i})) \\
&\textit{end}
\end{aligned}
$$

Fig. 3. The pseudo code for ordering

Note that for ZF or MMSE-THP, the system performance will be dominated by the signal component with the largest noise variance, and we can find the ordering algorithm in the minimax noise variance sense as [1].

3.5 Simulation and results

3.5.1 Perfect CSI

The mutual information for real x_k with i.i.d uniform distribution on the module interval [-t/2, t/2) is plotted in fig. 4, where the average transmitter energy is $t^2/12$. This figure also shows the mutual information curves for the upper (17) and lower (18) bounds for each sub channel. For comparison we also plotted the well-known AWGN channel capacity (with no ISI). Observe that the upper bound lies above the AWGN capacity and lower bound lies below this capacity (especially for high SNR values).

Figs. 5 and 6 give the performance comparison of the MMSE-GTHP with/without power loading (relation (24)) when 4QAM and 16QAM modulations are used, respectively. From these figures, it is clearly seen that the MMSE-GTHP with ordering can achieve better performance than the MMSE-GTHP with or without power allocation (4QAM or 16QAM,). When 4QAM modulation is used, at BER=2×10^{-4} we can observe that the MMSE-GTHP with power allocation achieves about 7dB gain, while this structure with power loading and ordering gives about 11.5dB gain. When 16QAM is used, at BER=5×10^{-4} the MMSE-GTHP with power allocation gives approximately 6.5dB gain, while this structure with power loading and ordering gives about 10dB gain. As can be seen from these figures, the performance of power loading is noticeable, especially when it combined with sub-channel ordering.

3.5.2 Imperfect CSI

For simulation purposes we have considered K=4 users. The entries of $\hat{\mathbf{H}}$ and $\Delta\mathbf{H}$ have been assumed to be zero mean i.i.d. complex Gaussian random variables, i.e., $\hat{\mathbf{H}} \sim CN(0,1)$

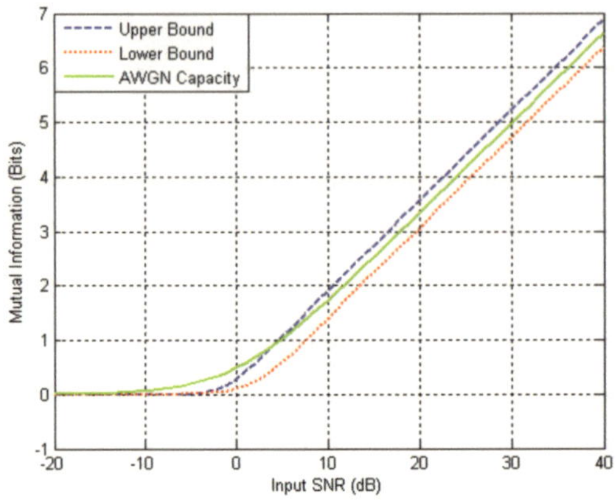

Fig. 4. Upper and lower bound of mutual information

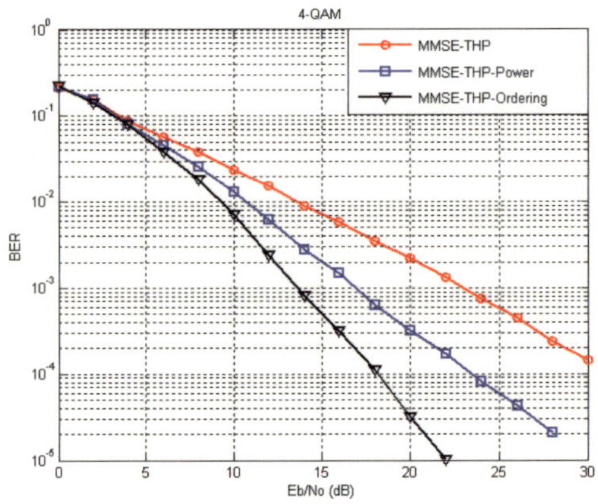

Fig. 5. Performance comparison of MMSE-GTHPwith power loading and ordering for 4QAM.

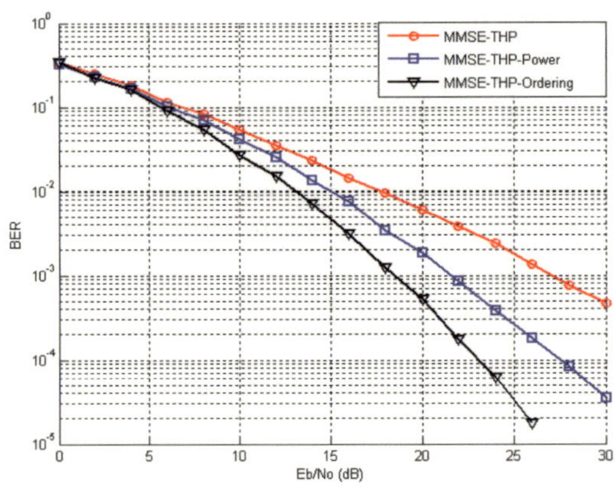

Fig. 6. Performance comparison of MMSE-GTHP. with power loading and ordering for 16QAM

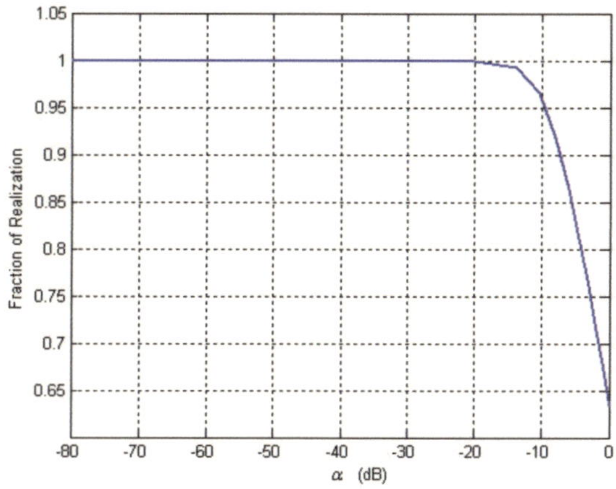

Fig. 7. Validation of approximation of α

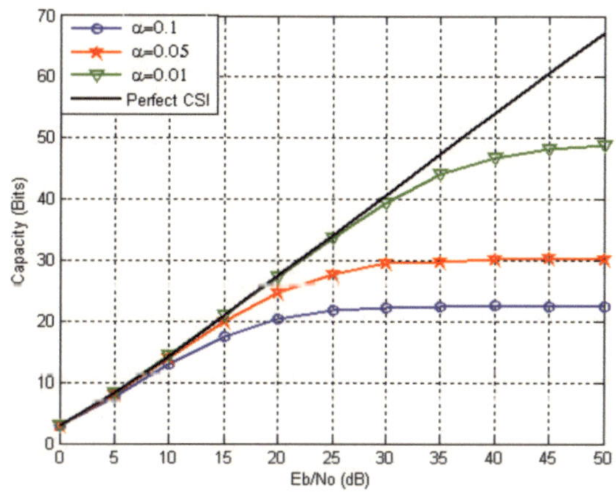

Fig. 8. Capacity with K=4 user and different value of α

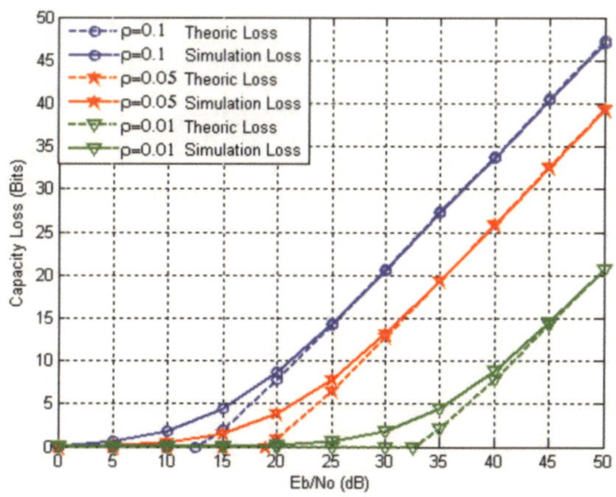

Fig. 9. Capacity loss for K=4 user

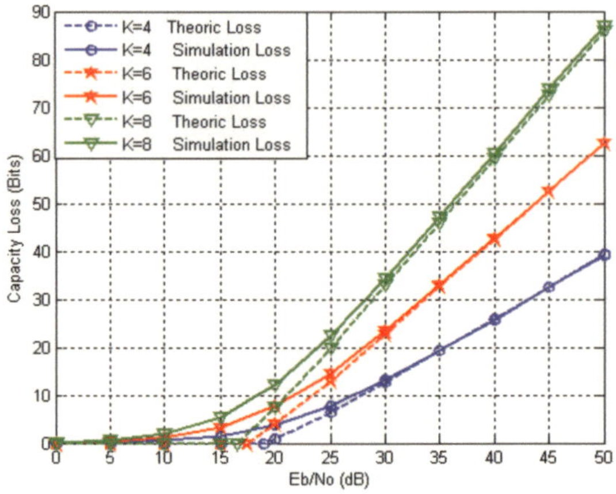

Fig. 10. Capacity loss for different user and $\rho = 0.05$

and $\Delta \boldsymbol{H} \sim CN(0,\rho^2)$. The validity of the approximations of $(\alpha p_j \le \sigma_n^2 ;\forall j)$ for $P_T / \sigma_n^2 = 16\ dB$ is shown in Fig. 7 by plotting the fraction of channel realizations in which the approximations are valid for different values of α . The simulation is done for more than 10^4 channel realization and the validity is calculated as the number of iteration where the inequality is valid. It can be seen that, for the particular values of the simulation parameters taken in this section, the capacity analysis is valid for values of α up to −10 dB. In Fig. 8, we have plotted the mean of the maximum achievable rates for the UPA scheme for different values of α . It can be seen that for values of $\alpha \le 0.01$ the capacity loss due to the presence of channel errors are negligible and by increasing the SNR value, the capacity increases up to a constant value. It means that by increasing the SNR value the capacity remains almost constant. Figs. 9 and 10 depict asymptotic capacity loss in a time varying channel for different values of ρ and different users, respectively. It can be seen that the attained aproximated bound is valid for variety of rang of the channel errors and number of users, especially for high SNR value (asymptotic case).

4. Robust MIMO-THP

4.1 Design criterion

The error, that is needed to be considered for the system illustrated in Fig. 1, should be the difference between the effective data vector $\boldsymbol{v}$ and the data vector entering the decision module $\boldsymbol{y}$, i.e.:

$$\boldsymbol{e} = \boldsymbol{y} - \boldsymbol{v} = [\boldsymbol{G}(\hat{\boldsymbol{H}} + \Delta\boldsymbol{H})\boldsymbol{F} - \boldsymbol{B}]\boldsymbol{x} + \tilde{\boldsymbol{n}} \tag{36}$$

The MMSE solution should minimize the error signal as:

$$\begin{cases} \arg\min\limits_{\boldsymbol{B},\boldsymbol{F},\boldsymbol{G}} E\left\| [\boldsymbol{G}(\hat{\boldsymbol{H}} + \Delta\boldsymbol{H})\boldsymbol{F} - \boldsymbol{B}]\boldsymbol{x} + \tilde{\boldsymbol{n}} \right\|^2 \\ s.t. \quad E\left\|\tilde{\boldsymbol{x}}\right\|^2 \le P_T \end{cases} \tag{37}$$

Instead of solving (37), it is easier to use the orthogonallity principle [1]. In this case, the MMSE solution should satisfy:

$$\mathrm{E}[\boldsymbol{e}\boldsymbol{r}^H]=0 \tag{38}$$

Thus, according to (36) we have:

$$\boldsymbol{G}\Phi_{rr}=\boldsymbol{B}\Phi_{xr} \tag{39}$$

The matrices Φ_{rr} and Φ_{xr} can be computed by using (36) as [13]:

$$\Phi_{rr}=\mathrm{E}[\boldsymbol{r}\boldsymbol{r}^H]=\sigma_x^2(\hat{\boldsymbol{H}}\hat{\boldsymbol{H}}^H+\zeta\boldsymbol{I}+\boldsymbol{C}_{\Delta H}) \tag{40.a}$$

$$\Phi_{xr}=\mathrm{E}[\boldsymbol{x}\boldsymbol{r}^H]=\sigma_x^2\boldsymbol{F}^H\hat{\boldsymbol{H}}^H \tag{40.b}$$

where $\zeta=\sigma_n^2/\sigma_x^2$. Substituting relation (40) for (39) and some manipulation, yielding [13]:

$$\boldsymbol{F}^H=\boldsymbol{B}^{-1}\boldsymbol{G}(\hat{\boldsymbol{H}}\hat{\boldsymbol{H}}^H+\zeta\boldsymbol{I}+\boldsymbol{C}_{\Delta H})\hat{\boldsymbol{H}}^{-H} \tag{41}$$

Since **F** is a unitary matrix [1]:

$$\boldsymbol{R}\boldsymbol{R}^H=(\hat{\boldsymbol{H}}\hat{\boldsymbol{H}}^H+\zeta\boldsymbol{I}+\boldsymbol{C}_{\Delta H})\hat{\boldsymbol{H}}^{-H}\hat{\boldsymbol{H}}^{-1}(\hat{\boldsymbol{H}}\hat{\boldsymbol{H}}^H+\zeta\boldsymbol{I}+\boldsymbol{C}_{\Delta H}) \tag{42}$$

where $\boldsymbol{R}=\boldsymbol{G}^{-1}\boldsymbol{B}$ is assumed. The matrix **R** can be found through Cholesky factorization of (42) and the matrices **G, B,** and **F** can be obtained as [13]:

$$\begin{aligned}&\boldsymbol{G}=\mathrm{diag}[r_{11}^{-1},\ldots,r_{KK}^{-1}]\\&\boldsymbol{B}=\boldsymbol{G}\boldsymbol{R}\\&\boldsymbol{F}=\hat{\boldsymbol{H}}^{-1}(\hat{\boldsymbol{H}}\hat{\boldsymbol{H}}^H+\zeta\boldsymbol{I}+\boldsymbol{C}_{\Delta H})\boldsymbol{R}^{-H}\end{aligned} \tag{43}$$

Using the **B, F** and **G** obtained in (43), the error covariance matrix can be computed as [13]:

$$\Phi_{ee}=E[\boldsymbol{e}\boldsymbol{e}^H]=\sigma_x^2\boldsymbol{G}(\zeta^2\hat{\boldsymbol{H}}^{-H}\hat{\boldsymbol{H}}^{-1}+\zeta\boldsymbol{I}+\boldsymbol{C}_{\Delta H})\boldsymbol{G}^H \tag{44}$$

It means that the error covariance matrix in imperfect CSI is the sum of covariance matrix in perfect CSI plus the term $\sigma_x^2\boldsymbol{G}\boldsymbol{C}_{\Delta H}\boldsymbol{G}^H$. This term tends to zero with low channel error assumption (perfect CSI), i.e.:

$$\Phi_{ee}=E[\boldsymbol{e}\boldsymbol{e}^H]=\sigma_n^2\boldsymbol{G}(\zeta\hat{\boldsymbol{H}}^{-H}\hat{\boldsymbol{H}}^{-1}+\boldsymbol{I})\boldsymbol{G}^H \tag{45}$$

In this case, **R** can be found through:

$$\boldsymbol{R}\boldsymbol{R}^H=(\hat{\boldsymbol{H}}\hat{\boldsymbol{H}}^H+\zeta\boldsymbol{I})\hat{\boldsymbol{H}}^{-H}\hat{\boldsymbol{H}}^{-1}(\hat{\boldsymbol{H}}\hat{\boldsymbol{H}}^H+\zeta\boldsymbol{I}) \tag{46}$$

and the matrices **G, B** and **F** can be computed as:

$$\begin{aligned}&\boldsymbol{G}=\mathrm{diag}[r_{11}^{-1},\ldots,r_{KK}^{-1}]\\&\boldsymbol{B}=\boldsymbol{G}\boldsymbol{R}\\&\boldsymbol{F}^H=\hat{\boldsymbol{H}}^{-1}(\hat{\boldsymbol{H}}\hat{\boldsymbol{H}}^H+\zeta\boldsymbol{I})\boldsymbol{R}^{-H}\end{aligned} \tag{47}$$

The above results (45-47) are the same as [12], where it is assumed that the perfect CSI is available. In this section, relations 45 to 47 are referred to as conventional optimization and relations (42-44) are referred to as robust optimization.

4.2 Robust optimization with channel estimation consideration

4.2.1 Channel estimation

The received signal at the base station during training period (uplink) , at time stand i, can be modeled as [13]:

$$y(i) = H^T a(i) + n(i) \tag{48}$$

During the training period of N symbols in uplink transmission, the received signal can be constructed as [13]:

$$y_s = sh_s + n \tag{49}$$

where $y_s = [y(0),...,y(N-1)]^T$, $n = [n(0),...,n(N-1)]^T$, $h_s = \mathrm{vec}[H^T]$, $s = [A(0),...,A(N-1)]^T$, and $A(i)$ can be constructed as block diagonal matrix with elements of $a(i)^T$. Based on the received signal in (49), the Best Linear Unbiased Estimator (BLUE) channel estimation can be performed as [20]:

$$\hat{h}_s = (s^H C_n^{-1} s)^{-1} s^H C_n^{-1} y_s = (s^H s)^{-1} s^H y_s = W y_s \tag{50}$$

with the covariance matrix of:

$$C_{\hat{h}_s} = (s^H C_n^{-1} s)^{-1} = \sigma_n^2 (s^H s)^{-1} \tag{51}$$

4.2.2 Improved robust optimization

In the robust optimization, only THP filters were optimized according to the MMSE criterion and the channel estimator was optimized separately from THP. Here, the above solution is extended to optimize THP filters together with the channel estimator conditioned on observed data. In this case, cost function should be optimized with respect to THP filters and the observed data. Thus, the goal is to optimize the precoder directly based on the available observation y_s.

Based on the linear model of (49), the conditional PDF $p_{h_s|y_s}(h_s \mid y_s)$ is a complex Gaussian process with moments $\mu_{h|y_s} = \mathrm{E}[h_s \mid y_s]$ and $C_{h|y_s} = E[(h_s - \mu_{h|y_s})(h_s - \mu_{h|y_s})^H \mid y_s]$ [18].

According to the Bayesian Gauss-Markov theorem, the Bayesian estimator can be written as [20]:

$$\mu_{h|y_s} = C_{h_s} s^H (s C_{h_s} s^H + \sigma_n^2 I)^{-1} y_s = W_s y_s \tag{52}$$

and the covariance matrix of channel estimator is:

$$C_{h|y_s} = C_{h_s} - W_s s C_{h_s} \tag{53}$$

where $C_{h_s} = E[h_s h_s^H]$. In order to optimize the THP filters, the cost function in the previous section should be modified conditional to the observed data, i.e.:

$$\Phi_{ee} = \mathrm{E}_{\mathrm{H}}[\boldsymbol{e}\boldsymbol{e}^{H} \mid \boldsymbol{y}_s] \tag{54}$$

By using the orthogonallity principle, the MMSE solution should be equivalent to:

$$\mathrm{E}_{\mathrm{H}}[\boldsymbol{e}\boldsymbol{r}^{H} \mid \boldsymbol{y}_s] = 0 \tag{55}$$

As relation (39) it is possible to write [13],

$$\boldsymbol{G}\Phi_{rr|y_s} = \boldsymbol{B}\Phi_{xr|y_s} \tag{56}$$

Like the sub-section 4.1, the matrix $\boldsymbol{R}$ can be found through Cholesky factorization of [13]:

$$\boldsymbol{R}\boldsymbol{R}^{H} = (\hat{\boldsymbol{H}}\hat{\boldsymbol{H}}^{H} + \zeta\boldsymbol{I} + \boldsymbol{C}_{h|y_s})\hat{\boldsymbol{H}}^{-H}\hat{\boldsymbol{H}}^{-1}(\hat{\boldsymbol{H}}\hat{\boldsymbol{H}}^{H} + \zeta\boldsymbol{I} + \boldsymbol{C}_{h|y_s}) \tag{57}$$

and matrices $\boldsymbol{G}$, $\boldsymbol{B}$ and $\boldsymbol{F}$ can be found as:

$$\begin{aligned} \boldsymbol{G} &= \mathrm{diag}[r_{11}^{-1},\ldots,r_{KK}^{-1}] \\ \boldsymbol{B} &= \boldsymbol{G}\boldsymbol{R} \\ \boldsymbol{F} &= \hat{\boldsymbol{H}}^{-1}(\hat{\boldsymbol{H}}\hat{\boldsymbol{H}}^{H} + \zeta\boldsymbol{I} + \boldsymbol{C}_{h|y_s})\boldsymbol{R}^{-H} \end{aligned} \tag{58}$$

In this case, the error covariance matrix has the form [13]:

$$\Phi_{ee} = \sigma_x^2\boldsymbol{G}(\zeta^2\hat{\boldsymbol{H}}^{-H}\hat{\boldsymbol{H}}^{-1} + \zeta\boldsymbol{I} + \boldsymbol{C}_{h|y_s})\boldsymbol{G}^{H} \tag{59}$$

It means that the improved robust optimization can be done by replacing $\boldsymbol{C}_{\Delta H}$ in the robust optimization with its equvalent, i.e. $\boldsymbol{C}_{h|y_s}$.

4.3 Power loading in imperfect CSI

In sub-section 3.4, we discussed about power loading of point-to-point MIMO-THP in perfect CSI. Now we develop this power loading in MIMO-BC-THP for imperfect case.

4.3.1 Optimal solution

It is easy to approximate the SER of each sub-streams for imperfect CSI as [13]:

$$SER \approx \sum_{k=1}^{K}(1-\frac{1}{\sqrt{M_k}})Q(\sqrt{\frac{3}{M_k-1}\frac{p_k}{\sigma_e^2}}) \tag{60}$$

so p_k is the power of transmitted symbols of kth user and from (59) we have,

$$\sigma_e^2 = [\Phi_{ee}]_{kk} = \frac{1}{|r_{kk}|^2}\left(\sigma_n^2 + \sum_{j=1}^{K} p_j|\delta_{kj}|^2 + \beta_k\right) \tag{61}$$

where $\delta_{ij} = [\Delta\boldsymbol{H}]_{ij}$, $\beta_k = \sigma_n^4\sum_{j=1}^{K}|\widetilde{h}_{kj}|^2 / p_j$, and $\widetilde{h}_{ij} = [\hat{\boldsymbol{H}}^{-H}\hat{\boldsymbol{H}}^{-1}]_{ij}$. Assuming small error in (61), i.e. $\alpha\, p_j \le \sigma_n^2\ ;\forall j$, σ_e^2 can be approximated as [13]:

$$\sigma_e^2 \approx \frac{1}{\left|r_{kk}\right|^2}\left(\sigma_n^2 + \alpha\, p_T + \beta_k\right) \tag{62}$$

Where, similar to previous section, the worst-case is assumed, i.e., $\alpha = \max_{i,j} |\delta_{ij}|^2 ;\ \forall i,j$. Thus power distribution that will minimize the average SER, when imperfect CSI is presented, can be found with the use of the Lagrange method as [13]:

$$L = \sum_{k=1}^{K}(1 - \frac{1}{\sqrt{M_k}})Q(\sqrt{p_k B_k}) - \lambda(K - \sum_{k=1}^{K} p_k) \tag{63}$$

where $B_k = \dfrac{3}{M_k - 1}\dfrac{\left|r_{kk}\right|^2}{\sigma_n^2 + \alpha\, p_T + \beta_k}$. Unfortunately, we did not find any explicit solution to solve (63). Therefore, some numerical or suboptimal solutions are necessary to solve it.

4.3.2 Suboptimal solution

In (63) since β_k is a function of p_k it is difficult to find explicit solution. One simple method to overcome this problem is that the initial power can be approximated as $p_j \approx P_T / K$, so that in this case β_k is not a function of p_k and,

$$\beta_k = \beta \approx \sigma_n^4 (p_T / K)\sum_{j=1}^{K}\left|\tilde{h}_{kj}\right|^2 \tag{64}$$

It means that the initial power distribution is assumed to be uniform, where the power distribution can be attained according to power allocation policies, i.e., relation (63). The simpler solution can be attained by distributing the power according to noise and channel error values, but without the interference term, i.e. neglecting the interference term and assuming $\beta = 0$. (These assumptions are only valid for the case of small error, i.e., $\beta \le \sigma_n^2 + \alpha\, p_T$.) In this case, the initial power is assigned as a uniform distribution with zero value. Solving $\partial L / \partial p_k = 0$ for p_k, yielding [13]:

$$p_k = \frac{1}{B_k} W(\frac{B_k}{A_k}\Lambda) \tag{65}$$

where $A_k = \dfrac{M_k}{(\sqrt{M_k} - 1)^2}\dfrac{1}{B_k}$, $B_k = \dfrac{3}{M_k - 1}\dfrac{\left|r_{kk}\right|^2}{\sigma_n^2 + \alpha\, p_T + \beta}$, $\Lambda = cte$, and $W(x)$ is the real valued Lambert W -function. The unique solution for $\boldsymbol{P} = [p_1, \ldots, p_K]$ can be found by the simple iterative procedure same as sub-section 3.2.2 [13]:

It should be noted that, the above power loading can be combined with conventional (equal to before section), robust, or improved robust optimization strategies. In the former cases, the power loadings are similar to conventional power loading in imperfect CSI but THP filters should be calculated according to robust/ improved robust optimization.

4.4 Simulation and results

For simulation purposes, K=4 user with 4-QAM signaling are assumed. The entries of $\hat{\boldsymbol{H}}$ and $\Delta\boldsymbol{H}$ have been assumed to be zero mean i.i.d. complex Gaussian random variables, i.e., $\hat{\boldsymbol{H}} \sim CN(0,1)$ and $\Delta\boldsymbol{H} \sim CN(0,\rho^2)$, respectively. In simulations, in order to compare our results more simply with other contributions, the BER is ploted instead of the SER.

Fig. 11 plots the mean BER versus Eb/No for Robust optimization together with conventional optimization for 4QAM modulation. It is observed that the robust optimization has better performance for all channel imperfection values, especially for high SNR. Fig. 12 compares the performance of the proposed improved robust optimization with conventional optimization. As can be realized, the proposed improved robust optimization algorithm substantially outperforms the conventional optimization, over the whole observation data lengths. In fact, the performance is noticeable for smaller Ns, where the channel estimator estimates the channel erroneously, especially for high SNR values.

In order to demonstrate the performance of power loading, for simplicity, the suboptimal solution is considered. The validity of the approximation of $\beta \le \sigma_n^2 + \alpha\, p_T$ is shown in Fig. 13 by plotting the fraction of channel realizations in which the approximations are valid for different values of ρ. It can be seen that for the values of SNR>20dB the approximation is

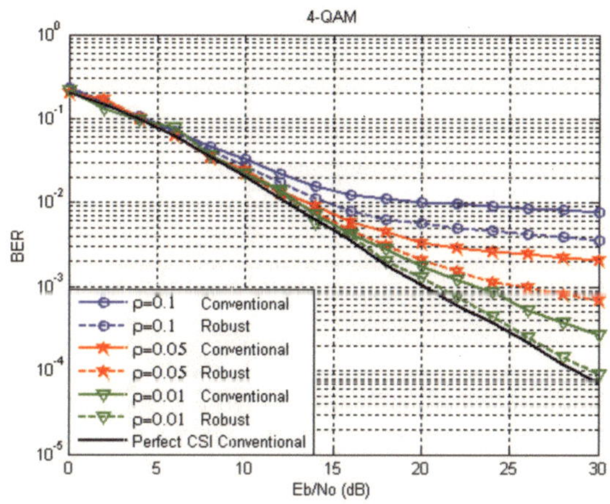

Fig. 11. Robust THP optimization performance

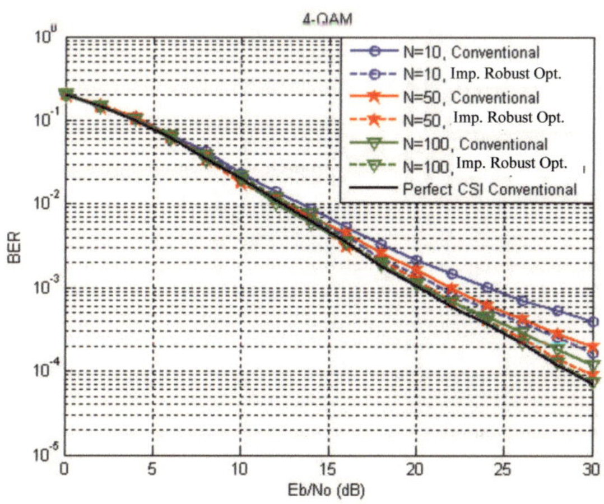

Fig. 12. Improved robust optimization performance with different N value

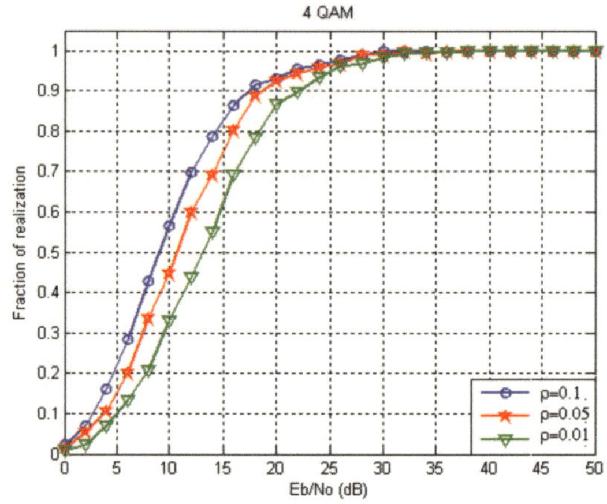

Fig. 13. Validation of approximating $\beta \le \sigma_n^2 + \alpha\, p_T$

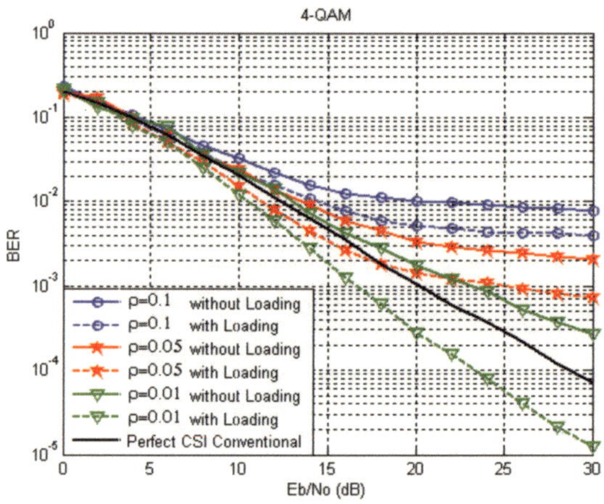

Fig. 14. Suboptimal power loading in conventional optimization

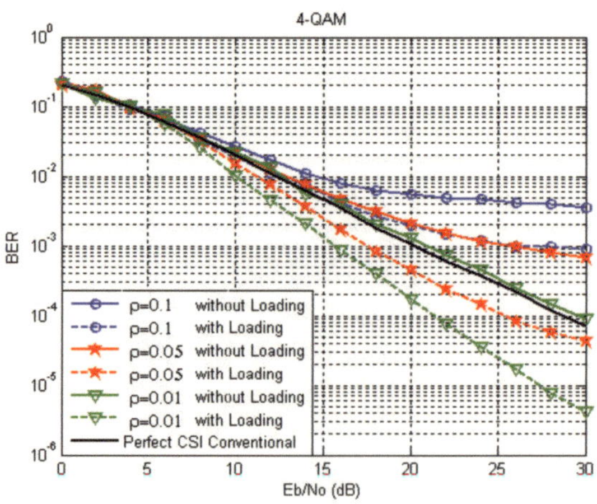

Fig. 15. Suboptimal power loading in robust optimization

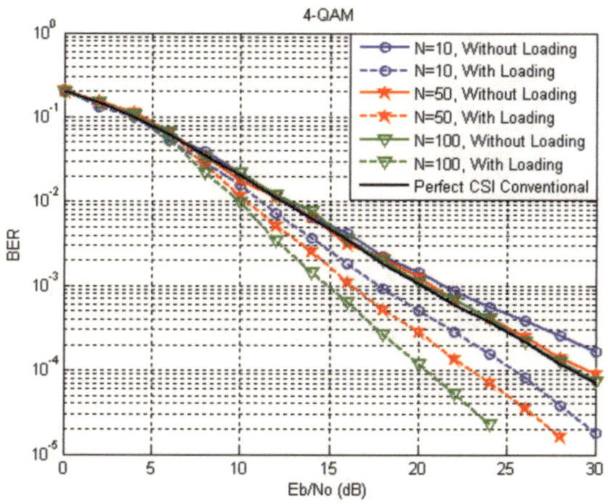

Fig. 16. Suboptimal power loading in improved robust optimization

valid. In figures 14-16 we can observe the performance of power loading for conventional, robust and improved robust optimizations, respectively, with the assumption that $\beta = 0$. As can be seen from these figures, the performance of power loading is noticeable, especially for low errors and high SNR values.

5. Joint channel estimation and spatial pre-equalization

Traditionally, channel estimation and pre-equalization are optimized separately and independently in which results to performance degradation. This loss may be causing some poor performance, especially, in erroneous conditions, as can be seen from previous section. In [21] Dietrich et. al proposed a new method for joint pilot symbol assisted channel estimation and equalization and applied it to the design of the space time decision feedback equalizer. Their research was developed in [13] for MIMO-BC-THP system with the assumption of the BLUE, as discussed in sub-section 4.2.2. The authors in [13] extracted the explicit solution for THP optimization with a good performance against to separate design of THP and channel estimator.

In this section, their work is extended as joint optimization in which the THP filters are optimized together with the channel estimation conditioned on observation data (with, approximately, the same order of complexity as a separate design). In the other words, in joint optimization, in contrast to separate optimization, the average cost function should be optimized with respect to THP filters and channel estimation, i.e. the expectation is taken with respect to the unknown channel parameters conditioned on the available observation data. It means that in contrast to improved robust optimization in which different channel estimation methods have to be investigated for a given optimized THP to find out the best combination, the best channel estimation can be chosen directly by minimizing the MMSE criterion. As a result, it will be shown that the joint optimization lead to a Linear MMSE (LMMSE) channel estimator and a new structure for THP filters based on the error covariance matrix of the channel estimator.

5.1 Channel estimation

By using the Bayesian Gauss-Markov theorem, the Bayesian LMMSE estimator can be obtained for linear model of (49) [20]:

$$\hat{\boldsymbol{h}} = \mathrm{E}[\boldsymbol{h} \mid \boldsymbol{y}_s] = \boldsymbol{C}_{h_s} \boldsymbol{s}^H (\boldsymbol{s}\boldsymbol{C}_{h_s} \boldsymbol{s}^H + \sigma_n^2 \boldsymbol{I})^{-1} \boldsymbol{y}_s = \boldsymbol{W}_s \boldsymbol{y}_s \tag{66}$$

and

$$\boldsymbol{C}_{h|y_s} = \boldsymbol{C}_{h_s} - \boldsymbol{W}_s \boldsymbol{s} \boldsymbol{C}_{h_s} \tag{67}$$

where $\hat{\boldsymbol{h}}$ indicate the estimation of $\boldsymbol{h}$ and:

$$\begin{aligned} \boldsymbol{C}_{h_s} &= \mathrm{E}[\boldsymbol{h}_s \boldsymbol{h}_s^H] \\ \boldsymbol{W}_s &= \boldsymbol{C}_{h_s} \boldsymbol{s}^H (\boldsymbol{s}\boldsymbol{C}_{h_s} \boldsymbol{s}^H + \sigma_n^2 \boldsymbol{I})^{-1} \end{aligned} \tag{68}$$

5.2 Joint optimization

In conventional THP optimization the error, by using the orthogonality principle of (38), the THP can be optimized in perfect CSI as [11] or in imperfect CSI as [13] in which a specific channel estimator (i.e., BLUE) is assumed and THP is optimized according to this estimator structure (i.e. improved robust optimization). In general aspect, this is not a desired method because it is necessary to select different channel estimators and optimize the THP filters to find the best combination where it is cumbersome work. Nevertheless, in the joint optimization the best channel estimator is determined in which the THP filters and channel estimation can be optimized jointly without any trial method. In this case, since the training sequence and $\boldsymbol{y}_s$ are given, the channel can be modelled as a random variable from point of view of the receiver. Thus, the cost function in (38) is a random variable and should be considered as (55), i.e.:

$$E[\boldsymbol{e}\boldsymbol{r}^H \mid \boldsymbol{y}_s] = 0 \tag{69}$$

where the expectation is taken with respect to the unknown channel parameters. The above equation can be written simpler as:

$$\boldsymbol{G}\Phi_{rr|y_s} = \boldsymbol{B}\Phi_{xr|y_s} \tag{70}$$

where the matrices $\Phi_{rr|y_s}$ and $\Phi_{xr|y_s}$ can be computed by using (1) as [22]:

$$\Phi_{rr|y_s} = \mathrm{E}[\boldsymbol{r}\boldsymbol{r}^H \mid \boldsymbol{y}_s] = \mathrm{E}[(\boldsymbol{H}\boldsymbol{F}\boldsymbol{x} + \boldsymbol{n})(\boldsymbol{H}\boldsymbol{F}\boldsymbol{x} + \boldsymbol{n})^H \mid \boldsymbol{y}_s] = \boldsymbol{V} + \zeta \boldsymbol{I} \tag{71.a}$$

$$\Phi_{xr|y_s} = \mathrm{E}[\boldsymbol{x}\boldsymbol{r}^H \mid \boldsymbol{y}_s] = \mathrm{E}[\boldsymbol{x}(\boldsymbol{H}\boldsymbol{F}\boldsymbol{x} + \boldsymbol{n})^H \mid \boldsymbol{y}_s] = \sigma_x^2 \boldsymbol{F}^H \hat{\boldsymbol{H}}^H \tag{71.b}$$

where $\zeta = \sigma_n^2 / \sigma_x^2$ and $\boldsymbol{V} = \mathrm{E}[\boldsymbol{H}\boldsymbol{H}^H \mid \boldsymbol{y}_s]$. In order to find a closed form solution for THP filters, it is needed to calculate the conditional mean estimate of $\boldsymbol{T} = \boldsymbol{H}\boldsymbol{H}^H$ over observed data, i.e., $\boldsymbol{y}_s$. The matrix $\boldsymbol{T}$ is well known as Gramian matrix where its probability distribution is a Wishart distribution [1]. In order to calculate $\boldsymbol{V}$, consider the cost function as:

$$J = \left\| \hat{\boldsymbol{T}} - \boldsymbol{T} \right\|_F^2 \tag{72}$$

where the lower index stands for Frobenius norm and $\hat{\boldsymbol{T}}$ is a nonlinear function of $\boldsymbol{y}_s$ where should be determined. The minimization of (72) lead to a non-linear conditional mean estimator as [22]:

$$\hat{T} = \mathrm{E}[T \mid y_s] = \mathrm{E}[HH^H \mid y_s] = \sum_{k=1}^{K} \mathrm{E}[h_k h_k^H \mid y_s] \tag{73}$$

where $H = [h_1, ..., h_K]$. It is possible to consider each expression in the summation of (73) as [24]:

$$\mathrm{E}[h_k h_k^H \mid y_s] = \mathrm{E}[h_k \mid y_s]\mathrm{E}[h_k \mid y_s]^H + C_{h_k|y_s} = \hat{h}_k \hat{h}_k^H + C_{h_k|y_s} \tag{74}$$

where

$$C_{h_k|y_s} = \mathrm{E}[(h_k - \hat{h}_k)(h_k - \hat{h}_k)^H \mid y_s] \tag{75}$$

Since the error $h_k - \hat{h}_k$ is statically independent from the observation data, we have:

$$C_{h_k|y_s} = \mathrm{E}[(h_k - \hat{h}_k)(h_k - \hat{h}_k)^H] \tag{76}$$

By substituting the relations (74) and (76) in (73) and rearrange the resultant sub-matrices in its original matrix form, we have [22],

$$\hat{T} = \hat{H}\hat{H}^H + C_{H|y_s} \tag{77}$$

where,

$$C_{H|y_s} = C_{h_s} - W_s S C_{h_s} \tag{78}$$

The matrices W_s and C_{h_s} are the same as (68) where is used in Bayesian LMMSE channel estimator. On the other hand, it is possible to show that $\hat{h} = \mathrm{E}[h \mid y_s] = W_s y_s$ [20], i.e., this joint optimization lead to a Bayesian LMMSE channel estimator (in the joint optimization, the explicit channel estimation is not needed). In this case, the matrices $\Phi_{rr|y_s}$ in (71.a) can be obtained as [22]:

$$\Phi_{rr|y_s} = \hat{H}\hat{H}^H + C_{H|y_s} + \zeta I \tag{79}$$

Substitution relations (71.b) and (79) in (70) and by some manipulating, the matrix R can be found through the Cholesky factorization of [22]:

$$RR^H = (\hat{H}\hat{H}^H + \zeta I + C_{H|y_s})\hat{H}^{-H}\hat{H}^{-1}(\hat{H}\hat{H}^H + \zeta I + C_{H|y_s}) \tag{80}$$

and matrices G, B and F can be found as:

$$\begin{aligned} G &= \mathrm{diag}[r_{11}^{-1}, ..., r_{KK}^{-1}] \\ B &= GR \\ F^H &= \hat{H}^{-1}(\hat{H}\hat{H}^H + \zeta I + C_{H|y_s})R^{-H} \end{aligned} \tag{81}$$

In this case, the error covariance matrix can be computed as [22]:

$$\Phi_{ee} = \sigma_x^2 G(\zeta^2 \hat{H}^{-H}\hat{H}^{-1} + \zeta I + C_{H|y_s})G^H \tag{82}$$

Note that if the perfect CSI is assumed, i.e. $\boldsymbol{C}_{H|y_s} = 0$, the relations (80-82) are the same as conventional THP optimization where denoted in [12] and here it is referred to as conventional optimization.

5.3 Simulation and results

For simulation purposes, K=4 user with 4-QAM signalling is assumed. The entries of $\boldsymbol{H}$ is assumed to be zero mean i.i.d. complex Gaussian random variables, i.e., $\boldsymbol{H} \sim CN(0,1)$. Fig. 16 compares the performance of the proposed joint optimization with conventional optimization. As can be realized, the proposed joint optimization algorithm substantially outperforms the conventional optimization, over the whole observation data lengths. In fact, the performance is noticeable for smaller Ns, where the channel estimator estimates the channel erroneously, especially for high SNR values. In order to observe the detailed results more precisely, some of the main part of Fig. 17 is reproduced again in Fig. 18.

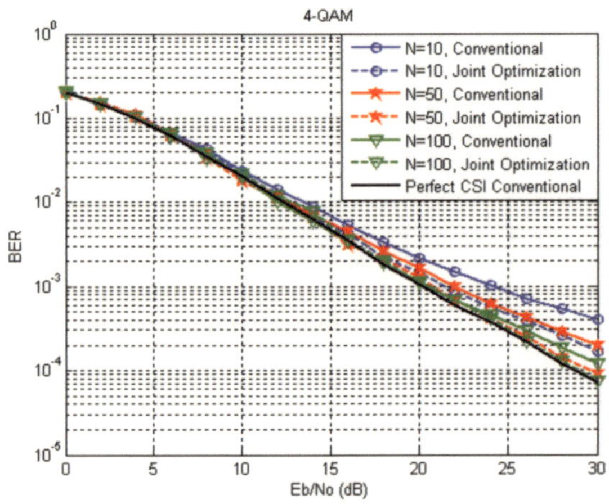

Fig. 17. Joint optimization performance with different observation data length

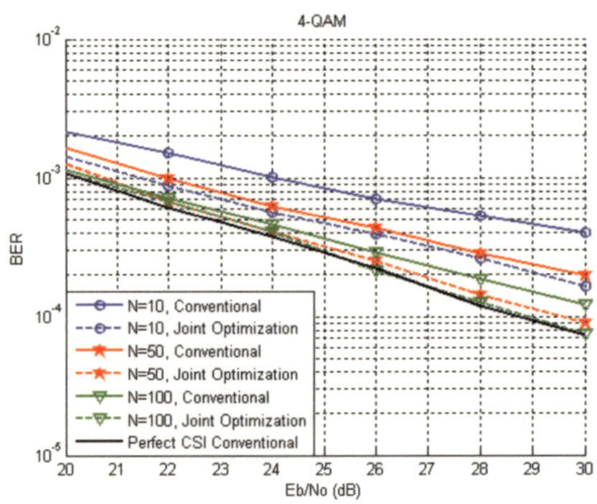

Fig. 18. A detailed reproduction of Fig. 17

6. THP optimization in correlated channel

Mobile channels suffer from multi path fading phenomenon so that it is necessary to adapt the transmitter resources (such as power and rate) with channel characteristics in order to achieve channel capacity. The capacity achieving transmitter adaptation strategy depends on how much CSI is available at the transmitter and the receiver. The use of channel feedback from receiver to transmitter is a standard method in precoding systems, as discussed in previous sections. While knowledge of the channel at the transmitter tends to good performance in precoding systems over slowly time-varying channels, the generation of reliable channel feedback is complicated in fast time-varying channels. This fast time variations impose a significant challenge for precoding design.

In order to increase the THP performance in fast time-varying channels, the THP should be optimized in each (or approximately some) symbol time. On the other hand, the generation of reliable channel feedback in each symbol time is complicated in fast time-varying channel and can be led to high bandwidth overhead. From practical implementation point of view, most of the THPs have been designed assuming that the wireless channel can be regarded as constant over a block of data. In mobile applications where the channel is time-varying, the assumption that the channel is constant over some periods only holds approximately and will affects the THP performance that are designed based on this assumption. Hence, a judicious and innovative THP system design that takes the time-varying nature of the channel into account is the key to solve the above problems.

Based on previous section on precoding and also works denoted in [25], this section extend and unify the THP concepts to time-varying MIMO channels, which will allow us to improve mobile systems performance, as well as to provide guidelines for future precoder designs employing low feedback overhead. A first-order Auto-Regressive (AR) model is used to characterize the channel coefficients that vary from symbol to symbol. Although, traditional THP with a perfect channel estimation in each symbol time has an advantage over MIMO channels, new THP design provides a significant advantage over correlated MIMO channels where the volume of feedback dominate to $1/N$ (N is data block length).

6.1 Correlated channel model

In time-varying channel, it is assumed that the perfect CSI in known at the beginning of each data block, but not during the block. It is desired to optimize THP at time t with the assumption that the outdated CSI, $H_{t-\tau}$, is available. $\boldsymbol{H}_{t-\tau}$ corresponds to the channel state τ seconds earlier (i.e. the beginning of block) where $\tau = lT_s$; $l \le N$. l and T_s denote as number of symbol and symbol time, respectively. Under the assumption of rich scattering, the MIMO channel matrix $\boldsymbol{H}_t$ can be modeled as a complex matrix whose entries are i.i.d. zero-mean complex Gaussian random variables with common variance σ_h^2, i.e. $\boldsymbol{H}_t \sim CN(0, \sigma_h^2 \boldsymbol{I})$ [25]. $\boldsymbol{H}_t$ and $\boldsymbol{H}_{t-\tau}$ are correlated realizations of the latter channel distribution. Thus, given the outdated CSI, $\boldsymbol{H}_{t-\tau}$, we can characterize the unknown current CSI, $\boldsymbol{H}_t$, using the conditional CSI model introduced in [25], as follows:

$$\boldsymbol{H}_t \sim CN\left(\rho_t \boldsymbol{H}_{t-\tau}, \sigma_h^2\left(1-\rho_t^{\,2}\right)\boldsymbol{I}\right) \tag{83}$$

where ρ_t is the common time-correlation of the i.i.d. time-varying MIMO channel coefficients, defined as:

$$\rho_t = E\{[\boldsymbol{H}_t]_{i,j}[\boldsymbol{H}_{t-\tau}]_{i,j}^H\}/\sigma_h^2 = R(\tau) \tag{84}$$

where $R(\tau)$ depending on the channel time-variation model.

6.2 THP optimization

The error that is needed to be considered for the system illustrated in Fig. 1, should be the difference between the effective data vector, $\boldsymbol{v}_t$, and the data vector entering the decision module, $\boldsymbol{y}_t$, i.e.:

$$\boldsymbol{e}_t = \boldsymbol{y}_t - \boldsymbol{v}_t = [\boldsymbol{G}_t\boldsymbol{H}_t\boldsymbol{F}_t - \boldsymbol{B}_t]\boldsymbol{x}_t + \tilde{\boldsymbol{n}}_t \tag{85}$$

where $\tilde{\boldsymbol{n}}_t = \boldsymbol{G}_t\boldsymbol{n}_t$. The MMSE solution should minimize the error signal as:

$$\begin{cases} \arg\min\limits_{\boldsymbol{B},\boldsymbol{F},\boldsymbol{G}} E_{H_t|\boldsymbol{H}_{t-\tau}}\left\{E_{a,x}\left\|[\boldsymbol{G}_t\boldsymbol{H}_t\boldsymbol{F}_t - \boldsymbol{B}_t]\boldsymbol{x}_t + \tilde{\boldsymbol{n}}_t\right\|^2\right\} \\ s.t. \quad E\left\|\tilde{\boldsymbol{x}}_t\right\|^2 \le P_T \end{cases} \tag{86}$$

where, P_T is total available power at transmitter. Similar to previous section, instead of solving (86), it is easier to use the orthogonallity principle [13]. In this case, the MMSE solution should satisfy [25]:

$$E_{H_t|\boldsymbol{H}_{t-\tau}}\left\{E_{a,x}[\boldsymbol{e}_t\boldsymbol{r}_t^H]\right\} = 0 \tag{87}$$

Thus:

$$\boldsymbol{G}_t\Phi_{rr,t} = \boldsymbol{B}_t\Phi_{xr,t} \tag{88}$$

The matrix $\Phi_{rr,t}$ can be computed by using (1) as:

$$\Phi_{rr,t} = E_{H_t|\boldsymbol{H}_{t-\tau}}\left\{E_{a,x}[\boldsymbol{r}_t\boldsymbol{r}_t^H]\right\} = E_{H_t|\boldsymbol{H}_{t-\tau}}[\sigma_x^2\boldsymbol{H}_t\boldsymbol{H}_t^H + \sigma_n^2\boldsymbol{I}] \tag{89}$$

The matrix $\boldsymbol{H}_t\boldsymbol{H}_t^H$ is well known as Gramian matrix where its probability distribution is a Wishart distribution. Calculation of the conditional expectation, as required in (89), seems to be difficult. We consider an approximate solution to solve it. To do this, by using the channel distribution of (83), we first instantiate the true channel $\boldsymbol{H}_t$ as $\boldsymbol{H}_t = \overline{\boldsymbol{H}}_t + \Delta_t$, where $\overline{\boldsymbol{H}}_t = \rho_t\boldsymbol{H}_{t-\tau}$ and Δ_t is the $CN(0,\sigma_h^2(1-\rho_t^2)\boldsymbol{I})$ -distributed uncertainty on the true channel given the outdated CSI [25]. Under the assumption of isotropic scattering and moving terminal, this model describes the time-correlation function as $R(\tau) = J_0(2\pi f_D\tau)$, where J_0 is the zero-th order Bessel function of the first kind and f_D is maximum Doppler frequency [26]. Then, we use the following expression as statistical model for the time variations of the channel [25]:

$$\boldsymbol{H}_t = \rho_t\boldsymbol{H}_{t-\tau} + \sigma_h\sqrt{1-\rho_t^2}\,\boldsymbol{E}_t \tag{90}$$

where ρ_t denotes the correlation coefficient between the time instants $t-\tau$ and t, and $\boldsymbol{E}_t$ is a circularly symmetric complex Gaussian matrix with i.i.d. entries, i.e. $\boldsymbol{E}_t \sim CN(0,\boldsymbol{I})$. In this case [25],

$$\Phi_{rr,t} = E_{H_t|H_{t-\tau}}[\sigma_x^2(\overline{H}_t + \Delta_t)(\overline{H}_t + \Delta_t)^H + \sigma_n^2 I] = \sigma_x^2 \overline{H}_t \overline{H}_t^H + C_{\Delta,t} + \sigma_n^2 I \quad (91)$$

where $C_{\Delta,t} = \sigma_h^2(1-\rho_t^2)I$. The matrix $\Phi_{xr,t}$ in (88) can be computed as [25]:

$$\begin{aligned}\Phi_{xr,t} &= \mathrm{E}[x_t r_t^H] = E_{H_t|H_{t-\tau}}\{E_{a,x}[x_t r_t^H]\} = E_{H_t|H_{t-\tau}}\{E_{a,x}[x_t x_t^H F_t^H H_t^H]\} \\ &= E_{H_t|H_{t-\tau}}[\sigma_x^2 F_t^H(\overline{H}_t + \Delta_t)^H] = \sigma_x^2 F_t^H \overline{H}_t^H\end{aligned} \quad (92)$$

Substituting relations (89) and (92) in (88) and after some manipulations lead to [25]:

$$F_t^H = B_t^{-1} G_t (\overline{H}_t \overline{H}_t^H + \zeta I + C_{\Delta,t}) \overline{H}_t^{-H} \quad (93)$$

where $\zeta = \sigma_n^2 / \sigma_x^2$. Since F_t is unitary matrix, we have:

$$R_t R_t^H = (\overline{H}_t \overline{H}_t^H + \zeta I + C_{\Delta,t}) \overline{H}_t^{-H} \overline{H}_t^{-1} (\overline{H}_t \overline{H}_t^H + \zeta I + C_{\Delta,t}) \quad (94)$$

where $R_t = G_t^{-1} B_t$ is assumed. The matrix R_t can be found through Cholesky factorization of (94) and the matrices G_t, B_t, and F_t can be found as:

$$\begin{aligned}G_t &= \mathrm{diag}[r_{11}^{-1}, \ldots, r_{KK}^{-1}] \\ B_t &= G_t R_t \\ F_t &= (\rho_t H_{t-\tau})^{-1}(\rho_t^2 H_{t-\tau} H_{t-\tau}^H + \sigma_h^2(1-\rho_t^2) I + \zeta I) R_t^{-H}\end{aligned} \quad (95)$$

with B_t, F_t and G_t that found in (95), the error covariance matrix can be computed as [25]:

$$\Phi_{ee,t} = E[e_t e_t^H] = \sigma_x^2 G_t (\zeta^2 \rho_t^2 H_{t-\tau}^{-H} H_{t-\tau}^{-1} + \zeta I + \sigma_h^2 (1-\rho_t^2) I) G_t^H \quad (96)$$

In situation where the channel is assumed to be quasi-static (i.e. $\rho_t \to 1$), the relations (94-96) tends to what is considered as conventional THP in relations (45-47) and also in previous section.

6.3 Simulation and results

In this section, we illustrate the improvement, in term of average BER, that our proposed MMSE-THP design offers over conventional THP designs that assumes the channel is constant over *N* symbol time. In order to do that, we use the well-known Jakes model [27] to instantiate a realistic outdated CSI model based on (83). In our simulations, we use the normalized Doppler frequency $f_D T = 0.001$, $f_D T = 0.005$, and $f_D T = 0.01$ according to slow, medium, and fast fading, respectively [28]. We further consider the case of a (4, 4) MIMO set-up with 4QAM-modulated data streams.

Figures (19a) to (19c) plot the average BER performances for *N*=5, 10, 20, respectively. Clearly, our robust MMSE-THP design exhibits a lower average BER performance when compared to the conventional design. More specifically, from the figures, it can be observed that the proposed method have more advantage in fast fading over the state-of-the-art design for shorter *N* where the channel correlation is noticeable.

Figures (20a) to (20c) plot the average BER performances for slow, medium, and fast fading, respectively. Observe that for slow fading case, our proposed method has good performance over all values of *N* while in the fast fading case; the performance is noticeable for shorter *N*s.

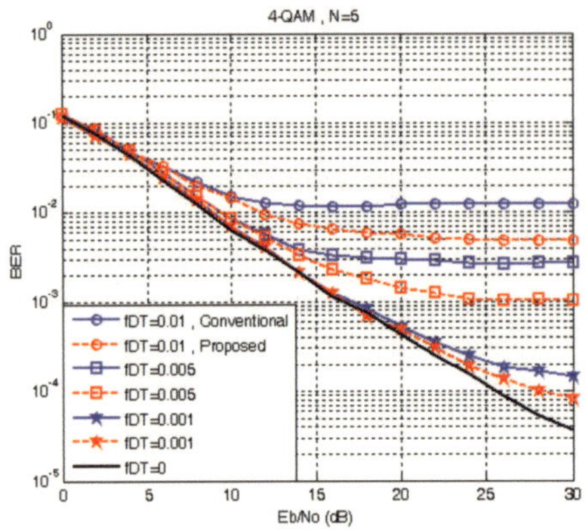

(a) N=5

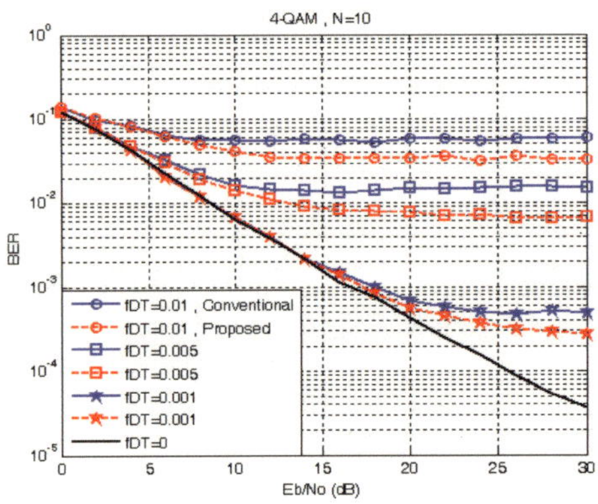

(b) N=10

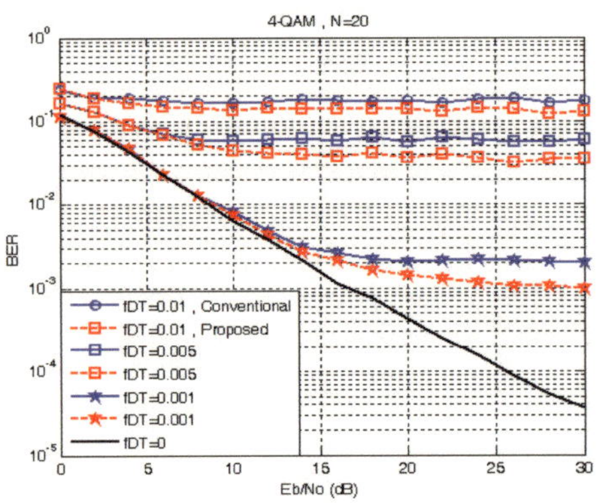

(c) N=20

Fig. 19. BER performance for different N values

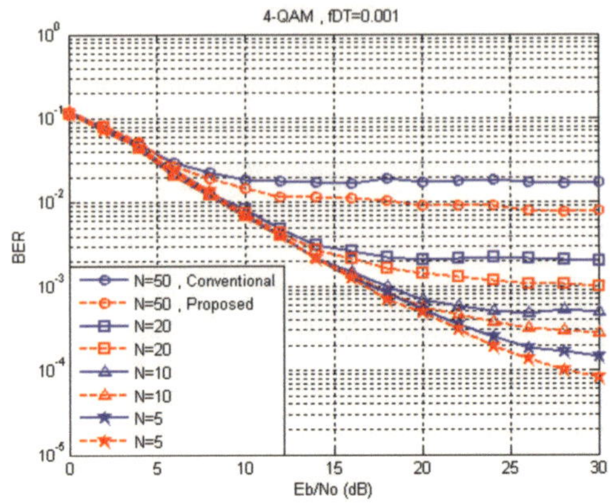

(a) $f_D T = 0.001$

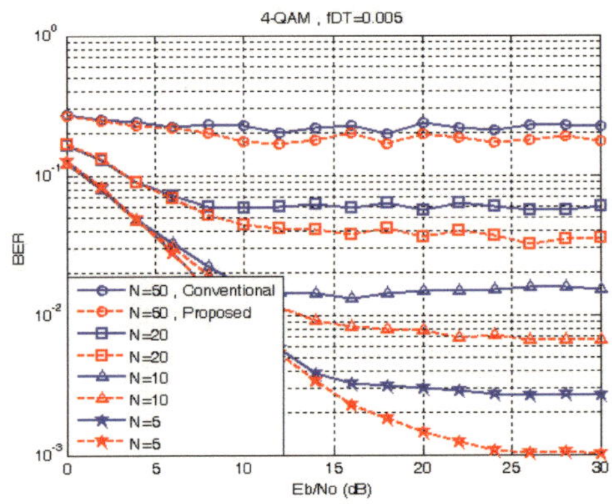

(b) $f_D T = 0.005$

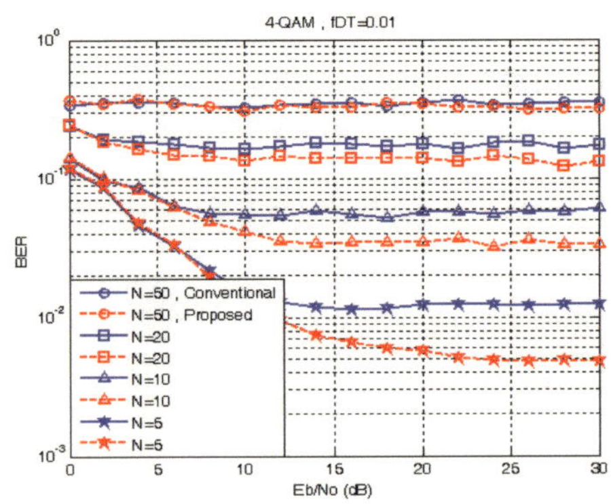

(c) $f_D T = 0.01$

Fig. 20. BER performance for different Doppler values

7. References

[1] H. Khaleghi Bizaki, "Precoding and Blind/Semi-blind Estimation in MIMO Fading Channels", PhD Thesis, Iran University of Science and Technology (IUSTz), Winter 2008.

[2] R. F. H. Fischer, C. Windpassinger, a. Lampe, and J. B. Huber, "Space-Time Transmission using Tomlinson-Harashima Precoding ", In Proceedings of 4th ITG Conference on Source and Channel Coding, pp. 139-147, Berlin, January 2002.

[3] C. Windpassinger, "Detection and Precoding for Multiple Input Multiple Output Channels", PhD Thesis, Erlangen, 2004.

[4] M. H. M. Costa, "Writing on Dirty Paper", IEEE Transactions on Information Theory, vol. IT-29, No. 3, May 1983.

[5] R. F. H. Fischer, C. Windpassinger, a. Lampe, and J. B. Huber, " MIMO Precoding for Decentralized Receivers"

[6] G. H. Golub and C. F. VanLoan, "Matrix Computations", The Johns Hopkins University Press, Baltimore, MD, USA, 3rd edition, 1996.

[7] M. Tomilson, "New Automatic Equalizer Employing Modulo Arithmetic", Electronic Letters, pp. 138-139, March 1971.

[8] H. Harashima, Miyakawa, "Matched –Transmission Technique for Channels with Intersymbole Interference", IEEE Journal on Communications, pp. 774-780, Aug. 1972.

[9] U. Erez, S. Shamai, and R. Zamir, "Capacity and Lattice Strategies for Cancelling Known Interference", In Proceeding of International Symposium on Information Theory abd Its Applications, Honolulu, Hi, USA, Nov. 2000.

[10] Q. Zhou, H. Dai, and H. Zhang, "Joint Tominson-Harashima Precoding and Scheduling for Multiuser MIMO with Imperfect Feedback", IEEE Wireless Comm. and Networking Conf. (WCNC), Vol. 3, pp: 1233-1238, 2006.

[11] H. Khaleghi Bizaki and A. Falahati, "Power Loading by Minimizing the Average Symbol Error Rate on MIMO-THP Systems", The 9th Int. Conf. on Advanced Comm. Technology (ICACT), Vol. 2, pp: 1323-1326, Feb. 2007

[12] J. Lin, W. A. Krzymein, "Improved Tomlinson Harashima Precoding for the Downlink of Multiple Antenna Multi-User Systems", IEEE Wireless Comm. and Networking Conf. (WCNC), pp: 466-472, March 2006.

[13] H. Khaleghi Bizaki, A. Falahati, "Tomlinson Harashima Precoder with Imperfect Channel State Information ", IET Communication Journal, Volume 2, Issue 1,Page(s):151 - 158, January 2008.

[14] T. Hunziker, D. Dahlhaus, "Optimal Power Adaptation for OFDM Systems with Ideal Bit-Interleaving and Hard-Decision Decoding", IEEE International Conference on Communications (ICC), vol. 5, pp:3392-3397, 2003.

[15] R. D. Wesel, J. Cioffi, "Achievable Rates for Tomlinson-Harashima Precoding", IEEE Transaction on Information Theory, vol. 44, No. 2, March 1998.

[16] M. Payaro, A. P. Neira, M. A. Lagunas, "Achievable Rates for Generalized Spatial Tomlonson-Harashima Precoding in MIMO Systems", IEEE Vehicular Technology Conference (VTC), vol. 4, pp: 2462 - 2466, Fall 2004.

[17] Bizaki, H.K.; Falahati, A., "Achievable Rates and Power Loading in MIMO-THP Systems ", 3rd International Conference on Information and Communication Technologies (ICTTA),Page(s): 1 - 7, 2008

[18] Payaro, M., Neira, A.P., and Lagunas, M.A., "Robustness evaluation of uniform power allocation with antenna selection for spatial Tomlinson-Harashima precoding", IEEE Int. Conf. Acoustics, Speech, and Signal Processing (ICASSP 2005), Philadelphia, (USA), 18–23 March 2005.

[19] H. Khaleghi Bizaki, "Channel Imperfection Effects on THP Performance in a Slowly Time Varying MIMO Channels", IEEE WCNIS2010-Wireless Communication, Conference date: 25-27 June 2010

[20] Kay, S.M.: 'Fundamentals of statistical signal processing: estimation theory' (Prentice-Hall, 1993)

[21] Dietrich, F.A., Joham, M., and Utschick, W., "Joint optimization of pilot assisted channel estimation and equalization applied to space-time decision feedback equalization", Int. Conf. on Communication (ICC), 2005, Vol. 4, pp. 2162–2167

[22] H. Khaleghi Bizaki and A. Falahati, "Joint Channel Estimation and Spatial Pre-Equalization in MIMO Systems ", IET Electronics Letters, Vol. 43, Issue 24, Nov. 2007

[23] Bizaki, H.Khaleghi, and Falahati, A., "Tomlinson-Harashima precoding with imperfect channel side information", 9th International Conference on Advanced Communication Technology (ICACT), Korea, 2007, Vol. 2, pp. 987–991

[24] Dietrich, F.A., Hoffman, F., and Utschick,W.: 'Conditional mean estimator for the Gramian matrix of complex gaussian random variables'. IEEE Int. Conf. on Acoustics, Speech, and Signal Processing (ICASSP), Philadelphia, Pennsylvania, USA, 2005, Vol. 3, pp. 1137–1140

[25] H. Khaleghi Bizaki, "Tomlinson-Harashima Precoding Optimization over Correlated MIMO Channels", IEEE WCNIS2010-Wireless Communication, Conference date: 25-27 June 2010.

[26] N. Khaled, G. Leus, C. Desset and H. De Man, "A Robust Joint Linear Precoder and Decoder MMSE Design for Slowly Time- Varying MIMO Channels", IEEE International Conference on Acoustics, Speech, and Signal Processing (ICASSP), vol. 4, pp: 485- 488, 2004.

[27] M. Patzold, "Mobile Fading Channels--Modeling, Analysis & Simulation", John Wiley, 2002.

[28] P. M. Castro, L. Castedo and J. Miguez, "Adaptive Precoding in MIMO Wireless Communication Systems Using Blind Channel Precoding Over Frequency Selective Fading Channels", IEEE 13th Workshop on Statistical Signal Processing, pp: 173 - 178, 2005.

10

Analysis and Design of Tomlinson-Harashima Precoding for Multiuser MIMO Systems

Xiang Chen, Min Huang, Ming Zhao, Shidong Zhou and Jing Wang
Research Institute of Information Technology, Tsinghua University
Beijing, China

1. Introduction

The multiuser multiple-input-multiple-output (MIMO) downlink has attracted great research interests because of its potential of increasing the system capacity(Caire & Shamai, 2003; Vishwanath et al., 2003; Viswanath & Tse, 2003; Weingarten et al., 2006). Many transmitter precoding schemes have been reported in order to mitigate the cochannel interference (CCI) as well as exploiting the spatial multiplexing of the multiuser MIMO downlink. Tomlinson-Harashima precoding (THP) has become a promising scheme since the successive interference pre-cancelation structure makes THP outperform linear precoding schemes (Choi & Murch, 2004; Zhang et al., 2005) with only a small increase in complexity. Many THP schemes based on different criteria have been reported in the literature(Doostnejad et al., 2005; Joham et al., 2004; Mezghani et al., 2006; Schubert & Shi, 2005; Stankovic & Haardt, 2005; Windpassinger et al., 2004), in which one is the zero-forcing (ZF) criterion and the other is the minimum mean square error (MMSE) criterion. This chapter will consider the above two criteria based THP schemes' analysis and design, respectively.
For the ZF-THP scheme, initial research mainly focuses on the scenarios that each receiver is equipped with a single antenna (Windpassinger et al., 2004), where there exists only the transmit diversity, but without any receive diversity. Presently, the receive diversity due to multiple antennas at each receiver is taken into account (Stankovic & Haardt, 2005; Wang et al., 2006).In these literatures, it is commonly assumed that the total number of receive antennas is less than or equal to that of transmit antennas. Under this assumption, firstly the layers are divided into groups which correspond to different users, and then the dominant eigenmode transmission is performed for each group. Hereby, this kind of schemes is regarded as *per-user* processing. Actually, it is more common in the cellular multiuser downlink systems that the number of users is not less than that of transmit antennas at the base station (BS), which is investigated as the generalized case with THP in this chapter. In order to avoid complicated user selection and concentrate on the essential of transceiver filters design, our consideration is limited into a unique case that the number of users equals the number of transmit antennas, denoted as M. Besides, it is assumed that the channels of these M users have the same large-scale power attenuation.[1] In this case a so-called *per-layer* processing can be applied by the regulation that each user be provided with only one

[1] In practice, when the number of users is large enough, we can find M users whose large-scale power attenuations are approximately equal by scheduling.

subchannel and all the M users be served simultaneously. Based on the criterion of maximum system sum-capacity, two per-layer joint transmit and receive filters design schemes can be employed which apply receive antenna beamforming (RAB) and receive antenna selection (RAS), respectively. Through a theorem and two corollaries, the differences of the equivalent channel gains and capacities between these two schemes are developed. Theoretical analysis and simulation results indicate that compared with linear-ZF and per-user processing, these per-layer schemes can achieve better rate region and sum-capacity performance.

For the MMSE-THP scheme, we address the problem of joint transceiver design under both perfect and imperfect channel state information (CSI). The authors in (Joham et al., 2004) designed THP based on the MMSE criterion for the MISO system where the users are restricted to use a common scalar receiving weight. This restriction was relaxed in (Schubert & Shi, 2005), i.e., the users may use different scalar receiving weights, where the authors used the MSE duality between the uplink and downlink and an exhaustive search method to tackle the problem. The problem of joint THP transceiver design for multiuser MIMO systems has been studied in (Doostnejad et al., 2005) based on the MMSE criterion. However, a per-user power constraint is imposed, which may not be reasonable in the downlink. Morevoer, only the inter-user interference is pre-canceled nonlinearly, whereas the data streams of the same user are linearly precoded. The work of (Doostnejad et al., 2005) has been improved in (Mezghani et al., 2006) under a total transmit power constraint, where the users apply the MSE dualtiy between the uplink and downlink and the *projected gradient* algorithm to calculate the solution iteratively. Again, only the inter-user interference is pre-subtracted.

The above schemes have a common assumption that the BS, has perfect CSI. In a realistic scenario, however, the CSI is generally imperfect due to limited number of training symbols or channel time-variations. Therefore, the robust transceiver design which takes into account the uncertainties of CSI at the transmitter (CSIT) is required. Several robust schemes have been proposed for THP in the multiuser MISO downlink, which can be classified into the worst-case approach (Payaro et al., 2007; Shenouda & Davidson, 2007) and the stochastic approach (Dietrich et al., 2007; Shenouda & Davidson, 2007). The worst-case approach optimizes the worst system performance for any channel error in a predefined uncertainty region. In (Payaro et al., 2007) a robust power allocation scheme for THP was proposed, which maximizes the achievable rates for the worst-case errors in the CSI in the small errors regime. The authors of (Shenouda & Davidson, 2007) designed the THP transmitter to minimize the worst-case MSE over all admissible channel uncertainties subject to power constraints on each antenna, or a total power constraint. On the other hand, the stochastic approach optimizes a statistical measure of the system performance assuming that the statistics of the uncertainty is known. A robust nonlinear transmit zero-forcing filter with THP was presented in (Hunger et al., 2004) using a conditional-expectation approach, and has been extended lately in (Shenouda & Davidson, 2007) by relaxing the zero-forcing constraint and using the MMSE criterion. The problem of combined optimization of channel estimation and THP was considered in (Dietrich et al., 2007) and a conditional-expectation approach is adopted to solve the problem. All the above robust schemes are designed for the MISO downlink where each user has only one single antenna.

In this chapter for the MMSE scheme, we propose novel joint THP transceiver designs for the multiuser MIMO downlink with both perfect and imperfect CSIT. The transmitter performs nonlinear stream-wise (both inter-user and intra-user) interference pre-cancelation. We first consider the transceiver optimization problem under perfect CSIT and formulate it as minimizing the total mean square error (T-MSE) of the downlink (Zhang et al., 2005)

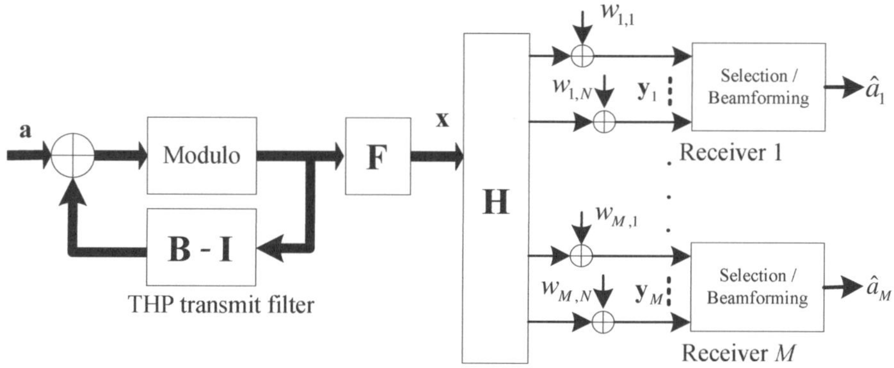

Fig. 1. Block diagram of ZF-THP in multiuser MIMO downlink system.

under a total transmit power constraint. Under the optimization criterion of minimizing the T-MSE, the stream-wise interference pre-cancelation structure is superior to the structure of inter-user only interference pre-cancelation combined with intra-user linear precoding adopted in (Doostnejad et al., 2005) and (Mezghani et al., 2006), which has already been proven to be true in a particular case, i.e., the single-user MIMO case (Shenouda & Davidson, 2008). By some convex analysis of the optimization problem, we find the necessary conditions for the optimal solution, by which the optimal transmitter and receivers are inter-dependent. We extend the iterative algorithm developed in (Zhang et al., 2005) to handle our problem. Although the iterative algorithm does not assure to converge to the globally optimal solution, it is guaranteed to converge to a locally optimal solution. Then, we make an extension of our design under perfect CSIT to the imperfect CSIT case which leads to a robust transceiver design against the channel uncertainty. The robust optimization problem is mathematically formulated as minimizing the expectation of the T-MSE conditioned on the channel estimates at the BS under a total transmit power constraint. An iterative optimization algorithm similar to its perfect CSIT counterpart can also be applied. Extensive simulation results are presented to illustrate the efficacy of our proposed schemes and their superiority over existing MMSE-based THP schemes.

The organization of the rest of this chapter is as follows. In Section 2, the system models for the multiuser MIMO downlink with THP established. In Section 3, two per-layer ZF-THP schemes are proposed and the analysis of the equivalent channel gains is given. In Section 4, the problem of the MMSE-THP design and analysis under both perfect and imperfect CSI is addressed. Simulation results are presented in Section 5. Section 6 concludes the chapter.

2. System models of multiuser MIMO downlink with THP

In this section, we will consider two system models for ZF-THP and MMSE-THP schemes, respectively.

2.1 System model for ZF-THP scheme

As mentioned in Section 1, for ZF-THP scheme, we consider the unique case that the number of users equals the number of transmit antennas at BS, denoted as M. Therein, each user is equipped with N receive antennas, as shown in Fig. 1. Perfect CSI is assumed at the transmitter (Windpassinger et al., 2004).

THP transmit filter group consists of a forward filter **F**, a backward filter **B**, and a modulo operator (Windpassinger et al., 2004). The transmit data symbol is denoted by the $M \times 1$ vector $\mathbf{a}$. After $\mathbf{a}$ passes through the THP transmit filter, the precoded symbol, which is denoted by the $M \times 1$ vector $\mathbf{x}$, is generated. It is assumed that the channel is flat fading. Denote the MIMO channel of user k by an $N \times M$ matrix $\mathbf{H}_k$. Each entry in $\mathbf{H}_k$ satisfies zero-mean unit-variance complex-Gaussian distribution, denoted by $CN(0,1)$. Through the channels, each user's $N \times 1$ received signal vector is

$$\mathbf{y}_k = \mathbf{H}_k\mathbf{x} + \mathbf{w}_k, \quad k = 1, 2, \cdots, M. \tag{1}$$

Therein the noise $\mathbf{w}_k$ is an $N \times 1$ vector, whose entries are independent and identically distributed (i.i.d.) random variables with the distribution $CN(0, \sigma_n^2)$.
Under the regulation that only one sub-channel be allocated to a user and all the M users be served simultaneously, every user's receive filter is a $1 \times N$ row vector, denoted by $\mathbf{r}_k$. For normalization, we assume $\|\mathbf{r}_k\|_2^2 = 1$, where $\|\cdot\|_2$ stands for the Euclidean norm of a vector. Thus, the detected signal can be expressed as

$$\hat{a}_k = \mathbf{r}_k\mathbf{H}_k\mathbf{x} + \mathbf{r}_k\mathbf{w}_k = \widetilde{\mathbf{h}}_k\mathbf{x} + \mathbf{r}_k\mathbf{w}_k, \quad k = 1, 2, \cdots, M, \tag{2}$$

where $\widetilde{\mathbf{h}}_k \triangleq \mathbf{r}_k\mathbf{H}_k$ is the equivalent channel row vector of user k. Construct the entire equivalent channel as $\widetilde{\mathbf{H}} \triangleq \left[\widetilde{\mathbf{h}}_1^H \quad \widetilde{\mathbf{h}}_2^H \quad \cdots \quad \widetilde{\mathbf{h}}_M^H\right]^H$.

2.2 System model for MMSE-THP scheme

Different from the above system model for ZF-THP scheme, we consider a more generalized model for MMSE-THP scheme, in which the number of users is not necessarily equal to that of transmit antennas. Therein, the BS is with M transmit antennas and K users are with N_k receive antennas at the kth user, $k = 1, \ldots, K$ (see Fig. 2). Let $\mathbf{H}_k \in \mathbb{C}^{N_k \times M}$ denote the channel between the BS and the kth user. The vector $\mathbf{d}_k \in \mathbb{C}^{L_k \times 1}$ represents the transmitted data vector for user k, where each entry belongs to the interval $[-\tau/2, \tau/2) + j \cdot [-\tau/2, \tau/2)$ (τ is the *modulo* base of THP as introduced later) and L_k is the number of data streams transmitted for user k. The data vectors are stacked into $\mathbf{d} \triangleq [\mathbf{d}_1^T \ \mathbf{d}_2^T \ \ldots \ \mathbf{d}_K^T]^T$, which is first reordered by a permutation matrix $\mathbf{\Pi} \in \mathbb{C}^{L \times L}$ ($\mathbf{\Pi}\mathbf{\Pi}^T = \mathbf{\Pi}^T\mathbf{\Pi} = \mathbf{I}_L$, $L \triangleq \sum_{k=1}^{K} L_k$) and then successively precoded using THP (see Fig. 2). The feedback matrix $\mathbf{F} \in \mathbb{C}^{L \times L}$ is a lower triangular matrix with zero diagonal. The structure of $\mathbf{F}$ enables inter-stream interference pre-cancelation and is different from the one used in (Mezghani et al., 2006) which only enables inter-user interference pre-cancelation. The *modulo* device performs a mod τ operation to avoid transmit power enhancement. Each entry of the output $\mathbf{w}$ of the *modulo* device is constrained in the interval $[-\tau/2, \tau/2) + j \cdot [-\tau/2, \tau/2)$. A common assumption in the literature is that the entries of $\mathbf{w}$ are uniformly distributed with unit variance (i.e., $\tau = \sqrt{6}$) and are mutually uncorrelated. Then $\mathbf{w}$ is linearly precoded by a feedforward matrix $\mathbf{P} \in \mathbb{C}^{M \times L}$ and transmitted over the downlink channel to the K users.
At the kth receiver, a decoding matrix $\mathbf{G}_k \in \mathbb{C}^{L_k \times N_k}$ and a *modulo* device are employed to estimate the data vector $\mathbf{d}_k$. Denote the estimate of $\mathbf{d}_k$ by $\tilde{\mathbf{d}}_k$, then it is given by

$$\tilde{\mathbf{d}}_k = (\mathbf{G}_k\mathbf{H}_k\mathbf{P}\mathbf{w} + \mathbf{G}_k\mathbf{n}_k) \bmod \tau, \tag{3}$$

in which $\mathbf{n}_k \in \mathbb{C}^{N_k \times 1}$ is the additive Gaussian noise vector at user k with zero mean and covariance matrix $\mathcal{E}\{\mathbf{n}_k\mathbf{n}_k^H\} = \sigma_{n,k}^2\mathbf{I}_{N_k}$. We assume that there is a total power constraint P_T at the BS so that tr $\left(\mathbf{P}^H\mathbf{P}\right) = P_T$.

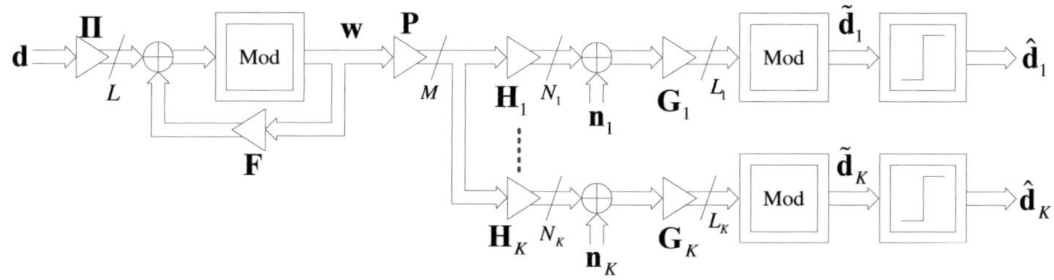

Fig. 2. Block diagram of MMSE-THP in multiuser MIMO downlink system.

3. Per-layer ZF-THP design and analysis

In this section, we will firstly propose two *per-layer* ZF-THP schemes for multiuser MIMO downlinks based on the system model in Fig. 1.

3.1 Capacity analysis for ZF-THP

Perform QR factorization to the conjugate transpose of the equivalent channel matrix $\widetilde{\mathbf{H}}$ in (2) in Section 2.1. This generates

$$\widetilde{\mathbf{H}} = \mathbf{S}\mathbf{F}^H, \tag{4}$$

where $\mathbf{F}$ is a unitary matrix and $\mathbf{S}$ is a lower-triangular matrix. In (Windpassinger et al., 2004), it is given that without account of the precoding loss (Yu et al., 2005), the sum-capacity of all layers is equivalent to

$$C^{sum} = \sum_{l=1}^{M} C_l = \sum_{l=1}^{M} \log\left(1 + \frac{\sigma_{x,l}^2}{\sigma_n^2} |s_{ll}|^2\right), \tag{5}$$

where $\sigma_{x,l}^2$ is the signal power in layer l. $|s_{ll}|^2$ can be interpreted as the equivalent channel gain of the lth layer.

If all the entries in $\widetilde{\mathbf{H}}$ have the distribution $CN(0,1)$, $|s_{ll}|^2$ is a random variable with the chi-square distribution of $2(M-l+1)$ degrees of freedom (Windpassinger et al., 2004). Nevertheless, when the total number of receive antennas is more than that of transmit antennas, the distribution of entries in $\widetilde{\mathbf{H}}$ will change with different precessing methods for receive antennas.

With the assumption that the channels of all the users have the same power attenuation, serving all the M users simultaneously means that the obtained receive diversity gain, which is defined by the negative slope of the outage probability versus signal-to-noise ratio (SNR) curve on a log-log scale (Tse & Viswanath, 2005), can be scaled by MN. In comparison, in per-user processing only $\lceil \frac{M}{N} \rceil$ users are served at one time, so the obtained receive diversity gain is scaled by M. Therefore, the strategy of serving all the M users simultaneously leads to N times larger receive diversity gain, which implies that each user should be provided with only one subchannel.

3.2 Per-layer transmit and receive filters design

From (2), the equivalent channel matrix $\widetilde{\mathbf{H}}$ is derived from the receive filters $\{\mathbf{r}_l, l = 1, \cdots, M\}$. Due to the channel matrix trianglization in (4), the higher layers will interfere with, but not be influenced by the lower ones. Denote the mapping $f_l : \bar{\mathbf{h}}_l = f_l(\{\mathbf{r}_p, p = 1, \cdots, l\})$, then $|s_{ll}|^2 = \|\bar{\mathbf{h}}_l\|_2^2 = \|f_l(\{\mathbf{r}_p\})\|_2^2$.

So the optimal $\{\mathbf{r}_l, l = 1, \cdots, M\}$ that maximizes the sum-capacity can be expressed as

$$\{\mathbf{r}_l^{opt}\} = \arg\max_{\{\mathbf{r}_l\}} \sum_{l=1}^{M} \log\left(1 + \frac{\sigma_{x,l}^2}{\sigma_n^2}\|f_l(\{\mathbf{r}_p\})\|_2^2\right), \quad \text{s. t.} \quad \|\mathbf{r}_l\|_2^2 = 1, \quad l = 1, \cdots, M. \tag{6}$$

According to (6), the design of one layer should take into account of its impact upon all the lower layers, and for each layer except the last one, there are multiple candidate users. So the solution of this optimization problem is very complicated. To make it practical, we employ a suboptimal approach, which conducts a per-layer optimization from high to low and converts the global optimization (6) into a series of greedy optimization as follows.

$$\mathbf{r}_l^{opt} = \arg\max_{\mathbf{r}_l} \log\left(1 + \frac{\sigma_{x,l}^2}{\sigma_n^2}\|f_l(\{\mathbf{r}_p\})\|_2^2\right), \quad \text{s. t.} \quad \|\mathbf{r}_l\|_2^2 = 1, \quad l = 1, \cdots, M. \tag{7}$$

When processing one layer, say layer l, we disregard its impact upon other layers and just maximize the power in $\bar{\mathbf{h}}_l$. Specifically, we suppose all the rest users as candidates, and generate their own receive filters according to some proper criterion. Thus, for layer l and user k, the equivalent channel row vector, denoted by $\tilde{\mathbf{h}}_{l,k}^{equi}$, can be obtained. Here, $\bar{\mathbf{H}}_{l-1}^{null}$ represents the subspace orthogonal to that spanned by $\{\tilde{\mathbf{h}}_p^H, p = 1, \cdots, l-1\}$, and the projection power of $\tilde{\mathbf{h}}_{l,k}^{equi,H}$ onto $\bar{\mathbf{H}}_{l-1}^{null}$ is interpreted as user k's *residual* channel gain in layer l. Then, the user with the largest residual channel gain is selected and placed into layer l. In this way, all the users can be arranged into the sequence of layers and $\{\bar{\mathbf{h}}_l, l = 1, \cdots, M\}$ can be obtained sequentially. Within this per-layer approach, the key is how to design the receive filters. For layer l and user k, we denote $\bar{\mathbf{H}}_k^{(l)}$ as the projection of $\mathbf{H}_k^H$ onto $\bar{\mathbf{H}}_{l-1}^{null}$, then the optimal receive filter $\mathbf{r}_{k,l}$ can be obtained by

$$\mathbf{r}_{l,k}^{opt} = \arg\max_{\mathbf{r}} \|\mathbf{r}(\bar{\mathbf{H}}_k^{(l)})^H\|_2^2, \quad \text{s. t.} \quad \|\mathbf{r}\|_2^2 = 1. \tag{8}$$

The solution of this maximization problem can be given by the theory of Rayleigh quotient (Horn & Johnson, 1985). That is, $\mathbf{r}_{l,k}^{opt}$ is the conjugate transpose of the eigenvector corresponding to the maximum eigenvalue of the matrix $(\bar{\mathbf{H}}_k^{(l)})^H \bar{\mathbf{H}}_k^{(l)}$. In essential, this processing method aims to maximize power gain and diversity gain of each layer through the design of receive antenna beamforming (RAB). The per-layer RAB scheme is summarized in Table 1-a. Therein EVD$(\cdot)$ returns the set of eigenvalues and eigenvectors, and Householder$(\cdot)$ returns the Householder matrix. $\mathbf{I}_N$ stands for an $N \times N$ identity matrix.
By this scheme, the user ordering $\{\pi_l\}$, the receive filters $\{\hat{\mathbf{r}}_l\}$, and the transmit filter $\hat{\mathbf{F}} = \mathbf{F}^{(M)} \cdots \mathbf{F}^{(1)}$ are all generated. However, the operations of eigenvalue decomposition (EVD) still consume a certain complexity. To further reduce the complexity and employ less analog chains at the receivers (Gorokhov et al., 2003), RAB can be replaced by receive antenna selection (RAS). Specifically, for a layer and a candidate user, instead of computing the the eigenvector, we just select the receive antenna whose equivalent channel vector has the maximum Euclidean norm, as shown in Table 1-b.
Remarks:

- For each layer, the aim of the receive filter design is to adjust the weights of receive antennas to maximize the power in the equivalent channel vector's component orthogonal to the higher layers' dimensions (i.e., $\|\bar{\mathbf{h}}_l\|_2^2$), but not the power in the equivalent channel vector itself (i.e., $\|\tilde{\mathbf{h}}_l\|_2^2$).

(a) The scheme of per-layer RAB	(b) The scheme of per-layer RAS
Given all user's $N \times M$ *channel matrices* $\mathbf{H}_k$,	*Given all user's* $N \times M$ *channel matrices* $\mathbf{H}_k$,
$k = 1, 2, \cdots, M$.	$k = 1, 2, \cdots, M$.
Initialization: The candidate user set	*Initialization: The candidate user set*
$\Phi = \{1, 2, \cdots, M\}$,	$\Phi = \{1, 2, \cdots, M\}$,
$\mathbf{F}^{(0)} = \mathbf{I}_M, \bar{\mathbf{H}}_k^{(0)} = \mathbf{H}_k^H, k = 1, 2, \cdots, M$.	$\mathbf{F}^{(0)} = \mathbf{I}_M, \bar{\mathbf{H}}_k^{(0)} = \mathbf{H}_k^H, k = 1, 2, \cdots, M$.
For the layer index $l : 1 \rightarrow M$	*For the layer index* $l : 1 \rightarrow M$
For the user index $k \in \Phi$	*For the user index* $k \in \Phi$
$\bar{\mathbf{H}}_k^{(l)} = \mathbf{F}^{(l-1)} \bar{\mathbf{H}}_k^{(l-1)}$	$\bar{\mathbf{H}}_k^{(l)} = \mathbf{F}^{(l-1)} \bar{\mathbf{H}}_k^{(l-1)}$
$\bar{\mathbf{H}}_k^{(l),proj}$ *is comprised by the*	$\bar{\mathbf{H}}_k^{(l),proj}$ *is comprised by the*
l*th to* M*th rows of* $\bar{\mathbf{H}}_k^{(l)}$	l*th to* M*th rows of* $\bar{\mathbf{H}}_k^{(l)}$
$\{[\lambda_n \quad \mathbf{u}_n], n = 1, \cdots, N\} =$	p_n *is the Euclidean norm of*
$\mathrm{EVD}((\bar{\mathbf{H}}_k^{(l),proj})^H \bar{\mathbf{H}}_k^{(l),proj})$	*the* n*th column of* $\bar{\mathbf{H}}_k^{(l),proj}$
$n_{max} = \arg\max_n \{\lambda_n\}$	$n_{max} = \arg\max_n \{p_n\}$
$\mathbf{r}_k = \mathbf{u}_{n_{max}}^H$	$\mathbf{r}_k$ *is a* $1 \times N$ *vector,* $[0, \cdots, 0, \star]$
$\lambda^{(k)} = \lambda_{n_{max}}$	$p^{(k)} = p_{n_{max}}$
end	*end*
$\widetilde{k} = \arg\max_{k \in \Phi} \{\lambda^{(k)}\}$	$\widetilde{k} = \arg\max_{k \in \Phi} \{p^{(k)}\}$
$\Phi = \Phi \backslash \{\widetilde{k}\}$	$\Phi = \Phi \backslash \{\widetilde{k}\}$
$\pi_l = \widetilde{k}$	$\pi_l = \widetilde{k}$
$\hat{\mathbf{r}}_l = \mathbf{r}_{\widetilde{k}}$	$\hat{\mathbf{r}}_l = \mathbf{r}_{\widetilde{k}}$
$\bar{\mathbf{h}}_l$ *is comprised by the* l*th to* M*th rows of* $\bar{\mathbf{H}}_{\widetilde{k}}^{(l)} \mathbf{r}_{\widetilde{k}}^H$	$\bar{\mathbf{h}}_l$ *is comprised by* l*th to* M*th rows of* $\bar{\mathbf{H}}_{\widetilde{k}}^{(l)} \mathbf{r}_{\widetilde{k}}^H$
$\mathbf{F}^{(l)} = \left[\begin{array}{c\|c} \mathbf{I}_{l-1} & \mathbf{0}_{(l-1)\times(M-l+1)} \\ \hline \mathbf{0}_{(M-l+1)\times(l-1)} & \mathrm{Householder}(\bar{\mathbf{h}}_l) \end{array}\right]$	$\mathbf{F}^{(l)} = \left[\begin{array}{c\|c} \mathbf{I}_{l-1} & \mathbf{0}_{(l-1)\times(M-l+1)} \\ \hline \mathbf{0}_{(M-l+1)\times(l-1)} & \mathrm{Householder}(\bar{\mathbf{h}}_l) \end{array}\right]$
end	*end*

Table 1. The schemes of per-layer RAB and per-layer RAS.

- In the successive mechanism of THP, the higher a layer, the less it costs for the interference suppression. In the per-layer schemes, the users with large residual channel gains are placed into the high layers. In this way, the power wasted in the interference suppression can be decreased, but the power contributing to the sum-capacity can be increased.
- As a suboptimal solution of (8), per-layer RAS is inferior to per-layer RAB. However, for the sake of practice, in per-layer RAS only the indexes of the selected antennas should be informed to the receivers, but in per-layer RAB, the counterparts are the designed receive filter weights.

3.3 Comparison between per-layer RAB and RAS

Here, we do not order the users and consider the lth layer's projected channel matrix $(\bar{\mathbf{H}}_k^{(l),proj})^H, \forall k$, whose entries have i.i.d. $CN(0,1)$ distribution. Using per-layer RAB, the equivalent channel gain is the square of its maximum singular value, while using per-layer RAS, the equivalent channel gain is the square of its maximum row vector's Euclidean norm. We denote these two kinds of channel gains by $\delta^2_{RAB}(l)$ and $\delta^2_{RAS}(l)$, respectively. With the decrease of l, the relative difference between $\delta^2_{RAB}(l)$ and $\delta^2_{RAS}(l)$ tends to decrease, which can be deduced below.

Theorem 1. *Given a $2 \times n$ matrix **A** such that all its entries have i.i.d. $CN(0,1)$ distribution. Denote the eigenvalues of $\mathbf{AA}^H$ as λ'_i, $i = 1,2$. Let $\lambda_1 \triangleq \max_i\{\lambda'_i\}$, $\lambda_2 \triangleq \min_i\{\lambda'_i\}$, and denote $\Delta\lambda \triangleq \lambda_1 - \lambda_2$. Then with $n \to \infty$, the ratio $E(\Delta\lambda)/E(\lambda_1) \to 0$.*

Proof: By the bidiagonalization (Th. 3.4 in (Wang et al., 2006)), **A** is unitarily similar to a lower-triangular matrix Λ, where

$$\Lambda \triangleq \begin{bmatrix} x_{2n} & 0 & \cdots 0 \\ y_2 & x_{2(n-1)} & \cdots 0 \end{bmatrix}, \tag{9}$$

$x^2_{2n}, x^2_{2(n-1)}$ and y_2^2 are independent chi-square distributed random variables with the degrees of freedom $2n$, $2(n-1)$ and 2, respectively. Then,

$$\mathbf{AA}^H = \Lambda\Lambda^H = \begin{bmatrix} x^2_{2n} & x_{2n}y_2 \\ x_{2n}y_2 & x^2_{2(n-1)} + y_2^2 \end{bmatrix}. \tag{10}$$

Therein we denote a new chi-square random variable $y^2_{2n} \triangleq x^2_{2(n-1)} + y_2^2$. Further, the eigenvalues of $\mathbf{AA}^H$ can be obtained as

$$\lambda_{1,2} = \frac{x^2_{2n} + y^2_{2n}}{2} \pm \sqrt{x^2_{2n}y_2^2 + (\frac{x^2_{2n} + y^2_{2n}}{2})^2 - x^2_{2n}y^2_{2n}}. \tag{11}$$

Substitute $x^2_{2n} + y^2_{2n}$ by x^2_{4n}, then

$$\Delta\lambda = 2\sqrt{x^2_{2n}y_2^2 + (\frac{x^2_{4n}}{2})^2 - x^2_{2n}y^2_{2n}}. \tag{12}$$

$$E(\Delta\lambda) \leq \sqrt{E((\Delta\lambda)^2)} = 2\sqrt{E(x^2_{2n}y_2^2) + E\Big((\frac{x^2_{4n}}{2})^2\Big) - E(x^2_{2n}y^2_{2n})}. \tag{13}$$

According to the character of chi-square distribution (Horn & Johnson, 1985), we have

$$E(x^2_{2n}y_2^2) = E(x^2_{2n})E(y_2^2) = 2n \cdot 2 = 4n, \tag{14}$$

$$E\Big((\frac{x^2_{4n}}{2})^2\Big) = \frac{4n(4n+2)}{4} = 4n^2 + 2n, \tag{15}$$

$$E(x^2_{2n}y^2_{2n}) = E(x^2_{2n})E(y^2_{2n}) = 2n \cdot 2n = 4n^2. \tag{16}$$

Plunging to (13),

$$E(\Delta\lambda) \leq 2\sqrt{6n}. \tag{17}$$

On the other hand,

$$E(\lambda_1) \geq E\left(\frac{x^2_{2n} + y^2_{2n}}{2} + \sqrt{(\frac{x^2_{2n} + y^2_{2n}}{2})^2 - x^2_{2n}y^2_{2n}}\right) = E(x^2_{2n}) = 2n. \tag{18}$$

So $E(\Delta\lambda)/E(\lambda_1) \leq \sqrt{6}n^{-1/2}$.
With $n \to \infty$, $\sqrt{6}n^{-1/2} \to 0$. Considering $E(\Delta\lambda)/E(\lambda_1) \geq 0$, it can be concluded that with $n \to \infty$, $E(\Delta\lambda)/E(\lambda_1) \to 0$.

■

Corollary 1. *Given the same conditions as Theorem 1. Additionally denote the row vectors of* $\mathbf{A}$ *as* $\mathbf{a}_i$, $i = 1,2$. *Then with* $n \to \infty$, *the ratio* $E(\lambda_1 - \max_i \|\mathbf{a}_i\|_2^2)/E(\lambda_1) \to 0$.

Proof: By the character of Rayleigh quotient (Horn & Johnson, 1985), $\forall r \in \mathbb{C}^{1\times 2}$, $rr^H = 1$, the maximum and minimum values of $r\mathbf{A}\mathbf{A}^H r^H$ are λ_1 and λ_2, respectively. Let $r' = [0 \quad 1]$ or $r' = [1 \quad 0]$, then the value of $r'\mathbf{A}\mathbf{A}^H r'^H$ is surely between λ_1 and λ_2. Obviously, $r'\mathbf{A}\mathbf{A}^H r'^H$ is equivalent to $\|\mathbf{a}_i\|_2^2$, $i = 1,2$. Thus,

$$0 \leq \lambda_1 - \max_i \|\mathbf{a}_i\|_2^2 \leq \Delta\lambda. \tag{19}$$

According to Theorem 1, with $n \to \infty$, the ratio $E(\lambda_1 - \max_i \|\mathbf{a}_i\|_2^2)/E(\lambda_1) \to 0$.

■

According to Corollary 1, when the number of transmit antennas K tends to be infinite, from the bottom layer ($l = K$) to the top layer ($l = 1$), the relative difference between $\delta_{RAB}(l)$ and $\delta_{RAS}(l)$, denoted by $\Delta G_l \triangleq (\delta_{RAB}(l) - \delta_{RAS}(l))/\delta_{RAB}(l)$, tends to decrease until zero.
Next, consider the capacity of each layer in both per-layer RAB and per-layer RAS. Denote the capacities of layer l in these two schemes as C_l^{RAB} and C_l^{RAS}, respectively, and denote their difference as $\Delta C_l \triangleq C_l^{RAB} - C_l^{RAS}$. Let $n = K - l + 1$, which means the degree of freedom in layer l, then ΔC_l can be rewritten as $\Delta C(n)$. Denote $\gamma \triangleq \sigma_x^2/\sigma_n^2$. In the medium and high (SNR) scenarios, the characteristic of $\Delta C(n)$ is described in the following corollary.

Corollary 2. *Given the same conditions as in Corollary 1,*

$$E(\Delta C(n)) = E\Big(\log(1 + \gamma\lambda_1(n)) - \log(1 + \gamma \max_i \|\mathbf{a}_i(n)\|_2^2)\Big). \tag{20}$$

In the medium and high SNR scenarios, with $n \to \infty$, *it holds that* $E(\Delta C(n)) \to 0$.

The details of the proof of Corollary 2 are omitted due to page limit[2]. Then, we consider the low SNR scenarios, where $\gamma \to 0$,

$$E(\Delta C(n)) \approx E(\gamma\lambda_1(n) - \gamma \max_i \|\mathbf{a}_i(n)\|_2^2) \leq E(\gamma\Delta\lambda(n)). \tag{21}$$

From (17), it can be inferred that in the low SNR scenarios $E(\Delta C(n))$ increases with n. This trend is opposite to that in the medium and high SNR scenarios.
Theorem 1 and its corollaries indicate the case of a two-row matrix, which corresponds to the scenarios with two receive antennas at each receiver. Thus, a conclusion can be drawn that when the number of transmit antennas increases infinitely, both ΔG_l and ΔC_l (at medium and high SNR) in those high layers will asymptotically tend to zero. This implies that in the case of a large number of transmit antennas, for those higher layers, whether applying RAB or RAS, the differences of channel gains could be approximately negligible, but RAS consumes much less complexity.

[2] Please refer to (Huang et al., 2010).

Scheme	Layer 1	Layer 2
per-layer RAB	$\widetilde{\lambda}^{max,2}_{M-2l+2}$	$\widetilde{\lambda}^{max,2}_{M-2l+1}$
per-layer RAS	$\widetilde{p}^{max,2}_{M-2l+2}$	$\widetilde{p}^{max,2}_{M-2l+1}$
per-user	$\widetilde{\lambda}^{max,2}_{M-2l+2}$	$\widetilde{\lambda}^{min,2}_{M-2l+2}$

Table 2. The equivalent channel gains of a unit in per-layer RAB, per-layer RAS and per-user processing.

3.4 Comparison with per-user processing

We still consider the case of two receive antennas for each receiver. In the per-user processing, each receiver owns a group of two adjacent layers, represented by a $2 \times M$ channel matrix. The channel matrices of lower groups should be orthogonally projected onto those of higher groups. Hence the equivalent channel matrix for group l, which includes layers $2l-1$ and $2l$, is a $2 \times (M-2l+2)$ matrix, and all its entries can be assumed to be i.i.d. $CN(0,1)$ random variables. Here, each group of two adjacent layers is interpreted as a basic unit. For the lth unit, the two equivalent channel gains are the squares of singular values of a $2 \times (M-2l+2)$ matrix, denoted by $\widetilde{\lambda}^{max,2}_{M-2l+2}$ and $\widetilde{\lambda}^{min,2}_{M-2l+2}$, respectively.

Accordingly, we also bind every two adjacent layers as a unit in both per-layer RAB and per-layer RAS schemes. In this way, for the lth unit, in per-layer RAB, one equivalent channel gain equals the square of the maximum singular value of a $2 \times (M-2l+2)$ matrix, and the other equals the square of the maximum singular value of a $2 \times (M-2l+1)$ matrix, denoted by $\widetilde{\lambda}^{max,2}_{M-2l+2}$ and $\widetilde{\lambda}^{max,2}_{M-2l+1}$, respectively; while in per-layer RAS, they are the squares of the maximum row-norm of a $2 \times (M-2l+2)$ matrix and the maximum row-norm of a $2 \times (M-2l+1)$ matrix, denoted by $\widetilde{p}^{max,2}_{M-2l+2}$ and $\widetilde{p}^{max,2}_{M-2l+1}$, respectively. The equivalent channel gains of a unit in these three schemes are summarized in Table 2.

Remarks:

- In layer 1, per-layer RAB and per-user processing have the same equivalent channel gain, which is larger than that of per-layer RAS.
- In layer 2, the chances of exploiting the maximum singular value or the maximum row-norm still exist in per-layer RAB and per-layer RAS; while in per-user processing only the minimum singular value is used, and hence, the diversity gain is lost.

Based on the above observations, per-layer RAB outperforms the other two schemes evidently. But the relation between per-layer RAS and per-user processing is indistinct. We analyze two extreme cases: with very low SNR, where the maximum sum-capacity is approximately achieved by allocating all the signal power into the best layer, or with very high SNR, where by allocating the power into all the layers averagely (Tse & Viswanath, 2005).

It can be derived from (19) that $E(\widetilde{p}^{max}_{K-2l+2}) < E(\widetilde{\lambda}^{max,2}_{K-2l+2})$, thus, at low SNR, per-layer RAS has smaller sum-capacity than per-user processing.

At very high SNR, the capacity depends on the product of the channel gains of two layers. Let $n = K - 2l + 2$, then the lower bounds can be developed in **Appendix A** that for per-user processing $E(\widetilde{\lambda}^{max,2}_n \widetilde{\lambda}^{min,2}_n) \geq 4n^2 - 6n$ and for per-layer RAS $E(\widetilde{p}^{max,2}_n \widetilde{p}^{max,2}_{n-1}) \geq 4n^2 - 4n$. Though the tightness of these two lower bounds are not proved, the advantage of per-layer RAS over per-user processing at high SNR can be additionally validated by the simulation results in Subsection 5.1.

4. Stream-wise MMSE-THP design and analysis

In this section, we firstly propose our joint THP transceiver design under perfect CSIT using a minimum total mean square error (MT-MSE) criterion. Using convex analysis for the optimization problem we derive the necessary conditions for the optimal transceiver in Subsection 4.1.1. Then the iterative algorithm proposed in (Zhang et al., 2005) is extended in Subsection 4.1.2 to obtain a locally optimal transceiver. Furthermore, we introduce a robust THP transceiver design for the multiuser MIMO downlink in Subsection 4.2, which is more effective against the uncertainty in the CSIT than the above simple solution. The robust optimization problem is mathematically formulated as minimizing the expectation of the T-MSE conditioned on the channel estimates at the BS under a total transmit power constraint. Then the iterative algorithm proposed in Subsection 4.1 is applied to solve the problem.

4.1 Transceiver optimization under perfect CSIT

4.1.1 Problem reformulation

Our design is based on the linear representation (Joham et al., 2004) (see Fig. 3) of the system in Fig. 2, where the *modulo* devices at the transmitter and receivers are replaced by the additive vector $\mathbf{a} \triangleq [\mathbf{a}_1^T \ \mathbf{a}_2^T \ \ldots \ \mathbf{a}_K^T]^T$ and $-\tilde{\mathbf{a}}_k$, $k = 1, ..., K$, where $\mathbf{a} \in \tau\left(\mathbb{Z}^{L\times 1} + j \cdot \mathbb{Z}^{L\times 1}\right)$ and $\tilde{\mathbf{a}}_k \in \tau\left(\mathbb{Z}^{L_k\times 1} + j \cdot \mathbb{Z}^{L_k\times 1}\right)$. The vectors $\mathbf{a}$ and $\tilde{\mathbf{a}}_k$ are chosen to make the same $\mathbf{w}$ and $\tilde{\mathbf{d}}_k$ as the *modulo* devices at the transmitter and receivers output respectively.

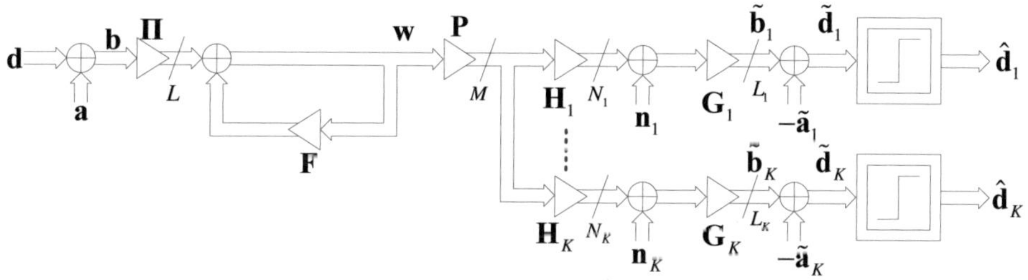

Fig. 3. Equivalent linear representation of THP in Fig. 2.

Define $\mathbf{b}_k \triangleq \mathbf{d}_k + \mathbf{a}_k$ and $\tilde{\mathbf{b}}_k \triangleq \tilde{\mathbf{d}}_k + \tilde{\mathbf{a}}_k$ and stack them into $\mathbf{b} \triangleq [\mathbf{b}_1^T \ \ldots \ \mathbf{b}_K^T]^T$ and $\tilde{\mathbf{b}} \triangleq [\tilde{\mathbf{b}}_1^T \ \ldots \ \tilde{\mathbf{b}}_K^T]^T$. Let $\mathbf{H} \triangleq \left[\mathbf{H}_1^T \ldots \mathbf{H}_K^T\right]^T$, $\mathbf{n} \triangleq \left[\mathbf{n}_1^T \ldots \mathbf{n}_K^T\right]^T$ and $\mathbf{G} \triangleq \text{blockdiag}(\mathbf{G}_1, ..., \mathbf{G}_K)$, then from Fig. 3 we have

$$\mathbf{\Pi b} + \mathbf{Fw} = \mathbf{w} \Rightarrow \mathbf{b} = \mathbf{\Pi}^T(\mathbf{I}_L - \mathbf{F})\mathbf{w} \tag{22}$$

and

$$\tilde{\mathbf{b}} = \mathbf{GHPw} + \mathbf{Gn}. \tag{23}$$

We consider the MSE between $\mathbf{b}$ and $\tilde{\mathbf{b}}$ rather than $\mathbf{d}$ and $\tilde{\mathbf{d}}$ in order to bypass the impact of the *modulo* operations and define it as the total MSE (T-MSE) of the downlink, which is written as follows:

$$\begin{aligned} \text{T-MSE} &= \mathcal{E}_{\mathbf{w,n}}\left\{\left\|\tilde{\mathbf{b}} - \mathbf{b}\right\|_2^2\right\} = \mathcal{E}_{\mathbf{w,n}}\left\{\left\|\left(\mathbf{GHP} - \mathbf{\Pi}^T(\mathbf{I}_L - \mathbf{F})\right)\mathbf{w} + \mathbf{Gn}\right\|_2^2\right\} \\ &= \left\|\mathbf{GHP} - \mathbf{\Pi}^T(\mathbf{I}_L - \mathbf{F})\right\|_F^2 + \text{tr}(\mathbf{G}^H\mathbf{G}\mathbf{\Sigma_n}), \end{aligned} \tag{24}$$

where $\mathbf{\Sigma_n} \triangleq \mathcal{E}\{\mathbf{nn}^H\} = \text{blockdiag}\left(\sigma_{n,1}^2 \mathbf{I}_{N_1}, ..., \sigma_{n,K}^2 \mathbf{I}_{N_K}\right)$.

So our transceiver design problem is to find a set $\left\{\mathbf{\Pi},\ \mathbf{F},\ \mathbf{P},\ \{\mathbf{G}_k\}_{k=1}^K\right\}$ that minimizes the T-MSE defined in (24) under a total transmit power constraint. Mathematically it can be formulated as follows:

$$\begin{aligned} \min_{\mathbf{\Pi},\ \mathbf{F},\ \mathbf{P},\ \{\mathbf{G}_k\}_{k=1}^K} \quad & \text{T-MSE} \\ \text{s.t.} \quad & \text{tr}\left(\mathbf{P}^H\mathbf{P}\right) = P_T, \\ & [\mathbf{F}]_{m,n} = 0,\ \forall\, 1 \leqslant m,n \leqslant L \text{ and } m \leqslant n. \end{aligned} \tag{25}$$

4.1.2 Iterative algorithm

In this subsection, through some analysis, we find the necessary conditions for the optimal $\mathbf{\Pi}$, $\mathbf{F}$, $\mathbf{P}$ and $\{\mathbf{G}_k\}_{k=1}^K$, which form an inter-dependence among them. This kind of inter-dependence leads to an iterative algorithm similar to the one proposed in (Zhang et al., 2005). In each iteration, we first determine the suboptimal reordering matrix $\mathbf{\Pi}$ and update $\mathbf{P}$ and $\mathbf{F}$ using the updated $\{\mathbf{G}_k\}_{k=1}^K$ in the last iteration, then update $\{\mathbf{G}_k\}_{k=1}^K$ using the above updated $\mathbf{\Pi}$, $\mathbf{P}$ and $\mathbf{F}$.

For ease of derivation, we introduce two new matrix variables $\mathbf{T} \triangleq \beta^{-1}\mathbf{P}$ and $\mathbf{R} \triangleq \beta\mathbf{G}$ to replace $\mathbf{P}$ and $\mathbf{G}$, where β is a positive real number. Then (24) is rewritten as

$$\text{T-MSE} = \left\|\mathbf{RHT} - \mathbf{\Pi}^T(\mathbf{I}_L - \mathbf{F})\right\|_F^2 + \beta^{-2} \cdot \text{tr}(\mathbf{R}^H\mathbf{R}\mathbf{\Sigma_n}). \tag{26}$$

Moreover, using the total power constraint in (25) we obtain

$$\beta = P_T^{\frac{1}{2}} \left(\text{tr}(\mathbf{TT}^H)\right)^{-\frac{1}{2}}. \tag{27}$$

Note that $\mathbf{F}$ only appears in the first term of (26). We expand the first term of (26) as follows:

$$\begin{aligned} & \left\|\mathbf{RHT} - \mathbf{\Pi}^T(\mathbf{I}_L - \mathbf{F})\right\|_F^2 \\ = \ & \|\mathbf{\Pi RHT} - (\mathbf{I}_L - \mathbf{F})\|_F^2 \quad (\text{for } \mathbf{\Pi}^T\mathbf{\Pi} = \mathbf{\Pi}\mathbf{\Pi}^T = \mathbf{I}_L) && (28) \\ = \ & \sum_{i=1}^{L} \left\| \begin{bmatrix} \mathbf{A}_i \\ \mathbf{B}_i \end{bmatrix} \mathbf{t}_i - \mathbf{e}_i + \mathbf{f}_i \right\|_2^2, && (29) \end{aligned}$$

where

$$\begin{bmatrix} \mathbf{A}_i \\ \mathbf{B}_i \end{bmatrix} = \mathbf{\Pi RH},\ \mathbf{A}_i \in \mathbb{C}^{i \times M},\ \mathbf{B}_i \in \mathbb{C}^{(L-i)\times M}, \tag{30}$$

$\mathbf{t}_i$, $\mathbf{e}_i$ and $\mathbf{f}_i$ are the ith columns of $\mathbf{T}$, $\mathbf{I}_L$ and $\mathbf{F}$ respectively. The equality in (28) follows from the fact that the Frobenius norm of a matrix remains constant after the multiplication of a unitary matrix (Horn & Johnson, 1985). For fixed $\mathbf{\Pi}$, $\mathbf{T}$ and $\mathbf{R}$, each term in the summation in (29) can be minimized separately. With the lower triangular and zero diagonal structure of $\mathbf{F}$, the optimal $\mathbf{f}_i$ that minimizes the ith term of (29) is easily computed as:

$$\mathbf{f}_i = -\begin{bmatrix} \mathbf{0}_{i\times M} \\ \mathbf{B}_i \end{bmatrix} \mathbf{t}_i,\ i = 1, \ldots, L. \tag{31}$$

By substituting (27) and (31) into (26), we rewrite the T-MSE as:

$$\text{T-MSE} = \sum_{i=1}^{L} \left\| \begin{bmatrix} \mathbf{A}_i \\ \mathbf{0} \end{bmatrix} \mathbf{t}_i - \mathbf{e}_i \right\|_2^2 + \xi \sum_{i=1}^{L} \text{tr}(\mathbf{t}_i^H \mathbf{t}_i), \tag{32}$$

where $\xi \triangleq P_T^{-1} \sum_{k=1}^{K} \sigma_{n,k}^2 \text{tr}\left(\mathbf{R}_k^H \mathbf{R}_k\right)$ is a nonnegative real number, i.e., $\xi \geq 0$.
For fixed $\mathbf{\Pi}$ and $\mathbf{R}$, the optimization problem in (25) can be reformulated as:

$$\min_{\mathbf{T}} g(\mathbf{T}) \triangleq \sum_{i=1}^{L} \left\| \begin{bmatrix} \mathbf{A}_i \\ \mathbf{0} \end{bmatrix} \mathbf{t}_i - \mathbf{e}_i \right\|_2^2 + \xi \sum_{i=1}^{L} \text{tr}(\mathbf{t}_i^H \mathbf{t}_i). \tag{33}$$

Notice that by the introduction of β the power constraint in the original optimization problem has been absorbed into the objective function, so (33) is an unconstraint optimization problem. The Hessian matrix of (32) with respect to $\mathbf{t}_i$ is calculated and shown below:

$$\nabla_{\mathbf{t}_i^T} \left[\nabla_{\mathbf{t}_i^*} g(\mathbf{T}) \right] = \mathbf{A}_i^H \mathbf{A}_i + \xi \mathbf{I}_M \succeq \mathbf{0}. \tag{34}$$

The Hessian matrix in (34) being positive semidefinite indicates that $g(\mathbf{T})$ is convex respect to $\mathbf{t}_i$. Then the optimal $\mathbf{t}_i$ is derived by calculating the first order derivative with respect to $\mathbf{t}_i^*$ and setting it to zero, i.e.,

$$\frac{\partial g(\mathbf{T})}{\partial \mathbf{t}_i^*} = \mathbf{A}_i^H \mathbf{A}_i \mathbf{t}_i - \begin{bmatrix} \mathbf{A}_i \\ \mathbf{0} \end{bmatrix}^H \mathbf{e}_i + \xi \mathbf{t}_i = \mathbf{0} \Rightarrow \mathbf{t}_i = \left(\mathbf{A}_i^H \mathbf{A}_i + \xi \mathbf{I}_M \right)^{-1} \begin{bmatrix} \mathbf{A}_i^H & \mathbf{0} \end{bmatrix} \mathbf{e}_i. \tag{35}$$

Now we consider the problem of the optimal ordering, i.e., the optimal $\mathbf{\Pi}$. (32) can be rewritten as:

$$\text{T-MSE} = \sum_{i=1}^{L} \left(\mathbf{t}_i^H \left(\mathbf{A}_i^H \mathbf{A}_i + \xi \mathbf{I}_M \right) \mathbf{t}_i - \mathbf{t}_i^H \begin{bmatrix} \mathbf{A}_i^H & \mathbf{0} \end{bmatrix} \mathbf{e}_i - \mathbf{e}_i^H \begin{bmatrix} \mathbf{A}_i \\ \mathbf{0} \end{bmatrix} \mathbf{t}_i + 1 \right). \tag{36}$$

Substituting (35) into (36) and after some algebraic manipulations, we rewrite the T-MSE as

$$\text{T-MSE} = L - \sum_{i=1}^{L} \text{tr} \left(\mathbf{e}_i^H \begin{bmatrix} \mathbf{A}_i \\ \mathbf{0} \end{bmatrix} \left(\mathbf{A}_i^H \mathbf{A}_i + \xi \mathbf{I}_M \right)^{-1} \begin{bmatrix} \mathbf{A}_i^H & \mathbf{0} \end{bmatrix} \mathbf{e}_i \right). \tag{37}$$

The T-MSE in (37) is a function of $\mathbf{\Pi}$ for fixed $\mathbf{R}$. An exhaustive search is needed to find the optimal reordering matrix that minimizes (37). To avoid the high complexity of this global optimal approach, we adopt a suboptimal successive reordering algorithm that only maximizes one term of the summation in (37) and starts from the L-th term till the 1st term. The maximization of the ith term determines the ith row of $\mathbf{\Pi}$. The procedure of the reordering algorithm is listed in Table 3.
Till now we have found the suboptimal $\mathbf{\Pi}$, the optimal $\mathbf{F}$, $\mathbf{T}$ and β for fixed $\mathbf{R}$. Next we calculate the optimal $\mathbf{R}$ under fixed $\mathbf{\Pi}$, $\mathbf{F}$ and $\mathbf{T}$.
The T-MSE in (26) can be expanded as the summation of the K users' MSEs, and the MSE for the kth user is written as follows:

$$\text{MSE}_k = \mathcal{E}_{\mathbf{w,n}} \left\{ \left\| \tilde{\mathbf{b}}_k - \mathbf{b}_k \right\|_2^2 \right\} = \left\| \mathbf{R}_k \mathbf{H}_k \mathbf{T} - \mathbf{E}_k^T \mathbf{\Pi}^T (\mathbf{I}_L - \mathbf{F}) \right\|_F^2 + \zeta_k \cdot \text{tr}(\mathbf{R}_k^H \mathbf{R}_k), \tag{38}$$

in which $\mathbf{E}_k \triangleq \left[\mathbf{e}_{\sum_{i=1}^{k-1} L_i + 1}, \ \ldots, \ \mathbf{e}_{\sum_{i=1}^{k} L_i} \right]$ and $\zeta_k \triangleq P_T^{-1} \sigma_{n,k}^2 \text{tr}(\mathbf{T}^H \mathbf{T}) \geq 0$.

Initialization:
$\mathbf{A} = \mathbf{RH}, \mathbf{\Pi} = \mathbf{0}_{L\times L}, \xi = P_T^{-1}\sum_{k=1}^{K}\sigma_{n,k}^2 \mathrm{tr}\left(\mathbf{R}_k^H\mathbf{R}_k\right).$
For $i = L : -1 : 1,$
$\mathbf{M} = \mathbf{A}\left(\mathbf{A}^H\mathbf{A} + \xi\mathbf{I}_M\right)^{-1}\mathbf{A}^H.$
$l^* = \max_{1\le l\le L} [\mathbf{M}]_{l,l}.$
The ith row $\mathbf{ß}_i$ of $\mathbf{\Pi}$ is obtained as: $\mathbf{ß}_i = \mathbf{e}_{l^*}^T.$
Then set the entries of the l^*th row of $\mathbf{A}$ to zeros.
end

Table 3. The suboptimal ordering algorithm for THP.

Since $\mathbf{R}_k$ is only related to MSE_k, the Hessian matrix of T-MSE with respect to $\mathbf{R}_k$ is equal to that of MSE_k, which is calculated as

$$\frac{\partial}{\partial(\mathrm{vec}(\mathbf{R}_k))^T}\left(\frac{\partial \text{T-MSE}}{\partial(\mathrm{vec}(\mathbf{R}_k))^*}\right) = \frac{\partial}{\partial(\mathrm{vec}(\mathbf{R}_k))^T}\left(\frac{\partial \mathrm{MSE}_k}{\partial(\mathrm{vec}(\mathbf{R}_k))^*}\right)$$
$$= \left(\mathbf{H}_k\mathbf{T}\mathbf{T}^H\mathbf{H}_k^H + \zeta_k\mathbf{I}_{N_k}\right)^T \otimes \mathbf{I}_{L_k} \succeq \mathbf{0}. \tag{39}$$

The Hessian matrix in (39) being positive semidefinite indicates that the T-MSE is also convex with respect to $\mathbf{R}_k$. Then the optimal $\mathbf{R}_k$ is calculated in the same way as (35) :

$$\frac{\partial \text{T-MSE}}{\partial \mathbf{R}_k^*} = \frac{\partial \mathrm{MSE}_k}{\partial \mathbf{R}_k^*} = \mathbf{R}_k\mathbf{H}_k\mathbf{T}\mathbf{T}^H\mathbf{H}_k^H - \mathbf{E}_k^T\mathbf{\Pi}^T(\mathbf{I}_L - \mathbf{F})\mathbf{T}^H\mathbf{H}_k^H + \zeta_k\mathbf{R}_k = \mathbf{0}.$$
$$\Rightarrow \mathbf{R}_k = \mathbf{E}_k^T\mathbf{\Pi}^T(\mathbf{I}_L - \mathbf{F})\mathbf{T}^H\mathbf{H}_k^H\left(\mathbf{H}_k\mathbf{T}\mathbf{T}^H\mathbf{H}_k^H + \zeta_k\mathbf{I}_{N_k}\right)^{-1}. \tag{40}$$

As the inter-dependence among the optimal $\mathbf{\Pi}$, $\mathbf{F}$, $\mathbf{T}$, β and $\{\mathbf{R}_k\}_{k=1}^K$ has been found, we now summarize our iterative algorithm in Table 4, where the notations with the superscript $(\cdot)^{(n)}$ denote the related variables in the nth iteration.

The convergence of our proposed iterative algorithm can be guaranteed. The proof of convergence is in **Appendix B**.

4.2 Robust optimization of transceivers under imperfect CSIT

4.2.1 Channel uncertainty model

We consider a TDD system where the BS estimates the CSI using the training sequences in the uplink. The maximum-likelihood estimate of the actual channel matrix $\mathbf{H}_k$ can be modeled as (Hassibi & Hochwald, 2003) $\widehat{\mathbf{H}}_k = \mathbf{H}_k + \Delta\mathbf{H}_k$, where $\Delta\mathbf{H}_k$ denotes the error matrix whose entries are i.i.d. complex Gaussian distributed with zero mean and variance $\sigma_{e,k}^2$. $\Delta\mathbf{H}_k$ is statistically independent of $\mathbf{H}_k$. According to (Kay, 1993), the distribution of $\mathbf{H}_k$ conditioned on $\widehat{\mathbf{H}}_k$ is Gaussian and can be expressed as

$$\mathbf{H}_k \mid \widehat{\mathbf{H}}_k = \rho_k\widehat{\mathbf{H}}_k + \Delta\widetilde{\mathbf{H}}_k, \tag{41}$$

where $\rho_k = \sigma_{h,k}^2/(\sigma_{h,k}^2 + \sigma_{e,k}^2)$ and the entries of $\Delta\widetilde{\mathbf{H}}_k$ are i.i.d. complex Gaussian distributed with zero mean and variance $\tilde{\sigma}_k^2 = \sigma_{e,k}^2\sigma_{h,k}^2/(\sigma_{h,k}^2 + \sigma_{e,k}^2)$. We assume that the information of $\widehat{\mathbf{H}}_k$, $\sigma_{h,k}^2$ and $\sigma_{e,k}^2$, $k = 1, \ldots, K$ is known at the BS.

Note that the channel uncertainty caused by the slow time-variations of the channel can also be modeled in the same manner as (41) except that ρ_k has a different relationship with $\tilde{\sigma}_k^2$ (Khaled et al., 2004).

Step (1) Set the iteration number $n = 0$ and initialize $\mathbf{\Pi}^{(0)} = \mathbf{I}_L$ and $\mathbf{R}_k^{(0)} = \mathbf{U}_k^H$, where $\mathbf{U}_k$ comprises the L_k dominant left singular vectors of $\mathbf{H}_k$.

Step (2) Set $n = n + 1$. Calculate the reordering matrix $\widetilde{\mathbf{\Pi}}$ using the algorithm described in Table 3 and $\mathbf{R}^{(n-1)}$.
Calculate (37) using $\widetilde{\mathbf{\Pi}}$ and $\mathbf{R}^{(n-1)}$ and denote the result as $\widetilde{C}$.

Calculate (37) using $\mathbf{\Pi}^{(n-1)}$ and $\mathbf{R}^{(n-1)}$ and denote the result as C.
If $\widetilde{C} \leqslant C$
$\mathbf{\Pi}^{(n)} = \widetilde{\mathbf{\Pi}}$,
else
$\mathbf{\Pi}^{(n)} = \mathbf{\Pi}^{(n-1)}$.
end
Update transmit processing:

$$\mathbf{t}_i^{(n)} = \left(\mathbf{A}_i^{(n),H}\mathbf{A}_i^{(n)} + \xi\mathbf{I}_M\right)^{-1} \left[\mathbf{A}_i^{(n),H} \; \mathbf{0}\right] \mathbf{e}_i,$$

$$\mathbf{f}_i^{(n)} = -\begin{bmatrix} \mathbf{0} \\ \mathbf{B}_i^{(n)} \end{bmatrix} \mathbf{t}_i^{(n)}, i = 1, \ldots, L,$$

and $\beta^{(n)} = P_T^{\frac{1}{2}} \left(\mathrm{tr}(\mathbf{T}^{(n)}\mathbf{T}^{(n),H})\right)^{-\frac{1}{2}}$,

where

$$\begin{bmatrix} \mathbf{A}_i^{(n)} \\ \mathbf{B}_i^{(n)} \end{bmatrix} = \mathbf{\Pi}^{(n)}\mathbf{R}^{(n-1)}\mathbf{H},\ \mathbf{A}_i^{(n)} \in \mathbb{C}^{i\times M},\ \mathbf{B}_i^{(n)} \in \mathbb{C}^{(L-i)\times M},$$

and $\xi = P_T^{-1} \sum_{k=1}^{K} \sigma_{n,k}^2 \mathrm{tr}\left(\mathbf{R}_k^{(n-1),H}\mathbf{R}_k^{(n-1)}\right)$.

Update receiver processing:

$$\mathbf{R}_k^{(n)} = \mathbf{E}_k^T \mathbf{\Pi}^{(n),T}(\mathbf{I}_L - \mathbf{F}^{(n)})\mathbf{T}^{(n),H}\mathbf{H}_k^H \cdot \left(\mathbf{H}_k\mathbf{T}^{(n)}\mathbf{T}^{(n),H}\mathbf{H}_k^H + \zeta_k\mathbf{I}_{N_k}\right)^{-1}, k = 1, \ldots, K,$$

where $\zeta_k = P_T^{-1}\sigma_{n,k}^2 \mathrm{tr}(\mathbf{T}^{(n),H}\mathbf{T}^{(n)})$.

Step (3) If $\| \mathbf{R}_k^{(n)} - \mathbf{R}_k^{(n-1)} \|_F^2 \geq \epsilon$, $\exists\, k \in \{1, \ldots, K\}$, then go to **Step (2)**. Otherwise, stop the iteration and the solution is given by $\mathbf{\Pi} = \mathbf{\Pi}^{(n)}$, $\mathbf{P} = \beta^{(n)}\mathbf{T}^{(n)}$, $\mathbf{F} = \mathbf{F}^{(n)}$, $\beta = \beta^{(n)}$ and $\mathbf{G}_k = \left(\beta^{(n)}\right)^{-1} \mathbf{R}_k^{(n)}$.

Table 4. The iterative algorithm for joint THP transceiver design.

4.2.2 Robust optimization problem formulation and iterative algorithm

When only the channel estimates $\widehat{\mathbf{H}}_k$, $k = 1, \ldots, K$ are available at the BS, the definition of T-MSE in (26) and MSE_k in (38) cannot be directly applied to the transceiver design. Instead, the expectation of MSE conditioned on $\widehat{\mathbf{H}}_k$ is an applicable performance measure and provides the robustness against the channel uncertainties in an average manner (Dietrich et al., 2007; Shenouda & Davidson, 2007). By (38), the conditional expectation of MSE of user k is expressed as

$$\mathcal{E}_{\mathbf{H}_k|\widehat{\mathbf{H}}_k}\{\mathrm{MSE}_k\} = \mathcal{E}_{\mathbf{H}_k|\widehat{\mathbf{H}}_k}\left\{\left\|\mathbf{R}_k\mathbf{H}_k\mathbf{T} - \mathbf{E}_k^T\mathbf{\Pi}^T(\mathbf{I}_L - \mathbf{F})\right\|_F^2 + \zeta_k \cdot \mathrm{tr}(\mathbf{R}_k^H\mathbf{R}_k)\right\}$$

$$= \mathrm{tr}\Bigg(\mathbf{R}_k \cdot \mathcal{E}_{\mathbf{H}_k|\widehat{\mathbf{H}}_k}\left\{\mathbf{H}_k\mathbf{T}\mathbf{T}^H\mathbf{H}_k^H\right\} \cdot \mathbf{R}_k^H - \mathbf{R}_k \cdot \mathcal{E}_{\mathbf{H}_k|\widehat{\mathbf{H}}_k}\{\mathbf{H}_k\} \cdot \mathbf{T}(\mathbf{E}_k^T\mathbf{\Pi}^T(\mathbf{I}_L - \mathbf{F}))^H$$

$$-\mathbf{E}_k^T\mathbf{\Pi}^T(\mathbf{I}_L - \mathbf{F})\mathbf{T}^H \cdot \mathcal{E}_{\mathbf{H}_k|\widehat{\mathbf{H}}_k}\left\{\mathbf{H}_k^H\right\} \cdot \mathbf{R}_k^H + \mathbf{E}_k^T\mathbf{\Pi}^T(\mathbf{I}_L - \mathbf{F})\left(\mathbf{E}_k^T\mathbf{\Pi}^T(\mathbf{I}_L - \mathbf{F})\right)^H\Bigg)$$

$$+ \zeta_k \cdot \mathrm{tr}(\mathbf{R}_k^H\mathbf{R}_k). \tag{42}$$

Using (41) it can be easily verified that

$$\mathcal{E}_{\mathbf{H}_k|\widehat{\mathbf{H}}_k}\{\mathbf{H}_k\} = \rho_k\widehat{\mathbf{H}}_k, \tag{43}$$

$$\mathcal{E}_{\mathbf{H}_k|\widehat{\mathbf{H}}_k}\left\{\mathbf{H}_k^H\right\} = \rho_k\widehat{\mathbf{H}}_k^H. \tag{44}$$

Next we calculate the quadratic term $\mathbf{Q}_k \triangleq \mathcal{E}_{\mathbf{H}_k|\widehat{\mathbf{H}}_k}\{\mathbf{H}_k\mathbf{T}\mathbf{T}^H\mathbf{H}_k^H\}$. Let $\mathbf{H}_k = \left[\mathbf{h}_{k,1}^T\ \mathbf{h}_{k,2}^T \ldots \mathbf{h}_{k,N_k}^T\right]^T$, where $\mathbf{h}_{k,m}$ is the mth row of $\mathbf{H}_k$. Then the element at the mth row and nth column of $\mathbf{Q}_k$ can be written as

$$\begin{aligned}[\mathbf{Q}_k]_{m,n} &= \mathcal{E}_{\mathbf{H}_k|\widehat{\mathbf{H}}_k}\left\{\mathbf{h}_{k,m}\mathbf{T}\mathbf{T}^H\mathbf{h}_{k,n}^H\right\} = \mathcal{E}_{\mathbf{H}_k|\widehat{\mathbf{H}}_k}\left\{\mathrm{tr}\left(\mathbf{h}_{k,m}\mathbf{T}\mathbf{T}^H\mathbf{h}_{k,n}^H\right)\right\} \\ &= \mathcal{E}_{\mathbf{H}_k|\widehat{\mathbf{H}}_k}\left\{\mathrm{tr}\left(\mathbf{h}_{k,n}^H\mathbf{h}_{k,m}\mathbf{T}\mathbf{T}^H\right)\right\} = \mathrm{tr}\left(\mathcal{E}_{\mathbf{H}_k|\widehat{\mathbf{H}}_k}\left\{\mathbf{h}_{k,n}^H\mathbf{h}_{k,m}\right\}\mathbf{T}\mathbf{T}^H\right).\end{aligned} \tag{45}$$

According to (41), we have

$$\mathcal{E}_{\mathbf{H}_k|\widehat{\mathbf{H}}_k}\left\{\mathbf{h}_{k,n}^H\mathbf{h}_{k,m}\right\} = \rho_k^2\hat{\mathbf{h}}_{k,n}^H\hat{\mathbf{h}}_{k,m} + \mathcal{E}\left\{\Delta\tilde{\mathbf{h}}_{k,n}^H\Delta\tilde{\mathbf{h}}_{k,m}\right\}, \tag{46}$$

where

$$\mathcal{E}\left\{\Delta\tilde{\mathbf{h}}_{k,n}^H\Delta\tilde{\mathbf{h}}_{k,m}\right\} = \begin{cases}\tilde{\sigma}_k^2\mathbf{I}_M, & m = n\\ 0. & m \neq n\end{cases}$$

Therefore,

$$[\mathbf{Q}_k]_{m,n} = \begin{cases}\rho_k^2\hat{\mathbf{h}}_{k,m}\mathbf{T}\mathbf{T}^H\hat{\mathbf{h}}_{k,n}^H + \tilde{\sigma}_k^2\mathrm{tr}\left(\mathbf{T}\mathbf{T}^H\right), & m = n\\ \rho_k^2\hat{\mathbf{h}}_{k,m}\mathbf{T}\mathbf{T}^H\hat{\mathbf{h}}_{k,n}^H. & m \neq n\end{cases}$$

Finally, $\mathbf{Q}_k$ can be expressed as

$$\mathbf{Q}_k = \rho_k^2\widehat{\mathbf{H}}_k\mathbf{T}\mathbf{T}^H\widehat{\mathbf{H}}_k^H + \tilde{\sigma}_k^2\mathrm{tr}\left(\mathbf{T}\mathbf{T}^H\right)\mathbf{I}_{N_k}. \tag{47}$$

By substituting (43), (44) and (47) into (42), we can obtain the explicit expression of $\mathcal{E}_{\mathbf{H}_k|\widehat{\mathbf{H}}_k}\{\mathrm{MSE}_k\}$ as follows:

$$\mathcal{E}_{\mathbf{H}_k|\widehat{\mathbf{H}}_k}\{\mathrm{MSE}_k\} = \left\|\rho_k\mathbf{R}_k\widehat{\mathbf{H}}_k\mathbf{T} - \mathbf{E}_k^T\mathbf{\Pi}^T(\mathbf{I}_L - \mathbf{F})\right\|_F^2 + \tilde{\zeta}_k \cdot \mathrm{tr}(\mathbf{R}_k^H\mathbf{R}_k), \tag{48}$$

in which $\tilde{\zeta}_k \triangleq (P_T^{-1}\sigma_{n,k}^2 + \tilde{\sigma}_k^2) \cdot \mathrm{tr}(\mathbf{T}^H\mathbf{T})$. Then the conditional expectation of the T-MSE can be expressed as:

$$\mathcal{E}_{\mathbf{H}|\widehat{\mathbf{H}}}\{\text{T-MSE}\} = \sum_{k=1}^{K} \mathcal{E}_{\mathbf{H}_k|\widehat{\mathbf{H}}_k}\{\text{MSE}_k\} = \left\|\mathbf{R}\widehat{\mathbf{H}}\mathbf{T} - \mathbf{\Pi}^T(\mathbf{I}_L - \mathbf{F})\right\|_F^2 + \tilde{\xi} \cdot \mathrm{tr}(\mathbf{T}^H\mathbf{T}), \tag{49}$$

where $\widehat{\mathbf{H}} \triangleq \left[\rho_1\widehat{\mathbf{H}}_1^T \ldots \rho_K\widehat{\mathbf{H}}_K^T\right]^T$ and $\tilde{\xi} \triangleq \sum_{k=1}^K \left(P_T^{-1}\sigma_{n,k}^2 + \tilde{\sigma}_k^2\right) \mathrm{tr}\left(\mathbf{R}_k^H\mathbf{R}_k\right)$.
Finally, the robust transceiver optimization problem is mathematically formulated as below:

$$\begin{aligned} &\min_{\mathbf{\Pi},\, \mathbf{F},\, \mathbf{T},\, \{\mathbf{R}_k\}_{k=1}^K} \mathcal{E}_{\mathbf{H}|\widehat{\mathbf{H}}}\{\text{T-MSE}\} \\ &\text{s.t.} \quad [\mathbf{F}]_{m,n} = 0,\ \forall\, 1 \leqslant m,n \leqslant L \text{ and } m \leqslant n. \end{aligned} \tag{50}$$

Following the derivation of transceiver design under perfect CSIT, we can easily obtain the necessary conditions for the optimal robust transceiver under imperfect CSIT as follows. For fixed $\mathbf{\Pi}$ and $\mathbf{R}$, the optimal $\mathbf{f}_i$ and $\mathbf{t}_i$ are listed below:

$$\mathbf{f}_i = -\begin{bmatrix} \mathbf{0}_{i\times M} \\ \widehat{\mathbf{B}}_i \end{bmatrix} \mathbf{t}_i, \tag{51}$$

$$\mathbf{t}_i = \left(\widehat{\mathbf{A}}_i^H\widehat{\mathbf{A}}_i + \tilde{\xi}\mathbf{I}_M\right)^{-1} \left[\widehat{\mathbf{A}}_i^H\ \mathbf{0}\right]\mathbf{e}_i,\ i = 1,\ldots,L, \tag{52}$$

where

$$\begin{bmatrix} \widehat{\mathbf{A}}_i \\ \widehat{\mathbf{B}}_i \end{bmatrix} = \mathbf{\Pi}\mathbf{R}\widehat{\mathbf{H}},\ \widehat{\mathbf{A}}_i \in \mathbb{C}^{i\times M},\ \widehat{\mathbf{B}}_i \in \mathbb{C}^{(L-i)\times M}. \tag{53}$$

By substituting (51) and (52) into (49), we reformulate $\mathcal{E}_{\mathbf{H}|\widehat{\mathbf{H}}}\{\text{T-MSE}\}$ as:

$$\mathcal{E}_{\mathbf{H}|\widehat{\mathbf{H}}}\{\text{T-MSE}\} = L - \sum_{i=1}^{L} \mathrm{tr}\left(\mathbf{e}_i^H \begin{bmatrix} \widehat{\mathbf{A}}_i \\ \mathbf{0} \end{bmatrix} \left(\widehat{\mathbf{A}}_i^H\widehat{\mathbf{A}}_i + \tilde{\xi}\mathbf{I}_M\right)^{-1} \left[\widehat{\mathbf{A}}_i^H\ \mathbf{0}\right]\mathbf{e}_i\right). \tag{54}$$

Then the suboptimal successive ordering algorithm presented in Table 3 can be applied with $\mathbf{H}$ replaced by $\widehat{\mathbf{H}}$ and the expression of ξ replaced by $\tilde{\xi} = \sum_{k=1}^K \left(P_T^{-1}\sigma_{n,k}^2 + \tilde{\sigma}_k^2\right) \mathrm{tr}\left(\mathbf{R}_k^H\mathbf{R}_k\right)$.
For fixed $\mathbf{\Pi}$, $\mathbf{F}$ and $\mathbf{T}$, the optimal $\mathbf{R}_k$ is given by

$$\mathbf{R}_k = \rho_k\mathbf{E}_k^T\mathbf{\Pi}^T(\mathbf{I}_L - \mathbf{F})\mathbf{T}^H\mathbf{H}_k^H\left(\rho_k^2\mathbf{H}_k\mathbf{T}\mathbf{T}^H\mathbf{H}_k^H + \tilde{\zeta}_k\mathbf{I}_{N_k}\right)^{-1}. \tag{55}$$

Now that we have found the inter-dependence among $\mathbf{\Pi}$, $\mathbf{F}$, $\mathbf{T}$ and $\{\mathbf{R}_k\}_{k=1}^K$ as shown in (51)-(55), the thread of the iterative algorithm proposed in Subsection 4.1 can be again adopted here to compute the robust transceiver. In each iteration, we first determine the suboptimal reordering matrix $\mathbf{\Pi}$ and update $\mathbf{T}$ and $\mathbf{F}$ using $\{\mathbf{R}_k\}_{k=1}^K$ updated in the last iteration, then update $\{\mathbf{R}_k\}_{k=1}^K$ using the above updated $\mathbf{\Pi}$, $\mathbf{T}$ and $\mathbf{F}$ until convergence. The formulation of the algorithm is similar to that described in Table 4 except for some notation and expression changes so we do not list the details of the modified algorithm.
The convergence of the iterative algorithm is also guaranteed. The proof of convergence is also similar to that in **Appendix B**.

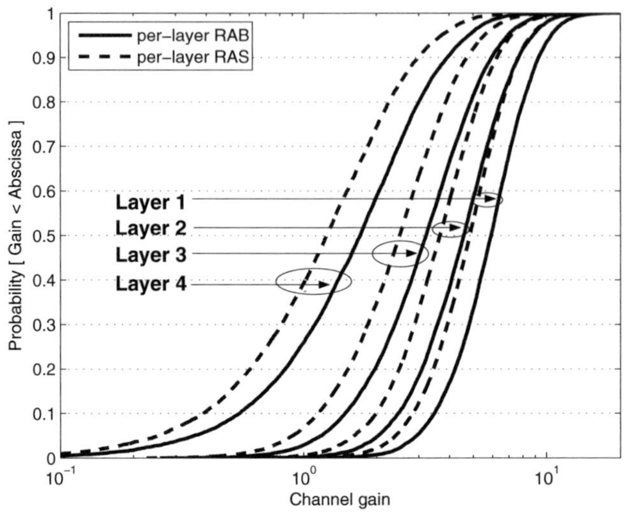

Fig. 4. The cumulative distribution function of equivalent channel gains of four layers applying per-layer RAB and RAS in $(2 \times 4) \times 4$ systems.

5. Simulation results

In this section, we will present some simulation results to verify the effectiveness of our proposed per-layer ZF-THP schemes and stream-wise MMSE-THP schemes, respectively.

5.1 Results of per-layer ZF-THP

We consider the system settings of one transmitter with M antennas and M receivers each with N antennas, denoted by $(N \times M) \times M$. For comparison, we also simulate the performance of the parallel linear-ZF precoding where each user combines its receive antennas according to the eigenmode, and the performance of the per-user processing which selects $\lceil \frac{M}{N} \rceil$ users with the largest channel gains and allocates each of them with N adjacent layers.

Fig. 4 and Fig. 5 compare the equivalent channel gains and capacities respectively of the four layers by per-layer RAB and RAS with the equal power allocation among layers. Here, we place users into layers according to their indexes, i.e., $\pi_l = l, l = 1, \cdots, 4$. It can be found that for the layers from 4 to 1, the relative differences of channel gains between these two schemes gradually decrease, though the rates of decreasing get slower. At the same time, the order of layers according to the differences of channel capacities between per-layer RAB and per-layer RAS exhibits two results. When SNR<3dB, the capacity difference is the largest in layer 1, and the smallest in layer 4. On the contrary, when SNR>7dB, the order from large to small is layer 4 to layer 1. These results are consistent with Corollary 1 and 2.

In Fig. 6, the rate regions of the $(2 \times 2) \times 2$ systems are considered, whose boundaries are generated by averaging channel realizations. For THP, the rate regions are asymmetric, which means the higher layer has the larger capacity than the lower one. With the ideal power allocation among layers, the order of the maximum sum-capacities from large to small is RAB, RAS, per-user, and linear-ZF.

In Fig. 7, the ergodic sum-capacity of the $(2 \times 4) \times 4$ system is evaluated. Here, in order to achieve the potential sum-capacity, the water-filling power allocation (Tse & Viswanath, 2005) among layers is applied. Among these four curves, the advantage of THP over linear-ZF, and

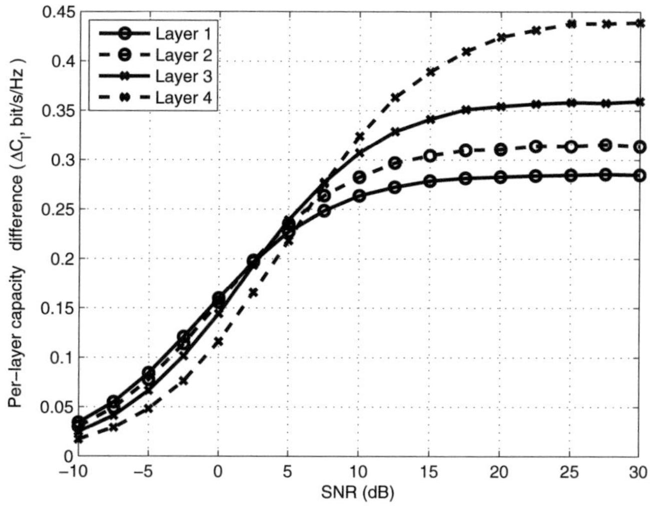

Fig. 5. The channel capacity differences of four layers between per-layer RAB and per-layer RAS in $(2 \times 4) \times 4$ systems.

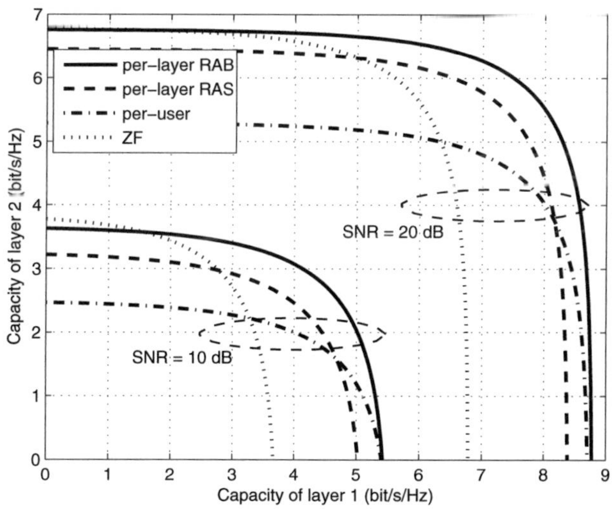

Fig. 6. The rate regions in $(2 \times 2) \times 2$ systems.

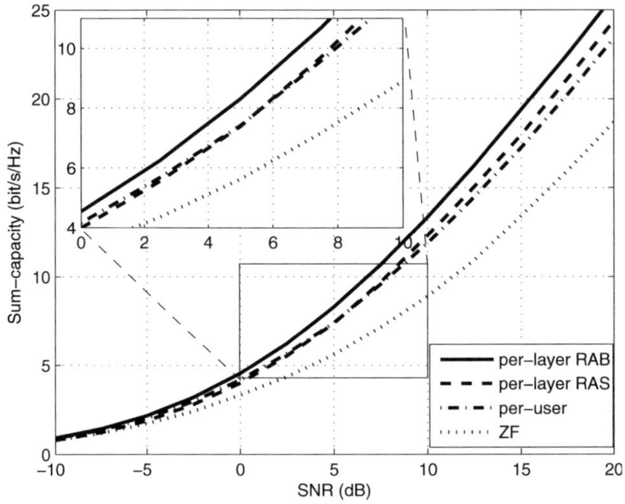

Fig. 7. The ergodic sum-capacities in $(2 \times 4) \times 4$ systems. The water-filling power allocation is applied.

the advantage of RAB over RAS and per-user are obvious; the curves of RAS and per-user are intersected at SNR$\approx$5dB, i.e., when SNR$<$5dB, per-user outperforms RAS, but when SNR$>$5dB, it is the opposite.
To sum up, the essential of these comparison results among THP schemes is as follows.

- With the larger power gain and diversity gain, per-layer beamforming outperforms per-user processing.
- Per-layer RAS and per-user processing actually exploit the same number of receive antennas, though for the former, these antennas belong to K users, while for the latter, they belong to $\lceil \frac{K}{N} \rceil$ users. In the low SNR scenarios, per-user method can obtain larger power gains due to its dominant intra-user processing and less inter-user interference suppression. However, in the high SNR scenarios, this effect of power gains makes trivial contribution to the system sum-capacity, but the larger multiuser and multi-antenna selection diversity gain in per-layer RAS turns to the dominant factor.

5.2 Results of stream-wise MMSE-THP

In this subsection, some results are presented to show the performance superiority of our proposed joint stream-wise THP transceiver designs in comparison with some existing THP schemes. The illustration is divided into two parts. The first part illustrates the performance under perfect CSIT and the second part focuses on the performance under imperfect CSIT. We assume quasi-static i.i.d. Rayleigh flat fading channel with unit channel variance between each transmit antenna at the BS and each receive antenna at each user. We also assume $\sigma_{n,k}^2 = 1$, $k = 1, \ldots, K$. The signal-to-noise ratio (SNR) in the following figures is defined as SNR $\triangleq 10 \cdot \log_{10} P_T$. QPSK or 16-QAM modulations are used in the simulations. We set the convergence threshold in the iterative algorithms $\epsilon = 10^{-5}$.

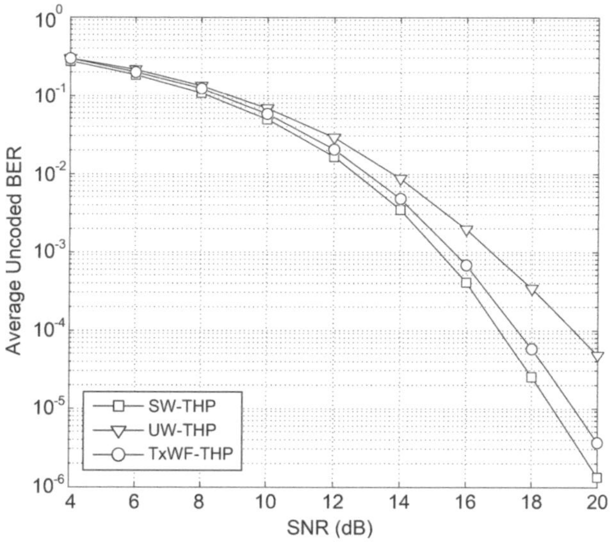

Fig. 8. Performance comparison of different schemes with $M = 6,\ K = 3,\ N_k = 2,\ L_k = 2,\ \forall k$ and QPSK.

5.2.1 Performance under perfect CSIT

We examine the performance of our proposed joint THP transceiver design under perfect CSIT in comparison with some existing *MMSE-based* THP schemes.

Figs. 8-9 compare our proposed *stream-wise* THP transceiver (denote as "SW-THP") with the *user-wise* THP transceiver in (Mezghani et al., 2006) (denoted as "UW-THP") and "TxWF-THP" in (Joham et al., 2004) in terms of average uncoded bit error rate (BER). In Fig. 8, we set $M = 6,\ K = 3,\ N_k = N = 2,\ L_k = L = 2,\ \forall k$ and use QPSK. For "TxWF-THP" we assume that the 2 antennas at the same receiver are decentralized, i.e., there are 6 *virtual* users in the system. From the simulation results we can see that our scheme clearly outperforms the other two schemes. The superiority over "UW-THP" comes from that our scheme performs stream-wise interference pre-cancelation while "UW-THP" only enables inter-user interference pre-cancelation and multiple streams of the same user are linearly precoded. Moreover, our scheme outperforms "TxWF-THP" because in our scheme the received signals from multiple antennas of one user can be jointly processed while in "TxWF-THP", the receive antennas of the same user are assumed to be decentralized and a common scaling factor is imposed for each single receive antenna, which is apparently suboptimal.

In Fig. 9, we increase N by 1 and keep the other parameters unchanged. Since $N > L$, "TxWF-THP" cannot be directly applied. Here we simply select $\mathbf{U}_k^H$ as the receiver for user k, where $\mathbf{U}_k$ comprises the L_k dominant left singular vectors of $\mathbf{H}_k$, then apply $\mathbf{U}_k^H \mathbf{H}_k$ as the equivalent $L_k \times M$ channel matrix, which is suitable for implementation of "TxWF-THP". It is shown in this figure that our scheme still performs best.

It is an interesting phenomenon that the comparison results of "UW-THP" and "TxWF-THP" in Fig. 8 and 9 are just opposite, so are those for 16-QAM[3]. This can be explained that for Fig. 8, more interferences are pre-canceled in "TxWF-THP" than in "UW-THP". For Fig. 9, however,

[3] Due to page limit, we don't show the results of 16-QAM here. Please refer to (Miao et al., 2009).

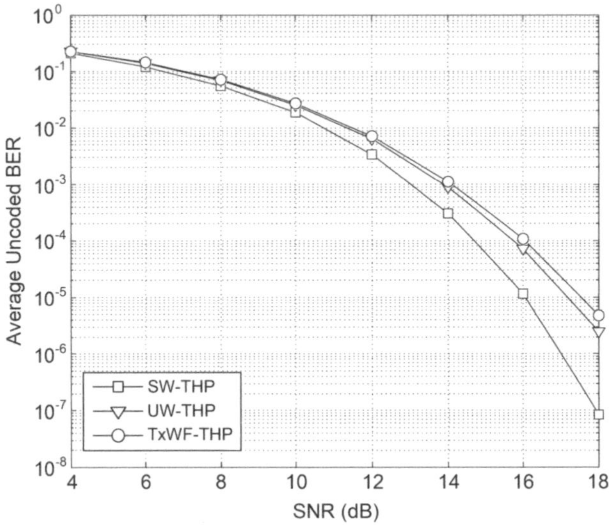

Fig. 9. Performance comparison of different schemes with $M = 6,\ K = 3,\ N_k = 3,\ L_k = 2,\ \forall k$ and QPSK.

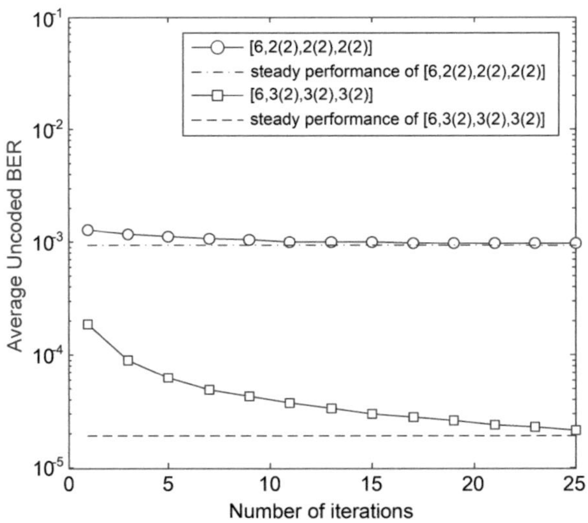

Fig. 10. verage uncoded 16-QAM BER performance of SW-THP under different number of iterations at SNR = 22dB.

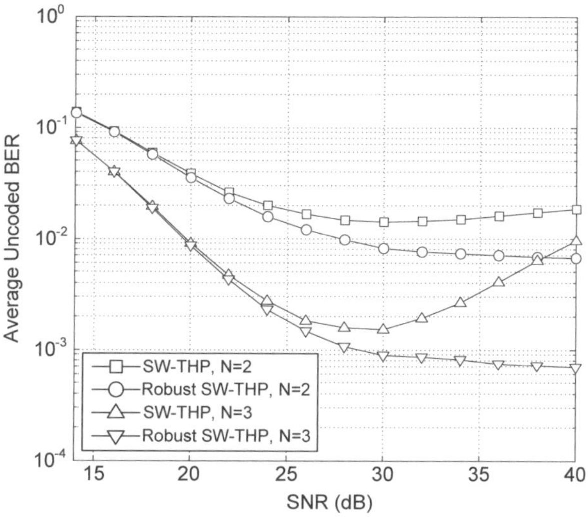

Fig. 11. Performance comparison of non-robust and robust SW-THP with $M = 6,\ K = 3,\ L = 2,\ \sigma_e^2 = 0.01$ and 16-QAM.

the additional receive antenna at each user provides more diversity for each data stream in "UW-THP" through joint transceiver design while the simple and suboptimal receiver structure we have adopted for "TxWF-THP" restricts its performance improvement.

We also present some simulation results to verify the convergence of our proposed iterative algorithm. We use the notation $[M, N_1(L_1), \ldots, N_K(L_K)]$ to denote a K-user MIMO system that has M antennas at the BS and N_k antennas at the kth receiver to support L_k data streams for user k, $k = 1, ..., K$. Fig. 10 shows the average uncoded BER of 16-QAM versus the number of iterations for our scheme under two system configurations at a fixed SNR = 22dB. The dashed lines denote the steady state performance. It can be seen that after about 10 iterations (for $[6, 2(2), 2(2), 2(2)]$) or 25 iterations (for $[6, 3(2), 3(2), 3(2)]$) the BER performance is quite close to the steady state performance. Similar results hold for QPSK, thus are not shown here for brevity.

5.2.2 Performance under imperfect CSIT

In this subsection, we examine the performance of SW-THP introduced in Subsection 4.1 and the robust version of SW-THP introduced in Subsection 4.2 under imperfect CSIT.

Fig. 11 compare "SW-THP" and "Robust SW-THP" for $M = 6,\ K = 3,\ L_k = 2,\ N_k = N = 2$ and 3 (the parameters are consistent with those in Subsection 5.1). We assume $\sigma_{e,k}^2 = \sigma_e^2 = 0.01,\ k = 1, \ldots, K$ for 16-QAM. It can be seen that the non-robust and robust schemes have close performance at low SNR since the noise is dominant in this regime. However, the gap between the two schemes enlarges as the SNR increases because the channel uncertainty gradually dominates and the efficacy of the robust scheme shows up.

In Fig. 12 we test the performance of "SW-THP" and "Robust SW-THP" under different channel errors. The antenna configurations are the same as above, i.e., $M = 6,\ K = 3,\ L_k = 2,\ N_k = N = 2$ and 3. We fix SNR $= 40$ dB, and change σ_e^2 from 0.01 to 0.1 for 16-QAM in Fig. 12. The simulated curves reveal that "Robust SW-THP" always holds a considerable superiority over the non-robust "SW-THP" as the channel error increases.

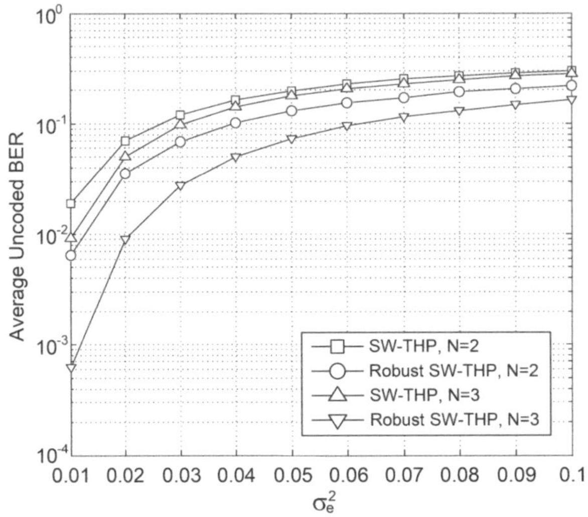

Fig. 12. Performance comparison of non-robust and robust SW-THP with $M = 6,\ K = 3,\ L = 2$, SNR = 40dB and 16-QAM for different channel errors.

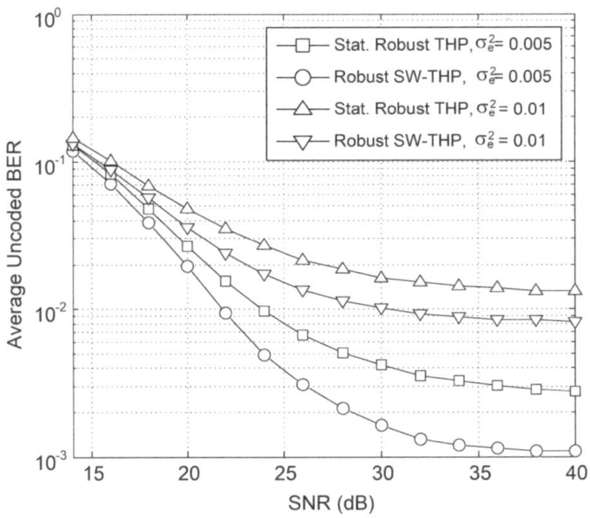

Fig. 13. Performance comparison of robust SW-THP and Stat. Robust THP in (Shenouda & Davidson, 2007) with $M = 4,\ K = 2,\ N = 2,\ L = 2$ and 16-QAM.

We also compare our robust THP scheme with "Statistically Robust Tomlinson-Harashima precoding" (denoted as "Stat. Robust THP") reported lately in (Shenouda & Davidson, 2007). The results are shown in Fig. 13. We choose $M = 4$, $K = 2$, $L_k = L = 2$, $N_k = N = 2$. For 16-QAM we set $\sigma_{e,k}^2 = \sigma_e^2 = 0.005$ and 0.01. Since "Stat. Robust THP" is designed only for the MISO downlink, it cannot be directly applied here. So for "Stat. Robust THP", we assume that the 2 antennas at the same receiver are decentralized, i.e., there are 4 *virtual* users in the system. From Fig. 13 we can find out that "Robust SW-THP" outperforms "Stat. Robust THP" in the entire SNR regime and also has a much lower BER floor at high SNR. The advantage of our scheme comes from two aspects. First, we jointly optimize the transmitter, the receivers and the precoding order with consideration of the channel errors. In contrast, only the transmitter of "Stat. Robust THP" is robustly designed and the receivers and the precoding order are determined as if the estimated CSI was the true one. Second, the received signals from the two antennas of the same user can be jointly processed using "Robust SW-THP" but can only be processed independently by "Stat. Robust THP".

6. Conclusions

In this chapter, the multiuser MIMO downlink systems with THP based on ZF and MMSE criteria have been investigated, respectively.

For the ZF-THP schemes, based on the criterion of maximum system sum-capacity, two per-layer joint transmit and receive filters design schemes have been studied, which apply RAB and RAS, respectively. Furthermore, the comparison between equivalent channel gains and capacities within these two per-layer schemes have been proved, including the trend with the layers from low to high and the trend with the increase of the number of transmit antennas to infinite.

For the MMSE-THP schemes, we have jointly designed the THP transceiver under both perfect and imperfect CSIT. First, we deal with the perfect CSIT case. Based on the equivalent linear representation of the THP downlink, we formulate a T-MSE minimization problem under the total power constraint. After some analysis, an iterative algorithm is proposed to obtain a locally optimal solution. Next, we consider the imperfect CSIT case. By minimizing the conditional-expectation of the T-MSE, we extend the algorithm derived under perfect CSIT to the robust version that is more effective against the channel uncertainty. The two characteristics of our schemes, i.e., the stream-wise interference pre-cancelation and the joint design of transceiver, ensure its performance superiority over the existing MMSE-based THP schmes, which has been verified by the simulation results.[4]

7. Appendix

7.1 Appendix A: Lower bounds for per-user processing and per-layer RAS

Derived from the proof of Theorem 1, $E(\widetilde{\lambda}_n^{max,2} - \widetilde{\lambda}_n^{min,2})^2 = 24n$. Besides,

$$E(\widetilde{\lambda}_n^{max,2} + \widetilde{\lambda}_n^{min,2})^2 \geq \left[E(\widetilde{\lambda}_n^{max,2} + \widetilde{\lambda}_n^{min,2})\right]^2 = (4n)^2. \tag{56}$$

[4] This work is partially supported by National Basic Research Program (2007CB310608), National Science and Technology Pillar Program (2008BAH30B09), National Natural Science Foundation of China (60832008), National Major Project (2008ZX03003-004), China 863 Project (2009AA011501) and PCSIRT.

For per-user processing,

$$E(\widetilde{\lambda}_n^{max,2}\widetilde{\lambda}_n^{min,2}) = \frac{1}{4}\left[E(\widetilde{\lambda}_n^{max,2}+\widetilde{\lambda}_n^{min,2})^2 - E(\widetilde{\lambda}_n^{max,2}-\widetilde{\lambda}_n^{min,2})^2\right] \geq 4n^2-6n. \tag{57}$$

For per-layer RAS, due to the operations of maximization,

$$E(\widetilde{p}_n^{max,2}\widetilde{p}_{n-1}^{max,2}) \geq E(x_{2n}^2 x_{2n-2}^2) = 2n\cdot(2n-2) = 4n^2-4n, \tag{58}$$

where x_{2n}^2 and x_{2n-2}^2 stand for two chi-square distributed random variables with $2n$ and $2n-2$ degrees of freedom, respectively.

7.2 Appendix B: Proof of convergence of the proposed iterative algorithm

We use the notation $f\left(\mathbf{\Pi}^{(n)},\mathbf{F}^{(n)},\mathbf{T}^{(n)},\mathbf{R}^{(n)}\right)$ to represent the T-MSE calculated after the nth iteration of our iterative algorithm. If we fix $\mathbf{\Pi}^{(n)}$ and $\mathbf{R}^{(n)}$, and update $\mathbf{F}^{(n)}$ and $\mathbf{T}^{(n)}$ using the optimal conditions derived in (31) and (35), then due to the convexity of T-MSE with respect to $\mathbf{t}_i$, we have the following inequality:

$$f\left(\mathbf{\Pi}^{(n)},\mathbf{F}^{(n)},\mathbf{T}^{(n)},\mathbf{R}^{(n)}\right) \geqslant f\left(\mathbf{\Pi}^{(n)},\mathbf{F}',\mathbf{T}',\mathbf{R}^{(n)}\right), \tag{59}$$

where $\mathbf{F}'$ and $\mathbf{T}'$ are the updated matrices using $\mathbf{\Pi}^{(n)}$ and $\mathbf{R}^{(n)}$.
We define $h\left(\mathbf{\Pi},\mathbf{R}\right) \triangleq \min_{\mathbf{F},\mathbf{T}} f\left(\mathbf{\Pi},\mathbf{F},\mathbf{T},\mathbf{R}\right)$. Obviously $h\left(\mathbf{\Pi},\mathbf{R}\right)$ is equal to (37). From the definition of h we know that

$$h\left(\mathbf{\Pi}^{(n)},\mathbf{R}^{(n)}\right) = f\left(\mathbf{\Pi}^{(n)},\mathbf{F}',\mathbf{T}',\mathbf{R}^{(n)}\right). \tag{60}$$

We denote the permutation matrix calculated using $\mathbf{R}^{(n)}$ and the suboptimal ordering algorithm in Table 3 as $\mathbf{\Pi}'$. As stated in Table 4, if $\mathbf{\Pi}'$ leads to a smaller value of (37) than $\mathbf{\Pi}^{(n)}$, then $\mathbf{\Pi}^{(n+1)} = \mathbf{\Pi}'$. Otherwise, $\mathbf{\Pi}^{(n+1)} = \mathbf{\Pi}^{(n)}$. Then the following inequality holds:

$$h\left(\mathbf{\Pi}^{(n)},\mathbf{R}^{(n)}\right) \geqslant h\left(\mathbf{\Pi}^{(n+1)},\mathbf{R}^{(n)}\right). \tag{61}$$

From the definition of h and iteration procedure, we have:

$$h\left(\mathbf{\Pi}^{(n+1)},\mathbf{R}^{(n)}\right) = f\left(\mathbf{\Pi}^{(n+1)},\mathbf{F}^{(n+1)},\mathbf{T}^{(n+1)},\mathbf{R}^{(n)}\right). \tag{62}$$

Since f is also convex with respect to $\mathbf{R}_k$, the following inequality holds:

$$f\left(\mathbf{\Pi}^{(n+1)},\mathbf{F}^{(n+1)},\mathbf{T}^{(n+1)},\mathbf{R}^{(n)}\right) \geqslant f\left(\mathbf{\Pi}^{(n+1)},\mathbf{F}^{(n+1)},\mathbf{T}^{(n+1)},\mathbf{R}^{(n+1)}\right)$$

Combing (59)-(63) we finally get:

$$f\left(\mathbf{\Pi}^{(n+1)},\mathbf{F}^{(n+1)},\mathbf{T}^{(n+1)},\mathbf{R}^{(n+1)}\right) \leqslant f\left(\mathbf{\Pi}^{(n)},\mathbf{F}^{(n)},\mathbf{T}^{(n)},\mathbf{R}^{(n)}\right). \tag{63}$$

The inequality in (63) indicates that during the iterations of our proposed iterative algorithm, the T-MSE is decreasing. Moreover, T-MSE is obviously lower bounded by 0. Therefore, the iterative algorithm guarantees the convergence.
Thus, we have completed the proof. ■

8. References

Caire, G. & Shamai, S. (2003). On the achievable throughput of a multi-antenna gaussian broadcast channel, *IEEE Trans. Inform. Theory* Vol.49(No.7): 1691–1706.

Choi, R. & Murch, R. (2004). A transmit pre-processing technique for multiuser mimo systems: a decomposition approach, *IEEE Trans. Wireless Commun.* Vol.3(No.1): 20–24.

Dietrich, F., Breun, P. & Utschick, W. (2007). Robust tomlinson-harashima precoding for the wireless broadcast channel, *IEEE Trans. Sig. Process.* Vol.55(No.2): 631–644.

Doostnejad, R., Lim, T. & Sousa, E. (2005). Joint precoding and beamforming design for the downlink in a multiuser mimo system, *Proc. Conf. on Wireless and Mobile Computing, Networking and Communications 2005*, Montreal, Canada, pp. 153–159.

Gorokhov, A., Gore, D. & Paulraj, A. (2003). Receive antenna selection for mimo flat-fading channels: theory and algorithms, *IEEE Trans. Information Theory* Vol.49(No.10): 2687–2696.

Hassibi, B. & Hochwald, B. (2003). How much training is needed in multiple-antenna wireless links?, *IEEE Trans. Inform. Theory* Vol.49(No.4): 951–963.

Horn, R. & Johnson, C. (1985). *Matrix Analysis*, Cambridge University Press, Cambridge, UK.

Huang, M., Chen, X., Zhou, S. & Wang, J. (2010). Per-layer transmit and receive filters design for tomlinson−harashima precoding in multiuser mimo systems, *submitted to Science in China Series−F* .

Hunger, R., Dietrich, F., Joham, M. & Utschick, W. (2004). Robust transmit zero-forcing filters, *Proc. ITG Workshop Smart Antennas*, Munich, Germany, pp. 130–137.

Joham, M., Brehmer, J. & Utschick, W. (2004). Mmse approaches to multiuser spatio-temporal tomlinson-harashima precoding, *Proc. 5th Int. ITG Conf. on Source and Channel Coding*, Erlangen, Germany, pp. 1117–1120.

Kay, S. (1993). *Fundamentals of Statistical Signal Processing, Volume I: Estimation Theory*, Prentice Hall PTR, London, UK.

Khaled, N., Leus, G., Desset, C. & Man, H. (2004). A robust joint linear precoder and decoder mmse design for slowly time-varying mimo channels, *Proc. IEEE Int. Conf. Acoustics, Speech, and Signal Processing*, Montreal, Canada, pp. 485–488.

Mezghani, A., Hunger, R., Joham, M. & Utschick, W. (2006). Iterative thp transceiver optimization for multi-user mimo systems based on weighted sum-mse minimization, *Proc. IEEE 7th Workshop on Signal Processing Advances in Wireless Communications*, Cannes, France, pp. 1–5.

Miao, W., Chen, X., Zhou, S. & Wang, J. (2009). Joint stream-wise thp transceiver design for the multiuser mimo downlink, *IEICE Transactions on Communications* Vol.E92-B(No.1): 209–218.

Payaro, M., Pascual-Iserte, A., Perez-Neira, A. & Lagunas, M. (2007). Robust design of spatial tomlinson-harashima precoding in the presence of errors in the csi, *IEEE Trans. Wireless Commun.* Vol.6(No.7): 2396–2401.

Schubert, M. & Shi, S. (2005). Mmse transmit optimization with interference pre-compensation, *Proc. Vehicular Technology Conf. 2005-Spring*, Stockholm, Sweden, pp. 845–849.

Shenouda, M. & Davidson, T. (2007). Tomlinson-harashima precoding for broadcast channels with uncertainty, *IEEE J. Sel. Area. Comm.* Vol.25(No.7): 1380–1389.

Shenouda, M. & Davidson, T. (2008). A framework for designing mimo systems with decision feedback equalization or tomlinson-harashima precoding, *IEEE J. Sel. Area. Comm.* Vol.26(No.2): 401–411.

Stankovic, V. & Haardt, M. (2005). Successive optimization tomlinson-harashima precoding *sothp* for multi-user mimo systems, *Proc. IEEE Int. Conf. Acoustics, Speech, and Signal Processing 2005*, Philadelphia, USA, pp. 1117–1120.

Tse, D. & Viswanath, P. (2005). *Fundamentals of wireless communication*, Cambridge University Press, Cambridge, UK.

Vishwanath, S., Jindal, N. & Goldsmith, A. (2003). Duality, achievable rates and sum capacity of gaussian mimo broadcast channels, *IEEE Trans. Inform. Theory* Vol.49(No.10): 2658–2668.

Viswanath, P. & Tse, D. (2003). Sum capacity of the vector gaussian broadcast channel and uplink-downlink duality, *IEEE Trans. Inform. Theory* Vol.49(No.8): 1912–1921.

Wang, D., Jorswieck, E., Sezgin, A. & Costa, E. (2006). Joint tomlinson-harashima precoding with diversity techniques for multiuser mimo system, *Proc. IEEE VTC 2005Spring*, Stockholm, Sweden, pp. 1017–1021.

Weingarten, H., Steinberg, Y. & Shamai, S. (2006). The capacity region of the gaussian multiple-input multiple-output broadcast hannel, *IEEE Trans. Inform. Theory* Vol.52(No.9): 3936–3964.

Windpassinger, C., Fischer, R., Vencel, T. & Huber, J. (2004). Precoding in multiantenna and multiuser communications, *IEEE Trans. Wireless Commun.* Vol.3(No.4): 1305–1316.

Yu, W., Varodayan, D. & Cioffi, J. (2005). Trellis and convolutional precoding for transmitter-based interference presubtraction, *IEEE Trans. Commun.* Vol.53(No.7): 1220–1230.

Zhang, J., Wu, Y., Zhou, S. & Wang, J. (2005). Joint linear transmitter and receiver design for the downlink of multiuser mimo systems, *IEEE Commun. Letters* Vol.9(No.11): 991–993.

11

Iterative Optimization Algorithms to Determine Transmit and Receive Weights for MIMO Systems

Osamu Muta, Takayuki Tominaga, Daiki Fujii, and Yoshihiko Akaiwa
Kyushu University
Japan

1. Introduction

As a technology to realize high data rates and high capacity in wireless communication systems, Multiple-Input Multiple-Output (MIMO) system has received increasing attention. In MIMO systems, high spectral efficiency is achieved by spatially multiplexing multiple data streams at the same time and frequency [1].

The MIMO system with space division multiplexing (SDM) technique is categorized into two cases, i.e., no channel state information (CSI) is needed at the transmitter or CSI is exploited in both transmitter and receiver. As a method used in the former systems, spatial filtering and maximum likelihood detection (MLD) are known [1], where the received SDM signal is de-multiplexed with signal processing at the receiver. One of the latter MIMO systems is called the Eigenbeam SDM (E-SDM) [1][2], where data streams are transmitted through multiple orthogonal eigenpath channels between the transmitter and the receiver. Thus, the E-SDM system with power control based on the water-filling theorem [3] improves the MIMO channel capacity, provided that accurate CSI is known to the transmitter and the receiver. Therefore, it is expected that the E-SDM system achieves significant increase of spectral efficiency. In the E-SDM system, it is important to find optimum transmit and receive weights for maximizing its capacity. Such the optimum weights are determined based on eigenvector of $\mathbf{H}^H\mathbf{H}$, where $\mathbf{H}$ denotes channel matrix and suffix H denotes complex conjugate transpose. As a method to find these eigenvectors, eigenvalue decomposition (EVD) of $\mathbf{H}^H\mathbf{H}$ or singular value decomposition (SVD) of $\mathbf{H}$ is well-known. Generally, SVD or EVD requires matrix decomposition operation based on QR decomposition.

MIMO techniques can be used for multiple access systems where multiple signals are sent from multiple terminals at the same time and same frequency, i.e., Space Division Multiple Access (SDMA) or multi-user MIMO (MU-MIMO) [6]-[9]. When such a multi-antenna system is used at a transmitter, the transmit weights are optimized under the constraint of total transmit power [5]-[9]. However, the maximum transmit power for each antenna element in SDMA systems is not restricted in general assumptions. Therefore, in the worst case, an amplifier whose maximum output power is the same as total transmit power is needed for each antenna element; these amplifiers cause an increase in cost. From this point of view, it is desirable to use a reasonable (i.e., low cost) power amplifier for each antenna element, where per-antenna transmit power is limited within a permissible output power.

To meet this requirement, it is necessary to determine weight coefficients so that the transmit power for each antenna is limited below a given threshold.

In Ref.[10], a method to maximize transmission rate in eigenbeam MIMO-OFDM system under constraint of the maximum transmit power for an antenna has been reported, where the weights are determined by considering only the suppression of inter-stream interference (i.e., the optimum weights are first determined without considering the constraint of per-antenna power, and then the total transmit power is normalized to meet the power constraint). However, this method does not optimize weight coefficients in presence of noise and interference. To find the optimum weights under per-antenna power constraint, these two factors (inter-stream interference and signal-to-noise power ratio) have to be taken into consideration simultaneously.

In this paper, first we propose an iterative optimization algorithm to find optimum transmit and receive weights in an E-SDM system, where the transmitter is equipped with a virtual MIMO channel and virtual receiver to obtain the optimum transmitter weight. The transmitter estimates the optimum transmitter weights by minimizing the error signal at the virtual receiver. Second, we propose an optimization method of transmit and receive weights under constraints of both total transmit power and the maximum transmit power for an antenna element in MU-MIMO systems, where the transmit weights are optimized by minimizing the mean square error of the received signal to obtain the minimum bit error rate (BER) under the per-antenna power constraint, provided that the knowledge of channel state information (CSI) and the receive signal to noise power ratio (SNR) is given. In our study, we solve this optimization problem by transforming the above constrained minimization problem to non-constrained one by using the *Extended Interior Penalty Function (EIPF) Method* [11]. After descriptions of the weight optimization methods, BER and signal-to-noise and interference power ratio (SINR) performance of MIMO systems are evaluated by computer simulation.

2. A least mean square based algorithm to determine the transmit and receive weights in Eigen-beam SDM

2.1 Eigen-beam SDM in MIMO systems

Figure 1 shows a MIMO system model considered in this paper, where N_t and N_r stand for the number of transmit and receive antenna elements, respectively. $\mathbf{W}_t$ denotes $N_t \times N_s$ transmitter weight matrix whose row vectors are given as eigenvector of channel autocorrelation matrix $\mathbf{H}^H\mathbf{H}$, where N_s is the number of data streams. $\mathbf{W}_r$ denotes $N_s \times N_r$ receiver weight matrix. $\mathbf{H}$ is $N_r \times N_t$ channel matrix. To achieve the maximum capacity, the receive weight matrix $\mathbf{W}_r$ is determined as

$$\mathbf{W}_r = \mathbf{W}_t^H \mathbf{H}^H \tag{1}$$

When the transmit and receive data stream vectors are defined as $\mathbf{s} = (s_1, s_2, \cdots, s_{Ns})^T$ and $\mathbf{s}_o = (s_{o1}, s_{o2}, \cdots, s_{oNs})^T$, respectively, the received data stream in E-SDM system is given as

$$\mathbf{s}_o = \mathbf{W}_r \mathbf{H} \mathbf{W}_t s + \mathbf{W}_r \mathbf{n} = \mathbf{W}_t^H \mathbf{H}^H \mathbf{H} \mathbf{W}_t s + \mathbf{W}_t^H \mathbf{H}^H \mathbf{n} \tag{2}$$

where $\mathbf{n} = (n_1, n_2, \cdots, n_{Nr})^T$ is noise signal vector.

2.2 Iterative optimization of transmit- and receive-weights in E-SDM

a. System Description

Figure 2 shows a block diagram of an E-SDM system using the proposed LMS based algorithm, where it is assumed that the transmitter is equipped with a virtual MIMO channel and virtual receiver. Figure 3 shows transmission frame structure assumed in this paper, where transmission frame is composed of pilot and data symbols. Pilot symbols are used for weight determination at the receiver. In this paper, for simplicity, we assume that channel state information is perfectly estimated at the receiver and correctly informed to the transmitter by a feedback channel.

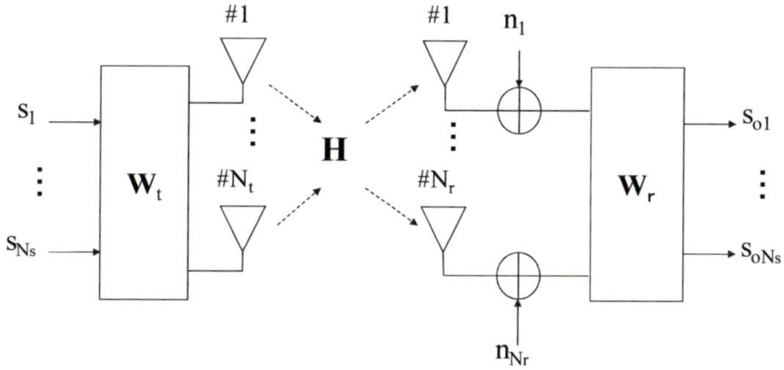

Fig. 1. MIMO System Model

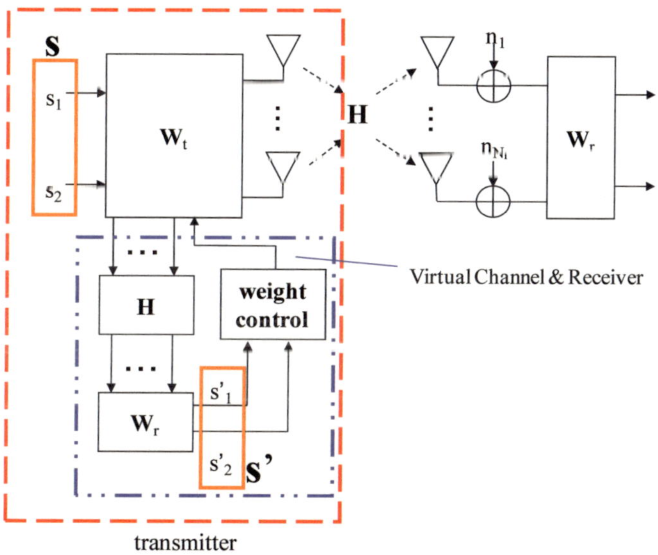

Fig. 2. E-SDM system with iterative weight optimization

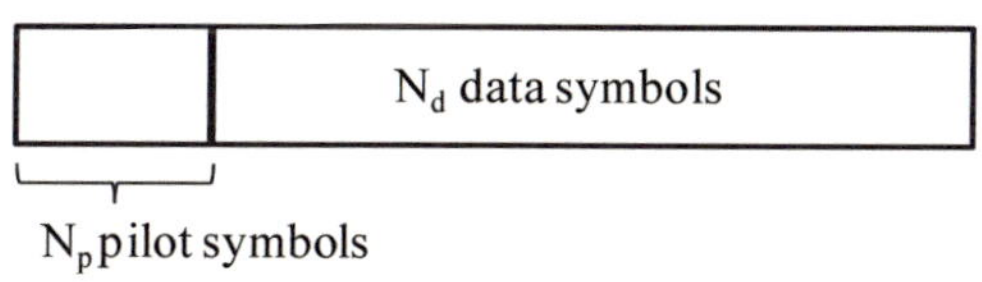

Fig. 3. Frame format

The optimum weight matrices are obtained by minimizing the error signal attributable to inter-stream interference and noise at the receiver side, i.e., the error signal is defined as the difference between transmit and receive signal vectors. This means that, in E-SDM system using the proposed algorithm, weight optimization cannot be performed at the transmitter. To solve this problem, we employ a virtual MIMO channel and virtual receiver on the transmitter side as shown in Fig.2. The received signal at the virtual receiver is expressed as

$$\mathbf{s}'_o = \mathbf{W}_r \mathbf{H} \mathbf{W}_t s = \mathbf{W}_t^H \mathbf{H}^H \mathbf{H} \mathbf{W}_t s \tag{3}$$

where $\mathbf{s}'_o = (s'_{o1}, s'_{o2}, \cdots, s'_{oNs})$ and s'_{oi} is i-th receive stream at the virtual receiver. After determining optimum transmitter weight matrix, the weighted data steam is transmitted to MIMO channel. At the receiver, optimum receiver weight matrix is calculated by observing the pilot symbols. It is noteworthy that the receiver can find optimum receive weight by minimizing the error signal at the receiver, if the optimum transmit weight is multiplied at the transmitter.

b. Iterative Algorithm to Determine the Transmit and Receive Weights

The detailed algorithm to determine optimum weights in the proposed method is explained as follows. For simplicity of discussion, it is assumed that channel matrix $\boldsymbol{H}$ is known to the transmitter. From the relation of Eq.(1), it can be seen that the maximum capacity in E-SDM system is achieved by constructing the matrix $\mathbf{W}_t$ whose row vectors are given as eigenvectors of $\boldsymbol{H}^H\boldsymbol{H}$. Therefore, in the proposed method, eigenvector of channel matrix is sequentially obtained by using a recursive calculation such as *least mean square* (LMS) algorithm. In the following discussion, we consider 2×2 MIMO system for simplicity, i.e., two eigenpaths exist. The detailed expression of the received signal in 2×2 MIMO system can be given as

$$\begin{bmatrix} s_{o1} \\ s_{o2} \end{bmatrix} = \begin{bmatrix} w_{t11}^* & w_{t21}^* \\ w_{t12}^* & w_{t22}^* \end{bmatrix} \mathbf{H}^H \mathbf{H} \begin{bmatrix} w_{t11} & w_{t12} \\ w_{t21} & w_{t22} \end{bmatrix} \begin{bmatrix} s_1 \\ s_2 \end{bmatrix} = \begin{bmatrix} \mathbf{w}_{t1}^H \\ \mathbf{w}_{t2}^H \end{bmatrix} \mathbf{H}^H \mathbf{H} \begin{bmatrix} \mathbf{w}_{t1} & \mathbf{w}_{t2} \end{bmatrix} \begin{bmatrix} s_1 \\ s_2 \end{bmatrix} \tag{4}$$

where $\mathbf{w}_{t1} = (w_{t11}, w_{t21})^T$ and $\mathbf{w}_{t2} = (w_{t12}, w_{t22})^T$ denote column vectors of weight matrix, i.e., the transmit weight vectors for data streams of s_1 and s_2. It is noteworthy that the discussion for 2×2 MIMO system can be easily extended to the case of arbitrary number of transmit and receive antennas as explained later.

First, we consider the optimization of the first weight vector $\mathbf{w}_{t1}$ corresponding to data stream s_1. The first received data stream in E-SDM system is given as

$$s_{o1} = \mathbf{w}_{t1}^H \mathbf{H}^H \mathbf{H} \mathbf{w}_{t1} s_1 \tag{5}$$

where the effect of noise is neglected here. The above equation suggests that the condition for orthogonal multiplexing of data streams in E-SDM system is given as $\mathbf{w}_{t1}^H \mathbf{H}^H \mathbf{H} \mathbf{w}_{t1} = 1$, i.e., when this condition is satisfied, $\mathbf{w}_{t1}$ becomes one of eigenvectors of $\mathbf{H}^H\mathbf{H}$. Thus, the error signal e_1 corresponding to the first data stream is defined as

$$e_1 = s_1 - s_{o1} \tag{6}$$

In this case, the error signal defined in Eq.(6) cannot be obtained at the transmitter. Therefore, by substituting s_{o1} for the first virtual received stream s'_{o1} in Fig.2, the error signal is modified to

$$e_1 = s_1 - s'_{o1} = s_1 - \mathbf{w}_{t1}^H \mathbf{H}^H \mathbf{H} \mathbf{w}_{t1} s_1 \tag{7}$$

Thus, the mean square error is obtained as

$$E\left[|e_1|^2\right] = E\left[|s_1|^2\right] - 2\mathbf{w}_{t1}^H E\left[\mathbf{H}^H\mathbf{H}\right]\mathbf{w}_{t1} E\left[|s_1|^2\right] + \left(\mathbf{w}_{t1}^H E\left[\mathbf{H}^H\mathbf{H}\right]\mathbf{w}_{t1}\right)^2 E\left[|s_1|^2\right] \tag{8}$$

In Eq.(8), we can see that local minimum value does not exist and therefore optimum solution is obtained with a simple iterative algorithm such as LMS method, since Eq.(8) is the fourth order equation with respect to the weight vector $\mathbf{w}_{t1}$ and the first, second and third terms of right side in Eq.(8) are the zero-th, second and fourth order expressions with respect to $\mathbf{w}_{t1}$, respectively.
By differentiating Eq. (8) with respect to $\mathbf{w}_{t1}$, we can obtain

$$\nabla\mathbf{w}_{t1} E\left[|e_1|^2\right] = -4E\left[\mathbf{H}^H\mathbf{H}\right]\mathbf{w}_{t1} E\left[|s_1|^2\right] + 4E\left[\mathbf{H}^H\mathbf{H}\right]\mathbf{w}_{t1}\mathbf{w}_{t1}^H E\left[\mathbf{H}^H\mathbf{H}\right]\mathbf{w}_{t1} \tag{9}$$

where $\nabla\mathbf{w} = \frac{\partial}{\partial \mathbf{w}_x} - j\frac{\partial}{\partial \mathbf{w}_y}$ ($\mathbf{w} = \mathbf{w}_x + j\mathbf{w}_y$). Thus, the recursive equation to obtain the first weight vector is given as

$$\mathbf{w}_{t1}(m+1) = \mathbf{w}_{t1}(m) - \frac{\mu}{4}\nabla\mathbf{w}_{t1} E\left[|e_1|^2\right] \tag{10}$$

In this paper, to achieve fast convergence time, we employ the normalized LMS algorithm[4]. Hence, after substituting Eq.(9) for the above equation and expectation operation is removed, Eq.(10) is reduced to

$$\mathbf{w}_{t1}(m+1) = \mathbf{w}_{t1}(m) + \frac{\mu}{|\mathbf{r}_1(m)|^2}\mathbf{H}^H\mathbf{r}_1(m)e_1^*(m) \tag{11}$$

where m is an integer number corresponding to the number of iterations in the LMS algorithm and μ denotes step size. $\mathbf{r}_1(m)$ is the received signal given by $\mathbf{r}_1(m)=\mathbf{H}\mathbf{w}_{t1}(m)s_1$. After the first weight vector is determined, we consider optimization of the second weight vector $\mathbf{w}_{t2}$ corresponding to data stream s_2. Similarly in the first case, the error signal for the second data stream is defined as

$$e_2 = s_2 - \mathbf{w}_{t2}^H \mathbf{H}^H \mathbf{H} \mathbf{w}_{t2} s_2 - \mathbf{w}_{t2}^H \mathbf{H}^H \mathbf{H} \mathbf{w}_{t1} s_1 \tag{12}$$

where $\mathbf{w}_{t1}$ is set to the optimum value obtained in the first case in Eq. (11). In Eq.(12), the second and third terms in right hand side of this equation mean that "condition where the second eigenvector exists" and the third term means "condition where a target vector $\mathbf{w}_{t2}$ is orthogonal to the first eigenvector $\mathbf{w}_{t1}$". Hence, if $e_2=0$, we can obtain the second eigenvector $\mathbf{w}_{t2}$. Thus, mean square error of the error signal e_2 is given as

$$\begin{aligned} E\left[|e_2|^2\right] &= E\left[|s_2|^2\right] - 2\mathbf{w}_{t2}^H E\left[\mathbf{H}^H\mathbf{H}\right]\mathbf{w}_{t2} E\left[|s_2|^2\right] + \left(\mathbf{w}_{t2}^H E\left[\mathbf{H}^H\mathbf{H}\right]\mathbf{w}_{t2}\right)^2 E\left[|s_2|^2\right] \\ &+ \mathbf{w}_{t2}^H E\left[\mathbf{H}^H\mathbf{H}\right]\mathbf{w}_{t1}\mathbf{w}_{t1}^H E\left[\mathbf{H}^H\mathbf{H}\right]\mathbf{w}_{t2}^H E\left[|s_1|^2\right] \end{aligned} \tag{13}$$

The above equation implies that local minimum solution does not exist and the optimum solution with minimum square error is definitely determined as well as in Eq. (8). Thus, by differentiating this equation respect to $\mathbf{w}_{t2}$, we can obtain

$$\begin{aligned}\nabla\mathbf{w}_{t2}E\left[|e_2|^2\right] &= -4E\left[\mathbf{H}^H\mathbf{H}\right]\mathbf{w}_{t2}E\left[|s_2|^2\right] + 4E\left[\mathbf{H}^H\mathbf{H}\right]\mathbf{w}_{t2}\mathbf{w}_{t2}^H E\left[\mathbf{H}^H\mathbf{H}\right]\mathbf{w}_{t2} \\ &+2E\left[\mathbf{H}^H\mathbf{H}\right]\mathbf{w}_{t1}\mathbf{w}_{t1}^H E\left[\mathbf{H}^H\mathbf{H}\right]\mathbf{w}_{t2}E\left[|s_1|^2\right]\end{aligned} \tag{14}$$

After substituting this equation for Eq.(10) and removing the expectation operation, Eq.(10) is reduced to

$$\mathbf{w}_{t2}(m+1) = \mathbf{w}_{t2}(m) + \frac{\mu}{|\mathbf{r}_2(m)|^2}\left(\mathbf{H}^H\mathbf{r}_2(m)e_2^*(m) - \frac{1}{2}\mathbf{H}^H\mathbf{w}_{t1}\mathbf{w}_{t1}^H\mathbf{H}\mathbf{w}_{t2}(m)|s_1|^2\right) \tag{15}$$

where $\mathbf{r}_2(m)=\mathbf{H}(\mathbf{w}_{t1}(m)s_1+\mathbf{w}_{t2}(m)s_2)$. The optimum weight matrix $\mathbf{W}_t$ is obtained by updating weight vectors of these two recursive equations, i.e., Eqs. (11) and (15).

The above discussion on 2×2 MIMO system is easily extended to $N_t\times 2$ or $2\times N_r$ MIMO system, i.e., for $N_t\times 2$ MIMO system, the received signal at the virtual receiver can be given as

$$\begin{bmatrix} s'_{o1} \\ s'_{o2}\end{bmatrix} = \begin{bmatrix} w_{t11}^* & \cdots & w_{tN_t1}^* \\ w_{t12}^* & \cdots & w_{tN_t2}^*\end{bmatrix}\mathbf{H}^H\mathbf{H}\begin{bmatrix} w_{t11} & w_{t12} \\ \vdots & \vdots \\ w_{tN_t1} & w_{tN_t2}\end{bmatrix}\begin{bmatrix} s_1 \\ s_2\end{bmatrix} = \begin{bmatrix}\mathbf{w}_{t1}^H \\ \mathbf{w}_{t2}^H\end{bmatrix}\mathbf{H}^H\mathbf{H}\begin{bmatrix}\mathbf{w}_{t1} & \mathbf{w}_{t2}\end{bmatrix}\begin{bmatrix} s_1 \\ s_2\end{bmatrix} \tag{16}$$

where $\mathbf{w}_{t1}=(w_{t11},\cdots,w_{tNt1})^T$ and $\mathbf{w}_{t2}=(w_{t12},\cdots,w_{tNt2})^T$. From this equation, it is clear that optimum weight matrixes for $N_t\times 2$ MIMO system are obtained by the same way as 2×2 MIMO case, since channel autocorrelation matrix $\mathbf{H}^H\mathbf{H}$ is given as $N_t\times N_t$ matrix. For case of $2\times N_t$ MIMO system, since the autocorrelation matrix $\mathbf{H}^H\mathbf{H}$ is given as 2×2 matrix, the same discussion as 2×2 MIMO case can be applied.

In addition, the proposed method can be applied to case where the rank of channel matrix is more than two, e.g., when the rank of channel matrix is 3, optimum weight matrix is obtained by minimizing the error function defined so that the third weight vector $\mathbf{w}_{t3}$ is orthogonal to both the first and second weight vectors of $\mathbf{w}_{t1}$ and $\mathbf{w}_{t2}$, where the weight vectors obtained in the previous calculation, i.e., $\mathbf{w}_{t1}$ and $\mathbf{w}_{t2}$, are used as the fixed vectors in this case. Thus, it is obvious that this discussion can be extended to case of channel matrix with the rank of more than 3.

In the proposed method, the parameter convergence speed depends on initial values of weight coefficients. When continuous data transmission is assumed, the convergence time becomes faster by employing weight vectors in last data frame as initial parameters in current recursive calculation.

2.3 Simulation results

We evaluate the performance of a MIMO system using the proposed algorithm by computer simulation. For comparison purpose, obtained eigenvalues, bit error rate (BER) and capacity performance of the E-SDM systems using the proposed algorithm are compared to cases with SVD. Simulation parameters are summarized in Table 1. QPSK with coherent detection is employed as modulation/demodulation scheme. Propagation model is flat uncorrelated quasistatic Rayleigh fading, where we assume that there is no correlation between paths. In the iterative calculation, an initial value of weight vector is set to $(1, 0, 0, \cdots, 0)^T$ for both $\mathbf{w}_{t1}$ and $\mathbf{w}_{t2}$. The step size of μ is set to 0.01 for $\mathbf{w}_{t1}$ and 0.0001 for $\mathbf{w}_{t2}$, respectively. A frame structure consisting of 57 pilot and 182 data symbols in Fig.3 is employed. For simplicity, we assume that channel parameters are perfectly estimated at the receiver and sent back to the transmitter side in this paper.

Number of users	1
Number of data streams	1, 2
(Number of the transmit antennas × Number of the receive antennas)	(2×2), (3×2), (4×2), (2×3), (2×4)
Data modulation /demodulation	QPSK / Coherent detection
Angular spread (Tx & Rx Station)	360°
Propagation model	Flat uncorrelated quasistatic Ralyleight fading

Table 1. Simulation parameters

Figure 4 shows the first and second eigenvalues measured by the proposed method as a function of the frame number in 2×2 MIMO system, where these eigenvalues are obtained by using channel matrix and the transmit and receive weights determined by the proposed algorithm. Figure 4 also shows eigenvalues determined by the SVD method. In Fig. 4, although the first eigenvalue obtained by the proposed method occasionally takes slightly smaller value than that of SVD, the proposed method finds almost the same eigenvectors as the theoretical value obtained by SVD.

Figure 5 shows BER performance of N_tx2 MIMO diversity system using the maximum ratio combining (MRC) as a function of transmit signal to noise power ratio, where average gain of channel is unity. Figure 6 also shows BER performance of 2xN_r MIMO MRC diversity system. In Figs. 5 and 6, the data stream is transmitted by the first eigenpath. Therefore, it can be seen that both methods (LMS, SVD) achieve almost the same BER performance. This result suggests that the eigenvector corresponding to the highest eigenvalue is correctly detected as the first weight vector, i.e., the first eigenpath. It can be also qualitatively explained that the highest eivenvalue is first found as the most dominant parameter determining the error signal.

Figures 7 and 8 show BER performance of N_t×2 and 2×N_r E-MIMO, respectively. The number of data streams is set to two, since the rank of channel matrix is two. Based on the BER minimization criterion [1], the achievable BER is minimized by multiplying the transmit signal by the inverse of the corresponding eigenvalue at the transmitter. In Figs. 7 and 8, we can see that both methods (LMS and SVD) achieve almost the same BER performance.

Figures 9 and 10 show the MIMO channel capacity in case of two data streams. In this paper, for simplicity, MIMO channel capacity is defined as the sum of each eigenpath channel capacity which is calculated based on Shannon channel capacity in AWGN channel [3];

$$C = \log_2 (1+SNR) \quad [\text{bit/s/Hz}] \tag{17}$$

The transmit power allocation for each eigenpath is determined based on the water-filling theorem [3]. In Figs.9 and 10, it can be seen that the E-SDM system with the proposed method achieves the same channel capacity as that of the ideal one (SVD).

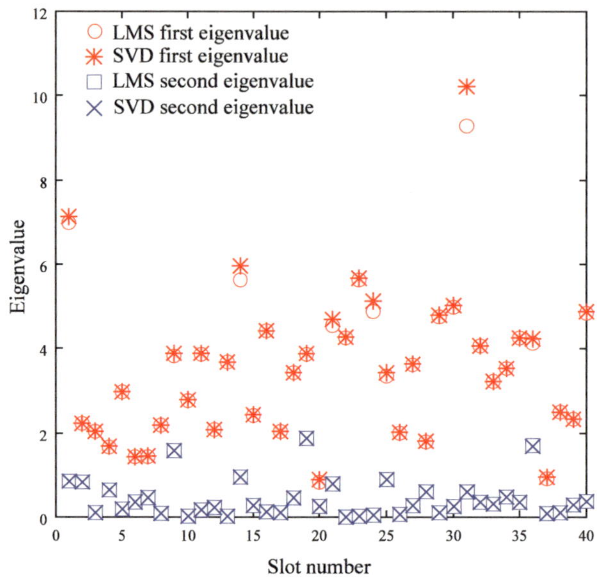

Fig. 4. Measured eigenvalues

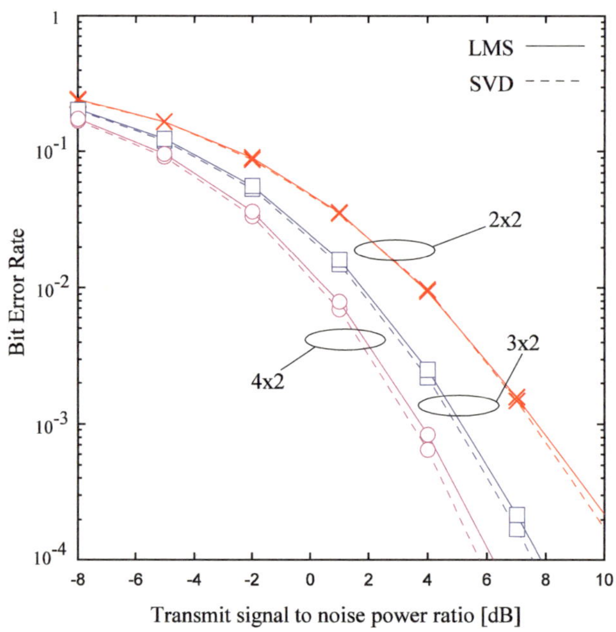

Fig. 5. Bit error rate performance (1 data stream, $N_t \times 2$)

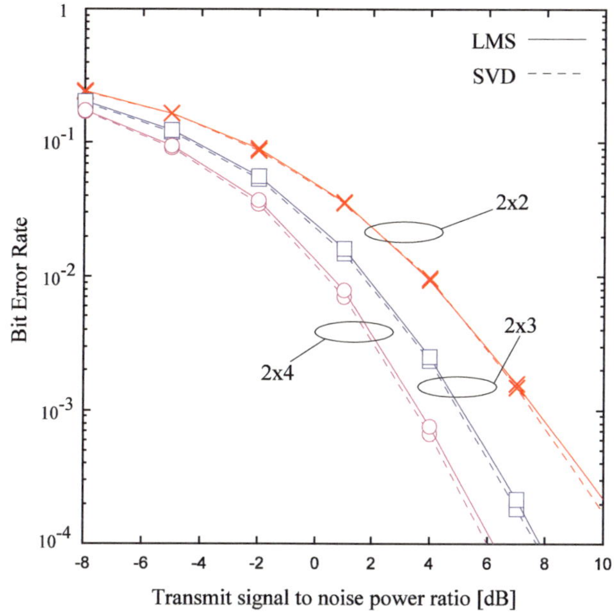

Fig. 6. Bit error rate performance (1 data stream, $2 \times N_r$)

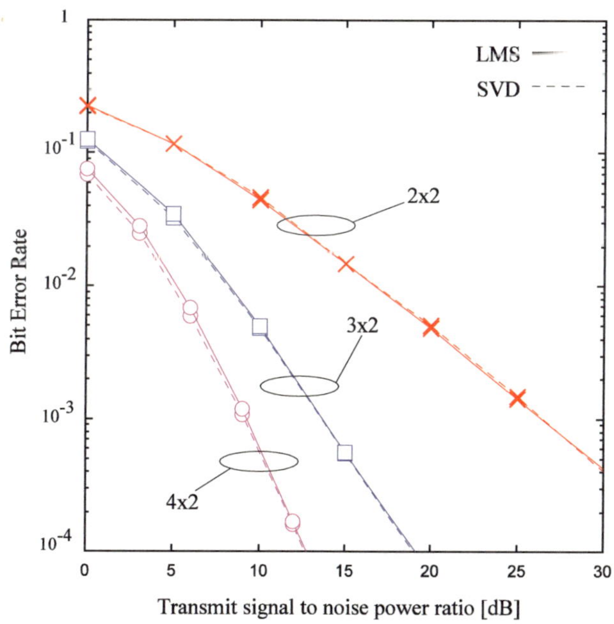

Fig. 7. Bit error rate performance (2 data stream, $N_t \times 2$)

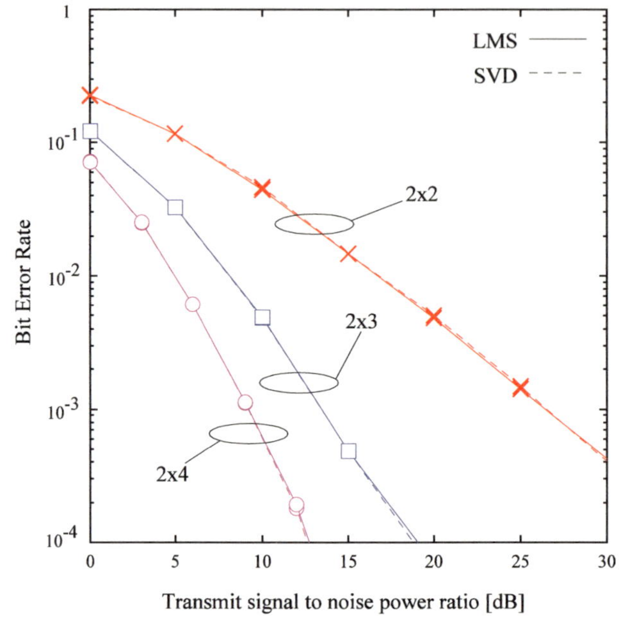

Fig. 8. Bit error rate performance (2 data stream, $2 \times N_r$)

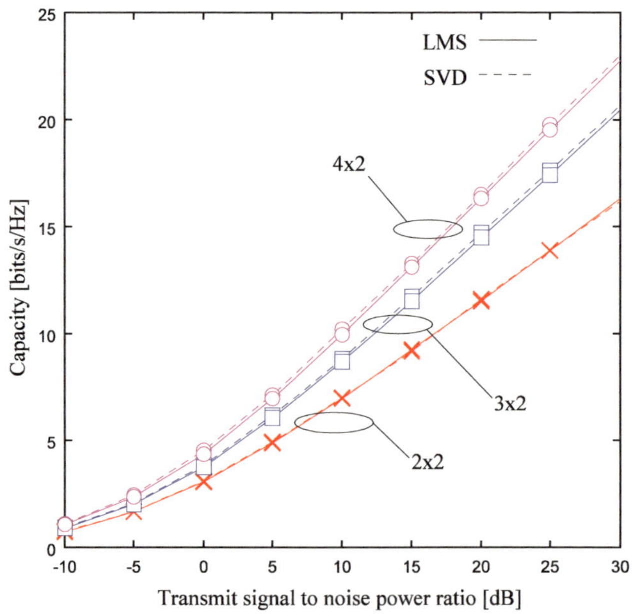

Fig. 9. Channel capacity performance ($N_t \times 2$)

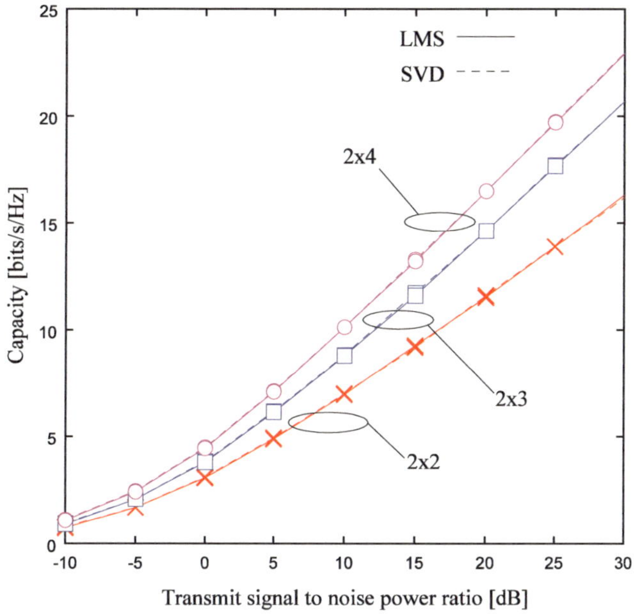

Fig. 10. Channel capacity performance (2×N$_r$)

3. Iterative optimization of the transmitter weights under constraint of the maximum transmit power for an antenna element in MIMO systems

3.1 System model

Figure 11 shows MU-MIMO system considered in this paper, where K antenna elements and single antenna element are equipped at the Base Station (BS) and Mobile Station (MS), respectively. Single antenna is assumed for each Mobile Station (MS). The number of users in SDMA is N. The receive signal at receive antenna $Y=[y_1, \cdots, y_N]^T$ is expressed as

$$\mathbf{Y} = \mathbf{W}_r^H \mathbf{H} \mathbf{W}_t^H \mathbf{X} + \mathbf{W}_r^H \mathbf{n} \tag{18}$$

where superscript ^{T} and superscript ^{H} denote transpose and Hermitian transpose, respectively. **H** is N×K complex channel metrics, $\mathbf{W}_t$ is N×K complex transmit weight matrices, $\mathbf{W}_r=\mathrm{diag}(w_1, \cdots, w_N)$ is receive weight metrics, $\mathbf{X}=[x_1, \cdots, x_N]^T$ is transmit signal, and $\mu=[n_1, \cdots, n_N]^T$ is noise signal. The average power of transmit signal is unity (i.e., $E[x_i^2]=1$), where E[ ] denotes ensemble average operation) and there is no correlation between each user signal (i.e., $E[x_{i1}\, x_{i2}]=0$), the condition to keep the total average transmit power to be less than or equal to P_{th} is given as

$$\sum_{i=1}^{N}\sum_{j=1}^{K}\left|w_{ij}\right|^2 \le P_{th} \tag{19}$$

where w_{ij} denotes the transmit weight of antenna #j for user #i. Then, the condition to constrain the average transmit power per each antenna to be less than or equal to p_{th} is given as

$$\sum_{i=1}^{N} \left| w_{ij} \right|^2 \le p_{th} \quad \forall j \quad (1 \le j \le K) \tag{20}$$

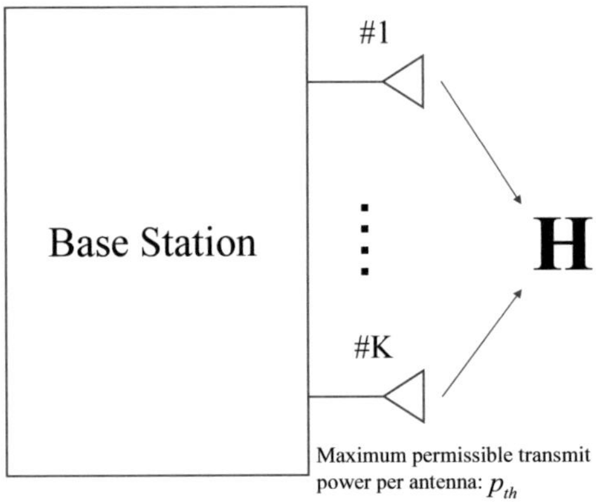

Fig. 11. MU-MIMO Systems

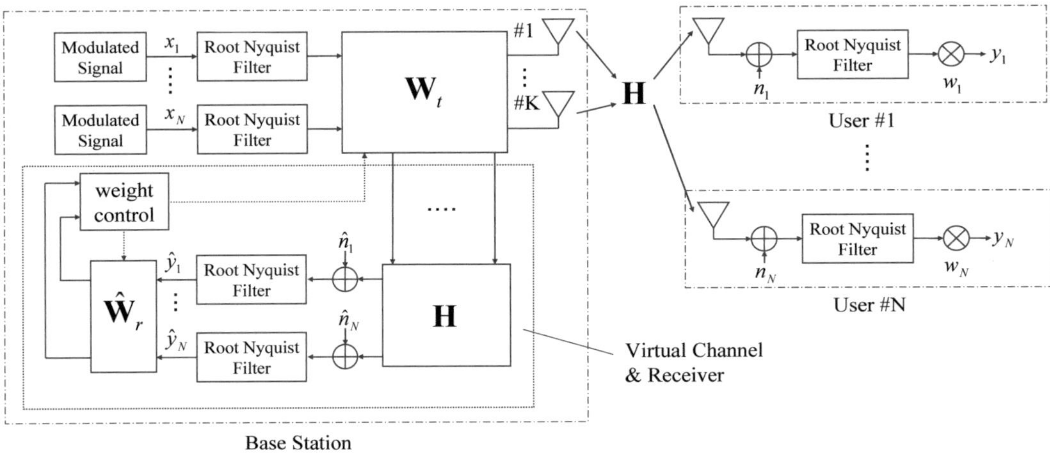

Fig. 12. System configurations

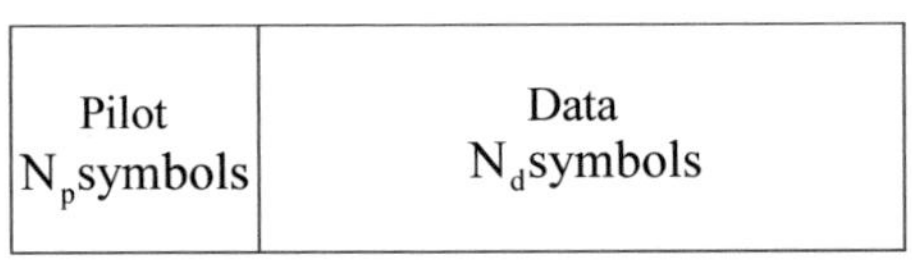

Fig. 13. Frame format

3.2 Transmitter and receiver model

Figure 12 shows the system configuration of the transmitter and receiver in MU-MIMO system considered in this paper, where the number of transmit antennas and the number of receive antennas are K and 1, respectively. A virtual channel and virtual receiver are equipped with the transmitter to estimate mean square error at the receiver side, where

$\hat{\mathbf{W}}_r$ =diag($\hat{w}_1, \cdots, \hat{w}_N$) and $\hat{\mathbf{n}}$ =[$\hat{n}_1, \ldots, \hat{n}_N$]^T denote the virtual receive weight and the virtual noise, respectively. We assume that the average power of additive white Gaussian noise (AWGN) is known to the transmitter, i.e., we assume $E\left[\hat{n}_i^2\right]$=E[$n_i^2$]. Then, the receive signal at the virtual receiver $\hat{\mathbf{Y}}$ is given as

$$\hat{\mathbf{Y}} = \hat{\mathbf{W}}_r^H \mathbf{H} \mathbf{W}_t^H \mathbf{X} + \hat{\mathbf{W}}_r^H \hat{\mathbf{n}} \tag{21}$$

The transmit weights are optimized by minimizing the error signal between transmit and receive signals at the virtual receiver under constraints given as Eqs.(19) and (20). Figure 13 shows a frame format assumed in this paper, where each frame consists of N_p pilot symbols and N_d data symbols. Pilot symbols are known and used for optimizing the receive weights on the receiver side.

3.3 Weight optimization

a. Problem Formulation

The transmit weights are optimized by minimizing the mean square error between transmit and receive signals at the virtual receiver under constraint given as Eqs. (19) and (20). From Eq.(21), the error signal between transmit signal **X** and receive signal at the virtual receiver $\hat{\mathbf{Y}}$ is given as

$$e = \mathbf{X} - \hat{\mathbf{Y}} = \mathbf{X} - \hat{\mathbf{W}}_r^H \mathbf{H} \mathbf{W}_t^H \mathbf{X} - \hat{\mathbf{W}}_r^H \hat{\mathbf{n}} \tag{22}$$

where **e**=[$e_1, \ldots, e_N$]. From Eqs.(19) and (20), the problem to minimize the mean square error under two constraints can be formulated as the following constrained minimizing problem;

$$\text{Minimize} \quad E\left[\|e(\mathbf{W})\|^2\right]$$

$$\text{Subject to} \quad g(\mathbf{W}) = \sum_{i=1}^{N}\sum_{j=1}^{K}\left|w_{ij}\right|^2 - P_{th} \le 0 \tag{23}$$

$$h_j(\mathbf{W}) = \sum_{i=1}^{N}\left|w_{ij}\right|^2 - p_{th} \le 0 \qquad \forall j$$

where $\|\cdot\|$ denotes vector norm. **W** is N×(N+K) complex matrix defined as **W**=[$\mathbf{W}_t$, $\hat{\mathbf{W}}_r$].

b. A EIPF based Approach for Weight Optimization

By introducing the extended interior penalty function (EIPF) method into the problem shown in Eq.(23), this problem can be transformed into the following non-constrained minimizing problem [11];

$$\text{Minimize} \quad E\left[\|e(\mathbf{W})\|^2\right] + r\{\Phi(\mathbf{W}) + \Psi(\mathbf{W})\}$$

$$\text{Subject to} \quad \Phi(\mathbf{W}) = \begin{cases} -\frac{1}{g(\mathbf{W})} & \text{if } g(\mathbf{W}) \le \varepsilon \\ -\frac{2\varepsilon - g(\mathbf{W})}{\varepsilon^2} & \text{if } g(\mathbf{W}) > \varepsilon \end{cases}$$

$$\Psi(\mathbf{W}) = \sum_{j=1}^{K} \psi_j(\mathbf{W})$$

$$\psi_j(\mathbf{W}) = \begin{cases} -\frac{1}{h_j(\mathbf{W})} & \text{if } h_j(\mathbf{W}) \le \varepsilon \\ -\frac{2\varepsilon - h_j(\mathbf{W})}{\varepsilon^2} & \text{if } h_j(\mathbf{W}) > \varepsilon \end{cases}$$

Here, ε(<0) and r(>0) denote the design parameters for non-constrained problem. In Eq.(24), $\Phi(\mathbf{W})$ and $\Psi(\mathbf{W})$ increase rapidly as approaches to the boundary. When g(**W**) = ε and $h_j(\mathbf{W})=\varepsilon$, the continuity of $\Phi(\mathbf{W})$ and $\Psi(\mathbf{W})$ is guaranteed as well as derivatives of these two functions. Thus, Eq. (24) can be minimized by using the Steepest Descent method; **W** is updated as

$$\mathbf{W}(m+1) = \mathbf{W}(m) - \mu\nabla_{\mathbf{W}}\left(E\left[\|e(\mathbf{W})\|^2 \right] + r\{\Phi(\mathbf{W}) + \Psi(\mathbf{W})\} \right) \tag{28}$$

where μ is a step size to adjust the updating speed. $\nabla_{\mathbf{W}}$ denotes a gradient with respect to **W**, which is defined as

$$\nabla_{\mathbf{W}} = \left[\begin{array}{ccc|ccc} \frac{\partial}{\partial w_{11}} & \cdots & \frac{\partial}{\partial w_{1K}} & \frac{\partial}{\partial \hat{w}_1} & & \mathbf{0} \\ \vdots & \ddots & \vdots & & \ddots & \\ \frac{\partial}{\partial w_{N1}} & \cdots & \frac{\partial}{\partial w_{NK}} & \mathbf{0} & & \frac{\partial}{\partial \hat{w}_N} \end{array} \right] \tag{29}$$

where j denotes an imaginary unit and

$$\frac{\partial}{\partial w_{ij}} = \frac{\partial}{\partial\{\mathrm{Re}(w_{ij})\}} + j\frac{\partial}{\partial\{\mathrm{Im}(w_{ij})\}}, \tag{30}$$

$$\frac{\partial}{\partial \hat{w}_i} = \frac{\partial}{\partial\{\mathrm{Re}(\hat{w}_i)\}} + j\frac{\partial}{\partial\{\mathrm{Im}(\hat{w}_i)\}}, \tag{31}$$

When **W** is updated as in Eq. (28) at every symbols, Eq. (28) can be reduced to

$$\mathbf{W}(m+1) = \mathbf{W}(m) - \mu\nabla_{\mathbf{W}}\left[\|\mathbf{e}(\mathbf{W})\|^2 + r\{\Phi(\mathbf{W}) + \Psi(\mathbf{W})\} \right]. \tag{32}$$

3.4 Performance evaluation

Performance of MU-MIMO system using the considered algorithm is evaluated by computer simulation. Simulation parameters are shown in Table 2. As a channel model, we consider a set of 8 plane waves transmitted in random direction within the angle range of 12 degrees at the BS. Each of the plane waves has constant amplitude and takes the random phase distributed from 0 to 2π. All users are randomly distributed with a uniform distribution in a range of the coverage area of a BS. Channel states and distribution of users change independently at every frame. Transmit weights are determined with recursive calculation given in Eq.(32). Receive weights are determined by observing the pilot symbols. The upper limit of the average transmit power for an antenna element normalized by the upper limit of the total transmit power is denoted as

$$\gamma = \frac{p_{th}}{P_{th}}, \tag{33}$$

where

$$\frac{1}{K} \leq \gamma \leq 1 \tag{34}$$

In Eq.(34), γ=1 corresponds to the case without constraint of per-antenna transmit power. The minimum value of γ is 1/K which corresponds to, the strictest case where per-antenna transmit power is limited within the minimum value. The maximum permissible power per user (P_{th}/N) to noise power ratio is defined as

$$\mathrm{SNR}_{\max} = \frac{P_{th} / N}{E[|n_i|^2]} \tag{35}$$

where $E[n_i^2]$ denotes the average noise power corresponding to the user #i.

Channel Model	Flat uncorrelated quasistatic Rayeigh fading
Modulation Method	QPSK
Number of Pilot Symbols (N_p)	34 [symbols/frame]
Number of Data Symbols (N_d)	460 [symbols/frame]
Average propagation loss	0 [dB] (Except for Figs.20 and 21)
Antenna element spacing	5.25λ

Table 2. Simulation Parameters

Figures 14(a) and (b) show complementary cumulative distribution function (CCDF) of average transmit power of transmit signal measured at every frames with respect to antenna #1. The number of transmit antennas is set to 4 and 8, respectively. The number of users is 2. The maximum permissible transmit power is set to P_{th}=1.0, and average noise power is set to E $[n_i^2]$=0.1. From these figures, we can see that transmit power of the signal at antenna #1 can be kept below p_{th}.

Figures 15 and 16 show the received SINR as a function of γ, where SNR_{max} is set to 10 dB. Note that SINR is the same as SNR when the number of users is 1. In these figures, we can see that the degradation in SINR at γ=1/K is about 0.5dB and 0.6~1.0dB for K=4 and 8 as compared with the case of γ =1. It is shown that SINR is slightly degraded when $\gamma \leq 0.4$ and $\gamma \leq 0.3$ for K=4 and K=8, respectively. This is because the probability that transmit power of the signal at a certain antenna element exceeds γ becomes low as γ increases. The received SINR is degraded as the number of users increases, because the diversity effect is reduced attributable to the decrease of a degree of freedom on the number of antennas.

Figures 17 and 18 show BER performance as a function of SNR_{max}, where the number of users is set to 1~3 for K=4 in Fig.17, and set to 3 for K=8 in Fig.18. In these figures, we can see that, when the maximum per-antenna transmit power is limited to 1/K, BER performances is degraded by about 0.7~0.8 dB at BER=10^{-2} as compared with case of γ=1.

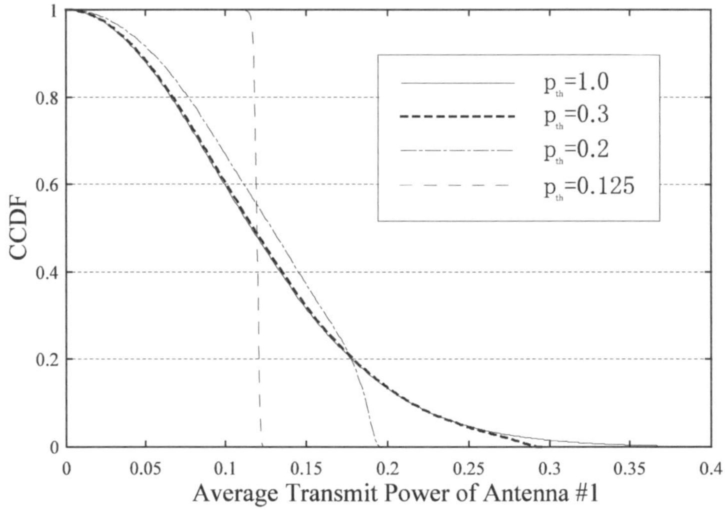

(a) K=4, N=2

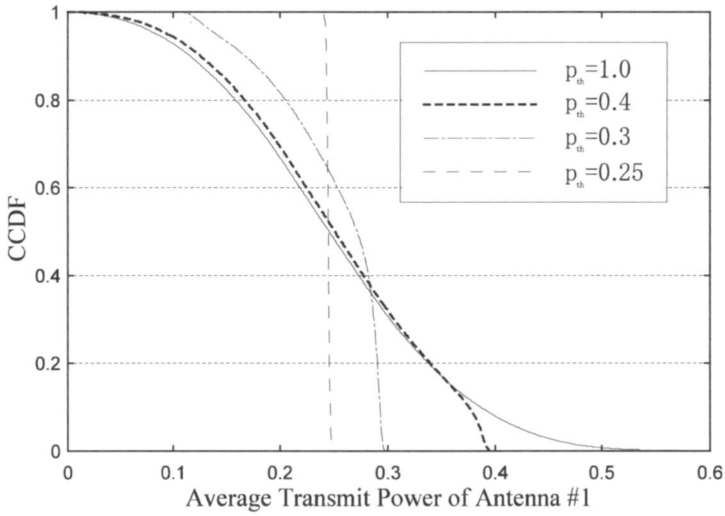

(b) K=8, N=2

Fig. 14. CCDF of average transmit power of the signal measured at every frames with respect to antenna #1

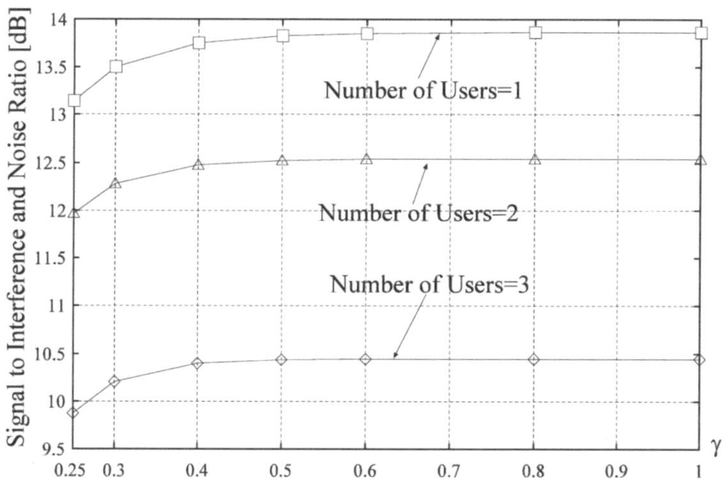

Fig. 15. SINR vs. γ (K=4, SNR_{max}=10dB)

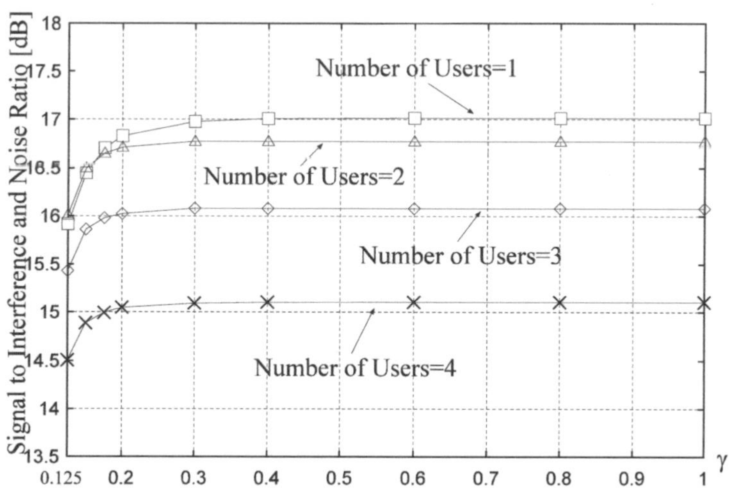

Fig. 16. SINR vs. γ (K=8, SNR_{max}=10dB)

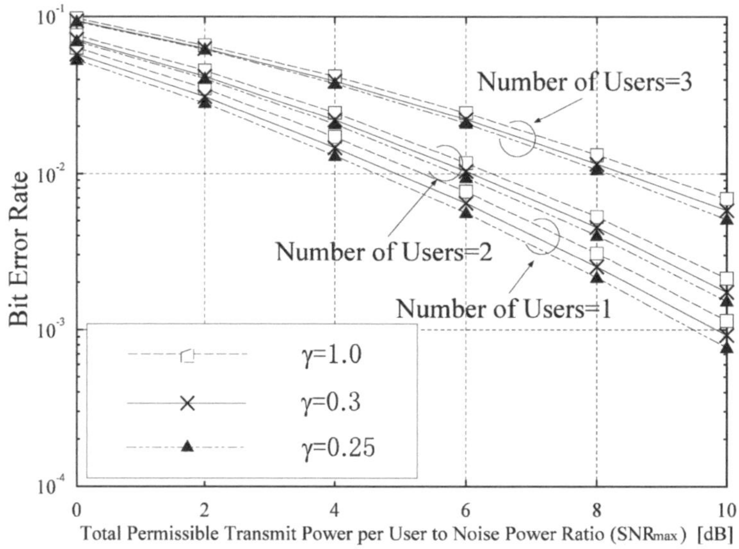

Fig. 17. Bit Error Rate Performance (K=4)

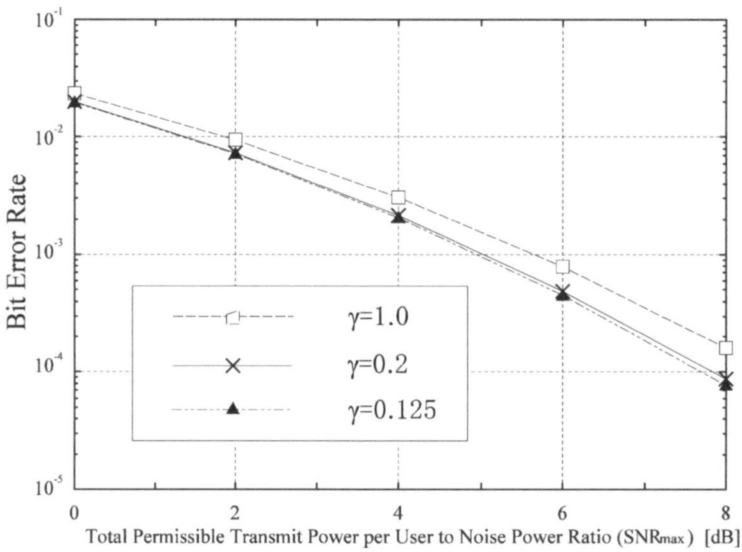

Fig. 18. Bit Error Rate Performance (K=8, N=3)

4. Conclusion

We proposed optimization algorithms of transmit and receive weights for MIMO systems, where the transmitter is equipped with a virtual MIMO channel and virtual receiver to calculate the transmitter weight. First, we proposed an iterative optimization of transmit and receive weights for E-SDM systems, where a least mean square algorithm is used to determine the weight coefficients. The proposed method can be easily extended to the case of E-SDM in MIMO system with arbitrary number of transmit and receive antennas. Second, we proposed a weight optimization method of MIMO systems under constraints of the total transmit power for all antenna elements and the maximum transmit power for an antenna element. The performance of the proposed method is evaluated for QPSK signal in MU-MIMO system with K antenna elements on the transmitter side and single antenna element at the receive side. It is clarified that the degradation of received SINR attributable to constraint of per antenna power is 0.5~1.0 dB in case where the maximum transmit power for an antenna element is limited to 1/K for the number of antenna of K=4 and 8. These results mean that the proposed optimization algorithm enables to use a low cost power amplifier at base stations in MIMO systems.

5. References

[1] T. Ohgane, T. Nishimura, & Y. Ogawa. Applications of Space Division Multiplexing and Those Performance in a MIMO Channel, *IEICE Transactions on Communications*, vol.E88-B, no.5, pp.1843-1851, May. 2005.

[2] G. Lebrun, J. Gao, & M. Faulkner. MIMO Transmission Over a Time-Varying Channel Using SVD, *IEEE Transactions on Wireless Communications*, vol. 4, No.2, pp. 757--764, March 2005.

[3] J. G. Proakis. Digital Communications, Fourth Edition, *McGraw-Hill*, 2001.

[4] S. Haykin. Adaptive Filter Theory, Fourth Edition, *Prentice Hall*, 2002.

[5] H. Yoshinaga, M.Taromaru, & Y.Akaiwa. Performance of Adaptive Array Antenna with Widely Spaced Antenna Elements, *Proceedings of the IEEE Vehicular Technology Conference Fall'99*, pp.72-76, Sept. 1999.

[6] T. Nishimura, Y. Takatori, T. Ohgane, Y. Ogawa, & K. Cho, Transmit Nullforming for a MIMO/SDMA Downlink with Receive Antenna Selection, *Proceedings of the IEEE VTC Vehicular Technology Conference Fall'02*, pp.190-194, Sept. 2002.

[7] Y. Kishiyama, T. Nishimura, T. Ohgane, Y. Ogawa, & Y. Doi. Weight Estimation for Downlink Null Steering in a TDD/SDMA System, *Proceedings of the IEEE VTC Vehicular Technology Conference Spring'00*, pp.346-350, May 2000.

[8] Y. Doi, Tadayuki Ito, J. Kitakado, T. Miyata, S. Nakao, T. Ohgane, & Y. Ogawa. The SDMA/TDD Base Station for PHS Mobile Communication, *Proceedings of the IEEE Vehicular Technology Conference Spring'02*, pp.1074-1078, May 2002.

[9] T. Nishimura, T. Ohgane, Y. Ogawa, Y. Doi, & J. Kitakado. Downlink Beamforming Performance for an SDMA Terminal with Joint Detection, *Proceedings of the IEEE Vehicular Technology Conference Fall'01*, pp.1538-1542, Oct. 2001.

[10] B. S. Krongold. Optimal MIMO-OFDM Loading with Power-Constrained Antennas, *Proceedings of the IEEE PIMRC'06*, Sept. 2006.

[11] S. S. Rao. Engineering Optimization, Theory and Practice, 3rd Edition, *Wiley-Interscience*, 1996.

12

Beamforming Based on Finite-Rate Feedback

Pengcheng Zhu[1], Lan Tang[2], Yan Wang[3], Xiaohu You[4]
[1,3,4]*National Mobile Communications Research Laboratory*
Southeast University, Nanjing, 210096
[4]*School of Electrical Science and Engineering*
Nanjing University, Nanjing, 210093
P. R. China

1. Introduction

Multiple-input multi-output (MIMO), emerged as one of the most significant breakthroughs in wireless communications theory over the last two decades, is considered as a key to meeting the increasing demands for high data rates and mass wireless access services over a limited spectrum bandwidth. Transmit beamforming with receive combining is a low-complexity technique to exploit the benefits of MIMO wireless systems. It has received much interest over the last few years, because it provides substantial performance improvement without sophisticated signal processing. In order to enable the beamforming operation, either full or partial channel state information (CSI) has to be furnished to the transmitter. With full CSI, the optimal transmit beamforming scheme is maximum ratio transmission (MRT) [Dighe et al. (2003a)], where the principal right singular vector of the channel matrix is used as the beamforming vector. In Rayleigh fading, exact expressions for the symbol error rate (SER) of MRT were derived in [Dighe et al. (2003a,b)], and the asymptotic error performance was studied in [Zhou & Dai (2006)].
However, in certain application scenarios, e.g. frequency division duplex (FDD) systems, CSI is not usually available at the transmitter. To cope with the lack of CSI, a beamforming scheme based on finite-rate feedback has been proposed in the literature, where the CSI is quantized at the receiver and fed back to the transmitter. This scheme has been adopted in current 3GPP specifications. Under the assumption of independent block-fading and the assumption of delay- and error-free feedback, the design and performance analysis of quantized beamforming systems have been well investigated. Different beamformer design methods were developed in [Mukkavilli et al. (2003); Love & Heath (2003); Xia & Giannakis (2006)]. In multiple-input single-output (MISO) cases, lower bounds to the outage probability and symbol error rate (SER) were derived in [Mukkavilli et al. (2003)] and [Zhou et al. (2005)], respectively. In MIMO cases, the average receive signal-to-noise ratio (SNR) and outage probability were studied in [Mondal & Heath (2006)]. Analytical results showed that full diversity order can be achieved by a well-designed beamformer [Love & Heath (2005)].
This chapter highlights recent advances in beamforming based on finite-rate feedback from a communication-theoretic perspective. We first study the SER performance when the feedback link is delay- and error-free. Then non-ideal factors in the feedback link are investigated, and countermeasures are proposed to compensate the performance degradation due to non-ideal feedback.

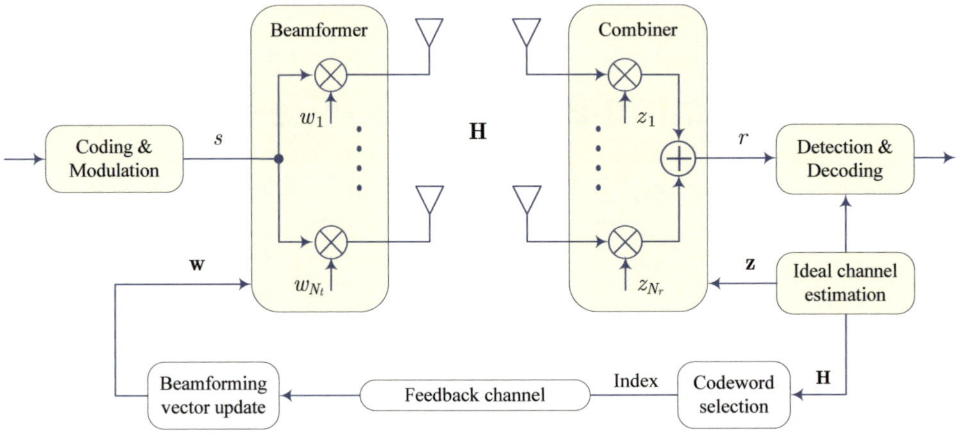

Fig. 1. A beamforming system based on finite-rate feedback.

The notations used in this chapter are conformed to the following convention. Bold upper and lower case letters are used to denote matrices and column vectors, respectively. $(\cdot)^T$, and $(\cdot)^*$ refer to transpose and conjugate transpose, respectively. $\|\cdot\|$ and $\|\cdot\|_F$ stand for vector 2-norm and matrix Frobenius norm, respectively. $\mathbf{I}_N$ refers to the $N \times N$ identity matrix. $\mathcal{CN}(\mu, \sigma^2)$ stands for the circularly symmetric complex Gaussian distribution with mean μ and covariance σ^2. Pr and $\mathbb{E}$ denote the probability and expectation operators, respectively.

2. An upper bound on the SER

In this section, we evaluate the SER of a beamforming system based on finite-rate feedback. Assuming a delay- and error-free feedback link, we derive an upper bound on the average SER, and prove that the bound is asymptotically tight in high SNR regions.

2.1 System model

Consider an MIMO system with N_t transmit and N_r receive antennas. The wireless channel is assumed to be frequency-flat, and is modeled as an $N_r \times N_t$ random matrix $\mathbf{H}$. The (m, n)-th entry of the channel matrix, $h_{m,n}$, denotes the fading coefficient between the n-th transmit antenna and the m-th receive antenna. We assume an independent and identically distributed (i.i.d.) Rayleigh fading. Then the fading coefficient $h_{m,n}$'s are independent of each other and distributed as $\mathcal{CN}(0, 1)$.

The system adopts transmit beamforming with receive combining, as shown in Figure 1. At the transmitter, the information-bearing symbol $s \in \mathbb{C}$ is weighted by a beamforming vector, and transmitted simultaneously from all antennas. Then the $N_t \times 1$ transmitted signal vector is given by $\mathbf{x} = \mathbf{w}\, s$, where $\mathbf{w} = [w_1, \cdots, w_{N_t}]^T$ denotes the beamforming vector. The beamforming vector is a unit-norm vector, satisfying $\|\mathbf{w}\| = 1$. At the receiver, the $N_r \times 1$ received signal vector can be expressed as

$$\mathbf{y} = \mathbf{H}\mathbf{w}\, s + \boldsymbol{\eta}, \tag{1}$$

where $\boldsymbol{\eta}$ refers to the noise vector with independent $\mathcal{CN}(0, N_0)$ entries. Assuming that the receiver knows the channel $\mathbf{H}$ and the beamforming vector $\mathbf{w}$, it performs receive combining to the received signal, using the maximum ratio combining (MRC) vector [Love & Heath (2003)]

$$\mathbf{z} = \frac{\mathbf{Hw}}{\|\mathbf{Hw}\|}.$$

The signal $r \in \mathbb{C}$ after receive combining can be written as

$$r = \mathbf{z}^*\mathbf{y} = \mathbf{z}^*\mathbf{Hw}\, s + \mathbf{z}^*\boldsymbol{\eta}, \tag{2}$$

and the corresponding instantaneous *receive SNR* is given by

$$\gamma = \gamma_S \|\mathbf{Hw}\|^2, \tag{3}$$

where

$$\gamma_S \triangleq \mathbb{E}(|s|^2)/N_0 \tag{4}$$

is the average *symbol SNR*.

In a beamforming system based on finite-rate feedback, the beamforming vector $\mathbf{w}$ is restricted to lie in a finite set (codebook) that is known to both the transmitter and receiver. The codebook, denoted as $\mathcal{C}$, is designed in advance and consists of $N_c = 2^B$ unit-norm codewords $\mathcal{C} = \{\mathbf{c}_1, \cdots, \mathbf{c}_{N_c}\}$. The receiver selects the favorable codeword from the codebook according to

$$\mathbf{c}_{\text{opt}} = \arg\max_{\mathbf{c}\in\mathcal{C}} \|\mathbf{Hc}\|^2. \tag{5}$$

If the codeword $\mathbf{c}_k$ is selected ($\mathbf{c}_{\text{opt}} = \mathbf{c}_k$), its index k is sent to the transmitter via a feedback link, requiring B bits each time. In this section, we focus on the case of delay- and error-free feedback. The transmitter obtains the index of the optimal codeword, and uses the codeword as beamforming vector. Then we have

$$\mathbf{w} = \mathbf{c}_{\text{opt}} \quad \text{(delay- and error-free feedback).} \tag{6}$$

As in [Love & Heath (2005)], we assume the beamforming codewords $\{\mathbf{c}_1, \cdots, \mathbf{c}_{N_c}\}$ span $\mathbb{C}^{N_t}$. This property guarantees the soundness of several steps in the following derivation. We note that it is a mild condition. To our knowledge, all well-designed codebooks, e.g. those in [Love & Heath (2003); Xia & Giannakis (2006)] and 3GPP specifications, satisfy this condition.

2.2 SER analysis

For notation brevity, phase-shift keying (PSK) signals are assumed in the derivations. Conditioned on the instantaneous SNR γ, the SER of M-ary PSK can be expressed as [Simon & Alouini (2005), Eq.(8.23)]

$$P_E = \frac{1}{\pi}\int_0^{\frac{(M-1)\pi}{M}} \exp\Big(-\frac{g_{\text{PSK}}\gamma}{\sin^2\theta}\Big)\, d\theta, \tag{7}$$

where $g_{\text{PSK}} = \sin^2(\pi/M)$ is a constellation-dependent constant. Applying (6) and (5) to (3), the instantaneous SNR of the beamforming system is given by

$$\gamma = \gamma_S\|\mathbf{Hw}\|^2 = \gamma_S\|\mathbf{Hc}_{\text{opt}}\|^2 = \gamma_S \max_{\mathbf{c}\in\mathcal{C}} \|\mathbf{Hc}\|^2 \tag{8}$$

Then substituting (8) into (7) and taking expectation, the average SER of the beamforming system can be written as

$$\overline{P}_E \triangleq \mathbb{E}\, P_E = \frac{1}{\pi}\int_0^{\frac{(M-1)\pi}{M}} \mathbb{E}\exp\Big(-\frac{g_{\text{PSK}}\gamma_S}{\sin^2\theta}\max_{\mathbf{c}\in\mathcal{C}}\|\mathbf{Hc}\|^2\Big)\, d\theta, \tag{9}$$

where the expectation is respect to the channel matrix $\mathbf{H}$.
To find an upper bound on the average SER, we first study the expectation term in the right-hand-side of (9), as shown in the following lemma.

Lemma 1. *Let* $\mathbf{H}$ *be an* $N_r \times N_t$ *random matrix with i.i.d.* $\mathcal{CN}(0,1)$ *entries, and define*

$$\tilde{\mathbf{H}} \triangleq \mathbf{H}/\|\mathbf{H}\|_F .$$

Then for a given beamforming codebook $\mathcal{C}$, *and a given* $t \geq 0$, *we have*

$$\mathbb{E} \exp\Big(-t \cdot \max_{\mathbf{c}\in\mathcal{C}} \|\mathbf{Hc}\|^2\Big) \leq (1 + t \cdot g_{\text{CB}})^{-N_t N_r}, \tag{10}$$

where

$$g_{\text{CB}} \triangleq \Big[\mathbb{E}\Big\{\big(\max_{\mathbf{c}\in\mathcal{C}} \|\tilde{\mathbf{H}}\mathbf{c}\|^2\big)^{-N_t N_r}\Big\}\Big]^{-\frac{1}{N_t N_r}} \tag{11}$$

is a codebook-dependent parameter.

Proof: Since the channel matrix $\mathbf{H}$ has i.i.d. $\mathcal{CN}(0,1)$ entries, $\|\mathbf{H}\|_F^2$ is chi-square distributed and independent of $\tilde{\mathbf{H}}$. The moment generating function of $\|\mathbf{H}\|_F^2$ is given by

$$\mathbb{E} \exp\Big(-s\|\mathbf{H}\|_F^2\Big) = (1+s)^{-N_t N_r}. \tag{12}$$

Therefore we have

$$\begin{aligned} \mathbb{E} \exp\big(-t \cdot \max_{\mathbf{c}\in\mathcal{C}} \|\mathbf{Hc}\|^2\big) &= \mathbb{E}\Big\{\mathbb{E}\Big[\exp\Big(-t \cdot \|\mathbf{H}\|_F^2 \cdot \max_{\mathbf{c}\in\mathcal{C}} \|\tilde{\mathbf{H}}\mathbf{c}\|^2\Big)\Big|\tilde{\mathbf{H}}\Big]\Big\} \\ &= \mathbb{E}\Big\{\big(1 + t \cdot \max_{\mathbf{c}\in\mathcal{C}} \|\tilde{\mathbf{H}}\mathbf{c}\|^2\big)^{-N_t N_r}\Big\}. \end{aligned} \tag{13}$$

Notice that when $x, t > 0$,

$$f(x) = \Big(1 + t \cdot x^{-\frac{1}{N_t N_r}}\Big)^{-N_t N_r}$$

is a concave function with respect to x. We can apply Jensen's inequality to the right-hand-side of (13), and obtain

$$\begin{aligned} \mathbb{E}\Big\{\big(1 + t \cdot \max_{\mathbf{c}\in\mathcal{C}} \|\tilde{\mathbf{H}}\mathbf{c}\|^2\big)^{-N_t N_r}\Big\} &= \mathbb{E}\Big\{\Big(1 + t \cdot \Big[\underbrace{\big(\max_{\mathbf{c}\in\mathcal{C}} \|\tilde{\mathbf{H}}\mathbf{c}\|^2\big)^{-N_t N_r}}_{\text{treated as a r. v.}}\Big]^{-\frac{1}{N_t N_r}}\Big)^{-N_t N_r}\Big\} \\ &\leq \Big(1 + t \cdot \Big[\mathbb{E}\Big\{\big(\max_{\mathbf{c}\in\mathcal{C}} \|\tilde{\mathbf{H}}\mathbf{c}\|^2\big)^{-N_t N_r}\Big\}\Big]^{-\frac{1}{N_t N_r}}\Big)^{-N_t N_r}. \end{aligned} \tag{14}$$

Substituting this into (13), we reach the desired result (10). □

In Lemma 1, the definition of g_{CB} is not given in a closed-form. Its value has to be evaluated numerically. In the following lemma, we will further study the parameter, and derive a closed-form approximation for it.

Lemma 2. *The parameter* g_{CB} *satisfies* $0 \leq g_{\text{CB}} \leq 1$. *Moreover, for a well-designed codebook, it can be approximated by*

$$g_{\text{CB}}^{-N_t N_r} \simeq N_c C_1 (N_t N_r - 1)!\,(N_t - 1) \sum_{n=0}^{N_t - 2} \binom{N_t - 2}{n} \frac{(-1)^n \big(C_2^{-(N_t N_r - 1 - n)} - 1\big)}{N_t N_r - 1 - n}, \tag{15}$$

where

$$C_1 \triangleq N_t N_r \prod_{n=1}^{\min(N_t,N_r)} \frac{[\min(N_t, N_r) - n]!}{(N_t + N_r - n)!} \tag{16}$$

and

$$C_2 \triangleq 1 - (1/N_c)^{\frac{1}{N_t-1}}. \tag{17}$$

Proof: We first prove $0 \leq g_{\mathrm{CB}} \leq 1$. It is clear from the definition that $g_{\mathrm{CB}} \geq 0$. To see $g_{\mathrm{CB}} \leq 1$, consider the following equality

$$\begin{aligned} \lim_{t\to\infty} t^{N_tN_r} \mathbb{E}\Big\{\big(1 + t \cdot \max_{\mathbf{c}\in\mathcal{C}} \|\tilde{\mathbf{H}}\mathbf{c}\|^2\big)^{-N_tN_r}\Big\} &= \mathbb{E}\Big\{\lim_{t\to\infty} t^{N_tN_r}\big(1 + t \cdot \max_{\mathbf{c}\in\mathcal{C}} \|\tilde{\mathbf{H}}\mathbf{c}\|^2\big)^{-N_tN_r}\Big\} \\ &= \mathbb{E}\Big\{\big(\max_{\mathbf{c}\in\mathcal{C}} \|\tilde{\mathbf{H}}\mathbf{c}\|^2\big)^{-N_tN_r}\Big\}. \end{aligned} \tag{18}$$

By the definition (11), (18) implies that

$$g_{\mathrm{CB}}^{-N_tN_r} = \lim_{t\to\infty} t^{N_tN_r} \mathbb{E}\Big\{\big(1 + t \cdot \max_{\mathbf{c}\in\mathcal{C}} \|\tilde{\mathbf{H}}\mathbf{c}\|^2\big)^{-N_tN_r}\Big\} \tag{19}$$

Notice that $\|\tilde{\mathbf{H}}\|_{\mathrm{F}} = 1$ and $\|\mathbf{c}\| = 1, \forall \mathbf{c} \in \mathcal{C}$. Therefore

$$\max_{\mathbf{c}\in\mathcal{C}} \|\tilde{\mathbf{H}}\mathbf{c}\|^2 \leq 1$$

and

$$\mathbb{E}\Big\{\big(1 + t \cdot \max_{\mathbf{c}\in\mathcal{C}} \|\tilde{\mathbf{H}}\mathbf{c}\|^2\big)^{-N_tN_r}\Big\} \geq (1+t)^{-N_tN_r}.$$

So we have

$$\begin{aligned} g_{\mathrm{CB}}^{-N_tN_r} &= \lim_{t\to\infty} t^{N_tN_r} \mathbb{E}\Big\{\big(1 + t \cdot \max_{\mathbf{c}\in\mathcal{C}} \|\tilde{\mathbf{H}}\mathbf{c}\|^2\big)^{-N_tN_r}\Big\} \\ &\geq \lim_{t\to\infty} t^{N_tN_r}(1+t)^{-N_tN_r} \\ &= 1, \end{aligned} \tag{20}$$

which implies $\beta_{\mathcal{C}} \leq 1$.

To obtain (15), recall Equation (13) in the proof of Lemma 1. Substituting it into (19) yields

$$g_{\mathrm{CB}}^{-N_tN_r} = \lim_{t\to\infty} t^{N_tN_r}\, \mathbb{E} \exp\big(-t \cdot \max_{\mathbf{c}\in\mathcal{C}} \|\mathbf{H}\mathbf{c}\|^2\big). \tag{21}$$

Let the eigen-decomposition of the channel be denoted as

$$\mathbf{H}^*\mathbf{H} = \big[\mathbf{u}_1, \cdots, \mathbf{u}_{N_t}\big] \begin{bmatrix} \lambda_1 & & \\ & \ddots & \\ & & \lambda_{N_t} \end{bmatrix} \begin{bmatrix} \mathbf{u}_1^* \\ \vdots \\ \mathbf{u}_{N_t}^* \end{bmatrix} \tag{22}$$

where $\lambda_1 \geq \cdots \geq \lambda_{N_t} \geq 0$ and $\mathbf{u}_1, \cdots, \mathbf{u}_{N_t}$ denote the eigenvalues and the eigenvectors, respectively. We have the following inequality

$$\|\mathbf{H}\mathbf{c}\|^2 = \mathbf{c}^*\mathbf{H}^*\mathbf{H}\mathbf{c} = \sum_{n=1}^{N_t} \lambda_n |\mathbf{u}_n^*\mathbf{c}|^2 \geq \lambda_1 |\mathbf{u}_1^*\mathbf{c}|^2. \tag{23}$$

Substituting (23) into (21), one obtains

$$g_{\mathrm{CB}}^{-N_tN_r} \le \lim_{t\to\infty} t^{N_tN_r}\, \mathbb{E}\, \exp\Big(-t\lambda_1 \cdot \max_{\mathbf{c}\in\mathcal{C}} |\mathbf{u}_1^*\mathbf{c}|^2 \Big). \tag{24}$$

In an i.i.d. Rayleigh fading scenario, $\mathbf{H}^*\mathbf{H}$ is Wishart distributed. The eigenvalue and eigenvector of a Wishart matrix are independent of each other. So (24) can be expressed as

$$\begin{aligned} g_{\mathrm{CB}}^{-N_tN_r} &\le \lim_{t\to\infty} t^{N_tN_r} \int \Big[\int_0^\infty e^{-txz} p_{\lambda_1}(x)\,\mathrm{d}x \Big] \mathrm{d}\Big(\Pr\Big\{ \max_{\mathbf{c}\in\mathcal{C}} |\mathbf{u}_1^*\mathbf{c}|^2 < z \Big\} \Big) \\ &= \int \Big[\lim_{t\to\infty} t^{N_tN_r} \int_0^\infty e^{-txz} p_{\lambda_1}(x)\,\mathrm{d}x \Big] \mathrm{d}\Big(\Pr\Big\{ \max_{\mathbf{c}\in\mathcal{C}} |\mathbf{u}_1^*\mathbf{c}|^2 < z \Big\} \Big) \end{aligned} \tag{25}$$

Since $\mathbf{H}^*\mathbf{H}$ is Wishart distributed, the probability density function (PDF) of its largest eigenvalue λ_1 has the asymptotic property [Zhou & Dai (2006)]

$$p_{\lambda_1}(x) = C_1\, x^{N_tN_r-1} + o(x^{N_tN_r-1}), \quad x \to 0^+, \tag{26}$$

where C_1 is defined in (16), and $o(x^{N_tN_r-1})$ stands for a function $a(x)$ satisfying $\lim_{x\to 0^+} a(x)/x^{N_tN_r-1} = 0$. Then we have

$$\begin{aligned} \lim_{t\to\infty} t^{N_tN_r} \int_0^\infty e^{-txz} p_{\lambda_1}(x)\,\mathrm{d}x &= \lim_{t\to\infty} t^{N_tN_r} \int_0^\infty e^{-yz} p_{\lambda_1}(y/t)\,\mathrm{d}(y/t) \\ &= \int_0^\infty e^{-yz} \Big[\lim_{t\to\infty} t^{N_tN_r-1} p_{\lambda_1}(y/t) \Big] \mathrm{d}y \\ &= \int_0^\infty e^{-yz} C_1 y^{N_tN_r-1} \mathrm{d}y \\ &= C_1 (N_tN_r-1)!\, z^{-N_tN_r} \end{aligned} \tag{27}$$

Substituting (27) into (25) yields

$$g_{\mathrm{CB}}^{-N_tN_r} \le C_1(N_tN_r-1)! \int z^{-N_tN_r} \mathrm{d}\Big(\Pr\Big\{ \max_{\mathbf{c}\in\mathcal{C}} |\mathbf{u}_1^*\mathbf{c}|^2 < z \Big\} \Big). \tag{28}$$

For a well-designed codebook, the Voronoi cells of the codewords can be approximated by 'spherical caps', which leads to a very tight bound [Zhou et al. (2005)]

$$\Pr\Big\{ \max_{\mathbf{c}\in\mathcal{C}} |\mathbf{u}_1^*\mathbf{c}|^2 < z \Big\} \ge 1 - N_c(1-z)^{N_t-1}, \ \ C_2 \le z \le 1, \tag{29}$$

where C_2 is defined in (17). The inequalities (28) and (29) are both tight. We now treat then as approximations, and substitute (29) into (28). After performing the integration, the right-hand-side of (15) is obtained. □

We then apply the results in Lemma 1 and 2 to the SER analysis. Setting $t = g_{\mathrm{PSK}}\gamma_S / \sin^2\theta$ and after some manipulations, (10) becomes

$$\mathbb{E}\, \exp\Big(-\frac{g_{\mathrm{PSK}}\gamma_S}{\sin^2\theta} \max_{\mathbf{c}\in\mathcal{C}} \|\mathbf{Hc}\|^2 \Big) \le \Big(\frac{\sin^2\theta}{g_{\mathrm{PSK}} g_{\mathrm{CB}} \gamma_S + \sin^2\theta} \Big)^{N_tN_r}. \tag{30}$$

Substituting this into (9), we obtain an upper bound on the average SER of M-ary PSK signal

$$\overline{P}_E \le \overline{P}_E^{\mathrm{ub}} = \frac{1}{\pi} \int_0^{\frac{(M-1)\pi}{M}} \Big(\frac{\sin^2\theta}{g_{\mathrm{PSK}} g_{\mathrm{CB}} \gamma_S + \sin^2\theta} \Big)^{N_tN_r} \mathrm{d}\theta, \tag{31}$$

which is the main result of this section.

At last, we give two remarks on the SER bound.

Remark 1 (Asymptotic behavior). The upper bound has the merit of being asymptotically tight. In fact, at high SNR, we have

$$
\begin{aligned}
G &= \lim_{\gamma_S \to \infty} \gamma_S^{N_t N_r} \overline{P}_E \\
&\overset{(9)}{=} \frac{1}{\pi} \lim_{\gamma_S \to \infty} \gamma_S^{N_t N_r} \int_0^{\frac{(M-1)\pi}{M}} \mathbb{E} \exp\left(-\frac{g_{\text{PSK}}\gamma_S}{\sin^2\theta} \max_{\mathbf{c}\in\mathcal{C}} \|\mathbf{Hc}\|^2\right) d\theta \\
&= \frac{1}{\pi} \int_0^{\frac{(M-1)\pi}{M}} \left(\frac{\sin^2\theta}{g_{\text{PSK}}}\right)^{N_t N_r} \left[\lim_{\gamma_S \to \infty} \left(\frac{g_{\text{PSK}}\gamma_S}{\sin^2\theta}\right)^{N_t N_r} \mathbb{E} \exp\left(-\frac{g_{\text{PSK}}\gamma_S}{\sin^2\theta} \max_{\mathbf{c}\in\mathcal{C}} \|\mathbf{Hc}\|^2\right)\right] d\theta \\
&\overset{(21)}{=} \frac{1}{\pi} \int_0^{\frac{(M-1)\pi}{M}} \left(\frac{g_{\text{PSK}}\gamma_S}{\sin^2\theta}\right)^{N_t N_r} (g_{\text{CB}})^{-N_t N_r} d\theta \\
&= \frac{1}{\pi} \int_0^{\frac{(M-1)\pi}{M}} \frac{(\sin\theta)^{2N_t N_r}}{(g_{\text{PSK}} g_{\text{CB}})^{N_t N_r}} d\theta \,.
\end{aligned} \tag{32}
$$

This equation shows that as $\gamma_S \to \infty$, $\overline{P}_E$ decreases as $G\gamma_S^{-N_t N_r} + o(\gamma_S^{-N_t N_r})$. $G^{\frac{-1}{N_t N_r}}$ is usually referred to as the *coding gain*. On the other hand, it is easily verified that

$$
\lim_{\gamma_S \to \infty} \gamma_S^{N_t N_r} \overline{P}_E^{\text{ub}} = G \,.
$$

Therefore, (31) is asymptotically tight at high SNR. Moreover, when $\gamma_S = 0$, both sides of (31) are equal to $(M-1)/M$. So the bound holds with equality. This guarantees the tightness of the bound at low SNR.

Remark 2 (Extension to other constellations). For brevity, we have assumed a phase-shift keying (PSK) signal in the derivation of the SER bound. However, the same procedure can be applied to other 2-D constellations. For example, the SER of square quadrature amplitude modulation (QAM), conditioned on the instantaneous SNR γ, can be expressed as [Simon & Alouini (2005)]

$$
P_{E,\text{QAM}} = \frac{4(\sqrt{M}-1)}{\pi M} \int_0^{\pi/4} \exp\left(-\frac{g_{\text{QAM}}\gamma}{\sin^2\theta}\right) d\theta + \frac{4(\sqrt{M}-1)}{\pi\sqrt{M}} \int_{\pi/4}^{\pi/2} \exp\left(-\frac{g_{\text{QAM}}\gamma}{\sin^2\theta}\right) d\theta,
$$

where M is the constellation size, and $g_{\text{QAM}} = 1.5/(M-1)$. This equation has a similar form to (7). Using the procedure of deriving (31), we can obtain an upper bound on the average SER of QAM

$$
\begin{aligned}
\overline{P}_{E,\text{QAM}} = \mathbb{E}P_{E,\text{QAM}} \leq{}& \frac{4(\sqrt{M}-1)}{\pi M} \int_0^{\pi/4} \left(\frac{\sin^2\theta}{g_{\text{QAM}} g_{\text{CB}}\gamma_S + \sin^2\theta}\right)^{N_t N_r} d\theta \\
&+ \frac{4(\sqrt{M}-1)}{\pi\sqrt{M}} \int_{\pi/4}^{\pi/2} \left(\frac{\sin^2\theta}{g_{\text{QAM}} g_{\text{CB}}\gamma_S + \sin^2\theta}\right)^{N_t N_r} d\theta \,.
\end{aligned} \tag{33}
$$

2.3 Extension to correlated rayleigh fading

In a correlated Rayleigh fading channel, the system model is the same as in Section 2.1, except that the channel matrix is modeled as

$$\text{vec}(\mathbf{H}) = \mathbf{\Phi}\mathbf{h}_w, \tag{34}$$

where $\mathbf{h}_w$ refers to an $N_tN_r \times 1$ random vector with independent $\mathcal{CN}(0,1)$ entries; $\mathbf{\Phi}$ is an $N_tN_r \times N_tN_r$ positive definite matrix; and $\text{vec}(\mathbf{H})$ denotes the $N_tN_r \times 1$ vector of stacked columns of $\mathbf{H}$. $\mathbf{\Phi}^2$ $(= \mathbf{\Phi}\mathbf{\Phi}^*)$ is usually called the channel correlation matrix.

The idea used in Section 2.2 can be extended to correlated Rayleigh fading scenarios. For M-ary PSK signals, we can prove that the average SER is upper bounded by

$$\overline{P}_E \le \frac{1}{\pi}\int_0^{\frac{(M-1)\pi}{M}} \left(\frac{\sin^2\theta}{g_{\text{PSK}}g_{\text{CB}}g_{\text{Cor}}\gamma_S + \sin^2\theta}\right)^{N_tN_r} d\theta, \tag{35}$$

where

$$g_{\text{Cor}} \triangleq \left[\det(\mathbf{\Phi}^2)\right]^{\frac{1}{N_tN_r}}$$

is a parameter depends on the channel correlation matrix. The proof of this bound is out of the scope of this book. Interesting readers are referred to [Zhu et al. (2010)] for detailed derivations.

The bound (35) is asymptotically tight at high and low SNRs [Zhu et al. (2010)]. However, at medium SNR, the tightness of the bound is not guaranteed because it does not fully reflect the effect of channel correlation. Based on extensive simulations, we propose the following conjectured SER formula

$$\overline{P}_E \overset{\text{conjectured}}{\le} \frac{1}{\pi}\int_0^{\frac{(M-1)\pi}{M}} \prod_{i=1}^{N_tN_r} \frac{\sin^2\theta}{g_{\text{PSK}}g_{\text{CB}}\gamma_S\lambda_{\mathbf{\Phi}^2;i} + \sin^2\theta} d\theta, \tag{36}$$

where $\lambda_{\mathbf{\Phi}^2;i}$ denotes the i-th eigenvalue of $\mathbf{\Phi}^2$. We have not been able to prove the conjecture as yet. Some discussion in support of it is presented in [Zhu et al. (2010)].

2.4 Numerical results

Simulations are carried out for 2Tx-2Rx and 4Tx-2Rx antenna configurations. QPSK and 16-QAM constellations are used in the simulations. The 2Tx-2Rx system uses the 2-bit Grassmannian codebook ([Love & Heath (2003)]-TABLE II), and the 4Tx-2Rx system adopts the 4-bit codebook in 3GPP specification ([3GPP TS 36.211 (2009)]-Table 6.3.4.2.3-2).

Figure 2 and 3 show the average SER in uncorrelated Rayleigh fading. The SER bounds (31) (33) are tight in these figures.

We also consider a correlated Rayleigh fading channel. The channel correlation matrix $\mathbf{\Phi}^2$ is generated according to the 802.11n model D [Erceg et al. (2004)]. We assume uniform linear arrays with 0.5-wavelength adjacent antenna spacing, as in [Erceg et al. (2004), Section 7]. Figure 4 and 5 plot the average SER in this fading environment. In both figures, the gap between the simulation result and the bound (35) is no more than 2 dB. The conjectured SER formula (36) is even tighter than the bound.

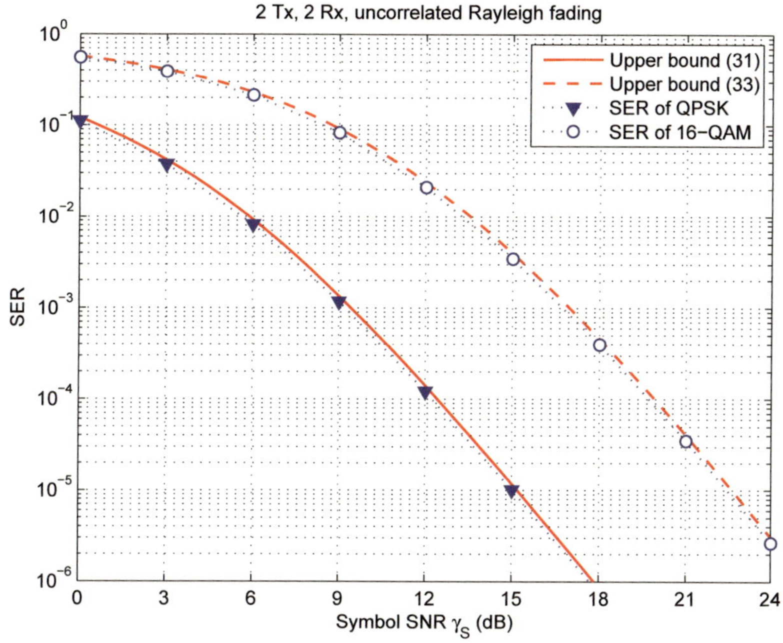

Fig. 2. SER of the 2Tx-2Rx beamforming system in Rayleigh fading environment.

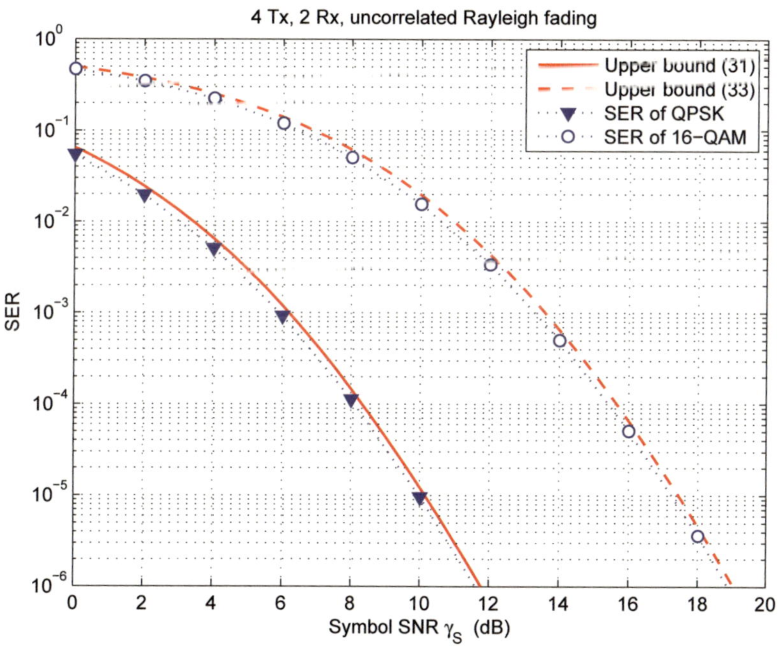

Fig. 3. SER of the 4Tx-2Rx beamforming system in Rayleigh fading environment.

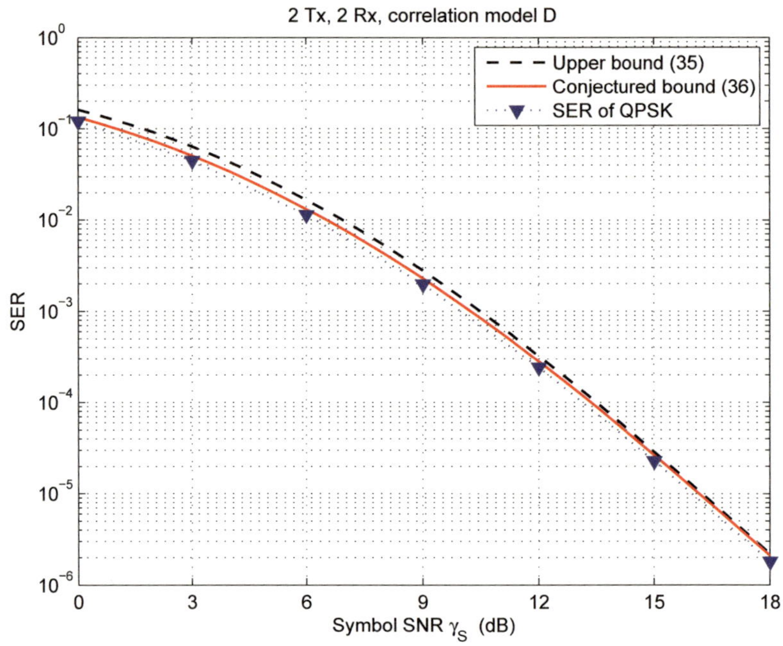

Fig. 4. SER of the 2Tx-2Rx beamforming system in correlated Rayleigh fading environment.

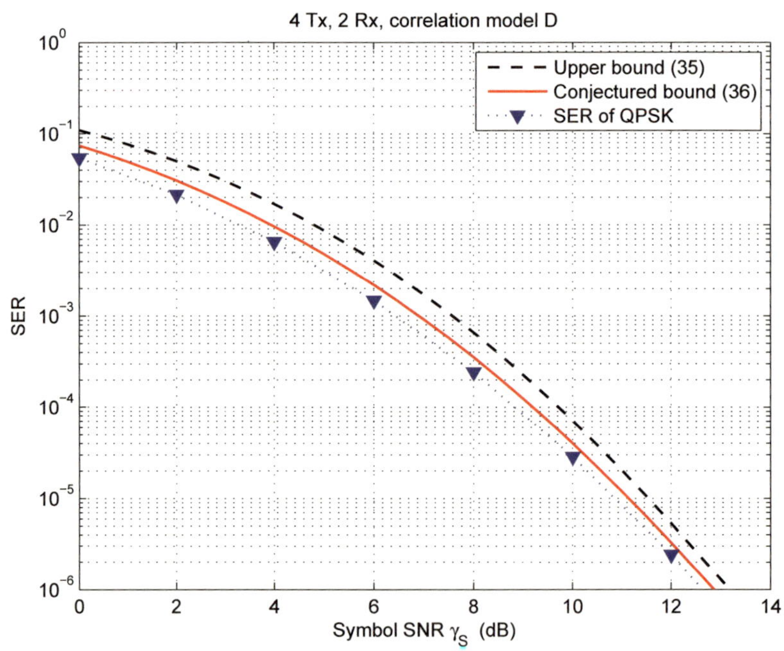

Fig. 5. SER of the 4Tx-2Rx beamforming system in correlated Rayleigh fading environment.

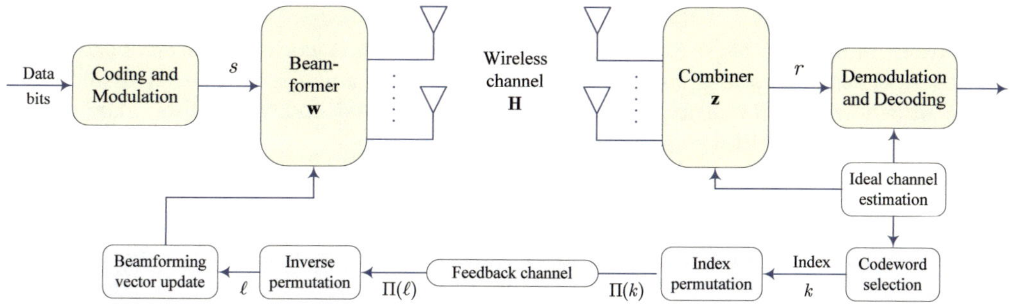

Fig. 6. A finite-rate beamforming system with index permutation.

3. Effect of feedback error and index assignment

In Section 2, the feedback link is assumed to be error-free, but feedback error is inevitable in practice. In this section, we study a finite-rate beamforming system with feedback error. It is shown that feedback error deteriorates not only the array gain but also the diversity gain. To mitigate the effect of feedback error, IA technique is adopted, which is popular in conventional VQ designs (see [Zeger & Gersho (1990)] [Ben-David & Malah (2005)] and references therein). IA technique is preferable to other error-protection methods, e.g. error-control coding, because it requires neither additional feedback bits nor additional signal processing, i.e. it is *redundancy-free*.

3.1 System model

The finite-rate beamforming system with feedback error is illustrated in Figure 6. The forward part of this system is similar to that in Section 2.1. The wireless channel is modeled as an $N_r \times N_t$ random matrix with i.i.d. $\mathcal{CN}(0,1)$ entries. The input-output of the forward part is given by

$$r = \mathbf{z}^* \mathbf{H} \mathbf{w}\, s + \mathbf{z}^* \boldsymbol{\eta}, \tag{37}$$

where

- $\mathbf{H}$ denotes the $N_r \times N_t$ channel matrix. Assuming i.i.d. Rayleigh fading, the entries of $\mathbf{H}$ are independent $\mathcal{CN}(0,1)$ random variables;
- $s \in \mathbb{C}$ is the information-bearing symbol;
- $\mathbf{w} \in \mathbb{C}^{N_t}$ stands for the unit-norm beamforming vector;
- $\mathbf{z} = \mathbf{H}\mathbf{w}/\|\mathbf{H}\mathbf{w}\|$ is the MRC combining vector;
- $\boldsymbol{\eta} \in \mathbb{C}^{N_r}$ refers to the noise vector with independent $\mathcal{CN}(0, N_0)$ entries;
- $r \in \mathbb{C}$ denotes the signal after receive combining.

The instantaneous receive SNR in (37) is given by

$$\gamma = \gamma_S \|\mathbf{H}\mathbf{w}\|^2, \tag{38}$$

where

$$\gamma_S \triangleq \mathbb{E}(|s|^2)/N_0 \tag{39}$$

is the average symbol SNR.

In the system, the beamforming vector $\mathbf{w}$ is determined by feedback information. The receiver conveys the feedback information to the transmitter via a low-rate feedback link, which consists of the five blocks at the bottom of Figure 6. The 'Index permutation' and 'Inverse permutation' blocks are used to cope with feedback error. A codebook $\mathcal{C} = \{\mathbf{c}_1, \cdots, \mathbf{c}_{N_c}\}$ is designed in advance and stored at both the transmitter and the receiver. The codewords $\mathbf{c}_k$'s are unit-norm vector. The receiver selects the optimal codeword that maximize the instantaneous receive SNR, i.e.

$$\mathbf{c}_{\text{opt}} = \arg\max_{\mathbf{c}\in\mathcal{C}} \|\mathbf{H}\mathbf{c}\|^2. \tag{40}$$

If the codeword $\mathbf{c}_k$ is selected ($\mathbf{c}_{\text{opt}} = \mathbf{c}_k$), its index k is fed into the 'Index permutation' block, which performs permutation Π on this index and outputs $\Pi(k)$. The permutation Π is an invertible (one-to-one and onto) operator from the index set $\{1, \cdots, N_c\}$ to itself. For each index k, Π uniquely maps it to another index $\Pi(k) \in \{1, \cdots, N_c\}$. Given the codebook cardinality N_c, there are $N_c!$ permutations [Ben-David & Malah (2005)]. For example, when $N_c = 3$, one possible permutation is to map $1 \to 3$, $2 \to 1$, and $3 \to 2$, respectively.
The permutated index $\Pi(k)$ is then sent to the transmitter via the 'feedback channel', which is modeled as a discrete memoryless channel (DMC) with transition probability

$$T[i,j] = \Pr\{\text{DMC output is } j \,|\, \text{DMC input is } i\}, \qquad i,j = 1, \cdots, N_c. \tag{41}$$

Due to possible feedback error, the feedback channel does not always output the correct information. Supposing that the output of the feedback channel is $\Pi(\ell)$, the transmitter performs the inverse-permutation Π^{-1} on $\Pi(\ell)$ and obtains the index ℓ. The corresponding codeword $\mathbf{c}_\ell$ is used to update the beamforming vector. Conditioned on the optimal codeword $\mathbf{c}_{\text{opt}} = \mathbf{c}_k$, the probability that the transmitter uses $\mathbf{c}_\ell$ as the beamforming vector is given by

$$\begin{aligned}\Pr(\mathbf{w} = \mathbf{c}_\ell|\mathbf{c}_{\text{opt}} = \mathbf{c}_k) &= \Pr\left(\text{DMC output is } \Pi(\ell) \,|\, \text{DMC input is } \Pi(k)\right) \\ &= T[\Pi(k), \Pi(\ell)], \qquad k,\ell = 1, \cdots, N_c.\end{aligned} \tag{42}$$

Feedback error will deteriorate the performance of beamforming. In the following, we will quantify the effect of feedback error on the diversity gain and array gain.

3.2 The diversity gain

Diversity gain refers to the slope of SER-vs-SNR curve (on a log-log scale) as SNR approaches infinity. With error-free feedback, a well-designed beamformer can provide full diversity gain $N_t \times N_r$ if the codebook cardinality $N_c \geq N_t$ [Love & Heath (2005)]. However, the diversity gain may decrease to N_r due to feedback error, as shown in the following lemma.

Lemma 3. *For the beamforming system described in Section 3.1, the diversity gain equals to N_r, if the transition probability of the DMC feedback channel satisfies*

$$T[i,j] \geq T_{\min} > 0, \qquad i,j = 1, \cdots, N_c. \tag{43}$$

Proof: According to [Tse & Viswanath (2005)], the diversity gain is equal to

$$\lim_{\gamma_S \to \infty} -\frac{\log P_{\text{out}}}{\log \gamma_S}, \tag{44}$$

where P_{out} denotes the outage probability. For the beamforming system based on finite-rate feedback, the outage probability is given by [Mukkavilli et al. (2003); Mondal & Heath (2006)]

$$\begin{aligned}P_{\text{out}} &= \Pr\left(\log_2(1 + \gamma_S\|\mathbf{H}\mathbf{w}\|^2) < R\right) \\ &= \Pr\left(\|\mathbf{H}\mathbf{w}\|^2 < (2^R - 1)/\gamma_S\right),\end{aligned} \tag{45}$$

where R denotes the desired transmission rate. By the law of total probability, the right-hand-side of (45) can be expanded to give

$$\begin{aligned}
P_{\text{out}} &= \sum_{k=1}^{N_c}\sum_{\ell=1}^{N_c} \Pr\left(\|\mathbf{H}\mathbf{c}_\ell\|^2 < \frac{2^R-1}{\gamma_S} \,\middle|\, \mathbf{c}_{\text{opt}} = \mathbf{c}_k \right) \Pr\left(\mathbf{w} = \mathbf{c}_\ell\,,\, \mathbf{c}_{\text{opt}} = \mathbf{c}_k\right) \\
&\overset{(42)}{=} \sum_{k=1}^{N_c}\sum_{\ell=1}^{N_c} \Pr\left(\|\mathbf{H}\mathbf{c}_\ell\|^2 < \frac{2^R-1}{\gamma_S} \,\middle|\, \mathbf{c}_{\text{opt}} = \mathbf{c}_k \right) T[\Pi(k),\Pi(\ell)]\Pr(\mathbf{c}_{\text{opt}} = \mathbf{c}_k) \\
&\overset{(43)}{\geq} T_{\min} \sum_{\ell=1}^{N_c}\sum_{k=1}^{N_c} \Pr\left(\|\mathbf{H}\mathbf{c}_\ell\|^2 < \frac{2^R-1}{\gamma_S} \,\middle|\, \mathbf{c}_{\text{opt}} = \mathbf{c}_k \right) \Pr(\mathbf{c}_{\text{opt}} = \mathbf{c}_k) \\
&= T_{\min} \sum_{\ell=1}^{N_c} \Pr\left(\|\mathbf{H}\mathbf{c}_\ell\|^2 < \frac{2^R-1}{\gamma_S} \right). \qquad (46)
\end{aligned}$$

Similarly, because $T[\Pi(k),\Pi(\ell)] \leq 1$ holds trivially, we have

$$P_{\text{out}} \leq \sum_{l=1}^{N_c} \Pr\left(\|\mathbf{H}\mathbf{c}_\ell\|^2 < \frac{2^R-1}{\gamma_S} \right). \qquad (47)$$

Since the codeword $\mathbf{c}_\ell$ is deterministic and unit-norm, $\mathbf{H}\mathbf{c}_\ell$ is a Gaussian distributed random vector with zero mean and covariance $\mathbf{I}_{N_r}$. So $\|\mathbf{H}\mathbf{c}_\ell\|^2$ has a central chi-square distribution with $2N_r$ degrees of freedom. Hence

$$\Pr\left\{ \|\mathbf{H}\mathbf{c}_\ell\|^2 < \frac{2^R-1}{\gamma_S} \right\} = \underbrace{1 - \exp\left(-\frac{2^R-1}{\gamma_S}\right) \sum_{m=0}^{N_r-1} \frac{1}{m!} \left(\frac{2^R-1}{\gamma_S}\right)^m}_{\triangleq\, f(\gamma_S)}, \quad \forall \ell. \qquad (48)$$

Substituting this equation into (46) and (47), one obtains

$$T_{\min} N_c f(\gamma_S) \leq P_{\text{out}} \leq N_c f(\gamma_S).$$

Then, using the squeezing theorem, we get the diversity gain

$$\lim_{\gamma_S\to\infty} -\frac{\log P_{\text{out}}}{\log \gamma_S} = \lim_{\gamma_S\to\infty} -\frac{\log f(\gamma_S)}{\log \gamma_S} = N_r\,, \qquad (49)$$

where the last equality can be derived by repeatedly applying L'Hospital's rule. This is the desired result. □

Two remarks about Lemma 3 are in order.

Remark 1 (BSC). The constraint (43) is satisfied by many DMC's. For example, binary symmetric channel (BSC) is an important DMC, whose transition probability is

$$T_{\text{BSC}}[i,j] = p^{\,d_{\text{H}}(i-1,j-1)}(1-p)^{B-d_{\text{H}}(i-1,j-1)} \qquad i,j = 1,\cdots,N_c, \qquad (50)$$

where p is a parameter of the BSC; $N_c = 2^B$; and $d_{\text{H}}(i-1,j-1)$ denotes the Hamming distance between the binary representations of $i-1$ and $j-1$. The BSC satisfies (43) if $p > 0$. Hence, a beamforming system based on finite-rate feedback can only achieve a diversity gain of N_r, if the feedback channel is a BSC.

Remark 2(Comparison with STBC). With error-free feedback, finite-rate beamforming outperforms space-time block coding (STBC), because beamforming provides not only diversity gain but also array gain. However, this conclusion should be reconsidered if there exists feedback error. According to Lemma 1, a beamforming system based on finite-rate feedback may suffer from a large diversity gain loss due to feedback error. So at sufficiently high SNR, the performance of beamforming will be worse than that of STBC. The comparison at low-to-medium SNR is of practical importance, but out of the scope of this book.

3.3 The array gain

The array gain is defined as the ratio of the average receive SNR $\mathbb{E}\gamma$ and the symbol SNR γ_S. It reflects the increase in average receive SNR that arises from the coherent combining effect of multiple antennas.

Consider the case that the receiver selects $\mathbf{c}_k$ as the optimal codeword, but the transmitter uses $\mathbf{c}_\ell$ as the beamforming vector due to feedback error. The average receive SNR conditioned on this case is

$$\gamma_S \, \mathbb{E}\big(\|\mathbf{H}\mathbf{c}_\ell\|^2 \,\big|\, \mathbf{c}_{\text{opt}} = \mathbf{c}_k\big) = \gamma_S \, \mathbf{c}_\ell^* \, \mathbb{E}(\mathbf{H}^*\mathbf{H}|\mathbf{c}_{\text{opt}} = \mathbf{c}_k)\, \mathbf{c}_\ell \, .$$

Therefore, by the law of total expectation, the array gain can be written as

$$\begin{aligned}\frac{\mathbb{E}(\gamma)}{\gamma_S} &= \sum_{k=1}^{N_c}\sum_{\ell=1}^{N_c} \mathbf{c}_\ell^* \, \mathbb{E}\left(\mathbf{H}^*\mathbf{H} \,\big|\, \mathbf{c}_{\text{opt}} = \mathbf{c}_k\right) \mathbf{c}_\ell \, \Pr\left(\mathbf{w} = \mathbf{c}_\ell \, , \, \mathbf{c}_{\text{opt}} = \mathbf{c}_k\right) \\ &\overset{(42)}{=} \sum_{k=1}^{N_c}\sum_{\ell=1}^{N_c} \mathbf{c}_\ell^* \, \mathbb{E}\left(\mathbf{H}^*\mathbf{H} \,\big|\, \mathbf{c}_{\text{opt}} = \mathbf{c}_k\right) \mathbf{c}_\ell \, T[\Pi(k), \Pi(l)] \, \Pr(\mathbf{c}_{\text{opt}} = \mathbf{c}_k). \end{aligned} \quad (51)$$

In the right-hand-side of (51), the accurate values of $\mathbb{E}\left(\mathbf{H}^*\mathbf{H}|\mathbf{c}_{\text{opt}} = \mathbf{c}_k\right)$ and $\Pr(\mathbf{c}_{\text{opt}} = \mathbf{c}_k)$ are hard to obtain, but their approximations can be derived as follows.

Since the channel matrix $\mathbf{H}$ has independent $\mathcal{CN}(0,1)$ entries, $\mathbf{H}^*\mathbf{H}$ is Wishart distributed. Its eigen-decomposition is denoted as

$$\mathbf{H}^*\mathbf{H} = \left[\mathbf{u}_1, \cdots, \mathbf{u}_{N_t}\right] \begin{bmatrix} \lambda_1 & & \\ & \ddots & \\ & & \lambda_{N_t} \end{bmatrix} \begin{bmatrix} \mathbf{u}_1^* \\ \vdots \\ \mathbf{u}_{N_t}^* \end{bmatrix} = \sum_{n=1}^{N_t} \lambda_n \mathbf{u}_n \mathbf{u}_n^* \quad (52)$$

where $\lambda_1 \geq \cdots \geq \lambda_{N_t} \geq 0$ and $\mathbf{u}_1, \cdots, \mathbf{u}_{N_t}$ denote the eigenvalues and the eigenvectors, respectively. The distribution of nonzero eigenvalues is known. The eigen matrix $\mathbf{U} = \left[\mathbf{u}_1, \cdots, \mathbf{u}_{N_t}\right]$ is uniformly distributed on the group of $N_t \times N_t$ unitary matrices and *independent of the eigenvalues* [Love & Heath (2003), Lemma 1].

If the feedback channel is perfect, the transmitter uses the eigenvector $\mathbf{u}_1$ as the beamforming vector, which is called maximum ratio transmission (MRT). But this is not the case in practice, where the quantized information — the optimal codeword $\mathbf{c}_{\text{opt}}$ — is fed back to the transmitter. Then, one would expect that the ideal feedback information $\mathbf{u}_1$ and the quantized version $\mathbf{c}_{\text{opt}}$ are 'close'. Since both $\mathbf{u}_1$ and $\mathbf{c}_{\text{opt}}$ belong to the unit hypersphere

$$\Omega^{N_t} \triangleq \{\mathbf{x} \in \mathbb{C}^{N_t} : \mathbf{x}^*\mathbf{x} = 1\}, \quad (53)$$

a suitable measure of their 'closeness' is the chordal distance, defined as

$$d_c(\mathbf{x}_1, \mathbf{x}_2) = \sqrt{1 - |\mathbf{x}_1^*\mathbf{x}_2|^2}, \qquad \mathbf{x}_1, \mathbf{x}_2 \in \Omega^{N_t}. \quad (54)$$

Now define a spherical cap centered at the codeword $\mathbf{c}_k$ as

$$\mathcal{S}(\mathbf{c}_k) = \left\{\mathbf{x} \in \Omega^{N_t} : d_c(\mathbf{x}, \mathbf{c}_k) < \sqrt{\alpha}\right\}$$
$$= \left\{\mathbf{x} \in \Omega^{N_t} : |\mathbf{x}^*\mathbf{c}_k|^2 > 1 - \alpha\right\}, \quad k = 1, \cdots, N_c. \tag{55}$$

We use the event $\{\mathbf{u}_1 \in \mathcal{S}(\mathbf{c}_k)\}$ to approximate the event $\{\mathbf{c}_{\text{opt}} = \mathbf{c}_k\}$, which leads to the following approximations

$$\mathbb{E}\left(\mathbf{H}^*\mathbf{H} \,|\, \mathbf{c}_{\text{opt}} = \mathbf{c}_k\right) \simeq \mathbb{E}\left\{\mathbf{H}^*\mathbf{H} \,|\, \mathbf{u}_1 \in \mathcal{S}(\mathbf{c}_k)\right\} \overset{(52)}{=} \sum_{n=1}^{N_t} \mathbb{E}(\lambda_n)\, \mathbb{E}\{\mathbf{u}_n\mathbf{u}_n^* \,|\, \mathbf{u}_1 \in \mathcal{S}(\mathbf{c}_k)\}, \tag{56}$$

$$\Pr(\mathbf{c}_{\text{opt}} = \mathbf{c}_k) \simeq \Pr\left\{\mathbf{u}_1 \in \mathcal{S}(\mathbf{c}_k)\right\}. \tag{57}$$

In order to simplify the right-hand-side of (56), we derive the following result.

Lemma 4. *If* $\mathbf{U} = [\mathbf{u}_1, \cdots, \mathbf{u}_{N_t}]$ *is uniformly distributed on the group of* $N_t \times N_t$ *unitary matrices, then*

$$\mathbb{E}\left\{\mathbf{u}_1\mathbf{u}_1^* \,|\, \mathbf{u}_1 \in \mathcal{S}(\mathbf{c}_k)\right\} = \frac{\alpha}{N_t}\mathbf{I}_{N_t} + (1-\alpha)\,\mathbf{c}_k\mathbf{c}_k^*, \quad k = 1, \cdots, N_c; \tag{58}$$

$$\mathbb{E}\left\{\mathbf{u}_n\mathbf{u}_n^* \,|\, \mathbf{u}_1 \in \mathcal{S}(\mathbf{c}_k)\right\} = \frac{N_t - \alpha}{N_t(N_t-1)}\,\mathbf{I}_{N_t} - \frac{1-\alpha}{N_t - 1}\,\mathbf{c}_k\mathbf{c}_k^*,$$
$$n = 2, \cdots, N_t,\; k = 1, \cdots, N_c. \tag{59}$$

Proof: Since $\mathbf{U}$ is uniformly distributed, $\mathbf{u}_1$ is uniformly distributed on the unit hypersphere Ω^{N_t}. Conditioned on a particular realization of $\mathbf{u}_1$, $\mathbf{u}_n, n = 2, \cdots, N_t$, is uniformly distributed on the set [Marzetta & Hochwald (1999)]

$$\mathcal{O}(\mathbf{u}_1) = \{\mathbf{u}_n \in \Omega^{N_t} : \mathbf{u}_n^*\mathbf{u}_1 = 0\}.$$

Therefore

$$\mathbb{E}\left\{\mathbf{u}_1\mathbf{u}_1^* \,|\, \mathbf{u}_1 \in \mathcal{S}(\mathbf{c}_k)\right\} = \int_{\mathcal{S}(\mathbf{c}_k)} C_0\, \mathbf{u}_1\mathbf{u}_1^* d\mathbf{u}_1, \tag{60}$$

$$\mathbb{E}\left\{\mathbf{u}_n\mathbf{u}_n^* \,|\, \mathbf{u}_1 \in \mathcal{S}(\mathbf{c}_k)\right\} = \int_{\mathcal{S}(\mathbf{c}_k)} C_0 \left(\int_{\mathcal{O}(\mathbf{u}_1)} C_1\, \mathbf{u}_n\mathbf{u}_n^* d\mathbf{u}_n\right) d\mathbf{u}_1, \quad n = 2, \cdots, N_t, \tag{61}$$

where

$$C_0 = \frac{(N_t-1)!}{2\pi^{N_t}\alpha^{N_t-1}}, \qquad C_1 = \frac{(N_t-2)!}{2\pi^{N_t-1}}, \tag{62}$$

such that

$$\int_{\mathcal{S}(\mathbf{c}_k)} C_0\, d\mathbf{u}_1 = 1, \qquad \int_{\mathcal{O}(\mathbf{u}_1)} C_1\, d\mathbf{u}_n = 1.$$

The calculation of (60). Let $\mathbf{\Theta} = [\mathbf{c}_k, \mathbf{\Theta}_0]$ be a unitary matrix, where $\mathbf{\Theta}_0$ is chosen arbitrarily with the constraint that $\mathbf{\Theta}$ is unitary. Using the transformation $\mathbf{v} = \mathbf{\Theta}^*\mathbf{u}_1$, $\mathcal{S}(\mathbf{c}_k)$ is rotated to

$$\mathcal{S}(\mathbf{e}_1) = \{\mathbf{x} \in \Omega^{N_t} : |\mathbf{x}^*\mathbf{e}_1|^2 > 1 - \alpha\},$$

where $\mathbf{e}_1 \triangleq [1, 0, \cdots, 0]^T$. The Jacobian of this transformation is 1, because $\boldsymbol{\Theta}$ is unitary. Hence, applying the transformation $\mathbf{v} = \boldsymbol{\Theta}^* \mathbf{u}_1$ to the right-hand-side of (60) gives

$$\mathbb{E}\left\{\mathbf{u}_1\mathbf{u}_1^* \mid \mathbf{u}_1 \in \mathcal{S}(\mathbf{c}_k)\right\} = \int_{\mathcal{S}(\mathbf{e}_1)} C_0 \boldsymbol{\Theta}\mathbf{v}\mathbf{v}^*\boldsymbol{\Theta}^* d\mathbf{v} = \boldsymbol{\Theta}\left(C_0 \int_{\mathcal{S}(\mathbf{e}_1)} \mathbf{v}\mathbf{v}^* d\mathbf{v}\right)\boldsymbol{\Theta}^*. \tag{63}$$

The surface integration $\int_{\mathcal{S}(\mathbf{e}_1)} \mathbf{v}\mathbf{v}^* d\mathbf{v}$ can be converted to $2N_t - 1$ dimensional multiple integration [Fleming (1977)]. Let

$$\mathcal{G} = \left\{\mathbf{y} \in \mathbb{R}^{2N_t - 1} : -\pi < y_1 < \pi,\ y_2^2 + \cdots + y_{2N_t-1}^2 < \alpha\right\}. \tag{64}$$

Construct the transformation $\mathbf{g} : \mathcal{G} \to \mathcal{S}(\mathbf{e}_1)$

$$\begin{cases} \Re(v_1) = g_1(\mathbf{y}) = \sqrt{1 - y_2^2 - \cdots - y_{2N_t-1}^2}\ \cos y_1, \\ \Im(v_1) = g_2(\mathbf{y}) = \sqrt{1 - y_2^2 - \cdots - y_{2N_t-1}^2}\ \sin y_1, \\ \Re(v_n) = g_{2n-1}(\mathbf{y}) = y_{2n-2}, \qquad n = 2, \cdots, N_t, \\ \Im(v_n) = g_{2n}(\mathbf{y}) = y_{2n-1}, \qquad n = 2, \cdots, N_t, \end{cases} \tag{65}$$

where $\Re(v_n)$ and $\Im(v_n)$ denote the real and imaginary parts of the nth entry of $\mathbf{v}$ respectively. It can be verified that

$$\mathcal{J}\mathbf{g} \triangleq \sqrt{\det\left[(D\mathbf{g})^T (D\mathbf{g})\right]} = 1, \tag{66}$$

where $D\mathbf{g}$ denotes the $2N_t \times (2N_t - 1)$ differential matrix. The (i, j)-th entry of $D\mathbf{g}$ is given by

$$[D\mathbf{g}]_{i,j} = \frac{\partial g_i}{\partial y_j}, \quad i = 1, \cdots, 2N_t,\ j = 1, \cdots, 2N_t - 1.$$

Then, under the transformation $\mathbf{v} = \mathbf{g}(\mathbf{y})$,

$$\begin{aligned} \left[\int_{\mathcal{S}(\mathbf{e}_1)} \mathbf{v}\mathbf{v}^* d\mathbf{v}\right]_{n,\ell} &= \int_{\mathcal{S}(\mathbf{e}_1)} \left[\Re(v_n)\Re(v_\ell) + \Im(v_n)\Im(v_\ell)\right] + j\left[\Im(v_n)\Re(v_\ell) - \Re(v_n)\Im(v_\ell)\right] d\mathbf{v} \\ &= \int_{\mathcal{G}} \left(\left[g_{2n-1}\, g_{2l-1} + g_{2n}\, g_{2l}\right] + j\left[g_{2n}\, g_{2l-1} - g_{2n-1}\, g_{2l}\right]\right) \mathcal{J}\mathbf{g}\, d\mathbf{y} \\ &= \int_{\mathcal{G}} g_{2n-1}\, g_{2l-1}\, d\mathbf{y} + \int_{\mathcal{G}} g_{2n}\, g_{2l}\, d\mathbf{y} + j\left(\int_{\mathcal{G}} g_{2n}\, g_{2l-1}\, d\mathbf{y} - \int_{\mathcal{G}} g_{2n-1}\, g_{2l}\, d\mathbf{y}\right) \end{aligned}$$

The calculation of these integrations is straightforward, e.g.

$$\int_{\mathcal{G}} g_3\, g_3 d\mathbf{y} = \int_{\mathcal{G}} y_2^2\, d\mathbf{y} = \int_{-\pi}^{\pi} dy_1 \int_{y_2^2 + \cdots + y_{2N_t-1}^2 < \alpha} y_2^2\, dy_2 \cdots dy_{2N_t-1} = \frac{\pi\alpha^{N_t}}{N_t!}.$$

Collecting the results, one obtains

$$C_0 \int_{\mathcal{S}(\mathbf{e}_1)} \mathbf{v}\mathbf{v}^* d\mathbf{v} = \frac{\alpha}{N_t} \mathbf{I}_{N_t} + (1 - \alpha)\, \mathbf{e}_1\mathbf{e}_1^*. \tag{67}$$

Substituting (67) into (63), we get (58).

The calculation of (61) . The derivation of the inner integration in the right-hand-side of (61) is along the line of the calculation of (60). So it is omitted here. The result is

$$\int_{\mathcal{O}(\mathbf{u}_1)} C_1 \mathbf{u}_n \mathbf{u}_n^* \, d\mathbf{u}_n = \frac{1}{N_t - 1} \left(\mathbf{I}_{N_t} - \mathbf{u}_1 \mathbf{u}_1^* \right). \tag{68}$$

Substituting this into (61) yields

$$\begin{aligned}\mathbb{E}\left\{\mathbf{u}_n \mathbf{u}_n^* \mid \mathbf{u}_1 \in \mathcal{S}(\mathbf{c}_k)\right\} &= \frac{C_0}{N_t - 1} \int_{\mathcal{S}(\mathbf{c}_k)} \mathbf{I}_{N_t} d\mathbf{u}_1 - \frac{C_0}{N_t - 1} \int_{\mathcal{S}(\mathbf{c}_k)} \mathbf{u}_1 \mathbf{u}_1^* d\mathbf{u}_1 \\ &= \frac{N_t - \alpha}{N_t(N_t - 1)} \mathbf{I}_{N_t} - \frac{1 - \alpha}{N_t - 1} \mathbf{c}_k \mathbf{c}_k^*,\end{aligned}$$

which is the desired result (59). □

Applying Lemma 4 to (56) and using the fact

$$\sum_{n=1}^{N_t} \mathbb{E}(\lambda_n) = \mathbb{E}(\|\mathbf{H}\|_{\mathrm{F}}^2) = N_t N_r,$$

we obtain

$$\mathbb{E}\left(\mathbf{H}^*\mathbf{H} \mid \mathbf{c}_{\mathrm{opt}} = \mathbf{c}_k\right) \simeq \frac{N_t(1-\alpha)}{N_t - 1}\Big(\mathbb{E}(\lambda_1) - N_r\Big)\mathbf{c}_k \mathbf{c}_k^* + \left(\frac{N_r(N_t - \alpha)}{N_t - 1} - \frac{1-\alpha}{N_t - 1}\mathbb{E}(\lambda_1)\right)\mathbf{I}_{N_t}$$
$$k = 1, \cdots, N_c. \tag{69}$$

Moreover, according to the results in [Mukkavilli et al. (2003)], the right-hand-side of (57) can be simplified as

$$\Pr(\mathbf{c}_{\mathrm{opt}} = \mathbf{c}_k) \simeq \Pr\left\{\mathbf{u}_1 \in \mathcal{S}(\mathbf{c}_k)\right\} = \alpha^{N_t - 1}, \quad k = 1, \cdots, N_c. \tag{70}$$

The calculation is straightforward, and omitted here. To consist with

$$\sum_{k=1}^{N_c} \Pr(\mathbf{c}_{\mathrm{opt}} = \mathbf{c}_k) = 1,$$

we set

$$\alpha = (1/N_c)^{\frac{1}{N_t - 1}}.$$

Substituting (69) and (70) into (51), the array gain of the beamforming system can be approximated by

$$\begin{aligned}\frac{\mathbb{E}(\gamma)}{\gamma_S} \simeq& \frac{N_r(N_t - \alpha)}{N_t - 1} - \frac{1-\alpha}{N_t - 1}\mathbb{E}(\lambda_1) \\ &+ \frac{N_t(1-\alpha)}{N_c(N_t - 1)}\Big(\mathbb{E}(\lambda_1) - N_r\Big) \sum_{k=1}^{N_c}\sum_{\ell=1}^{N_c} |\mathbf{c}_k^* \mathbf{c}_\ell|^2 \, T[\Pi(k), \Pi(l)],\end{aligned} \tag{71}$$

which is the main result about the array gain.
We note that, in general, $\mathbb{E}(\lambda_1)$ can be obtained by numerical integration or simulation. It has closed-form expressions in some cases. In MISO systems, $\mathbf{H}$ reduces to a vector $\mathbf{h}$. Therefore

$$\mathbb{E}(\lambda_1) = \mathbb{E}(\|\mathbf{h}\|^2) = N_t, \qquad \text{when} \quad N_r = 1. \tag{72}$$

If $\min(N_t, N_r) = 2$, a closed form expression of $\mathbb{E}(\lambda_1)$ is derived in [Kang & Alouini (2004)]. As a special case, when the feedback is error-free, we have $T[i, j] = \delta_{i,j}$. In this case, (71) reduces to the result in [Mondal & Heath (2006)].

3.4 The index assignment

In the beamforming system, the adopted IA scheme Π affects the behavior of the feedback of the codeword index, which in turn impacts on the overall system performance. In this section, we focus on the design of the IA scheme, using the array gain (or equivalently, the average receive SNR) as a design metric.
According to (71), an IA scheme Π that maximizes the array gain can be obtained by solving

$$\begin{aligned} \underset{\Pi}{\text{maximize}} : \quad & \sum_{k=1}^{N_c} \sum_{\ell=1}^{N_c} |\mathbf{c}_k^* \mathbf{c}_\ell|^2 \, T[\Pi(k), \Pi(\ell)] \\ \text{subject to} : \quad & \Pi \text{ is a permutation} \end{aligned} \tag{73}$$

Maximization of the cost function in (73) over all of the $N_c!$ possible permutations is a special case of the quadratic assignment problem (QAP), and is known to be NP-complete. If N_c is small, it can be solved by brute-force search. But for a large codebook, e.g. $N_c \geq 16$, brute-force search is prohibitive since $16! > 10^{13}$, and suboptimal methods have been proposed in the literature [Zeger & Gersho (1990); Ben-David & Malah (2005)].
BSC is an important case of DMC. The optimization problem (73) can be simplified in the case of BSC feedback channel. In practice, feedback errors seldom occur, and the effect of multiple bit errors can be neglected. So the transition probability (50) can be approximated by

$$T_{\text{BSC}}[i,j] \simeq T_{\text{AP}}[i,j] = \begin{cases} (1-p)^B, & d_{\text{H}}(i-1, j-1) = 0; \\ p(1-p)^{B-1}, & d_{\text{H}}(i-1, j-1) = 1; \\ 0, & d_{\text{H}}(i-1, j-1) \geq 2, \end{cases} \tag{74}$$

and (73) can be simplified to

$$\begin{aligned} \underset{\Pi}{\text{maximize}} : \quad & \sum_{k=1}^{N_c} \sum_{\ell=1}^{N_c} |\mathbf{c}_k^* \mathbf{c}_\ell|^2 \, T_{\text{AP}}[\Pi(k), \Pi(\ell)] \\ \text{subject to} : \quad & \Pi \text{ is a permutation} \end{aligned} \tag{75}$$

The advantage of (75) is two-fold. 1) The computational complexity of the cost function is reduced, since most of T_{AP}'s are zero. 2) Its solution doesn't depend on the parameter p of the BSC [Zeger & Gersho (1990)]. Once we solve (75) for a particular value of p, the solution is valid for other values.

3.5 Numerical results

Simulations are carried out in (2,1,8) and (4,2,64) systems, where (N_t, N_r, N_c) denotes a system with N_t transmit antennas, N_r receive antennas, and codebook cardinality N_c. A BSC feedback channel is adopted in the simulations, Codebooks are downloaded from [Love's webpage (2006)]. We design good IA's for these codebooks by solving the simplified problem (75). To study the worst-case performance, bad IA's are also designed by minimizing the cost function of (75). The IA's of the (2,1,8) system are obtained by brute-force search, and shown in Table 1. The IA's of the (4,2,64) system are designed using the binary switching algorithm [Zeger & Gersho (1990)].
From the IA results in Table 1, we can gain an insight into how the good IA improves the system performance. For example, the first codeword is closest (with respect to the chordal distance) to the last codeword. The Hamming distance between their original indexes (1 and 8 respectively) is 3, while the Hamming distance between their good indexes (7 and

Codeword	Original IA, k	Good IA, $\Pi_{\text{good}}(k)$	Bad IA, $\Pi_{\text{bad}}(k)$
$[0.8393 - j\,0.2939,\ -0.1677 + j\,0.4256]^T$	1	7	1
$[-0.3427 + j\,0.9161,\ 0.0498 + j\,0.2019]^T$	2	1	4
$[-0.2065 + j\,0.3371,\ 0.9166 + j\,0.0600]^T$	3	6	2
$[0.3478 - j\,0.3351,\ 0.2584 + j\,0.8366]^T$	4	8	8
$[0.1049 + j\,0.6820,\ 0.6537 + j\,0.3106]^T$	5	2	5
$[0.0347 - j\,0.2716,\ 0.0935 - j\,0.9572]^T$	6	4	3
$[-0.7457 + j\,0.1181, -0.4553 - j\,0.4719]^T$	7	3	6
$[-0.7983 + j\,0.3232,\ 0.5000 + j\,0.0906]^T$	8	5	7

Table 1. Index assignment schemes for a 3-bit codebook

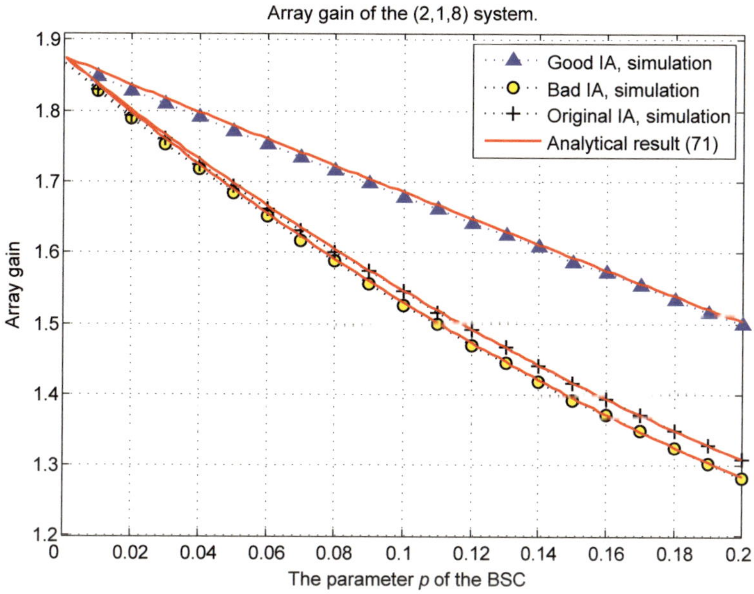

Fig. 7. The array gain of the (2,1,8) system.

5 respectively) is 1. Similarly, the second codeword is closest to the fifth codeword. The Hamming distance between their original indexes (2 and 5) is 2, and that between their good indexes (1 and 2) is 1. It is shown from these examples that the good IA scheme has assigned close indexes (in term of Hamming distance) to close codewords (in term of chordal distance). Figure 7 and 8 show the variation of the array gain with the increase of p. In the simulations, the SNR is fixed at $\gamma_S = 10$ dB. As shown in both figures, the good IA always outperforms the bad IA and the original IA. Furthermore, we can see that the analytical result (71) is tight. Figure 9 and 10 depict the SER simulation results. QPSK and 16-QAM modulations are used in (2,1,8) and (4,2,64) systems respectively. In the simulations, thirty-six symbols are transmitted in a block, and ideal coherent detection is adopted. In Figure 9, the SER of 'good IA' is much lower than that of 'bad IA', and approaches that of 'error-free feedback'. In Figure 10, the good IA outperforms the bad IA, though it cannot reach the performance of error-free feedback. In

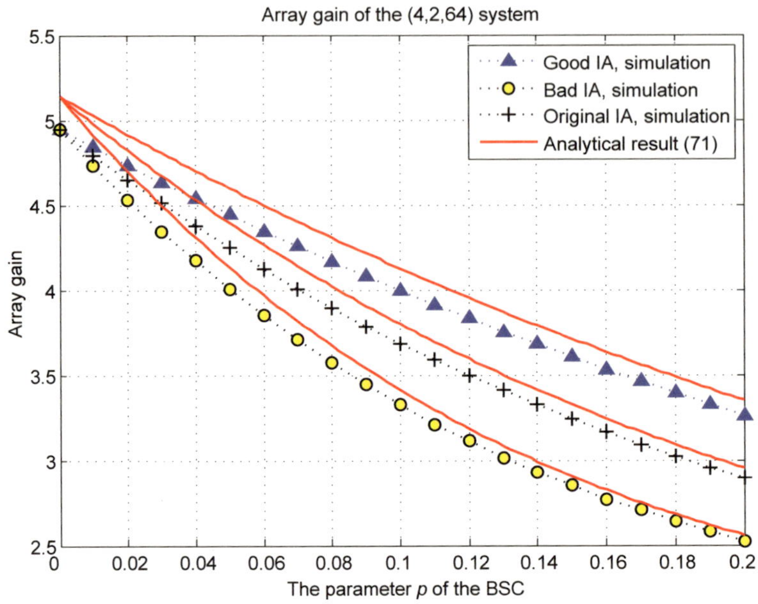

Fig. 8. The array gain of the (4,2,64) system.

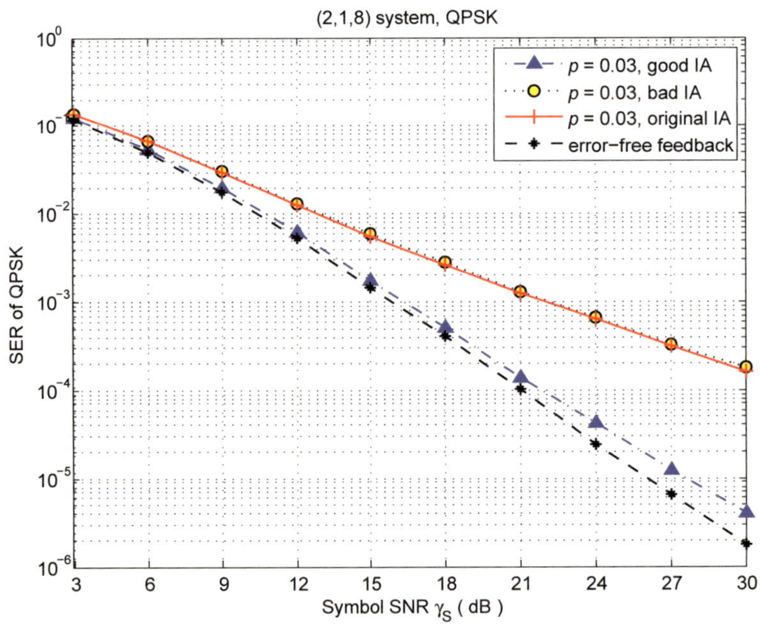

Fig. 9. SER of QPSK in the (2,1,8) system.

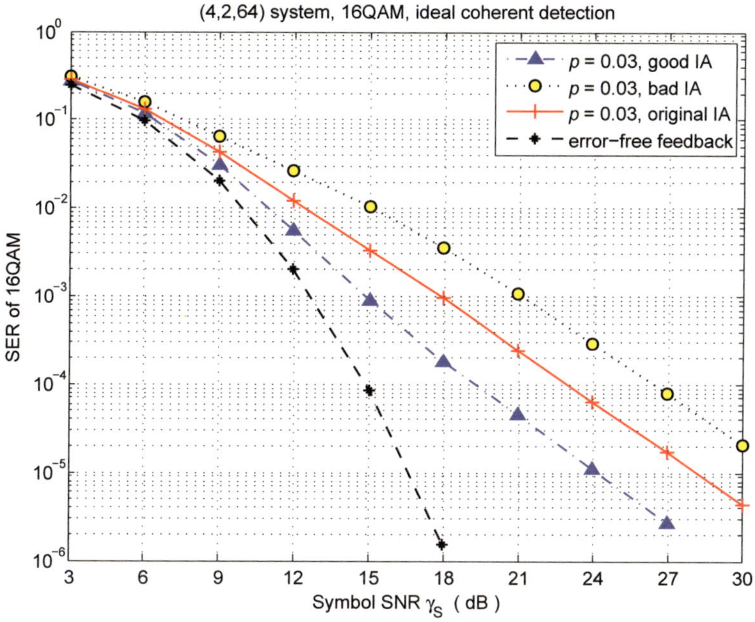

Fig. 10. SER of 16-QAM in the (4,2,64) system.

this case, IA technique shows its limitation. Other technique, such as error-control coding and automatic repeat-request (ARQ), is required to further compensate the performance degradation caused by feedback errors.

Moreover, it is observed from Figure 9 and 10 that the diversity gain (the slope of SER curve at high SNR) fits well with the conclusion of Lemma 3, except that the diversity gain of (2,1,8) system with good IA is not clear because the SNR is not large enough.

4. Effect of feedback delay and channel prediction

Feedback delay is harmful to a beamforming system based on finite-rate feedback, resulting in significant capacity loss [Huang et al. (2006)] and error-performance degradation [Ma & Zhang (2007)]. As an effective countermeasure against the feedback delay, channel prediction has been used in beamforming systems [Zhou & Giannakis (2004); Ma et al. (2008)]. For example, the minimum mean square error (MMSE) channel predictor was used in the predictive feedback scheme proposed in [11], where the authors analyzed the joint effect of imperfect channel estimation and feedback delay on the capacity of a beamforming system based on finite-rate feedback.

4.1 System model

Consider a beamforming system with N_t transmit and N_r receive antennas, as shown in Figure 11. The wireless channel is frequency-flat, and its temporal variation is slow enough to be considered quasi-static within some time interval T_B (called a block). Let $h_{m,n}(j)$ denote the fading coefficient of the (m,n)th channel branch (between transmit antenna n and receive antenna m) at block j. The fading coefficients are assumed to be zero-mean and jointly Gaussian. The channel is spatially white and temporally correlated according to Jakes' model,

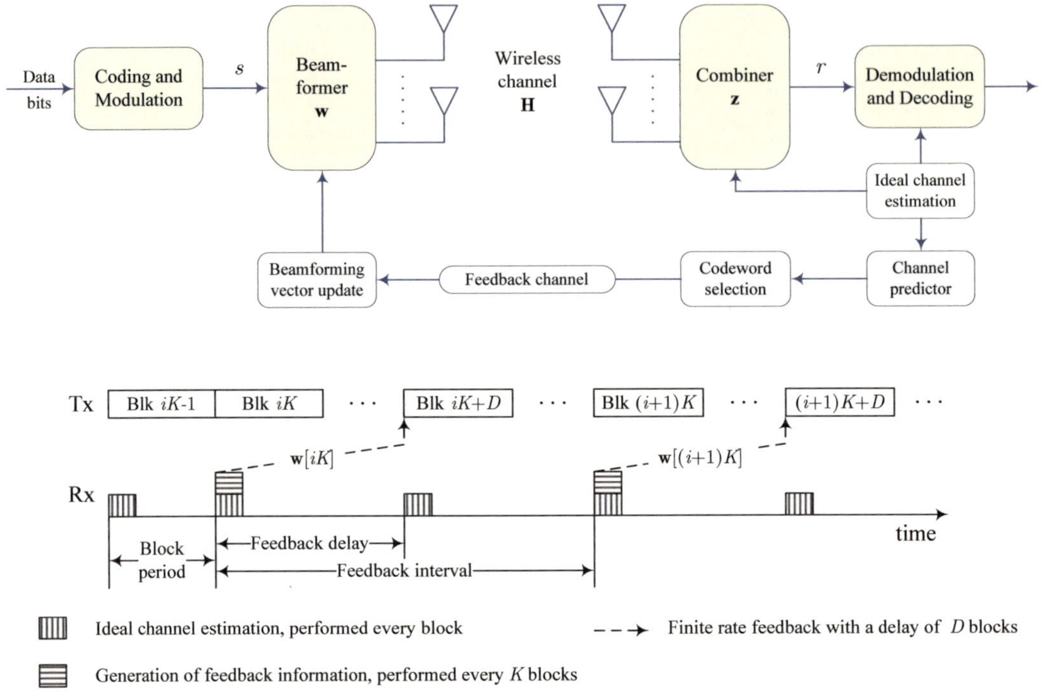

Fig. 11. A finite-rate beamforming system with channel prediction. (Top: system model. Bottom: frame structure.)

that is,

$$\mathbb{E}\{h_{m,n}(j)h^*_{m',n'}(j')\} = \begin{cases} J_0(2\pi f_d T_B|j-j'|), & m=m' \text{ and } n=n', \\ 0, & \text{else}, \end{cases} \tag{76}$$

where f_d denotes the Doppler spread; and $J_0(\cdot)$ is the zeroth-order Bessel function of the first kind. The channel coefficients at block j are collected into the $N_r \times N_t$ matrix $\mathbf{H}(j)$. We further assume that the receiver performs channel estimation at every block, and the estimation is perfect.

The frame structure is illustrated in the bottom of Figure 11. The system adopts a codebook based feedback scheme. A B-bit beamforming codebook $\mathcal{C} = \{\mathbf{c}_1, \cdots, \mathbf{c}_{N_c}\}$ is designed in advance, where $\mathbf{c}_1, \cdots, \mathbf{c}_{N_c} \in \mathbb{C}^{N_t}$ are *unit-norm* codewords and $N_c = 2^B$ In each feedback, the receiver selects a favorable codeword from the codebook and feeds back a B-bit codeword index to the transmitter. The feedback link is error-free, but has a delay of D blocks. Moreover, the feedback is carried out every K blocks, so that the average feedback rate $B/(KT_B)$ bps satisfies the feedback rate constraint.

The process of the ith feedback is described as follows. At block iK, the receiver predicts the MIMO channel, using a Pth order linear predictor. The predictor buffers the latest $P+1$ channel realizations $\mathbf{H}(iK), \cdots, \mathbf{H}(iK-P)$, and calculates the channel prediction according to

$$\widehat{\mathbf{H}}[iK] = \sum_{p=0}^{P} a^*(p)\mathbf{H}(iK-p), \tag{77}$$

where $\mathbf{a} \triangleq [a(0), \cdots, a(P)]^T$ is the predictor coefficient vector. The prediction is passed to the quantizer, which selects a favorable codeword from the beamforming codebook

$$\mathbf{w}[iK] = \arg\max_{\mathbf{c}\in\mathcal{C}} \|\widehat{\mathbf{H}}[iK]\mathbf{c}\|^2. \tag{78}$$

The index of $\mathbf{w}[iK]$ is sent to the transmitter via the feedback link. After a delay of D blocks, the transmitter obtains this index and updates the beamforming vector at block $iK + D$.
As shown in Figure 11, the beamforming vector $\mathbf{w}[iK]$ is used from block $iK + D$ to block $iK + D + K - 1$. In block $iK + k$, the $N_r \times 1$ received signal vector can be expressed as

$$\mathbf{y}(iK+k) = \mathbf{H}(iK+k)\mathbf{w}[iK]s(iK+k) + \boldsymbol{\eta}(iK+k),$$

where $s(iK + k) \in \mathbb{C}$ is the data symbol; and $\boldsymbol{\eta}(iK + k)$ denotes the noise vector with i.i.d. $\mathcal{CN}(0, 1)$ entries. The receiver performs maximum ratio combining on the received signal with the combining vector

$$\mathbf{z}(iK+k) = \frac{\mathbf{H}(iK+k)\mathbf{w}[iK]}{\|\mathbf{H}(iK+k)\mathbf{w}[iK]\|^2}.$$

Then the post-combining signal is given by

$$\begin{aligned} r(iK+k) &= \mathbf{z}^H(iK+k)\,\mathbf{y}(iK+k) \\ &= s(iK+k) + \frac{\mathbf{w}^*[iK]\mathbf{H}^*(iK+k)}{\|\mathbf{H}(iK+k)\mathbf{w}[iK]\|^2}\boldsymbol{\eta}(iK+k). \end{aligned} \tag{79}$$

We assume no temporal power control is used, i.e., the average symbol SNR

$$\gamma_S \triangleq \mathbb{E}\{|s(iK+k)|^2\}/N_0$$

is time invariant. So the receive SNR of the post-combining signal is given by

$$\gamma(iK+k) = \gamma_S\|\mathbf{H}(iK+k)\mathbf{w}[iK]\|^2, \quad k = D, \cdots, D+K-1. \tag{80}$$

4.2 SER analysis

We assume PSK signal in the analysis. Conditioned on the instantaneous SNR, the symbol error probability of M-ary PSK is given by Equation (7) shown at the beginning of Section 2.2. Therefore, according to (80), the SER at block $iK + k, k = D, \cdots, D + K - 1$, can be written as

$$\begin{aligned} P_e(iK+k) &= \frac{1}{\pi}\mathbb{E}\left\{\int_0^{\frac{(M-1)\pi}{M}} \exp\left(-\frac{g_{\text{PSK}}\gamma(iK+k)}{\sin^2\theta}\right) d\theta\right\}. \\ &= \frac{1}{\pi}\int_0^{\frac{(M-1)\pi}{M}} \mathbb{E}\exp\left(-\frac{g_{\text{PSK}}\gamma_S}{\sin^2\theta}\|\mathbf{H}(iK+k)\mathbf{w}[iK]\|^2\right) d\theta. \end{aligned} \tag{81}$$

where $g_{\text{PSK}} = \sin^2(\pi/M)$ depends on the constellation size.
According to (77) (78), the beamforming vector $\mathbf{w}[iK]$ is determined by the channel matrices $\mathbf{H}(iK - p), p = 0, \cdots, P$. Then $P_e(iK + k)$ depends on the joint distribution of $\mathbf{H}(iK + k)$ and $\mathbf{H}(iK - p), p = 0, \cdots, P$. Since this joint distribution is related to k and independent of i, so is

$P_e(iK+k)$. That is, $P_e(iK+k)$ is periodic with period K. Therefore we can confine our attention to just one period, and calculate the average SER of the system according to

$$\begin{aligned}\overline{P}_e &= \frac{1}{K}\sum_{k=D}^{D+K-1} P_e(iK+k)\big|_{i=0} \\ &= \frac{1}{\pi K}\sum_{k=D}^{D+K-1}\frac{1}{\pi}\int_0^{\frac{(M-1)\pi}{M}} \mathbb{E}\exp\left(-\frac{g\gamma_S}{\sin^2\theta}\left\|\mathbf{H}(k)\mathbf{w}[0]\right\|^2\right)d\theta \end{aligned} \tag{82}$$

Since the random variables $\mathbf{H}(k)$ and $\mathbf{w}[0]$ in (82) are both related to the channel prediction $\widehat{\mathbf{H}}[0]$, some relationship can be established between the expectation in (82) and the statistics of the channel prediction, as shown in the following lemma.

Lemma 5. *Given $t \geq 0$ and $k = D, \cdots, D+K-1$, we have*

$$\mathbb{E}\exp\left(-t\left\|\mathbf{H}(k)\mathbf{w}[0]\right\|^2\right) = \left(\frac{1}{1+t\beta(k)}\right)^{N_r}\mathbb{E}\exp\left(-\frac{t\left[1-\beta(k)\right]}{1+t\beta(k)}\max_{\mathbf{c}\in\mathcal{C}}\left\|\sigma_{\text{Prd}}^{-1}\widehat{\mathbf{H}}[0]\mathbf{c}\right\|^2\right), \tag{83}$$

where the parameters

$$\beta(k) \triangleq 1 - \frac{\left|\sum_{p=0}^{P} a(p)J_0(2\pi f_d T_B|p+k|)\right|^2}{\sum_{p=0}^{P}\sum_{q=0}^{P} a^*(p)a(q)J_0(2\pi f_d T_B|p-q|)}, \tag{84}$$

$$\sigma_{\text{Prd}} \triangleq \sqrt{\sum_{p=0}^{P}\sum_{q=0}^{P} a^*(p)a(q)J_0(2\pi f_d T_B|p-q|)}, \tag{85}$$

depends on the predictor coefficients $a(0), \cdots, a(P)$.

Proof: By the law of total expectation, the left-hand-side of (83) can be written as

$$\mathbb{E}\exp\left(-t\left\|\mathbf{H}(k)\mathbf{w}[0]\right\|^2\right) = \mathbb{E}\left\{\mathbb{E}\left[\exp\left(-t\left\|\mathbf{H}(k)\mathbf{w}[0]\right\|^2\right)\middle|\widehat{\mathbf{H}}[0]\right]\right\}. \tag{86}$$

The inner expectation in the right-hand-side of (25) depends on the conditional distribution of $(\mathbf{H}(k)|\widehat{\mathbf{H}}[0])$, which can be obtained as follows. For the (m,n)th channel branch, the channel prediction is given by $\hat{h}_{m,n}[0] = \sum_p a^*(p)h_{m,n}(-p)$. Since the channel coefficients are jointly Gaussian, the conditional distribution of $(h_{m,n}(k)|\hat{h}_{m,n}[0])$ must be Gaussian. This type of conditional distribution has been well studied in the literature. Applying the existing result to our case, we can easily obtain the conditional mean and variance, and write

$$\left(h_{m,n}(k)\,\middle|\,\hat{h}_{m,n}[0]\right) \sim \mathcal{CN}\left(\rho(k)\hat{h}_{m,n}[0], \beta(k)\right), \quad m=1,\cdots,N_r,\ n=1,\cdots,N_t, \tag{87}$$

where $\beta(k)$ is given by (84) and

$$\rho(k) \triangleq \frac{\sum_{p=0}^{P} a(p)J_0(2\pi f_d T_B|p+k|)}{\sum_{p=0}^{P}\sum_{q=0}^{P} a^*(p)a(q)J_0(2\pi f_d T_B|p-q|)}. \tag{88}$$

Since no spatial correlation exists, the result of (87) implies that $(\mathbf{H}(k)|\widehat{\mathbf{H}}[0])$ is matrix-Gaussian distributed with mean $\rho(k)\widehat{\mathbf{H}}[0]$ and covariance $\beta(k)\,\mathbf{I}_{N_r}\otimes\mathbf{I}_{N_t}$. To proceed, notice that $\mathbf{w}[0]$ is a *unit-norm* codeword determined by $\widehat{\mathbf{H}}[0]$. Then conditioned on $\widehat{\mathbf{H}}[0]$, $\mathbf{H}(k)\mathbf{w}[0]$ is a Gaussian random vector with mean $\rho(k)\widehat{\mathbf{H}}[0]\mathbf{w}[0]$ and covariance $\beta(k)\,\mathbf{I}_{N_r}$. So the inner expectation in the right-hand-side of (86) can be derived as [Zhou & Giannakis (2004), Eq.(36)]

$$\mathbb{E}\Big[\exp\left(-t\,\|\mathbf{H}(k)\mathbf{w}[0]\|^2\right)\Big|\,\widehat{\mathbf{H}}[0]\Big] = \left(\frac{1}{1+t\,\beta(k)}\right)^{N_r}\exp\left(-\frac{t\,\rho^2(k)\left\|\widehat{\mathbf{H}}[0]\mathbf{w}[0]\right\|^2}{1+t\,\beta(k)}\right). \tag{89}$$

Substituting (89) into (86) and noticing

$$\left\|\widehat{\mathbf{H}}[0]\mathbf{w}[0]\right\|^2 = \max_{\mathbf{c}\in\mathcal{C}}\left\|\widehat{\mathbf{H}}[0]\mathbf{c}\right\|^2, \qquad \rho^2(k) = \sigma_{\mathrm{Prd}}^{-2}\left[1-\beta(k)\right],$$

we reach the desired result of (83). □

Applying (83) to (82), the average SER can be written as

$$\begin{aligned}\overline{P}_e = {} & \frac{1}{\pi K}\sum_{k=D}^{D+K-1}\int_0^{\frac{(M-1)\pi}{M}}\left(\frac{\sin^2\theta}{\sin^2\theta+g_{\mathrm{PSK}}\gamma_S\beta(k)}\right)^{N_r} \\ & \times\mathbb{E}\exp\left(-\frac{g_{\mathrm{PSK}}\gamma_S[1-\beta(k)]}{\sin^2\theta+g_{\mathrm{PSK}}\gamma_S\beta(k)}\max_{\mathbf{c}\in\mathcal{C}}\left\|\sigma_{\mathrm{Prd}}^{-1}\widehat{\mathbf{H}}[0]\mathbf{c}\right\|^2\right)\,d\theta\,. \end{aligned}\tag{90}$$

In (90), we treat $\sigma_{\mathrm{Prd}}^{-1}\widehat{\mathbf{H}}[0]$ as a random matrix, whose distribution can be obtained as follows. The entries of $\widehat{\mathbf{H}}[0]$ must be Gaussian and independent of each other, because the predictor is linear and the channel is spatially white. Using (76) (77), it is can be verified that all entries of $\widehat{\mathbf{H}}[0]$ have zero mean and variance σ_{Prd}^2. This implies that $\sigma_{\mathrm{Prd}}^{-1}\widehat{\mathbf{H}}[0]$ is an $N_r\times N_t$ random matrix with i.i.d. $\mathcal{CN}(0,1)$ entries. Therefore Lemma 1 in Section 2.2 can be applied to the right-hand-side of (90). After some manipulations, we obtain the following upper bound on the average SER

$$\overline{P}_e \le \frac{1}{\pi K}\sum_{k=D}^{D+K-1}\int_0^{\frac{(M-1)\pi}{M}}\frac{\sin^{2N_r}\theta\left[\sin^2\theta+g_{\mathrm{PSK}}\gamma_S\beta(k)\right]^{(N_t-1)N_r}}{\left[\sin^2\theta+g_{\mathrm{PSK}}\gamma_S\beta(k)+g_{\mathrm{PSK}}g_{\mathrm{CB}}\gamma_S(1-\beta(k))\right]^{N_tN_r}}\,d\theta\,, \tag{91}$$

which is the main result of the SER analysis. Note that the value of the codebook-dependent parameter g_{CB} can be calculated numerically according to (11), or using the closed-form approximation (15).

4.3 Predictor design

We will utilize the SER bound (91) to a design good predictor in terms of error performance. But the integral in the bound complicates the optimization problem. To make the problem tractable, we focus on high SNR regime. When γ_S is large, the "$\sin^2\theta$" in the denominator of (91) can be omitted. and the following inequality is obtained

$$\begin{aligned}&\frac{\sin^{2N_r}\theta\left[g_{\mathrm{PSK}}\gamma_S\beta(k)\right]^{(N_t-1)N_r}}{\left[g_{\mathrm{PSK}}\gamma_S\beta(k)+g_{\mathrm{PSK}}g_{\mathrm{CB}}\gamma_S(1-\beta(k))\right]^{N_tN_r}} \\ \ge\ & \frac{\sin^{2N_r}\theta\left[\sin^2\theta+g_{\mathrm{PSK}}\gamma_S\beta(k)\right]^{(N_t-1)N_r}}{\left[\sin^2\theta+g_{\mathrm{PSK}}\gamma_S\beta(k)+g_{\mathrm{PSK}}g_{\mathrm{CB}}\gamma_S(1-\beta(k))\right]^{N_tN_r}}\end{aligned}$$

Therefore, the average SER can be further upper bounded by

$$\overline{P}_e \leq \frac{1}{\pi K (g_{\mathrm{PSK}} \gamma_S)^{N_r}} \left[\int_0^{\frac{(M-1)\pi}{M}} \sin^{2N_r} \theta \, \mathrm{d}\theta \right] \sum_{k=D}^{D+K-1} \frac{[\beta(k)]^{(N_t-1)N_r}}{[\beta(k) + g_{\mathrm{CB}}(1-\beta(k))]^{N_t N_r}} \,. \tag{92}$$

It is observed that the SER decreases at a speed of $\gamma_S^{-N_r}$, i.e., the diversity order is N_r. Since full diversity order $N_t N_r$ can be achieved by finite-rate beamforming without feedback delay [Love & Heath (2005)], the diversity from transmit antennas is lost due to feedback delay. Based on (92), the predictor design can be formulated as a minimization problem. To get a concise description, we collect the predictor coefficients into a vector $\mathbf{a} \triangleq [a(0), \cdots, a(P)]^T$, and define the following notations

- The temporal correlation matrix $\mathbf{R} \in \mathbb{C}^{(P+1)\times(P+1)}$, with the (p,q)-th entry

$$[\mathbf{R}]_{p,q} \triangleq J_0(2\pi f_d T_B |p-q|), \quad p,q = 0,1,\cdots,P. \tag{93}$$

- The cross-correlation vector $\mathbf{r}(k) \in \mathbb{C}^{(P+1)}$ with lag k, $k = D, \cdots, D+K-1$. The p-th entry of $\mathbf{r}(k)$ is

$$[\mathbf{r}(k)]_p \triangleq J_0(2\pi f_d T_B |p+k|), \quad p = 0,1,\cdots,P. \tag{94}$$

In the right-hand-side of (92), only the parameter $\beta(k)$ is related to the predictor. So only the summation needs to be considered in the predictor design. Since

$$\beta(k) = 1 - |\mathbf{r}^*(k)\,\mathbf{a}|^2 / \mathbf{a}^*\mathbf{R}\mathbf{a}$$

by definition, the summation in (92) can be expressed as

$$\sum_{k=D}^{D+K-1} \frac{\left[\mathbf{a}^*\mathbf{R}\mathbf{a}\right]^{N_r} \left[\mathbf{a}^*\mathbf{R}\mathbf{a} - |\mathbf{r}^*(k)\,\mathbf{a}|^2\right]^{(N_t-1)N_r}}{\left[\mathbf{a}^*\mathbf{R}\mathbf{a} - (1-g_{\mathrm{CB}})\,|\mathbf{r}^*(k)\,\mathbf{a}|^2\right]^{N_t N_r}} \,, \tag{95}$$

which is used as the metric of the predictor design. Because a coefficient vector $\mathbf{a}$ and its "normalized" version $\mathbf{a}/\sqrt{\mathbf{a}^*\mathbf{R}\mathbf{a}} \in \{\mathbf{x} : \mathbf{x}^*\mathbf{R}\mathbf{x} = 1\}$ yield the same value of (95), we impose the constraint $\mathbf{a}^*\mathbf{R}\mathbf{a} = 1$, and propose the following predictor design criterion

$$\begin{aligned} \underset{\mathbf{a}}{\text{minimize}} \,:\, & \sum_{k=D}^{D+K-1} \frac{\left[1 - |\mathbf{r}^*(k)\,\mathbf{a}|^2\right]^{(N_t-1)N_r}}{\left[1 - (1-g_{\mathrm{CB}})\,|\mathbf{r}^*(k)\mathbf{a}|^2\right]^{N_t N_r}} \\ \text{subject to} \,:\, & \mathbf{a}^*\mathbf{R}\mathbf{a} = 1 \,. \end{aligned} \tag{96}$$

The closed-form solution to (96) is hard to obtain. So we have to use numerical methods to solve it. Since the predictor can be designed offline, the computational complexity is affordable. However, there are still some issues that need to be addressed.

a) The minimization problem (96) is not always convex. (The convexity of the cost function depends on the values of $N_t, N_r, g_{\mathrm{CB}}$.) For a non-convex problem, most optimization algorithms can only guarantee convergence to a local minimum, i.e., a suboptimal solution.

b) An initial value for numerical methods can be obtained in a heuristic way. In the summation in (96), the term with index $k = D$ is a decreasing function of $|\mathbf{r}^*(D)\,\mathbf{a}|^2$. The predictor that maximizes $|\mathbf{r}^*(k)\,\mathbf{a}|^2$ under the constraint $\mathbf{a}^*\mathbf{R}\mathbf{a} = 1$ is given by

$$\mathbf{a}_{\mathrm{ini}} = \left[\mathbf{r}^*(D)\mathbf{R}^{-1}\mathbf{r}(D)\right]^{-\frac{1}{2}} \mathbf{R}^{-1}\mathbf{r}(D) \,. \tag{97}$$

This initial value gives satisfactory results in our simulations.

	3Tx-2Rx system	2Tx-2Rx system
Antenna Configuration	$N_t = 3, N_r = 2$	$N_t = N_r = 2$
Codebook	3 bit ($N_c = 8$)	2 bit ($N_c = 4$)
Channel correlation	Jakes' model	
Carrier frequency	1.8 GHz	
Block period (T_B)	$1/3$ ms	
Terminal speed	60 km/h	30 — 120 km/h
Doppler spread (f_d)	100 Hz	50 — 200 Hz
Feedback interval (K)	6 blocks	4 blocks
Feedback delay (D)	6 blocks	2, 4, 6 blocks
Predictor order (P)	2, 8, 16, 86	8
Modulation	QPSK	
Average symbol SNR	0 — 30 dB	15 dB

Table 2. System parameters in numerical simulations.

c) The temporal correlation matrix $\mathbf{R}$ (defined in (93)) is ill-conditioned (see [Baddour & Beaulieu (2005), Section III-B] and references therein). To avoid possible numerical instability, one may add a small perturbation, namely, replacing $\mathbf{R}$ by $\mathbf{R} + \epsilon\mathbf{I}$, where ϵ is a small positive number.

4.4 Numerical results

Numerical results are presented for a 3Tx-2Rx system and a 2Tx-2Rx system, as specified in Table 2. In both systems, the average feedback rate B/KT_B is fixed at 1.5 kbps. This feedback rate is adopted in 3GPP specifications. Moreover, we use the beamforming codebooks listed in [Love & Heath (2003)].

Predictors are designed by solving (96). Specifically, we use the `fmincon` function in MATLAB and the initial value given in (97). To avoid possible numerical instability, the correlation matrix $\mathbf{R}$ is replaced by $\mathbf{R} + (1.6 \times 10^{-3})\mathbf{I}$. We also consider the case of *delayed feedback* [Huang et al. (2006); Ma & Zhang (2007); Ma et al. (2008)]. This case is equivalent to a trivial zeroth order channel predictor with a single coefficient $a(0) = 1$. The results in Section 4.2 and 4.3 are applicable to this case (setting $P = 0$ and $a(0) = 1$).

Figure 12 illustrates the SER of the 3Tx-2Rx system for different symbol SNR's and predictor orders. We can see that compared with delayed feedback, even the 2nd order predictor improves the error performance considerably. The SER decreases with the increase of the predictor order, but the decrease of SER is no longer remarkable when $P > 16$. It is also observed that the analytical result (91) is tight, and the diversity gain is 2, as discussed in Section IV-A.

Figure 13 depicts the SER of the 2Tx-2Rx system for different terminal speeds and feedback delays. It is shown that the system with channel prediction is more robust to the movement of the terminal. However, the SER is sensitive to the value of the feedback delay D both in the case of delayed feedback and in the case of predicted feedback.

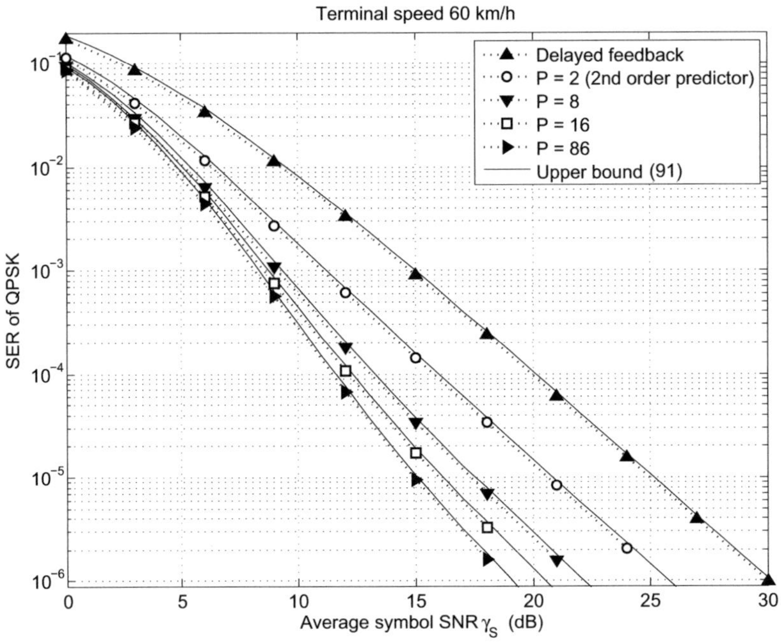

Fig. 12. SER performance of the 3Tx-2Rx system.

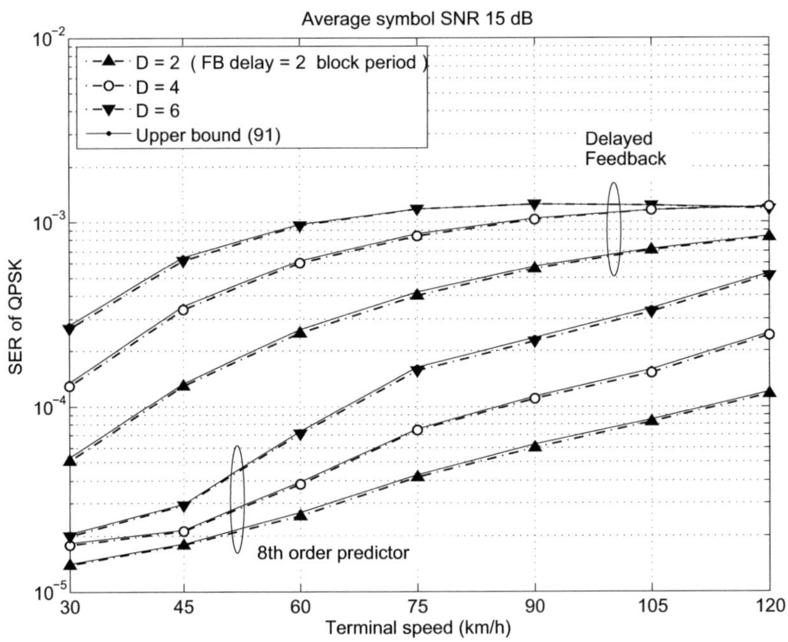

Fig. 13. SER performance of the 2Tx-2Rx system.

5. Conclusion

This chapter highlighted recent advances in beamforming based on finite-rate feedback from a communication-theoretic perspective. We first studied the SER performance of the finite-rate beamforming system. In a Rayleigh fading environment, an upper bound on the average SER was derived, which has a simple structure and is tight at all SNRs. Secondly, feedback error problem in a finite-rate beamforming system was investigated. The effect of feedback error on the diversity and array gains was quantified analytically. Index assignment technique was introduced to provide a redundancy-free protection against feedback error, and a good index assignment scheme was designed to maximize the array gain. Thirdly, we investigated the effect of feedback delay on finite-rate beamforming, and proposed to use a channel prediction scheme to compensate the performance degradation. For a beamforming system with delayed finite-rate feedback, an upper bound on the SER was derived, and a predictor was designed to provide good SER performance. The derived analytical results and the proposed performance enhancement schemes in this section were all verified by simulations under typical system configurations.

6. References

P. A. Dighe, R. K. Mallik, and S. S. Jamuar, "Analysis of transmit-receive diversity in Rayleigh fading," *IEEE Trans. Commun.*, vol. 51, no. 4, pp. 694-703, Apr. 2003.

P. A. Dighe, R. K. Mallik, and S. S. Jamuar, "Analysis of *K*-transmit dual-receive diversity with cochannel interferers over a Rayleigh fading channel," *Wireless Personal Communications*, vol. 25, no. 2, pp. 87-100, 2003.

Q. Zhou and H. Dai, "Asymptotic analysis in MIMO MRT/MRC systems," *EURASIP J. Wireless Commun. Net.*, 2006.

M. Kang, and M.-S. Alouini, "A comparative study on the performance of MIMO MRC systems with and without cochannel interference," *IEEE Trans. Commun.*, vol.52, no.8, pp.1417-1425, Aug. 2004.

K. K. Mukkavilli, A. Sabharwal, B. Aazhang, and E. Erkip, "On beamforming with finite rate feedback in multiple-antenna systems," *IEEE Trans. Inform. Theory*, vol. 49, no. 10, pp. 2562-2579, Oct. 2003.

D. J. Love, and R. W. Heath, Jr., "Grassmannian beamforming for multiple-input multiple-output wireless systems," *IEEE Trans. Inform. Theory*, vol. 49, no. 10, pp. 2735-2747, Oct. 2003.

D. J. Love, and R. W. Heath, Jr., "Necessary and sufficient conditions for full diversity order in correlated Rayleigh fading beamforming and combining systems," *IEEE Trans. Wireless Commun.*, vol. 4, pp. 20-23, Jan. 2005.

P. Xia, and G. B. Giannakis, "Design and analysis of transmit-beamforming based on limited-rate feedback," *IEEE Trans. Signal Process.*, vol. 54, pp. 1853-1862, May 2006.

S. Zhou, and G. B. Giannakis, "How accurate channel prediction needs to be for transmit-beamforming with adaptive modulation over Rayleigh MIMO channels?" *IEEE Trans. Wireless Commun.*, vol. 3, no. 4, pp. 1285-1294, July 2004.

S. Zhou, Z. Wang, and G. B. Giannakis, "Quantifying the power loss when transmit beamforming relies on finite-rate feedback," *IEEE Trans. Wireless Commun.*, vol. 4, pp. 1948-1957, July. 2005.

B. Mondal, and R. W. Heath, Jr., "Performance analysis of quantized beamforming MIMO systems," *IEEE Trans. Signal Process.*, vol. 54, no. 12, pp. 4753-4766, Dec. 2006.

K. Huang, B. Mondal, R. W. Heath, Jr., and J. G. Andrews, "Effect of feedback delay on multi-antenna limited feedback for temporally-correlated channels," in *Proc. Globecom*, San Francisco, CA, Nov.27-Dec.1, 2006.

Y. Ma, and D. Zhang, "Error Rate of Transmit Beamforming with Delayed and Limited Feedback," in *Proc. Globecom*, Washington, DC, Nov. 26-30, 2007.

Y. Ma, A. Leith, and R. Schober, "Predictive feedback for transmit beamforming with delayed feedback and channel estimation errors," in *Proc. ICC*, Beijing, China, May 19-23, 2008.

K. Zeger, and A. Gersho, "Pseudo-Gray coding," *IEEE Trans. Commun.*, vol.38, no.12, pp.2147-2158, Dec. 1990.

G. Ben-David, and D. Malah, "Bounds on the performance of vector-quantizers under channel errors," *IEEE Trans. Inform. Theory*, vol.51, no.6, pp.2227-2235, June 2005.

T. L. Marzetta, and B. M. Hochwald, "Capacity of a mobile multiple-antenna communication link in Rayleigh flat fading," *IEEE Trans. Inform. Theory*, vol.45, no.1, pp.139-157, Jan. 1999.

K. E. Baddour, and N. C. Beaulieu, "Autoregressive Modeling for Fading Channel Simulation," *IEEE Trans. Wireless Commun.*, vol. 4, no. 4, pp. 1650-1662, July 2005.

P. Zhu, L. Tang, Y. Wang, and X. You, "Index Assignment for Quantized Beamforming MIMO Systems," *IEEE Trans. Wireless Commun.*, vol. 7, no. 8, pp. 2917 - 2922, Aug. 2008.

P. Zhu, L. Tang, Y. Wang, and X. You, "Quantized beamforming with channel prediction," *IEEE Trans. Wireless Commun.*, vol. 8, no. 11, pp. 5377 - 5382, Nov. 2009.

P. Zhu, L. Tang, Y. Wang, and X. You, "An upper bound on the SER of transmit beamforming in correlated Rayleigh fading," *IEEE Trans. Commun.*, vol. 58, no. 2, pp. 457 - 462, Feb. 2010.

M. K. Simon, and M. -S. Alouini, *Digital communication over fading channels*, 2nd Ed., Wiley, 2005.

D. Tse, and P. Viswanath, *Fundamentals of wireless communication*, Cambridege University Press, 2005.

W. Fleming, *Functions of several variables*, 2nd Ed., Spinger-Verlag, 1977.

D. J. Love. (2006, Nov.) Personal Webpage on Grassmannian Subspace Packing. [Online]. Available: http://cobweb.ecn.purdue.edu/~djlove/ grass.html

Third Generation Partnership Project (3GPP), "Physical channels and modulation," 3GPP TS 36.211 V8.2.0, Mar. 2008.

V. Erceg, *et al.*, "TGn channel models," IEEE 802.11-03/940r4, May 2004.

13

Computationally Efficient Symbol Detection Schemes in Multi-Device STBC-MIMO Systems

Daniel C. Lee and Muhammad Naeem
Simon Fraser University, School of Engineering Science, Burnaby, BC, Canada

1. Introduction

The multi input multi output (MIMO) communication system has significantly higher channel capacity than the Single-Input-Single-Output (SISO) system for the same total transmission power and bandwidth (Foschini et al., 1998 & Telatar, 1999). It is known that the use of Space Time Block Code (STBC) can realize the increased capacity of MIMO systems and thus improve data throughput and spectral efficiency (Tarokh et al., 1998). In this work, we focus on the system that comprises one receiving station and multiple transmitting devices (e.g., uplinks in cellular systems). The receiver's front end has multiple antennas, and each transmitting device has multiple transmit antennas. The system discussed in this chapter is illustrated in Fig. 1. A Space Time Block Code (STBC) is used in each transmitting device, and joint signal detection is performed at the receiver. We refer to such a system as Multi-Device (MD) STBC-MIMO system. Generally in a MD-STBC-MIMO system, the number of receive antennas is typically smaller than the number of all transmit antennas used by all transmitting devices in the system. An example of MD-STBC-MIMO would be the uplink multiple access communication system, where the number of receive antennas at the base station or the access point is smaller than the total number of transmit antennas at the mobile devices.

In this work, we address symbol detection in Multi-Device (MD) STBC-MIMO systems. As will be discussed in section 3, the maximum a posteriori (MAP) detection, which reduced to the Maximum Likelihood (ML) detection in the case of a priori equally likely symbol blocks, minimizes the probability of detection error, and thus is optimal. However, a computationally efficient algorithm for achieving MAP or ML detection is not known. Some studies with sphere decoding (SD) algorithms exhibit that their expected computational complexity grows polynomially with the problem size, say m, up to some value of m for the cases of small constellation sizes (Vikalo et al., 2005), but it grows exponentially for the cases of large constellation sizes. Also, for some sphere decoding algorithms, operation at a low SNR requires inordinately high computation, although operation at a high SNR is efficient. In any case, an algorithm with polynomial growth of expected complexity for all values of the problem size, m, has not yet been found. In fact, Jalden et al. (2005) shows that even the expected computational complexity of the sphere decoding grows exponentially with the problem size in MIMO communication systems.

In this work, we present two evolutionary optimization methods, Biogeography-Based Optimization (BBO) and Estimation of Distribution Algorithm (EDA) to solve the problem

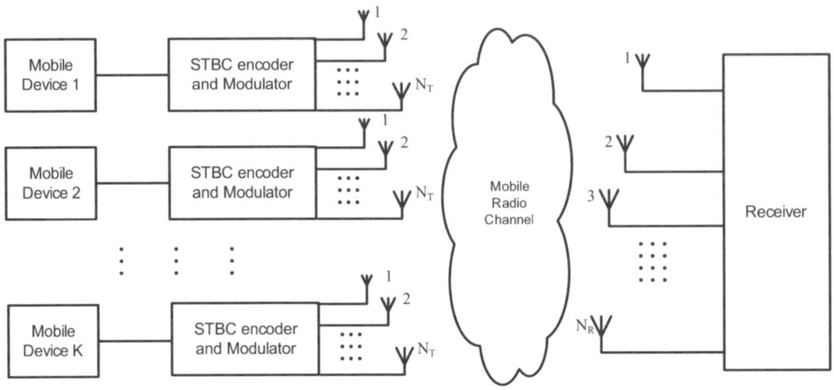

Fig. 1. A block diagram of a MD-STBC-MIMO system

of detection in a MD-STBC-MIMO system. Our simulation results show that BBO and EDA can meet the best known detector (i.e., sphere decoder) with less complexity and has better performance than other methods such as Minimum Mean Square Error (MMSE), Vertical Bell Laboratories Layered Space-Time (V-BLAST) (Wolniansky et al. ,1998), Semi-Definite Relaxation (SDR) (Nekuii et al. 2008), and genetic algorithm (GA).

2. System model

Fig. 1 shows an MD-STBC-MIMO system with one receiver and multiple transmit devices. Each of the K mobile devices (information senders) has N_T transmit antennas that apply STBC, and the receiver's front end has N_R receive antennas. The multiple devices in the system can cause co-channel interference. An IQ-modulation scheme (e.g. MPSK, M-QAM, etc.) maps source information into complex numbers. Even if transmit devices each employ an orthogonal space-time code, orthogonality among their signals cannot be guaranteed due to the absence of coding across different mobile devices.

First, let us consider the case of single mobile device; i.e. K=1. The mobile device transmits its signals through N_T transmit antennas, and the receiver has N_R antennas. We denote by T the number of time slots in the space-time code block. We assume that the channel is quasi-static; i.e., the channel gain remains constant during each time block of data. We also assume that the channel gain at each time block is known to the receiver. This assumption is often used in the literature and is reasonable if training or pilot signals are used. A complex $N_T \times N_R$ matrix H represents the MIMO channel and another complex $T \times N_T$ matrix S represents the input signal in the space-time code block. The relationship between the input and output signal is

$$\tilde{Y} = SH + \tilde{Z} \quad (1)$$

where $\tilde{Y}$ is the $T \times N_R$ complex output matrix, and $\tilde{Z}$ represents the additive white noise matrix.

In analyzing the system with linear dispersion space-time coding, the relation between the input and output of the channel is often expressed in another form (Hassibi et al., 2002) than (1). We now briefly describe this alternative form. An input signal, denoted by matrix S, in the space time block code (STBC) can be expressed as:

$$S = \sum_{q=1}^{Q} \left[\left(\alpha_q + j\beta_q \right) C_q + \left(\alpha_q - j\beta_q \right) D_q \right] \tag{2}$$

where Q is the number of symbols conveyed in a space time code block, and $a_q + j\beta_q$, q=1,…,Q are complex numbers that represent the Q symbols. (Note that a_q and β_q denote the real and imaginary parts of a symbol.) Then, the Q symbols can be represented as a $2Q$-dimensional real-valued row vector χ, where components of χ are constituted by a_q and β_q, q=1,…, Q. The real and imaginary parts of matrix $\tilde{Y}$'s components can be arranged as a $2TN_R$ -dimensional real-valued row vector y. In this alternative form, χ and y are arranged in such a way that their relation is expressed as:

$$y = \chi \Omega + Z \tag{3}$$

where $2Q \times 2TN_R$ real-valued matrix Ω is derived from the component of matrices H, C_q, D_q, q=1,…,Q, and Z is the $2TN_R$ -dimensional real-valued vector representing noise.
In the case of multiple mobile devices, equation (1) is naturally generalized to

$$\tilde{Y} = \sum_{k=1}^{K} S_k H_k + \tilde{Z} \tag{4}$$

where the $T \times N_T$- complex matrix S_k is the input signal from mobile device k, and the $N_T \times N_R$-complex matrix H_k represents the channel from the kth device to the receiver. Correspondingly, (3) is naturally generalized to

$$y = \begin{bmatrix} \chi_1 & \chi_2 & \cdots & \chi_K \end{bmatrix} \begin{bmatrix} \Omega_1 \\ \Omega_2 \\ \vdots \\ \Omega_K \end{bmatrix} + Z \tag{5}$$

where χ_k is a $2Q_k$-dimensional real-valued row vector that represents the Q_k complex symbols sent from mobile device k in a space time code block. Note that (5) can model the case in which different mobile devices use different code rates Q_k/T and different space time codes. We denote by $N_S = \sum_{k=1}^{K} Q_k$ the total number of symbols (from all mobile devices) transmitted in a space-time coded block through all of their transmit antennas.

3. Signal detection

The detector at the receiver has to choose from M^{N_S} possible sequences of symbols transmitted in a space-time code block, where M is the size of the symbol constellation associated with the modulation scheme. ML detection is known to yield the lowest symbol error probability in the case of a priori equally likely symbols. ML detection chooses transmitted symbols $\left[\chi_1, \chi_2, \cdots, \chi_K\right]$ that maximize $P\left(y \middle| \chi_1, \chi_2, \cdots, \chi_K\right)$. In the case of additive white Gaussian noise, Z, the ML detection is reduced to choosing the vector $\left[\chi_1, \chi_2, \cdots, \chi_K\right]$, from M^{N_S} possibilities, that has the shortest Euclidean distance:

$$\left\| y - \sum_{k=1}^{K} \chi_k \Omega_k \right\| \tag{6}$$

Defining $b = \log_2 M$, and M as the size of the symbol constellation, the ML detection scheme can be implemented by searching through all $M^{N_S} = 2^{bN_S}$ possible symbol sequences. Performing such an exhaustive search to find the minimum of (6) is computationally inefficient, especially for large N_S. Computational complexity increases exponentially with $N_S = \sum_{k=1}^{K} Q_k$, the number of bits per symbol, transmit antennas per device, and the number of mobile devices (K). High-speed communication requirements demand a low-complexity detection scheme. For low-complexity near-optimal detection, in this chapter we apply two population-based evolutionary algorithms, biogeography-based optimization (BBO) and estimation-of-distribution algorithm (EDA), to this MD-STBC-MIMO detection problem. The MD-STBC-MIMO detection problem is converted into a discrete optimization problem that searches the space of $M^{N_S} = 2^{bN_S}$ symbol combinations. Namely, the receiver's MD-STBC-MIMO detection problem is to find the value of $\left[\chi_1, \chi_2, \cdots, \chi_K\right]$ that minimizes cost function $\left\| y - \sum_{k=1}^{K} \chi_k \Omega_k \right\|$ for a received signal y and the known channel condition $\Omega_1, \Omega_2, \cdots, \Omega_K$. In section 4, we describe how we can apply BBO and EDA to the MD-STBC-MIMO signal detection problem.

4. BBO and EDA algorithms

Population-based evolutionary algorithms (EAs) in general have been often used to solve difficult optimization problems. Candidate solutions to an optimization problem are represented as individuals in the population. Most of the evolutionary algorithms are inspired by the theory of biological evolution (e.g., selection, crossover, mutation, recombination, and reproduction). In EAs the objective function value of a candidate solution can be considered as the fitness of the individual in the concept of natural selection. For the MD-STBC-MIMO detection problem, expression (6) can be used as the fitness function, where the smaller value of (6) means the better fitness. If each candidate solution being represented as a binary string, the search space is $I_s = \{0,1\}^n$ where $n = N_S \log_2 M$. The MD-STBC-MIMO detection problem can be converted into a binary optimization problem by associating M^{N_S} symbols with 2^{bN_S} bit strings. Each of the solutions has length, $n = N_S \log_2 M$ bits. In this section, we present a MD-STBC-MIMO detector that utilizes BBO-based and EDA-based evolutionary algorithms.

4.1 BBO

Biogeography-based optimization (Simon, 2008) is a population-based, stochastic global optimization EA, which is based on the mathematics of biogeography theory. Biogeography is the study of the geographical distributions of biological organisms. Mathematical models of biogeography describe how species migrate from one island to another, how new species arise, and how species become extinct.

Consider an optimization problem:

$$\begin{array}{ll} \max_x & F(x) \\ \text{subject to} & x \in X \end{array} \tag{7}$$

where $x \equiv (x_1, x_2, \ldots x_m)^T$ is a vector and X is a constraint set. In the original BBO, each candidate solution is represented by a vector variable of the optimization problem. In the context of evolutionary algorithms, a candidate solution is also referred to as an "individual," and a group of candidate solutions is referred to as a "population" of individuals. In BBO, each individual (candidate solution to an optimization problem) is analogically considered as an island (habitat) in Biogeography. The fitness value, $F(x)$, of each individual x corresponds to the Habitat Suitability Index (HSI) of an island in Biogeography. In Biogeography, features that affect HSI include vegetation, rainfall, topographic diversity, temperature, etc., and these features are characterized by variables that are called Suitability Index Variables (SIVs). As mentioned earlier, a candidate solution $x \equiv (x_1, x_2, \ldots x_m)^T$ in optimization problem (7) analogically corresponds to an island (a habitat) in Biogeography. Then, components $x_1, x_2, \ldots x_m$, of x correspond to its SIVs, and F(x) correspond to the HSI of island x. We will often use these terminologies to refer to a candidate solution $x \equiv (x_1, x_2, \ldots x_m)^T$, its components, $x_1, x_2, \ldots x_m$, , and the fitness value of a candidate solution x.

A good solution indicates an island with a high HSI, which are well suited as habitats for biological species. An island with a high HSI tends to have a large number of species, while an island with a low HSI tends to have a small number of species. An island with a high HSI tends to have a low immigration rate because it is already saturated with species (Simon, 2008). Also, many species emigrate from high-HSI islands to nearby habitats, as animals ride flotsam, fly or swim to neighboring islands. These habitats are said to have high emigration rates. Suppose there are several candidate solutions to a problem. A good solution is analogous to an island with a high HSI, while a poor solution is analogous to an island with a low HSI. High-HSI solutions are more likely to share their features (SIVs) with other habitats, which is analogous to emigration. Low HSI habitats tend to accept features of other solutions, which is analogous to immigration. Through this kind of probabilistic evolution, biogeography-based optimization searches for a good solution to an optimization problem.

We denote by λ and μ the immigration rate and emigration rate, respectively. Immigration rate λ and emigration rate μ are functions of the island's HSI (or equivalently, the number of species), S, on the island. An island has the maximum possible immigration rate I when there is no species in the island. As the number of species increases and the island becomes more crowded, the immigration rate decreases because fewer species can successfully survive immigration. The immigration rate is zero if the island has the largest possible number of species that the island can support. Similarly, when there is no species in the island, no species can emigrate from it, so its emigration rate is zero. The maximum emigration rate E occurs when the island contains the largest possible number of species. In short λ is a non-increasing function of HSI S, and is a non-decreasing function of S. Simple examples would be the linear functions illustrated in Figs. 2a and 2b, which can be mathematically expressed as

$$\lambda(S) = I\left(1 - \frac{S}{S_{\max}}\right), \quad \text{for } 0 \le S \le S_{\max} \tag{8}$$

$$\mu(S) = E\frac{S}{S_{\max}}, \quad \text{for } 0 \le S \le S_{\max} \tag{9}$$

Parameters and notations used in BBO are summarized in table 1, and the step-by-step algorithm of BBO used for the purpose of this chapter is presented in Fig. 3. We use the notation p_i to denote the ith population member (island) and $p_i(s)$ to denote the sth SIV of the ith population member.

λ_i	Immigration rate into the ith island in the population
μ_i	Emigration rate from the ith island in the population
S_{max}	The maximum number of species that a habitat can support
p	The population size (the number of islands in the population)
p_i	The ith island in the population
$p_i(s)$	The sth SIV of the ith population member (island)
q	The probability of mutation
I	The maximum possible immigration rate
E	The maximum possible emigration rate
g	The number of iterations used as termination condition

Table 1. Parameters and notation of BBO

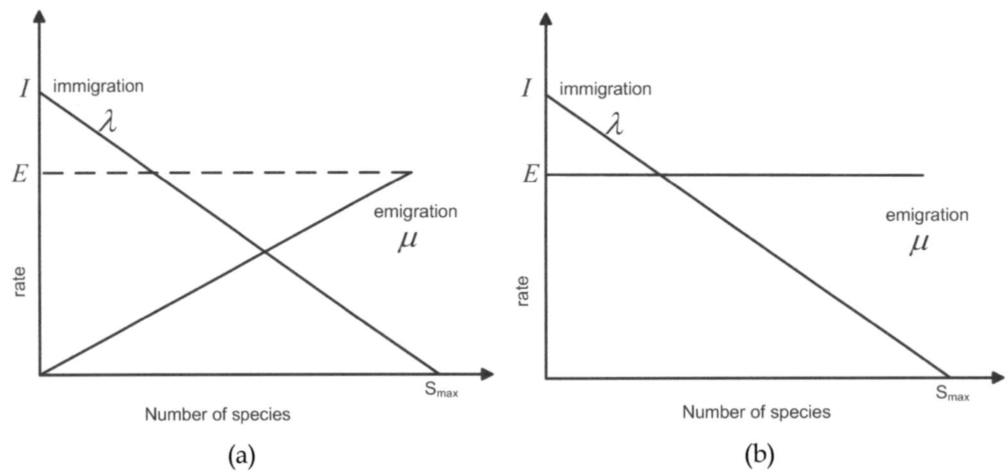

Fig. 2. (a) Typical migration Model used for BBO (b) Low- complexity migration Model for BBO

Various algorithms have been developed that use different migration schemes (Simon, 2008). The migration algorithm we use is basically a simplified version of the partial immigration-based BBO. We used a linear, decreasing λ (immigration rate) curve with a maximum of I and a constant μ (emigration rate) equal to E, as illustrated in Fig. 2b, in order to reduce computation. This constant emigration rate reduces the process of selecting the island that will emigrate to each island that is decided to accept immigration. Our preliminary results indicated that this computational simplification resulted in BER comparable to the implementation of (Simon, 2008).

```
For each iteration g
  For each island p_i
    For each SIV s
      With probability λ_i , p_i(s) is decided to accept immigration;
      If p_i(s) is decided to accept immigration, then
        Select one island, p_j, that emigrates to p_i from the rest of the population with
          equally likely probability;
          Assign p_i(s) = p_j(s) (p_j(s) emigrates into p_i(s);
      End If
    Next SIV
  NEXT Island

  For each island p_i
    For each SIV s
      With probability q,  p_i(s) is decided to mutate;
      If p_i(s) is decide to mutate, then
        Replace p_i(s) with a randomly generated SIV;
      End If
    Next SIV
  NEXT island

  For each island p_i
    Calculate HIS
  NEXT island

Sort population
NEXT generation
```

Fig. 3. One generation of BBO pseudo code

BBO takes the advantage of mutation. In each island in each generation, each SIV mutates with a particular probability. Simon (2008) introduces a procedure of assigning different probabilities of mutation. His mutation approach tends to increase diversity among the population. The aim of this scheme is to make an island with low HSI more likely to mutate its SIVs. However, we use a less complex approach of equal probability of mutation, q for all islands and all SIVs. Our results indicated that this equal mutation probability did not much influence the BER performance in our symbol detection problem.

4.2 EDA

Unlike other population-based evolutionary algorithms such as the genetic algorithm, in EDA the individuals are generated without the crossover and mutation operators. Instead, in EDA, a new population is generated based on a probability distribution, which is estimated from the best-selected individuals of the previous iterations (Larrañaga et al., 2001).
In general, conventional EDAs can be characterized and described by parameters and notations $(I_s, F, \Delta_l, \eta_l, p_s, \Gamma, I_{Ter})$, where

1. I_s denotes the space of all potential solutions (entire search space of individuals).
2. F denotes the fitness function.
3. Δ_l denotes the population (the set of individuals) at the l_{th} iteration.
4. η_l denotes the set of best candidate solutions selected from set Δ_l at the l_{th} iteration.
5. p_s is the selection probability. The EDA selects $p_s|\Delta_l|$ individuals from set Δ_l to make up the set η_l.
6. We denote by Γ_l the distribution estimated from η_l (the set of selected candidate solutions) at each iteration.
7. I_{Ter} is the maximum number of iterations.

In conventional EDAs each individual can be designated by a binary string of length n (n-dimensional binary vector). We denote by a binary row vector $X = (x_1, x_2, \cdots, x_n)$, $x_i \in \{0,1\}$ as an individual. In each iteration, an EDA maintains a population of individuals. We denote by $|\Delta_l|$ the number of individuals in population Δ_l. Population Δ_l can be specified by the following matrix

$$\Lambda = \begin{pmatrix} X^1 \\ X^2 \\ \vdots \\ X^{|\Delta_l|} \end{pmatrix} = \begin{pmatrix} x_1^1 & x_2^1 & \vdots & x_n^1 \\ x_1^2 & x_2^2 & \vdots & x_n^2 \\ \cdots & \cdots & \cdots & \cdots \\ x_1^{|\Delta_l|} & x_2^{|\Delta_l|} & \vdots & x_n^{|\Delta_l|} \end{pmatrix} \tag{10}$$

where superscript j in the row vector $X^j = (x_1^j, x_2^j, x_3^j, ..., x_n^j)$ indexes an individual in the population. A typical EDA applied to the MD-STBC-MIMO detection problem can be described in the following steps:

Step 0: *Generate an initial population Δ_0. The initial population ($|\Delta_0|$ individuals) is typically obtained by sampling according the uniform (equally likely) distribution (Larrañaga et al., 2001):*

$$p(\theta_1, \theta_2, \cdots, \theta_n) = \prod_{i=1}^{n} p_i(\theta_i), \qquad p_i(\theta_i = 1) = p_i(\theta_i = 0) = 0.5, \; i = 1,2,...,n. \tag{11}$$

(In accordance with Eqn. (11), in a typical EDA the joint probability distribution from which the individuals are sampled is factorized as a product of n univariate marginal probability distributions, each following a Bernoulli distribution with parameter value equal to 0.5.

For iterations l = 1,2, ..., follow Step 1 through Step 6:

Step 1: *Evaluate the individuals in the current population Δ_{l-1} according to the fitness function F. Sort the candidate solutions (individuals in the current population) according to their fitness orders.*

Step 2: *If the best candidate solution satisfies the convergence criterion[1] or if the number of iterations exceeds its limit I_{Ter}, then terminate; otherwise go to step 3.*

Step 3: *Select the best $p_s|\Delta_{l-1}| = |\eta_{l-1}|$ candidate solutions (individuals) from the current population Δ_{l-1}. This selection is accomplished according to the sorted candidate solutions.*

---

1 A simple example of the convergence criterion would be to terminate the algorithm if there is no improvement of the fitness value in the iteration.

Step 4: *Estimate the probability distribution* $p(\theta_1,\theta_2,\cdots,\theta_n)$ *on the basis of* $|\eta_{l-1}|$ *best candidate solutions. We denote this estimation by*

$$\Gamma_{l-1} \equiv P(\theta_1,\theta_2,\cdots,\theta_n\,|\,\eta_{l-1}) \tag{12}$$

Step 5: *Generate new* $|\Delta_l|-|\eta_{l-1}|$ *individuals on the basis of this new estimated probability distribution* Γ_{l-1}. *Combine these newly generated* $|\Delta_l|-|\eta_{l-1}|$ *individuals with members of* η_{l-1} *to form a new population* Δ_l.

Step 6: *Go to step 1 and repeat the steps.*

We followed the steps of the above pseudo code for our EDA implementation for the MD-STBC-MIMO detecton problem. In our experimentation, for estimation (12), we used the simple scheme of estimating the marginal distributions separately and using the product form

$$\begin{aligned}\Gamma_{l-1} = p(\theta_1,\theta_2,\cdots,\theta_n\,|\,\eta_{l-1}) &= \prod_{i=1}^{n} p_i(\theta_i\,|\,\eta_{l-1}) \\ &= \prod_{i=1}^{n}\left(\frac{\sum_{j=1}^{|\eta_{l-1}|}\delta\left(x_i^j=\theta_i\,|\,\eta_{l-1}\right)}{|\eta_{l-1}|}\right)\end{aligned} \tag{13}$$

in order to generate the samples in the next iteration (generation), where δ is an indicator function for the individual indexed by j in the set η_{l-1}.

$$\delta\left(x_i^j=\theta\,|\,\eta_{l-1}\right)=\begin{cases}1 & if\ x_i^j=\theta\\ 0 & otherwise\end{cases} \tag{14}$$

The use of a product-form distribution as in (13) is a part of our heuristic presented in this chapter. In fact, the statistics of the candidate solutions in η_{l-1} may show correlations among the variables $x_1,x_2,\ldots,x_n$. From these statistics, we could construct an empirical distribution (12) that captures correlations among variables, but that procedure would increase computational complexity. In the future, we will study to find how much performance improvement can be made by using such expensive procedures. Product-form distributions like (13) in EDA should not be discredited a priori because the benefit of searching variable correlations could, under particular circumstances, remain unclear (Platel et al.,2005).

A typical EDA can get stuck in a local optimum due to premature convergence of the probability distributions, or can be slowed down due to non-convergence of the probability distributions. We present a method of avoiding these two problems. Our approach is to apply a threshold on estimated parameters of the distributions. During the execution of a typical EDA, some of the estimated probabilities $P(\theta_i=1\,|\,\eta_{l-1}),\ldots,i=1,2,\ldots,$ or n, may become 0 or become very close to 0 at an early stage of the execution (at a small value of iteration count l). In that case, the algorithm is not likely to explore the candidate solutions with $x_i=1$ during the rest of the execution. In order to thwart such premature convergence, we present an idea of adjusting the estimated probabilities $P(\theta_i=1\,|\,\eta_{l-1})$, i = 1, 2, .., n after estimating these at each iteration. The adjustment in general can be regarded as a mapping from set of n-tuples

$$\Pi \equiv \{(P(\theta_1 = 1 | \eta_{l-1}), P(\theta_2 = 1 | \eta_{l-1}), ..., P(\theta_n = 1 | \eta_{l-1})) \mid 0 \le P(\theta_i = 1 | \eta_{l-1}) \le 1, i = 1, ..., n\}$$

to set Π itself. First, we address the problem that a marginal probability value, in the estimated distribution, prematurely converges to 1. To avoid this, we set some thresholds $0.5 < \gamma_1, \gamma_2, ..., \gamma_n < 1$. At any iteration, if the probability value $P(\theta_i = 1 | \eta_{l-1})$, i= 1, 2, ..., n , is greater than γ_i, we set that value to γ_i –i.e., we set $P(\theta_i = 1 | \eta_{l-1}) = \gamma_i$. This way, some degree of randomness remains in the algorithm until the termination criterion is satisfied. A simpler application of this idea is to set the same threshold $\gamma = \gamma_1 = \gamma_2 = ... = \gamma_n$. We can similarly address the problem that a probability value prematurely converges to 0. We define thresholds $0 < \tilde{\gamma}_1, \tilde{\gamma}_2, ..., \tilde{\gamma}_n < 0.5$. At any iteration, if the estimated probability value $P(\theta_i = 1 | \eta_{l-1})$, i = 1, 2, .., n, is less than $\tilde{\gamma}_i$, we set $P(\theta_i = 1 | \eta_{l-1}) = \tilde{\gamma}_i$, so that some degree of randomness remains in the algorithm until the termination criterion is satisfied. A simpler application of this idea is to set the same threshold $\tilde{\gamma} = \tilde{\gamma}_1 = \tilde{\gamma}_2 = ... = \tilde{\gamma}_n$.

5. Computational complexity

A motivation for applying the proposed near-optimal algorithms to a MD-STBC-MIMO problem is their low computational complexity. In this section, the computational complexity of BBO and EDA for MD-STBC-MIMO symbol detection is compared with that of MMSE, SD, SDR, V-BLAST, GA and the exhaustive search. The computational complexity of exhaustive search (an implementation of the ML detector) is $O\left(M^{N_S}\right)$ or $O\left(2^n\right)$, where $N_S = \sum_{k=1}^{K} Q_k$, $n \equiv N_S \log_2 M$, so an exhaustive search is usually impractical for real-time operations of symbol detection. A number of suboptimal detection schemes with better computational complexity have been presented in the literature.

The worst-case complexity of SD is exponential, and its expected complexity depends on the problem size and the SNR (Hassibi & Vikalo, 2005). The expected complexity of SD is roughly $O\left(N_S^3\right)$ at high SNRs (Hassibi & Vikalo, 2005) and $O\left(N_S^6\right)$ at low SNRs (Damen et al., 2000). MMSE is one of the sub-optimal detectors that involve inverting a matrix, and its computational complexity is $O\left(N_S^3\right)$ (Comaniciu et al.,2005). In V-BLAST, each iteration requires either a Zero-Forcing (ZF) or MMSE operation, and the number of iterations is equal to the total number of transmitted antennas. If the total number of transmit antennas in the system is equal to the number of receive antennas, then the complexity is $O\left(N_S^3\right)$. If the total number of transmit antennas in the system is greater than the number of receive antennas, then the complexity is $O\left(N_S^4\right)$ (Hassibi, 2000). The computational complexity of SDR (Kisialiou et al., 2005) is $O\left(N_S^{3.5}\right)$.

Typically, the time complexities of population-based algorithms are analyzed in terms of the number of function evaluations. The number of function evaluations in GA grows on the order of $O(gp)$, where g is the number of iterations (generations) and p is the population size. The number of function evaluations both in BBO and EDA also grows on the order of $O(gp)$. Now, let us consider the number of operations required for each function evaluation. Each function evaluation computes expression (6). Recall that the dimensions of Ω_k, χ_k, and y are $2Q_k \times 2TN_R$, $1 \times 2Q_k$, and $1 \times 2TN_R$, respectively. Computing the term $\sum_{k=1}^{K} \chi_k \Omega_k$ requires about $4TN_R\left(\sum_{k=1}^{K} Q_k\right) \approx 4N_s TN_R$ operations. Computing the norm square of the difference between two $2TN_R$-dimensional vectors as in $y - \sum_{k=1}^{K} \chi_k \Omega_k$ requires about $6TN_R$ arithmetic operations. Therefore, the total number of arithmetic operations to evaluate (6) is on the order of $4N_s TN_R + 6TN_R$. Combining the complexities of these procedures, we can say that the evaluation of functon (6) requires $O(N_s TN_R p)$ operations in each generation (iteration), where p is the population size. In each generation of GA, crossover and mutation procedures are performed, and these procedures take on the average $O(pN_s \log_2 M) = O(pN_s)$ computer operations, where $N_S \log_2 M$ is the length of the binary string that represents a chromosome[2]. As can be deduced from the pseudo code in Fig. 3, the average number of operations for migration and mutation in each generation (iteration) of BBO is also on the order of $O(pN_s \log_2 M)$. The average number of operations to estimate the probability distribution in each iteration of EDA is on the order of $O(N_s p \log_2 M)$. The sorting of p individuals (chromosomes) in GA, EDA and BBO can be performed in $O(p \log p)$. Therefore, the number of operations in each generation of GA, EDA and BBO can be expressed as $O(N_s TN_R p + N_s p + p \log p) \approx O(N_s TN_R p + p \log p)$.

6. Simulation results

In this section, we present the simulation results of the proposed BBO-based and EDA-based detection schemes and their comparison with other detection techniques applied to the MD-STBC-MIMO system. The system model used in our simulations is depicted in Fig. 1. The channels are assumed to be quasi-static, and different channels in the MD-STBC-MIMO system are assumed to be statistically independent. In all our simulations, we used the 4−QAM modulation. Each point in the plots of Figs. 4−10 is a value averaged over multiple independent simulation runs. In each simulation run, the set of transmitted signals $[\chi_1 \ \ \chi_2 \ \ \cdots \ \ \chi_K]$, channel matrices $[\Omega_1 \ \ \Omega_2 \ \ \cdots \ \ \Omega_K]^T$ and noise Z are generated randomly and independently. The main objective is to find the vector $[\chi_1 \ \ \chi_2 \ \ \cdots \ \ \chi_K]$ that minimizes (6). This experimental setup enables us to compare different algorithms in terms of the performance averaged over different channel and noise realizations. Some BBO and EDA parameters are kept constant through all simulations, such as $I = 1$, $q = 0.1$, $p_s = 0.5$. As mentioned in section 4.1, for BBO we assumed constant emigration rate over all islands. In each simulation experiment, we set the BBO, EDA, and GA to have the same population size and number of iterations for fair comparison.

---

[2] In this discussion, we fix M, the constellation size of the modulation.

The simulation results in Figs. 4 through 7 show the BER performance comparison between MMSE, V-BLAST, SDR (Luo, 2010), SD (Hassibi & Vikalo, 2005), GA, EDA and BBO detectors. The MD-STBC-MIMO system configuration, (K,N_T,N_R,M), is set to (4,2,6,4), (5,2,8,4), (6,2,10,4) and (3,4,4,4,) for Figs. 4, 5, 6 and 7, respectively. The Figs. 4, 5, and 6 use orthogonal space time coding scheme (Alamouti, 1998). Fig. 7 uses a non-orthogonal space time coding scheme (Boariu et al., 2003). BBO, EDA and GA parameters, (g, p), which denote the number of iterations and the population size are set to (60,100), (100,100), (100,120) and (100,150) for Figs. 4, 5, 6 and 7, respectively. For these Figs., the total number, N_S, of symbols transmitted from all users are set 8, 10, 12, and 12, respectively. Thus, our BBO, EDA and GA experiments have search spaces of 4^8, 4^{10}, 4^{12} possible solutions. Figs. 4, 5, 6 and 7 indicate that in the MD-STBC-MIMO system, EDA and BBO have significantly better BER performance than MMSE, SDR, V-BLAST, and GA. BBO and EDA outperform other suboptimal detection methods in all four Figs. and can achieve performance close to the sphere decoding[3] detector. While BBO and EDA have performances close to that of the sphere decoding, they require much less computation. Table 2 shows the average number of operations required by SD, EDA and BBO for Figs. 4 to 7. From Table 2 we can observe that the average number of operations required by EDA or BBO is much smaller than the sphere decoder.

	Average number of operations			
	Fig. 4	Fig. 5	Fig. 6	Fig. 7
SD	16,777,216	64,000,000	19,110,2976	19,110,2976
EDA or BBO	20,4000	42,0000	60,0950	752640
$\frac{\text{Ave.Number of EDA or BBO operations}}{\text{Ave.Number of SD operations}}$	0.0120	0.0065	0.0031	0.0039

Table 2. Computation complexity comparison

Intuitively, for any population-based evolutionary algorithm, the larger population size and the larger number of iterations will produce the better results. However, the larger population size and larger number of iterations both result in the more computational load. Fig. 8 shows how the BER performance of BBO, EDA, and GA imporves as the number of iterations increases. The BER results of other schemes (SD, V-BLAST, SDR, ZF, MMSE) and the number of iterations (generations) in BBO, EDA, and GA are irrelevant, so we indicated the BER results of SD, V-BLAST, SDR, ZF, and MMSE as horizontal lines in Fig. 8. For this experimentation, we used MD-STBC-MIMO system configuration (K,N_T,N_R,M) = (4,2,6,4) and an Alamouti space time block code. We assumed quasi-static channels and fixed the SNR to 8 dB. Fig. 8 shows that the BER perfromance of BBO and EDA with the population size fixed to 100 approaches the BER perfromance of the sphere decoder in less than 100

[3] Due to heavy computational load of performing the Maximum Likelihood (ML) detection, which is optimal, it is difficult to compare performance with the ML detector. We used sphere decoding as the benchmark.

iterations while the GA performance is nowhere close to that of the sphere decoder even with 100 iterations.

Fig. 9 shows how the BER performance of BBO depends on the population size (p) and the number (g) of iterations. Thus, Fig. 9 exhibits the tradeoff between the population size and the number of iterations required to achieve a desired BER in BBO. The MD-STBC-MIMO system configuration was set (K,N_T,N_R,M) = (4,2,4,4), and we used an Alamouti code. We assumed quasi-static channels and fixed the SNR to 8 dB. This tradeoff result is useful from the system design point of view. Increase in the population size and increase in the number of iterations each improves performance. While the increase in the population requires more memory module of the hardware, the increase in the number of iterations require fast processing in order to finish computation within a specified time. The tradeoff results such as Fig. 9 can guide how to set the population size and the number of iterations on the basis of hardware to be used. For example, if the hardware has high processing capabilities and small memory space, then we can set the population size low and run more iterations of BBO. Vice versa, if the hardware has large memory space and a slow processor, we can set the population size large and run a smaller number of iterations to get the same BER performance. We observe results similar to Fig. 9 for EDA.

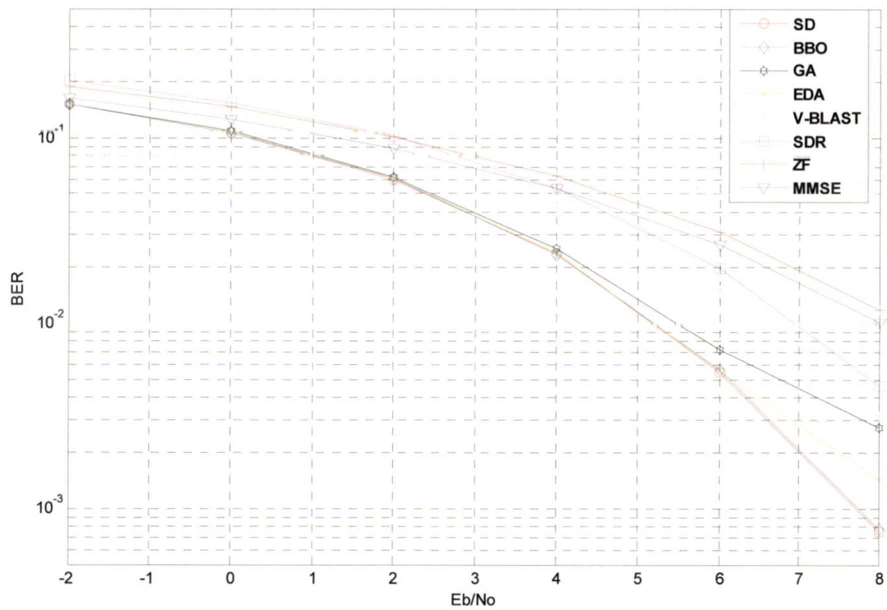

Fig. 4. BER performance comparison for (K,N_T,N_R,M) = (4,2,6,4) over quasi-static fading channel. The (g,p) for GA, BBO and EDA is set to (60,100).

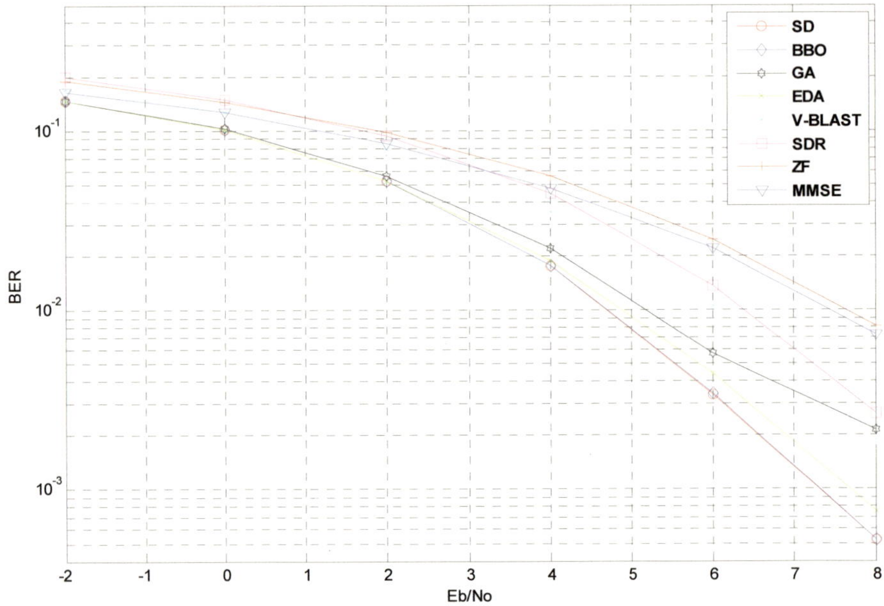

Fig. 5. BER performance comparison for (K,N_T,N_R,M) = (5,2,8,4) over quasi-static fading channel. The (g,p) for GA, BBO and EDA is set to (100,100).

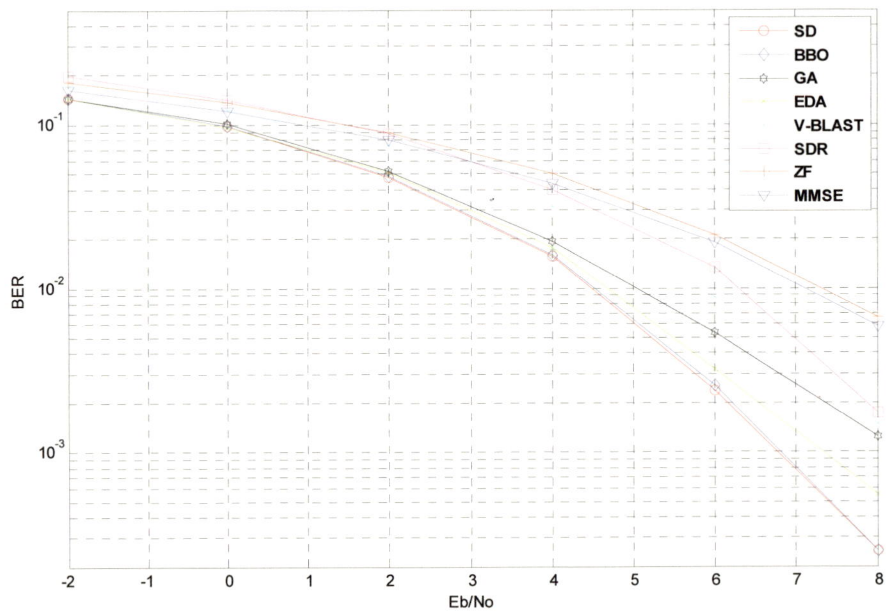

Fig. 6. BER performance comparison for (K,N_T,N_R,M) = (6,2,10,4) over quasi-static fading channel. The (g,p) for GA, BBO and EDA is set to (100,120).

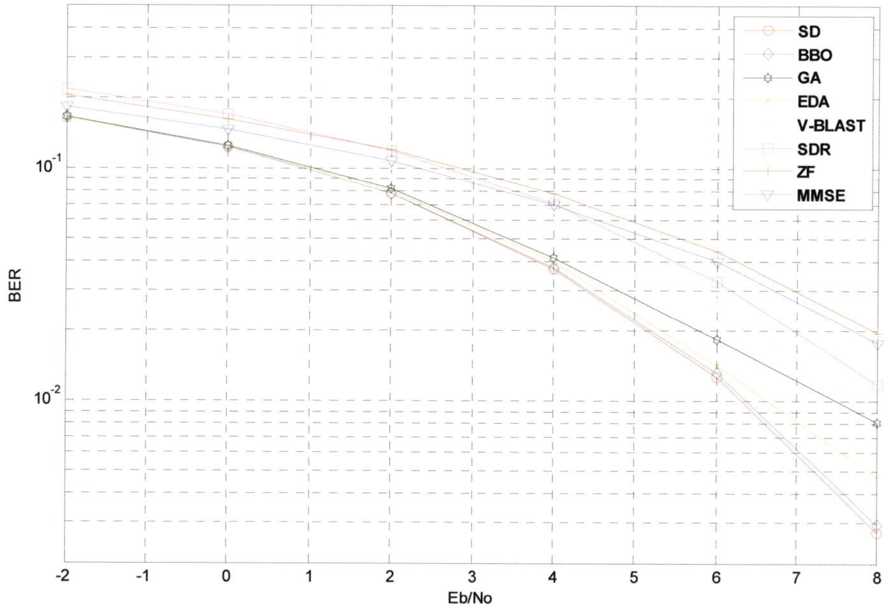

Fig. 7. BER performance comparison for $(K,N_T,N_R,M) = (3,4,4,4)$ over quasi-static fading channel. The (g,p) for GA, BBO and EDA is set to (100,150).

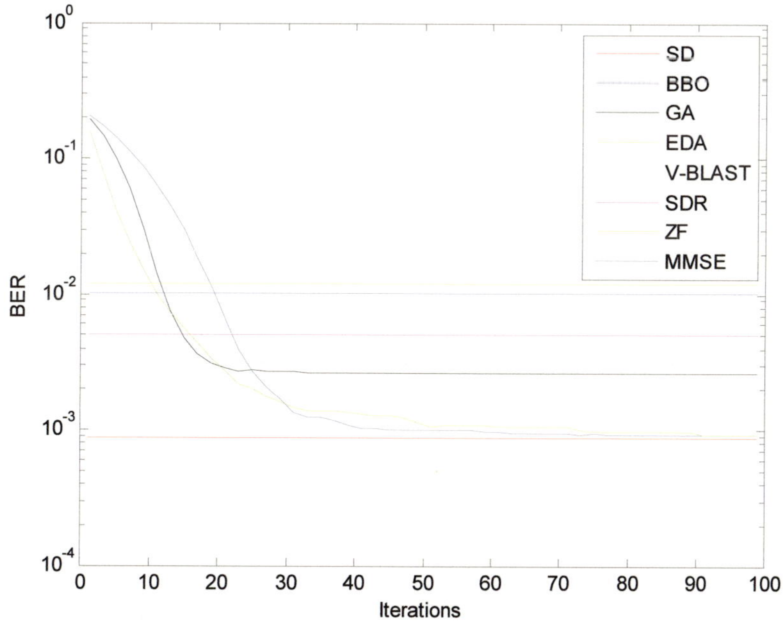

Fig. 8. How BER improves with the number of iterations in BBO, EDA, and GA. $(K,N_T,N_R,M) = (4,2,6,4)$. The population size is fixed to 100.

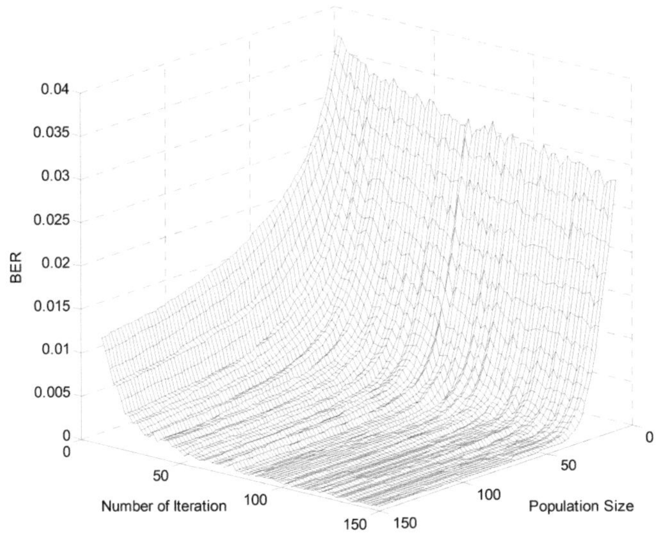

Fig. 9. BER dependency on the population size and the number of iterations in BBO. $(K,N_T,N_R,M) = (4,2,4,4)$

7. Conclusions

In this chapter, we proposed two population-based evolutionary algortihms, BBO and EDA, for symbol detection in the Multi-Device (MD) Space-Time Block Coded (STBC) Multi Input Multi Output (MIMO) Communication System. The proposed BBO and EDA algortihms have low complexity as compared with the sphere decoding algorithm, which is the best known algorithm for STBC decoding. Thus, the proposed BBO and EDA algorithms are suitable for high-speed real-time communications. In addition, compared to other Evolutionary Algorithms like GA, BBO and EDA algorithms show significantly better performance for the MD-STBC-MIMO detection. The proposed algorithms also show consistently better BER performance-complexity trade-offs in comparison to existing algorithms. Moreover, we believe that BBO and EDA for MD-STBC MIMO symbol detection still have room to improve further in terms of performance-complexity trade-offs. For example, we believe that BBO can be further improved by adjusting migration and mutation procedures.

8. References

Alamouti, S. M. (1998). A Simple Transmit Diversity Technique for Wireless Communications. *IEEE journal on select areas in communications*, Vol. 16, No. 8, pp. 1451-1458.

Boariu, A. and Ionescu, D.M. (2003) A class of nonorthogonal rate one space-time block codes with controlled interference. *IEEE Trans. Wireless Communication*, Vol. 2, No. 2 pp. 270-276.

Comaniciu, C.; Mandayam, N.B. & Poor, H.V. (2005) *Wireless Networks: Multiuser Detection in Cross-Layer Design*, Springer, NY.
Damen, O.; Chkeif, A. & Belfiore, J. C. (2000). Lattice code decoder for space-time codes. *IEEE Communications Lett.* Vol. 4, No. 5, pp. 161–163.
Foschini, G. J. & Gans, M. J. (1998). On limits of wireless communications in a fading environment when using multiple antennas. *Wireless Personal Communications*, Vol. 6, pp. 311-335.
Goldberg, D. (1989). *Genetic Algorithms in Search, Optimization, and Machine Learning*, Addison-Wesley.
Hassibi. B. (2000) An efficient square-root algorithm for BLAST. *Proceedings of IEEE ICASSP* pp. 737-740.
Hassibi, B. & Hochwald, B.M. (2002). High-rate codes that are linear in space and time. *IEEE Trans. Information Theory*, Vol.48, No.7, pp. 1804–1824.
Hassibi . B. & Vikalo, H. (2005). On the sphere decoding algorithm: Part I, the expected complexity. *IEEE Trans. Signal Processing*, Vol. 53, No. 8, pp. 2806-2818.
Jalden, J. & Ottersten, B. (2005). On the complexity of sphere decoding in digital Communications. *IEEE Trans. on Signal Processing*, Vol. 53, No 4, pp. 1474-14844.
Kisialiou, M. & Luo, Z.-Q. (2005). Performance analysis of quasi-maximum-likelihood detector based on semi-definite programming. *Proc. IEEE Int. Conf. Acoust., Speech, Signal Process.*, Vol. 3, pp. 433-436.
Larrañaga, P. & A Lozano, J. (2001) *Estimation of Distribution Algorithms: A New Tool for Evolutionary Computation*. Kluwer Academic Publishers.
Luo, Z. -Q. (2010) http://www.ece.umn.edu/~luozq/software/sw_about.html.
Nekuii, M.; Kisialiou, M.; Davidson, T.N. & Luo, Z. Q. (2008). Efficient Soft Demodulation of MIMO QPSK via Semidefinite Relaxation. *Proceedings of IEEE International Conference on Acoustics, Speech, and Signal Processing*, pp. 2665-2668.
Platel, M.D.; Schliebs, S. & Kasabov, N. (2009) Quantum-Inspired Evolutionary Algorithm: A Multimodel EDA. *IEEE Transactions on Evolutionary Computation*, Vol. 13, No. 6, pp. 1218–1232.
Simon, D. (2008). Biogeography-based optimization. *IEEE Transactions on Evolutionary Computation* pp. 702-713.
Tarokh, V.; Seshadri N., & Calderbank A. R.(1998). Space-time codes for high data rate wireless communications: performance criterion and code construction. *IEEE Trans. Information Theory*, Vol. 44, pp. 744-765.
Telatar, E. (1999). Capacity of Multi-antenna Gaussian Channels. *European Trans. on Telecomm*. vol. 10, pp. 569-709, Nov. 1999.
Verdú, S. (1986). Minimum Probability of Error for Asynchronous Gaussian Multiple Access Channels. *IEEE Trans. Information Theory*, Vol. 32, pp. 85-96.
Verdú S. (1989). Computational Complexity of Optimum Multiuser Detection," Algorithmica, Vol.4, No.3, pp. 303-312.
Vikalo, H. & Hassibi, B.(2005).On the sphere decoding algorithm: Part II, generalizations, second-order statistics and applications to communications. *IEEE Trans. on Signal Processing*, Nol 53, No. 8, pp. 2819-2834.
Wolniansky, P.W.; Foschini, G.J.; Golden, G.D. & R.A. Valenzuela (1998). V-BLAST: an architecture for realizing very high data rates over the rich-scattering wireless channel. *Proc. ISSSE-98*, Pisa, Italy. pp. 87-91.

Wei, H. & Hanzo, L. (2004). Reduced-complexity near-optimum genetic algorithm assisted multiuser detection for synchronous multicarrier cdma. *Proceedings of IEEE 59th Vehicular Technology Conference*, Vol. 3, pp. 1717 - 1721.

Part 4

Application and Case Studies

14

Analysis and Mitigation of Phase Noise in Centralized/De-centralized MIMO Systems*

Wei Zhang[1], Xiujun Zhang[2], Shidong Zhou[3] and Jing Wang[4]
[1]*Harbin Engineering University*
[2,3,4]*Tsinghua University*
P.R.China

1. Introduction

The multiple-input multiple-output (MIMO) techniques are anticipated to be widely employed in future wireless networks to address the ever-increasing capacity demands. A MIMO channel, typically modeled as a matrix with independent and identically distributed (i.i.d.) complex Gaussian entries, provides multiple spatial dimensions for communications. At high signal-to-noise-ratio (SNR), Shannon capacity can increase linearly with the minimum number of transmit and receive antennas. However, achieving these dramatic capacity gains in practice, especially for outdoor deployment, could be problematic. One severe problem is the rank deficiency and ill-conditionness of the MIMO channel matrix H. This is mainly caused by the spatial correlation due to the scattering environment and the antenna configurations, and sometimes by the "keyhole" effect even though the fading is essentially uncorrelated on each end of the channel. Therefore, the MIMO capacity may be greatly reduced and adding more (co-located) antennas only wastes resources.

The distributed antenna is an efficient way to solve this problem. In the downlink of a distributed antenna system, antennas allocated at several base stations (BS) constitute an antenna pool and can be used with a suitable MIMO transmission mode. Compared with a conventional centralized MIMO system, the distributed (de-centralized) MIMO system possesses many advantages, such as higher system capacity, better handoff performance, lower outage probability and higher data rates, especially at cell boundaries. However, it is also easier to be affected by phase noise (PN).

Oscillator noise stems from oscillator inaccuracies in both the transmitter and the receiver, and manifests itself in the baseband as additional phase and amplitude modulation of the received samples. The influence of oscillator noise on the signal depends on the noise characteristics of the oscillators in the system and on the signal bandwidth. It is generally split in amplitude noise and phase noise (PN). The influence of the amplitude noise on the data samples is often neglected (Robins, 1982). The phase noise contribution of both the transmitter and receiver can be viewed as an additional multiplicative effect of the radio channel, like fast and slow fading.

*The chapter is supported by the Fundamental Research Funds for the Central Universities in China and the China Postdoctoral Science Foundation funded project.

In centralized MIMO systems, the presence of phase noise is an important limiting factor for a broadband system, i.e., a MIMO-OFDM system's performance, and depends on the quality and the operating conditions of the system's RF hardware. In conventional mobile radio systems around a carrier frequency of 2GHz, the phase noise constitutes typically no severe limitation, however, when the carrier frequency goes up to a higher frequency, such as 15GHz, its effects were less negligible and hence had to be investigated in more depth.
In distributed MIMO systems, system performance are limited by independent PN at local oscillators of N_t transmit antennas and N_r receive antennas. Some MIMO techniques, such as precoding, will suffer from independent PN as the orthogonality among sub-data-streams is destroyed. This involves with both narrowband and broadband MIMO systems.
This chapter will deal with the phase noise in centralized and decentralized (distributed) MIMO systems. The organization of this chapter is as follows.
Section 2 presents an overview of the phase noise effect and modeling of phase noise. PN can be modeled as a sampled *Wiener Process* for a free-running oscillator, and an *Ornstein-Uhlenbeck* Process (a Gaussian process with zero-mean and bounded variance) for a phase-locked loop (PLL) frequency synthesizer. In the following sections, we will consider effects of either/both of these two types of phase noise.
Section 3 addresses the influence of independent PN in centralized and distributed MIMO systems. We concentrate on narrowband systems applying linear precoding. In such a scenario, independent PN gives rise to SNR loss at the receiver and induces interference among sub-streams, and hence degrades the system performance. Then we discuss the feasible synchronization schemes for frequency and phase synchronization.
Simulation results are then given in Section 4 and finally Section 5 discusses the future research directions and concludes the chapter.

2. Phase noise model

In practice, a local oscillator (LO) can be realized with a free-running oscillator or with a PLL synthesizer. To characterize carrier imperfections, we use a random carrier time shift $\alpha(t)$ rather than the phase shift $\theta(t)$. A noisy LO output signal is then described as $x(t + \alpha(t))$(Demir, 1998; Demir et al., 2000; Mehrotra, 1999; Petrovic et al., 2007). The phase and time shift at the carrier frequency f_c are related by $\theta(t) = 2\pi f_c \alpha(t)$. To avoid confusion, in the following $\alpha(t)$ can also be called phase shift, phase perturbation, or phase noise.

2.1 Free-running oscillators

The main sources of oscillator noise are shot noise, thermal noise and flicker noise (also called $1/f$ noise). The former two noise sources can be characterized as white Gaussian noise. The $1/f$ noise is really a nonstationary process because it blows up at $f = 0$. For analytical simplicity, a cut-off frequency is introduced, below which the spectrum deviates from $1/f$ and attains a finite value at $f = 0$, and thus the $1/f$ noise can be modeled as a stationary and colored stochastic process (Demir, 1998). The power in such $1/f$ noise source is concentrated at low frequencies, frequencies close to the carrier frequency.
Furthermore, the spectrum of the oscillator with colored noise sources has the shape of a Lorentzian around the carrier, and away from the carrier, the colored noise sources contribute terms that have a frequency dependence as $1/f^2$ multiplied with the spectral density of the colored noise source, whereas the white noise sources contribute a term that has a $1/f^2$

frequency dependence at remote frequencies. Unfortunately, up to now the $1/f$ noise cannot be described in time domain with a mathematical form, for this reason in the following we only consider white Gaussian noise source.
If the noise source is white Gaussian, the time shift $\alpha(t)$ can be approximately described as a *Wiener* or *Brownian motion* process:

$$\alpha(t) = \sqrt{c}B(t) \tag{1}$$

where c is a parameter describing the oscillator quality, and $B(t)$ stands for a standard Wiener process, i.e., an accumulated Gaussian random variable (r.v.) of unit variance. Therefore, $B(t)$ has the following properties:

1. $B(0) \sim \mathcal{N}(0,1)$;
2. $B(t_1) - B(t_2) \propto \sqrt{|t_2 - t_1|} \times \mathcal{N}(0,1)$;
3. the variance of $B(t)$ increases linearly with time, i.e. $\sigma_B^2(t) = t$.

As such the variance of $\alpha(t)$ also increases linearly with time, i.e., $\sigma_\alpha^2(t) = ct$. In addition, the autocorrelation function of $\alpha(t)$ is given by

$$E[\alpha(t)\alpha(t+\tau)] = c\min(t, t+\tau) \tag{2}$$

Consider the relationship between Wiener process and normal distribution. We have

$$\alpha(t) = \sqrt{c}\int_0^t \xi(t')dt' \tag{3}$$

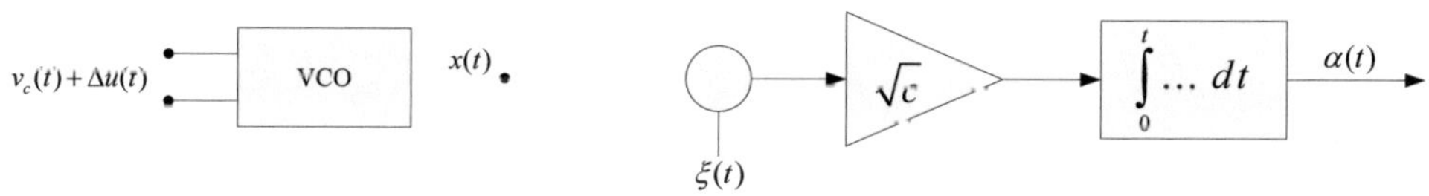

Fig. 1. Equivelent mathematical phase noise model of a free-running oscillator

Phase noise of both free-running oscillators and open-loop VCO can be described directly with (3), as shown in Fig. 1.
The constant c cannot be directly obtained in practice. To evaluate the performance of an oscillator, we are usually interested in the power spectrum density (PSD) around the first harmonic, i.e. PSD for f around f_c. Let f_m denote the offset frequency related to carrier frequency f_c. For $0 \leq f_m << f_c$, the *single-sideband* phase noise PSD $\mathcal{L}(f_m)$ (in dBc/Hz) can be approximated as (Mehrotra, 1999)

$$\mathcal{L}(f_m) \approx 10\lg\left(\frac{f_c^2 c}{\pi^2 f_c^4 c^2 + f_m^2}\right) \tag{4}$$

For $\pi f_c^2 c << f_m << f_c$, $\mathcal{L}(f_m)$ can be approximated by

$$\mathcal{L}(f_m) \approx 10\lg\left(\left(\frac{f_c}{f_m}\right)^2 c\right) \tag{5}$$

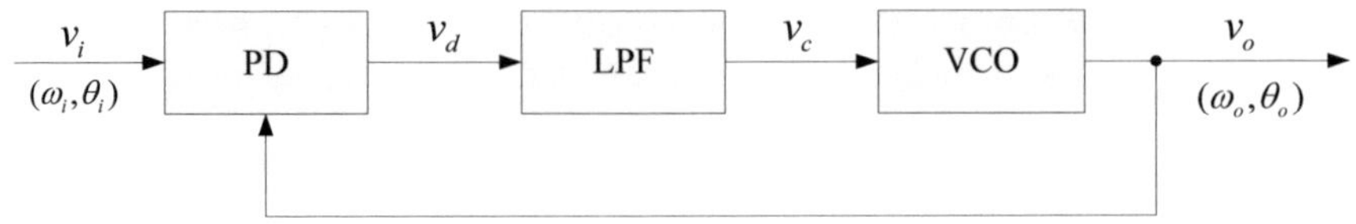

Fig. 2. PLL block diagram

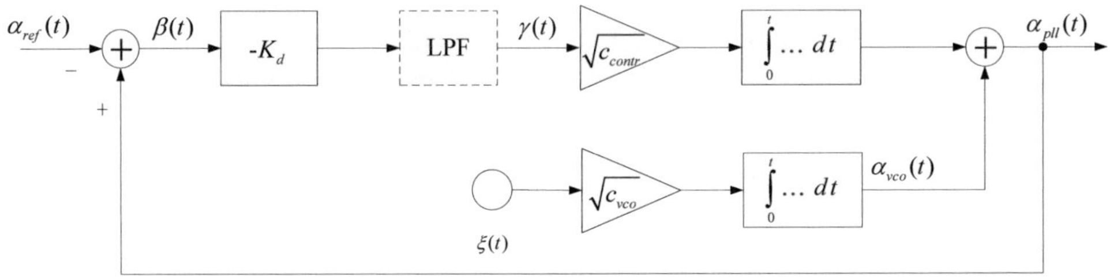

Fig. 3. Equivalent mathematical model of PLL phase noise

i.e., $\mathcal{L}(f_m)$ and c are related by $\mathcal{L}(f_m) \approx 10\log 10(c{f_c}^2/f_m^2)$. This approximation holds for $f_m \gg$ 3dB bandwidth of phase noise (denoted by $\Delta f_{3\text{dB}}$). As $\theta(t) = 2\pi f_s \alpha(t)$, Wiener process $\theta(t)$ can also be defined as $\sigma_\theta^2(t) = 4\pi\Delta f_{3\text{dB}}t$, and

$$\sigma_\theta^2(t) = 4\pi^2 f_s^2 \sigma_\alpha^2 = 4\pi^2 f_s^2 ct = 4\pi\Delta f_{3\text{dB}}t \tag{6}$$

the constant c is therefore related to $\Delta f_{3\text{dB}}$ by $c = \Delta f_{3\text{dB}}/\pi f_c^2$ (Petrovic et al., 2007). The time-sampling of the Wiener process $\alpha(t)$, $\alpha[n]$, may ba represented in an integral-sum form

$$\begin{aligned}\alpha[n] = \alpha(n\cdot\Delta t) &= \sqrt{c}\sum_{i=0}^{n-1}\int_0^{\Delta t}\xi(t' + i\cdot\Delta t)dt' \\ &= \sqrt{c}\sum_{i=0}^{n-1} W_i(\Delta t) \\ &= \sqrt{c\cdot\Delta t}\sum_{i=0}^{n-1} w[i]\end{aligned} \tag{7}$$

where $W_i(\Delta t) \sim \mathcal{N}(0, \Delta t)$, and $w[i]$ is a Gaussian random variable, $w[i] \sim \mathcal{N}(0,1)$. The variance of $\alpha(n\Delta t)$ increases linearly with time, i.e., $\sigma_\alpha^2(n\Delta t) = cn\Delta t$.

2.2 Phase-locked loop frequency synthesizer

Consider the basic PLL circuit in Fig. 2. Let $\alpha_{ref}(t)$, $\alpha_{vco}(t)$ and $\alpha_{pll}(t)$ denote the PN of the reference signal, the PN of the open-loop VCO and the PN at the output of PLL circuit, respectively. $\alpha_{ref}(t)$ and $\alpha_{vco}(t)$ can also be modeled as Wiener process/one-dimensional Brown motion process. The equivalent phase noise model of a PLL is shown in Fig. 3. The input of the phase detector (PD) is given by

$$\beta(t) = \alpha_{pll}(t) - \alpha_{ref}(t) \tag{8}$$

$\beta(t)$ is one component of a multidimensional *Ornstein-Uhlenbeck* process (Mehrotra, 2002). The correlation properties of $\beta(t)$ and $\alpha_{ref}(t)$ are

$$E\left[\beta(t_1)\alpha_{ref}(t_2)\right] = \sum_{i=1}^{n_0} \mu_i e^{\lambda_i \min\{0, t_2 - t_1\}} \tag{9}$$

$$E\left[\beta(t_1)\beta(t_2)\right] = \sum_{i=1}^{n_0} \nu_i e^{-\lambda_i |t_2 - t_1|} \tag{10}$$

where $n_0 = 1 + o_{lpf}$, o_{lpf} is the order of the low-pass filter. The calculation of parameters μ_i, ν_i and λ_i depends on the PLL circuit. The variance of the phase noise at the PLL output is then obtained by

$$\begin{aligned}
\sigma^2_{\alpha_{pll}}(t) &= E[\alpha^2_{pll}(t)] - \left(E[\alpha_{pll}(t)]\right)^2 \\
&= E[\alpha^2_{pll}(t)] \\
&= E[(\alpha_{ref}(t) + \beta(t))^2] \\
&= E[\alpha^2_{ref}(t) + \beta^2(t) + 2\alpha_{ref}(t)\beta(t)] \\
&= c_{ref}t + \sum_{i=1}^{n_0} (2\mu_i + \nu_i)
\end{aligned} \tag{11}$$

The random differential equation of $\beta(t)$

$$\begin{aligned}
\dot{\beta}(t) &= \dot{\alpha}_{pll}(t) - \dot{\alpha}_{ref}(t) \\
&= \sqrt{c_{contr}}\gamma(t) + \sqrt{c_{vco}}\xi_{vco}(t) - \sqrt{c_{ref}}\xi_{ref}(t)
\end{aligned} \tag{12}$$

For the reference signal $\alpha_{ref}(t)$ the PLL acts as a low-pass filter, whereas for the VCO signal the PLL acts like a high-pass filter. The 3dB corner frequency for both filters is identical: $\omega_{G_{pll}} = 2\pi f_{G_{pll}}$. To minimize the phase noise at the PLL output, the corner frequency has to be chosen as high as possible. On the other hand, to keep the system stable, $\omega_{G_{pll}}$ should not be higher than 1/10 of the crystal frequency. The constants c_{vco} at the control node of the VCO can be calculated by $\sqrt{c_{vco}} = \omega_{G_{pll}}/K_d$, where K_d is the given phase detector constant. Equation (12) is then rewritten as

$$\beta(t) = \sqrt{c_{contr}} \int_0^t \gamma(t')dt' + \sqrt{c_{vco}} \int_0^t \xi_{vco}(t')dt' - \sqrt{c_{ref}} \int_0^t \xi_{ref}(t')dt' \tag{13}$$

Sampling $\beta(t)$ with a frequency $1/\Delta t$ and after some deviations, the discrete time model follows as

$$\beta[n] = \beta[n-1] + \Delta t \left(\sqrt{c_{contr}}\gamma[n-1] + \sqrt{c_{vco}} w_{vco} - \sqrt{c_{ref}} w_{ref}\right) \tag{14}$$

It can be shown that $\alpha_{pll}(t)$ is asymptotically a zero-mean Gaussian process (Mehrotra, 2002), and the incremental random process $\alpha_{pll}(t) - \alpha_{pll}(t+\tau)$ has the mean

$$\mu(t) = \lim_{t\to\infty} (\alpha_{pll}(t) - \alpha_{pll}(t+\tau)) = 0 \tag{15}$$

and the variance

$$\sigma(t,\tau) = \lim_{t\to\infty} E\left\{(\alpha_{pll}(t) - \alpha_{pll}(t+\tau))^2\right\}$$
$$= c_{ref}|\tau| + 2\sum_{i=1}^{n_o} (\mu_i + \nu_i)\left(1 - e^{-\lambda_i|\tau|}\right) \quad (16)$$

which is independent of t. Therefore, the cumulative phase-noise increment between samples l and k of the received signal, $\Delta\theta_{kl} = 2\pi f_c(\alpha_{pll}(k\Delta t) - \alpha_{pll}(l\Delta t))$, has the characteristic function (Petrovic et al., 2007)

$$E\left\{e^{j\Delta\theta_{kl}}\right\} = e^{-\frac{\omega_0^2}{2}\left(c_{ref}|k-l|\Delta t + 2\sum_{i=1}^{no}(\mu_i+\nu_i)\left(1-e^{-\lambda_i|k-l|\Delta t}\right)\right)} \quad (17)$$

Algorithms for the calculation of PLL phase noise PSD are also given in literature (Mehrotra, 2002).

3. Phase noise in MIMO systems

3.1 MIMO system model

Consider a narrowband MIMO system with N_T transmit antennas and N_R receive antennas, as shown in Fig. 4. The system transmission model is expressed as (Tse & Viswanath, 2004)

$$\mathsf{y} = \mathsf{H}\mathsf{x} + \mathsf{w} \quad (18)$$

where $\mathsf{x} = [x_0, x_1, \ldots, x_{N_T-1}]^T$, $\mathsf{x} \in \mathcal{C}^{N_T}$ and $\mathsf{y} = [y_0, y_1, \ldots, y_{N_R-1}]^T$, $\mathsf{y} \in \mathcal{C}^{N_R}$ represent transmit signal vector and receive signal vector at a symbol time, respectively (the time index is dropped for simplicity). $\mathsf{H} = [h_{ij}]_{N_R\times N_T}$, $\mathsf{H} \in \mathcal{C}^{N_R\times N_T}$ denotes channel matrix, where h_{ij} represents the channel gain from transmitted antenna j to received antenna i, and is an independent and identically distributed complex Gaussian random variable of zaro mean and unity variance, i.e., $h_{ij} \sim \mathcal{CN}(0,1)$. $|h_{ij}|$ is Rayleigh distributed and $|h_{ij}|^2$ is exponentially distributed with a unity mean. $\mathsf{w} = [w_0, w_1, \ldots, w_{N_R-1}]^T$ denotes an N_r-dimensional complex Gaussian random vector with zero mean and variance N_0, i.e. $\mathsf{w} \sim \mathcal{CN}(0, N_0\mathsf{I}_{N_R})$, where I is identity matrix.

3.1.1 Precoding based on singular value decomposition

Assuming the channel is known at the transmitter, the *singular value decomposition* (SVD) of the channel matrix H

$$\mathsf{H} = \mathsf{U}\Lambda\mathsf{V}^H \quad (19)$$

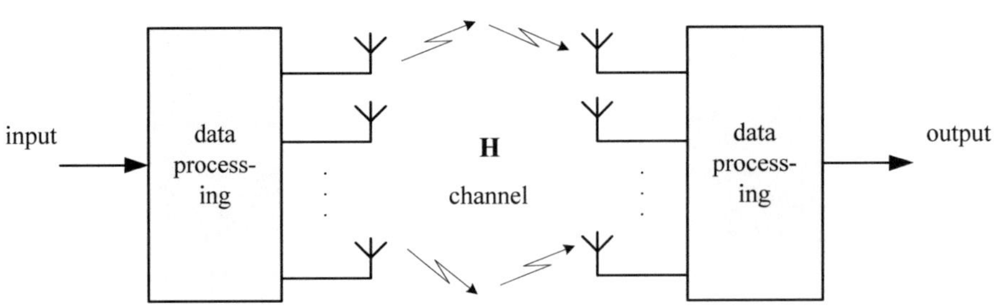

Fig. 4. MIMO system

where

$$\Lambda = \begin{pmatrix} \Lambda_1 & 0 \\ 0 & 0 \end{pmatrix} \quad (20)$$

and the diagonal elements $\Lambda_1 = \text{diag}\{[\lambda_1, \lambda_2, \ldots, \lambda_r]\}$ are the ordered *singular values* of the matrix H, where

$$\lambda_1 \geq \lambda_2 \geq \ldots \geq \lambda_r > 0, \qquad r = \text{rank}(\mathsf{H}) \quad (21)$$

and the rank of H, $r \leq \min(N_T, N_R)$. In (19), U and V are $N_R \times N_R$ and $N_T \times N_T$ unitary matrices, respectively, and satisfy $\mathsf{U}\mathsf{U}^H = \mathsf{I}_{N_R}$ and $\mathsf{V}\mathsf{V}^H = \mathsf{I}_{N_T}$, $\mathsf{U} \in \mathcal{C}^{N_R \times N_R}$, $\mathsf{V} \in \mathcal{C}^{N_T \times N_T}$. Since

$$\mathsf{H}\mathsf{H}^H = \mathsf{U}\Lambda\Lambda^T\mathsf{U}^H \quad (22)$$

The squared singular values λ_i^2 are the eigenvalues of the matrix $\mathsf{H}\mathsf{H}^H$ and also of $\mathsf{H}^H\mathsf{H}$. The SVD can be rewritten as

$$\mathsf{H} = \sum_{i=1}^{r} \lambda_i \mathsf{u}_i \mathsf{v}_i^H \quad (23)$$

i.e., the sum of rank one matrices $\lambda_i \mathsf{u}_i \mathsf{v}_i^H$'s. It can be seen that the rank of H is precisely the number of non zero singular values.
If we define

$$\begin{aligned} \tilde{\mathsf{y}} &:= \mathsf{U}^H\mathsf{y} \\ \tilde{\mathsf{x}} &:= \mathsf{V}^H\mathsf{x} \\ \tilde{\mathsf{w}} &:= \mathsf{U}^H\mathsf{w} \end{aligned} \quad (24)$$

then we can rewrite the channel (18) as

$$\tilde{\mathsf{y}} = \mathsf{U}^H\mathsf{H}\mathsf{V}\tilde{\mathsf{x}} + \tilde{\mathsf{w}} = \Lambda\tilde{\mathsf{x}} + \tilde{\mathsf{w}} \quad (25)$$

where $\tilde{\mathsf{w}} \sim \mathcal{CN}(0, N_0\mathsf{I}_{N_R})$ has the same distribution as w and $\|\tilde{\mathsf{x}}\|^2 = \|\mathsf{x}\|^2$. Thus, the energy is preserved and we have an equivalent representation as a parallel of Gaussian channels (as shown in Fig. 5)

$$\tilde{y}_i = \lambda_i\tilde{x}_i + \tilde{w}_i, \qquad i = 1, 2, \ldots, r. \quad (26)$$

where $x_i \sim \mathcal{CN}(0, \sigma_x^2)$. The data transmitted on such a Gaussian channel construct a *data sub stream*. If the power is equally assigned to all transmit antennas and H is the

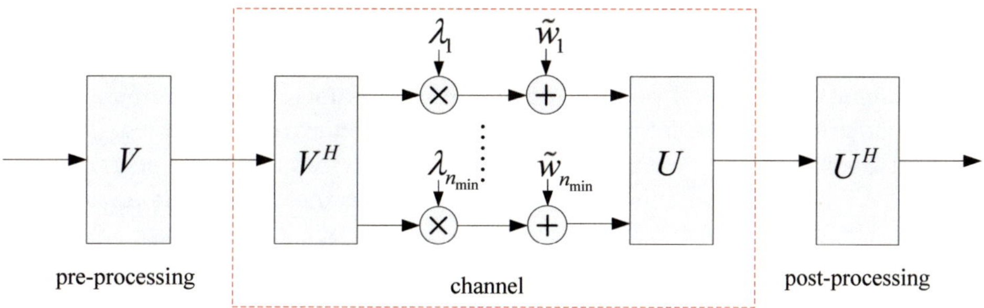

Fig. 5. Singular value decomposition of a MIMO channel

time-invariant deterministic channel, then

$$\sum_{i=1}^{r} \lambda_i^2 = \mathrm{Tr}[\mathrm{HH}^H] = \sum_{i,j} |h_{ij}|^2. \tag{27}$$

If H is Rayleigh fading channel, and $h_{ij} \sim \mathcal{CN}(0,1)$, then

$$E\left[\sum_{i=1}^{r} \lambda_i^2\right] = E\left[\mathrm{Tr}[\mathrm{HH}^H]\right] = \sum_{i,j} E\left[|h_{ij}|^2\right] = N_R \times N_T. \tag{28}$$

It follows that

$$E[\lambda_i^2] = N_R N_T / r. \tag{29}$$

3.1.2 MIMO system with phase noise

In decentralized (distributed) MIMO systems, antennas are placed in different positions in a cell, and can be seen as being mutually independent. Signals received by different antennas are transmitted to a central processing unit via optical fiber. To minimize the cost, the RF unit and the antenna(s) should be placed together. When multiple remote antennas form a virtual cell, RF signals of different antennas have independent clocks. Given a MIMO system using precoding based on SVD, in the following we will investigate the impacts of the dependent and independent clocks on system performance.

Consider the narrowband MIMO system in Fig. 4. When there exists phase noise, we rewrite (25) as

$$\tilde{\mathrm{y}} = \mathrm{U}^H \Theta \mathrm{H} \Phi \mathrm{V} \tilde{\mathrm{x}} + \tilde{\mathrm{w}} \tag{30}$$

where $\Phi = \mathrm{diag}\{[e^{j\phi_1}, e^{j\phi_2}, \ldots, e^{j\phi_{N_T}}]\}$ and $\Theta = \mathrm{diag}\{[e^{j\theta_1}, e^{j\theta_2}, \ldots, e^{j\theta_{N_R}}]\}$. $[\phi_1, \phi_2, \ldots, \phi_{N_T}]$ denote the phase noise related to different transmit antennas, whereas $[\theta_1, \theta_2, \ldots, \theta_{N_R}]$ denote the phase noise related to different receive antennas.

3.2 Phase noise in centralized MIMO systems

In a centralized MIMO system, multiple transmit/receive antennas are placed together with the same transmitter/receiver and so we have $\phi_1 = \phi_2 = \ldots = \phi_{N_T} = \phi$ and $\theta_1 = \theta_2 = \ldots = \theta_{N_R} = \theta$. That is to say, at any instant phase noises at transmitter(s)/receiver(s) are fully correlated, but ϕ and θ are statistically independent. Therefore, (30) can be simplified to be

$$\tilde{\mathrm{y}} = e^{j(\phi+\theta)} \Lambda \tilde{\mathrm{x}} + \tilde{\mathrm{w}} \tag{31}$$

This is equivalent to the case of point-to-point communications: in transmitters, phase noise can result in adjacent-channel interference and modulation errors; in receivers, phase noise can result in demodulation errors and degraded sensitivity and dynamic range, such as the rotation of signal constellation, and further degrade the system bit error rate (BER) performance. The specification, calculation, and reduction of oscillator noise are therefore of great importance in wireless system design (Rohde & Netwkirk, 2000).

3.3 Phase noise in decentralized MIMO systems

3.3.1 Identical transmitter clock, independent receiver clocks

Consider an uplink transmission in a distributed MIMO system, where all transmit antennas share the clock and distributed receive antennas have independent clocks. We have then $\phi_1 = \phi_2 = \ldots = \phi_{N_T} = \phi$ and $\theta_1 \neq \theta_2 \neq \ldots \neq \theta_{N_R}$, i.e., phase noises of different RF units are mutually independent. Equation (30) can be simplified to be

$$\begin{aligned}\tilde{\mathsf{y}} &= e^{j\phi}\mathsf{U}^H\Theta\mathsf{H}\mathsf{V}\tilde{\mathsf{x}} + \tilde{\mathsf{w}} \\ &= e^{j\phi}\mathsf{U}^H\Theta\mathsf{U}\Lambda\tilde{\mathsf{x}} + \tilde{\mathsf{w}}\end{aligned} \tag{32}$$

Precoding can be regarded as a generalization of beamforming to support multi-layer transmission in multi-antenna wireless communications. Independent receiver phase noises not only reduce the signal power, but also result in interference among sub-streams. Assuming $r = N_R \leq N_T$, the received signals of the ith data sub-stream are given by

$$\tilde{y}_i = \lambda_i \tilde{x}_i \sum_{l=1}^{N_R} |u_{li}|^2 e^{j(\theta_l+\phi)} + \sum_{k=1,k\neq i}^{r} \lambda_k \tilde{x}_k \sum_{l=1}^{N_R} u_{li}^* u_{lk} e^{j(\theta_l+\phi)} + \tilde{w}_i, \tag{33}$$

where the first term on the right hand side (RHS) of (33) is the desired signal, the second term is the interference introduced by phase noises. Since matrix U is unitary, $h_{ij} \sim \mathcal{CN}(0,1)$ and h_{ij} is independent of phase noise, it follows that the element $|u_{li}|^2$ is independent and identically distributed random variable, and satisfies

$$\begin{aligned}&\sum_{l=1}^{N_R} |u_{li}|^2 = 1 \\ &\sum_{l=1}^{N_R} u_{lk}^* u_{lj} = 0, \quad \text{for } j \neq k\end{aligned} \tag{34}$$

According to the law of large numbers,

$$N_R \times \frac{1}{N_R} \sum_{l=1}^{N_R} |u_{li}|^2 e^{j(\theta_l+\phi)} \longrightarrow N_R \times E\left[|u_{li}|^2 (\cos(\theta_l+\phi) + j\sin(\theta_l+\phi))\right] \tag{35}$$

Assume θ_l to be symmetrically distributed about 0, then $E[\sin\theta_l] = 0$. Further we have

$$\begin{aligned}E[\cos(\theta_l+\phi)] &= E[\cos\theta_l\cos\phi + \sin\theta_l\sin\phi] = E[\cos\theta_l]\cos\phi \\ E[\sin(\theta_l+\phi)] &= E[\sin\theta_l\cos\phi + \cos\theta_l\sin\phi] = E[\cos\theta_l]\sin\phi\end{aligned} \tag{36}$$

and from (34) we get

$$\sum_{l=1}^{N_R} |u_{li}|^2 = 1 \quad \Longrightarrow \quad \frac{1}{N_R}\sum_{l=1}^{N_R} |u_{li}|^2 = \frac{1}{N_R} \tag{37}$$

when $N_R \longrightarrow \infty$,

$$\frac{1}{N_R}\sum_{l=1}^{N_R} |u_{li}|^2 = \frac{1}{N_R} \quad \Longrightarrow \quad E[|u_{li}|^2] = \frac{1}{N_R} \tag{38}$$

The RHS of (35) is then simplified to be

$$\begin{aligned} & N_R \times E\left[|u_{li}|^2(\cos(\theta_l+\phi)+j\sin(\theta_l+\phi))\right] \\ &= N_R \times E[|u_{li}|^2]E[\cos(\theta_l+\phi)+j\sin(\theta_l+\phi)] \\ &= E[\cos\theta_l]e^{j\phi} \end{aligned} \tag{39}$$

Therefore,

$$\left\| \sum_{l=1}^{N_R} |u_{li}|^2 e^{j(\theta_l+\phi)} \right\|^2 = (E[\cos\theta_l])^2 \tag{40}$$

The power of the desired signal

$$P_{R,i} = \left\| \lambda_i \tilde{x}_i \sum_{l=1}^{N_R} |u_{li}|^2 e^{j(\theta_l+\phi)} \right\|^2 = \lambda_i^2 |\tilde{x}|^2 (E[\cos\theta_l])^2 \tag{41}$$

Consider only the part of $\mathsf{U}^H\Theta\mathsf{U}$ in (32). Denote $\mathsf{C} = \mathsf{U}^H\Theta\mathsf{U}$. Since $\mathsf{U}^H\mathsf{U} = \mathsf{U}\mathsf{U}^H = \mathsf{I}$ and $|e^{\theta_i}| = 1$, the diagonal entries of Θ, $e^{j\theta_1}, e^{j\theta_2}, \ldots, e^{j\theta_{N_R}}$, are eigenvalues of C, and the eigenvalue of $\mathsf{C}\mathsf{C}^H$ is 1. That is to say, on either of parallel Gaussian channels the sum of the interference power to other subchannels and desired signals equals to 1 (the square of λ is not counted).Then the average power of the desired signals in (33) is given by

$$\begin{aligned} E[P_{R,i}] &= E\left[\lambda_i^2 \tilde{x}_i \tilde{x}_i^* \sum_{l=1}^{N_R} |u_{li}|^2 e^{j(\theta_l+\phi)} \sum_{m=1}^{N_R} |u_{mi}|^2 e^{-j(\theta_m+\phi)}\right] \\ &= E[\lambda_i^2]E[|x_i|^2]\left(\frac{1}{N_R} + \frac{N_R(N_R-1)}{2} E\left[|u_{li}|^2|u_{mi}|^2 \cdot 2\Re\left(e^{j(\theta_l-\theta_m)}\right)\right]\right) \\ &= \frac{N_T N_R P}{r^2}\left(\frac{1}{N_R} + \frac{N_R-1}{N_R} E\left[\cos(\theta_l-\theta_m)\right]\right) \\ &= \frac{N_T N_R P}{r^2}\left(\frac{1}{N_R} + \frac{N_R-1}{N_R} E\left[\cos\theta_l\cos\theta_m - \sin\theta_l\sin\theta_m\right]\right) \\ &= \frac{N_T N_R P}{r^2}\left(\frac{1}{N_R} + \frac{N_R-1}{N_R} E[\cos\theta_l]E[\cos\theta_m]\right) \\ &= \frac{N_T P}{r^2}\left(1 + (N_R-1)(E[\cos\theta_l])^2\right) \end{aligned} \tag{42}$$

where $E[|x|^2] = P/r$ is assumed, the average channel gain $E[\lambda_k^2] = E[\lambda_i^2] = N_R N_T / r$, and θ_l is symmetrically distributed about 0. The average interference power from a parallel channel to other channels is given by

$$E[P_{I,i}] = E[\lambda_i^2]E[|x|^2] - E[P_R] = \frac{N_T P}{r^2}(N_R-1)\left(1-(E[\cos\theta_l])^2\right) \tag{43}$$

Define $\gamma_0 = P/N_0$. The signal-to-interference-and-noise-ratio (SINR) is then calculated by

$$SINR_i = \frac{E[P_{R,i}]}{E[P_{I,i}]+N_0} = \frac{1+(N_R-1)(E[\cos\theta_l])^2}{(N_R-1)\left(1-(E[\cos\theta_l])^2\right) + r^2/(N_T\gamma_0)} \tag{44}$$

It is shown that the independent phase noises worsen the received SNR and degraded the overall performance of the communication systems.

3.3.2 Identical receiver clock, independent transmitter clocks

Consider a downlink scenario where one mobile terminal receive signals from multiple cell antennas. Now $\theta_1 = \theta_2 = \ldots = \theta_{N_R} = \theta$, and without the loss of generality, assuming $\phi_1 \neq \phi_2 \neq \ldots \neq \phi_{N_T}$, (30) is simplified to be

$$\begin{aligned}\tilde{\mathsf{y}} &= e^{j\theta}\mathsf{U}^H\mathsf{H}\Phi\mathsf{V}\tilde{\mathsf{x}} + \tilde{\mathsf{w}} \\ &= e^{j\theta}\Lambda\mathsf{V}^H\Phi\mathsf{V}\tilde{\mathsf{x}} + \tilde{\mathsf{w}}\end{aligned} \tag{45}$$

Similarly, the signal received on ith parallel channel is expressed as

$$\tilde{y}_i = \lambda_i\tilde{x}_i\sum_{l=1}^{N_T}|v_{li}|^2 e^{j(\phi_l+\theta)} + \lambda_i\sum_{k=1,k\neq i}^{r}\tilde{x}_k\sum_{l=1}^{N_T}v_{li}^*v_{lk}e^{j(\phi_l+\theta)} + \tilde{w}_i, \tag{46}$$

Also we get

$$\begin{aligned}E[P_{R,i}] &= \frac{E[\lambda_i^2]E[|x|^2]}{N_T}\left(1+(N_T-1)(E[\cos\phi_l])^2\right) \\ &= \frac{N_R P}{r^2}\left(1+(N_T-1)(E[\cos\phi_l])^2\right)\end{aligned} \tag{47}$$

and

$$SINR = \frac{E[P_{R,i}]}{E[P_{I,i}]+N_0} = \frac{1+(N_T-1)(E[\cos\phi_l])^2}{(N_T-1)\left(1-(E[\cos\phi_l])^2\right)+r^2/(N_R\gamma_0)} \tag{48}$$

3.3.3 Independent transmitter/receiver clocks

From the analysis above, we can find out that with the increase of phase noise, the SINR degrades rapidly. Therefore, for a more generalized scenario where RF units of all transmit/receive antennas have independent clocks, only the research on the condition that $\theta << 1$ and $\phi << 1$ is meaningful. Under these assumptions, we have $e^{j\theta_l} \approx 1 + j\theta_l$ and $e^{j\phi_m} \approx 1 + j\phi_m$, and (30) can be simplified to be

$$\tilde{\mathsf{y}} \approx \Lambda\tilde{\mathsf{x}} + j\mathsf{U}^H\bar{\Theta}\mathsf{H}\mathsf{V}\tilde{\mathsf{x}} + j\mathsf{U}^H\mathsf{H}\bar{\Phi}\mathsf{V}\tilde{\mathsf{x}} - \mathsf{U}^H\bar{\Theta}\mathsf{H}\bar{\Phi}\mathsf{V}\tilde{\mathsf{x}} + \tilde{\mathsf{w}} \tag{49}$$

where $\bar{\Phi} = \mathrm{diag}\{[\phi_1,\phi_2,\ldots,\phi_{N_T}]\}$, $\bar{\Theta} = \mathrm{diag}\{[\theta_1,\theta_2,\ldots,\theta_{N_R}]\}$. When $\theta_l \longrightarrow 0$ and $\phi_l \longrightarrow 0$, the fourth term on the RHS of (49) is infinitesimal of higher order and can be ignored. The second term is only related to receiver phase noise and has the autocorrelation matrix

$$\begin{aligned}&E\left[j^2\mathsf{U}^H\bar{\Theta}\mathsf{H}\mathsf{V}\tilde{\mathsf{x}}\tilde{\mathsf{x}}^H\mathsf{V}^H\mathsf{H}^H\bar{\Theta}^T\mathsf{U}\right] \\ &= -\mathsf{U}^H E\left[\bar{\Theta}\mathsf{H}\mathsf{H}^H\bar{\Theta}^T\right]\mathsf{U}\cdot E[|x|^2] \\ &= -\mathsf{U}^H E\left[\bar{\Theta}[\mathsf{h}_{1\cdot}^T,\mathsf{h}_{2\cdot}^T,\ldots,\mathsf{h}_{N_R\cdot}^T]^T[\mathsf{h}_{1\cdot}^H,\mathsf{h}_{2\cdot}^H,\ldots,\mathsf{h}_{N_R\cdot}^H]\bar{\Theta}^T\right]\mathsf{U}\cdot E[|x|^2]\end{aligned} \tag{50}$$

where $E[|x|^2] = P/r$, and the channel matrix is assumed to be known, $\mathsf{H} = [\mathsf{h}_{1\cdot}^T,\mathsf{h}_{2\cdot}^T,\ldots,\mathsf{h}_{N_R\cdot}^T]^T$. Since phase noises are mutually independent, i.e., for $i \neq j$, $E[\theta_i\theta_j^*] = 0$, the further simplification is given by

$$\begin{aligned}&-\mathsf{U}^H E\left[\bar{\Theta}[\mathsf{h}_{1\cdot}^T,\mathsf{h}_{2\cdot}^T,\ldots,\mathsf{h}_{N_R\cdot}^T]^T[\mathsf{h}_{1\cdot}^H,\mathsf{h}_{2\cdot}^H,\ldots,\mathsf{h}_{N_R\cdot}^H]\bar{\Theta}^T\right]\mathsf{U} \\ &= -\mathsf{U}^H\mathrm{diag}\left\{E\left[||\mathsf{h}_{1\cdot}||^2\right]\sigma_{\theta_1}^2, E\left[||\mathsf{h}_{2\cdot}||^2\right]\sigma_{\theta_2}^2,\ldots,E\left[||\mathsf{h}_{N_R\cdot}||^2\right]\sigma_{\theta_{N_R}}^2\right\}\mathsf{U}\end{aligned} \tag{51}$$

where $\sigma_{\theta_l}^2 = E[\theta_l^2]$ is noise variance. Since U is unitary matrix, the noise power introduced by receiver phase noise is given by

$$\mathrm{tr}\left\{\mathsf{U}^H \mathrm{diag}\left\{E\left[||\mathsf{h}_{1\cdot}||^2\right]\sigma_{\theta_1}^2, E\left[||\mathsf{h}_{2\cdot}||^2\right]\sigma_{\theta_2}^2, \ldots, E\left[||\mathsf{h}_{N_R\cdot}||^2\right]\sigma_{\theta_{N_R}}^2\right\}\mathsf{U}\right\}\cdot E[|x|^2] = \frac{P}{r}\sum_{l=1}^{N_R} E\left[||\mathsf{h}_l||^2\right]\sigma_{\theta_l}^2 \tag{52}$$

If the channel is independent and identically distributed and Rayleigh faded, and $h_{ij} \sim \mathcal{CN}(0,1)$, then $E\left[|h_{ij}|^2\right] = 1$, and (52) is rewritten as

$$\mathrm{tr}\left\{\mathsf{U}^H \mathrm{diag}\left\{E\left[||\mathsf{h}_{1\cdot}||^2\right]\sigma_{\theta_1}^2, E\left[||\mathsf{h}_{2\cdot}||^2\right]\sigma_{\theta_2}^2, \ldots, E\left[||\mathsf{h}_{N_R\cdot}||^2\right]\sigma_{\theta_{N_R}}^2\right\}\mathsf{U}\right\}\cdot E[|x|^2] = \frac{N_T P}{r}\sum_{l=1}^{N_R}\sigma_{\theta_l}^2 \tag{53}$$

The third term in (49) related only to the transmitter phase noise, whose autocorrelation matrix is calculated by

$$\begin{aligned} &E[j^2\mathsf{U}^H\mathsf{H}\bar{\Phi}\mathsf{V}\tilde{\mathsf{x}}\tilde{\mathsf{x}}^H\mathsf{V}^H\bar{\Phi}^T\mathsf{H}^H\mathsf{U}] \\ &= -\mathsf{U}^H E[\mathsf{H}\bar{\Phi}\bar{\Phi}^T\mathsf{H}^H]\mathsf{U}\cdot E[|x|^2] \\ &= -\mathsf{U}^H E[[\mathsf{h}_{\cdot 1}\phi_1^2, \mathsf{h}_{\cdot 2}\phi_2^2, \ldots, \mathsf{h}_{\cdot N_T}\phi_{N_T}^2][\mathsf{h}_{\cdot 1}, \mathsf{h}_{\cdot 2}, \ldots, \mathsf{h}_{\cdot N_T}]^H]\mathsf{U}\cdot E[|x|^2] \end{aligned} \tag{54}$$

For Rayleigh fading channels

$$\mathrm{tr}\left\{\mathsf{U}^H E[[\mathsf{h}_{\cdot 1}\phi_1^2, \mathsf{h}_{\cdot 2}\phi_2^2, \ldots, \mathsf{h}_{\cdot \mathsf{N}_T}\phi_{N_T}^2][\mathsf{h}_{\cdot 1}, \mathsf{h}_{\cdot 2}, \ldots, \mathsf{h}_{\cdot \mathsf{N}_T}]^H]\mathsf{U}\right\}\cdot E[|x|^2] = N_R\sum_{l=1}^{N_T}\sigma_{\phi_l}^2\cdot E[|x|^2] \tag{55}$$

Then the average SINR on the receiving side

$$\begin{aligned} SINR &= \frac{E[P_R]}{E[P_I] + N_0} \\ &\approx \frac{\sum_{i=1}^r E[\lambda_i^2] - N_T\sum_{l=1}^{N_R}\sigma_{\theta_l}^2 - N_R\sum_{l=1}^{N_T}\sigma_{\phi_l}^2}{N_T\sum_{l=1}^{N_R}\sigma_{\theta_l}^2 + N_R\sum_{l=1}^{N_T}\sigma_{\phi_l}^2 + N_0/E[|x|^2]} \\ &= \frac{1 - \frac{1}{N_R}\sum_{l=1}^{N_R}\sigma_{\theta_l}^2 - \frac{1}{N_T}\sum_{l=1}^{N_T}\sigma_{\phi_l}^2}{\frac{1}{N_R}\sum_{l=1}^{N_R}\sigma_{\theta_l}^2 + \frac{1}{N_T}\sum_{l=1}^{N_T}\sigma_{\phi_l}^2 + N_0/(E[|x|^2]N_R N_T)} \\ &\approx \frac{1 - E[\sigma_{\theta_l}^2] - E[\sigma_{\phi_l}^2]}{E[\sigma_{\theta_l}^2] + E[\sigma_{\phi_l}^2] + r/(\gamma_0 N_R N_T)} \end{aligned} \tag{56}$$

where $\gamma_0 = P/N_0$. It follows that the average SINR is subject to the average of the variance of phase noises at both transmitters and receivers.

3.4 Discuss on synchronization

The suppression of phase noise should be done by two ways: on one hand, making the noise as low as possible during system design; on the other hand, suppressing and eliminating the effects of phase noise by means of algorithms.

The MIMO system we studied here is a narrowband system. When receive beamforming is used, the RF signal from the transmitter can be used as a reference signal in order to realize the synchronization in frequency and phase; whereas when only transmit beamforming is used, frequency synchronization can be realized by synchronizing all transmitter carrier frequencies with receiver, or using an extra transmitter as reference frequency. In the latter case, a time-division periodical synchronization scheme is required in order to correct the phase error induced by propagation. However, phase noise during the signal transmission can not be eliminated, especially when joint transmit/receive beamforming (linear precoding) is applied. Therefore, minimizing the noise during circuit design is a better way to solve this problem.

In a distributed MIMO system with independent transmitter and receiver clocks, a common independent RF reference signal is necessary. The meaningful phase noise estimation and suppression can only be made for broadband MIMO systems.

4. Simulation results

Referring to (6), the variance of Wiener phase noise will increase linearly with time. Assuming phase noises are independent and identically distributed, carrier frequency $f_c = 2$GHz, and the constant of an open-loop VCO $c_{vco} = 10^{-19}$, the variation of phase noise variance σ_θ^2 with time is shown in Fig. 6 for a free-running oscillator. At $t = 0.5$ms, $\sigma_\theta^2 \approx 0.01$, and if the average of $E[\sigma_{\theta_l}^2] \approx 0.01$, the corresponding SINR will be less than 20dB. It should be emphasized that different from application of single-layer beamforming (for point to point communication), application of precoding demands on electronic devices with a better noise characteristics. It is because independent phase noises result in not only the reduction of signal's active power, but also the interference among parallel channels.

Fig. 7 describes the phase noise characteristics of a PLL synthesizer. Within a time duration less than 100ms, phase noise at the PLL output is determined by $\beta(t)$, which is one component of the variance bounded multi-dimension Ornstein-Uhlenbeck process. Therefore, the variance of $\alpha_p ll(t)$ approximates a constant. In a relative large observation time range, $\alpha_p ll(t)$ will depend on the phase noise of the reference signal $\alpha_p ll(t)$. Furthermore, with a VCO of the same quality, PLL output signal has a much better noise characteristic, and better performance can be obtained by the system.

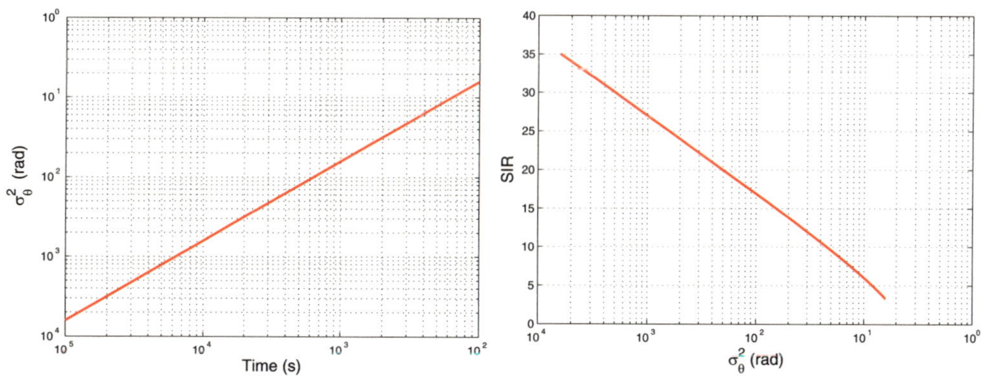

(a) Wiener noise: noise power increase linearly with time

(b) Wiener noise: SINR decreases with the increased noise power

Fig. 6. Simulation for a free-running oscillator

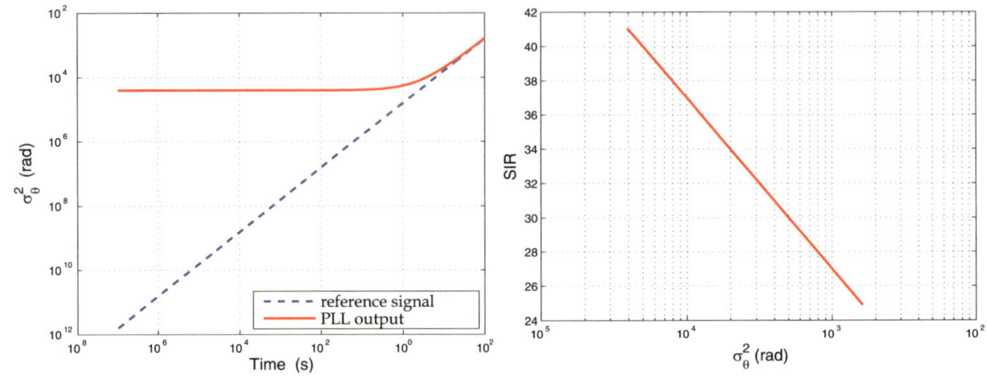

(a) PLL phase noise: noise power varies with time, for $c_{ref} = 10^{-25}$, $c_{vco} = 10^{-19}$

(b) PLL phase noise: SINR decreases with the increased noise power

Fig. 7. Simulation for a PLL synthesizer

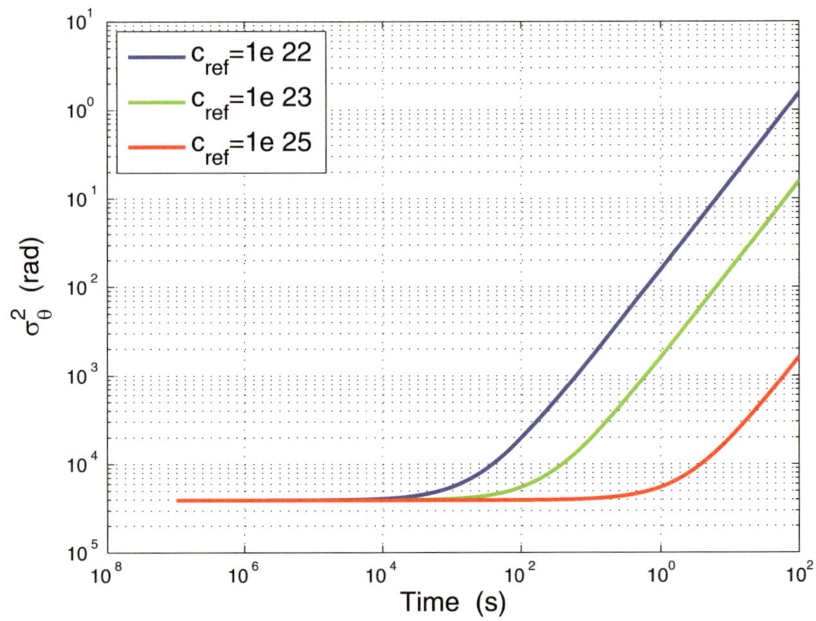

Fig. 8. Comparison of phase noise variance of PLL with difference reference crystal

Fig. 8 compares the phase noise variance of PLL synthesizers using reference clocks of different quality. Obviously, the better the quality of the clock, the better the stability of the PLL is.

In the literature (Mudumbai et al., 2007) the influence of independent phase noises on a distributed wireless sensor network using transmit beamforming has been studied. Application of transmit/receive beamforming can bring the potential power gain, but requires the carrier synchronization in phase and frequency. In addition, if single-layer beamforming is used(transmit/receive beamforming), independent phase noises at transmitter/receiver will reduce the power gain, but no interference is introduced; whereas with precoding (joint transmit/receive beamforming), extra interference exits among sub-streams and further worsens the receive SNR.

5. Conclusions and future research

In this chapter, we have investigated the effects of phase noise in centralized and distributed narrowband MIMO systems, and discussed the feasibility of phase and frequency synchronization. In centralized narrowband MIMO systems, the effects of phase noise is similar to that in simple point-to-point communications, except that phase noise may affect the accuracy of channel estimation and thus the channel capacity is overestimated (Taparugssanagorn & Ylitalo, 2009).

The simulation results have shown that in distributed MIMO systems, independent phase noises not only give rise to the reduction of power gain provided by transmit/receive beamforming, but also induce the interference among parallel channels provided by precoding. The signal-to-interference ratio decreases rapidly when the average interference power increases. Therefore, in decentralized MIMO systems more precise devices are demanded, which contradicts the low cost we expected. Furthermore, phase synchronization in a distributed narrowband system raises the overhead cost. Even so, during data transmission phase noise can not be corrected.

For a broadband system, overhead for phase noise estimation and mitigation is also a challenge. It is a hot topic at the moment as well as in the future.

6. References

Bittner, S., Rave, W. & Fettweis, G. (2007). Phase noise suppression in ofdm with spatial multiplexing, *VTC-2007 Spring*, pp. 1826–1830.

Demir, A. (1998). Phase noise in oscillators:dae's and colored noise sources, *Proc. IEEE/ACM Int. Conf. CAD*, pp. 170–177.

Demir, A., Mehrotra, A. & Roychowdhury, J. (2000). Phase noise in oscillators: A unifying theory and numerical methods for characterisation, *IEEE Trans. on Circuits Syst.* 47(5): 655–674.

Kim, Y. W. & Yu, J. D. (2008). Phase noise model of single loop frequency synthesizer, *IEEE Trans. on Broadcasting* 54(1): 112–119.

Leeson, D. B. (1966). A simple model of feedback oscillator noise spectrum, *Proc. of the IEEE* 54(2): 329–330.

Liu, P., Wu, S. & Bar-Ness, Y. (2006). A phase noise mitigation scheme for mimo wlanss with spatially correlated and imperfectly estimated channels, *IEEE Commun. Letters* 10(3): 141–143.

Mehrotra, A. (1999). *Simulation and modeling techniques for noise in radio frequency integrated circuits*, PhD thesis, Univ. California, Berkeley.

Mehrotra, A. (2002). Noise analysis of phase-locked loops, *IEEE Trans. on Circuits Syst.* 49(9): 1309–1316.

Mudumbai, R., Barriac, G. & Madhow, U. (2007). On the feasibility of distributed beamforming in wireless networks, *IEEE Trans. on Wireless Commun.* 6(4): 1754 – 1763.

Petrovic, A., Wolfgang, R. & Fettweis, G. (2007). Effects of phase noise on ofdm systems with and without pll: characterization and compensation, *IEEE Trans. on Commun.* 55(8): 1607–1616.

Robins, W. P. (1982). *Phase Noise in Signal Sources (Theory and Applications)*, Peter Peregrinus Ltd., London.

Rohde, U. L. & Netwkirk (2000). *RF/Microwave Circuit Design for Wireless Applications*, John Wiley & Sons. Inc.

Schenk, T. C. W., Tao, X.-J., Smulders, P. F. M. & Fledderus, E. R. (2004). Influence and suppression of phase noise in multi-antenna ofdm, *VTC-2004 Fall*, Vol. 2, pp. 1443–1447.

Taparugssanagorn, A. & Ylitalo, J. (2009). Characteristics of short-term phase noise of mimo channel sounding and its effect on capacity estimation, *IEEE Trans. on Instru. & measul.* 58(1): 196–201.

Tse, D. & Viswanath, P. (2004). *Fundamentals of Wireless Communication*, Cambridge University Press.

Zhang, W. (2010). *Phase Noise in MIMO systems*, PhD thesis, Tsinghua University, Beijing.

15

A List Based Detection Technique for MIMO Systems

SaiRamesh Nammi
Ericsson Inc.

1. Introduction

It is well known that multiple transmit and receive antennas can significantly increase the data carrying capacity of wireless systems (1) - (3). Usually a multiple-input multiple-output (MIMO) system requires a channel code to approach capacity (4). However, for such MIMO systems, the optimal joint maximum-likelihood (ML) detection using exhaustive search is too complex and is almost impossible. Motivated by turbo decoding (5), iterative detectors and decoders, treating the channel code as the outer code and the space time mapper as the inner code, have been commonly employed in the literature. Although the iterative decoder structures are extensively investigated in the code theory literature, the design and role of the detector structure still need investigation. In this context, the authors in (6), (9) studied the iterative algorithm for MIMO systems using optimum soft-input/soft-output ML MIMO detector. Unfortunately the optimal MIMO detector complexity increases exponentially with the number of transmit antennas or/and the number of bits per constellation point.

In this chapter, we propose a novel threshold list subset detector (T-LSS) that extends the List subset detector for iterative turbo MIMO systems (20). List based techniques have also been considered by other researchers in various contexts. In (15), a list-sequential detector, based on a modified stack algorithm, is discussed for MIMO systems. Detection over multiple input multiple output channels for uncoded system using Chase (25) type of algorithm has been proposed in (17), and for multiuser detection in spread spectrum systems in (28). The T-LSS detector generates lists of candidate codewords based on the subset sum algorithm (22). The T-LSS detector do not require any search radius as in LSD rather depends on the threshold value. Second, the lists of codewords are generated at each iteration of the receiver by taking into account the *a priori* information fed back from the channel decoder. Third, the candidate codeword selection in T-LSS does not directly involve channel estimation as in LSD. Our results show that for small average list sizes, the performance of T-LSS is almost same as full *A Posteriori Probability* (APP) detection. Moreover, the average complexity of T-LSS is relatively small. Using Extrinsic Information Transfer (EXIT) chart analysis, we can design the optimal list size per iteration. Our results indicate that we can reduce the complexity significantly for an iterative MIMO system by using T-LSS in the first iteration and using very low list sizes optimized using EXIT chart.

The rest of the chapter is organized as follows. Section II describes the system model. Section III describes the T-LSS detector structure in the iterative framework. Performance analysis of T-LSS is given in Section IV and Section V discusses the implementation complexity. Selection

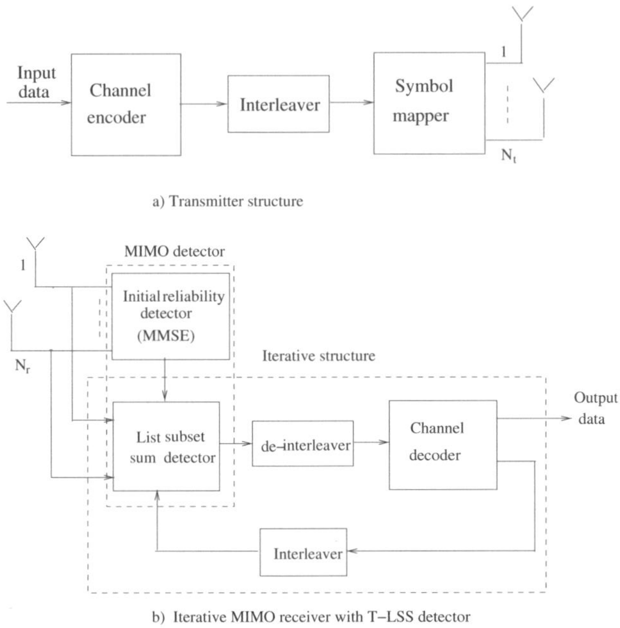

Fig. 1. Block diagram of turbo MIMO system.

of list sizes using EXIT charts is described in Section VI. Section VII discusses numerical results of T-LSS. Finally, conclusions are presented in Section VIII.
Notation: Bold upper case (lower case) letters denote matrices (vectors). The notations $(\cdot)^T$, $(\cdot)^H$, $(\cdot)^*$, $E[\cdot]$, $\text{Re}\{\cdot\}$ and $\text{Im}\{\cdot\}$ denote transpose, Hermitian, conjugation, expectation, real and imaginary parts respectively. A^c is the complement of A, $||\cdot||$ is Euclidian norm, and $\binom{m}{k}$ denotes $m!/(m-k)!k!$.

2. System description

Consider a MIMO system with N_t transmit antennas and N_r receive antennas. At the transmitter side, the input data bits are encoded by a channel code, randomly interleaved and then mapped to modulation symbols before being transmitted by the antennas. Denoting a block of information bits by the vector $\mathbf{d}$ and the transfer function of the channel code by $\mathbf{G}$, the output codeword can be written as $\tilde{\mathbf{c}} = \mathbf{G}\mathbf{d}$, and $\mathbf{c} = \Pi(\tilde{\mathbf{c}})$ represents the interleaved sequence of coded bits. The symbol mapper converts this bit sequence into an M-ary symbol sequence as follows. Assume that $\mathbf{c}$ has LN_tM_c elements, where M_c is the number of bits per constellation point and L is a non-negative integer. The symbol mapper first partitions $\mathbf{c}$ into L subvectors, each of length N_tM_c, to form an $N_tM_c \times L$ matrix, $\mathbf{C} = [\mathbf{c}^1, \mathbf{c}^2 ..., \mathbf{c}^L]$, where $\mathbf{c}^l = [c_{(l-1)N_tM_c+1}, c_{(l-1)N_tM_c+2}, ..., c_{lN_tM_c}]^T$, $1 \leq l \leq L$, and c_i is the i-th element of $\mathbf{c}$. Each subvector $\mathbf{c}^l$ is then mapped to the symbol vector $\mathbf{x}^l = [x_1^l, x_2^l, ..., x_{N_t}^l]^T$ through a unique predetermined bit mapping scheme $x_i^l = F[c_{(l-1)N_tM_c+(i-1)M_c+1}, ..., c_{(l-1)N_tM_c+iM_c}]$, $1 \leq i \leq N_t$. We call each sub vector $\mathbf{c}^l$ a code bit vector. The elements of the symbol vector $\mathbf{x}^l$ are then transmitted by the N_t transmit antennas.

Ignoring the superscript (l) for simplicity, the received $N_r \times 1$ vector $\mathbf{r}$ due to the transmission of the $N_t \times 1$ symbol vector $\mathbf{x}^l$ can be written as

$$\mathbf{r} = \mathbf{H}\mathbf{x} + \mathbf{n} \tag{1}$$

where $\mathbf{H}$ is a $N_r \times N_t$ complex MIMO channel matrix, and $\mathbf{n}$ is complex additive noise vector whose elements are independent, complex-valued Gaussian variables with zero mean and variance σ^2 per each dimension.

3. Soft MIMO detectors

3.1 MIMO detector background

In the literature, different optimal and suboptimal structures have been studied. The suboptimal detectors can be classified into linear and nonlinear detectors. Linear detectors include zero-forcing (ZF) and minimum mean-square error (MMSE) detectors, and the nonlinear receivers include decision feedback, nulling-cancelling and variants relying on successive interference cancellation. These suboptimal detectors are easy to implement but their bit error rate (BER) performance is significantly inferior to that of the optimum MIMO detector (10). The authors in (12), (13) studied nulling-cancelling detector with iterative processing.

There has also been considerable progress in using low complexity (near) ML detection based on lattice decoding. In (19), a near optimal detection method is proposed using sphere-decoder (SD) concept, which has low complexity at high SNR. The work in (4) extends the sphere detector algorithm of (19) to coded MIMO systems with complex constellations. It is shown that for good choices of the initial radius and for sufficient list sizes, list sphere decoder (LSD) in a concatenated system can achieve a bit error rate (BER) performance close to capacity limits. However, the worst case complexity of LSD can be very high and the expected complexity is polynomial with the number of transmit antennas (4). Besides, the LSD does not exploit channel decoder information in generating the list of candidate codewords. Based on semi-definite relaxation (SDR) of the ML problem, the authors in (14) proposed a soft Quasi ML detection for MIMO systems.

3.2 Conventional soft MIMO detector

A typical iterative receiver structure consists of an inner MIMO detector and an outer channel decoder. The MIMO detector calculates the log likelihood ratio (LLR) $L_D(c_i|\mathbf{r}) = \log(P(c_i = 1|\mathbf{r})/P(c_i = 0|\mathbf{r}))$, for each bit $c_i, 1 \le i \le N_t M_c$, and sends the extrinsic LLR values $L_{\text{ext}}(c_i|\mathbf{r}) = L_D(c_i|\mathbf{r}) - L_A(c_i)$ to the channel decoder, where $L_A(c_i) = \log(P(c_i = 1)/P(c_i = 0))$ is the *a priori* information. A direct computation of detector extrinsic LLR is performed as (9)

$$L_{\text{ext}}(c_i|\mathbf{r}) = \log \frac{\sum\limits_{\mathbf{c}^l \in \text{Ł} \bigcap c_{i,1}} p\,(\mathbf{r}|\mathbf{c}^l).\exp\,(\frac{1}{2}{\mathbf{c}^l}^T_{[i]}\mathbf{L}_{A,[i]})}{\sum\limits_{\mathbf{c}^l \in \text{Ł} \bigcap c_{i,0}} p\,(\mathbf{r}|\mathbf{c}^l).\exp\,(\frac{1}{2}{\mathbf{c}^l}^T_{[i]}\mathbf{L}_{A,[i]})} \tag{2}$$

where Ł is a set containing all possible code bit vectors and is called the list. The set Ł $\bigcap c_{i,1}$ contains code bit vectors in the list with $c_i = 1$, $(l-1)N_t M_c + 1 \le i \le l N_t M_c$, $\mathbf{L}_{A,[i]}$ denotes the vector of all *a priori* values by omitting the bit i, ${\mathbf{c}^l}^T_{[i]}$ denotes the coded bit vector $\mathbf{c}^l$ by removing the i-th bit, and $p\,(\mathbf{r}|\mathbf{c}^l)$ is the conditional probability density function (pdf).

From (2), it is clear that an exhaustive search over all possible bit vectors requires computation of (2) over a list Ł containing $2^{N_t M_c}$ candidate code bit vectors. This provides MAP detection but its complexity is too high for large $N_t M_c$. In order to reduce complexity, the authors in (4) extend the SD concept for coded turbo MIMO systems by generating a list of code bit vectors from the detector to the decoder. Although the complexity is reduced, the list generation process in (4) does not utilize the information fed back from the channel decoder. Accordingly, the work in (34) modifies the SD algorithm with a Finkle-Pohst MAP search, which is repeated for each iteration. In general, the SD based approach relies on a search radius that sometimes involves trial and error (4). Furthermore, many floating point operations, including QR decompositions or Cholesky decompositions, are required in generating the list of candidate code bit vectors. Therefore, we propose a different soft MIMO detector structure that generates lists requiring only a few floating point operations and mostly comparisons. Besides, instead of using a search radius, we use alternative parameters which are easier to characterize.

3.3 Proposed soft MIMO detector

In our novel receiver structure, we reduce the list size significantly so that the computation of (2) is performed only for a few candidate bit words. Toward this end, we propose to optimize the list from iteration to iteration. Thus, for the k-th iteration between the detector and the decoder, $L_{\text{ext}}^{(k)}(c_i|\mathbf{r})$ is computed as

$$L_{\text{ext}}^{(k)}(c_i|\mathbf{r}) = \log \frac{\sum\limits_{\mathbf{c}^l \in Ł^{(k)} \bigcap c_{i,1}} p\,(\mathbf{r}|\mathbf{c}^l).\exp\,(\frac{1}{2}\mathbf{c}^{l\,T}_{[i]}\mathbf{L}_{A,[i]})}{\sum\limits_{\mathbf{c}^l \in Ł^{(k)} \bigcap c_{i,0}} p\,(\mathbf{r}|\mathbf{c}^l).\exp\,(\frac{1}{2}\mathbf{c}^{l\,T}_{[i]}\mathbf{L}_{A,[i]})} \tag{3}$$

where $Ł^{(k)}$ is the list of code bit vectors in the k-th iteration. Note that unlike 2, $Ł^{(k)}$ does not usually contain all possible code bit vectors. The detector constructs the list based on the LLR information $\{\alpha_i\}$ for each bit. In the first iteration, we use an initial reliability detector to generate $\{\alpha_i\}$. For subsequent iterations, $\{\alpha_i\}$ are obtained from the channel decoder. Thus, our receiver consists of three components : (1) an initial reliability detector, (2) a list subset sum detector and (3) a channel decoder as shown in Fig. 1. The initial reliability detector and the list subset detector form the proposed MIMO detector.

3.3.1 Initial reliability detector

The initial reliability detector is used only during the first iteration to assist in the generation of the list $Ł^{(1)}$. It consists of a linear minimum mean squared error (MMSE) detector. Such detectors are known to have information lossless property (24). Other detectors namely zero forcing (ZF), zero forcing with successive interference cancellation (ZF-SC), minimum mean squared error with successive interference cancellation (MMSE-SC) can also be used to obtain the initial reliability of bits. Let $\mathbf{y} = \mathbf{P}^H\mathbf{r}$ be the $N_t \times 1$ vector output of MMSE detector, where $\mathbf{P} = (\mathbf{H}\mathbf{R}_x\mathbf{H}^H + \sigma^2\mathbf{I})^{-1}\mathbf{H}$ is the MMSE detector coefficients (10), $\mathbf{R}_x = E[\mathbf{x}\mathbf{x}^H]$, and $\mathbf{I}$ is the identity matrix of size $N_r \times N_r$. Assuming MMSE output as Gaussian (21), the MMSE output is used to obtain the initial reliability of the $N_t M_c$ bits. The reliability of the coded bit c_i is measured by the magnitude of the log-likelihood ratio, $\log(P(c_i = 1|y_t)/P(c_i = 0|y_t))$, where $t = \lceil i/M_c \rceil$. As an example, for quadrature phase-shift keying (QPSK) with constellation points $(1+j)/\sqrt{2}$, $(-1+j)/\sqrt{2}$, $(-1-j)/\sqrt{2}$, $(1-j)/\sqrt{2}$ representing bit pairs (0,0), (0,1), (1,1), and (1,0) respectively, $\mathrm{Im}\{y_t\}$ and $\mathrm{Re}\{y_t\}$ give the reliabilities of the first and the second bit respectively (32). Let us denote the reliability of the coded bits by $\{\alpha_i\}$ and let α_i^h be the hard decision of α_i.

3.3.2 List subset sum detector

The list subset sum detector takes reliability values $\{\alpha_i\}$ as input soft information to produce a list $Ł^{(k)}$ of most likely code bit vectors. This list is used to produce extrinsic LLR $L_{\text{ext}}(c_i|\mathbf{r})$ for each bit c_i. It consists of two parts : a list generator and an extrinsic LLR calculator.

List generator: The motivation of list based algorithms is that flipping of α_i^h at a few bit positions may produce the MAP code bit vector in $Ł^{(k)}$ of 3 with high probability. We use the MAP metric of the transmitted vector $\mathbf{c}$, denoted by $J(\mathbf{c})$ as (27)

$$J(\mathbf{c}) = \sum_{i \in S_c} |\alpha_i| - \sum_{i \in S_c^n} |\alpha_i| = \sum_{i \in U} |\alpha_i| - 2 \sum_{i \in S_c^n} |\alpha_i| \tag{4}$$

where $U = \{1, 2, ..., N_t M_c\}$ is the set of bit positions, $S_c = \{i | c_i = \alpha_i^h, i \in U\}$ and $S_c^n = \{i | c_i \neq \alpha_i^h, i \in U\}$. Since the first term in (4) is independent of the value of $\mathbf{c}$, $J(\mathbf{c})$ depends only on the second term $\sum_{i \in S_c^n} |\alpha_i|$. Hence we need to find set of vectors which minimizes the second term. i.e. the perturbation vectors should be designed in ascending order of the sum $S = \sum_{i \in S_c^n} |\alpha_i|$. That is the problem is equivalent to the subset-sum problem (22), where the objective is to find all subsets of a set of numbers that have sum less than or equal to a given number V_{th}. Let $Ł^{(k)}$ be the set of all perturbation vectors whose reliability sum S is below a certain threshold V_{th}, and $\gamma^{(k)}$ be the cardinality of $Ł^{(k)}$. The choice of V_{th} will be discussed in the next section.

Let us define a vector $\mathbf{p}_i$ as a zero element vector with a 1 at the i-th bit position. The weight of the vector $\{\alpha_i\}$ is defined as $w = \sum_{i=1}^{N_t M_c} |\alpha_i|$. The list is generated as follows :

Step 1: Sort $|\alpha_j|, j = 1, ..., N_t M_c$ in the ascending order. Set the size of the list $\Lambda = 1$, bit position $i = 1$, first term in the list $\mathbf{v}_{\text{old}}^1 = [0\,0\,0...0]$ and the weight $w_{\text{old}}^1 = 0$.
Step 2: For each vector in the list $m = 1$ to Λ, create $\mathbf{v}_{\text{new}}^m = \mathbf{v}_{\text{old}}^m + \mathbf{p}_i$, and modify the weight of each vector as $w_{\text{new}}^m = w_{\text{old}}^m + |\alpha_i|$.
Step 3: Merge lists $\mathbf{v}_{\text{new}}$ and $\mathbf{v}_{\text{old}}$ to form a new list $\mathbf{v}_{\text{new}}$.
Step 4: If $k = 1$ i.e., in the first iteration between the detector and the decoder, keep only the vectors in $\mathbf{v}_{\text{new}}$ that have a sum less than or equal to V_{th}.
Step 5: If $k > 1$, then sort vectors in $\mathbf{v}_{\text{new}}$ in ascending order according to their weights w_{new}. Truncate the size of $\mathbf{v}_{\text{new}}$ to $\Lambda^{(k)}$, if the size is greater than $\Lambda^{(k)}$, where $\Lambda^{(k)}$ is the maximum size of the list, and set the value of $\Lambda^{(k)}$ to the size of $\mathbf{v}_{\text{new}}$.
Step 6: Set $\mathbf{v}_{\text{old}} \leftarrow \mathbf{v}_{\text{new}}$.
Step 7: Set $i \leftarrow i + 1$. Repeat Step 2 to 7 if $i \leq N_t M_c$.
Step 8: Re-arrange the positions of the elements in each test vector in $\mathbf{v}_{\text{new}}$ back to their original positions before Step 1, and then add each vector to $\boldsymbol{\alpha}^h$ to generate the test list $Ł^{(k)}$.

Extrinsic LLR calculator : Once the list $Ł^{(k)}$ is available, the extrinsic LLR values $L_{\text{ext}}^{(k)}(c_i|\mathbf{r})$ are computed for each coded bit c_i. For higher number of transmit antennas, computation of (3) is very cumbersome, and max-log approximation is used as (4)

$$L_{\text{ext}}^{(k)}(c_i|\mathbf{r}) \approx \frac{1}{2} \max_{\mathbf{c}^l \in \text{Ł}^{(k)} \bigcap c_{i,+1}} \{-\frac{1}{\sigma^2}||\mathbf{r} - \mathbf{Hx}||^2 + \mathbf{c}^{l\,T}_{[i]}\mathbf{L}_{A,[i]}\}$$
$$-\frac{1}{2} \max_{\mathbf{c}^l \in \text{Ł}^{(k)} \bigcap c_{i,0}} \{-\frac{1}{\sigma^2}||\mathbf{r} - \mathbf{Hx}||^2 + \mathbf{c}^{l\,T}_{[i]}\mathbf{L}_{A,[i]}\} \quad (5)$$

After all the code bit vectors $\mathbf{c}^l$, $1 \leq l \leq L$, have been processed by the MIMO detector, the extrinsic LLR values associated with $\mathbf{c}$ are made available to the channel decoder as *a priori* information after de-interleaving:

$$L^A(\tilde{\mathbf{c}}) = \Pi^{-1}\{L_{\text{ext}}(\mathbf{c})\} \quad (6)$$

3.4 Channel decoder

The decoder considered in this chapter is a soft-input soft-output (SISO) convolutional decoder based on the BCJR algorithm (30). This decoder is based on trellis structure, where a branch metric computation uses the soft information $L^A(\tilde{\mathbf{c}})$ coming from the detector. Based on forward and reverse recursions of BCJR, the LLR for each coded and data bit is obtained. The *a priori* information $L^A(\tilde{c}_i)$ is removed from the LLR of each coded bit to produce extrinsic information $L^{ext}(\tilde{c}_i)$. Finally, the extrinsic information is fed back to the detector as *a priori* information after interleaving as

$$L_A(\mathbf{c}) = \Pi\{L^{\text{ext}}(\tilde{\mathbf{c}})\} \quad (7)$$

The *a priori* information $L_A c_i$ for each bit c_i is used by the MIMO detector as described in Section III A. The iteration between the detector and the decoder are repeated to decrease the BER. After the last iteration, the LLR of data bits are subjected to hard decision to produce final data decisions.

4. BER analysis of MIMO detector

In this section, we consider BER analysis of the MIMO detector. It is useful for two reasons. First, it directly provides analytical performance results of the proposed detector when used in an uncoded MIMO system. Second, we use this analysis to show selection of V_{th} used in the list generation algorithm. We follow a procedure similar to (32). However, our final results are different from (32).
We consider QPSK modulation. The probability of an error e in bit c_i can be written as

$$P_{b,i} = \sum_{u=0}^{1} P(e|c_i = u)P(c_i = u) \quad (8)$$

Assume bits 0 and 1 to be equally likely and $P(e|c_i = u)$ to be same for u = 0 and 1. We can then write $P_{b,i} = P(e|c_i = 0)$. When the MIMO detector uses V_{th} to control the list, we can write

$$P_{b,i} = P(e|c_i = 0, \alpha_i \leq V_{\text{th}})P(\alpha_i \leq V_{\text{th}}|c_i = 0)$$
$$+P(e|c_i = 0, \alpha_i > V_{\text{th}})P(\alpha_i > V_{\text{th}}|c_i = 0) \quad (9)$$

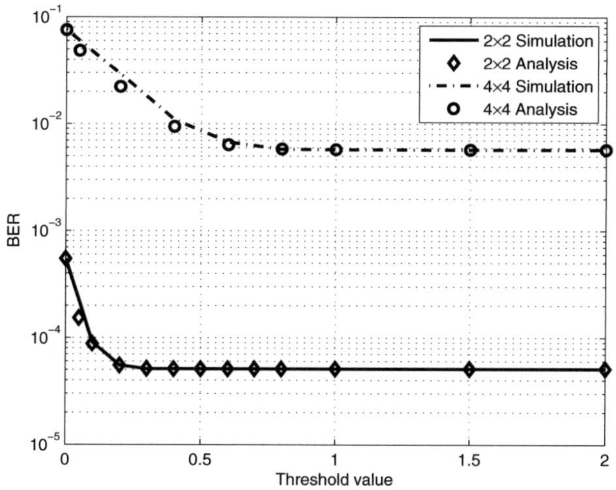

Fig. 2. Effects of V_{th} on the BER of T-LSS based MIMO system.

Let us denote $P(\alpha_i > V_{\text{th}}|c_i = 0)$ as $P_{\delta,i}$. When $\alpha_i > V_{\text{th}}$, the decision on the i-th bit is 1. Hence $P(e|c_i = 0, \alpha_i > V_{\text{th}})$ =1. We approximate $p(\alpha_i|c_i = 0)$ to be Gaussian with mean μ_i and variance σ_i^2. Thus, we get, $P_{\delta,i} = Q((V_{\text{th}} + |\mu_i|)/\sigma_i^2)$ where $Q(.)$ is the Gaussian tail probability (10). The evaluation of the term $P(e|c_i = 0, \alpha_i \leq V_{\text{th}})$ is difficult. However, it can be approximately evaluated as follows.

Consider asymptotically large V_{th}, i.e., $V_{\text{th}} \to \infty$. In this case, the list contains all possible $2^{N_t M_c}$ members. Obviously, in this case, $P(e|c_i = 0, \alpha_i \leq V_{\text{th}} \to P_{\text{ML}})$, where P_{ML} is the bit error probability for a maximum likelihood detector. Even for a moderately large V_{th}, since we are interested in the near ML region of operation, we can still write $P(e|c_i = 0, \alpha_i \leq V_{\text{th}}) \approx P_{\text{ML}}$. Therefore, from (9)

$$P_{b,i} \approx (1 - P_{\delta,i})P_{\text{ML}} + P_{\delta,i} \tag{10}$$

The average BER can be written as

$$P_b = \frac{1}{N_t M_c} \sum_{i=1}^{N_t M_c} P_{b,i} = P_{\text{ML}} + \frac{1 - P_{ML}}{N_t M_c} \sum_{i=1}^{N_t M_c} P_{\delta,i} \tag{11}$$

We have found this expression to provide accurate results in our extensive simulations. As an example, Figure 2 shows our results for various values of V_{th}. The channel is randomly generated and kept fixed at an SNR of 10 dB. For the 2×2 system, the rows of the channel matrix is are [1.3-j0.19 0.08-j0.16], and [0.84-j0.09 -0.85-j0.81] respectively. For the 4×4 system, the rows are [0.27-j0.23 0.35-j0.21 0.21+j1.51 0.91-j0.52], [0.82+j0.46 0.40+j0.50 -0.69+j0.024 0.35-j1.202], [1.16+j0.56 0.38-j0.03 0.47-j0.27 1.11+j1.25], and [-0.02-j0.14 -0.15-j0.36 0.69+j0.75 1.16+j0.24] respectively.

The selection of V_{th} is done as follows. Suppose we want P_b to be within a fraction of ϵ of P_{ML}, i.e., $P_b = (1 + \epsilon)P_{\text{ML}}$. From (11), we get

$$\sum_{i=1}^{N_t M_c} P_{\delta,i} = (\frac{\epsilon P_{\text{ML}}}{1 - P_{\text{ML}}}) N_t M_c \tag{12}$$

which gives

$$\sum_{i=1}^{N_t M_c} Q(\frac{V_{\text{th}} + |\mu_i|}{\sigma_i}) = (\frac{\epsilon P_{\text{ML}}}{1 - P_{\text{ML}}}) N_t M_c \tag{13}$$

Thus, V_{th} can be found by solving (13). The parameter P_{ML} is difficult to obtain (31). We propose to use interference-free error performance P_{IF} in place of P_{ML}. This is given by

$$P_{\text{IF}} = \frac{1}{N_t M_c} \sum_{i=1}^{N_t M_c} P_{\text{IF},i} = \frac{1}{N_t M_c} \sum_{i=1}^{N_t M_c} Q(\sqrt{2\gamma_i}) \tag{14}$$

where γ_i is the average received SNR for the i-th bit when only the symbol x_l, $l = \lceil i/M \rceil$, is present in the system and other symbols are set to zero. Since $P_{\text{IF}} \leq P_{\text{ML}}$, using P_{IF} for P_{ML} in (13) ensures an error performance better than or equal to $(1 + \epsilon)P_{\text{ML}}$. The parameters μ_i and σ_i are obtained as described in (32).

5. Complexity analysis

In this section we describe the exact number of computations needed for T-LSS detector. The complexity depends on
1. MMSE detector
2. List generator
3. Extrinsic LLR calculator

5.1 MMSE detector

Let N_v be the number of V_{th} values where (13) is evaluated. The calculation of V_{th} requires about $(N_v + 1)N_t M_c + 3$ multiplications and $(N_v + 1)(N_t M_c - 1)$ additions. the Q function values can be obtained from tables or through a combination of tables and a few floating point operations.
Total number of multiplications for computing $\mathbf{H}\mathbf{R}_x\mathbf{H}^H + \sigma^2\mathbf{I}$ is equal to $(1/2)N_r(N_r + 1)(N_t + 1)$ assuming $\mathbf{R}_x = \sigma_s^2\mathbf{I}$, where $\sigma_s^2 = E[|x_i|^2]$. The number of addition operations is equal to $(1/2)N_r(N_r + 1)(N_t - 1) + N_r$. The number of multiplications and additions for calculating $(\mathbf{H}\mathbf{R}_x\mathbf{H}^H + \sigma^2\mathbf{I})^{-1}\mathbf{H}$ is equal to $N_r^2 N_t$ and $(N_r - 1)N_r N_t$ respectively. The generation of MMSE detector coefficients requires matrix inversion of size $N_r \times N_r$. From (23), matrix inversion operation requires N_r^3 multiplications and $N_r^3 - 2N_r^2 + N_r$ additions. Since the MMSE detector gives reliability of received code vector of $N_t M_c$ bits, the total number of multiplications per bit is equal to $((1/2)N_r(N_r + 1)(N_t + 1) + N_r^3 + N_r^2 N_t + (N_v + 1)N_t M_c + 3)/N_t M_c$ and addition operations is approximately equal to $(N_r^3 - 2N_r^2 + 2N_r + (1/2)N_r(N_r + 1)(N_t - 1) + (N_r - 1)N_r N_t + (N_v - 1)(N_t M_c - 1))/N_t M_c$. Note that in the case of block fading channels, these operations need to be done only once for the whole block. Then the number of operations per bit will decrease depending on the block size.

5.2 List generator

The sorting operation in Step 1 is needed over $N_t M_c$ elements. Hence the total number of comparison operations are $O(N_t M_c \log_2(N_t M_c))$. This can usually be neglected compared to the other comparison operations. Step 2 requires $\sum_{j=1}^{N_t M_c} 2^{\min(j-1, \log_2 \gamma^{(k)})} = 2\gamma^{(k)} - 1 + (N_t M_c - \log_2 \gamma^{(k)})\gamma^{(k)}$ additions, since the list is restricted to a maximum size of $\gamma^{(k)}$, and until that size is reached, the list size doubles for every increase in i value in $\mathbf{p}_i$. In Step 4, we compare the sum with V_{th} for each member in the list. Since the number of members doubles for every increase in i value in $\mathbf{p}_i$ and half of the members in the list each time have already been compared, we have a worst case number of comparisons as $2\gamma^{(1)} - 1 + (N_t M_c - \log_2\gamma^{(1)})\gamma^{(1)}$. In Step 5, until i exceeds $\log_2\gamma^{(k)}$, only $(2^i - 1)$ comparisons are needed for a given i. This is due the special structure of the algorithm where the members after merging in Step 3 remain in the form of two sorted lists. Therefore, the cumulative number of comparisons till i exceeds $\log_2\gamma^{(k)}$ can be obtained by summing $(2^i - 1)$ over i. Defining $\gamma^{(k)'} = 2^n$, where $n = \lceil \log_2\gamma^{(k)} \rceil$, the number of comparisons is about $2\gamma^{(k)'} - \log_2\gamma^{(k)'}$. When $i > \log_2\gamma^{(k)}$ then the list doubles (after Step 3) for every increase in the i value, but members remain in two sorted lists. We thus need to select $\gamma^{(k)}$ elements from $2\gamma^{(k)}$ elements available in two sorted lists. Thus, $\gamma^{(k)}$ comparisons are needed for each i when $i > \log_2\gamma^{(k)}$. Thus, Step 5 requires $2\gamma^{(k)'} - \log_2\gamma^{(k)'} + (N_t M_c - \log_2\gamma^{(k)})\gamma^{(k)}$ comparisons. In summary, if η iterations are used, the total number of additions per bit is approximately $(1/(N_t M_c)) \sum_{k=1}^{\eta} (2\gamma^{(k)} - 1 + (N_t M_c - \log_2\gamma^{(k)})\gamma^{(k)})$. The total number of comparisons per bit is about $\eta \log_2(N_t M_c) + (1/(N_t M_c))(2\gamma^{(1)} - 1 + (N_t M_c - \log_2\gamma^{(1)})\gamma^{(1)}) + (1/(N_t M_c)) \sum_{k=2}^{\eta} (2\gamma^{(k)'} - \log_2\gamma^{(k)'} + (N_t M_c - \log_2\gamma^{(k)})\gamma^{(k)})$.

5.3 Extrinsic LLR calculator

The extrinsic LLR values are calculated using (5). We can see that $\mathbf{r} - \mathbf{Hx}$ requires $N_r N_t$ multiplications and $N_r N_t$ addition operations for each member in the list. The norm operation $||\mathbf{r} - \mathbf{Hx}||^2$ requires N_r multiplications and $N_r - 1$ additions. Multiplication by $1/\sigma^2$ requires 1 multiplication operation. The term $\mathbf{c'}_{[i]}^{T} \mathbf{L}_{A,[i]}$ require $N_t M_c - 2$ addition operations and the summation of the two terms require 1 addition operation. There are $\gamma^{(k)}$ comparison operations. Thus there are $\frac{1}{N_t M_c}(N_r N_t + N_r + 1)\sum_{k=1}^{\eta} \gamma^{(k)}$ multiplications per bit, $\frac{1}{N_t M_c}(N_r N_t + N_t M_c + N_r - 2)\sum_{k=1}^{\eta} \gamma^{(k)}$ additions per bit and $\frac{1}{N_t M_c}\sum_{k=1}^{\eta} \gamma^{(k)}$ comparisons per bit.

6. Iterative list detection for turbo MIMO systems

For MIMO systems concatenated with a channel code, the list size is controlled using V_{th} during the first iteration. Equation (5) is used to generate extrinsic LLR values that are passed to the channel decoder via the de-interleaver. Since $\mathbf{L}^{(k)}$ does not contain all the code bit vectors, it may so happen for a few bits that they are either 0 or 1 throughout the list. In these cases, we can set the LLR of that particular bit to some fixed value such as +8 or -8 depending on the bit reliability, or we can use the bit flipping algorithm as in (28). In bit flipping method, there is a list size penalty. Once we obtain the channel decoder output information of all the coded bits, they are passed through the interleaver to the MIMO detector. Since the channel decoder outputs are more reliable than the initial reliability detector, the detector uses them for subsequent iterations. In that case, the list size can simply be fixed at the value provided by the first iteration. Fixed list sizes have also been considered in (20). However, further reduction in list sizes can be obtained using an EXIT chart as described below.

6.1 List detector transfer curves

The mutual information transfer curve of List MIMO detector can be computed by assuming *a priori* information L from the channel decoder to be Gaussian distributed with variance σ_A^2 and mean $\sigma_A^2/2$. The measure of information is computed between the *a priori* LLR and the true transmitted bits as well as between extrinsic output LLR and the transmitted bits. The mutual information at the input of the MIMO detector is denoted as I_{A_1} and at the output as I_{E_2}. The mutual information between transmitted bit x and the LLR L is given by (35), (36)

$$I(B,L) = \frac{1}{2} \sum_{d_i=-1,1} \int_{-\infty}^{\infty} p(l|B=b) \times \log_2 \frac{2p(l|B=b)}{p(l|B=-1)+p(l|B=1)} dl \tag{15}$$

The measure of a-priori information is characterized by σ_A^2. The value σ_A for that particular measure of input mutual information I_A is calculated using the relationship

$$J(\sigma) = I_{A_1}(\sigma_A = \sigma) \tag{16}$$

where

$$J(\sigma_A) = 1 - \int_{-\infty}^{\infty} \frac{e^{-(l-\sigma_A^2/2)^2/2\sigma_A^2}}{\sqrt{2\pi\sigma_A^2}} \log_2(1+e^{-l})dl \tag{17}$$

For every given value of I_A, the MIMO detector is simulated to obtain histogram based pdf, which is used in 15 to compute the mutual output information I_{E_1}.

6.2 List optimization using mutual information approach

Since we would like to ensure convergence, the EXIT optimization is done under two worse scenarios: (1) A low SNR value is used so that the list size found from the chart also works at high SNR. In practice, charts pre-computed at several SNR values can be used. (2) The EXIT chart assumes that extrinsic LLR is used both in the list generator as well as in the extrinsic LLR calculator. This is required to simplify the EXIT plots. This, however, is a worse scenario compared to the actual algorithm where full LLR is used in the list generator while extrinsic LLR is used in the extrinsic LLR calculator.

Fig. 3 shows the EXIT characteristics for the MIMO detector using T-LSS and a rate 1/2 convolutional decoder with constraint length 7 with polynomials [147, 117]. From the numerical simulations, the average list size for T-LSS is found to be 35. This is the list size in the first iteration. For subsequent iterations, the figure shows that the list size can be reduced to 32 and 24, that can provide performance very close to a full list with 256 members.

Fig. 4 shows the EXIT characteristics for the case of $N_t > N_r$, In this case, the average list size for the T-LSS detector is 81. The list size can be reduced to 48 and then to 32 for the second and higher iterations. This can reduce numerical complexity significantly.

7. Numerical Results and Discussions

A MIMO system with N_t transmit and N_r receive antennas is considered. We use a complex Gaussian random MIMO channel. In our work, the channel changes for each code bit vector although our technique applies to block fading channels as well. The channel information is assumed to be known at the receiver only. QPSK modulation is considered for all the cases. Unless specified, ϵ is set to 0.01.

Fig 5 shows the BER plot of the MIMO detector (uncoded system) with symmetrical configurations for 2×2, 4×4, 6×6 and 8×8 systems. Full search maximum likelihood detector (MLD) results are also shown. It can be seen that with T-LSS we can reach the

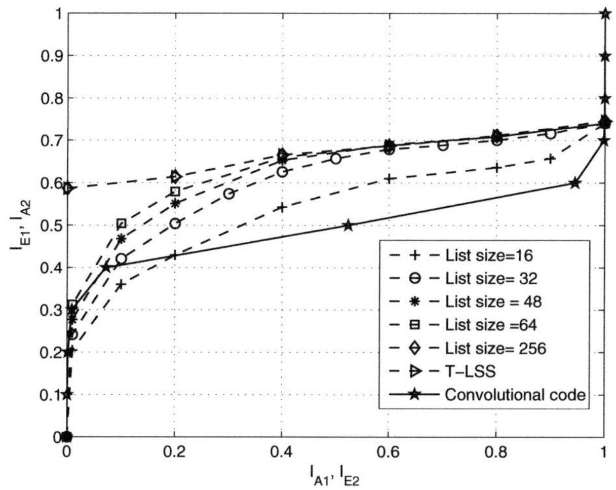

Fig. 3. EXIT characteristics for 4 × 4 MIMO system with LSS detector at E_b/N_0 = 4.0 dB. Note that average complexity for T-LSS is 35.

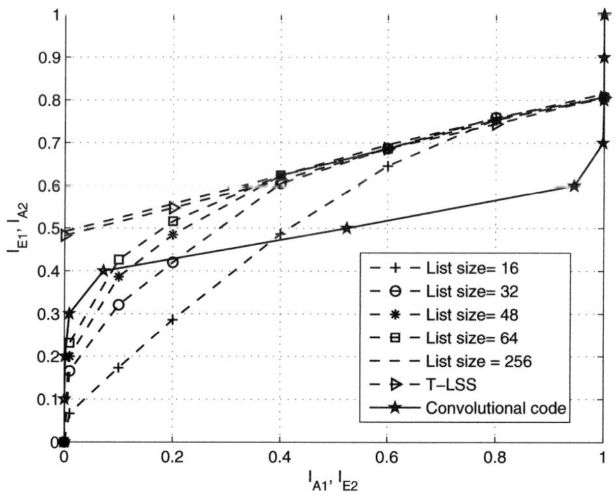

Fig. 4. EXIT characteristics for 4 × 2 MIMO system with T-LSS detector at E_b/N_0 = 6.0 dB. Note that average complexity for T-LSS is 81.

same performance as that of MLD. The average list S_{avg} required for T-LSS symmetrical configuration is shown in Fig 6. We observe that we can get significant reduction in the list size at high E_b/N_o. For example, for a 4×4 system at $E_b/N_o = 14dB$, the average list size equal to 18, as compared to 256 for MLD.

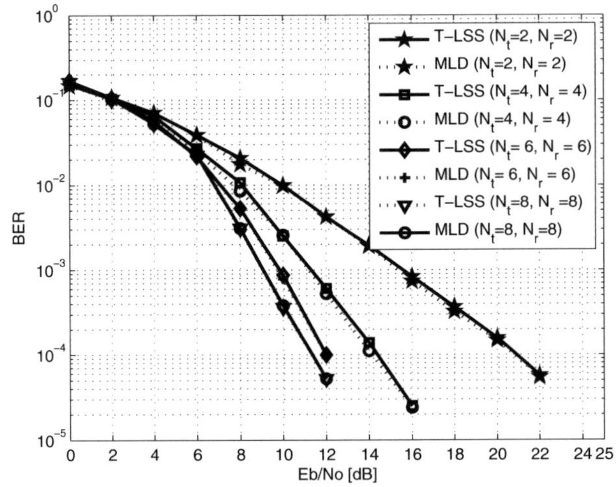

Fig. 5. BER performance of T-LSS for symmetrical configuration.

Fig 7 shows the BER results for a $4 \times N_r$ system for different values of N_r. We note that the T-LSS detector works for any of the combinations of N_t and N_r unlike the SD, or the original vertical Bell Labs layered space-time (V-BLAST) algorithm. Fig 8 shows the average list sizes of the T-LSS for these configurations. Reduction in the number of receiver antennas results in a loss in diversity and, therefore, an equivalent SNR loss. Therefore, the benefit does

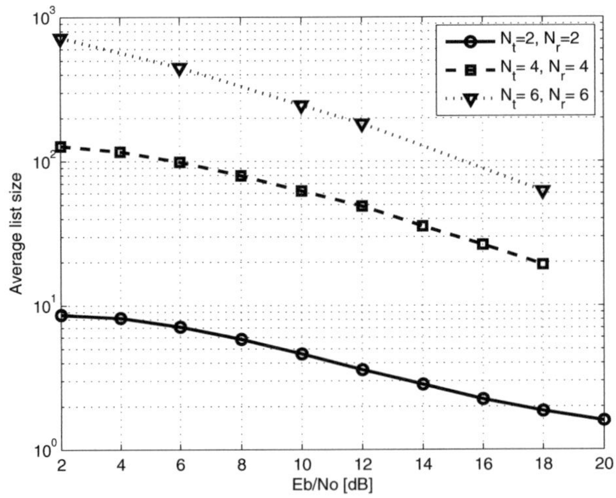

Fig. 6. Average list size per bit of T-LSS for symmetrical configuration.

not significantly improve with an increase in SNR in the low SNR region. Nevertheless, the average list size for the 4 × 2 system is reduced to 90 in contrast with the ML size of 256.

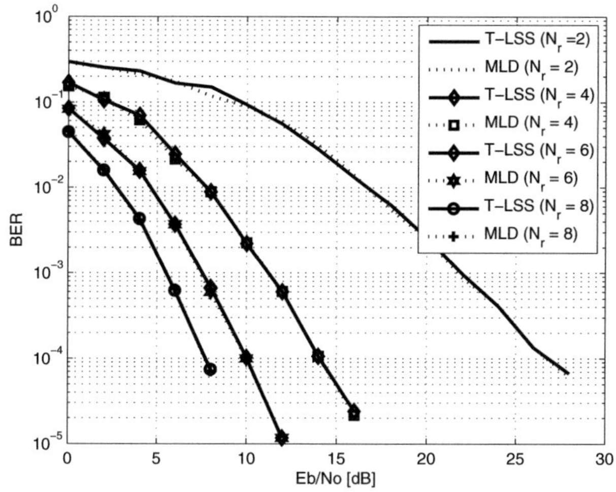

Fig. 7. BER performance of T-LSS for $4 \times N_r$ systems.

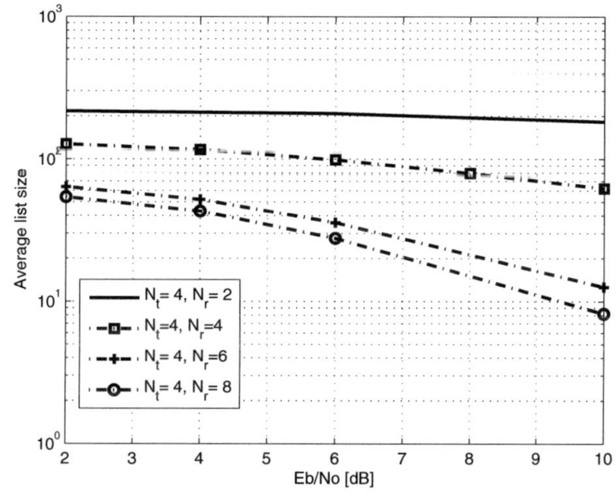

Fig. 8. Average list size per bit of T-LSS for $4 \times N_r$ systems.

The role of the parameter ϵ on the average list size for a 4×4 system is shown in Fig 9. Note that to get exact ML performance, we need a list size of 256 in this case. However, by relaxing the requirement to $(1+\epsilon)P_{\text{ML}}$, the list size is significantly reduced. Thus at a 15 dB SNR, we reduce the list size 17 times by targeting for a BER of $1.01P_{\text{ML}}$ instead of exact P_{ML}. The complexity reduction depends on the SNR.

Fig. 10 shows the BER plot for the iterative MIMO system. A convolutional code with polynomials as described in Section VI is considered. In the first iteration, the list is generated from the T-LSS algorithm. Subsequent iterations use $\gamma^{(k)}$ obtained from the EXIT chart at 4.0

dB (Fig 3). We observe that the BER performance of the proposed method closely matches with full search MLD where a list size of 256 is used. For the proposed method, using list sizes of 35, 32, 24 and 24 respectively in the four iterations, the number of multiplications per bit is about 420 using $N_v = 100$. For the full list MLD, the number of multiplications per bit is about 2688 that comes from the extrinsic LLR calculator. For larger number of antennas or constellation sizes, the potential benefit in complexity will be much more significant.

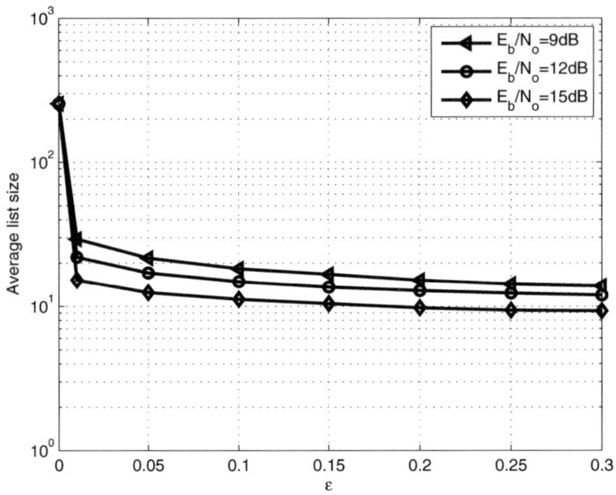

Fig. 9. Effects of ϵ on the complexity.

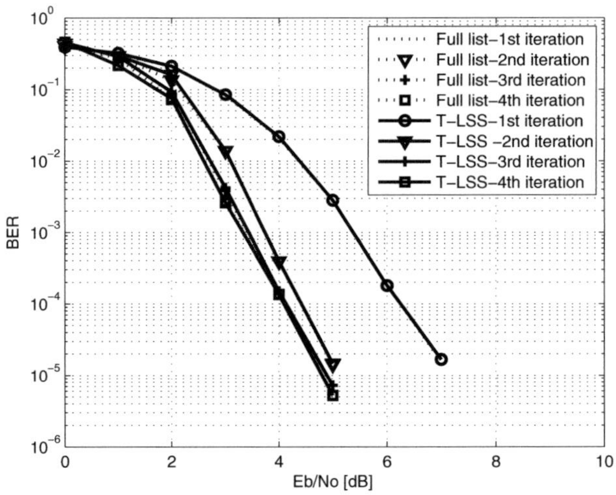

Fig. 10. BER performance of iterative T-LSS for 4×4 MIMO system.

8. Conclusions

A soft input soft output MIMO detector for turbo MIMO systems is presented. The detector searches over a list of candidate codewords to find the most likely codeword vector. The

BER performance of the detector is analyzed, and a method to choose the optimal list size is described. During the first iteration, the detector uses a threshold value to limit the number of list members, while subsequent iterations use list sizes obtained through EXIT analysis. The numerical complexity of the detector is analyzed. We show that the detector provides BER performance very close to the use of ML detector based receiver at significantly low complexity. Possible future work includes design of capacity approaching irregular LDPC codes based on the proposed T-LSS detector.

9. References

[1] I. E. Telatar, "Capcity of multi-antenna Gaussian channels," *Eur. Trans. Telecommun.*, vol. 10, pp. 585-595, Nov. 1999.

[2] G. D. Golden, G. J. Foschini, R. A. Valenzuela, and P. W. Wolnianski, "Detection algorithm and initial laboratory results using the V-BLAST space-time communication architecture," *Electron. Lett.*, vol. 35, pp. 14-15, Jan. 1999.

[3] G. J. Foschini, "Layered space-time architecture for wireless communication in a fading environment when using multi-element antennas," *Bell Labs. Tech. J.*, vol. 1, pp. 41-59, Feb. 1996.

[4] B. M. Hochwald and S. T. Brink, "Achieving near-capacity on a multiple-antenna channel," *IEEE Trans. Commun.*, vol. 51, pp. 389-399, Mar. 2003.

[5] C. Berrou, A. Glavieux, and P. Thitimajshima, "Near Shannon limit error correcting coding and decoding: turbo codes," *in Proc. IEEE ICC*, Geneva,Switzerland, pp. 1064-1070, May 1993.

[6] A. M. Tonello, "Space-time bit-interleaved coded modulation with an iterative decoing strategy," in *Proc. IEEE VTC'* 2000, pp. 473-478, Boston, May 2001.

[7] S. T. Brink, J. Speidel, and R. H. Yan, "Iterative demapping and decoding for multilevel modulation," in *Proc. IEEE GLOBECOM'* 1998, pp. 579-584, Sydney, Nov 1998.

[8] J. Hagenauer, E. Offer, and L. Papke, "Iterative decoding of binary block and convolutional codes," *IEEE Trans. Inform. Thoery*, vol. 42, pp.429-445, March 1996.

[9] A. V. Zelst, R. V. Nee, and G. Awater, "Turbo-BLAST and its performance," in *Proc. IEEE VTC'* 2001, pp. 1282-1286, Greece, May 2001.

[10] J. G. Proakis, *Digital Communications*, 4th ed. New York: McGraw-Hill, 2000.

[11] R. D. Maddock and A. H. Banihashemi, "Reliability-based coded modulation with low-density parity-check codes," *IEEE Trans. Commun.*, vol. 54, pp. 403-406, Mar. 2006.

[12] M. Sellathurai and S. Haykin, "Turbo-BLAST for wireless communications: theory and experiments, "*IEEE Trans. Signal Processing*, vol. 50, pp. 2538-2546, Oct. 2002.

[13] S. L. Ariyavisitakul, "Turbo space-time processing to improve wireless channel capacity," *IEEE Trans. Commun.*, vol. 48, pp. 1347-1359, Aug 2000.

[14] B. Steingrimsson, Z. Q. Luo and K. M. Wong, "Soft quasi maximum likelihood detection for multiple antenna wireless channels, "*IEEE Trans. Signal Processing*, vol. 51, pp. 2710-2719, Nov. 2003.

[15] S. Baro, J. Hagenauer, and M. Witzke, " Iterative detection of MIMO transmission using a list-sequential (LISS) detector," *Proc. ICC'*2004, Seattle, 2003.

[16] D. Chase, "A class of algorithms for decoding block codes with channel measurement information," *IEEE Trans. Inform. Theory*, vol. 18, pp. 170-182, Jan. 1972.

[17] D. Waters and J. R. Barry, "The Chase family of detection algorithms for multiple-input multiple-output channels," *Proc. IEEE GLOBECOM'*2004, Dallas, 2004.

[18] Z. Qin and K. C. Teh, "Iterative reduced-complexity multiuser detection based on Chase decoding for synchronous turbo-coded CDMA system," *IEEE J. Select. Areas Commun.*, vol. 24, pp. 200-208, Jan. 2006.

[19] E. Viterbo and J. Boutros, "A universal lattice code decoder for fading channels," *IEEE Trans. Inform. Theory*, vol. 45, pp. 1639-1642, July 1999.

[20] S. Nammi and D. K. Borah, "Iterative List-Subset Detectors for Turbo Product Coded MIMO Wireless Systems," *Proc. IEEE VTC'*2007, Dublin, 2007.

[21] H. V. Poor and S. Verdu, "Probability of error in MMSE multiuser detection," *IEEE Trans. Inform. Theory*, vol. 43, pp. 858-871, May 1997.

[22] T. H. Cormen, C. E. Leiserson, and R. L. Rivest, *Introduction to algorithms*, 4th ed. New York: McGraw-Hill, 1990.

[23] R. L. Burden, J. D. Faires, and A. C. Reynolds, *Numerical analysis*, 2nd ed. Boston: Prindle, Weber and Schmid, 1981.

[24] D. Tse and P. Viswanath, *Fundamentals of Wireless Communication*, Cambridge University Press, 2005.

[25] D. Chase, "A class of algorithms for decoding block codes with channel measurement information," *IEEE Trans. Inform. Theory*, vol. 18, pp. 170-182, Jan. 1972.

[26] R. Pyndiah, "Near-optimum decoding of product codes: Block turbo codes," *IEEE Trans. Commun.*, vol. 46, pp. 1003-1010, Aug 1998.

[27] T. Kaneko, T. Nishijima, H. Inazumi, and S. Hirasawa, "An efficient maximum-likelihood-decoding algorithm for linear block codes with algebraic decoder," *IEEE Trans. Inform. Theory*, vol. 40, pp. 320-327, March 1994.

[28] Z. Qin and K. C. Teh, "Iterative reduced-complexity multiuser detection based on chase decoding for synchronous turbo-aided CDMA system," *IEEE J. Select. Areas Commun.*, vol. 24, pp. 200-208, Jan. 2006.

[29] E. Zimmermann, W. Rave, and G. Fettweis, "On the complexity of sphere decoding," *Proc. WPMC'*2004, Italy, 2004.

[30] L. R. Bahl, J. Cocke, F. Jelinek, and J. Raviv, "Optimal decoding of linear codes for minimizing symbol error rate," *IEEE Trans. Inform. Theory*, vol. 20, pp. 284Ű287, March 1974.

[31] X. Zhu, and R. D. Murch, " Performance analysis of maximum likelihood detection in a MIMO antenna system," *IEEE Trans. Commun.*, vol. 50, pp. 187-191, Feb. 2002.

[32] S. Nammi and D. K. Borah, "List-based detection in fading channels with long intersymbol interference," *IEEE Trans. Wireless Commun.*, vol. 8, pp. 1276-1283, March 2009.

[33] K. R. Narayanan, X. Wang and G. Yue "Estimating the PDF of the SIC-MMSE equalizer output and its applications in designing LDPC codes with turbo equalization," *IEEE Trans. Wireless Commun.*, vol. 4, pp. 278-287, Jan. 2005.

[34] H. Vikalo, B. Hassibi and T. Kailath, "Iterative decoding of MIMO channels via modified sphere decoder," *IEEE Trans. Wireless Commun.*, vol. 3, pp. 2299-2311, 2004.

[35] S. T. Brink, G. Kramer, and A. Ashikhmin, "Design of low-density parity check codes for modulation and detection," *IEEE Trans. Commun.*, vol. 52, pp. 670-678, April 2004.

[36] S. T. Brink, "Convergence behavior of iteratively decoded parallel concatenated codes," *IEEE Trans. Commun.*, vol. 49, pp. 1727-1737, Oct. 2001.

16

Narrowband Interference Suppression in MIMO Systems

Vladimir Poulkov, Miglen Ovtcharov, Georgi Iliev and Zlatka Nikolova
Technical University of Sofia
Bulgaria

1. Introduction

Multiple Input Multiple Output (MIMO) communication technology has received recently significant attention triggered by the rapid development of Orthogonal Frequency Division Multiplexing (OFDM) broad-band wireless and high speed wireline communication systems. Such systems are very sensitive to Narrowband Interferences (NBI) due to their relatively low transmission power. This type of signal interference can be found in the new unlicensed frequency bands, e.g., the Industrial Scientific Medical (ISM) band, coming from systems such as Bluetooth or microwave ovens which interfere with OFDM based Wireless Local Area Networks (WLAN), like Hiperlan II. Other examples of NBI, typical for Digital Subscriber Line (DSL) communications, are strong Radio Frequency Interferences (RFI) from short-wave radio, Citizen's Band (CB) radio and amateur "ham" radio which interfere with Hybrid Fiber Coaxial (HFC) networks and high speed DSL. Ineffective shielding of a cable network may also cause the ingress from external home electrical devices, such as cordless phones, TVs and computers.

Let's consider a strong NBI signal which resides within the same frequency band as a wideband OFDM signal. In this case, severe Signal to Noise Ratio (SNR) degradation is likely to occur across all OFDM subcarriers due to spectral leakage of the NBI signal from block processing in the OFDM receiver. In Fig. 1, the magnitude-squared of the received signal samples $|X[k]|^2$ is plotted. For flat-fading channels it is straightforward to ascertain the fact that the highest peak corresponds to the subcarrier affected by NBI, as shown in Fig. 1.(a). Fig.1.(b) shows that in frequency-selective channels the OFDM signal presents a large dynamic range, and some signal peaks can have values close to the peak induced by NBI. This observation indicates that the detection of NBI becomes difficult in frequency-selective channels, especially if the Signal to Interference Ratio (SIR) is high (Giorgetti et al., 2005).

The presence of strong NBI causes nonlinear distortion in Automatic Gain Control and Analog to Digital Converter functional blocks as well as spectral leakage in the Discrete Fourier Transform (DFT) process. Many subcarriers close to the interference frequency will suffer serious Signal to Noise Ratio (SNR) degradation. Therefore, NBI suppression is of primary importance.

The issue of NBI suppression for OFDM systems has been studied extensively in recent years, and a number of general approaches are proposed. Various *Frequency Excision methods*, where the affected frequency bins of the OFDM symbol are excised or their usage avoided, have received specific attention. Often degradation in an OFDM based receiver is beyond the reach of the frequency excision method when the SIR is less than 0dB. In such cases other approaches, related to the *Cancellation techniques* that aim the elimination or mitigation of the effect of the NBI on the received OFDM signal, are recommended. *Linear and Nonlinear filtering methods* for NBI cancellation are also proposed.

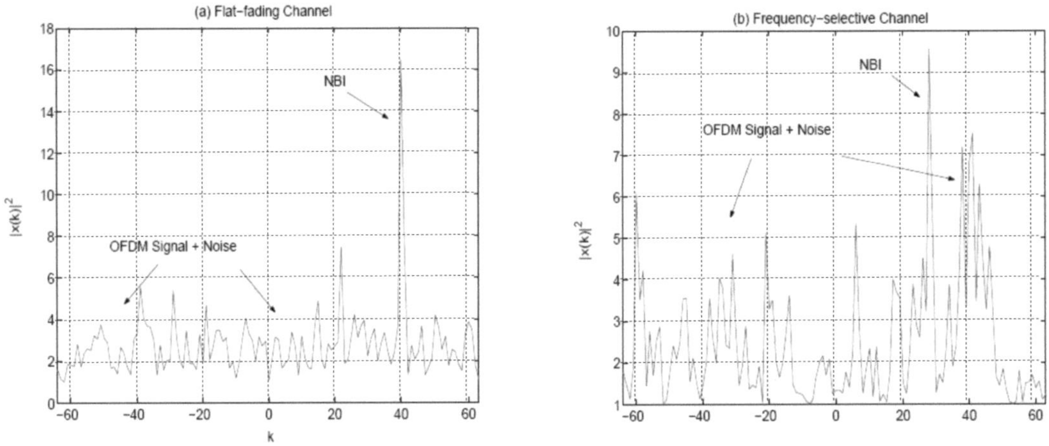

Fig. 1. The magnitude-squared of FFT bins: $|X[k]|^2$

Different specific approaches for NBI suppression are deployed in MIMO systems depending on the type of channel. For wireless channels one major group of methods is *Block Coding (BC): Space Time and Space Frequency*. It utilizes the properties of MIMO systems over fading channels. Other methods such as *adaptive beam-forming* of the MIMO antenna system and *Forward Precoding* at the transmission side are reviewed. In wire-line subscriber systems methods such as *Dynamic Spectrum Management (Level 3), Optimal Precoding* and *Multi-channel Adaptive Filtering* are applied.

In this chapter the major NBI suppression methods their performance, computational complexity and application to different types of MIMO communication systems are considered. Results of simulation experiments for wireless MIMO systems and for MIMO DSL channels are analyzed. It could be seen that the methods have different computational complexity and performance depending on the parameters of the MIMO system and the communication channels.

2. Overview of NBI suppression methods

2.1 Frequency excision

This approach uses the Fast Fourier Transform (FFT) based frequency-domain excision to remove the NBI. The DFT output of block of N_{FFT} samples, $r_{m,n}$ is given by:

$$r_{m,n} = \sum_{k=1}^{N_{FFT}} r_{m,k} e^{-j2\pi kn/N_{FFT}}, \quad k = 1,....N_{FFT} \tag{1}$$

where m is the number of the OFDM symbol, n is the number of the subcarrier of the OFDM symbol, $r_{m,k}$ is the digital complex baseband signal sample at the demodulator input for m.

In the frequency domain, the NBI manifests itself as a peak in the spectra. Interferences can be excised by comparing and limiting the magnitude of each frequency bin to a threshold. The following method is used to determine the threshold. The mean value of the amplitude of the frequency bins and its variance are computed (Jyh-Ching et al., 2004):

$$T_{\text{mean}} = \sum_{n=1}^{N_{FFT}} \frac{10\log_{10}\left|r_{m,n}\right|}{N_{FFT}} \tag{2}$$

$$T_{\text{var}} = \frac{1}{N_{FFT}}\left[\sum_{n=1}^{N_{FFT}} \left(10\log_{10}\left|r_{m,n}\right|\right)^2 - \frac{1}{N_{FFT}}\left(\sum_{n=1}^{N_{FFT}} \left(10\log_{10}\left|r_{m,n}\right|\right)\right)^2\right]. \tag{3}$$

The threshold is determined according to:

$$T_{\text{excision}} = T_{\text{mean}} - \alpha T_{\text{var}}^{1/2}. \tag{4}$$

The scale factor α in the above equation is adjusted to maintain the threshold at some value of the noise floor. Each frequency bin is compared to the threshold and, if it exceeds the threshold, its value is held at the threshold. After the application of Inverse Fast Fourier Transform (IFFT), the signal is much less contaminated with narrow band interferences.
This method is characterized with relative high efficiency due to the fact that the FFT/IFFT digital signal processing could be realized in real time. In addition, it is possible to excise multiple NBIs without any initial limiting conditions. Drawbacks of this method are the impossibility to fully suppress the interference signals because of the NBI interference with the information subcarriers thus violating their orthogonality and the strong dependence of the threshold on the type and conditions of the channel.

2.2 Frequency identification and cancellation

The discrete complex baseband signal for sample *k* at the input of an OFDM receiver can be expressed as (Baccareli et al., 2002):

$$r(k) = s_t(k) * h(k) + n(k) + i_n(k), \tag{5}$$

where $s_t(k)$ is the complex output signal for sample *k* of the transmitter, $h(k)$ is the complex impulse response for sample *k* of the channel, $n(k)$ is the complex Additive White Gaussian Noise (AWGN) for sample *k* and $i_n(k)$ is n-th complex single tone NBI for sample *k*. Initially the frequency position of the interference tone has to be estimated in order to approach the problem of NBI identification. After sampling the received signal $r(t)$ to obtain $r(k)$, FFT must be applied to this sequence via a Goertzel algorithm or using a butterfly lattice. It is important to note that appropriate setting of the sampling time *T* is a fundamental step; therefore, to increase the frequency resolution, an eight times oversampling in the frequency domain is proposed.
Assuming that the power spectrum properties of all signals constituting *r*(*k*) are known, the narrow band interfering signal can be modeled as a complex sinusoidal tone:

$$i_n(k) = \alpha_n e^{j(\widehat{\omega}_n k/T + \varphi_n)} = \alpha_n \cos(\widehat{\omega}_n k / T + \varphi_n) + j\alpha_n \sin(\widehat{\omega}_n k / T + \varphi_n). \tag{6}$$

$$i_n(k) = a_n \cos(\hat{\omega}_n k / T) + j b_n \sin(\hat{\omega}_n k / T) . \qquad (7)$$

Thus, it is clear that frequency domain processing is an appealing approach to estimate frequency, because the spectral properties of $r(k)$ are known.
This method is implemented in several steps. First the complex NBI frequency is estimated by finding the maximum amplitude in the oversampled signal spectrum:

$$\hat{\omega}_n = \arg \omega \in L_\infty \max(P_R(\omega)) . \qquad (8)$$

Then amplitude and phase estimation is done. A matrix form can be used to represent the sampled interference (Baccareli et al., 2002):

$$\mathbf{i_n} = \mathbf{M}\mathbf{x} . \qquad (9)$$

The matrix **M** is defined as:

$$\mathbf{M} = \begin{bmatrix} \cos(\hat{\omega}_n k_1 / T) & j\sin(\hat{\omega}_n k_1 / T) \\ \cos(\hat{\omega}_n k_2 / T) & j\sin(\hat{\omega}_n k_2 / T) \\ \ldots\ldots & \ldots\ldots \\ \cos(\hat{\omega}_n k_N / T) & j\sin(\hat{\omega}_n k_N / T) \end{bmatrix} . \qquad (10)$$

The vector **x** gathers the coefficients a_n and b_n:

$$\mathbf{x} = \begin{bmatrix} a_n \\ b_n \end{bmatrix} . \qquad (11)$$

And $\boldsymbol{i}_n$ is a column vector of the n-th complex NBI where N is the FFT length.

$$\mathbf{i}_n = \left[i(k_1) \quad i(k_2) \quad \cdots \quad i(k_N) \right]^T . \qquad (12)$$

Applying the Maximum Likelihood (ML) algorithm, the solution is given by (Iltis & Milstein, 1985):

$$\hat{\mathbf{x}} = \left(\mathbf{M}^{\mathrm{T}}\mathbf{M}\right)^{-1} \mathbf{M}^{\mathrm{T}}\mathbf{r} . \qquad (13)$$

Here $\mathbf{r}$ is the input signal vector and $\hat{\mathbf{x}}$ is the estimation of the complex amplitude of the n-th complex NBI. The next step is to estimate the exact values of the amplitude, frequency and phase of each of the harmonic interference signals. The algorithm uses the estimation results from the first step as initial conditions of a Normalized Least Squares (NLS) optimization procedure (Baccareli et al., 2002). The cancellation of the NBI is the last step. The identified frequency, amplitude and phase of the interference in the received signal are used for the synthesis of complex harmonic signals, which are subtracted from the received signal:

$$\overline{r}(k) = r(k) - \sum_{n=1}^{M} \alpha_n e^{j(\omega_n k/T + \varphi_n)} , \qquad (14)$$

where $\acute{r}(k)$ is the input signal for the OFDM demodulator after the NBI cancellation.

2.3 Complex adaptive narrowband filtering

A NBI occupies a much narrower frequency band with higher power spectral density compared with a wideband signal. It could be assumed that usually a wideband signal has autocorrelation properties quite similar to that of AWGN. Therefore, filtering in the frequency domain could be realized. In OFDM systems such linear filtering is performed at the input of the demodulator. An example of such application is using an adaptive complex variable filter section with independent tuning of the central frequency and the bandwidth (Iliev et al., 2004). The transform function of such a filter section is defined as:

$$H_{RR}(z)=H_{II}(z)=\hat{\beta}\frac{1+2\hat{\beta}cos\theta z^{-1}+(2\hat{\beta}-1)z^{-2}}{1+2(2\hat{\beta}-1)cos\theta z^{-1}+(2\hat{\beta}-1)^2 z^{-2}},$$
$$H_{RI}(z)=-H_{IR}(z)=\hat{\beta}\frac{2(1-\hat{\beta})\sin\theta z^{-1}}{1+2(2\hat{\beta}-1)\cos\theta z^{-1}+(2\hat{\beta}-1)^2 z^{-2}}, \quad (15)$$

$$\hat{\beta}=\beta+2\beta\gamma(\beta-1) \quad (16)$$

where the coefficients β and γ define the filter bandwidth and parameter θ controls the central pass-band frequency ω_0.

The proposed implementation has two very important advantages: first, an extremely low passband sensitivity increases the resistance to quantization effects; second, the central frequency and filter bandwidth can be independently controlled over a wide frequency range. Fig. 2 shows the adaptive complex notch/bandpass narrowband system based on a variable complex filter section. The band-pass variable complex filter section is tuned at the NBI frequency, thus the signal $y(n)$ represents the output of the filter which is in fact the NBI signal. Further, the NBI signal $y(n)$ is subtracted by the input complex signal $x(n)$ which is the additive sum of OFDM signal, NBI signal and AWGN. If perfect filtering is assumed, the result signal $e(n)$ is supposed to be NBI-free OFDM signal. Here $e_R(n)=x_R(n)-y_R(n)$ and $e_I(n)=x_I(n)-y_I(n)$ are the real and imaginary output of the band-stop (BS) filter. The cost function minimizes the power of the BS filter output signal $[e(n)e^*(n)]$, where $e(n)=e_R(n)+je_I(n)$.

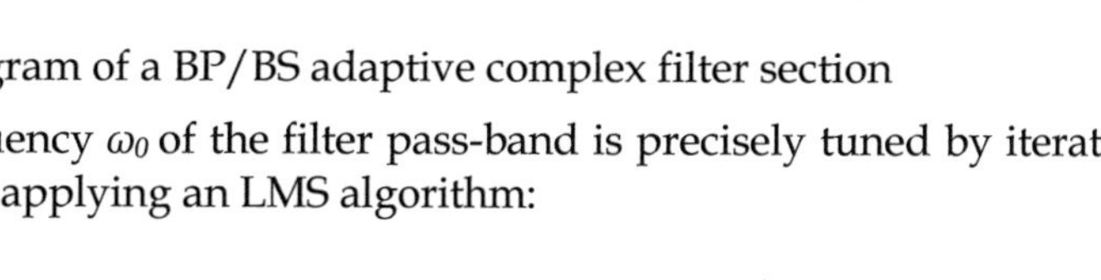

Fig. 2. Block-diagram of a BP/BS adaptive complex filter section

The central frequency ω_0 of the filter pass-band is precisely tuned by iterative update of the filter coefficients applying an LMS algorithm:

$$\theta(n+1)=\theta(n)+\mu\mathrm{Re}[e(n)y'^*(n)]. \quad (17)$$

Here μ is the step size controlling the speed of convergence, (*) denotes complex-conjugate, $y'(n)$ is the derivative of $y(n) = y_R(n) + jy_I(n)$ with respect to the coefficient subject of adaptation.

2.4 Nonlinear filtering

Nowadays communication systems work in a saturated and noisy electromagnetic spectrum. Many of the interference signals could be approximated as independent stationary random processes with Gaussian distribution, but quite a lot of the radio-frequency interferences do not fit into this category. A big variety of impulse noises, to which OFDM systems are quite sensitive, could not be considered as Gaussian and/or stationary random processes. The system model has to take that into account to realize an effective NBI suppression. Under the above conditions nonlinear filtering methods, such as the two considered below, offer an improvement in NBI suppression.

a. Nonlinear ACM filter for autoregressive interference

Lets consider the received sampled signal from equation (8). In complex form the narrowband interference $\{i_k\}$ is modeled as an autoregressive Gaussian process of order *p*, i.e., assuming a model of the form (Hsu & Giordano, 1978), (Iltis & Milstein, 1985):

$$i_k = \sum_{i=0}^{p} \phi_i i_{k-i} + e_k , \tag{18}$$

where $\{e_k\}$ is a white Gaussian process, and where the autoregressive parameters $\Phi_1, \Phi_2, \ldots, \Phi_p$ are known to the receiver. Under this model, the received signal has a state space representation as follows:

$$x_k = \Phi x_k - 1 + w_k , \tag{19}$$

$$z_k = Hx_k + s_k + n_k , \tag{20}$$

$$x_k = \begin{bmatrix} i_k & i_{k-1} & \cdots & i_{k-p+1} \end{bmatrix}^T , \; H_k = \begin{bmatrix} 1 & 0 \ldots 0 \end{bmatrix}^T , \; w_k = \begin{bmatrix} e_k & 0 \ldots . 0 \end{bmatrix}^T , \tag{21}$$

$$\Phi = \begin{bmatrix} \phi_1 & \phi_2 & \cdots & \phi_{p-1} & \phi_p \\ 1 & 0 & \cdots & 0 & 0 \\ \cdots & \cdots & \cdots & \cdots & \cdots \\ 0 & 0 & \cdots & 1 & 0 \end{bmatrix} . \tag{22}$$

The first component of the state vector x_k is the interference i_k. Hence, by estimating the state, an estimate of the interference which can be subtracted from the received signal to reject the interference, can be obtained.

In (Masreliez, 1975) an Approximate Conditional Mean (ACM) filter for estimating the state of a linear system with Gaussian noise and non-Gaussian measurement noise is developed. The nature of the nonlinearity is determined by the probability density of the observed noise. Under this assumption, the filtered estimate and its conditional covariance P_k and $\hat{x}_k$ are obtained recursively through the equations for state (23,24) and time (25) estimation:

$$P_k = M_k - M_k H^T G_k(z_k) H M_k\,, \tag{23}$$

$$\hat{x}_k = \bar{x}_k + M_k H^T g_k(z_k)\,, \tag{24}$$

$$M_{k+1} = \Phi P_k \Phi^T + Q_k\,; \quad \bar{x}_{k+1} = \Phi \hat{x}_k\,. \tag{25}$$

Here G_k and g_k are the nonlinear components due to the non-Gaussian noise, ε_k is the error signal and σ its variance defined by the following equations (Laster & Reed, 1997):

$$Q_k = \mathrm{E}\{\mathbf{w_k}\ \mathbf{w_k}^{\mathrm{T}}\}\,, \tag{26}$$

$$g_k(z_k) = \frac{1}{\sigma^2}\left[\varepsilon_k - \tanh\left(\frac{\varepsilon_k}{\sigma^2}\right)\right], \tag{27}$$

$$G_k(z_k) = \frac{1}{\sigma^2}\left[1 - \frac{1}{\sigma^2}\mathrm{sech}^2\left(\frac{\varepsilon_k}{\sigma^2}\right)\right], \tag{28}$$

$$\varepsilon_k = x_k - \mathbf{H}\bar{\mathbf{x}}_\mathbf{k}\,, \tag{29}$$

$$\sigma^2 = \sigma^2_n + \mathbf{H M_k H^T}\,, \tag{30}$$

The ACM filter here is seen to have a structure similar to that of Kalman - Bucy filter, as the time-update equations (23), (24) are identical. Without the nonlinear functions *tanh* and *sech* the recursive algorithm of Masreliez is reduced to the linear Kalman - Bucy algorithm (Sampei 1997). From equations (24), (27) and (29) follows, that with the ACM filter the decision feedback is realized via the function *tanh*, by correcting the measured value in the interval [-1, 1]. The behavior of the filter depends on the variance of the error signal. When the system is in a steady state, the variance is small and the nonlinear element in the feedback determines the behavior of the filter. When the variance is high, the feedback is in a linear working state and the behavior is similar to that of a linear Kalman-Bucy filter.

b. Adaptive nonlinear filter based on LMS algorithm

It has been shown in (Sampei 1997) that better interference rejection can be obtained by using one or two-sided interpolation filter. For such a filter the following Widrow Least Mean Squared (LMS) algorithm equations hold (Iltis & Milstein, 1985) (Johnson 1984):

$$X_k = \left[z_{k+N},\ z_{k+N-1}, \dots, z_{k+1}, z_{k-1}, \dots, z_{k-N}\right]^T, \tag{31}$$

$$\theta_k = \left[a_{-N}(k),\ a_{-N+1}(k),\ \dots, a_{-1}(k), a_1(k), \dots, a_N(k)\right]^T, \tag{32}$$

where $2N + 1$ is the length of the data window. In order to ensure independence of the parameter μ (for control of the stability and convergence of the algorithm) from the input signal level the equations are normalized as follows (Johnson 1984):

$$\theta_k = \theta_{k-1} + \frac{\mu_o}{r_k}\xi_k X_k\,, \tag{33}$$

Where r_K is the estimate of the power of the input signal determined as

$$r_k = r_{k-1} + \mu_o \left[\left| X_k \right|^2 - r_{k-1} \right]. \tag{34}$$

The output signal of the filter ε_K, which is the error signal is defined as:

$$\varepsilon k = zk - \hat{z}k, \tag{35}$$

where $\check{z}_K$ is the predicted value of the NBI on the basis of L samples from the input signal z_K. By transforming the prediction error a nonlinear transversal filter for the prediction of z_k is obtained. The differential equation of the filter is in the form (Johnson 1984):

$$\hat{z}_k = \sum_{i=0}^{L} a_i(k-1)\left[\hat{z}_{k-i} + \rho_{k-1}(\xi_{k-1}) \right], \tag{36}$$

$$\rho_k(\xi_k) = \xi_k - \tanh\left(\frac{\xi_k}{\sigma_k^2} \right). \tag{37}$$

Here ρ_k is a nonlinear function as the one used in equation (30) and σ_k variance of a Gaussian random variable.

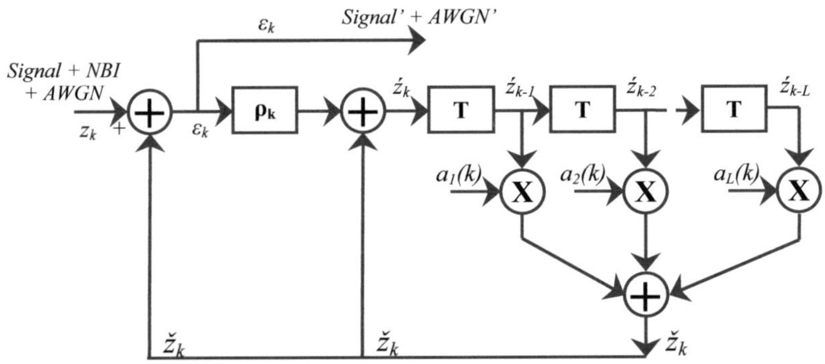

Fig. 3. One-sided Adaptive Nonlinear Filter Based on LMS Algorithm.

A block scheme of an one-sided nonlinear adaptive filter, based on LMS algorithm is shown on Fig.3. For the realization of the filter from equation (36) an estimation of σ_k and the tap weight coefficients $a_i(k)$, has to be obtained using LMS algorithm (Carlemalm et al., 2004).

3. NBI suppression in wireless MIMO OFDM systems

Following some of the specific methods for NBI suppression in MIMO communication systems are discussed, such as adaptive beamforming of the antenna system, forward precoding at the transmission side and Orthogonal Space-Time Block Coding.

3.1 Adaptive beamforming of the MIMO antenna system

One of the fastest developing methods for noise suppression in modern MIMO communication systems is employing adaptive antenna arrays in the transmit and receive side of the radio-communication system. A block scheme of an MIMO OFDM system with adaptive beamforming is shown in Fig.4, where R is the number of transmit antennas, M - number of receive antennas, N - number of subcarriers in the OFDM symbol, $s_k(t)$ $0 \le k < N$ is the k transmitted normalized information symbol from block 't' (Iserte et al., 2001).

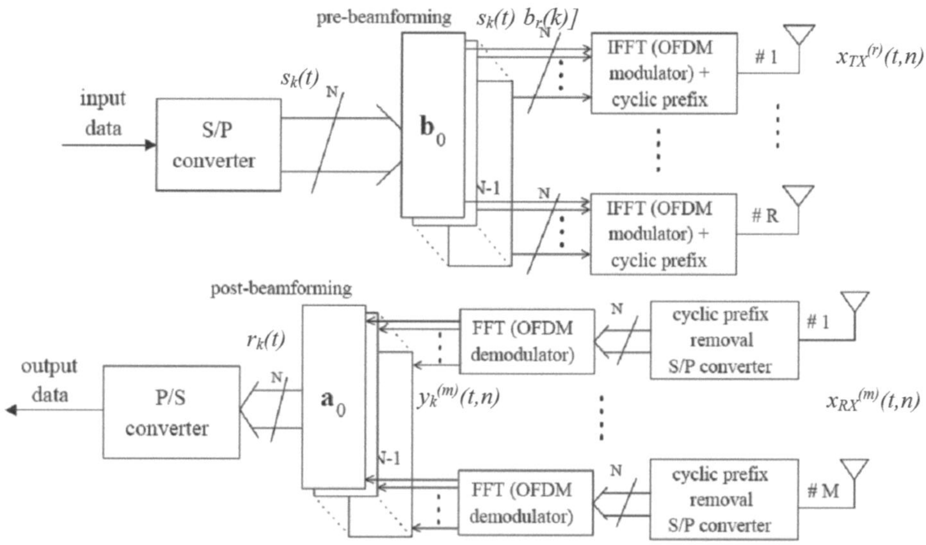

Fig. 4. Block diagram of MIMO OFDM communication system with adaptive beamforming.

The control of the antenna array in the transmit side is done via a number of vectors $\boldsymbol{b}_k = [b_1(k)\ b_2(k)...b_R(k)]^T$, with which the corresponding information symbols are multiplied before the process of OFDM modulation. In the receive side after OFDM demodulation the FFT demodulated information symbols $y_k^{(m)}(t)$ for each subcarrier k, are multiplied with a number of vectors for control of the antenna array $\boldsymbol{a}_k = [a_1(k)\ a_2(k)...a_M(k)]^T$ за $L \ge N_t$ (Thung et al., 2001):

$$\mathbf{r}_\mathbf{k}(t) = \mathbf{a}_\mathbf{k}^\mathbf{H}\mathbf{y}_\mathbf{k}(t) = \mathbf{a}_\mathbf{k}^\mathbf{H}\mathbf{H}(k)\mathbf{b}_\mathbf{k}s_k(t) + \mathbf{a}_\mathbf{k}^\mathbf{H}\mathbf{n}_\mathbf{k}(t), \tag{38}$$

where $\boldsymbol{H(k)}$ is the matrix of the frequency response of the MIMO radio-channel, $\boldsymbol{n}_k(t)$ is the vector sum of NBI and AWGN, $r_k(t)$ is the signal at the output of the control block of the antenna array of the receiver for the k-th carrier frequency. It is assumed that all the MIMO channel state information (CSI) is available at the transmitter, including the statistical characteristics of the narrowband and wideband noises. The maximum of the signal-to-noise plus NBI is obtained when the receiver has characteristics equal to a matched filter in relation to the communication channel (Wong et al., 2001), (Iserte et al., 2001):

$$SNIR_k\big|_{MAX} \Leftrightarrow \mathbf{a}_\mathbf{k} = \alpha_k\mathbf{R}_\mathbf{n}^{-1}(k)\mathbf{H}(k)\mathbf{b}_\mathbf{k}, \tag{39}$$

where $\mathbf{R}_\mathbf{n}(k) = E\left\{\mathbf{n}_\mathbf{k}(t)\mathbf{n}_\mathbf{k}^\mathbf{H}(t)\right\}$ is the covariance addition matrix of the NBI and AWGN.

In the real systems there are limits for the total emitted power from all of the antennas. For MIMO OFDM systems with adaptive beamforming of the transmission antenna system the emitted power from all of the antennas for the k-th subcarrier is proportional to $\|\mathbf{b}_k\|^2$. The goal of the optimization procedure is to find an optimal vector with emitted power weight coefficients - $\boldsymbol{b}_k$ for each antenna:

$$J(k)\big|_{MIN} = SNIR_k - \lambda(k)\left(\|\mathbf{b}_k\|^2 - |\beta(k)|^2\right), \tag{40}$$

where $\lambda(k)$ are the eigenvalues, $|\beta(k)|^2$ is limit of the emitted power from all antennas.

3.2 NBI suppression with low-complexity precoding at the transmit side

Methods based on forward precoding of the signal at the transmission side assume that the wireless channel characteristics (narrowband and wideband noises included) are known in advance at the transmission side. One recently proposed method of low computational complexity is precoding at the transmit side with maximizing the diversity of the received signals, Fig.5 (Rico et al., 2007).

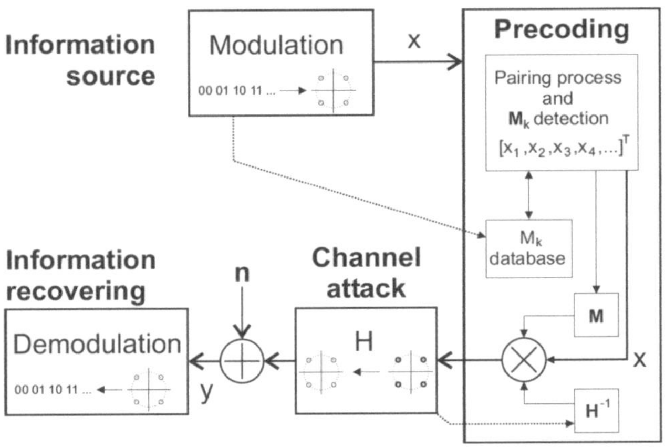

Fig. 5. Block diagram of MIMO OFDM communication system with forward precoding

This approach uses the information about the type of the transmitted symbols as criteria for the selection of optimal precoding matrix. The signal at the input of the receiver side is given as:

$$\mathbf{y} = \mathbf{H}\mathbf{T}\mathbf{x} + \mathbf{n} \tag{41}$$

where $\boldsymbol{y}$ is the vector $[n_r \ x \ 1]$ of the received signal, $\boldsymbol{x}$ - vector $[n_t \ x \ 1]$ of the transmitted signal, $\boldsymbol{H}$ – matrix $[n_t \ x \ n_r]$ of the communications channel, $\boldsymbol{T}$ – precoding matrix $[n_t \ x \ n_r]$, $\boldsymbol{n}$ – vector $[n_r \ x \ 1]$ of the complex additive noise signal at the input of the receiver. The precoding matrix of the maximum diversity $\boldsymbol{T}$ is used for maximizing the energy of the received signal and to suppress the interference signals. According (Rico et al., 2007), $\boldsymbol{T}$ is defined as:

$$\mathbf{T} = \frac{1}{|\mathbf{H}|}\begin{bmatrix} \sum_{i=1}^{j} H_{i,1}\mathbf{M}_{i,1} & \sum_{i=1}^{j} H_{i,1}\mathbf{M}_{i,2} & \cdots & \cdots & \sum_{i=1}^{j} H_{i,1}\mathbf{M}_{i,j} \\ \sum_{i=1}^{j} H_{i,2}\mathbf{M}_{i,1} & \sum_{i=1}^{j} H_{i,2}\mathbf{M}_{i,2} & \cdots & \cdots & \sum_{i=1}^{j} H_{i,2}\mathbf{M}_{i,j} \\ \vdots & \vdots & \ddots & & \vdots \\ \vdots & \vdots & & \ddots & \vdots \\ \sum_{i=1}^{j} H_{i,j}\mathbf{M}_{i,1} & \sum_{i=1}^{j} H_{i,j}\mathbf{M}_{i,2} & \cdots & \cdots & \sum_{i=1}^{j} H_{i,j}\mathbf{M}_{i,j} \end{bmatrix}, \quad (42)$$

where $\boldsymbol{M}_{i,j}$ is a sub-precoding matrix of identical dimensions to **T**, the function of which is maximizing the energy of the received signal by using apriority information of the transmitted symbols.

3.3 Space block coding

This type of block coding is used for improving the communications quality, including better interference suppression. Transmitted information is coded and transmitted via several antennas simultaneously to obtain maximum space diversity of the emitted signal. The main idea of this approach is that if the error probability of receiving a message by transmission over a wireless channel is p, with the simultaneous transmission of n orthogonal copies of the message over n independent wireless channels, the total error probability is p^n. Hence, the use of space block codes leads to relatively lower error probability or, under identical other conditions, to an increase of the channel throughput.

a. Orthogonal Space-Time Block Coding (OSTBC)

Alamouti (Alamouti, 1998) proposes an OSTBC transmit matrix for complex signals by using two transmit antennas $N_T = 2$. The block diagram of an OSTB precoder is shown on Fig.6. The proposed matrix is the only one which ensures maximum diversity equal to 2 in the transmit side with code rate $R=1$:

$$\mathbf{G}_2 = \begin{bmatrix} S_1 & -S_2{}^* \\ S_2 & S_1{}^* \end{bmatrix}, \quad (43)$$

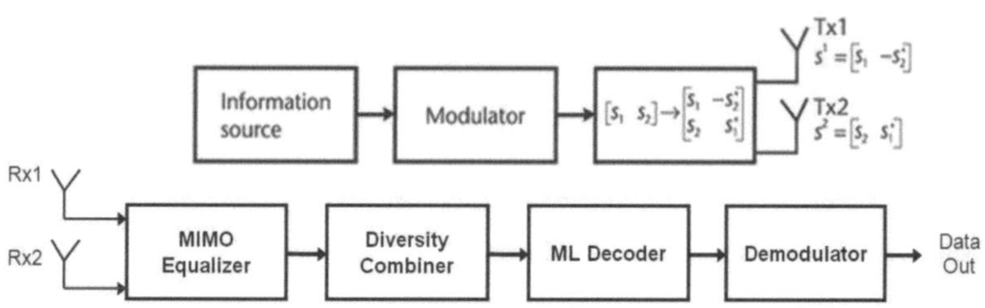

Fig. 6. Block diagram of an OSTBC MIMO OFDM system.

At the receiver side the decoding is performed using the method proposed by Tarokh (Tarokh, 1999), applying the maximum likelihood (ML) criteria. For N_T=2, R=1 and coding matrix $\mathbf{G}_2$, the decoded signal is described by the following equations:

$$\hat{S}_1 = \arg\min\left(\left|\left[\sum_{j=1}^{m}\left(r1^j h1,_j * +(r2^j)*h2,_j\right)\right]-s1\right|^2 + \left(-1+2\sum_{j=1}^{m}\sum_{i=1}^{2}\left|hi,_j\right|^2\right)\left|s1\right|^2\right), \quad (44)$$

$$\hat{S}_2 = \arg\min\left(\left|\left[\sum_{j=1}^{m}\left(r1^j h2,_j * -(r2^j)*h1,_j\right)\right]-s2\right|^2 + \left(-1+2\sum_{j=1}^{m}\sum_{i=1}^{2}\left|hi,_j\right|^2\right)\left|s2\right|^2\right), \quad (45)$$

where $\hat{S}_K$ for $k = 1,2$, is the estimation of the decoded signal for the k-th OFDM symbol at the input of the demodulator, $h_{i,j}$ - transmitting coefficient of the channel from the i-th TX antenna to j-th RX antenna, $r_P{}^j$ – received signal from the j-the RX antenna for the time interval of the OFDM symbol $P = 1,2$.

b. Space Frequency Block Code (SFBC)

There are many algorithms for the synthesis of Space-Frequency Block Codes. The algorithm described below is characterized with maximum of the diversity coefficient and maximal code rate $R = 1$ (Shao et al., 2003). Let's consider the radio-communications system shown on Fig.7, with M transmit and N receive antennas and number of subcarriers N_C, N_C>>M, N. The synthesis of the space-frequency coding matrix **C** passes through several stages. At the first stage, the vector of the input symbols **s** with dimension $[N_C\ x\ 1]$ is divided into groups of G vectors $\{\mathbf{s}_g\}$ with dimension $[ML\ x\ 1]$. At the second stage, each vector $\mathbf{s}_g$ is multiplied from the left with a coding rotational (CR) matrix $\boldsymbol{\Theta}$ (Xin et al., 2001) with dimension $[ML\ x\ ML]$. As a result of the multiplication a vector v_g with dimension $[ML\ x\ 1]$ is obtained. At the third stage each vector v_g is divided to L sub-vectors with dimension $[M\ x\ 1]$, after which each sub-vector is used for the synthesis of a diagonal sub-matrix $\mathbf{D}_{Sg,k}$. At the last stage, all GL sub-matrices from all G groups are interleaved, so that the space-frequency coding matrix **C** show on Fig.8 is obtained (Shao et al., 2003).

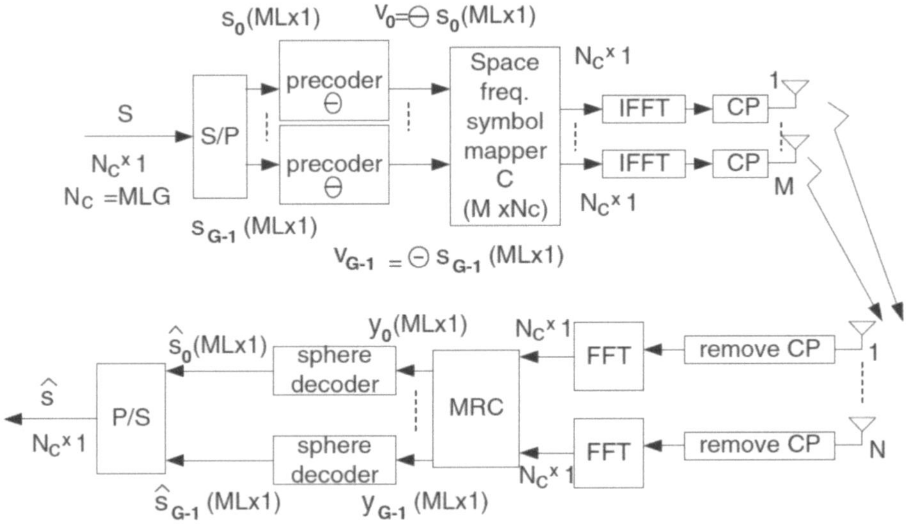

Fig. 7. Block diagram of a MIMO SFBC-OFDM system.

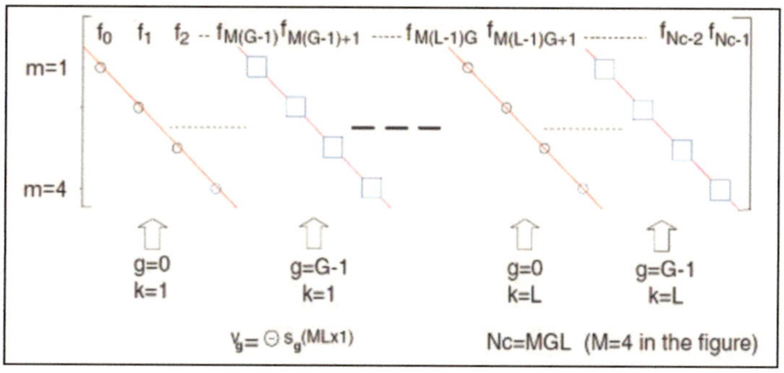

Fig. 8. Structure of the SFBC matrix **C**.

At the receive side the signal from each of the receive antennas is equalized, synchronized and transformed to baseband. After removing the guard interval, the OFDM symbols are demodulated via FFT and are fed into the space-frequency baseband decoder (SFBDC). After combining the information symbols from group *g* for all *N* receive antennas the result at the output of the combination scheme (MRC) is:

$$\mathbf{y}_g = \boldsymbol{\Sigma}_g^{-1/2}\boldsymbol{\Theta}\,\mathbf{s}_g + \boldsymbol{\eta}_g\,. \tag{46}$$

The decoding of the information symbols is performed by the method described in [459], by applying the maximum likelihood criteria:

$$\hat{\mathbf{s}}_g = \arg\min \left\|\mathbf{y}_g - \mathbf{H}\mathbf{s}_g\right\|^2, \tag{47}$$

where $\mathbf{s_g}$ is the transmitted information signal and $\mathbf{y_g}$ the decoded received signal at the output of the combination scheme (Shao et al, 2003).

4. RFI suppression in MIMO DSL DMT communication systems

Some of the specific methods for RFI suppression in digital subscriber lines such as optimal dynamic spectrum management of the signal (DSM Level 3), optimal precoding and multichannel block adaptive filtering will be considered below.

4.1 Optimal dynamic signal spectrum management (DSM Level 3)

Dynamic spectrum management is a new promising alternative, based on optimal dynamic management of the spectrum of each transmitter depending on the parameters of the specific wire-line (copper cable) channel. Major practical problem of the proposed algorithms for DSM is the extremely high computational complexity exponentially proportional to the number of tones in the discrete multi-tone (DMT) symbols.
If we consider a MIMO DSL DMT communication system with independent transmission channels, for each tone of the DMT symbol the following equation is valid:

$$\mathbf{y}_k = \mathbf{H}_k\mathbf{x}_k + \mathbf{n}_k\,, \tag{48}$$

where $\mathbf{y_k}$ is the vector of the received signals for tone *k*; $\mathbf{x_k}$ is the vector of the transmitted signals for tone *k*; $\mathbf{n_k}$ is the additive sum vector of the background AWGN, crosstalk noise between the twisted pairs and RFI for tone k; $\mathbf{H_k}$ is the *[NxN]* dimensional channel matrix, the diagonal elements of which are the transmission functions of the direct channels and the rest are the transmission functions of the crosstalk; N is the number of twisted pairs in the cable. The total transmission speed for subscriber n from the MIMO channel is:

$$R_n = \sum_{k=1}^{K} b_k^n \quad [bits] \tag{49}$$

where b_k^n is the number of transmitted bits in the QAM symbol for subscriber n and tone k. For a channel with two subscribers with limiting conditions for the maximal emitted power and in conformity with the PSD mask for each of the DSL transmitters, the spectrum optimization problem is the following (Cendrillon et al., 2003):

$$\max_{S1,S2} R_2 \quad for \quad R_1 \geq R_1^{target}, \tag{50}$$

$$\sum_{k=1}^{K} s_k^n \leq P_n, \quad n = 1,2\,. \tag{51}$$

The direct approach to solving the optimization problem (50-51) leads to an exponential increase of the computational complexity. The computational complexity could be reduced significantly by using a Lagrange unconditional optimization instead of the conditional optimization from (50-51):

$$L = \sum_{k=1}^{K} L_k + \lambda_1 P_1 + \lambda_2 P_2, \tag{52}$$

$$L_k = wb_k^1 + (1-w)b_k^2 - \lambda_1 s_k^1\left(b_k^1, b_k^2\right) - \lambda_2 s_k^2\left(b_k^1, b_k^2\right), \tag{53}$$

$$\max_{S1,S2} L\left(w, \lambda_1, \lambda_2, s_k^1, s_k^2\right), \tag{54}$$

where w is the Lagrange optimization constant. The limits for the maximal emitted power are imposed indirectly through the factors λ_1 and λ_2. The proposed method leads to the division of the global optimization problem (50-51) to K independent Lagrange optimization problems (53-54) with linear computational complexity (Cendrillon et al., 2003):

$$O(K) = K\left(b_{\max} + 1\right)^2. \tag{55}$$

4.2 Optimal precoding

Modern GDSL communication systems employ time, frequency, code and space multiplexing to achieve maximum transmission speed in a cable channel. One of the methods to achieve this goal is the design of an optimal linear precoder, that maximizes the total quantity of transmitted information under the condition of imposed limits for the Minimal Mean Square Error (MMSE).

Let us consider a complex model of a MIMO DSL channel (Perez-Cruz et al., 2008):

$$\mathbf{y} = \sqrt{snr}\mathbf{HPx} + \mathbf{w}\ , \tag{56}$$

where **y** is a n-dimensional vector of the received discrete signals; **x** is a m-dimensional vector of the transmitted discrete signals, which are independent random values with zero average and unit variance; **w** is a n-dimensional additive sum vector of the received signals of the background complex AWGN and complex RFI; **H** is the $[n\ x\ m]$ dimensional channel matrix, the diagonal elements of which are the transmission functions of the direct channels and the rest of the elements are the transmission functions of the crosstalk; snr is a scale factor function of the total transmitted power; **P** is a $[m\ x\ m]$ dimensional precoding matrix. The precoding matrix **P** is obtained as the result of a nonlinear optimization problem for maximizing the transmitted information and thus mitigating the RFI, under the following limiting conditions:

$$\max_{\mathbf{P}} I(\mathbf{x},\mathbf{y})\ , \tag{57}$$

$$Tr\left\{E[\mathbf{Pxx}^H\mathbf{P}^H]\right\} = Tr\left\{\mathbf{PP}^H\right\} \leq m\ . \tag{58}$$

The solution of the optimization procedure (57-58) is described through the equation:

$$\mathbf{P}^* = \lambda^{-1}\mathbf{H}^H\mathbf{HP}^*\mathbf{E}\ . \tag{59}$$

Here the parameters λ and the matrix of the Minimal Mean Square Error (MMSE) - **E** are determined by (Kay, 2008):

$$\lambda = \left\|\mathbf{H}^H\mathbf{HP}^*\mathbf{E}\right\| / \sqrt{m}\ , \tag{60}$$

$$\mathbf{E} = E\left[\left(\mathbf{x} - E[\mathbf{x}\mid\mathbf{y}]\right)\left(\mathbf{x} - E[\mathbf{x}\mid\mathbf{y}]\right)^H\right]. \tag{61}$$

For finding **P**, which is a global solution of the optimization problem (57-58), it is necessary to find the local solutions for the different values of the signal-to-noise ratio. Another computational problem is determining the MMSE matrix – **E**, which is of an exponential computational complexity. When the dimension of the **E** matrix is big, instead of direct computation for finding a representative estimation of the error matrix the *Monte Carlo* method is used. After this the precoding matrix **P** is obtained using the iteration procedure (Perez-Cruz et al., 2008):

$$\mathbf{P}_{k+1} = \lambda^{-1}\left(\mathbf{P}_k + \mu\, snr\, \mathbf{H}^H\mathbf{HP}_k\mathbf{E}\right), \tag{62}$$

$$Tr\left\{\mathbf{P}_{k+1}\mathbf{P}_{k+1}{}^H\right\} = m\ , \tag{63}$$

where μ is a constant and λ^{-1} is chosen according the condition (60).

4.3 Multichannel adaptive filtering

The algorithms for block-based multichannel transform domain adaptive filtering solve specific problems in relation with the space-time interferences between the input signals of the adaptive filter. Two major approaches for the suppression of the space-time interferences between the input signals are available: Frequency Domain Adaptive Filtering (FDAF) and Transform Domain Adaptive Filtering (TDAF).

a. Multichannel FDAF

This method is based on the block approach in solving the multichannel identification problem by forming each block from L consecutive samples from the error signal vector **e(n)**. The generalized description of a MIMO system leads to the solution of a system of equations, by assigning one equation to each filter output q (Spors et al., 2009):

$$\mathbf{e}_{\mathbf{q}}(n) = \mathbf{y}_{\mathbf{q}}(n) - \sum_{p=1}^{P} \hat{\mathbf{h}}_{\mathbf{p,q}}^{\mathbf{T}} \mathbf{x}_{\mathbf{p,q}}(n), \tag{63}$$

The resulting matrix is transformed via DFT to a diagonal matrix. The error signal block **e(m)** described in the time domain as a block with length L consecutive samples is:

$$\mathbf{e}(m) = [e(mL), e(mL+1), ..., e(mL+L-1)]^T, \tag{64}$$

where m is the index of the block. The input signal of the system $\boldsymbol{y}(m)$ in the time domain is: described in a similar way:

$$\mathbf{y}(m) = [y(mL), y(mL+1), ..., y(mL+L-1)]^T. \tag{65}$$

The algorithm requires DFT with dimension $2L$ which needs preliminary addition of zero samples to the error signal and the output signal of the filter. The input signals in the frequency domain are described as:

$$\mathbf{X}(m) = [\mathbf{X}_1(m), \mathbf{X}_2(m), ..., \mathbf{X}_P(m)]. \tag{66}$$

It can be shown that the resulting normal equation for the MIMO case can be decomposed into a series of independent multiple-input single-output (MISO) normal equations for each DSL channel. Hence, the consideration of a MISO system is sufficient in the context of this work. Under the above considerations a generalized FDAF algorithm for a MISO system could be synthesized in the following four equations (Spors et al., 2009):

$$\mathbf{S}(m) = \lambda \mathbf{S}(m-1) + (1-\lambda)\mathbf{X}^H(m)\mathbf{G}\mathbf{X}(m), \tag{67}$$

$$\mathbf{K}(m) = (1-\lambda)\mathbf{S}^{-1}(m)\mathbf{X}^H(m), \tag{68}$$

$$\mathbf{e}'(m) = \mathbf{y}'(m) - \mathbf{G}\mathbf{X}(m)\hat{\mathbf{h}}'(m-1), \tag{69}$$

$$\hat{\mathbf{h}}'(m) = \hat{\mathbf{h}}'(m-1) + \mathbf{G}\mathbf{K}(m)\mathbf{e}'(m), \tag{70}$$

where **S(m)** is the co-variation matrix, λ the optimization constant; $\hat{\mathbf{h}}(\mathbf{m})$ the vector with the estimations of the filter coefficients; **G** zero matrix for complementing the signal with zero samples; **K(m)** is the vector with Kalman amplification coefficients.

b. Block-based Multichannel TDAF

Transform-domain adaptive filtering (TDAF) is a technique that performs the filter adaptation in a transform domain. In the ideal case, the far-end signals will be decorrelated

by a suitably chosen transformation. The ideal transformation can be deduced from the covariance matrix $Rxx(n)$ and is data-dependent in general.
The spatio-temporal decoupling consists of two steps: (1) temporal decoupling using a discrete Fourier transform (DFT) based transformation and (2) a spatial decoupling using a unitary transform. In order to derive a block-based algorithm for multichannel TDAF both block-based FDAF and the concept of TDAF are combined in the following. The two stage approach to TDAF separates the temporal decoupling from the spatial decoupling. Hence, FDAF can be utilized for temporal decoupling combined with the concept of spatial decoupling from TDAF. For this purpose the eigenvalue decomposition of covariance matrix of TDAF is introduced into (Spors et al., 2009):

$$\underline{\mathbf{S}}(m) = \mathbf{A}\,\mathbf{U}(m)\mathbf{T}(m)\mathbf{U}^H(m)\mathbf{A}^T , \tag{71}$$

$$\underline{\mathbf{X}}(m) = \mathbf{X}(m)\mathbf{A}\,\mathbf{U}(m) , \tag{72}$$

where **T(m)** is transformed covariance matrix and **U(m)** is an unitary matrix. Furthermore introducing (72) into the error signal of FDAF (69) reads:

$$\underline{\mathbf{e}}'(m) = \underline{\mathbf{y}}'(m) - \mathbf{G}\underline{\mathbf{X}}(m)\mathbf{G}_U\hat{\underline{\mathbf{h}}}'(m-1) . \tag{73}$$

Finally, the coefficient update is introduced by equation (74) where G, G_U are constant matrixes.

$$\hat{\underline{\mathbf{h}}}'(m) = \mathbf{G}_U\hat{\underline{\mathbf{h}}}'(m-1) + \mathbf{G}\underline{\mathbf{K}}(m)\underline{\mathbf{e}}'(m) . \tag{74}$$

The algorithm defined by Eq. (67-74) constitutes a combination of the concepts of FDAF and TDAF. The decoupling of the covariance matrix is performed in a two-step approach. The DFT is used for temporal decoupling and an eigenvalue decomposition for spatial decoupling. The temporal decoupling is performed in a very efficient manner by applying FDAF. The required DFTs can be realized efficiently by the fast Fourier transform (FFT). For an exact spatial decoupling, an eigenvalue decomposition has to be performed. However, the derived formulation allows also the usage of generic MIMO filtering of the far-end signals with the potential of finding efficient approximations of the exact solution. Fig. 9 illustrates a block-diagram of the algorithm exploiting these structures.

5. Comparison of NBI suppression methods

A comparison of some of the above described NBI suppression methods based on simulation experiments is given below. Methods with relatively low computational complexity were selected for practical reasons. The bit error ratio (BER) as a function of the Signal to Interference Ratio (SIR) is estimated and the simulation results for different MIMO systems are presented.

5.1 Computational complexity of NBI suppression methods

Algorithm analysis (Knuth, 1997) stands for determination of the amount of resources (i.e. time and storage) necessary to execute it. Most algorithms are designed to work with inputs of arbitrary length. Usually the efficiency or running time of an algorithm is stated as a

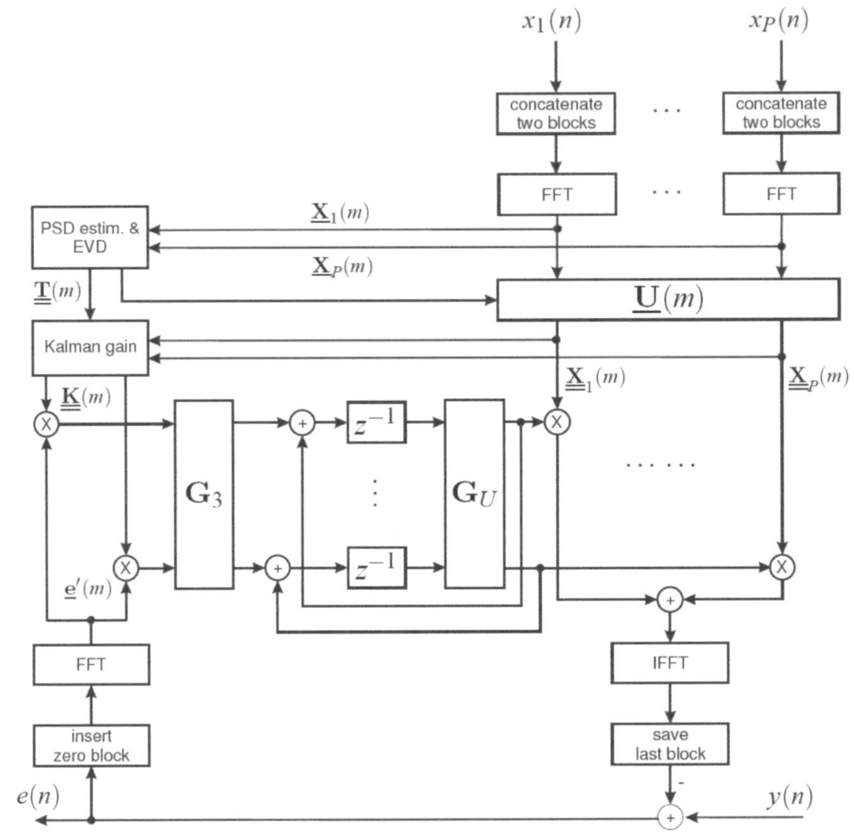

Fig. 9. Block diagram of TDAF algorithm.

function relating the input length to the number of steps (time complexity) or storage locations (space complexity). Algorithm analysis is an important part of a broader computational complexity theory, which provides theoretical estimates for the resources needed by any algorithm which solves a given computational problem. These estimates provide an insight into reasonable directions of search for efficient algorithms.

In theoretical analysis of algorithms it is common to estimate their complexity in the asymptotic sense, i.e., to estimate the complexity function for arbitrarily large input. Big O notation, omega notation and theta notation are used to this end. Usually asymptotic estimates are used because different implementations of the same algorithm may differ in efficiency. However the efficiencies of any two "reasonable" implementations of a given algorithm are related by a constant multiplicative factor called a *hidden constant*.

Time efficiency estimates depend on what we define to be a step. For the analysis to correspond usefully to the actual execution time, the time required to perform a step must be guaranteed to be upper bounded by a constant.

The computational complexity of some scalar and matrix operations is listed in Tables 1 and 2. Here $M(n)$ stands for the complexity of the chosen multiplication algorithm (Knuth,1998). Based on such estimations the computational complexity of major NBI suppression methods for wireless MIMO and MIMO GDSL systems are given in Tables 3 and 4. For the simulation experiments the methods with lowest computational complexity are chosen.

Function	Input	Algorithm	Complexity
Polynomial evaluation	One polynomial of degree n with fixed-size polynomial coefficients	Direct evaluation	$\Theta(n)$
		Horner's method	$\Theta(n)$
Polynomial GCD (over $Z[x]$ or $F[x]$)	Two polynomials of degree n with fixed-size polynomial coefficients	Euclidean algorithm	$O(n^2)$
		Fast Euclidean algorithm	$O(n (\log n)^2 \log\log n)$
DFT, IDFT, Circular Convolution	m numbers, n-digit each	Radix 2^K FFT	O(m log(m))

Table 1. Complexity of some basic algebra functions.

Function	Input	Output	Algorithm	Complexity
Matrix multiplication	Two $n \times n$-matrices	One $n \times n$-matrix	Schoolbook matrix multiplication	$O(n^3)$
			Strassen algorithm	$O(n^{2.807})$
			Coppersmith–Winograd algorithm	$O(n^{2.376})$
Matrix multiplication	One $n \times m$-matrix & One $m \times p$-matrix	One $n \times p$-matrix	Schoolbook matrix multiplication	$O(nmp)$
Matrix inversion	One $n \times n$-matrix	One $n \times n$-matrix	Gauss–Jordan elimination	$O(n^3)$
			Strassen algorithm	$O(n^{2.807})$
			Coppersmith–Winograd algorithm	$O(n^{2.376})$
Determinant	One $n \times n$-matrix	One number with at most $O(n \log n)$ bits	Laplace expansion	$O(n!)$
			LU decomposition	$O(n^3)$
			Bareiss algorithm	$O(n^3)$
			Fast matrix multiplication	$O(n^{2.376})$

Table 2. Complexity of some matrix algebra functions.

NBI Suppression Methods	Number of Additions	Number of Multiplications	Computational Complexity
Frequency Excision	$6MN$	$4MN\log(N)$	$\sim O(KMN\log(N))$
Adaptive Linear Filtering	KMN	$28MN+KMN^2$	$\sim O(KMN^2)$
Identification and Compensation	$2MN^2$	$M(N+2)N^3$	$\sim O(KMN^4)$
Adaptive Antenna Beamforming	NA*	NA*	$\sim O(KN(MS)^3)$
Optimal Transmitter Precoding	NA*	NA*	$\sim O(KN(MS)!)$
OSTBC M,S=2 M,S=3,4	$N(3MS-1)$ $N(6MS+M-1)$	$N(3MS+2)$ $N(7MS+6)$	$\sim O(KMSN)$
SFBC	NA*	NA*	$\sim O(KNG(MS)!)$

*Note: * number of additions and multiplications depends on the method used for matrix computation.*

Table 3. Computational complexity of NBI suppression methods in wireless MIMO systems.

RFI Suppression Methods	Number of Additions	Number of Multiplications	Computational Complexity
Frequency Excision	$MN+K$	$KMN\log(N)$	$\sim O(K2N\log(2N))$
Adaptive Linear Filtering	KMN	$KMN+MN^2$	$\sim O(K(2N)^2)$
Identification and Compensation	$2MN^2$	$KM(N-1)N^3$	$\sim O(K(2N)^4)$
Dynamic Spectrum Management Level 3	NA*	NA*	$\sim O(KNM^3)$
Optimal Precoding	NA*	NA*	$\sim O(KNM!)$
Block-based Multichannel FDAF	NA*	NA*	$\sim O(KNML!)$
Block-based Multichannel TDAF	NA*	NA*	$\sim O(KNML^4)$

*Note: * number of additions and multiplications depends on the method used for matrix computation.*

Table 4. Computational complexity of RFI suppression methods in MIMO GDSL system.

5.2 Performance of NBI suppression methods for wireless systems

To evaluate the performance of NBI suppression methods in wireless systems, simulations relative to complex baseband presentation are conducted, assuming standard MIMO OFDM receiver. The information source is modeled by a generator of uniformly distributed random integers based on the modified version of Marsaglia's "Subtract with borrow algorithm" (Tezuka et al., 1993). The method can generate all the double-precision values in the closed interval $[2^{-53}, 1-2^{-53}]$. Theoretically, it can generate over 2^{1492} values before repeating itself. The channel coder is implemented as a convolutional encoder with code rate is $Rc = 1/2$ and Viterbi hard threshold convolutional decoding. A block interleaver - deinterleaver is implemented with and algorithm that chooses a permutation table randomly, using the initial state input that is provided. The Orthogonal Space-Time Block Encoder (OSTBC) for complex signals is realized using the methods described in (Alamouti, 1998), (Bernguer & Dong, 2003). The number of transmit antennas NT and receive antennas NR can be set from 1 to 4. For 2x2 MIMO system, the code rate is Rc=1 whereas for 3x3 and 4x4 MIMO systems, the code rate is Rc=1/2. The digital modulator block is implemented as a bank of NT blocks each one performing 256-point IFFT. The OFDM symbol per each output channel consists of 128 data bins. Each OFDM data symbol can use different modulation formats. In the experiments Grey encoded 64-QAM modulation format is used for each transmitter. After the IFFT process, the prefix and suffix guard intervals are added. The MIMO wireless flat fading channel is realized as given in (Bernguer & Dong, 2003). Channel estimation based on optimal training preamble is adopted. In the OFDM block demodulator, the guard prefix and suffix intervals are removed per each channel and 256-point FFT is applied. Finally, OSTB decoding, 64-QAM demodulation and error correction decoding are made.

The excision method is applied to the MIMO OFDM signal with a complex NBI at each input of the receiver. The signal is converted into the frequency domain by applying an FFT at each input, oversampled by 8 and the noise peaks in the spectra are limited to the determined threshold. After conversion back in the time domain the signal is fed into the corresponding OFDM demodulator. For more precise frequency excision, FFT of higher order than the one in the demodulator is applied.

For the realization of the filtering method a complex Adaptive Notch Filter Bank (ANFB) is connected at the receiver's input. The filtering process is applied independently at each receiver input. The adaptation algorithm tunes the filters at each receiver input, in such a way that the central frequency and bandwidth match to the NBI signal spectrum. In the simulations, the central frequency of the notch filters is chosen to be equal to the NBI central frequency, while their bandwidth is equal to 20% of the bandwidth between two adjacent OFDM sub-carriers.

The frequency cancellation method is implemented in several stages as discussed in section 2.2. First, the complex NBI frequency is estimated by finding the maximum in the oversampled signal spectrum per channel. Using ML approach, per each input, the NBI amplitude and phase are estimated. Then a NLS optimization algorithm is realized, where precise estimation of the NBI's complex amplitude, phase and frequency are done.

Using the above general simulation model of an OSTBC MIMO system, different experiments are performed, estimating the bit error ratio (BER) as a function of the Signal to Interference Ratio (SIR). The NBI is modeled as a single complex tone, the frequency of which is located in the middle between two adjacent OFDM sub-carriers. The MIMO channels are subject to flat fading and background AWGN. The Signal to AWGN ratio at the input of the OFDM receiver is 15dB. For comparison in Fig.10 the results for simulation of a conventional Single Input Single Output (SISO) system employing different NBI suppression methods are presented: frequency excision, frequency cancellation and adaptive filtering. The experiments show that for high NBI, where the SIR is less than 0 dB, all three suppression methods lead to a significant improvement in performance. It is evident that the frequency cancellation method gives best performance, while filtering is better than the excision method.

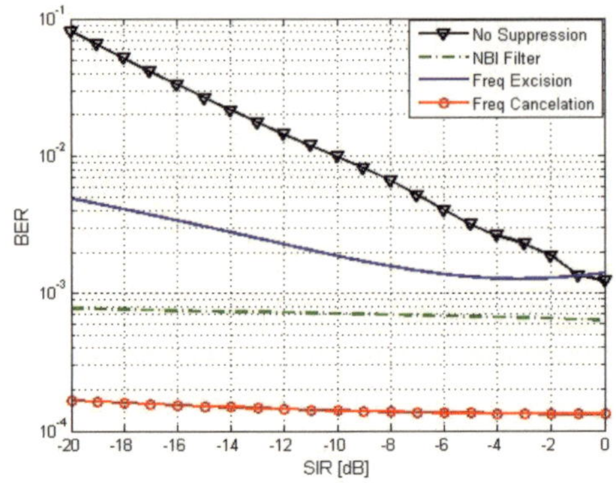

Fig. 10. BER as a function of SIR for SISO channel

In the case of 2x2 and 4x4 MIMO channels the results are presented in Fig.11 a,b. It could be seen that filtering with ANFB gives better performance for higher values of the antenna diversity in the OSTBC MIMO system. The frequency excision method manifests good performance for high SIR. It should be noted that the adaptive filtering and frequency cancellation schemes lead to a slight degradation of the overall performance when SIR>0, which is due to either an amplitude and phase distortion of the adaptive notch filters or to a wrong estimation of the NBI parameters during the identification process. The degradation could be reduced using higher-order notch filters or implementing more sophisticated identification algorithms. The degradation effect could be avoided by simply switching off the filtering when SIR~0. Such a scheme is easily realizable, as the amplitude of the NBI can be monitored at the bandpass output of the filters (Fig. 2).

The results have outlined the fact that the Frequency Identification and Cancellation method achieves the highest performance. However, the extremely high computational complexity limits its applications to the hardware resources. In this aspect, the Adaptive Notch Filter Bank offers a trade off between good performance and reasonable computational complexity. The frequency excision method shows relatively good results and his main advantage is its computational efficiency.

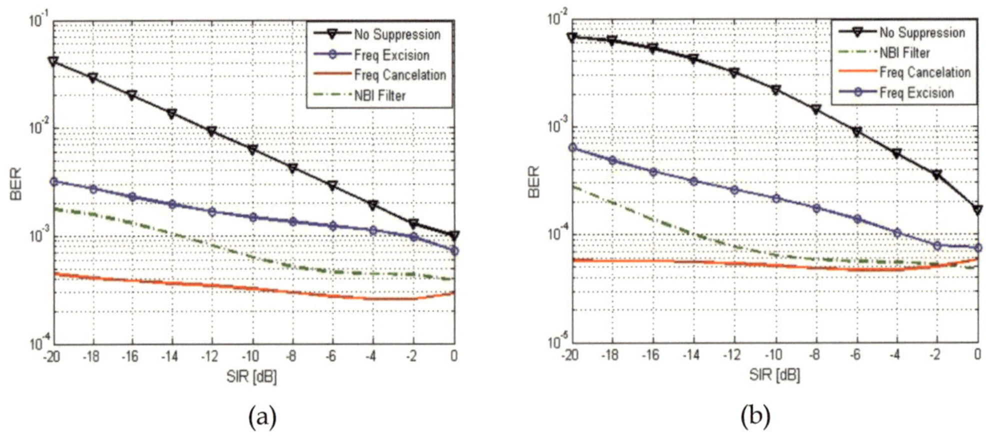

Fig. 11. BER as a function of SIR for 2x2 and 4x4 MIMO channels

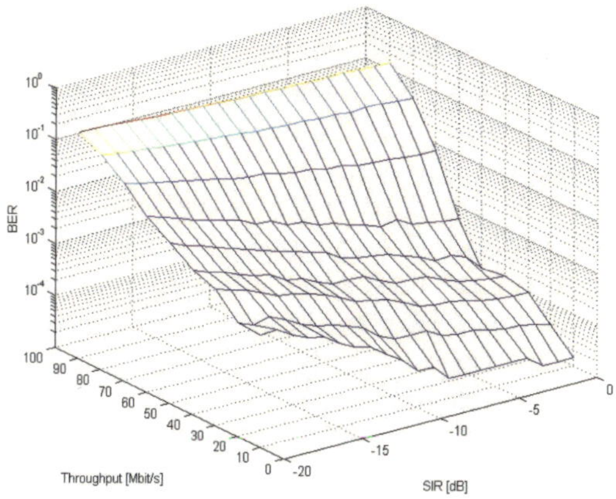

(a)

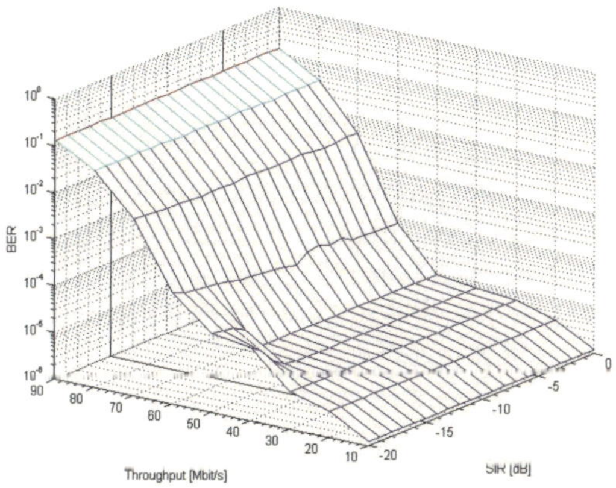

(b)

Fig. 12. BER, Throughput and SIR for 4x4 MIMO OSTBC single user communication system.

The relation between BER, Throughput and SIR for 4x4 MIMO OSTBC OFDM single user communication system in flat fading environment and constant AWGN level for SNR=20dB is presented on Fig.12 where (a) is without NBI excision and (b) with NBI excision.

5.3 RFI suppression performance in wire-line systems

To evaluate the performance of RFI suppression methods, simulations are conducted assuming typical MIMO DMT receiver. The implementation of the information source, channel encoder, interleaver - deinterleaver and optimal linear precoder for complex signals is the same as the one for wireless channels. For the simulations, the number of used cable pairs is: N=[1,2,3,4] and the corresponding number of MIMO inputs/outputs is: K=2N-1. The Digital Modulator is implemented as *8192-point* Inverse Fast Fourier Transform (IFFT) process. The DMT symbol consists of up to *4096 tones*. Each DMT data tone can use a different Grey encoded QAM modulation format depending on the Signal to Noise Ratio. After the IFFT process, the prefix and suffix guard intervals are added.

The MIMO wire-line channel model is realized in accordance with ITU-T Recommendations G.993.2 and G.996.1. The DSL cable test loop is modeled for common mode excitation, using the ABCD parameters block matrix. For experiments 26-gauge cable (*AWG 26*) is chosen. The background noise with Power Spectrum Density (PSD) at *-140 dBm/Hz* is modeled as Additive Complex Arbitrary White Gaussian Noise - Nc(0, 1). The model includes wide-scale frequency variations with standard statistics determined from measured actual Far End Crosstalk (FEXT) and Near End Crosstalk (NEXT) transfer functions.

In the DMT Demodulator the guard prefix and suffix intervals are removed and *8192-point* FFT is applied. Further, Frequency Domain Per-Tone Adaptive Channel Equalization and DMT demodulation are performed. Finally, OLP, *64-QAM* demodulation and Error Correction Decoding are implemented.

For the simulation of the Excision, Frequency Identification and Cancellation and Adaptive Filtering method in the MIMO GDSL the same principles as in wilreless system simulation are applied. The RFI is modeled as a complex single tone, the frequency of which is located in the middle between two adjacent DMT tones.

Using the above simulation model of MIMO Gigabit DSL system, different experiments have been performed, estimating BER as a function of the Signal to Interference Ratio (SIR). In respect to the number of used twisted pairs, four DSL systems are considered: Single Input Single Output (SISO) VDSL2 (1-pair), MIMO GDSL: 2, 3 and 4-pair. The DSL channels are subject to FEXT, NEXT and background AWGN with flat PSD at *-140 dBm/Hz*.

In case of 2 and 4-pair MIMO GDSL channel (Fig.13 a,b) the best results in terms of RFI filtering are obtained for the highest value of channel diversity: 4-pair Gigabit DSL MIMO system. The Frequency Excision Method manifests good performance for high SIR. It should be noted that the adaptive filtering scheme and frequency cancellation scheme lead to a slight degradation of the overall performance when SIR>0, which is due to the amplitude and phase distortion caused by the adaptive notch filters or due to a wrong estimation of RFI parameters during the identification procedure. The degradation could be reduced by implementing higher-order notch filters or by applying more sophisticated identification algorithms. Its negative effect can be avoided by switching off the filtering when SIR~0.

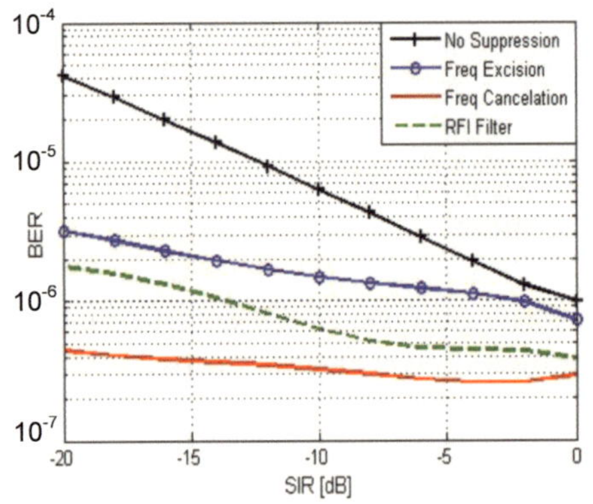

(a)

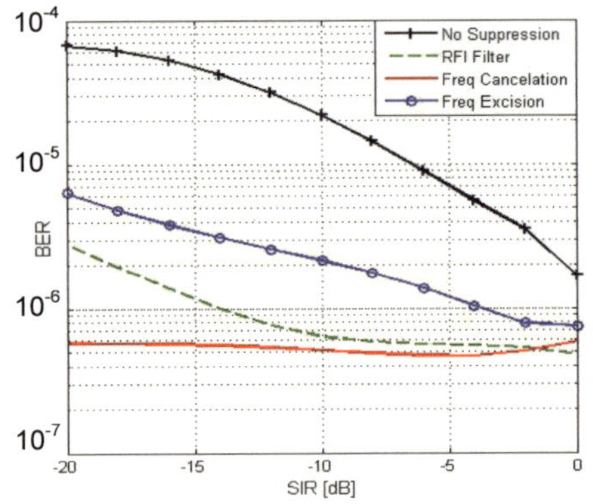

(b)

Fig. 13. BER as a function of SIR for 2-pair and 4-pair GDSL

The results have outlined the fact, that the Frequency Identification and Cancellation Method gives highest performance. However, the extremely high computational complexity limits its application up to the available hardware resources. In this aspect, the Complex Adaptive Notch Filter Bank turns to be the optimal narrowband interference suppression technique, offering the trade-off between excellent performance and low computational complexity. The Frequency Excision Method demonstrates reasonable performance as well as high computational efficiency. Figure 14 presents the relation between BER, Throughput and SNR for 4x4 MIMO GDSL DMT single user communication system for SIR=-10dB (Cable 24 AWG, L=300m.), where (a) is without RFI cancellation and (b) with RFI cancellation.

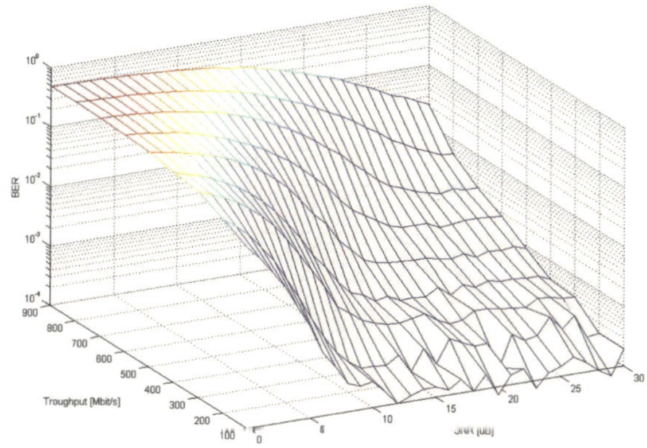

(a)

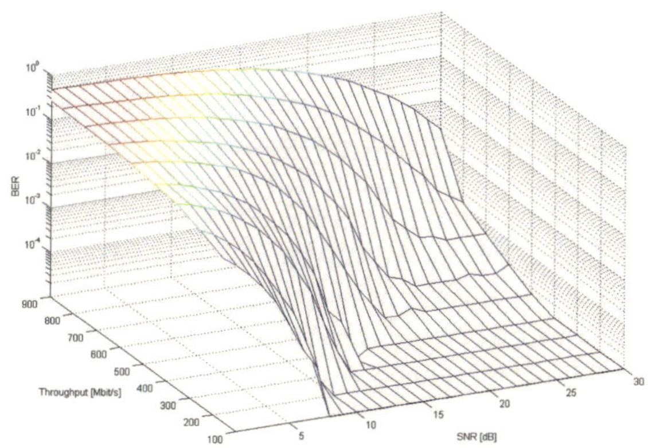

(b)

Fig. 14. Relation between BER, Throughput and SNR for 4x4 MIMO GDSL DMT single user communication system.

6. Conclusion

In this chapter some of the major NBI suppression methods were outlined. Their application in different types of wireless MIMO and MIMO DSL channels MIMO was considered. Using a simulation model the different methods were analyzed and compared in terms of computational complexity and error rate improvement. The authors consider such an approach to the comparison of narrowband interference suppression methods important for correct overall performance estimation. Each method has a different computational complexity and error rate performance depending on the type and parameters of the MIMO system and the communication channel. Choosing a NBI suppression method for practical implementation in most cases is a trade-off between good performance and lower computational complexity.

From the general methods for NBI suppression - Frequency Excision, Adaptive Filtering, Identification and Cancellation, the last gives best error rate performance. However, the extremely high computational complexity limits its application up to the available hardware resources. Complex Adaptive Filtering offers a trade-off between good error performance and lower computational complexity. The Frequency Excision Method demonstrates reasonable error performance as well as high computational efficiency.

The simulation results also show that the application of a general method for NBI suppression along with one of the specific methods that are used in wireless MIMO systems such as Orthogonal Space-Time Block Coding improves the error rate performance. Similar is the case in wire-line subscriber systems where other specific methods such as Dynamic Spectrum Management (Level 3), Optimal Precoding and Multi-channel Adaptive Filtering are used.

7. References

Alamouti S. (1998). "Simple Transmit Diversity Technique for Wireless Communications", *IEEE Journal Select. Areas Commun,*, Vol. 16, No. 8, October 1998, pp. 1451–1458.

Baccareli E.; Baggi M., Tagilione L. (2002). "A Novel Approach to In-Band Interference Mitigationin UltraWide Band Radio Systems, " *IEEE Conference on UltraWide Band Systems and Technologies*, 2002.

Bernguer I. & Dong X. (2003). "Space-Time Coding and Signal Processing for MIMO Communications", *J Comput. Sci &Technol.*, Vol. 18, No 6, pp. 689-702, Nov. 2003.

Carlemalm C.; Poor H., Logothetis A. (2004). "Suppression of multiple narrowband interferers in a spread-spectrum communication system", *IEEE J. Select. Areas Communications*, vol.3, No. 5, pp. 1431-1436, Sept 2004.

Cendrillon R.; Yu W., Moonen M., (2003). "Optimal Multi-user Spectrum Management for Digital Subscriber Lines", *IEEE Trans.Commun.*, 2003.

Giorgetti A.; Chiani M., Win M. (2005). "The Effect of Narrowband Interference on WidebandWireless Communication Systems," *IEEE Trans. Communications,* 53(12), pp. 2139–2149, December 2005.

Hsu F. & Giordano A. (1978). "Digital whitening techniques for improving spread spectrum communications performance in the presence of narrow-band jamming and interference," *IEEE Trans. on Communications,* COM-26, pp. 209–216, February 1978.

Iliev G.; Nikolova Z., Stoyanov G., Egiazarian K. (2004). "Efficient Design of Adaptive Complex Narrowband IIR Filters," *Proceedings of XII European Signal Processing Conference, EUSIPCO 2004,* Vienna, Austria, pp. 1597–1600, 6–10 September 2004.

Iltis R. & Milstein L. (1985). "An approximate statistical analysis of the Widrow LMSalgorithm with application to narrow-band interference rejection," *IEEE Trans. On Communications,* COM-33, pp. 121–130, February 1985.

Iserte A.; P´erez-Neira A., Hern´andez M. A. (2001). "Pre- and Post-Beamforming in MIMO Channels applied to HIPERLAN/2 and OFDM," in *IST Mobile Communic. Summit,* Barcelona, Sept. 2001.

ITU-T *(2006). Recommendation G.993.2 - 2006,* "Very High Speed Digital Subscriber Line 2 (VDSL 2)", February 2006.

ITU-T *(2006). Recommendation G.996.1,* "Test Procedures for Digital Subscriber Line (DSL) Transceivers", February 2006.

Johnson C. (1984). "Adaptive IIR filtering: Current results and open issues," *IEEE Trans. On Information Theory,* IT-30, pp. 237–250, March 1984.

Juang J.-C.; Chang C.-L., Tsai Y.-L. (2004). "An Interference Mitigation Approach Against Pseudolites," *Proceedings of the 2004 International Symposium on GNSS/GPS,* Sydney, Australia, 6–8 December 2004.

Kay S.; Fundamentals of Statistical Signal Processing: Estimation Theory. Englewood Cliffs, NJ, USA: Prentice - Hall, 1993.

Knuth D. (1997). *The Art of Computer Programming,* Volume 1: *Fundamental Algorithms,* ISBN 0-201-89683-4. Section 1.2.11: Asymptotic Representations, pp. 107–123. Third Edition. Addison-Wesley.

Knuth D. (1998). "Teach Calculus with Big O". *Notices of the American Mathematical Society* 45 (6): pp 687-693. (June/July 1998).

Laster J. & Reed J. (1997). "Interference Rejection in Digital Wireless Communications", *IEEE SIGNAL PROCESSING MAGAZINE,* MAY1997, pp.37-62.

Masreliez C. (1997). "Approximate non-Gaussian filtering with linear state and observation relations," *IEEE Trans. Automatic Control,* AC–20, pp. 107–110, February 1975.

Perez - Cruz F.; Rodrigues M., Verdu S. (2008). "Optimal Precoding for Digital Subscriber Lines", *International Conference on Communications (ICC),* 19-23 May 2008, Beijing (China).

Rico U.; Alsusa E., Masouros C. (2007). "A Simple Low-Complexity Precoding Technique for MIMO Systems", *IEEE Communications Society - WCNC 2007 proceedings.,* 1525-3511/07/2007.

Sampei S. (1997). *"Applications of Digital Wireless Technologies to Global Wireless Communications"*, Prentice Hall, 1997.

Shao L.; Roy S., Sandhu S. (2003). "Rate-one Space frequency block codes with maximum diversity gain for MIMO-OFDM", *IEEE Global Telecommunications Conference, 2003.(GLOBECOM '03),* Volume: 2, 1-5 Dec.2003 pp: 809 - 813.

Spors S.; Buchner H., Helwani K. (2009). „Block-Based Multichannel Transform-Domain Adaptive Filtering", *17th European Signal Processing Conference (EUSIPCO 2009), Glasgow, Scotland, August 24-28, 2009,* pp. 1735-1739.

Tarokh, V.; Jafarkhani H., Calderbank A. (1999). "Space-Time Blok Codes from Orthogonal Designs", *IEEE Trans. Inform. Theory,* Vol. 45, No. 5, July 1999, pp. 1456–1467.

Tezuka S.; L'Ecuyer P., Couture R. (1993). "On the lattice structure of the add-with-carry and subtract-with-borrow random number generators", *TOMACS,* vol. 3 , No 4, pp. 315 - 331, Oct. 1993.

Tung T.; Yao K., and Hudson R. (2001). "Channel Estimation and Adaptive Power Allocation for Performance and Capacity Improvement of Multiple-Antenna OFDM Systems," in *IEEE SPAWC,* Taiwan, March 2001, pp. 82–85.

Wong K.; Cheng R., Murch R. (2001). "Adaptive Antennas at the Mobile and Base Stations in an OFDM/TDMA System", *IEEE Trans. on Comm.,* vol. 49, pp.195–206, Jan. 2001.

Xin Y.; Wang Z. Giannakis G. (2001). Space-time constellation-rotating codes maximizing diversity and coding gains. *GLOBECOM* 2001.

17

Fundamentals for a Pragmatic MIMO Performance Evaluation

Shirook M. Ali
Advanced Technology, Research In Motion Limited
Canada

1. Introduction

The user's experience of a mobile communication system entirely depends on the performance of the bidirectional radio link between the base station (BS) and the handset. Each mobile network operator creates a system of linked BSs located to provide coverage of as large a physical area as possible. While the BS will typically be equipped with a relatively large high-gain antenna and a transmitter cable for delivering some tens of watts of radio frequency (RF) power, the handset relies on an antenna whose dimensions are severely constrained by those of the handset in which it is fitted, with typical maximum effective radiated power of 1 watt. With this said, the communication link quality, and hence the user's experience, can mainly get as good as the handset antenna performance.
Multiple Input Multiple Output (MIMO) has promised the potential for improved data rates and signal quality (Foschini, 1998). The promised improvement in the system's performance comes at the expense of added challenges, especially from the perspective of handset antenna design. An important aspect in the process of designing a multi-antenna handset is the proper evaluation of its performance. Although the antenna itself can be seen as a deterministic element, its performance becomes random once operated in a realistic, random, environment and in the presence of the operating user. The key questions in ensuring performance are: What are the parameters that need attention when designing the multi antenna system and how should the evaluation of this system be done?
In this chapter the author discusses some aspects that are fundamental for a pragmatic evaluation of multi-antenna system. This becomes especially important to realize at the early stages of a design process when the multi-antenna system is evaluated entirely or partially through simulations.

2. The antenna as part of the propagation environment and of the handset

Let's start our discussion by defining the antenna element as the transducer element that is designed to transmit or receive electromagnetic waves. In other words, an antenna converts electromagnetic waves into electrical currents and vice-versa. The interaction of the propagation channel environement with the antenna is through the antenna complex field radiation pattern, which in turn is scaled by the antenna efficiency. On the other hand, the interaction of the RF front-end in the handset with the antenna comes down to the amount of signal that goes through the antenna and the amount of signal that reflects back from it

into the RF front-end. Therefore, when designing a system, it is necessary to consider the antenna as an integrated part of both the propagation channel environment and of the handset RF front end. Consequently, its evaluation should be as such.
Fig. 1 shows an illustration of this intergation. The blue parts show the handset elements while the yellow part shows the propagation channel environement. The antenna part of the diagram is shaded to reflect its integration into both the channel and into the handset. The evaluation of the multi-antenna system throughout this chapter will be based on the assumption illustrated in this figure.

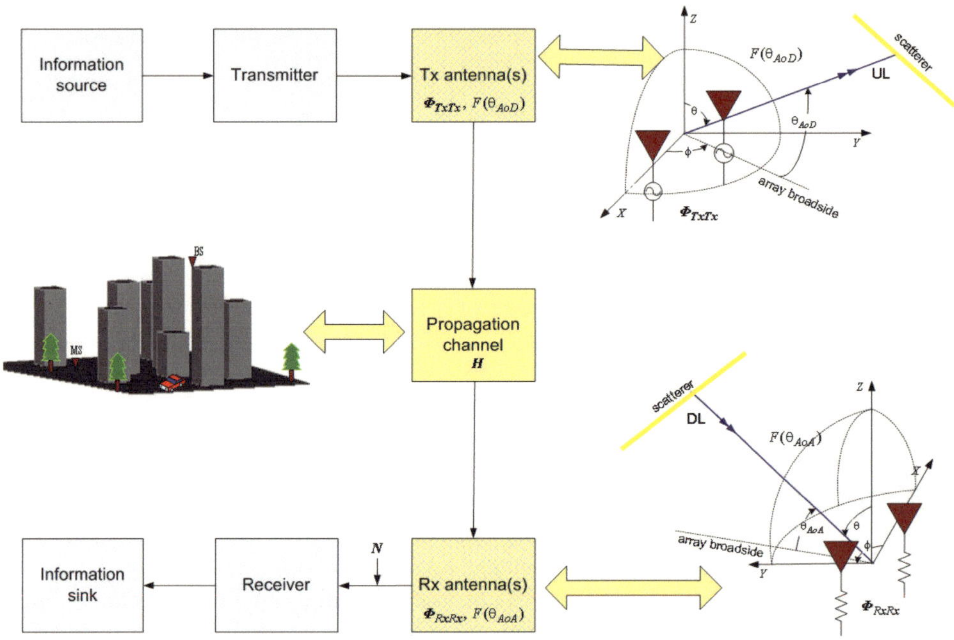

Fig. 1. Illustration of the antenna as a common element of the handset and of the propagation channel. The antenna can be tied to the channel through its coupling matrix $\boldsymbol{\Phi}$ and the field pattern F.

3. Evaluating the performance with partial Information

An advantages of evaluating an antenna system within a statistical channel model rather than a deterministic channel model is it allows for the incorporation of the random aspects that are likely to happen in a realistic communications situation (Ali,[1] 2008), (Ali[2], 2008), (Ali[1], 2009), (Kanj, 2008), (Kanj[1], 2008), (Lusina, 2008), (Lusina, 2009). As discussed later in the chapter, these random elements of the environment could facilitate knowledge of the parameter(s) that need attention during the design. In many evaluation scenarios reported in the literature, however, only part of the antenna information - the antenna information refers to the radiation patterns when discussing the interaction between the channel and the antenna - is included into the statistical channel model. The incorporation of partial information could very much lead to an inaccurate evaluation, as is illustrated below.
Fig. 2 shows the performance of a 2X2 PIFA-based antenna design when integrated into the spatial channel model (SCM), (Technical specifications group radio access network, 2006).

The channel settings are chosen as described in Table I. The effective radiation pattern of each PIFA is uploaded into the channel code in four different test scenarios: (i) using the default setting in the SCM, which assumes the upload of only one cut plane of the radiation pattern in the 0 deg evaluation; (ii) uploading five equally-spaced cut planes in the elevation; (iii) including eighteen equally-spaced in elevation cut planes; and (iv) uploading the whole 3D pattern, i.e., taking all 180 elevation cut planes. In the fourth test scenario (iv), a 3D-to-2D equivalent transformation is used (Kanj, 2009). The four test scenarios are simulated in five different orientations of the handset with respect to the servicing BS. The performance metrics are the correlation given in Fig. 2(a), the power transfer factor (Kanj, 2008) given in Fig. 2 (b), and the channel capacity given in Fig. 2(c).

Notice that the worst performance, i.e., higher correlation and lower power transfer and channel capacity, is seen when including the minimum antenna information with the single cut plane. The most optimistic performance is seen when more information is included with the five cut planes uploaded. After a convergence test, it is found that at least eighteen cut planes in the elevation plane are required to capture the majority of the antenna properties and have a performance equivalent to that when the whole 3D pattern is included. Although it may be argued that this performance is dependent on the particular antenna and scenario being examined, we would most likely come to the same conclusion that partial information would result in either a conservative or in an optimistic performance prediction. Therefore, the 3D antenna properties need to be included in its evaluation and not parts of it aforementioned.

Amongst the three performance metrics, the correlation metric seems to be the one that is the least affected by the partial antenna information. Therefore, at this point in our discussion, the channel power factor and the capacity seem to better reflect the antenna properties within the channel than the correlation values. The author will continue to examine the parameters that better describe the antenna performance as the chapter progresses; in an effort to answer the question: What parameters should be considered when designing a multi-antenna system?

Parameter	Setting
• Carrier frequency	2 GHz
• MS height	1.5 m
• BS height	32 m
• BS to MS distance	100 m
• Realizations	10,000
• Antennas characteristics	Equivalent 2D of the embedded patterns, effective gain
• Scenario	Urban macro-cell
• Usage scenarios	Free space (FS), Voice (HH), Data (D), Head (H)

Table I. Channel simulation parameters

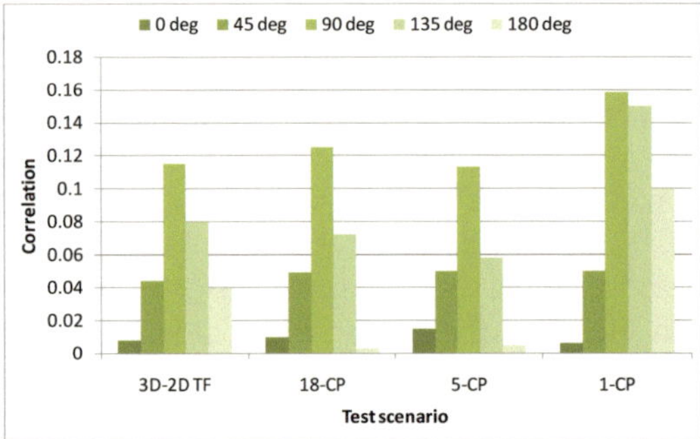

(a)

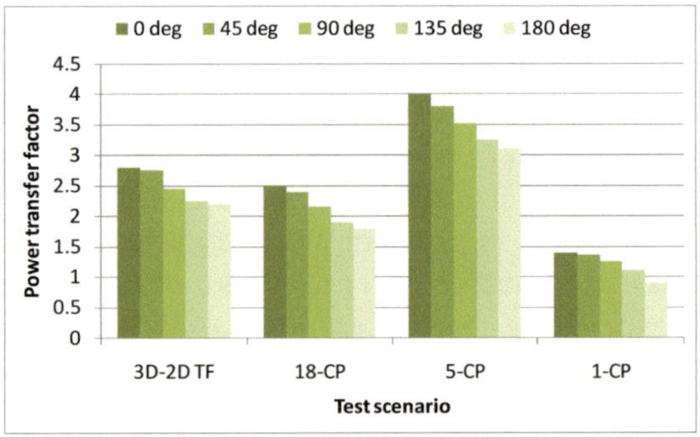

(b)

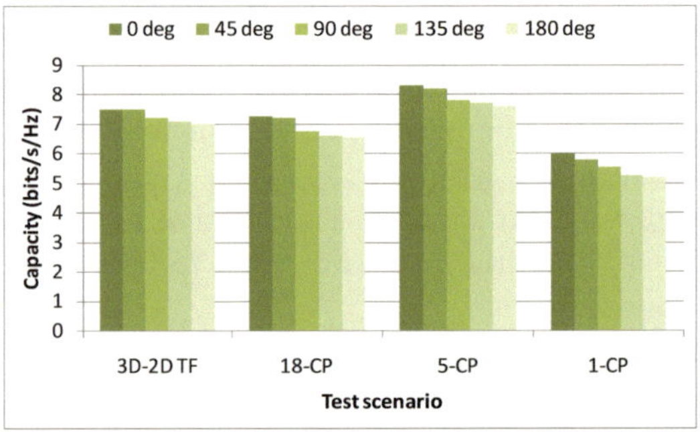

(c)

Fig. 2. Performance of a 2X2 antenna system when different percentage of the antenna information is incroporated into the channel: (a) correlation values, (b) power transfer factor, and (c) channel capacity

4. Propagation environment domination

As discussed in Section 2, the antenna is part of the communication channel and therefore the channel characteristics would significantly affect the overall multi-antenna performance. Even though the antenna itself is a deterministic element, its performance cannot be predicted in a generalized manner and any predictions would have to be made under certain assumptions of the propagation channel environment.

Let's investigate the propagation environment looking at the parameters that mainly affect the channel behavior, from an antenna stand point. These are the frequency, the scattering richness of the environment, the antenna polarization, and the presence of the user. For this, dipole antennas can be used where the channel matrix H, see Fig. 1, is constructed at three frequencies, 700 MHz, 950 MHz, and 1880 MHz, for the four investigated scenarios shown in Fig. 3. These are investigated in two cases. The first assumes a poor scattering environment and the second assumes a rich scattering environment, as shown in Fig. 3. The presence of the user's head is also noted in each of the described test scenarios. As the discussion is from the antenna view point, the un-normalized channel matrix is used in the analyses (Kanj, 2008), (Suvikunnas, 2007), (Pérez, 2003).

4.1 Channel capacity and power transfer

Channel capacity results for the described scenarios are given in Figs. 4-6, for both scattering environment assumptions. The first observation that can be made is that the channel capacities indicate a decline in the capacity as the operating frequency becomes higher when operating in a poor scattering environment. This is true whether the user is present or not. However, the same cannot be said for capacities in rich scattering environment scenarios, where a fluctuating channel capacity is measured. This can also be seen when considering the power transfer factor shown in Fig. 7. The highest power transfer is noticed at 700 MHz in Fig. 7(a) with the poor scattering environment while a fluctuating channel power is measured with the rich scattered environment; see Fig. 7(b).

Without loss of generality, with the channel being frequency-dependant one can expect that the performance of a system would vary with frequency. Therefore, there would be a frequency at which the performance trend of the multi-antenna system would go through a turning point. Let's try to apply this general statement on our defined test scenarios. Notice that the capacity at frequency 700 MHz shows that the performance is comparable whether the environment is a rich or a poor scattering one, and whether or not the user is present. The capacity is slightly higher with the rich scattering environment but the system performance is comparable. Also, the power transfer factor appears to be linearly associated with the channel capacity in the poor scattering environment. Therefore, at this frequency we can assume that the system performance is relatively independent of the scattering environment; i.e., the antenna system properties are dominant in the performance equation. On the other hand, as we go to higher frequencies, the channel power becomes less associated with the channel capacity in the rich scattering environment; therefore, the propagation environment characteristics are dominating the system performance. The turning frequency in the performance occurs at 1800 MHz in the poor scattering environment case, but occurs at a lower frequency (950 MHz) in the rich scattering case.

When the discussion comes to which polarization improves the multi-antenna system capacity for a spatial multiplexing MIMO system, by intuition, one might think of the cross-polarization choice. This has been indeed discussed widely in the literature (Chuang, 1994), (Ogawa, 2001), (Valenzuela, 2007). However, in many of these research investigations the user's presence, the environment scattering richness, and the operating frequency are not fully considered leading to conclusions that are based on partial information. From the defined experiments shown in Fig. 3 and the related discussion, it can be stated that a generalized statement on the best polarization choice cannot be made. At 700 MHz, the user appears as a smaller object to the propagating wave. Therefore, its effects are mainly shown in the power loss and do not disturb the propagating wave's characteristics much. Therefore, as the intuitive designer might plan, the x-polarized multi-antenna design is the better choice. This, on the other hand, becomes less true as the frequency of operation increases and as the channel scattering richness changes. As a matter of fact, the user presence becomes more effective as we operate higher in frequency where the channel properties become increasingly dominant. Therefore, a vertically polarization choice could yield higher channel capacity in the presence of the user when operating in a rich scattering environment. This effect could change significantly in a poorly scattered environment, as shown in Figs. 4-6 (a) and (b), respectively (Ali[2], 2009).

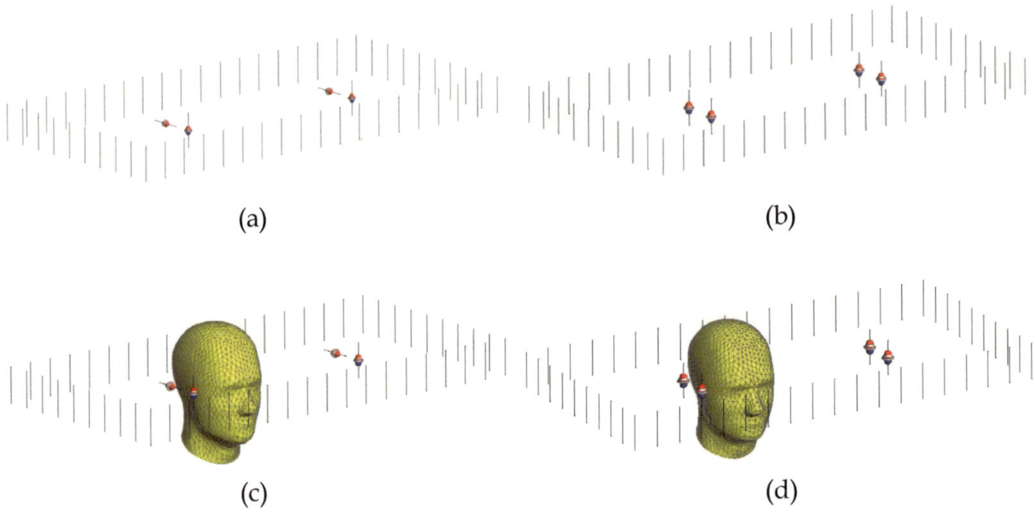

Fig. 3. Test scenarios for a 2×2 MIMO configuration: (a) vertically cross-polarized dipoles without the user head, (b) vertically co-polarized dipoles without the user head, (c) cross-polarized dipoles with the user head, and (d) co-polarized dipoles with the user head. Illustration not to scale.

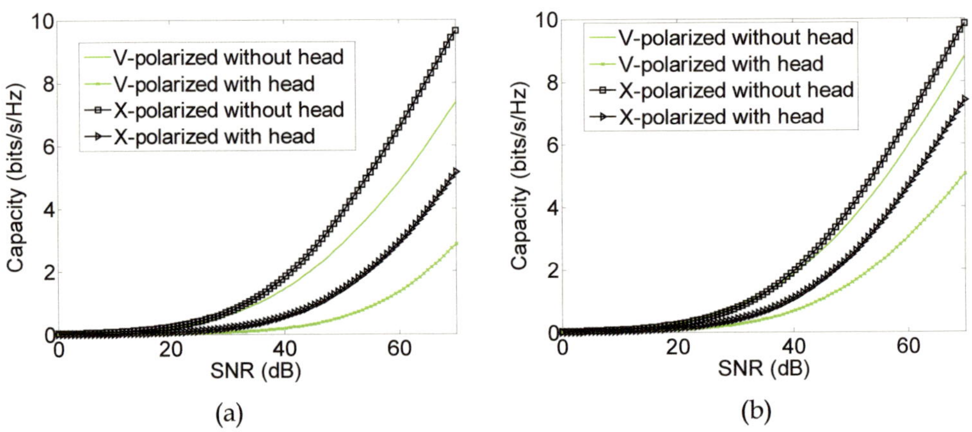

Fig. 4. Capacity for the different tested scenarios at 700 MHz: (a) poor scattering environment, and (b) rich scaterring environment

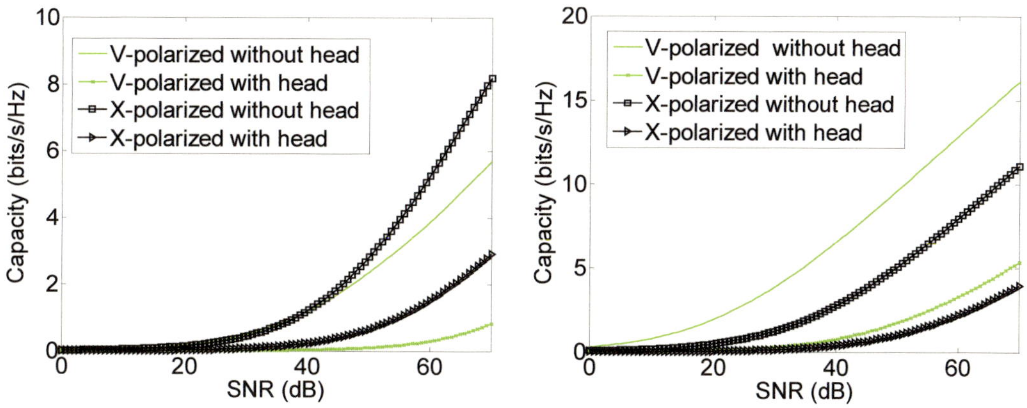

(a) (b)

Fig. 5. Capacity for the different tested scenarios at 950 MHz: (a) poor scattering environment, and (b) rich scaterring environment

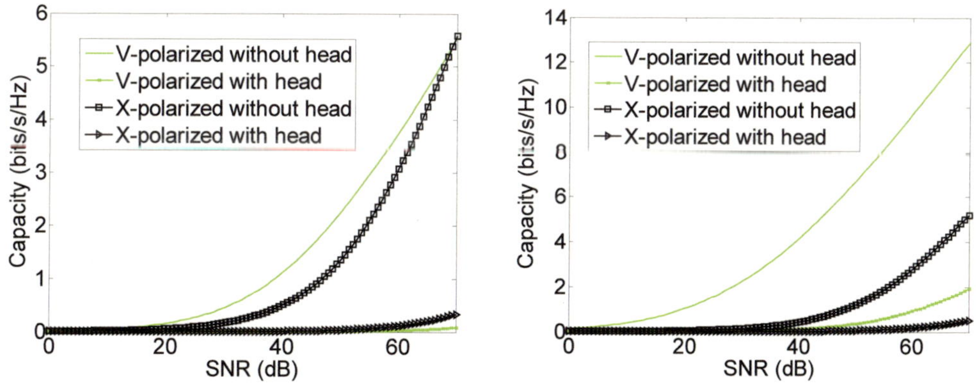

(a) (b)

Fig. 6. Capacity for the different tested scenarios at 1880 MHz: (a) poor scattering environment, and (b) rich scaterring environment

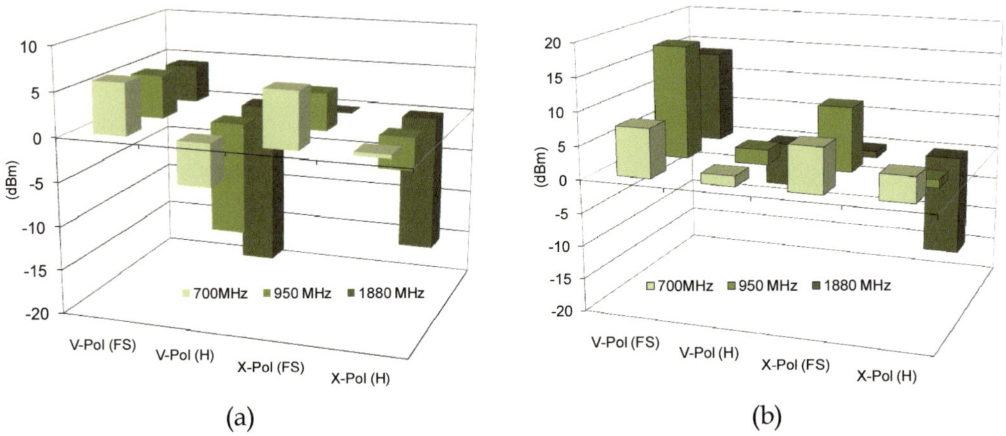

Fig. 7. Channel average power factor (dBm) at the three frequencies of operation for the four investigated scenarios in free space (FS) and with the user head (H) : (a) poor scattering environment, and (b) rich scaterring environment

4.2 Correlation

Since the environment is already included in the defined setting of Fig. 3, the correlation values computed here will be based on S-parameters measured at the terminals of each antenna (Blanch, 2003). Therefore, their values are expected to be, in general, low and reflect a deterministic evaluation. The involvement of the environment on the correlation would be included through the amount of signal reflected back into the system and the amount of signal leaked from one antenna terminal into the other antenna in a given handset configuration. These correlation values are shown in Fig. 8.

Examining the correlation values with the defined test scenarios comes with a general interesting observation. That is, regardless of the test scenario, the correlation values measure the highest at the lowest frequency of operation, 700 MHz. This can be expected due to the Omni-directional pattern type of the antennas at this frequency. The observation agrees well with the observation made before in Section 4.1 where the antenna properties were found to be dominating on the performance at this low frequency of operation and not the propagation environment.

Other observations can be made based on our defined test experiment but they cannot be said to be general. Looking at Figs. 8 (a) and (b), as the environment scattering richness changes from a poor scattering environment to a richer one, higher correlation values vary from the vertically polarized test scenario to the cross-polarized scenario in the absence of the user. The correlation values follow the same trend and alternate with frequency, as observed with the channel power values and capacity measurements from Section 4.1.

In the presence of the user, the lowest correlation values are seen in test scenario (c) where the antennas are cross polarized to each other, for both poor and rich scattering environments. This agrees well with the intuitive choice of choosing a cross polarization antenna configuration for improved system performance, which in this case is defined by the performance metric of the correlation values. However, the channel power factor and the capacity measurements given in Figs. 4-7 showed that this scenario performance is not necessarily the best. Hence, once again, evaluating the system performance having the correlation values as a main performance metric can lead to inconsistencies.

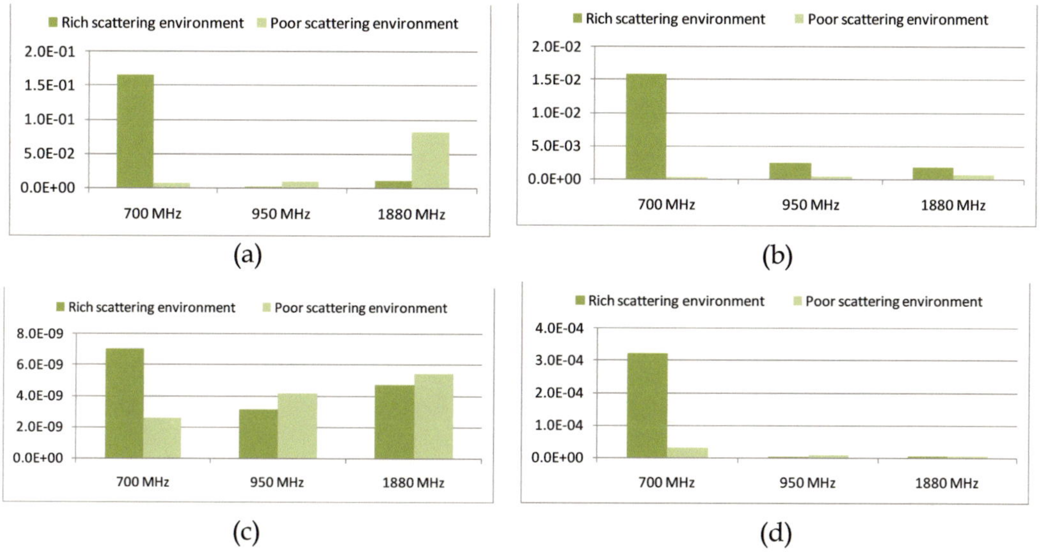

Fig. 8. Correlation values at different frequencies: (a) vertically co-polarized dipoles without the user head, (b) vertically co-polarized dipoles without the user head, (c) cross-polarized dipoles with the user head, and (d) cross-polarized dipoles with the user head. Calculations are based on (Blanch, 2003)

5. Handling of the antenna efficiency

A vital parameter in the communication system design is the antenna efficiency as this directly plays into the antenna link budget (Ali[3], 2008) (Aulraj, 2004), (3G Americas, 2010). The antenna efficiency describes how much of the signal provided to the antenna terminal is in fact being radiated to the destination. Of course, there is the orientation dependence factor that is described by the antenna directivity, but this quantity is tied to the efficiency and is scaled by it. The higher the antenna efficiency, the better the link margin, and the more tolerant the handset antenna is to changes in the environment. Additionally, there are other benefits from designing a high efficiency antenna as it would have a lower quality (Q-factor) value (Balanis, 1005) and hence can be expected to have less specific absorption rate (SAR) readings. This leads to the following question: How should the antenna efficiency be accounted for in the multi-antenna evaluation?

5.1 Evaluation setup

To investigate the proper handling of the antenna efficiency in multi-antenna system, the PIFA based antennas discribed in Section 3 are used agian here. The effects of the change in the efficiencies of these antennas on the link performance are emphasized through the presence of a user and in different usage scenarios. The main usage scenarios considered are: (i) Head only H; (ii) voice scenario with the user head and hand HH; and (iii) data scenario D with the user's two hands. The scenarios are shown in Fig. 9 where the phantom head and the hand models are used to simulate the user.

To reflect realistic usage, the handset and the user are factored into the propagation channel matrix, $\boldsymbol{H}$ as discribed in Section 2, see Fig. 1. The simulations are carried out having the discussion of Section 2 in mind where the channel matrix ***H*** is computed statistically and is modeled with the SCM for an urban macro-cell environment but with specific antenna properties at the handset. The antennas and the user are simulated through full-wave EM simulations that are performed with a 3D solver, FEKO (FEKO, 2006).

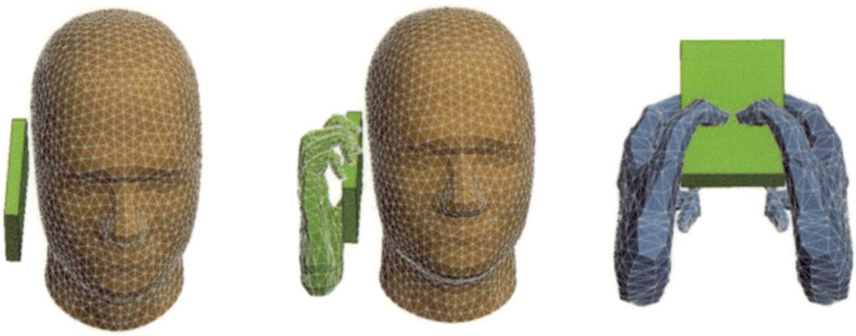

(a) (b) (c)

Fig. 9. Different usage scenarios: (a) Head only - H, (b) Voice with head and hand - HH, and (c) data with two hands – D

5.2 Antenna efficiency in the presence of the user

The antenna patterns and efficiency definitions are not simple enough so as to be derived directly from conventional pattern descriptions when the antenna is placed in the vicinity of or on a lossy medium. This is due to losses in the medium that cause the waves in the far-field to attenuate more quickly and finally to zero. The antenna efficiency is proportional to its gain (Balanis, 2005),

$$G(\theta,\phi) = \eta \cdot D(\theta,\phi) \, .$$

Here, η is the total efficiency factor and $D(\theta,\phi)$ is the antenna directivity, which is obtained from the antenna normalized power pattern that is observed in the far-field. Typically, the antenna efficiency cannot be accurately computed with full-wave electromagnetic simulators as the losses in the material and from mismatch are usually ignored. In fact, in many EM simulators, η is set by default to 1, which corresponds to 100% efficiency, i.e.

$$G(\theta,\phi) = D(\theta,\phi) \, .$$

For a more realistic presentation of the antenna performance into the channel for prorogation model analysis, the gain G should be taken as defined in the first equation where it is scaled by the antenna efficiency, and not as defined in the second equation. Therefore, in our analyses, the measured antenna efficiencies are used and these are given in Table II, where all losses are accounted for. As expected, the highest efficiency is measured in the FS scenario since there are no losses into the user's tissue or from losses as a result of reflections back into the transmitter RF circuitry. The largest loss in efficiency is seen in the voice scenario. Notice that as the near-field environment surrounding the antenna, in this case the user, becomes asymmetrical with respect to the antennas. This is reflected in the efficiencies of the two antennas where these become asymmetrical as well. The loss of efficiency, as will be discussed later, is reflected in the channel power transfer factor, and in turn, on the link throughput.

Scenario	Antenna 1	Antenna 2
Free Space	45	44.5
Head	35	34
Head & Hand	15	11
Data	23	22

Table II. Measured efficiency (%) in different usage scenarios

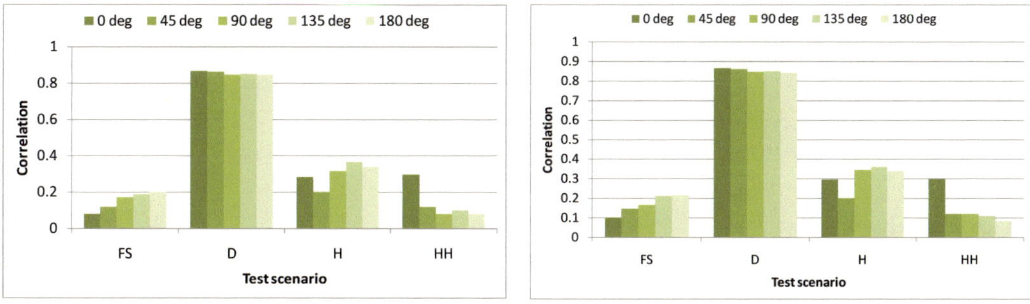

(a) (b)

Fig 10. Correlation values: (a) without including the antenna efficiency, and (b) including the antenna efficiency

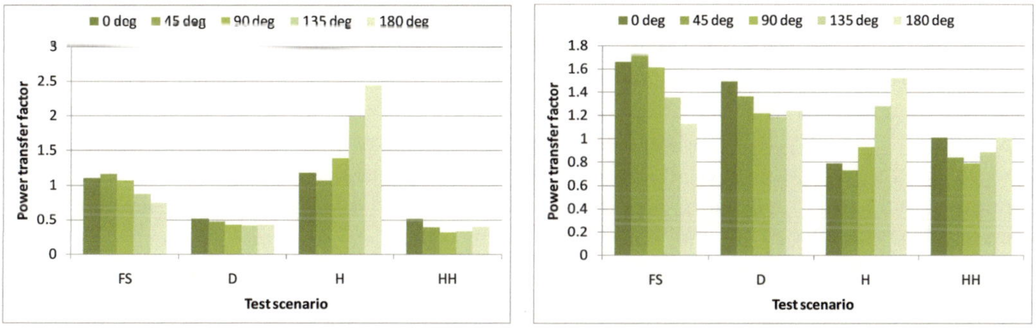

(a) (b)

Fig. 11. Channel power transferred in different usage scenarios at 20 dB SNR: (a) without including the antenna efficiency, and (b) including the antenna efficiency

The correlation values are shown in Fig. 10 for the different usage scenarios with and without including the antenna efficiencies at different oreintations of the user with respect to the servering BS. Notice that including the antenna efficiency did not have a major effect on the correlation values and in all user orientations. This is expected because the correlation is mainly defined by the normalized far-field radiation patterns of the two antennas. The performance of the correlation values can be predicted looking at the behavior of the radiation patterns, see for example Figs. 13(d)-16(d).

The channel power transfer factor with and without the antenna efficiency is given in Fig. 11. Two main observations can be made. The first is that power transfer factor is much less than the actual amount when the efficiency is not included. For example, in the H case and at user orientation 180°, the channel power is 1.5 dB which corresponds to a 0.5 dB decrease from the maximum and this comes from the antenna efficiency being at about 35%. However, it is not always the case that the reduction in the antenna efficiency can be linearly reflected into the channel power factor. As discussed in Section 4, it depends how dominant the propagation environement is and how the antenna radiation pattern interacts with the propagation envirnment . In some scenarios, such as the H scenario, the pattern shape would have a larger effect on the power especially since the pattern becomes directional due to the user's head and the performance becomes much more related to the user's orientation, see Fig. 14(d).
An important remark to make here is that accounting for the antenna efficiency cannot be considered as a linear scale factored into the SNR as many might think . It needs to be a weighted factor in accociation to the behavior of the antenna far-field radiation pattern, which ties into the user's orientation (Ali, 2010).

5.3 Link analysis

Lets analyze the effects of the antenna efficiency at the link level and investigate the performance of the system based on the antenna properties and the channel statistics. For this, an in-house MIMO-OFDM system based on 3GPP Release 8 standard (TSG-RAN TR 36.211, 2009) is used for the analysis. The channel matrices output from the integration of the EM and the SCM model, as discribed in Sections 2 and 3, are inputs into the link simulator. The evaluation is based on the throughput of the system in the different usage scenarios. Table III shows the simulation parameters of the LTE link simulator.
The link throughput in the different usage scenarios with and without the antenna efficiencies at 0° user orientation is given in Fig. 12. Notice that it takes 5 dBs more to achieve 1 Mbps moving from a FS scenario to the data scenario when efficiency is not included. In reality, however, this would take more than 15 dB as can be seen from the curves when efficiency is included. Furthermore, the difference in the performance is more pronounced in the variation of the efficiencies in the different usage scenarios; see Table II. Generally, including the antenna efficiency costs the system higher SNR to achieve the same performance compared to the case when the efficiency is ignored, becoming more pronounced with the higher SNR.

Parameter	Setting
• MIMO Mode	Spatial Multiplexing
• Antenna configuration	2×2
• Equalization	MMSE
• Channel Estimation	known
• Modulation	16QAM
• Coding rate	1/2
• HARQ	Enables (with one re-transmission)
• Correlation	Low

Table III. LTE Link simulation parameters

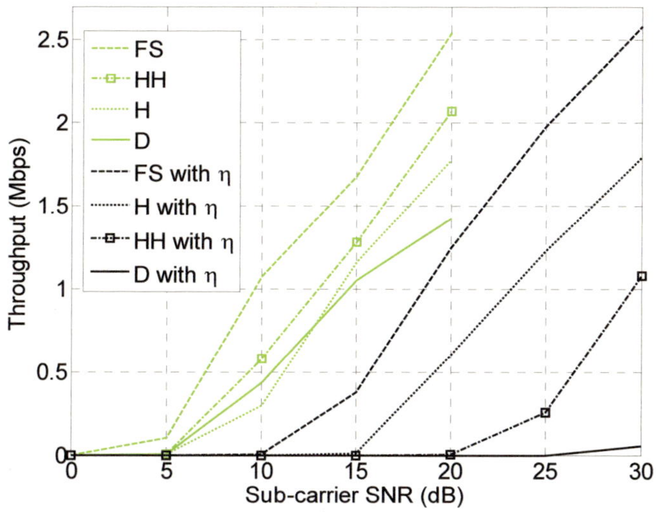

Fig. 12. Link throughput in the different usage scenarios with and without including the antenna efficiency and at 0° orientation

Notice that the link throughput showed a better performance of the voice scenario over the H scenario when the efficiency is ignored, which aligns well with the correlation results of Fig. 10(a). However, as we mentioned before, the correlation factor alone is not sufficient to accurately predict the performance. Therefore, the combined channel power in addition to including the antenna efficiencies is required. This is established in the black throughput curves in Fig. 12 where the H scenario shows improved performance relative to the HH scenario.

6. How deterministic is the multi-antenna design problem?

A handset antenna design can be thought of as a determanistic problem where its solution falls into the average of the general statistical problem. From this prespective, the mobile network operator would confine the handset manufacturer to an antenna performance with some minimum requirments taking into account their network and the physical area that it services.
The change in the antenna properties can be significant depending on the operating frequency and on the operational mode of the antenna. This is illustrated in Figs. 13-16 through the far-field radiation patterns. In all these simulations, the same multi-antenna configuration is used that is based on a typical PIFA design. Nevertheless, its radiation properties change significantly with the frequency and with the usage scenario. Assuming that the propagation environement changes slow enough allowing us to neglect its effects, the same antenna can easily have a performance that measures differently moving from one usage scenario to another, or from one frequency to another. Therefore, our one determanistic antenna design needs to solve the problem in multiple situtations.
Now including the propagation channel environement into the discussion, which, as explained earleir in the chapter, could in some scenarios be the dominating factor in the performance equation, the problem randomness becomes even more pronounced. With a randomly changing and dominant environment, we are faced with the folowing challenge: How well can we predict the performance of our multi-antenna system?

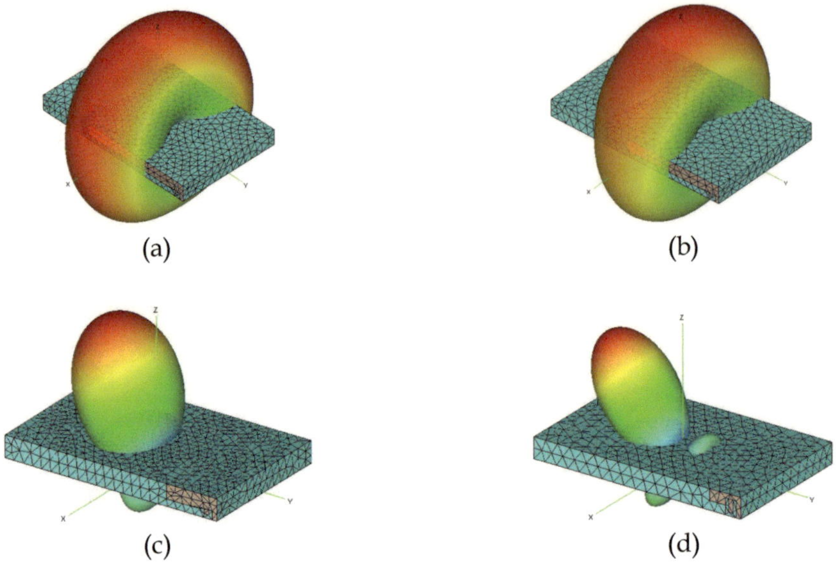

Fig. 13. The total radiation pattern in FS: (a) at 700 MHz, (b) at 950 MHz, (c) at 1880 MHz, and (d) at 2.6 GHz

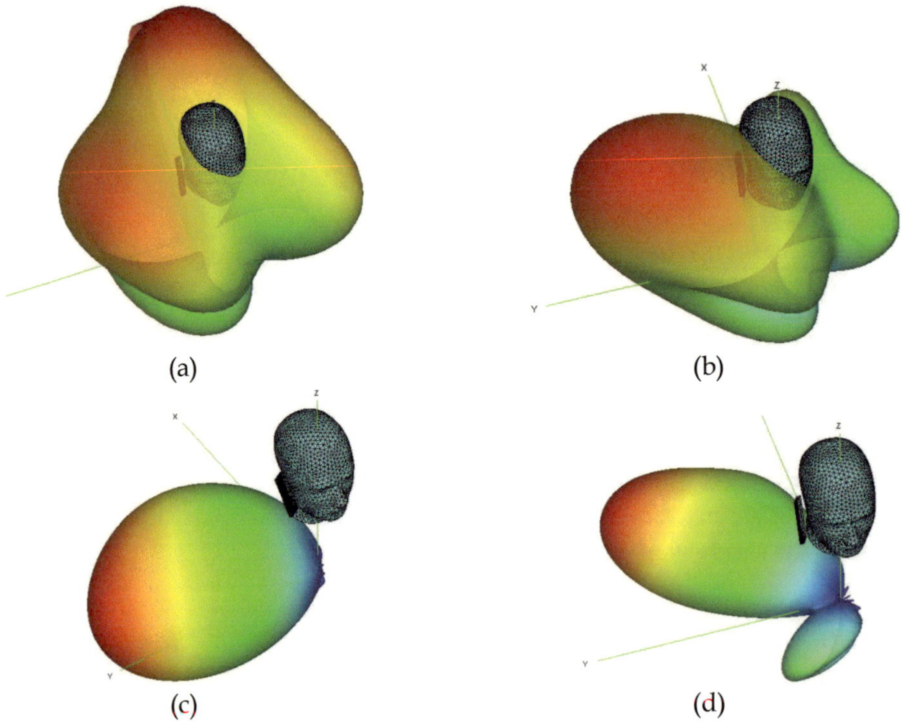

Fig. 14. The total radiation pattern in the presence of the user head: (a) at 700 MHz, (b) at 950 MHz, (c) at 1880 MHz, and (d) at 2.6 GHz

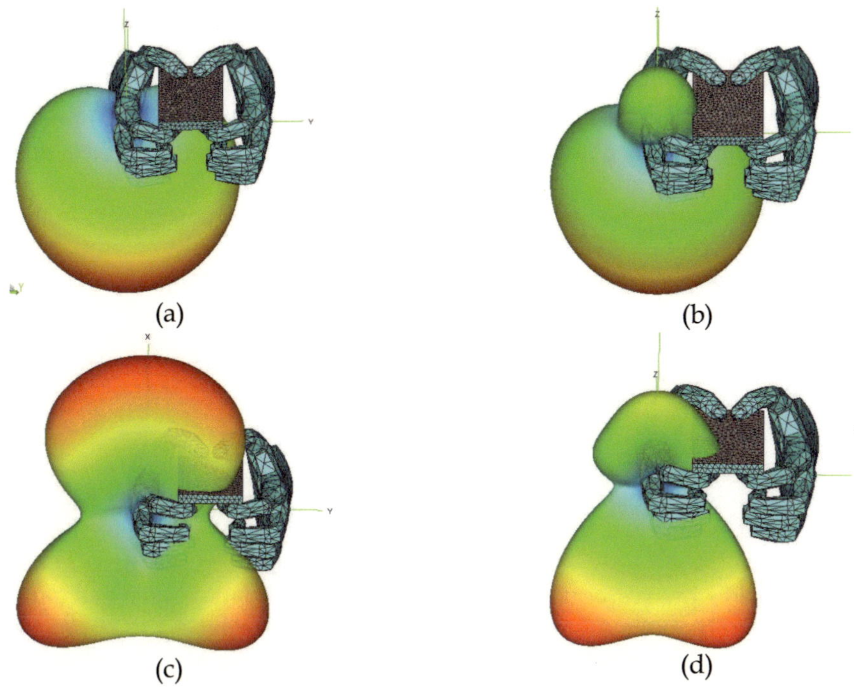

Fig. 15. The total radation pattern in the data position: (a) at 700 MHz, (b) at 950 MHz, (c) at 1880 MHz, and (d) at 2.6 GHz

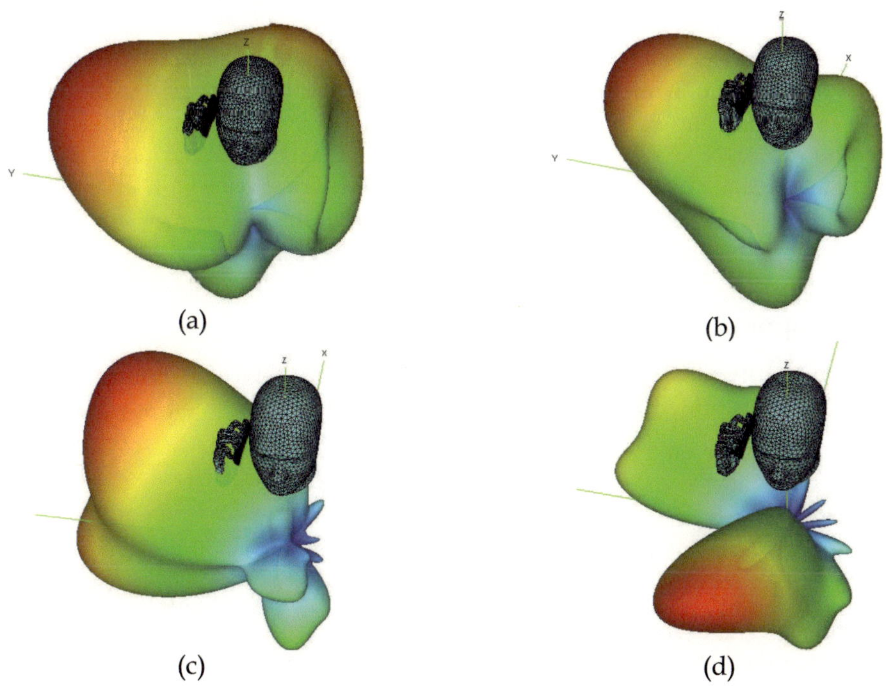

Fig. 16. The total radation pattern in the voice position: (a) at 700 MHz, (b) at 950 MHz, (c) at 1880 MHz, and (d) at 2.6 GHz

What we need to look at in the design of a MIMO antenna problem is the scalar parameters that better describe the antenna performance. In other words, since the radiation pattern can change significantly with the user and interact unpredictably with the channel, we cannot really count on it as a design target. Hence, quantities that by definition depend on the antenna radiation pattern, such as correlation; can become weak performance metrics in the evaluation. On the other hand, quantities such as the antenna efficiency and gain are to some extent controlled by the designer, of course, under the constraints of the laws of physics. Hence, the target should be to design a high efficiency multi-antenna system rather than a low correlation system and expect that the MIMO system performance would be good.

7. Final remarks

Ideally, the performance of a multi-antenna handset ought to be measured with the actual interaction of the antennas, the RF front ends, and the baseband processing elements. It also is useful to test the device with a performance metric or the figure of merit consistent with user operation and experience to provide a real-world metric. However, if the evaluation needs to take place at an early stage during the design process - where modelling and simulations are used - then careful measures need to be considered in the evaluation.

In evaluating a multi-antenna system it is important to include the complete antenna information, or enough of it, into the performance analysis. Incomplete information could very much lead to mistaken performance predictions. To this same point, the propagation channel environment can be a dominating factor in the performance equation and, therefore, its assumptions need to be laid out clearly as well. The performance in many cases would be scenario dependant.

A multi-antenna system designed with the antenna efficiency in mind could yield a system performance that is more likely to satisfy the performance requirements than a design that is based on weak or random performance metrics, such as correlation values. The design procedure could also be a simpler one. The key equation here becomes what would the target antenna efficiency be? And if known, how would replacing a high-efficiency antenna with two relatively lower efficiency antennas affect the multi-antenna system performance? Key questions from the author's view point yet to be answered.

8. Acknowledgment

The author would like to recognize and thank the researchers of Advanced Technology, Research In Motion Limited for the fruitful discussions and for their support during the writing of this chapter.

9. References

Ali[1], S.M., Kohandani, F. & Geyi, W. (2008). Macrocell MIMO system analysis using electromagnetic-based and measurement-based models, *IEEE/URSI Int. Symposium on Antennas and Propagation*

Ali[2], S.M., Kohandani, F., Qu, S.& Geyi, W. (2008). MIMO Link-level Simulations in an Urban Microcell Environment by an EM-based Model, *IEEE/URSI Int. Symposium on Antennas and Propagation*

Ali[3], S.M., Qu, S. & Kohandani, F. (2008). Capacity of an antenna system in the transmission mode with near-field effects, *24th Annual Review of Progress in Applied Computational Electromagnetics*, CD ROM

Ali[1], S.M., Qu, S., Kohandani, F. & Lusina, P. (2009). Deterministic-based and statistical-based channel models in the evaluation of single-user MIMO systems, *Antenna and Propagation Wireless Letters*, vol. 8, pp. 927 - 930

Ali[2], S.M. & Amineh, R.K. (2009). User effect analysis on polarization in MIMO systems, *IET Journal of Microwaves, Antennas and Propagation*, accepted

Ali, S.M., Mobasher, A. & Lusina, P. (2010). Antenna efficiency effect on MIMO link performance in the user's presence, accepted, *Vehicular Technology Conference*

Aulraj, A.J., Gore, D.A., Nabar, R.U. &Bolcskei, H. (2004) . An overview of MIMO communications–a key to Gigabyte wireless, *Proceedings of the IEEE*, vol. 92, no. 2

Balanis, C.A. (2005). *Antenna theory analysis and design*, 3rd ed., John Wiley & Sons, Inc., Hoboken, NJ, 2005

Blanch, J., Romeu, J. & Cordella, I. (2003). Exact representation of antenna system diversity performance from input parameter description, *Electronics Letters*, vol. 39, pp. 705-707

Chuang, H.R. (1994) . Human operator coupling effects on radiation characteristics of a portable communication dipole antenna, *IEEE Transactions on Antennas Propagation*, vol. 42, no. 4, pp. 556–560

FEKO® User Manual, Suite 5.3, (2006), EM Software & Systems-S.A. (Pty) Ltd, 32 Techno Lane, Technopark, Stellenbosch, 7600, South Africa

Foschini, G. J. & Gans, M.J. (1998). On limits of wireless communications in a fading environment when using multiple antennas, *Wireless. Personal Communication*, vol. 40, no. 6, pp. 311-335

Kanj[1], H., Ali, S.M. & Lusina, P. (2008). Effect of scattereres on multi-antenna systems, *24th Annual Review of Progress in Applied Computational Electromagnetics ACES*, CD ROM.

Kanj[2], H., Ali, S.M., Lusina, P. & Kohandani, F. (2008). A modelling approach for simulating compact MIMO systems, *European Microwave Week*

Kanj, H., Lusina, P., Ali, S.M., Kohandani, F. (2009). A 3D-to-2D transform algorithm for incorporating 3D antenna radiation patterns in 2D channel simulators, *IEEE Antenna and Propagation Wireless Letters*, vol. 8, pp. 815-818

Lusina, P., Kohandani, F. & Ali, S.M. (2009). Impact of MIMO channel models on outage capacity, *IEEE Radio and Wireless Symposium*

Lusina, P. Kohandani, F. & Ali, S.M. (2008). Antenna parameters effects on spatial channel models, *IET Communications*, vol. 3, issue 9, pp. 1463-1472.

Ogawa, K, Matsuyoshi, T. & Monma, K. (2001). An analysis of the performance of a handset diversity antenna influenced by the head, hand, and shoulder effects at 900 MHz: Part I - Effective gain characteristics, *IEEE Transactions on Vehicular Technology*, vol. 50, no. 3

Pérez,J., Ibáňez, J., Vielva, L. & Santamarĺa, I. (2004). Capacity estimation of polarization-diversity MIMO system in urban microcellular environments, *Proc. PIMRC*, vol. 4

Suvikunnas, P., Salo, J. & Vainikanianen, P. (2007). Impact of power normalization in experimental MIMO antenna performance studies, *IEEE Antennas and Wireless Propagation Letters*, vol. 6, pp. 43-46

Technical Specifications Group Radio Access Network (2006). Spatial channel model for multiple input multiple output (MIMO) simulations - TR25.996, *3GPP Partners.*

TSG-RAN TR 36.211 (2009). Evolved universal terrestrial radio access (E-UTRA); physical channels and modulation, Version 8.8.0

Valenzuela, J., Martinez, A. & Sanchez, D. (2007). Effects of user presence on receiving diversity and MIMO capacity for Rayleigh-fading channels, *IEEE Antennas and Propagation Letters*, vol. 5

3G Americas (2010). MIMO and smart antennas for 3G and 4G wireless systems: practical aspects and deployment considerations, Chapter 10

Part 5

Implementation and Experimental Evaluation

MIMO Beamforming Network Having Polarization Diversity

Kadir M.F. A., Suaidi M.K. and Aziz M. Z. A.
Universiti Teknikal Malaysia Melaka (UTeM)
Malaysia

1. Introduction

The technology of mobile communication and wireless local network (WLAN) are expending at a fast rate which is to assure that the end users reach a maximum data transfer and a get a better of quality service. The MIMO system is introduced to improve the communication system without having an additional transmit power or larger bandwidth, this because the MIMO system can utilize the multipath propagation.
In [Agilent, 2008], there are three categories of MIMO system. The first system is spatial diversity which can be obtain by spaced the antenna in a dense of multipath scattering environment. The second category is spatial multiplexing where an independent data is transmitted over different antennas. Third category is beamforming network. Generally the MIMO system can be divided into two parts, which the first part is the digital signal processing (DSP) and representing as **Part A** in Fig. 1. The second part is radio frequency (RF) device and representing as **Part B** in Fig. 1. (Jensen & Wallace, 2004).

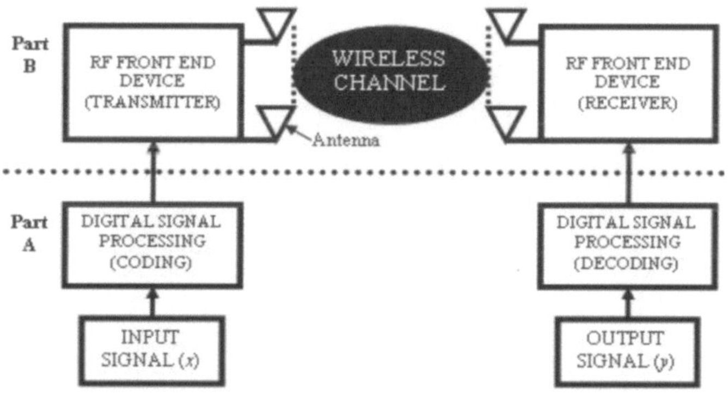

Fig. 1. Block diagram of MIMO systems

This paper is more focused to the RF section and antenna, the beamforming technique is used to act as MIMO system. The term beamforming related to the device in which energy radiated by an antenna that focused is focused along a specific direction in space (Mariadoss etc. al. 2005). So, there are several types of beamforming such as Blass matrix, Nolen matrix and Butler matrix (Desmond, 2001; Ahmad & Seman, 2005). The Butler matrix is used in this project because it simple circuit and easy to be fabricated (Mariadoss etc. al. 2005).

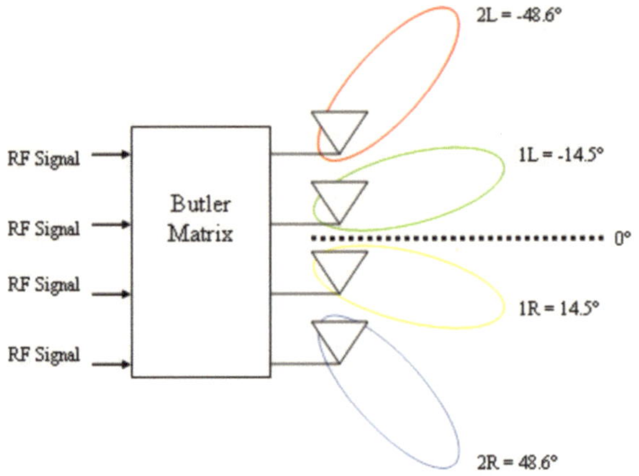

Fig. 2. An example of Butler matrix

Moreover, diversity technique was introduced in this measurement setup. The diversity is used to reduce the fading issues and multipath problems (Duman & Ghrayeb, 2007). By implementing diversity, the received power can be increased significantly, thus increasing the capacity of the system. Polarization diversity, angle diversity and pattern diversity are among the examples of diversity technique.

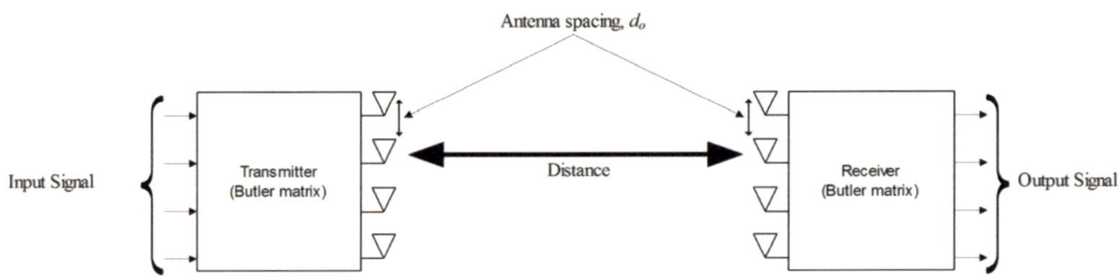

Fig. 3. The Concept of RF MIMO Front End Systems Measurement

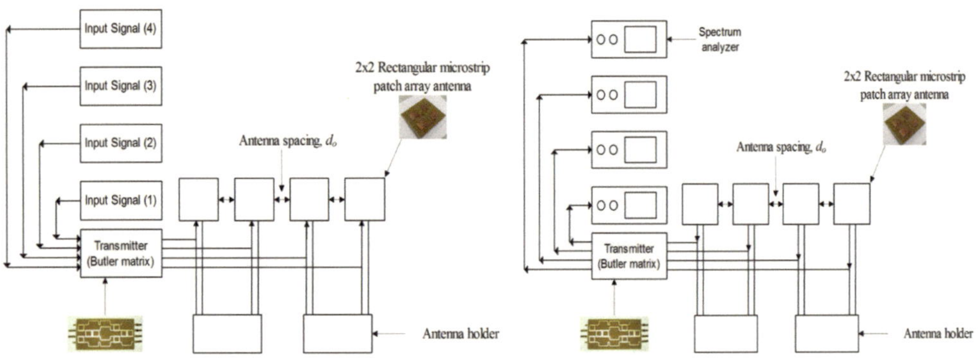

(a) Transmitter Setup (b) Receiver Setup

Fig. 4. The Measurement Setup at (a) Transmitter (b) Receiver

2. MIMO channel matrix

Each of antennas in MIMO systems has its own vector. The transmit antenna vector is represented as x and the receive antenna is represented as y vector. The MIMO channel input-output relationship can be expressed as:

$$y = H x + \eta \tag{1}$$

where the η is the noise and H is the MIMO channel matrix. The MIMO channel matrix size is depending on the number of transmit and receive antennas. The channel matrix size can be express as below:

$$H = \begin{bmatrix} \rho_{11} & \cdots & \rho_{14} \\ \rho_{21} & \cdots & \rho_{24} \\ \vdots & \ddots & \vdots \\ \rho_{41} & & \rho_{ij} \end{bmatrix} \quad M\,x\,N \tag{2}$$

where M was the number of transmit antenna and **N** was the number of receive antenna. The ρ_{ij} was represented as the correlation between transmit power and receive power where i^th^ was the input port and j^th^ was the output port signals. The ρ_{ij} components were direct depended on the physical characteristics of the propagation environment and also the structure of the antenna array (Kermoal et al., 2000)

Correlation coefficient was a statistic method to measure the correlation between two variables. The correlation coefficient can be calculated as:

$$\rho_{yx} = \frac{cov_{yx}}{\sigma_{yy}\sigma_{xx}} \tag{3}$$

The channel matrix, H will be analyzed for each type of antenna configuration which can be referred in Appendix. The eigenvalues can be calculated to estimate the receiving power for each eigen paths. The eigenvalues were calculated using eigen value decomposition (EVD) technique. The calculated eigenvalues were related with the MIMO average channel capacity by using the equation below (Hirayama etc. al. 2007):

$$C_{ave} = \left[\sum_{j=1}^{N} \log_2(1 + \lambda_j SNR / N) \right] \tag{4}$$

where N was the number of transmit because there was no setting channel state information (CSI) at transmitter. λ_j was the eigenvalues from the matrix, H while the noise power of -76dBm was obtain from measurement.

3. Results and discussion

3.1 Antenna separation comparison

Fig. 5. shows the average channel capacity for configuration A and B. The distance for this measurement is set for 12 meters. Its shows that the average channel capacity for antenna separation 12 cm are higher for both configurations. The average channel capacity difference for antenna spacing between 12 cm and 6 cm for configuration A is 2.2848 bps/Hz and configuration B is 2.473bps/Hz. For the entire antenna configuration can be referred to Appendix.

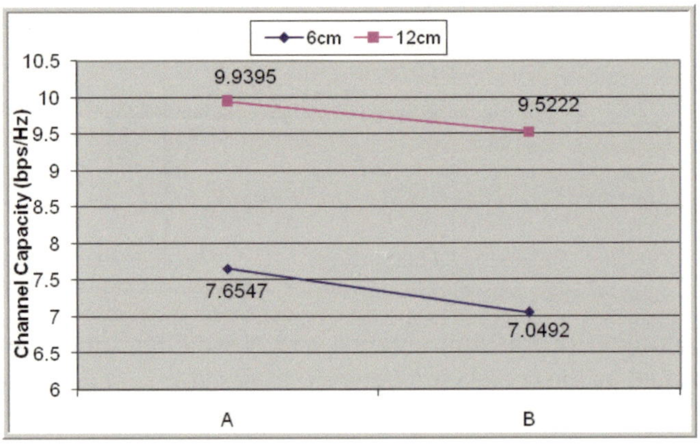

Fig. 5. Average Channel Capacity for Normal Configuration

In Fig. 6 (a)., the configuration E shows a higher average channel capacity is 9.193 bps/Hz with antenna separation is 12 cm and it is higher form others configuration. Fig. 6 (b). shows an average channel capacity for polarization diversity applied to the receiver side. Two configuration show high value which is indicated at configuration G and configuration I with antenna separation 12 cm.

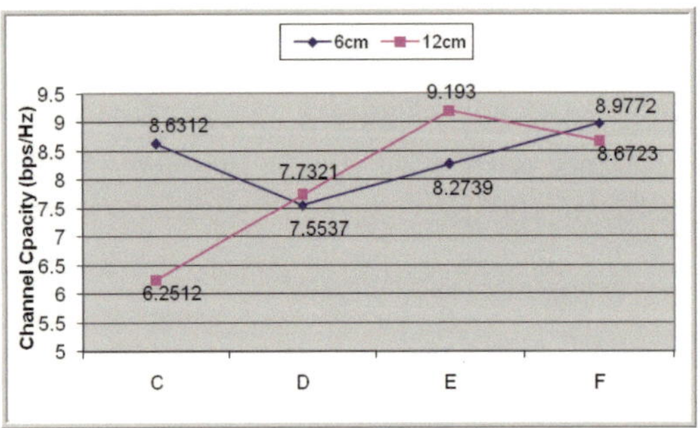

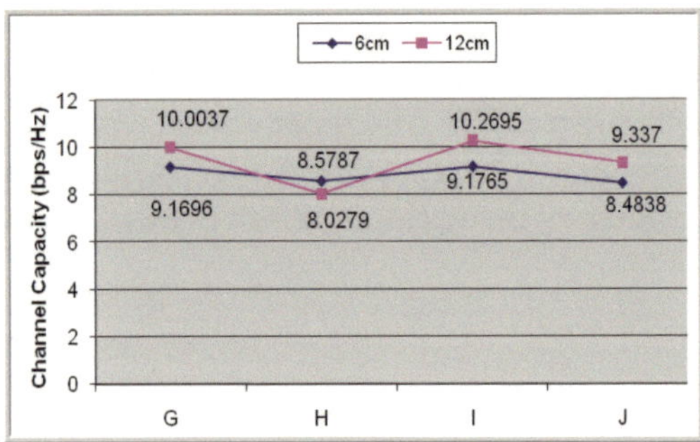

(a) (b)

Fig. 6. (a) Average Channel Capacity for Polarization Diversity at Transmitter Fig. 6 (b). Average Channel Capacity for Polarization Diversity at Receiver

Fig. 7. shows the average channel capacity for both sides polarization diversity. The higher values are indicated at configuration K, L and M. For L an M the antenna separation is 12 cm, meanwhile the configuration K antenna separation is 6 cm.

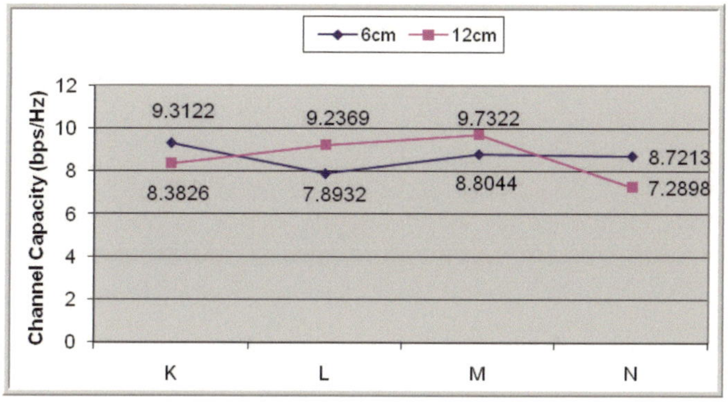

Fig. 7. Average Channel Capacity for Both Sides Polarization Diversity

4. Conclusion

As the conclusion, if the antenna spacing is double, the average channel capacity is increased. When polarization diversity is applied to the system, its show that average channel capacity is better than the normal configuration.

5. References

Agilent (2008). MIMO Channel Modelling and Emulation Test Challenges, *Application Note*, Agilent, pp. 1-60, Agilent Technologies, Inc, 5989-8973EN, USA

Ahmad, S.R. & Seman, F.C. (2005). 4-port Butler matrix for Switched Multibeam Antenna Array, *Proceedings of Asia-Pacific Conference on Applied Electromagnetics*, 0-7803-9431-3, Dec. 2005

Duman, T.M & Gharayeb, A. (2004). *Coding for MIMO Communication Systems, Wiley*, 978-0-470-02809-4, England

Hirayama, H.; Hayashi, A. & Kikuma, N.; Sakakibara. K. (2007). An Experimental Consideration On Spatial Correlation And Minimum Eigenvalue For MIMO Systems Using Polarization, *Proceedings of The Second European Conference On Antennas and Propagation*, pp. 1-5, 978-0-86341-842-6, Edinburgh, Nov. 2007

Jensen, M.A.; & Wallace, J.W. (2004). A Review of Antennas and Propagation for MIMO Wireless Communication, *IEEE Transaction on Antenna And Propagation*, Vol. 52, No. 11, November 2004 pp. 0018-926X, Nov. 2004

Kermoal, J.P.; Schumacher, L. & Pedersen, P.E. (2000). Experimental Investigation of Correlation Properties of MIMO Radio Channel for Indoor Picocell Scenarios, *Proceeding of IEEE Vehicular Technology Conference*, pp. 14-21, 0-7803-6507-0, Boston, MA, Sept. 2000, IEEE

Mariadoss, P.Q; Rahim, M.K.A. & Aziz, M.Z.A.A. (2005). Design and Implementation of A Compact Butler Matrix using Mitered Bends, *Proceeding of APMC 2005*, 0-7803-9433-X, Dec. 2005

19

VLSI Implementation of Least Square Channel Estimation and QPSK Modulation Technique for 2×2 MIMO System

Sudhakar Reddy Penubolu[1] and Ramachandra Reddy Gudheti[2]
[1]*Srikalahasteeswara Institute of Technology, J.N.T.University, Anantapur,*
[2]*Sri Venkateswara University, Tirupathi,*
India

1. Introduction

Future Wireless communication systems have to be designed to integrate features such as high data rates, high quality of service and multimedia in the existing communication framework. Increased demand in wireless communication systems has led to the demand for higher network capacity and performance. Higher bandwidth, optimized modulation offer practically limited potential to increase the spectral efficiency. Hence MIMO systems utilizes space multiplex by using array of antenna's for enhancing the efficiency at particular utilized bandwidth. MIMO use multiple inputs multiple outputs for a single channel. These systems are defined by spectral diversity and spatial multiplexing. MIMO describes the ways to send data from multiple users on the same frequency/time channel using multiple antennas at the transmitter and receiver end.

A transmitter/receiver system uses multiple antennas not only for transmitting data between corresponding antennas but also between adjacent antennas. The data is received in the form of MIMO Channel Matrix. A top level MIMO system (Joseph et al., 2007) is shown in Fig.1.1. MIMO system is used in many applications like WiMax, Wi-Fi, WLANs, and many more applications.

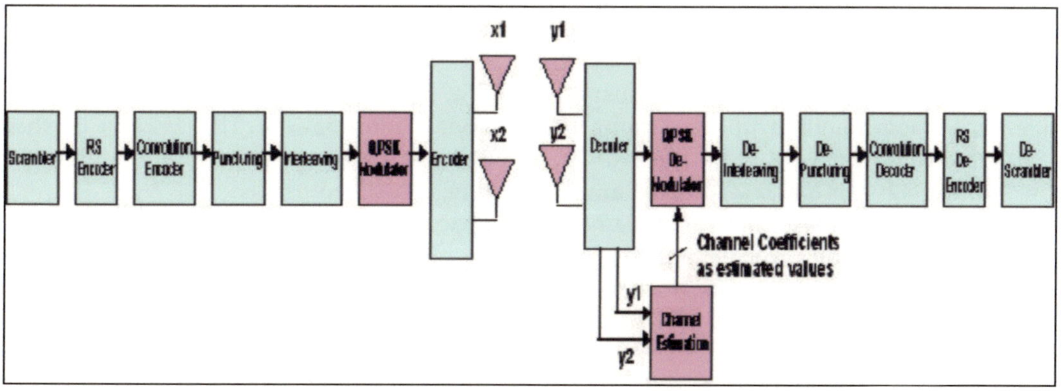

Fig. 1.1. MIMO System

In recent years the telecommunications industry has experiencing a tremendous growth in the area of wireless communication. This growth has been ignited the widespread popularity of mobile telephones and wireless computer networking. But there are limits to growth, and the radio spectrum used for wireless communications. Hence considerable effort has been invested in making more efficient use of the resources. Using the spectrum more efficiently caters for the ever increasing demand for faster communications since more bits per second can be transmitted using the same bandwidth. Recent major industry research focus in this area has been the use of multiple antennas for transmitting and receiving instead of the single antenna systems. The uses of multiple transmit and receive antennas, and efficient coding techniques could increase the performance of wireless communication systems. The implementation carried is for 2x2 MIMO systems with good channel estimation method and better coding techniques for better performance. The design is carried till the backend VLSI flow. The design is to implementing 2x2 MIMO systems with QPSK modulation technique and estimating channel coefficients using LS channel estimation method.

2. Back ground discussion

Due to demand in the wireless products, the high data rates, capacity, accuracy, with less hardware and the reliability of the system is a greater challenge. Hence numerous efficient channel estimation and modulation techniques are to be modified for greater performance of the overall system. In this, literature review is carried for different MIMO architectures for different modulation techniques and the channel estimation methods.

In digital communication system the channel accepts electrical/electromagnetic signals, and the resultant output is smeared or distorted version of the input due to the non-ideal nature of the communication channel. Also the information-bearing signal is corrupted by unpredictable noise. The smearing and noise introduce errors in the information being transmitted and limits the rate at which information is to be communicated from the source to the destination. The probability of incorrectly decoding a message symbol at the receiver is often used as a measure of performance of digital communication systems. The main function of modulator and the demodulator is to combat the degrading effects of the channel on the signal and maximize the information rate and accuracy.

In digital modulation, information signals to be modulated is digital. Therefore, digital information modulates an analog carrier and hence called as “digital modulation”. There are basically three types of digital modulation techniques i.e Amplitude Shift Keying (ASK), Phase Shift Keying (PSK) and Frequency Shift Keying (FSK). Phase Shift Keying is a digital modulation technique which is one most used. In PSK modulation, the phase of the carrier is altered in accordance with the input binary coded information. The PSK is further subdivided into BPSK, 8-PSK, 16-PSK, QPSK, DPSK. In binary phase shifting keying the transmitted signal is sinusoid of fixed amplitude .It has one fixed phase.

Chia-Liang Liu (Chia L.L, 1998) explained the impact of I/Q imbalance on QPSK-OFDM-QAM Detection. The in-phase and the quadrature phase components are very important component in QPSK. Any imbalance in these the performance of the whole system crashes down at the detectors. Either the phase or the amplitude imbalances lead to the introduction of ICI, there may be cross- talk between in-phase and quadrature channels.

Wout Joseph, et.al. (Joseph et al., 2007) proposed an algorithm for the performance study on IEEE 802.16-2004 for different conditions like for different channel models and MIMO system. Different channel models are compared by applying different diversity schemes. The goal is to analyze the performance of WiMax system. Performance study is carried for 2x2 and 2x1 MIMO systems. The configuration of WiMax transmitter and receiver are done for MIMO and SISO. Initially a model was implemented in MATLAB/Simulink.

Yantao Qiao, et al. (Yantao.Q et al., 2005), has made a research on an iterative algorithm of least square channel estimation in MIMO OFDM systems. The main objective of this paper is an iterative channel estimation algorithm for MIMO OFDM. Compared to common least square channel estimation, this has greatly improved estimation accuracy and, low pass filtering in time domain reduces AWGN and ICI significantly. MIMO-OFDM is used in mobile applications. Many techniques have been applied to the MIMO channel both in time and frequency domains which increases the capacity and increase the reliability of wireless link.

Cui.S et al., (Cui.S et al., 2004) proposed channel estimation in MIMO wireless communication systems plays a key role in the performance of the space time decoders that depends on the accuracy of the channel knowledge .In this paper the study of the performance of the MIMO channel estimation using training sequence is carried out. The Least Square, MMSE and new scaled least square approaches to the channel estimation are studied and the optional choice of the training signals is investigated for each of these techniques.

Sarod Yatawatta, et.al,(Yatawatta.S et al.,2005) proposed a solution for minimizing the energy spent on during the channel estimation when subjected to known error and delay when timing symbols are transmitted. The minimization of energy is carried by reducing the hardware, also by using a low rank equalization at the receiver. A scalar energy reduction at channel estimation is explained,

Benoit Le Saux, et.al,(Saux.B.L, 2006) proposed a MIMO system with OFDM has greater potential like reduction in inter-symbol interference, decrease in fading, increase in bandwidth, and increase in data rates. The performance of MIMO system degrades due to inaccurate channel estimation over frequency selective fast-varying channels. Pilot-aided turbo channel estimation improved by addition of linear algorithm in the iterative process is discussed. This improves the channel estimation and reduction in the use of number of pilots.

Changchuan Yin, et.al,(Changchuan.Y et al.,2006) presents the theoretical and simulation results for LS channel estimation in MIMO-OFDM based on 2 types of pilot structure i.e. block type pilot and the comb-type pilot. The CSI is obtained by blind channel estimation and pilot aided channel estimation. The pilot based channel estimation is based on LS or MMSE criterion. LS channel estimation is more practical then MMSE because LS does not require extra information about channel correlation and noise variance.

Riza Abdolee, et.al,(Reza A et al., 2007) proposed a method to reduce the computational complexity of channel estimation algorithm for MIMO-OFDM. Channel estimation suffer from high computational complexity. Solution for high efficient channel estimation and simplified computational complexity is stated. Normal matrix inversion is solved using QR decomposition method. This method is used to simplify LS method. The results show QR decomposition can greatly reduce the complexity of LS channel estimation.

Deseada Bellido, et.al,(Bellido et al., 2007) proposed LS channel estimation algorithm for MIMO-OFDM. This evaluation has been made using pilot design rules that guarantee a bounded error level for the estimation. This method is used for estimation of the channel matrix. Diversity technique is used for channel matrix. The capacity of channel depends on

the knowledge of the channel matrix gains at the transmitter and receiver. The estimation is carried out by transmission of set of pilot symbols known to transmitter and receiver.
Markus Myllyla, et.al,(Markus M et al., 2006) proposed a method for performance evaluation of FPGA implementation of a LMMSE based detector for radio channels. Complex part of high level 2x2 LMMSE detector blocks which is solved by CORDIC based architectures.
Sudhakar, et al (Sudhakar et al., 2009) proposed a design method and FPGA implementation of channel estimation and Modulation technique for MIMO system.

3. Problem definition/statement

While designing VLSI circuits for very efficient implementations the designer should consider algorithmic and hardware architectures trade-offs. Rapid prototyping for such applications will imply the short designs satisfying all the design constraints, such as timing and silicon area. The wireless systems are operated under harsh and challenging channel conditions. A 2x2 MIMO system is been designed, simulated, implemented, synthesized and physical design is carried in macro level to tape out the design.
The aim of this work is VLSI implementation of LS channel estimation method and QPSK modulation technique for 2×2 MIMO system. The design specifications are verified using MATLAB. The RTL coding is carried for the design to be implemented on Xilinx FPGA. Design synthesis and macro level physical designing is also carried out.
The main Objectives of the present work on MIMO Systems are

- To review literature on MIMO and its architecture, various channel estimation methods and QPSK modulation technique.
- To identify a channel estimation method for MIMO architecture.
- To model and simulate MIMO system with QPSK modulation technique.
- To implement and optimize MIMO System on FPGA.
- To verify MIMO system for its functionality.

The design methodology carried to meet the objectives is as follows. The flowchart is also shown in Fig.3.1 that shows the design methodology followed for developing 2x2 MIMO systems.

4. Least square channel estimation and QPSK modulation for 2x2 MIMO systems

The MIMO system has multiple transmitter antennas and multiple receive antennas so that the data is transmitted in parallel. This work demonstrates simple working of a 2x2 MIMO system carried till backend of the VLSI flow. The MIMO system is designed with least square channel estimation method and QPSK modulation technique.
In communication systems, channels are usually multi-path channels, which cause inter-symbol interference in the received signal. As discussed in literature review various detection algorithms offer a very good receiver performance and reduced computations. Channel estimators require the channel impulse response. The channel estimation is based on the known sequence of bits called training sequence which is unique for each transmitter. Here the known training sequence is transmitted so that the channel coefficients are obtained. There are different standards used for transmitting training sequence like IEEE 802.16 standard.

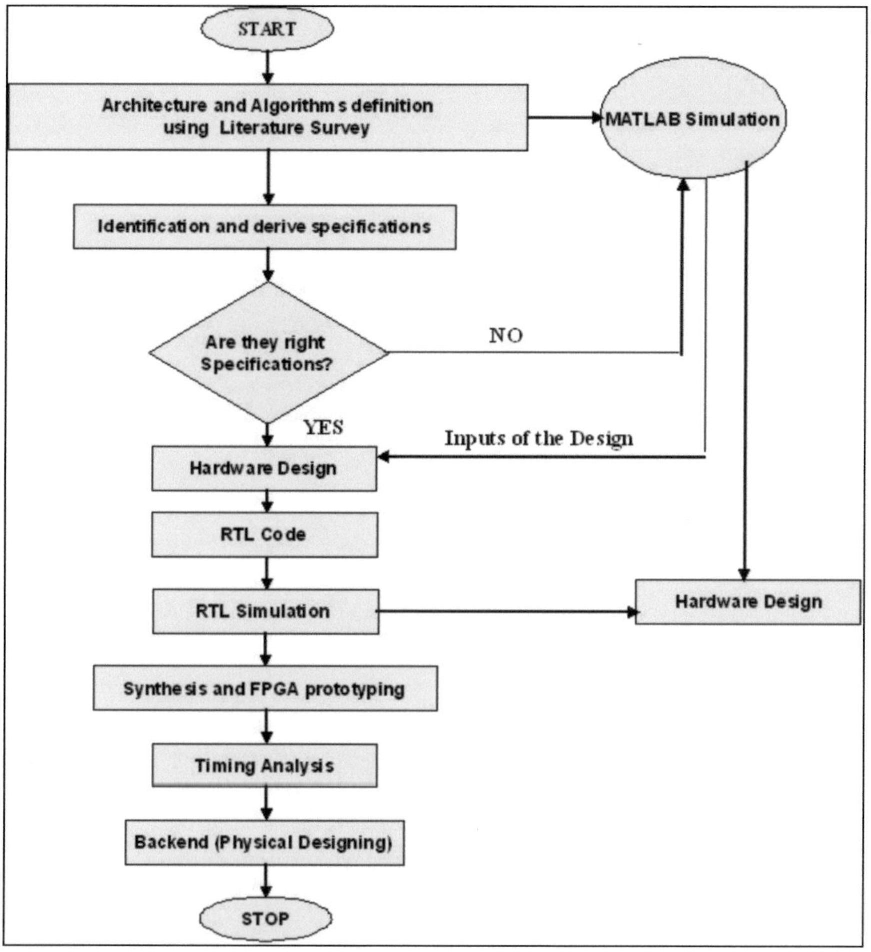

Fig. 3.1. Flow Chart showing methodology carried for Design

4.1 LS Channel estimator for 2x2 channel

2x2 joint channel estimation is considered for the MIMO design. The channel is considered to be ideal wherein noise is not taken in account. The sequence '*x*' is taken from 2 different transmitters and 4 channel paths are considered as shown in Fig.4.1. From each transmitter unique training sequences are transmitted in concatenation with the QPSK modulated data. The training sequence is transmitted for the identification of the transmitter from which the data is obtained at the receiver end. Here diversity technique is used but the Alamouti encoder and decoder is not used.

The 2x2 channel estimation block diagram is shown in Fig.4.1. Initially the training sequence is transmitted so that the channel coefficients are calculated. First the transmitter is made active and the training sequence is sent. The tx2 is made inactive i.e. nothing is sent from tx2. The h11 and h12 channel coefficients are obtained. Then the tx1 is made inactive and tr2 is sent through tx2. Then the h21 and h22 coefficients obtained by simplifying the equations. The data obtained at the receivers rx1 and rx2 are shown by the equations (1) and (2).

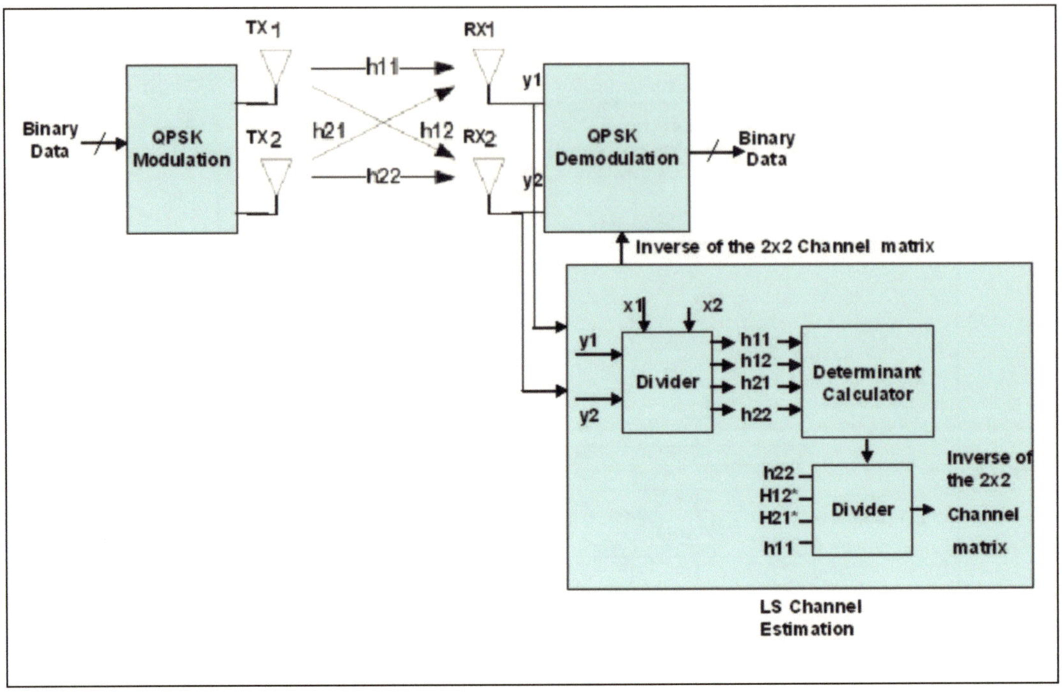

Fig. 4.1. QPSK modulator/demodulator with LS channel estimation

$$y1 = x1 * h11 + x2 * h21 \tag{1}$$

$$y2 = x1 * h12 + x2 * h22 \tag{2}$$

After which the transmitted bits are decoded by multiplying y with the inverse of the 2x2 channel matrix h.

$$h^{-1} = y/x \tag{3}$$

4.2 QPSK Modulation technique for 2x2 channel

Phase shift keying is a digital modulation technique; perhaps one of the most used in today's digital communication systems. In PSK modulation scheme, the phase of the carrier is altered in accordance with the input binary coded information. The PSK is further subdivided into BPSK, 8-PSK, 16-PSK, QPSK, DPSK. In binary phase shifting keying the transmitted signal is sinusoid of fixed amplitude .It has one fixed phase. The Fig 4.2 shows wave form of phase shift keying. The QPSK is a phase modulation scheme, used in constellation mapping. The constellation map of QPSK modulator is shown in Fig.4.3. Here the input bits stream is converted into complex stream using equation (4) and where I and Q both are in phase with I-out and Q-out respectively are shown in table 4.1. QPSK modulator accepts the binary bits as inputs consider as a symbol and converts them into complex value. QPSK takes only 4 symbols and generate its complex value in this fashion because the bit rate is ½.

$$D = (I + jQ)*K_{MOD} \text{ where } K_{MOD} = 1/1.414 \tag{4}$$

Input Bits	I-out	Q-out
00	-1	-1
01	-1	+1
10	+1	-1
11	+1	+1

Table 4.1. The Inputs and Outputs of QPSK Modulator

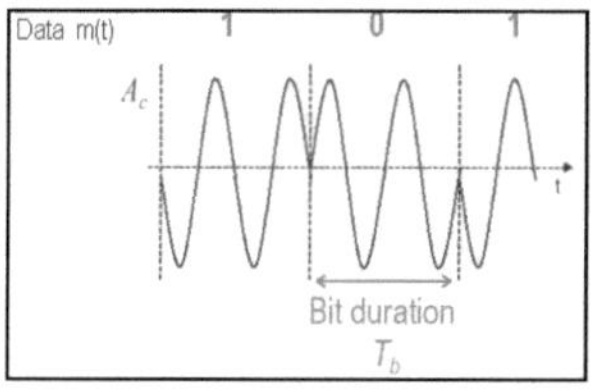

Fig. 4.2. Phase shift keying

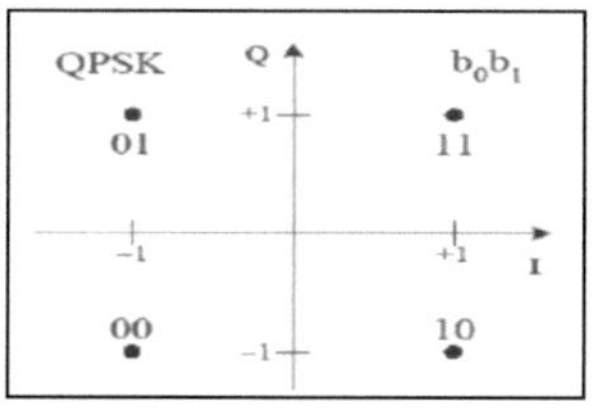

Fig. 4.3. QPSK Constellation Mapping

4.2 Working of 2x2 MIMO system

The 2x2 MIMO systems have different blocks like QPSK modulator, ideal channel, channel estimation block and QPSK demodulation. QPSK modulator takes the binary bits as input. Here the inputs bits are taken as symbols. For example 00, 01, 10, 11 are 4 inputs symbols for QPSK. The symbols are converted into complex values as shown in table 4.1. The complex output of the modulator is appended with the training sequence at 2 transmitters. Here the diversity technique is used to send the same data with different training sequences. The channel coefficients are multiplied with the modulated data. Here 4 paths are considered for 2x2 channels as shown in Fig.4.1.

The channel is an ideal channel i.e. the noise is not considered. The reliability is increased by employing the diversity technique i.e. transmitting the same information across multiple channels. The redundant data is transmitted so that the fading can be reduced. If one of the channel is not used or if the data is lost in space then the information/data can still be recovered from redundant transmission over the channels and hence the reliability of the communication system is improved.

The data transmitted from transmitter tx1 across the channels h11 and h12 and received by both the receiver's rx1 and rx2. Here the use of the available channels is done to increase the capacity and reliability. Then the received sequence is taken into the channel estimation block to find the estimated channel coefficients. The received data from LS channel estimator block are the given to the decision block in the QPSK demodulator block to get back the sent binary sequence. The transmitted sequence is then checked with the received sequence at the QPSK modulator and the Demodulator block respectively.

5. Simulation results

This section mainly deals with the simulation and VLSI implementation of 2x2 MIMO systems. The results are analyzed and compared with the MATLAB values.

5.1 MATLAB simulation results

The specifications of MIMO system from MATLAB are shown below.

1. 2x2 MIMO Technique: 2 transmitters and 2 Receiver antennas
2. QPSK modulation technique
3. Number of bits transmitted 64bits
4. Number of iterations = 100
5. Signal to noise ratio SNR = [0:3:9] i.e. 0, 3, 6 , 9 in dB

QPSK modulated output is shown in Fig.5.1. Input binary data is sent through QPSK modulator, the data taken into is in the form of symbols. The in-phase and quadrature waveform and the QPSK modulated data are shown in Fig.5.1. The bit error rate with respect to signal to noise ratio, the error rate goes on reducing as shown in Fig.5.2 (a). The error rate goes on reducing from 10^0 to 10^{-1} as the SNR is increased from 0 to 10dB. As the number of iterations increases the error at the output which is given as the difference between the input and the output signal becomes as zero. The error plot is shown in Fig.5.2 (b) is for 100 iterations. There is a mismatch at 2 points and the difference is plotted. These plots as shown in Fig.5.2 are for LS channel estimation method. As the number of iterations

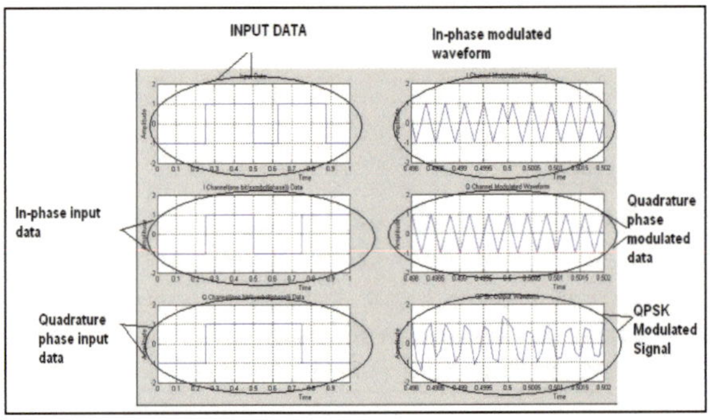

Fig. 5.1. QPSK modulator MATLAB outputs

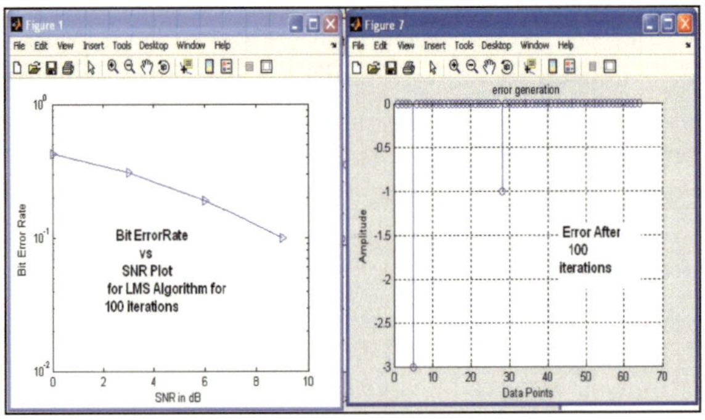

(a) (b)

Fig. 5.2. BER vs SNR plot and Error generation after 100 iterations

increase the error reduces so that the BER vs SNR ratio goes on reducing and becomes negligible. Whenever there is a mismatch between the input and the output sequence of MIMO system then the difference is shown as the error.
The plots shown in Fig.5.3 are the input to a system is a random data generated using 'randn' in MATLAB. The random data generated is fed to the QPSK modulator. The data get modulated and then the modulated output is sent through the QPSK demodulator. The demodulated output same as input as shown in Fig.5.3. The plots as shown in Fig.5.3 are from MATLAB. Here the input is taken as 64 symbols.

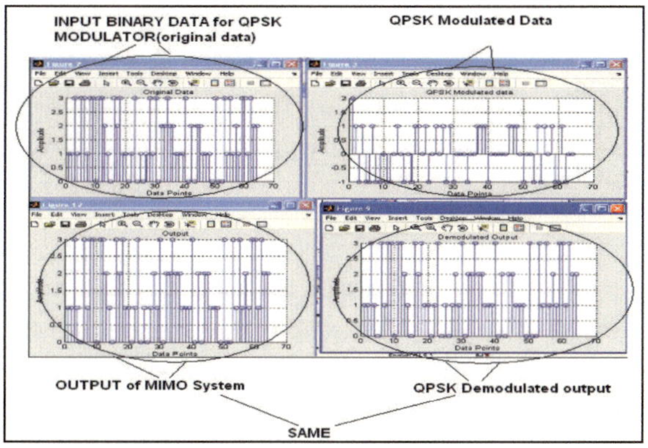

Fig. 5.3. The output of MIMO system and QPSK demodulated output

5.2 Modelsim simulation results

2x2 MIMO system is designed using Verilog-HDL and simulated in Modelsim. Each block is separately verified for its functionality. Each block is synthesized and the integrated output is also verified by implementing Xilinx FPGA. The top block of 2x2 MIMO system block diagram from Xilinx is shown in Fig.5.4. All the blocks in 2x2 MIMO system like QPSK modulator/demodulator, ideal channel and the LS channel estimation blocks are integrated and the connections are shown in Fig.5.4, with inputs and the outputs.
The output of integrated 2x2 MIMO system design is as shown in Fig.5.5. The waveform shows the 2x2 MIMO system output. Here the clock is applied and the reset is high then the

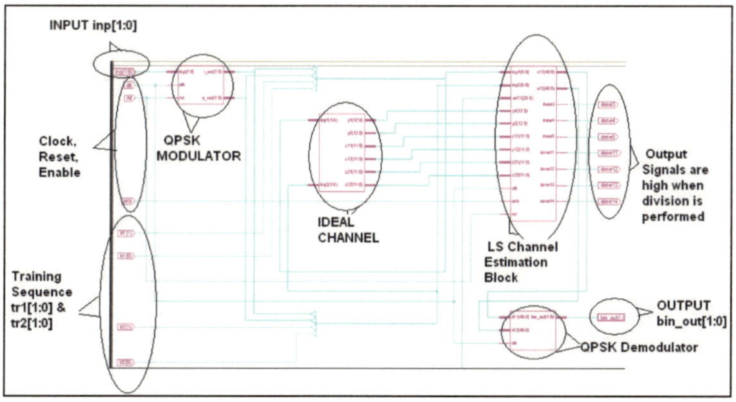

Fig. 5.4. 2x2 MIMO System in Xilinx

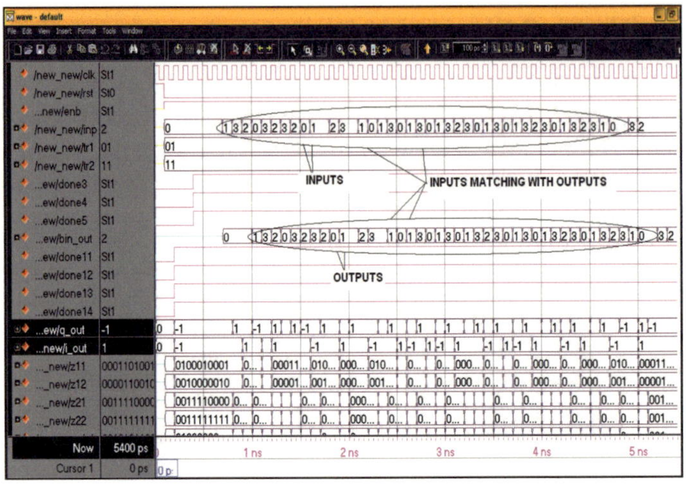

Fig. 5.5. The 2x2 MIMO System output

system is made reset. Then the reset signal is made low and the enable signal is made high. The inp[1:0] is the given the value as the required symbol to transmit. The training sequence tr1 is given as +1 and training sequence tr2 is given as -1.Then the output bin_out[1:0] is obtained as the input inp[1:0]. In Modelsim simulation the channel coefficients are represented 'z' in terms of 'h'.

5.3 VLSI implementation results

5.3.1 2x2 MIMO system

The design is implemented on Xilinx Virtex II-Pro board. The synthesis report is shown below.

Name of the device	Device utilization
Number of Slices	3999 out of 4928
Number of Slice Flip Flops	343 out of 9856
Number of 4 input LUTs	7603 out of 9856
Number of IOs	18
Number of bonded IOBs	18 out of 396
Number of MULT18X18s	11 out of 44
Number of GCLKs	1 out of 16

Table 5.1. Device utilization summary 2x2 MIMO system target device using 2vp7ff896-6

Timing parameters	Parameter values
Speed Grade	-6
Minimum period	137.548ns
Maximum Frequency	7.270MHz)
Minimum input arrival time before clock	39.872ns
Maximum output required time after clock	3.692ns
Maximum combinational path delay	No path found

Table 5.2. Timing Report of 2x2 MIMO systems

The device utilization report from Xilinx is shown above table 5.1. The timing report is shown below table 5.2 for the 2x2 MIMO system design. The device utilized is 2vp7ff896 with a speed grade of 6 and the slices taken are 3999 with a frequency of 7.2MHz.

5.3.2 QPSK modulator

The device utilization of QPSK modulator is shown in below table 5.3. It utilizes about 1 slice and its operating frequency is about 666.66MHz from DC. The timing report is shown table 5.4.

Name of the Device	Device utilization
Number of Slices	1 out of 4928
Number of Slice Flip Flops	4 out of 9856
Number of 4 input LUTs	2 out of 9856
Number of IOs	8
Number of bonded IOBs	8 out of 396
IOB Flip Flops	4
Number of GCLKs	1 out of 16

Table 5.3. Device utilization summary of QPSK Modulator using 2vp7ff896-6

Timing Parameter	Parameter values
Minimum input arrival time before clock	2.246ns
Maximum output required time after clock	3.615ns
Maximum combinational path delay	No path founds

Table 5.4. Timing Report of QPSK Modulator

5.3.2 Ideal channel

The device utilization of ideal channel is shown in below table 5.5. It utilizes about 10 slices and its operating frequency is about 400MHz from DC. This is a combinational block and hence has maximum combinational delay of 9.05ns. The timing report is shown table 5.6.

Name of the Device	Device utilization
Number of Slices	10 out of 4928
Number of 4 input LUTs	2 out of 9856
Number of IOs	86
Number of bonded IOBs	86 out of 396
Number of MULT18X18s	3 out of 44

Table 5.5. Device utilization summary of ideal channel using 2vp7ff896-6

Timing Parameter	Parameter values
Maximum combinational path delay	9.055ns

Table 5.6. Timing Report of ideal channel

The device utilization of LS Channel Estimation is shown in below table 5.7. It utilizes about 4321 slices and its operating frequency is about 37MHz from DC. The timing report is shown table 5.8.

Name of the Device	Device utilization
Number of Slices	4231 out of 4928
Number of Slice Flip Flops	509 out of 9856
Number of 4 input LUTs	7956 out of 9856
Number of IOs	218
Number of bonded IOBs	218 out of 396
IOB Flip Flops	4
Number of MULT18X18s	18 out of 44
Number of GCLKs	1 out of 16

Table 5.7. Device utilization summary of LS channel estimation using 2vp7ff896-6

Timing Parameter	Parameter values
Minimum period	142.444ns
Maximum Frequency	7.020MHz
Minimum input arrival time before clock	96.582ns
Maximum output required time after clock	9.734ns

Table 5.8. Timing Report of LS channel estimation

The device utilization of QPSK demodulator is shown in below table 5.9. It utilizes about 6 slices and its operating frequency is about 500MHz from DC. The timing report is shown table 5.10.

Name of the Device	Device utilization
Number of Slices	6 out of 4928
Number of 4 input LUTs	10 out of 9856
Number of IOs	101
Number of bonded IOBs	16 out of 396
IOB Flip Flops	2
Number of GCLKs	1 out of 16

Table 5.9. Device utilization summary of QPSK Demodulator

Timing Parameter	Parameter values
Minimum input arrival time before clock	3.708ns
Maximum output required time after clock	3.615ns

Table 5.10. Timing Report of QPSK Demodulator

The Table 5.11 shows the device utilization for each block in 2x2 MIMO systems.

5.4 Synthesis and timing verification of 2x2 MIMO system

The 2x2 MIMO system code is simulated in Modelsim and synthesized using Xilinx before giving to Design Compiler for timing analysis. Design compiler is tools which synthesize the design using the user given constraints, libraries and the RTL code. The clock to be specified as a constraint for which approximately the design work is determined using Xilinx. A clock of 50-60% from Xilinx is created for timing verification. Clock network delay or min latency of 25% of the clock, uncertainty for setup is given as 10% of the clock, for hold 5% of the

Block	QPSK Modulator	Ideal Channel	LS Channel Estimation	QPSK Demodulator	2x2 MIMO S/m	2x2 MIMO for 2vp30ff896 device
No.of Slices Out of 4928	1	10	4321	6	4231	1902 out of 13696
No.of FFs Out of 9856	4	-	509	-	343	273 out of 27392
No.of 4input LUT out of 9856	2	2	7956	10	7956	3595 out of 27392
No.of IOs Out of 39	8	86	218	101	18	27 out of 556
No. of Mux out of 44	-	3	18	-	11	8 out of 136
Frequency of operation in MHz	666	400	7	500	7.2	13

Table 5.11. Block wise Device utilization for 2x2 MIMO System

clock. An external delay is specified as 40-50% of clock specified. Then the compiled results are checked for the setup and hold violations i.e. the slack is met or not. If not met the constraints are again changed as per requirement. The design timing verification is checked again. The libraries used are cb13fs120_tsmc_max.db and cb13io320_tsmc_max.db. These libraries are set as link and target library.

Name of the Device	Value Before Ungrouping	Value After Ungrouping
Number of ports	18	18
Number of nets	170	27550
Number of cells	6	26251
Number of references	5	123
Combinational area	38888.750000	38888.750000
Non combinational area	1902.000000	1902.000000
Net Interconnect area	6541.367676	8807.866211
Total cell area	40790.750000	40790.750000
Total area	47332.117676	49598.616211

Table 5.12. Area report before and after Ungrouping

The total area is changed as the hierarchical design is ungrouped. The area is the constraint which is increased when the design is ungrouped. The area occupied before ungroup is 47332μm^2 and after ungroup the area is 49598μm^2. The area report is shown table 5.12.

The power is the main constraint in any design. The power obtained from DC is the approximate value. The total dynamic power is the power consumption of the design which is 2.9289mW. Here there are 2 path groups and the timing is met. There is no violation before ungrouping. For setup and hold the timings are met before and after ungrouping. The power report is shown table 5.13.

Type of the Power	Before Ungrouping	After Ungrouping
Cell Internal Power	1.8290 mW	1.8093 mW
Net Switching Power	1.0680 mW	1.1196 mW
Total Dynamic Power	2.8970 mW	2.9289 mW
Cell Leakage Power	224.1040 uW	224.1024 uW

Table 5.13. Power report

Cell	Reference	Library	Area Attributes
U3	bufbd1	cb13fs120_tsmc_max	1.000000
U4	bufbd1	cb13fs120_tsmc_max	1.000000
cha_est1	channel_est	40617.250000	h,n
cv1	channel_virtual	126.250000	h
qd41	qpsk_demod5_1	16.000000	h, n
qmm1	qpskmoddd_1	129.250000	h, n
Total cells		40790.750000	

Table 5.14. The Cell report

The cell report table 5.14 shows the total number of cell in the 2x2 MIMO system i.e. 40790 cells. After all timing verifications the net list, ddc file and constraint file is saved to carry out the timing verification in PT. Then after PT static timing verification is carried out. In PT we get the exact timing report for both setup and hold.

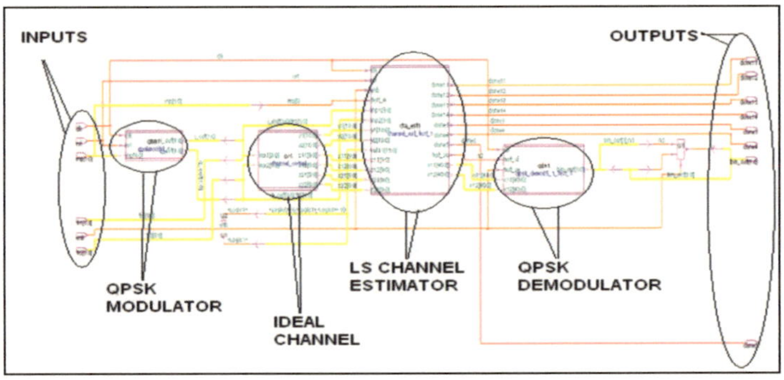

Fig. 5.6. The 2x2 MIMO system Schematic after DFT from DC

Parameters	Design Compiler	After Ungroup in DC	Primetime
Area	47332	49598	-
Maximum Timing	0.97	0.33	0.33
Minimum Timing	0	0.03	0.03
Dynamic Power	2.897mW	29289nW	-
Cell leakage power	224.10µW	224.1024 µW	-

Table 5.15. Area, Timing, Power Reports

The design for testability (DFT) is carried on 2x2 MIMO system. The Fig 5.6 shows schematic view of 2x2 MIMO system.The Table 5.15 shows the area, power and timing reports obtained from DC and PT for 2x2 MIMO system.

5.5 Physical design of 2x2 MIMO system

After synthesis, design is taken to the backend flow to tape out the design. This has inputs generated from the front end ASIC design flow. This involves the floor plan, power plan, timing setup, placement, clock tree synthesis, routing, design rule check, design for manufacturability and GDSII. The Fig.5.7 shows the physical design flow that is to be carried to tape-out the 2x2 MIMO systems.

Physical design flow is the back end flow in a chip design flow. Technology file (cb13_4m_tlu.tf) is a file to be specified when we create a design library. This consists of Layer Definitions, Via Definitions, Process Design Rules, TLU Parasitic capacitance models, Preferred routing directions and units.

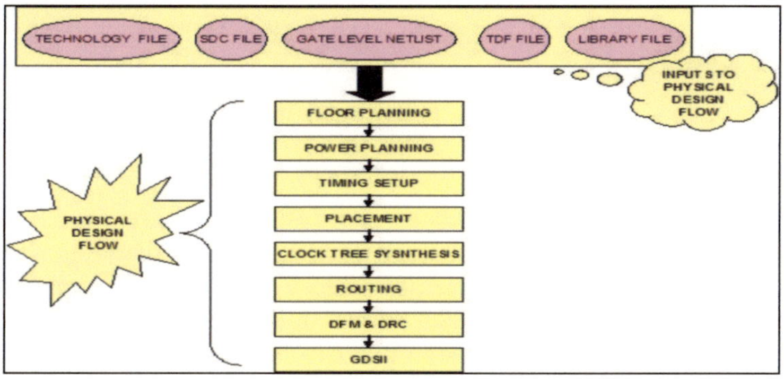

Fig. 5.7. Physical Design Flow

Design Libraries which utilizes the Synopsis database (Milky Way), which has .NETL specify the net views, .EXP specify expanded net list views, .CEL specify the cell views, .FRAM specify frame views and .TIM specify timing views. Gate-level net list consisting of inputs pads, output pads and their logical connectivity. SDC file consisting of the design constraints specified from Design compiler. The design is carried in Macro level. So TDF file is not required which consists of boundary and pads/pins information. Finally, the output of backend physical design flow is the routed net list, parasitic file and GDS. This GDS consists of static timing analysis, simulation, logic equivalence check, and physical verification for final verification. Here 0.13μm technology is used in Astro design flow. Astro is a physical design tool.

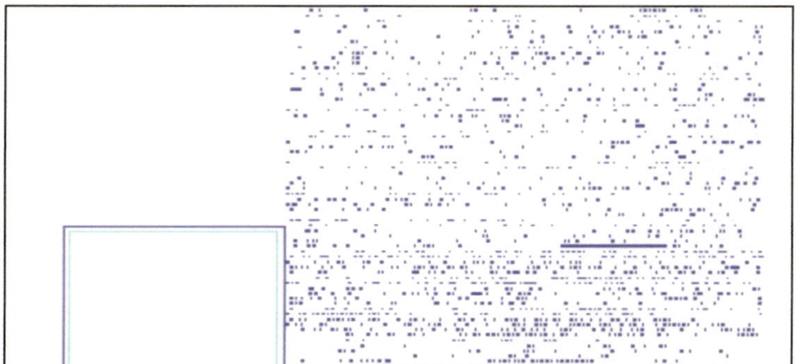

Fig. 5.8. Floor Planning of 2x2 MIMO System from Astro

Fig. 5.9. Standard Cell Placement during In-Placement in Astro

Floor planning is first important step in back end design flow. The Fig.5.8 shows the floor planning of the design. To have a good IC to be developed, floor planning should be perfect. This phase consisting of planning and sizing of blocks and interconnects. Inputs of Floor-plan define different control parameters like aspect ratio, core utilization.

Placement is the key step in physical design process. Poor placement consumes large areas hence performance gets reduced. Inputs to the placement are set of blocks, number of terminal for each block and net list. Main objectives of placement are to perform Standard cell placement, perform logic and placement optimization, to meet the timing depending on the placement and optimizing of the design and reducing congestion and crosstalk free placement. The Fig 5.9 shows the placement of cell during in-placement. Pre-placement optimization removes all wire load models effects to create a clean state. Perform zero wire load model optimization and logic remapping. This reduces the total cell area by gate downsizing and buffer remove which is introduced to remove the congestion i.e. collapsing high fan-out nets. Also meet the setup timing. The high fan-out buffers are first collapsed and then synthesized for quick placement as shown in Fig.5.10. Here the fan-outs considered are 20. Timing report is

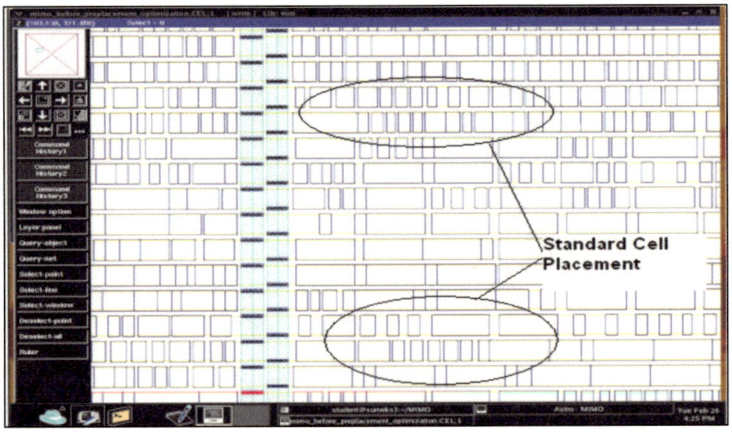

Fig. 5.10. Standard Cell Placement During Pre-Placement in Astro

checked for setup, hold, maximum capacitance, maximum fan-out and nets. The design is carried up to the placement of the cells. The clock tree synthesis and the routing is not carried for physical design. The DRC is also not checked for the design.

2x2 MIMO system results are explained in detail. The RTL code is synthesized implemented on Virtex2Pro Xilinx board. Then design timing verification is performed using DC and PT. The 2x2 MIMO system design is taken to physical design to tape-out.

6. Conclusion

The simple working of a MIMO system is carried till backend of the VLSI flow. The design is simulated in MATLAB to arrive at the specifications. The RTL code is successfully simulated in Modelsim. The design is synthesized and implemented on Virtex2Pro FPGA board. The synthesis and timing is verified and the timing is met for both setup and hold in DC and PT. DFT is also carried without timing violation. The top design takes about 3999 number of slices out of 4928 slices i.e. in Virtex2Pro the device selected is 2vp7ff896 at speed grade of 6 with operating frequency 7.27MHz and minimum period of 137.548ns. Timing verified is all met with positive slack with zero violations.

In the design the channel is considered to be ideal. In the future work the noise is to be added and the estimation of channel using different channel models is to be carried. Different channel estimation is to be simulated in MATLAB and then taken to the complete VLSI flow. The frequency of the design is to be optimized. The complete backend flow has to be completed till the tape-out of the design.

- Scope of the future work is to improve the design further for the noise to be included in the channel and use any improved matrix inversion technique for improving the design frequency of operation. This can be done using QRD algorithm.
- Iterative algorithm is applied for the design so as to reduce the effect of noise on data. The iterative algorithms like Recursive LS, Least Mean Square, MMSE, etc. Channel estimation and compensation for different channel models for delays is to be implemented.
- Fixed point implementation of the design is to be carried out by obtaining the floating-point values from MATLAB because the fixed- point representation will be more efficient.
- MIMO performance can be improved by using OFDM. By incorporating OFDM the performance of the overall system can be improved.

7. Reference

Joseph, W. Reynders, W. Debruyne, J. and Martens, L., "*Influence of Channel Models and MIMO on the Performance of a System based on IEEE 802.16*",Wireless Communications and Networking Conference, ISBN 1-4244-0659-5, pp.1826-1830, 11-15th March, 2007.

Simon S Haykin and Michael Moher,"*Modern Wireless Communication*, Second Edition, Prentice-Hall publication 2004, ISBN-13: 9780130224729.

George Tsoulos, "*MIMO System Technology For Wireless Communications*",Revised Edition, CRC Publisher, 2006.

Chia-Liang Lui, "*Impacts of I/Q Imbalance on QPSK -OFDM-QAM Detection*", IEEE Transactions on Consumer Electronics, vol. 44, no. 3, pp. (984-989), August, 1998.

Yantao Qiao, Songyu Yu, Pengcheng Su and Lijun Zhang, *"Research on an Iterative Algorithm of LS Channel Estimation in MIMO OFDM Systems"*, IEEE Transactions on Broadcasting, vol. 51, no. 1, pp. (149-153), March 2005.

S.Cui, A.J.Goldsmith and A.Bahai, *"Energy Efficiency in MIMO and cooperative MIMO techniques in Sensor Networks"*, IEEE Journal on Selective Areas in Communication, vol. 22, no. 6, pp. (1089-1098), August 2004.

Yatawatta,S, Petropulu,A.P and Graff,C.J, *"Energy efficient Channel Estimation in MIMO Systems"*,IEEE international conference on Acoustics,Speech and Signal processing, vol.4, no.1, pp.(317-320), 18-23rd March,2005.

Saux,B.L and Helard,M., *"Iterative Channel Estimation based on Linear Regression for a MIMO-OFDM System"*, IEEE interanational Conf. on Networking and Communications,vol.1, no.1, pp.(356-361),19-21st June, 2006.

Changchuan Yin, Jingyu Li, Xiaolin Hou and Guangxin Yue, *"Pilot Aided LS Channel Estimation in MIMO-OFDM Systems"* The 8th International Conference on Signal Processing,vol.3, pp.(16-20), 2006.

[Reza Abdotee, Tharek Abd.Rehman and Savia Mahdaliza Idrus, *"Computational Complexity Reduction for MIMO-OFDM Channel Estimation Algorithm "*, elecktika Journal of electrical engineering , vol.9,No.1, pp.(30-36) , 2007.

Bellido, Deseada, Entrambasaguas and Jose T. *"MSE evaluation at reception end in MIMO-OFDM systems using LS channel estimation"*, Waveform Diversity and Design Conference, vol.1, pp. (174-177), 4-8th June,2007.

Markus Myllyla, Markku Juntti, Matti Limingoja,Aaron Byman and Joseph,R. Cavallaro *"Performance Evaluation of Two LMMSE Detectors in a MIMO-OFDM Hardware Testbed"*, Conf.on Signals, systems and Computers, Fortieth Asilomar Pacific Grove, pp. (1161-1165), Oct29-Nov1 2006.

Ian Griffiths *"FPGA Implementation Of MIMO Wireless Communications System"*, University Of New Castle ,Australia, 1st November, 2005.

Bernard Sklar *"Digital Communications–Fundamentals and Application"* Published by PEARSON Education. Year 2003.

G.Proakis, Masoud Salehi *"Communication System Engineering"*, Second edition by John. Published by Pearson Education. 2002.

John G. Proakis, Dimitris G. Manolakis, *"Digital Signal Processing Principles Algorithms and Applications"*, 4th Edition, Boston: McGraw Hill.

Himanshu Bhatnagar - *"Advanced ASIC Chip Synthesis Using Synopsys® Design Compiler™ Physical Compiler™ and PrimeTime®"*, Kluwer academic publishers, Second edition - 2002.

Synopsys User guide for Design Compiler.

Synopsys User guide for Primetime.

Synopsys User guide for Astro.

David Garrett, Linda Davis, Stephen ten Brink, Bertrand Hochwald and Geoff Knagge, *"Silicon Complexity for Maximum Likelihood MIMO Detection Using Spherical Decoding"*, IEEE journal of Solid State Circuits, vol.39, no.9, pp. (1544-1552), September, 2004.

Yvo L.C. de Jong and Tricia J. Willink, *"Iterative Tree Search Detection for MIMO Wireless Systems"*, IEEE Transactions on Communications, vol. 53, no.6, pp. (930-935), June, 2005.

Sudhakar Reddy P, Ramachandra Reddy G, *"Design and FPGA Implementation of Channel estimation method and Modulation technique for MIMO system"*, European journal of scientific research, ISSN 1450-216X,Vol.25,No.2,pp.257-265, 2009.

20

Efficient Implementation of MIMO Decoders

Muhammad S. Khairy[1], Mohamed M. Abdallah[2] and S. E.-D. Habib[3]
[1]*Electrical Engineering and computer Science Dept.,*
Univ. of California, Irvine, Irvine, CA 92697-2625,
[2]*Electrical and Computer Eng. Dept., Texas A&M Univ. at Qatar, PO Box 23874, Doha,*
[3]*Electronics and Comm. Dept., Faculty of Eng., Cairo Univ., Giza 12613,*
[1]*USA*
[2]*Qatar*
[3]*Egypt*

1. Introduction

Multiple-Input Multiple-Output (MIMO) technology has emerged as a promising technology for achieving the high data rates of next generation wireless communication systems. MIMO systems improve either the bit-error rate (BER) performance by using spatial diversity or the date rate via spatial multiplexing. However, Maximum-likelihood (ML) detection for high order MIMO systems faces a major challenge in computational complexity that grows exponentially with the number of transmit and receive antennas. This limits the practicality of these systems from an implementation point of view, particularly for mobile battery-operated devices.

This reality motivated researchers to consider other suboptimal approaches for MIMO decoding, such as Zero Forcing (ZF), Minimum Mean Square Error (MMSE) and VBLAST (Proakis & Salehi, 1994; Guo & Nilsson, 2003; Myllyla et al., 2005). All of these suboptimal approaches vary in performance and complexity. Recently, the sphere decoding (SD) algorithm which is a tree-based search algorithm enabled the implementation of efficient MIMO decoders that achieve near MLD performance together with reduced complexity (Burg et al., 2005; Barbero et al., 2005; Khairy et al., 2009). Instead of the exhaustive search over all possible combinations of the transmitted symbols, the SD algorithm reduces the complexity by searching only over a finite number of symbols within a circle of radius R centered at the received vector. While the SD approach provides a near ML solution, the runtime measured by the required operations to find the optimum solution is highly dependent on the received signal to noise ratio and the channel conditions. Consequently, the SD algorithm experiences variable throughput problems as the search radius R for each symbol varies according to the noise levels and the channel coefficients. Moreover, the sequential search results in hardware implementations that are not fully pipelined.

To alleviate these problems, the fixed sphere decoding (FSD) algorithm was recently proposed (Barbero & Thompson, 2006 b). The FSD algorithm achieves a fixed throughput performance and enables fully-pipelined hardware by performing fixed number of operations per detected symbol, independent of the noise level. The FSD algorithm is performed in two stages where the first one is a pre-processing stage for the received

symbols that incorporates a QR decomposition of the channel matrix for the MIMO system. This pre-processing allows for employing a tree-based search algorithm at the second stage that finds the optimal symbol which has the minimum distance to the received symbol.
In this chapter, we present an efficient FPGA prototype of the FSD that implements an efficient QR decomposition of the channel matrix for a MIMO system using CORDIC array techniques followed by the FSD tree-based search algorithm. This prototype provides an improved hardware implementation in terms of area and throughput compared to up-to-date published results in the literature. The FPGA implementation is incorporated with a MATLAB simulation model of an OFDM based MIMO system to validate the hardware design.
As a practical application, we employed the FSD in the receiver of the IEEE802.16e WiMAX MIMO system. The IEEE 802.16e standard supports three types of MIMO space time codes (STC), referred to as matrix A, B, and C. These STC codes achieve different levels of throughput and diversity depending on the quality of the MIMO channels. We employed the FSD as a STC detector to decode the different STC schemes. In achieving that, the received signal is adaptively pre-processed according to the STC type prior feeding it to the FSD. At the end of the chapter, we present the performance of a two by two STC-WiMAX system with the FSD in terms of bit error rate and throughput. The results indicate that our FPGA implementation achieves throughput values that satisfy the requirements of the WiMAX standard.

2. System model

Consider the OFDMA-based 2x2 WiMAX MIMO system shown in Fig. 1. At the transmitter, the input data is first encoded then modulated using QAM modulation technique. The STC operates on input data symbols sequentially and distributes the data symbols to each antenna path. The WiMAX standard supports three STC referred to in the standard by matrix A, B and C, respectively. In particular, the STC matrix A achieves full diversity by employing the Alamouti coding, while the STC matrix B achieves full rate by employing spatial multiplexing, and the STC matrix C achieves full rate and diversity, by employing the Golden code. The transmitter encodes the incoming symbol streams into two spatial streams using one of the matrices A, B or C. The block of subcarrier mapping and PseudoRandom Binary Sequence (PRBS) function denotes data truncation or puncturing, pilot insertion, and IFFT input packing. Then IFFT is applied to generate the OFDMA symbol.
At the receiver, a FFT operation is first performed at each receive antenna, followed by subcarrier de-mapping and PRBS. The STC Decoder is implemented using the MIMO FSD decoder. The received signal, after performing the FFT operation, can be expressed as follows

$$\mathrm{r} = \mathrm{Hs} + \mathrm{n} \tag{1}$$

where s denotes the vector of transmitted symbols, n is the complex additive white Gaussian noise of zero mean and variance N_o, and r is the received vector. H denotes the 2 x 2 channel matrix where each element, h_{ij}, is modeled as Rayleigh fading channel. The channel matrix is assumed to be perfectly known at the receiver.
We develop a rapid prototype of the system model of interest whereby the FSD algorithm is implemented using Stratix II FPGA board and the rest of the system blocks are implemented using MATLAB. Such prototype allows for testing the WiMAX MIMO detectors for several OFDM systems with different FFT sizes.

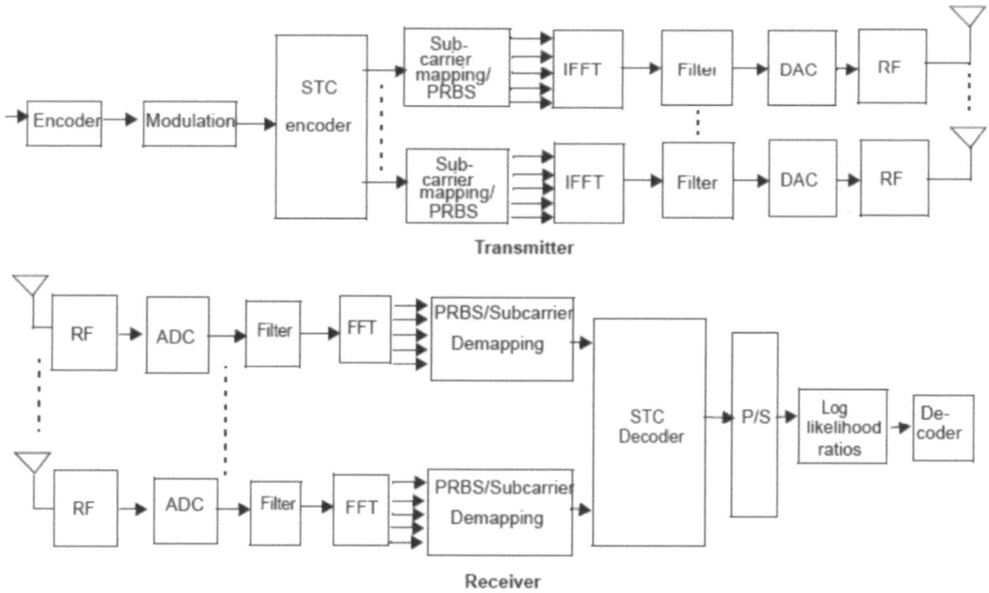

Fig. 1. WiMAX System Model

3. FSD algorithm

The Fixed Sphere Decoder searches over a fixed number of vectors within a circle of radius C centered at the received vector r. This results in fixed complexity and hence fixed throughput. The symbol with the minimum Euclidian distance to the received vector, $\hat{s}_{FSD}$, can be obtained as follows,

$$\hat{s}_{FSD} = \arg\{\min_{s \in S} \|r - Hs\|^2 \leq C^2\} \quad (2)$$

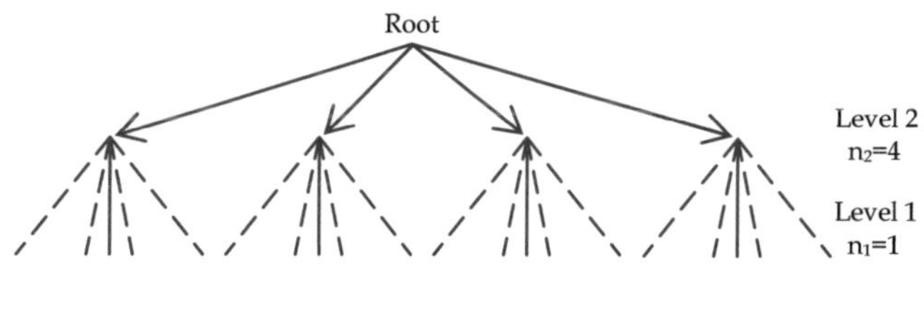

Fig. 2. Tree Search for 2x2 MIMO system and QPSK

The FSD search for the transmitted vectors can be viewed as a tree search of N levels. Fig. 2 shows the tree for 2x2 MIMO system with QPSK modulation. Starting from the root, there exists 4 branches corresponding to 4 QPSK symbols and for each node in the top level there exists another 4 different branches. In general, the search process starts from level N. The number of nodes considered per level is denoted as n_i. (Barbero & Thompson, 2006 b) showed that n_i during the tree search satisfies the following relation

$$E[n_N] \geq E[n_{N-1}] \geq \cdots \geq E[n_1]$$

where E[n_i] represents the expected number of visited nodes per level i. The numbers of visited nodes per level are pre-determined. All node symbols of level N are visited where only one node per level is chosen for the next levels. Given that the channel matrix H can be decomposed using QR to an upper triangular matrix R and a unitary matrix Q.

$$H = QR \tag{3}$$

where, $Q^HQ=QQ^H=I$. Since Q is a unitary matrix, the expression given in (1) can be rewritten as follows,

$$\|r - Hs\|^2 = \left\|Q^H(r - Hs)\right\|^2 = \|\hat{y} - Rs\|^2 \tag{4}$$

where $\hat{y} = Q^H r$

Hence the expression in (2) can be equivalently expressed as

$$\hat{s}_{FSD} = \arg\{\min_{s\in S}\|\hat{y} - Rs\|^2 \le C^2\} \tag{5}$$

The second norm can be expressed as sum of a partial Euclidian distance d_i where:

$$D = \|\hat{y} - Rs\|^2 = \sum_{i=1}^{N} d_i \tag{6}$$

The partial Euclidian distance d_i can be evaluated as:

$$d_i = |b_i - R_{ii}s_i|^2 \tag{7}$$

where $b_i = \hat{y}_i - \sum_{j=i+1}^{N} R_{ij}s_j$

We note that using the QR decomposition of the channel matrix H, rather than the Cholesky considered in (Barbero & Thompson, 2005 ; Barbero & Thompson, 2006 b), and subsequently casting the FSD by Eqns. (5-8) leads to improvements in hardware and throughput as will be proved in Section 6 and 7 of this chapter (Khairy et. al, 2009) . In the subsequent sections, we first present an overview of the QR decomposition and the implementation of the FSD, and then we present the decoding technique for the different STC matrices.

4. QR decomposition using CORDIC

The QR decomposition is done using CORDIC (Coordinate Rotation Digital Computer) (Volder, J.E, 1959). CORDIC is used to rotate 2-point vector (x, y) by angle ζ_v to a new vector with coordinates (x', y')

$$x' = \cos(\zeta_v) \times (x - y\tan\zeta_v) \tag{9a}$$

$$y' = \cos(\zeta_v) \times (y + x\tan\zeta_v) \tag{9b}$$

By selecting ζ_v as follows

$$\tan\zeta_v = 2^{-v}, \tag{10}$$

The trigonometric functions of equation 9 are then implemented via right-shifts and add/subtract operations only. The first step of the CORDIC algorithm is to express any arbitrary angle ξ by a sequence of forward or backward rotations obeying equation 10 as follows

$$\zeta = \sum_v \rho_v \zeta_v \quad (11)$$

where $\rho_v = \pm 1$. Throughout all these macro rotations, the multiplication by $\cos \zeta_v$ can be collected together into a single constant

$$K = \prod_v \cos \zeta_v \quad (12)$$

The CORDIC algorithm can be seen as composed of stages performing mini rotations followed by a gain correction. In general, at the mini rotation number i the new rotated vector can be expressed as:

$$y_{i+1} = y_i + \rho_i \, x_i \, 2^{-i} \quad (13a)$$

$$x_{i+1} = x_i - \rho_i \, y_i \, 2^{-i} \quad (13b)$$

$$\zeta_{i+1} = \zeta_i - \rho_i \times \tan^{-1} 2^{-i} \quad (13c)$$

The CORDIC method provides a pipelined implementation of the given rotations using only shifts and addition/subtractions. There exist two modes of CORDIC Units. The CORDIC vector mode rotates the input vector by the angle necessary to align it along the x-axis. The result of the vector mode is a rotation angle and a scaled magnitude of the original vector. The vectoring function works by minimizing the y component at each rotation. The second CORDIC mode is the rotate mode, which rotates the incoming vector by a certain angle. The QR is implemented using systolic array as shown in Fig.3. The QR operation can be accomplished by a sequence of circular row operations to nullify elements below the main diagonal of the matrix. The round boundary cell performs the vectoring operation, i.e., it computes the angles needed for annihilation of the incoming data samples. The boundary cell sends the angle values to the inner square cells in the same row. The inner square cell calculates the new rotated sample values based on the angle values given from the boundary cell.

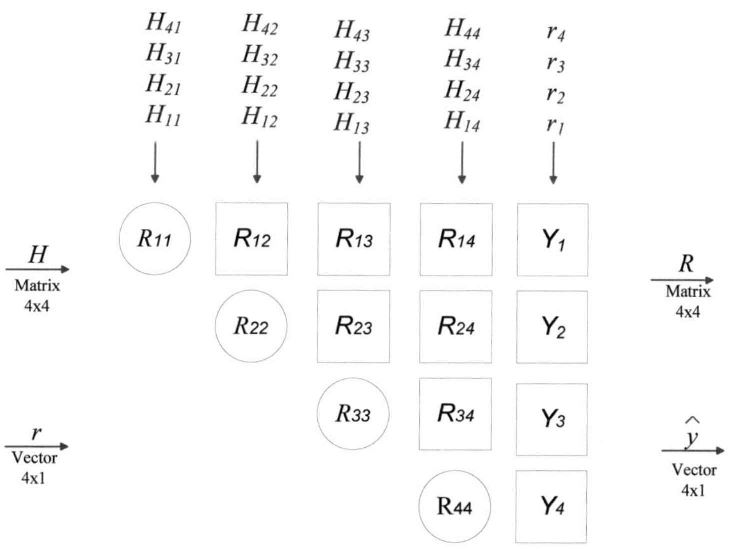

Fig. 3. QR Systolic Array for a 4x4 channel matrix

5. Architecture and FPGA implementation

In this section we consider the architecture and implementation of the FSD for a 4x4 un-coded MIMO system with 16-QAM modulation techniques. The FSD tree search has [16, 1, 1, 1] visited nodes per level as going from the root of the tree until the leaf respectively. Figure 4 shows the block diagram for the MIMO detector.

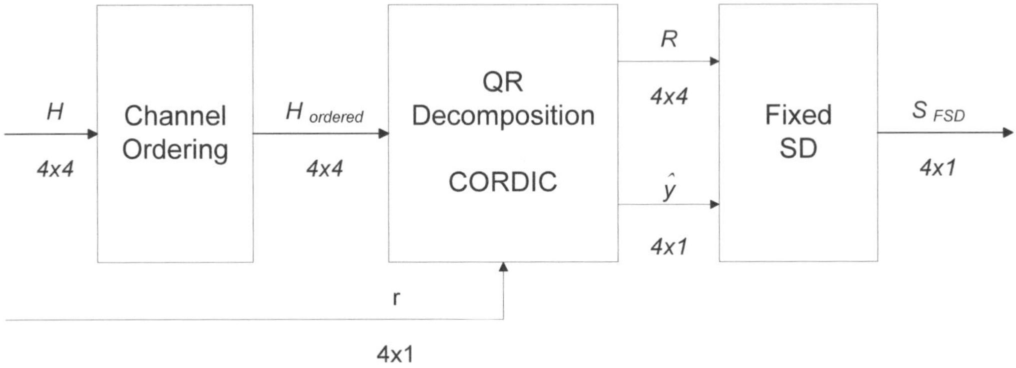

Fig. 4. FSD with QR and channel Ordering

The channel matrix H is assumed to be perfectly known at the receiver. The ordered channel matrix as well as the received vector is decomposed to upper triangular matrix R and a vector $Q^H r$ using the CORDIC QR decomposition.

5.1 Channel ordering

Channel ordering is performed using the columns of the channel matrix H. Channel columns are ordered in an iterative way according to the norm of each column. The received symbol with the largest noise amplification is selected for level N as this level considers all node symbols. For the remaining N-1 columns, we choose a channel column corresponding to the received symbols with the smallest noise amplification (Barbero & Thompson, 2006 b).

5.2 CORDIC systolic array

The basic CORDIC unit is implemented using only adder/subtract and shifter units. It can be adapted for vector mode and rotate mode by setting the sign of ρ_i.

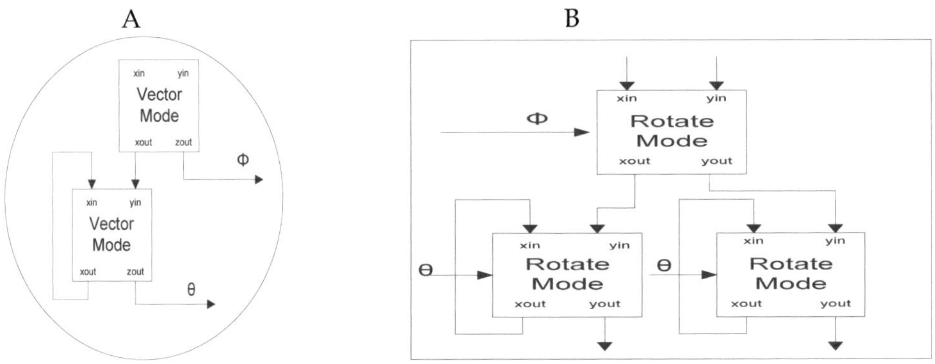

Fig. 5. (A) Vector mode CORDIC. (B) Rotate mode CORDIC.

For complex data, the boundary cell which performs vector mode consists of two real CORDIC blocks. The first one rotates the vector formed from the real and imaginary parts of the input, x, to set its imaginary part to zero producing an angle Φ. The second block computes the angle Θ needed to annihilate the rotated real part of the first CORDIC block against the stored one of the previous row.
The angle Φ calculated from the boundary cell is used to rotate both the real and imaginary part of the input vector. Both the rotated real and imaginary part, with the stored real and imaginary parts of the previous row, are rotated with angle Θ calculated from the boundary cell. On the other hand, the inner square cell which performs the complex rotate mode consists of three real CORDIC rotate blocks.

5.3 Fixed sphere decoder:

The search stages are composed of four pipelined levels and a decision unit. Considering 16-QAM modulation scheme, level 4 computes the 16 different Euclidian distances D_4. At levels 3 to 1 only the symbol with minimum d_i is chosen. The calculation of the 16 minimum chosen branch distances (the minimum branch distance for each node) at each level from 3 to 1 can be done fully parallel or sequential (Hess et al., 2007) or in between (Barbero & Thompson, 2006 b). We chose to divide them into 2 iterations each calculates 8 different distances. This approach increases the throughput of the system as we can detect a new MIMO symbol every 2 clock cycles.

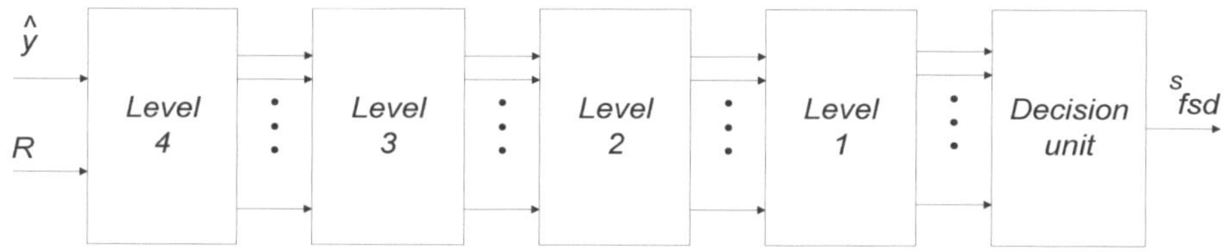

Fig. 6. Sphere Decoder Search stages

5.3.1 Level 4

Level 4 is the top level in tree after the root. At this level, the 16 different branch distances d_4 are calculated based on equation (7). The sphere ALU for the exhaustive search SD in (Barbero et al., 2006 b) which is used to calculate the branch distance is adapted and modified to calculate the Partial Euclidian distance of the FSD algorithm. For 16-QAM modulation, the partial Euclidian distance is calculated as follow

$$D_4 = d_4 = |b_4 - R_{44}s_4|^2 \tag{14a}$$

$$D_4 = R_{44}^2 \times |s_4|^2 + |b_4|^2 - 2 \times \mathrm{Re}\,\{s_4^* \times b_4\} \tag{14b}$$

The term $|b_4|^2$ is added to all the 16 partial distances. So, removing this term won't affect the calculation since we are looking for the symbol with the minimum Euclidian distance. The term $R_{44}^2|s_4|^2$ is equivalent to $2|R_{44}|^2$ or $10|R_{44}|^2$ or $18|R_{44}|^2$ since we adopt 16-QAM modulation scheme. This calculation can be done using shift and add as shown in the architecture of Fig. 7. Registers are inserted to divide the operation into two clock cycles.

5.3.2 Levels 3 to 1

Each node in the tree has 16 branches. We need only to find the symbol that minimizes the distance d_i.

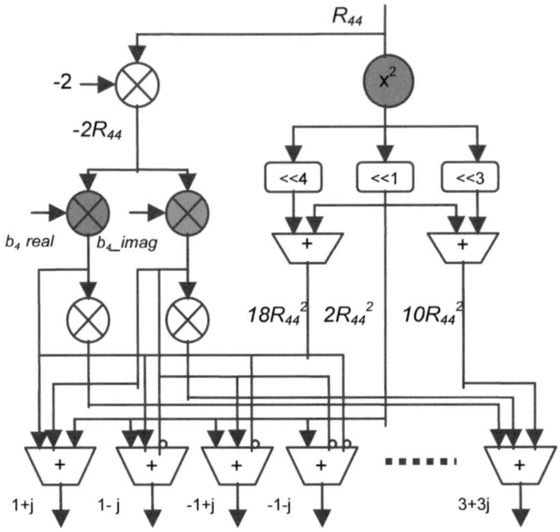

Fig. 7. Architecture of Level 4 FSD

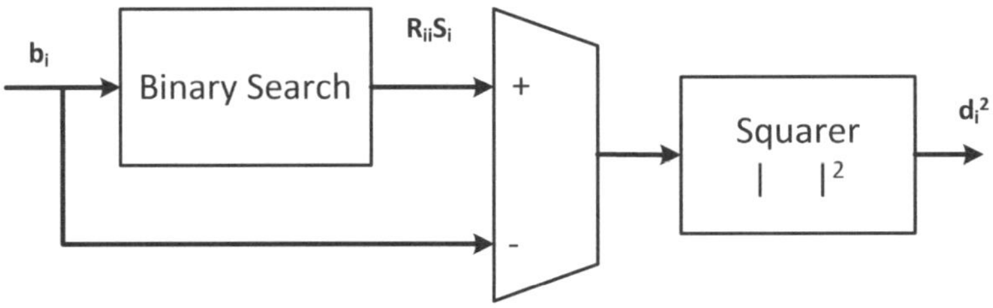

Fig. 8. Hardware Architecture for processing each node of Levels 3 to 1

$$d_i = |b_i - R_{ii}s_i|^2 \tag{15}$$

This is done by searching for the symbol s_i so that b_i is closest to $R_{ii}s_i$. This can be accomplished using binary search by comparing the real and imaginary of b_i with the boundaries of the used constellation.

The calculation of the distance d_i is performed in parallel for 8 branches out of 16. Therefore, 2 iterations are required to perform all distance calculation and only two clock cycles are required to perform the search of each stage. The calculation of b_i, using (8), requires the knowledge of the detected symbol of all the previous levels, $R_{ij}s_j$. Since we consider all the symbols in level 4, calculation of $R_{34}s_4$ can be done in parallel with level 4 as shown in figure 9. The 16 values of $R_{34}s_4$ are calculated over two clock cycles since we use two iterations over the parallel units for calculating the Euclidian distances.

For the case of b_2 and b_1, the calculation of $R_{ij}s_j$ can't be applied in parallel with the previous levels as only one symbol is considered for the previous levels 3 and 2. However, if we consider equation (8) we can find that $R_{i4}s_4$ can be pre-calculated in the same manner shown in Fig. 9 which reduces the hardware area. Also, to reduce the hardware complexity, the different values of $R_{ij}s_j$ are pre-calculated, and then, using a multiplexer for each branch distance unit we choose the one corresponding to the symbol s_j. The architecture used for calculating $R_{ij}s_j$ is shown in fig. 10. We use shifters and adders instead of multipliers so the hardware complexity is significantly reduced.

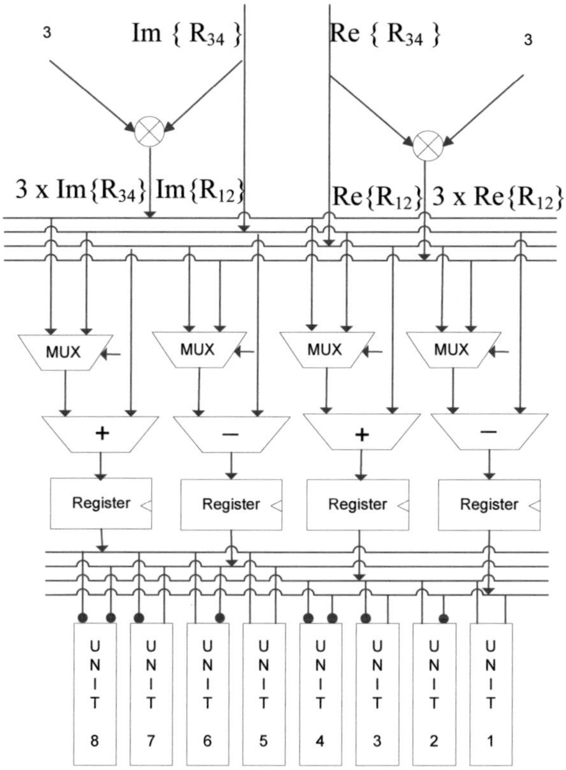

Fig. 9. Calculation of $R_{34}^{*} s_4$ for the 8 parallel units of Level 3

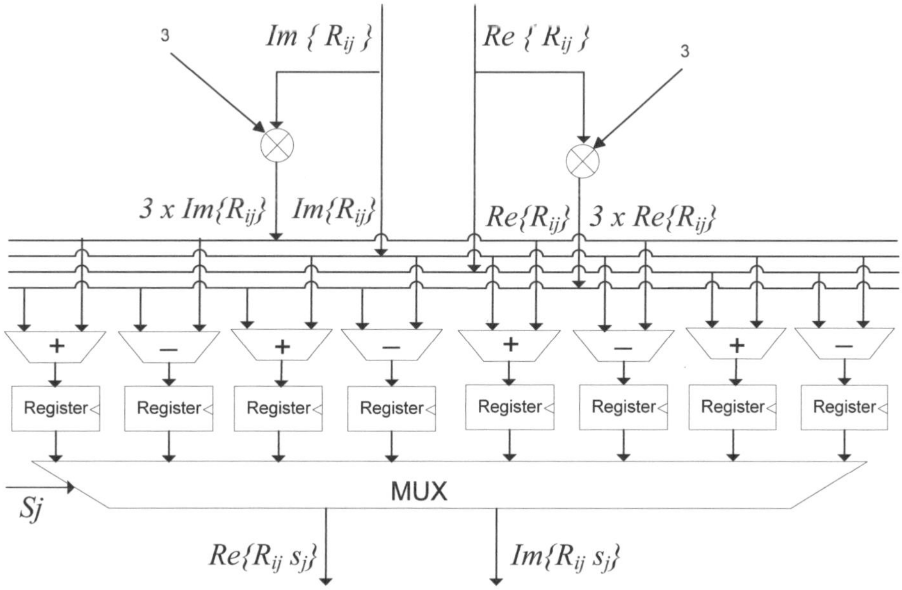

Fig. 10. Hardware Circuit for $R_{ij}S_j$ calculation

5.3.3 Decision unit

The decision unit adds the branch distances of level 4 to level 1 and then searches over the calculated 16 Euclidian distances to find the vector with the minimum distance. The transmitted vector with the minimum Euclidian distance is selected as the detected FSD symbol.

6. Symbol decoding

The WiMAX standard of the transmission scheme for two transmit and two receive antennas includes three matrices that define the transmission format with the row index indicating antenna number and column index indicating OFDMA symbol time (IEEE Standard 802.16e, 2005). One of the following three transmission matrices can be used.

6.1 Matrix A:

Matrix A achieves full diversity, and hence, improves the bit error rate given a rate one. This code is the Alamouti code.

$$A = \begin{bmatrix} s_i & -s_{i+1}^* \\ s_{i+1} & s_i^* \end{bmatrix} \tag{16}$$

The two received vector symbols at two successive time symbols can be written as:

$$\begin{bmatrix} r_0^{(0)} \\ r_1^{(0)} \end{bmatrix} = \begin{bmatrix} h_{00} & h_{01} \\ h_{10} & h_{11} \end{bmatrix} \times \begin{bmatrix} s_0 \\ s_1 \end{bmatrix}$$

$$\begin{bmatrix} r_0^{(1)} \\ r_1^{(1)} \end{bmatrix} = \begin{bmatrix} h_{00} & h_{01} \\ h_{10} & h_{11} \end{bmatrix} \times \begin{bmatrix} -s_1^* \\ s_0^* \end{bmatrix}$$

They can be combined into a matrix equation as follow:

$$\begin{bmatrix} r_0^{(0)} \\ r_1^{(0)} \\ r_0^{*(1)} \\ r_1^{*(1)} \end{bmatrix} = \begin{bmatrix} h_{00} & h_{01} \\ h_{10} & h_{11} \\ h_{01}^* & -h_{00}^* \\ h_{11}^* & -h_{10}^* \end{bmatrix} \times \begin{bmatrix} s_0 \\ s_1 \end{bmatrix} \tag{17}$$

resulting in a modified received vector $r'_{4\times1}$ and a channel matrix $H'_{4\times2}$. To use the sphere decoder, the channel matrix H must be first decomposed using the QR.

$$H'_{4\times2} = Q_{4\times4}\,R_{4\times2} = [Q'_{4\times2} \quad Q''_{4\times2}] \begin{bmatrix} R'_{2\times2} \\ 0_{2\times2} \end{bmatrix} \tag{18}$$

The sphere decoder is searching for the vector s that minimizes the following equations:

$$\hat{s}_{FSD} = \arg\left\{\min_{s\in S}\left\|r' - H's\right\|^2 \le C^2\right\} \tag{19}$$

$$\left\|r' - H's\right\|^2 = \left\|Q^H(r' - Hs)\right\|^2 = \left\|\begin{bmatrix} Q'^H_{2\times4} \\ Q''^H_{2\times4} \end{bmatrix} r' - \begin{bmatrix} R' \\ O \end{bmatrix} s\right\|^2 \tag{19a}$$

So,

$$\left\|r' - H's\right\|^2 = \|\hat{y} - R's\|^2 + \left\|Q''^{H}_{2\times4} \times r'\right\|^2 \quad (19b)$$

Since the second term doesn't depend on the symbol s, so equation 19 can be written as

$$\hat{s}_{FSD} = \arg\{\min_{s\in S}\|\hat{y} - R's\|^2 \leq C^2\} \quad (20)$$

where $\hat{y}_{2\times1} = Q'^{H}_{2\times4} \times r'_{4\times1}$

The Hardware architecture required to decode the symbols coded by Matrix A is shown Figure 11.

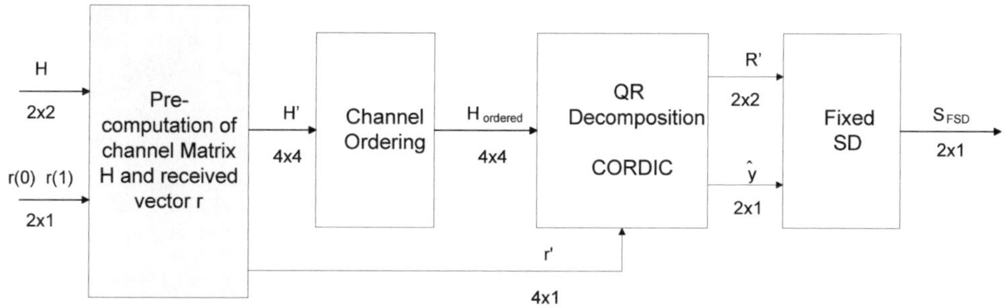

Fig. 11. Hardware Architecture for Matrix A decoding

The composed channel matrix H' is first ordered and then applied to the QR block. The CORDIC systolic array is fed by a non square matrix H' of dimension 4x2. By using the given rotation technique, the upper triangular matrix R and the vector y are obtained. The R Matrix has a special feature of having the off diagonal element R_{12} set to zero at the last step of the decomposition. Thus, R is a diagonal matrix. So the decoding process is simplified since each symbol can be decoded independently. For each stage we are looking for the symbol s_i that minimizes equation 7. This is accomplished by searching inside the scaled constellation, shown in figure 12, for the symbol s so that $R_{ii}s_i$ is closest to b_i.

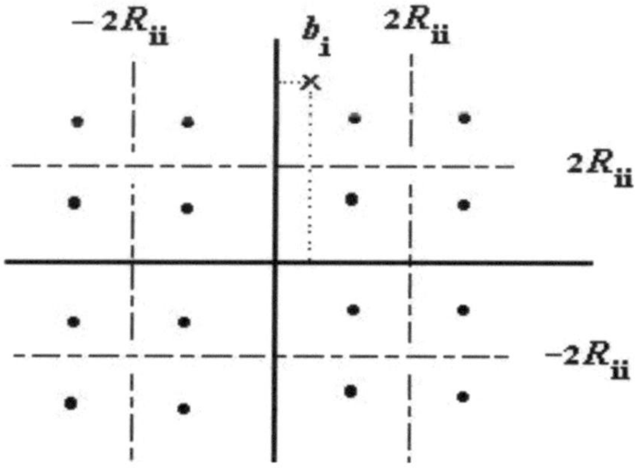

Fig. 12. 16-QAM Constellation

We compare the real and the imaginary value of b_i with the threshold levels 0, $2R_{ii}$ and $-2R_{ii}$ that define the boundaries of each symbol. This comparison is done using comparators and multiplexers as shown in figure 13.

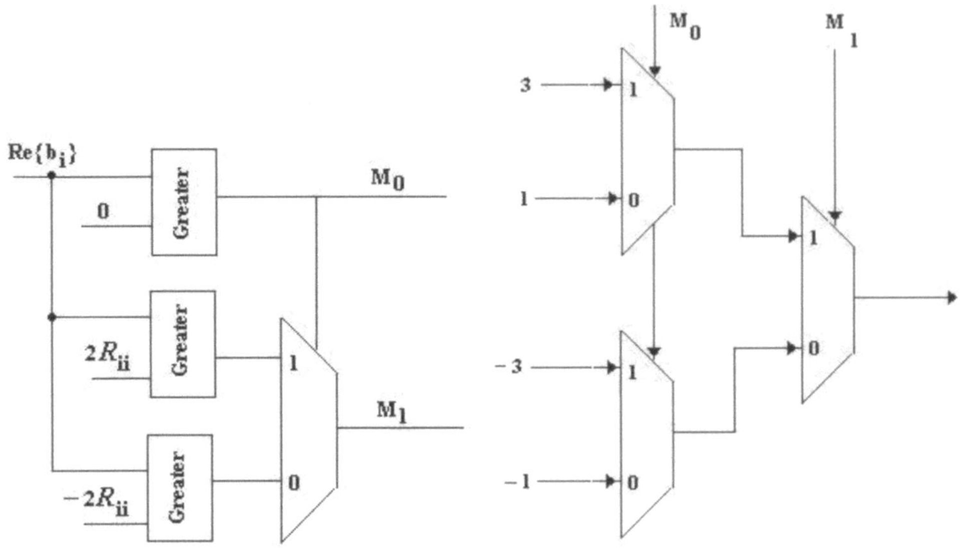

Fig. 13. Binary search Architecture

Therefore, there is no use for the tree search of the FSD. Only two parallel binary search blocks are required to find the transmitted symbols.

6.2 Matrix B:

Matrix B achieves spatial multiplexing with rate two employed for good channel conditions. It is similar to 2x2 un-coded MIMO system. Hence, it can be decoded directly using the traditional 2-level FSD. It doesn't require any additional adjustment over the usual FSD for the 2x2 MIMO systems.

$$B = \begin{bmatrix} s_i \\ s_{i+1} \end{bmatrix} \tag{21}$$

6.3 Matrix C:

Matrix C achieves full diversity and full rate. This code is called the golden code.

$$C = \frac{1}{\sqrt{1+p^2}} \times \begin{bmatrix} s_i + jp \times s_{i+3} & p \times s_{i+1} + s_{i+2} \\ s_{i+1} - p \times s_{i+2} & jp \times s_i + s_{i+3} \end{bmatrix}, p = \frac{-1+\sqrt{5}}{2} \tag{22}$$

The received vectors are:

$$\begin{bmatrix} r_0^{(0)} \\ r_1^{(0)} \end{bmatrix} = \begin{bmatrix} h_{00} & h_{01} \\ h_{10} & h_{11} \end{bmatrix} \times \begin{bmatrix} s_1 + jp \times s_4 \\ s_2 - p \times s_3 \end{bmatrix}$$

$$\begin{bmatrix} r_0^{(1)} \\ r_1^{(1)} \end{bmatrix} = \begin{bmatrix} h_{00} & h_{01} \\ h_{10} & h_{11} \end{bmatrix} \times \begin{bmatrix} p \times s_2 + s_3 \\ jp \times s_1 + s_4 \end{bmatrix}$$

They can be combined as:

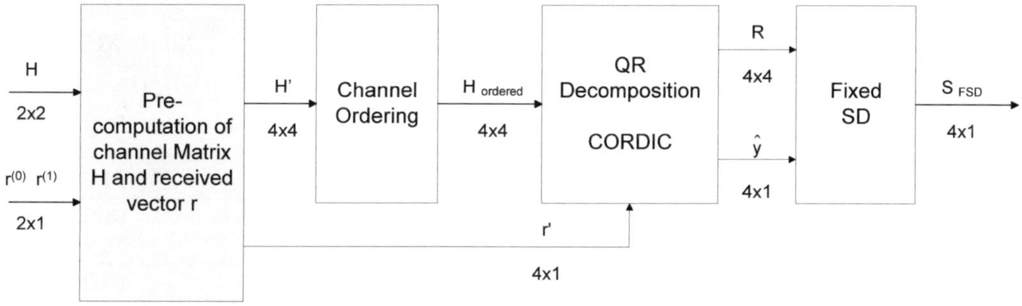

Fig. 14. Hardware Architecture for Matrix B decoding

$$\begin{bmatrix} r_0^{(0)} \\ r_1^{(0)} \\ r_0^{(1)} \\ r_1^{(1)} \end{bmatrix} = \begin{bmatrix} h_{00} & h_{01} & -ph_{10} & jph_{00} \\ h_{10} & h_{11} & -ph_{11} & jph_{10} \\ jph_{01} & ph_{00} & h_{00} & h_{01} \\ jph_{11} & ph_{10} & h_{10} & h_{11} \end{bmatrix} \times \begin{bmatrix} s_1 \\ s_2 \\ s_3 \\ s_4 \end{bmatrix} \tag{23}$$

$$r_{4\times1}{}' = H_{4\times4}{}' \times s_{4\times1} \tag{24}$$

Although the system is 2x2 but we should use 4 Levels FSD to be able to decode the four transmitted symbols. Pre-computations of the channel matrix H and the received vector r have to be done to adjust them into a form suitable for the FSD as explained by equations (23-24). A block diagram describing the decoding process is shown in figure 14.

The formed matrix H' and the vector r' are applied to the QR systolic structure to decompose them into the upper triangular matrix R and the vector y. The QR decomposition and FSD tree search is equivalent to the algorithms described in section 4 and 3 respectively.

7. Area report

The CORDIC algorithm is implemented with 9 iterations and 12 bit internal word length while a 16 bit word length is used for internal data width of FSD. ALTERA STRATIX II FPGA is used. The area report is given in Table I. Since, a new MIMO symbol is obtained every two clock cycles, the throughput for 4x4 system using 16-QAM modulation is calculated according to

$$\text{Throughput} = 4x4x\,(fclk/2) \tag{25}$$

Operation over a frequency of 100 MHz results in a throughput of 800 Mbps. Compared to (Barbero & Thompson, 2006 b), we have achieved the double the throughput of the presented FSD-1 and 33% increase in the throughput compared to the optimized FSD2.

Compared to the implementation of Ref. (Barbero & Thompson, 2006 b) given in Table I (using Vertix II Pro XC2VP70 FPGA), the number of multipliers are drastically reduced where only 2 multipliers (4 DSP Blocks) are required for the FSD and no multipliers are needed for the QR. Additionally, the proposed FSD requires less number of registers and Look-Up Tables LUT's.

	QR-CORDIC	4x4 FSD	FSD1/2[5]
Target FPGA	EP2S60F672C3	EP2S60F672C3	Xilinx XC2VP70
Combinational ALUTs/LUT	(5700) 11.78 %	(13,743) 28.2%	(16,119) 24%
Dedicated Logic Registers	(1520) 3.14%	(1412) 2.94%	15,332/16,855 23%/25.47%
DSP Block/Multiplier	-	4	(82) 25%
Memory bits/Block RAM	6080 < 1%	-	(82) 25%
Maximum clock freq.	160 MHz	102 MHz	100 MHz
Throughput	-	800 Mbps	400/600 Mbps

Table 1. Area Report and maximum clock frequency of 4x4 MIMO FSD

8. Simulation results

We developed a MATLAB Model for the FUSC WiMAX MIMO system with 2 transmitting antennas and 2 receiving antennas. A floating point simulation has been conducted for the QPSK modulation scheme. Fig. 15 shows bit error rate (BER) versus E_b/N_o for QPSK for the three different Matrices A, B and C. As seen, STC matrix A achieves the best BER but it only achieves data rate 1. STC matrix B has a data rate of 2 but its performance is about 6dB lower that Matrix A at 10^{-4} BER. Finally the STC matrix C has a full data rate of 2 and full diversity on the expense of higher bit error rate compared to the STC matrix A.

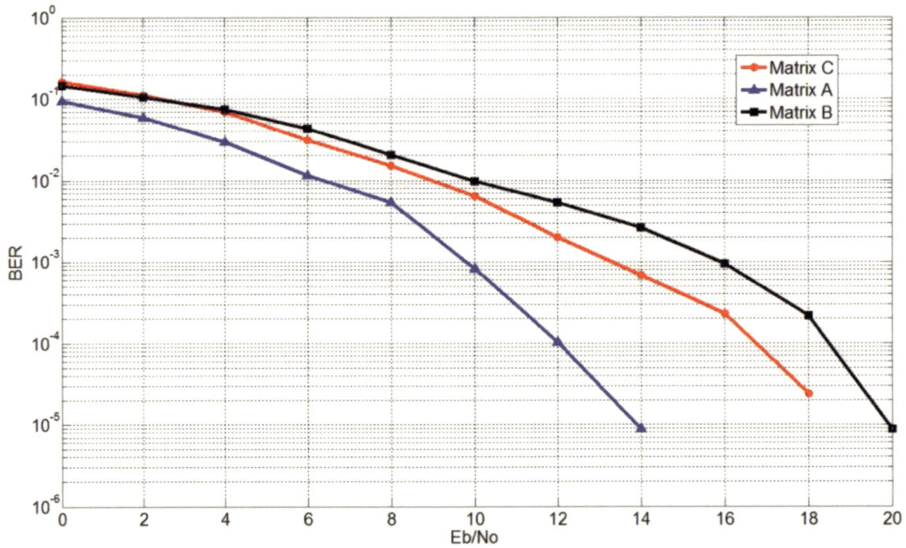

Fig. 15. Simulation Results Matrix A, B and C

Fig. 16 shows that the fixed-point BER performance of the fixed sphere decoder when employed to decode the symbols coded with Matrix A, B and C. The quantization effect has been simulated for 16, 14, 12 and 10 data bits.

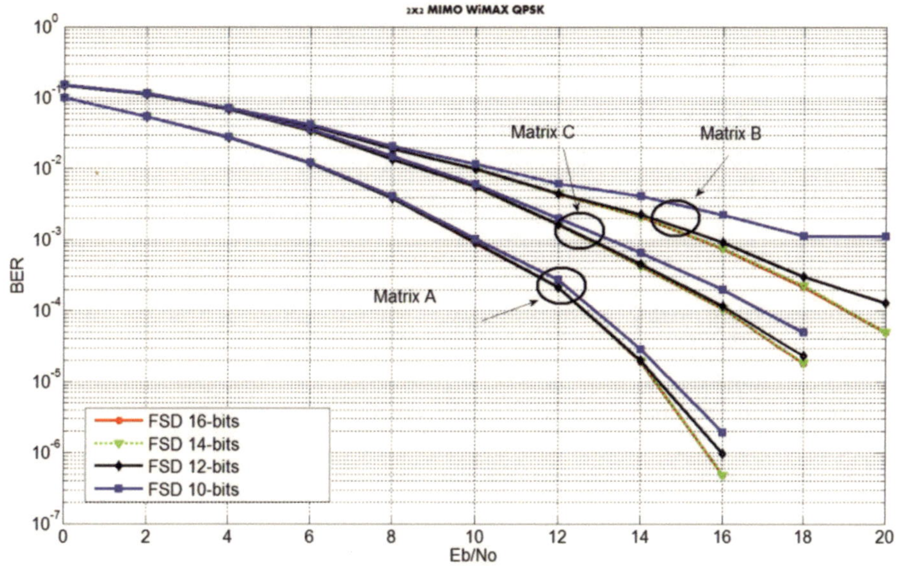

Fig. 16. Fixed Point Simulation of FSD for Matrix A, B and C

Finally the quantization effect of QR is taking into effect, where 12-bits per internal data word with 9 iterations have been used while a 16 bit data word is used for the FSD. Fig. 17 shows the degradation effect due to the quantization of the QR. This effect is apparent at high SNR where the noise effect is negligible compared to the quantization effect. Yet, this BER degradation at high SNR is less pronounced for the QR Decomposition than the same effect if Cholesky - based inversion of the H matrix is applied (Barbero & Thompson, 2006 a).

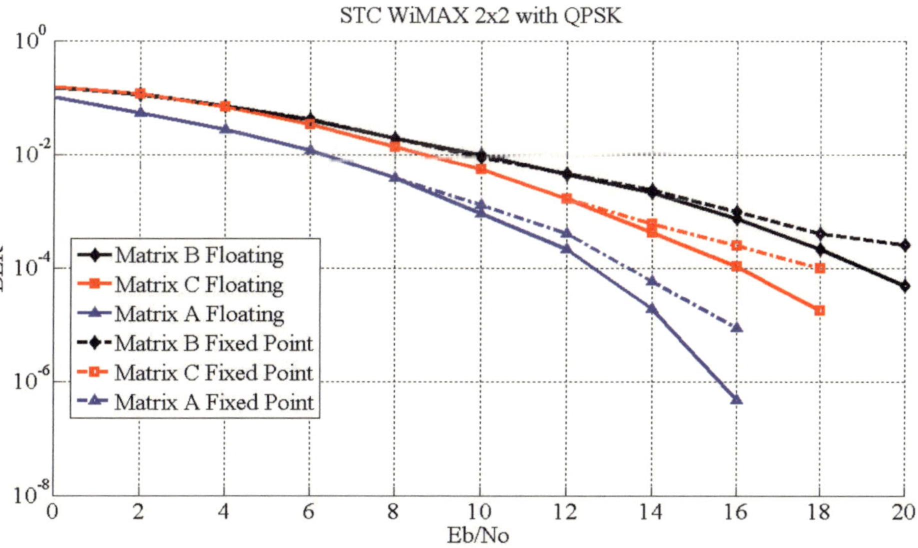

Fig. 17. QR quantization effect on Fixed Point Simulation

9. Conclusion

We have presented a FPGA architecture of the FSD as a MIMO decoder for the WiMAX system which achieves close-to ML BER performance. We have shown how the FSD is adaptively used to decode the three different STC. The presented FSD achieves fixed and higher throughput compared to (Barbero & Thompson, (2006) b) and uses less number of multipliers and LUTs.

10. References

"Air Interface for Fixed and Mobile Broadband Wireless Access Systems", IEEE Standard 802.16e, 2005

Barbero, L.G.; Thompson, J.S. (2005); "Rapid prototyping of the sphere decoder for MIMO systems," *in Proc. IEE/EURASIP Conference on DSP Enabled Radio (DSPeR '05)*, vol. 1, Southampton, UK, Sept. 2005, pp. 41–47

Barbero, L.G.; Thompson, J.S (2006) a, "Real Time implementation of a sphere decoder based MIMO Wireless system", *in EURASIP European Signal Processing Conference (EUSIPCO '06)*, Florence, Italy, Sep. 2006

Barbero, L.G.; Thompson, J.S (2006) b, "Rapid Prototyping of a Fixed-Throughput Sphere Decoder for MIMO Systems," *IEEE International Conference on Communication 2006*, vol. 7, pp. 3082-3087, June 2006.

Burg, A.; Borgmann, M.; Wenk, M.; Zellweger, M.; Fichtner, W.; Bolcskei, H. (2005). "VLSI implementation of MIMO detection using the sphere decoding algorithm," *IEEE Journal of Solid-State Circuits*, vol.40, no.7, pp. 1566- 1577, July 2005

Guo, Z. and Nilsson, P. (2003), "A VLSI implementation of MIMO detection for future wireless communications," in Proc. IEEE PIMRC'03, vol. 3, pp. 2852–2856, 2003.

Hess,C; Wenk, M. ; Burg, A.; Luethi, P. ; Studer, C.; Felber, N. and Fighter, N. (2007) " Reduced Complexity MIMO Detector with Close to ML Error Rate Performance ", *Proc. of ACM Great Lakes Symposium on VLSI (GLSVLSI)*, 200-203, Stresa-Lago Maggiore, Mar 2007

Khairy, M.S.; Abdallah, M.M.; Habib, S.E.-D. (2009), "Efficient FPGA implementation of MIMO decoder for mobile WiMAX system," in Proc. IEEE International Conference on Communications 2009 (ICC 2009), Dresden, Germany, June 2009.

Khairy, M.S.; Mehlfuhrer, C.; Rupp, M. (2010), "Boosting sphere decoding speed through Graphic Processing Units," 16th European Wireless Conference (EW), Lucca, Italy, 12-15 April 2010.

Myllyla, M.; Hintikka, J.-H.; Cavallaro, J.R.; Juntti, M.; Limingoja, M.; Byman, A (2005)., "Complexity Analysis of MMSE Detector Architecture for MIMO OFDM Systems," in Proceedings of the 2005 Asilomar conference, pp 75-81, Pacific Grove, CA, 2005.

Proakis, J; Salehi, M (1994); "Communication Systems Engineering," Prentice Hall PTR

Volder, J.E. (1959) "The CORDIC Trigonometric Computing Technique," *IRE Trans. on Electronic Computers*, vol. EC-8, no. 3, pp. 330–4, 1959.

21

MIMO System Implementation for WSN Using Xilinx Tools

Wael M El-Medany
University Of Bahrain
Bahrain

1. Introduction

Wireless sensor networks are one of the most rapidly evolving research and development fields for microelectronics. Their applications are countless, and the market potentials are huge. However, many technical hurdles have to be overcome to achieve a widespread diffusion of wireless sensor network technology [1].
This work presents the design and FPGA hardware implementation of a *Multiple-Input/Multiple-Output (MIMO)* system for Wireless Sensors Networks (WSN). This system will offer *more parallel channels between the sensor nodes and the base station at the same frequency band, thereby increasing spectral efficiency.* The hardware design of the MIMO wireless sensor network system has been described using VHDL (VHSIC Hardware Description Language). The design has been simulated and synthesized using Xilinx ISE 10.1i software tools, then tested in hardware level using Xilinx FPGA. The design offers remote monitoring system with MIMO wireless sensor network.
Sensor networks differ from traditional networks in several ways, sensor networks have severe energy constraints, redundant low data rate, and many-to-one flows. The end-to-end routing schemes that have been proposed in the literature for mobile ad-hoc networks are not appropriate under these settings [2]. A wireless sensor network has important applications such as remote environmental monitoring and target tracking. This has been enabled by the availability, particularly in recent years, of sensors that are smaller, cheaper, and intelligent. These sensors are equipped with wireless interfaces with which they can communicate with one another to form a network.
The design of a WSN depends significantly on the application, and it must consider factors such as the environment, the application's design objectives, cost, hardware, and system constraints [3]. WSNs have great potential for many applications in scenarios such as military target tracking and surveillance [4, 5], natural disaster relief [6], biomedical health monitoring [7, 8], and hazardous environment exploration and seismic sensing [9].
Mobile Ad-Hoc and Wireless Networks Evolving Mobile Network concept known as Multiple-Input/Multiple -Output, which has the potential to increase communications data rates by 10-20 times above current systems data rates. Such systems can use reprogrammable logic circuits, specifically Field Programmable Gate Arrays (FPGA), which offers a cost effective and high performance hardware alternative to Application Specific Integrated Circuits (ASIC) in low and mid volume products. Furthermore, FPGAs are

becoming important building blocks in embedded systems design [10]. Nowadays many system designs that used to be built in custom silicon VLSI [11] are now implemented in Field Programmable Gate Arrays. This is because of the high cost of building a mask production of a custom VLSI especially for small quantity [12]. In addition to the cost effective, FPGA gives large number of inputs and outputs. The problem with FPGA was that it is volatile; this means that once the power is switched off, the design will be erased, but this problem has been recently solved by the use of the first non-volatile FPGA that has been introduced by Xilinx in 2007, that is Xilinx Spartan 3 AN. The Spartan 3 AN FPGA has a flash memory built on the chip, which keep the design even when the power is off.

With the rapid development of computer technology, the monitored control design of the wireless sensor networks is becoming the core of the design for building automation system [13, 14]. Real-time monitoring provides reliable, timely information of petroleum product's status, important in taking decisions for petroleum production improvement. Evaluation of petroleum production systems is a time consuming and difficult process because it means performing visits to selected petroleum fields to be able to measure and register certain physical, chemical and biological characteristics of the petroleum production areas [15]. For remote monitoring, GSM network has been used for remote communications [16-19].

This research introduces a MIMO wireless sensor networks for a petroleum fields that has a large number of parallel channel offered by Xilinx FPGA for the communications with the wireless sensor nodes, in order to read the data simultaneously from the sensor nodes. The target technology is XC3S1400 AN-5fgg676, which has 676 pines on the package; the design has been synthesized and implemented using Xilinx ISE 10.1i.

2. MIMIO-WSN system architecture

A multiple-input / multiple-output wireless sensor networks system architecture comprising N wireless sensor node and transceiver systems, where N is (I*J) sensor node matrix. There are I-groups of sensor node, each group comprising J-sensor nodes, and each group is communicating with the base station through a communication channel, with a total of I-communication channel. There are I-sensor nodes that can communicate in parallel with the base station through the I-communication channel. At the same time there are I-transceiver antennas at the base station. Figure 1 shows a block diagram for the system architecture of the multiple-input / multiple-output wireless sensor networks. The main controller for the base station has targeted the Xilinx XC3S1400 AN-5fgg676 FPGA, which can be programmed with a design that contains up to 1.4 million of logic gates, with a package of 676 input/output pines.

Around 600 pines from the 676 pines of the FPGA package are available for user applications. Those large numbers of input/output pines are used to increase the number of parallel communication channels. For each channel there are two pines that are required for the communications, one for the data transmit and the other for the data receives. This means that we have up to 300 communications channel, and thereby the number of base station transceiver is also 300, which means that the (I) rows of the sensor node array can be increased up to 300. The VHDL top-level model of the main controller in the base station is shown in figure 2, where Rx1, Rx2,…., and Rx300 are the receiver end of the base station, and Tx1, Tx2,… , and Tx300 are the transmitting end. In accordance there are 300 UART

(Universal Asynchronous Receiver Transmitter) in the main controller of the base station. The UART has been designed using VHDL as a component, and implemented on the FPGA.

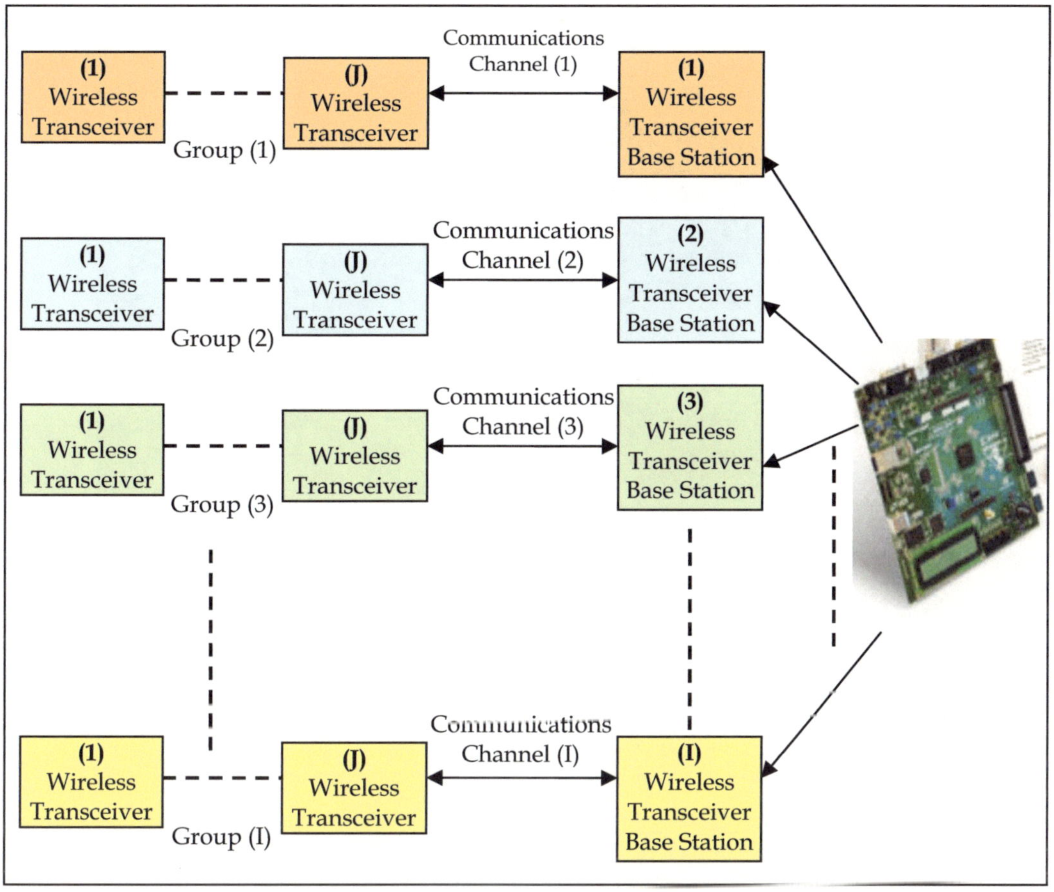

Fig. 1. IMIO-WSN System Architecture

A sensor node is a multi-functional unit performing many different tasks, from managing acquisition to handling communication protocol schedule and preparing data packets for transmission, after filtering, synchronizing and signal processing on data gathered from sensors. Thus, each sensor node requires processing and storage capabilities. Figure 3 shows a block diagram for the main components of the sensor node, there are six units in each of the sensor nodes. The most important unit of the sensor node is the processing unit, which has the main controller of the sensor node. The processing unit handle all the arithmetic and logical operations, receiving data from the ADC, storing data, sending data to the network protocol through the communications unit. The communication unit is a UART, and the sensing unit is changing according to the application, they have a variety of sensing devices, including security and pressure sensing devices and cameras.

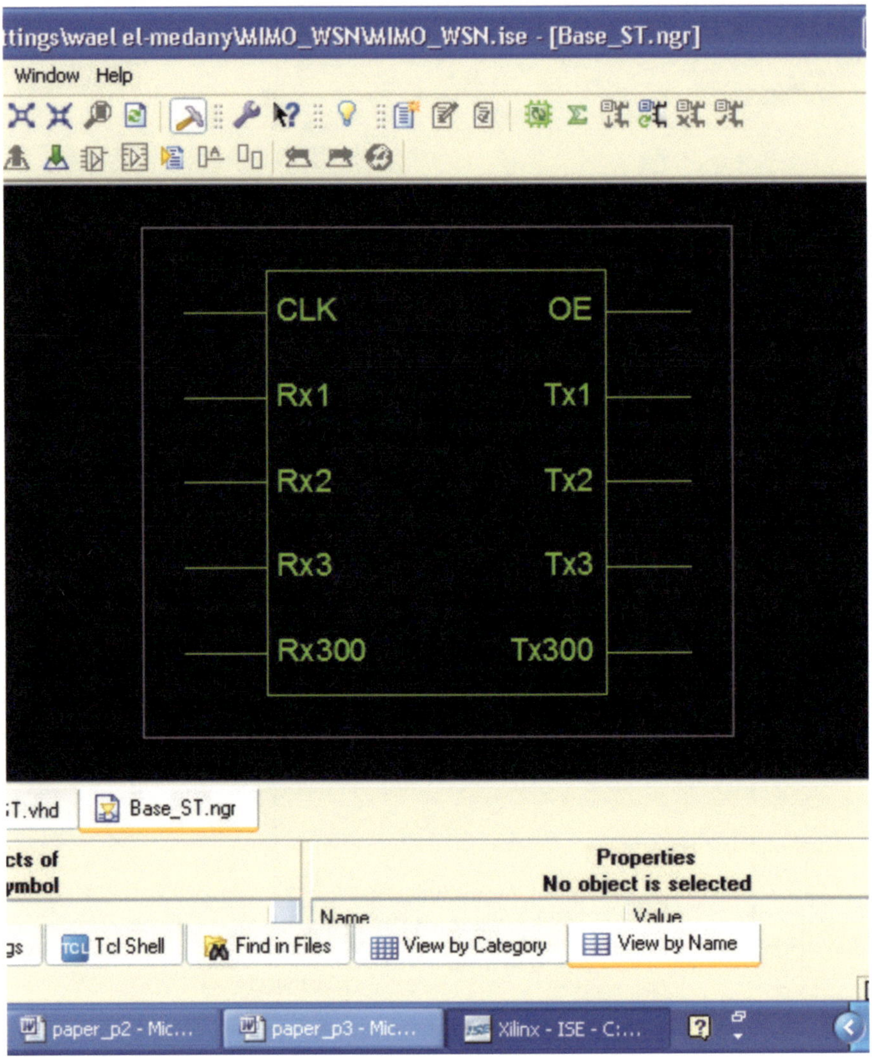

Fig. 2. VHDL Top-Level Design Model

3. Simulations results and efficiency

The VHDL models have been simulated functionally to verify the correctness of the behavioral description for the models. Figure 4 shows the simulation results for the processing unit in the sensor node. Where "clk" is the system clock input, "TS" is an input bit-vector of 8-bit which represents the inputs from the sensing unit. Where the analog value from the sensing unit has to be converted to digital value in binary representation using the ADC; this binary value is represented by "TS" input. This part of the sensor node has been tested using Spartan 3 starter kit from Xilinx, which has four seven segment digits built on the board, the four seven segments are common anodes, and the cathodes for the four seven segments are connected together in parallel. To display an output, you have to do a multiplexing between the four seven segments. The signal "SSCATH" in figure 4 is the

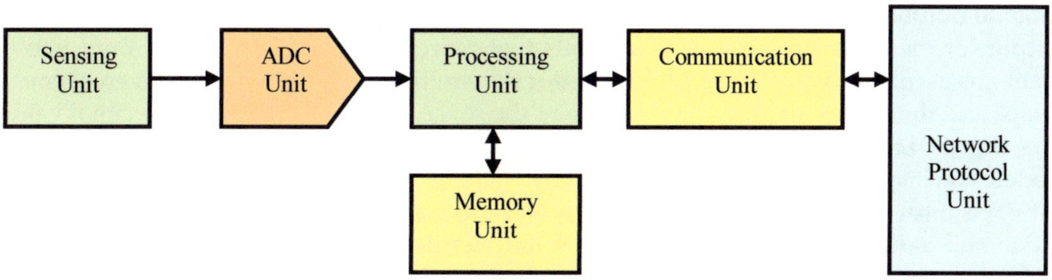

Fig. 3. Block diagram for the main components of the sensor node

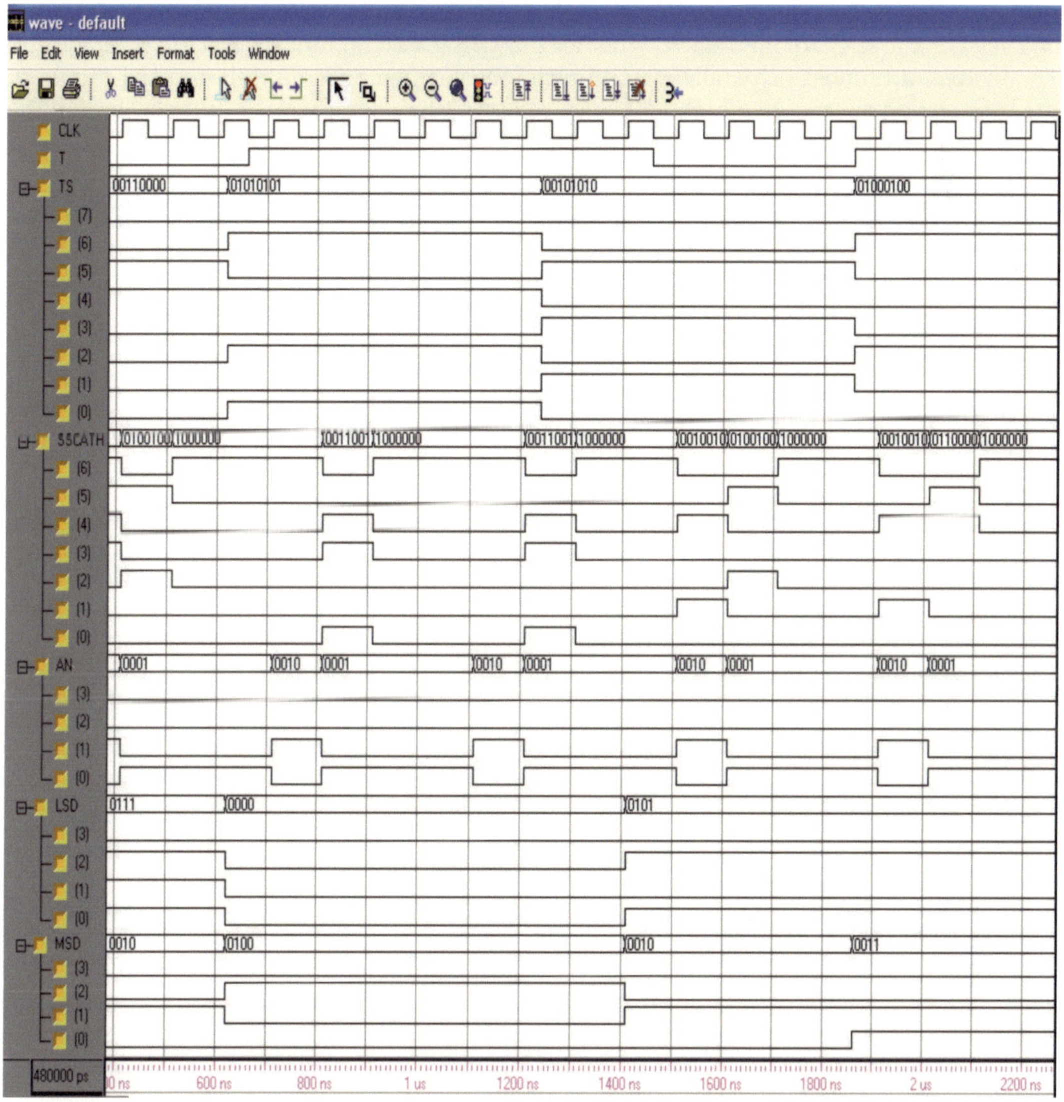

Fig. 4. Simulation results for the Processing Unit in the Sensor Node

cathode output for the seven segment display on the system board, and "AN" is the anode output for the seven segment, it is four bits that representing four digits, but only two of them are used. The values on "AN" represent the multiplexing between the seven segment digits, and the values on "SSCATH" represent the seven segment code for the decimal value. The signal "LSD" and "MSD" are internal signal that representing the two BCD (Binary Coded Decimal) digits of the senor reading. Where "LSD" is least significant digit and "MSD" is most significant digit. Signal "T" is an output signal that indicates the high sensing input and normal input. Figure 5 shows the simulation for the multiplexing between different sensor nodes in the base station. Where "TSC" and "HSC" are two 8-bit inputs from the analog to digital converter of the two Sensing Circuits, "Sel" is the selector input, and "Q" is an 8-bit output of the multiplexer. If (Sel='0') then (Q = HSC), and if (Sel='1') then (Q = TSC). The selector "Sel" is changing sequentially using binary counter. The MSB of TSC, HSC, and Q are on the left, whether the LSB (Least Significant Bit) are on the right. The increased number of parallel communications channel and multiplexing the inputs of the base station system give high efficiency to our system compared to other system in the literature, as well as reducing cost by using the FPGA devices.

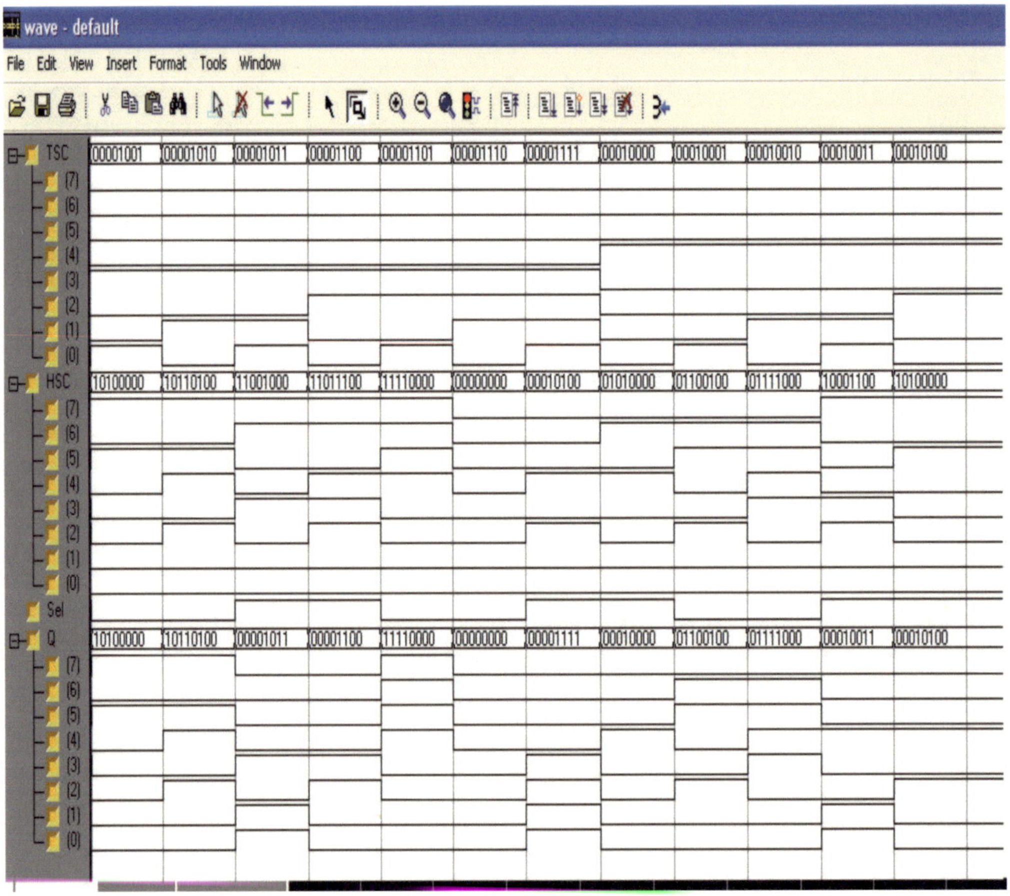

Fig. 5. Simulation results for the multiplexing between different sensor nodes in the Base Station

4. Conclusion

A hardware design and FPGA implementation of a Multiple-Input/Multiple-Output system for Wireless Sensors Networks has been introduced in this paper. The system has increased the number of parallel channels and hence the number of sensor nodes. The system uses the large numbers of Inputs/Outputs pines offered by the FPGA chip to increase the number of parallel channels between the base station and the sensor nodes. The system was designed using VHDL in a high level design method. All parts of the design have been tested in both simulation and hardware level. The implemented design targeted the non-volatile Xilinx XC3S1400 AN-5fgg676 FPGA for final prototype.

5. References

L. Benini, E. Farella, C. Guiducci, "Wireless sensor networks: Enabling technology for ambient intelligence", Microelectronics Journal 37, pp. 1639–1649, September 2006.

B. Krishnamachari, D. Estrin, S. Wicker, "Modeling Data-Centric Routing in Wireless Sensor Networks", IEEE INFOCOM 2002.

J. Yick, B. Mukherjee, D. Ghosal, "Wireless sensor network survey", Computer Networks 52, pp. 2292–2330, 2008.

G. Simon, M. Maroti, A. Ledeczi, G. Balogh, B. Kusy, A. Nadas, G. Pap, J. Sallai, K. Frampton, Sensor network-based countersniper system, in: Proceedings of the Second International Conference on Embedded Networked Sensor Systems (Sensys), Baltimore, MD, 2004.

J. Yick, B. Mukherjee, D. Ghosal, Analysis of a Prediction-based Mobility Adaptive Tracking Algorithm, in: Proceedings of the IEEE Second International Conference on Broadband Networks (BROADNETS), Boston, 2005.

M. Castillo-Effen, D.H. Quintela, R. Jordan, W. Westhoff, W. Moreno, Wireless sensor networks for flash flood alerting, in: Proceedings of the Fifth IEEE International Caracas Conference on Devices, Circuits, and Systems, Dominican Republic, 2004.

T. Gao, D. Greenspan, M. Welsh, R.R. Juang, A. Alm, Vital signs monitoring and patient tracking over a wireless network, in: Proceedings of the 27th IEEE EMBS Annual International Conference, 2005.

K. Lorincz, D. Malan, T.R.F. Fulford-Jones, A. Nawoj, A. Clavel, V. Shnayder, G. Mainland, M. Welsh, S. Moulton, Sensor networks for emergency response: challenges and opportunities, Pervasive Computing for First Response (Special Issue), IEEE Pervasive Computing, October–December 2004.

G. Wener-Allen, K. Lorincz, M. Ruiz, O. Marcillo, J. Johnson, J. Lees, M. Walsh, Deploying a wireless sensor network on an active volcano, Data-Driven Applications in Sensor Networks (Special Issue), IEEE Internet Computing, March/April 2006.

T. Wollinger, J. Guajardo, C. Paar, "Security on FPGAs: state-of-the-art implementations and attacks", ACM Trans Embedded Comp Syst, (3):534–74, 2004.

Jan M. Rabaey, Digital Integrated Circuits, A Design Perspective, Second Ed., Prentice Hall, 2003.

Design of a VLSI Integrated Circuit, IEEE, Piscataway, USA.

Chen Xi, Chang Jinzhao, Liu Junfeng, "The monitored control design of the central air-conditioning system", In Control & Measurment 2007, vol.6, no.1, pp. 245-7.

Jung-Ho Huh, M.J. Brandemuehl, "Optimization of air-conditioning system operating strategies for hot and humid climates" In Energy & Buildings 2008, vol.40, no.7, pp. 1202-13.

Wu Ming-fang; Tang De-dong, "Smart instrument for measuring petroleum product's humidity based on electromagnetism oscillation technique", In Instrument Techniques and Sensor 2008, no.4, pp. 16-18.

W M El-Medany, "FPGA Implementation for Humidity and Temperature Remote Sensing System", 14th Annual IEEE International Mixed- Signals, Sensors, and Systems Test Workshop, IMS3TW 08, Canada.

J. M. Jasso, G. O. Vargas, R. C. Miranda, E. V. Ramos, A. Z. Garrido, G. H. Ruiz, "FPGA-based real-time remote Monitoring system", Journal of Computers and Electronics in Agriculture, V 49, 2005, p 272–285.

GSM Based Remote Sensing and Control Systems Using FPGA, Wael M El-Medany, Mahmoud R El-Sabry, The IEEE International Conference on Computer & Communication Engineering, ICCCE'08, Kuala Lumpur, Malaysia, May 2008.

G. Aranguren, L. Nozal, A. Blazquez, and J. Arias, "Remote control of sensors and actuators by GSM", IEEE 28th Annual Conference of the Industrial Electronics Society IECON 02, vol. 3, 2002 p 2306 - 2310.

22

Experimental Evaluation of MIMO Coded Modulation Systems: are Space-Time Block Codes Really Necessary?

Francisco J. Vázquez Araújo[1], José A. García-Naya[1],Miguel González-López[1], Luis Castedo[1] and Javier Garcia-Frias[2]

[1]*University of A Coruña*
[2]*University of Delaware*
[1]*Spain*
[2]*USA*

1. Introduction

The use of multiple transmit and/or receive antennas, referred to as Multiple-Input Multiple-Output (MIMO) systems, is one of the most promising transmission techniques for achieving the high data rates demanded by the future wireless communication systems. This assertion relies on the theoretical and experimental evidence that the capacity of a MIMO system is considerably higher than that of a conventional single antenna system (Telatar, 1995).

Extracting the maximum capacity and diversity from the MIMO channel requires specific coding techniques that spread channel symbols over both spatial and temporal dimensions of the MIMO channel. The Alamouti code (Alamouti, 1998) is one of the most widely used Space-Time Block Codes (STBC) because of its low encoding and decoding complexity, and its ability to provide the maximum transmit diversity. For these reasons, it has been adopted by the IEEE 802.16-2009 standard (WiMAX) for wireless local and metropolitan area networks (*IEEE Standard for Local and Metropolitan Area Networks. Part 16: Air Interface for Fixed and Mobile Broadband Wireless Access Systems*, 2009), as well as in the recently approved IEEE 802.11n (WiFi) next-generation wireless standard for Local Area Networks (*IEEE Standard for Wireless LAN Medium Access Control (MAC) and PHYsical Layer (PHY) Specifications: Amendment: Medium Access Control (MAC) Enhancements for Higher Throughput*, 2009).

The utilization of the Alamouti code is limited to the case of two transmit antennas (i.e., $n_T = 2$) but it does not impose any constraint into the number of receive antennas (i.e., n_R). However, information-theoretic analysis show that the signal structure imposed by the Alamouti code reduces the capacity of the MIMO channel when there is more than one receiving antenna (Sandhu & Paulraj, 2000). In the particular case of 2×2 MIMO systems, this limitation is overcome with the utilization of the Golden code (Belfiore et al., 2005). The Golden code is another example of STBC and constitutes an appealing alternative to the Alamouti code since it does not suffer from capacity loss and exhibits a reasonable complexity cost.

In spite of their attractive properties, STBCs need an outer channel code to approach the theoretical capacity limit of a MIMO channel since STBCs provide little (or no) coding gain. Remarkable coding gains can be obtained if a capacity-approaching binary encoder, such as Turbo (Berrou et al., 1993) or Low-Density Parity Check (LDPC) (Gallager, 1963; MacKay, 1999), is employed. In this chapter, we focus on a particular subclass of LDPC codes known as Serially-Concatenated Low-Density Generator Matrix (SCLDGM) codes (Garcia-Frias & Zhong, 2003), whose performance is similar to that of general LDPC codes but with very low encoding complexity. Alternatively, Irregular Repeat-Accumulate (IRA) codes (Jin et al., 2000) can also be used (ten Brink & Kramer, 2003; Yue & Wang, 2005), but SCLDGM codes are preferable because their regular versions already approach the capacity limit.

Without using the aforementioned STBCs, the capacity of a MIMO channel can be approached for an arbitrary number of transmitting and receiving antennas by spatially multiplexing the output of a Bit-Interleaved Coded Modulation (BICM) scheme (Tonello, 2000; Zehavi, 1992), constructed with a properly designed capacity-approaching code, i.e. with a capacity-approaching code specifically designed to match the EXtrinsic Information Transfer (EXIT) (ten Brink, 2001) characteristic of the channel. The main difficulty when implementing BICM with spatial multiplexing is the complexity of the detection stage. In general, complexity of optimum detection (Log-Likelihood Ratio (LLR) computation) in BICM with spatial multiplexing is considerably higher than that in systems using STBCs, and it is only affordable for a moderate number of antennas and small constellation sizes.

When detection complexity grows excessively and the number of receiving antennas, n_R, is higher than or equal to the number of transmitting ones, n_T, there exist suboptimum methods for LLR computation with near-optimum performance, such as ML or Maximum *A Posteriori* (MAP) List Sphere Detection (LSD) (Hochwald & ten Brink, 2003; Vikalo et al., 2004) or Soft Interference Cancellation with MMSE filtering (SIC-MMSE) (Wang & Poor, 1999). However, when $n_T > n_R$, these high-performance suboptimum detectors cannot be utilized. The reason is that either the underlying system of equations is underdetermined or the decoding complexity grows exponentially with $n_T - n_R$. To overcome this limitation, high data rate linear STBCs have been proposed under the name of Linear Dispersion (LD) codes (Hassibi & Hochwald, 2002). In a sense, LD codes are an extension of V-BLAST for the case $n_T > n_R$. Since the use of LD codes modifies the EXIT characteristic of the resulting channel, it is necessary to specifically design codes matched to this new EXIT characteristic.

It is not clear in the literature which MIMO signaling scheme, i) concatenation of channel coding with an STBC or ii) BICM with spatial multiplexing, is better in terms of approaching capacity. For a 2×1 MIMO system, concatenation with the Alamouti code may appear preferable at a first glance since it employs simpler detectors and can approach capacity using conventional SISO optimized channel encoders. However, it is not clear whether this signaling technique is able to outperform BICM with spatial multiplexing when employing channel codes optimized for this specific MIMO configuration and each modulation format. For a 2×2 MIMO system, concatenating with a Golden code seems the best option to avoid the capacity loss introduced by the Alamouti code. Now, specific capacity approaching channel codes should be designed and, again, it is not known wether this concatenated scheme performs better than BICM with spatial multiplexing, or not. Finally, for a 3×1 MIMO system, either concatenation with a LD code or spatial multiplexing can be used. In each case, a specific channel code should be designed but the performance of these optimized coded modulation schemes is unknown

In this chapter we shed light into this controversy by comparing the performance of the above-mentioned MIMO scenarios when SCLDGM capacity approaching codes are employed. The data rate is two bits per channel use for the 2×1 and the 2×2 cases, and one bit per channel use for the 3×1 scenario. We specifically optimized regular SCLDGM codes for each system configuration using EXIT analysis techniques and assess its ability to approach the MIMO channel theoretical capacity limits.

A major contribution of this chapter is that performance evaluation is carried out not only over synthetically generated, spatially uncorrelated, Rayleigh distributed, flat-fading channels (ergodic and quasi-static) but also over realistic indoor scenarios. Although computer simulations are necessary and recommendable for wireless systems evaluation, they only reflect the simulated environment rather than the actual scenarios in which wireless systems operate. Channel models typically omit important issues such as quantization effects, power amplifier non-linearities, mutual antenna coupling, and phase noise. This is particularly important when dealing with MIMO channels since the scientific community has not reached a consensus on a reference channel model due to the extremely large number of parameters to be considered.

For the experimental evaluation of the MIMO coded modulation techniques we have used a MIMO hardware demonstrator developed at the University of A Coruña. The demonstrator has been constructed from Commercial Off-The-Shelf (COTS) modules, including the RF front-ends. We also developed a distributed multilayer software architecture necessary for the configuration and utilization of the hardware platform. Different experiments were carried out at the 5 GHz Industrial, Scientific and Medical (ISM) band considering different Tx/Rx locations and antenna positions. The results are presented in terms of Block Error Rate (BLER) versus E_b/N_0 at reception and are representative of the performance obtained over a typical indoor scenario.

The remainder of this chapter is organized as follows: Section 2. describes the different MIMO signaling techniques under consideration, namely, BICM with spatial multiplexing or in concatenation with STBCs (Alamouti, Golden or LD codes). Section 3. explains the utilization of SCLDGM codes in the schemes under consideration, and the optimization procedure. Section 4. presents the results of computer simulations assuming an identically and independently distributed (i.i.d.), spatially white, flat-fading Rayleigh MIMO channel (ergodic and quasi-static). These computer simulations corroborate the validity of the designed codes and show that spatial multiplexing and concatenation with a STBC yield the same performance. Section 5.describes the hardware and the experiments carried out to assess the performance of the MIMO signaling methods over a realistic indoor scenario. The results confirm those obtained by simulations: the performance of systems employing BICM with spatial multiplexing is the same as that when concatenating with a STBC. Finally, Section 6. provides the conclusions of this study.

2. Coded modulation for MIMO channels

2.1 Encoder

Figure 1 shows the block diagrams of the two basic MIMO coded modulation signaling methods considered in this chapter: Bit Interleaved Coded Modulation (BICM) with spatial multiplexing and channel coding concatenated with a STBC. We will assume that a stream of information bits $\mathbf{u} = [u_1, u_2, ..., u_K]$ inputs a rate $R_c = K/N$ temporal encoder (in our case an SCLDGM encoder) to produce a coded bit sequence $\mathbf{c} = [c_1, c_2, ..., c_N]$. This sequence is then Gray-mapped to a constellation carrying M_c bits per symbol, resulting in the sequence $\mathbf{s} = [s_1, s_2, ..., s_L]$, with $L = N/M_c$.

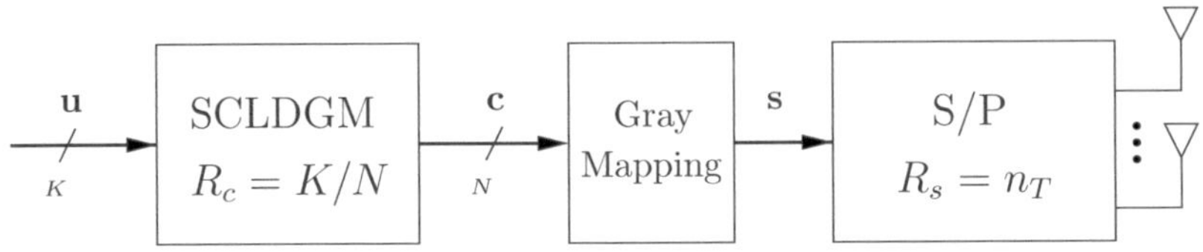

(a) BICM with spatial multiplexing

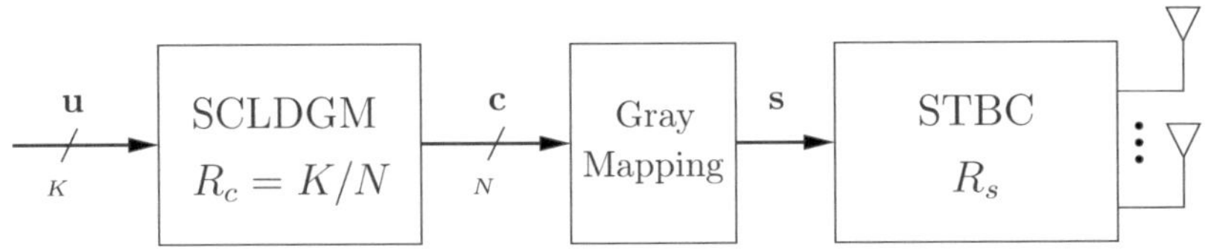

(b) Channel coding concatenated with STBC

Fig. 1. MIMO signaling schemes.

In the BICM scheme with spatial multiplexing (Fig. 1.a), the transmitted symbols s_k are serial-to-parallel (S/P) converted to produce the sequence of transmitted vectors $\mathbf{s}[k]$, $k = 1, 2, ..., L/R_s$, where $R_s = n_T$ is the spatial rate. BICM with spatial multiplexing is a good option for signaling over a MIMO channel in the general case of $n_T \times n_R$ with $n_T, n_R \geq 2$, since, as we will see bellow, it is capable of approaching the capacity limits. However, the complexity of optimum MAP detection, exponential in both n_T and M_c, constitutes an important limitation for a high number of transmitting antennas and/or modulation formats. In this case, it is possible to employ suboptimum detection methods such as LSD (List Sphere Detection) (Hochwald & ten Brink, 2003; Vikalo et al., 2004) or SIC-MMSE (Soft Interference Cancellation with MMSE filtering) (Wang & Poor, 1999). However, these methods either require the observations to be fully determined ($n_T \leq n_R$) or have an exponential complexity in $n_T - n_R$. This is an important limitation because $n_T > n_R$ occurs frequently in practice (for example, in the downlink of cellular communication systems).

When an STBC is concatenated after the constellation mapper (Fig. 1.b), the sequence $\mathbf{s}$ is partitioned into blocks of Q symbols. Each block is then encoded into an $n_T \times T$ symbol matrix, $\mathbf{S}[k]$, which is transmitted using T channel uses, resulting in a spatial rate $R_s = Q/T$. The mapping $[s_k, s_{k+1}, ..., s_{k+Q}] \rightarrow \mathbf{S}[k]_{n_T \times T}$ performed by the STBC greatly affects the features of the MIMO system: it may change its associated capacity, the attained diversity, and the complexity of the detection process, as well as the applicable detection methods.

Notice that, in spite of their differences, the two MIMO signaling schemes in Fig. 1 are closely related. Indeed, BICM with spatial multiplexing can be interpreted as the concatenation of a temporal encoder and a trivial STBC with $Q = n_T$ and $T = 1$.

2.2 Channel model

After transmission through the MIMO channel, the matrix of received vectors, $\mathbf{X}[k]$, is

$$\mathbf{X}[k] = \mathbf{H}[k]\mathbf{S}[k] + \mathbf{N}[k], \quad k = 1, 2, ..., L/R_s, \tag{1}$$

where $\mathbf{H}[k]$ is the $n_R \times n_T$ MIMO channel matrix and $\mathbf{N}[k] = [\mathbf{n}_1\ \mathbf{n}_2\ \cdots\ \mathbf{n}_T]$, where each column $\mathbf{n}_t$ contains independent AWGN samples. For the simulations in Section 4. we will assume a spatially uncorrelated, Rayleigh-fading MIMO channel where the elements in $\mathbf{H}[k]$ are distributed as $\mathcal{CN}(0,1)$. Under the ergodic assumption, the channel matrix changes each time a new vector of symbols is transmitted, whereas under the quasi-static assumption, it remains constant during the transmission of a whole codeword (i.e., $\mathbf{H}[k] \equiv \mathbf{H}, k = 1, 2, ..., L/R_s$). We assume in both cases that the channel changes in an independent fashion from one realization to the next.
For the experiments in Section 5, no assumption on the channel coefficients distribution is made. Since the delay spread in indoor channels is typically small, it is reasonable to assume that the flat-fading hypothesis holds true. Also, since neither the transmitter nor the receiver is moving, the channel behaves in a quasi-static manner and remains unchanged during the transmission of a data frame.

2.3 Decoder

In all the signaling methods we will assume that there is no Channel State Information (CSI) at the transmitter, while the receiver has perfect CSI. In the experiments in Section 5the channel is estimated in a previous step assuming that all transmitted symbols are known at the receiver, which yields to an almost perfect estimation of the channel.
Turbo-like receivers take as input the channel Log-Likelihood Ratios (LLRs), which have different expressions depending on the channel model and the detector type. For the sake of clarity, we will drop the time index k hereafter.
For BICM with spatial multiplexing, the information contained in the received vector $\mathbf{x}$ about one of its bits, $v_k = \pm 1$, is represented by the channel LLR, L_{ch}, which is computed by the optimum MIMO detector as

$$L_{\text{ch}} = \log \frac{P(\mathbf{x}|v_k = +1)}{P(\mathbf{x}|v_k = -1)} = \log \frac{P(v_k = +1|\mathbf{x})}{P(v_k = -1|\mathbf{x})} - \underbrace{\log \frac{P(v_k = +1)}{P(v_k = -1)}}_{L_k}$$

$$= \log \frac{\sum\limits_{\mathbf{s} \in \mathbf{S}_k^+} \exp\left(-\frac{\|\mathbf{x} - \mathbf{H}\mathbf{s}\|^2}{N_0} + \sum\limits_{i=1}^{n_T M_c} \frac{v_i L_i}{2} \right)}{\sum\limits_{\mathbf{s} \in \mathbf{S}_k^-} \exp\left(-\frac{\|\mathbf{x} - \mathbf{H}\mathbf{s}\|^2}{N_0} + \sum\limits_{i=1}^{n_T M_c} \frac{v_i L_i}{2} \right)} - L_k, \qquad (2)$$

where $\mathbf{S}_k^+$ and $\mathbf{S}_k^-$ represent the set of all transmitted symbol vectors $\mathbf{s}$ where bit $v_k = +1$ and $v_k = -1$, respectively. Note that the MIMO detector makes use of the bit Log *Prior* Ratio (LPRs), L_k, which turn outs to be the output messages from the channel decoder.
When the Alamouti STBC is used concatenated with a channel encoder, optimum detection can be performed in an independent fashion over the transmitted streams, thanks to the orthogonality of the effective channel matrix. At the receiver, multiplying the observations by the Hermitian of the channel matrix gives two new observations, x_1 and x_2, corresponding to an equivalent, spatially decoupled model. Then, optimum computation of the channel LLRs for each stream, x, can be realized as

$$L_{\text{ch}} = \log \frac{P(x|v_k = +1)}{P(x|v_k = -1)} = \log \frac{\sum_{s \in S_k^+} \exp\left(-\frac{\|x - \mathcal{F}s\|^2}{\mathcal{F}N_0} + \sum_{i=1}^{M_c} \frac{v_i L_i}{2} \right)}{\sum_{s \in S_k^-} \exp\left(-\frac{\|x - \mathcal{F}s\|^2}{\mathcal{F}N_0} + \sum_{i=1}^{M_c} \frac{v_i L_i}{2} \right)} - L_k, \quad (3)$$

where $\mathcal{F} = \|\mathbf{H}\|_F^2$ is the squared Frobenius norm of the channel matrix.
Optimum LLR computation for LD-coded MIMO systems can be carried out in a completely analogous way to (2) just by considering the resulting equivalent channel model. Regarding detection, the Golden code is also a linear code and, thus, the same equivalent observation model as that of linear dispersion codes applies. In any case, this different equivalent channel model has to be taken into account when performing code design.

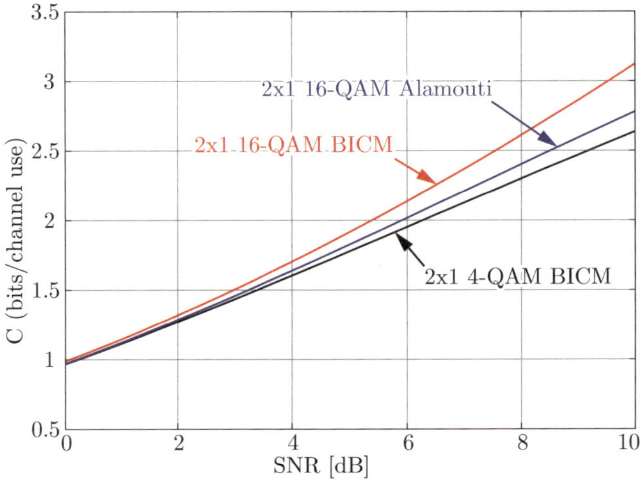

Fig. 2. Constrained capacity of 2×1 BICM with spatial multiplexing and in concatenation with Alamouti STBC in the region of interest.

2.4 Constrained system capacity

Figures 2, 3 and 5 plot the constrained capacities of the different MIMO coded modulation signaling schemes previously described for the cases of 2×1 (see Fig. 2), 2×2 (see Fig. 3) and 3×1 (see Fig. 5), respectively, over the spatially white, Rayleigh-distributed MIMO channel. The target rates are 2 bits/channel use for 2×1 and 2×2, and 1 bit/channel use for 3×1. We use the term constrained capacity to refer to the channel capacity when the transmitter is constrained to use a specific modulation format (4-QAM, 16-QAM, etc.). This constrained capacity is calculated by measuring the mutual information between the output L-values from the detector, L_{ch}, and their associated coded bits.
When $n_T = 2$, the utilization of the Alamouti code as an inner code is very attractive because it allows for spatially decoupling the ML detection, notably simplifying the overall decoding procedure. The price to be paid is the spatial rate consumed by the Alamouti code, $R_s = 1$, which forces the utilization of a higher order modulation to compensate for the rate loss (16-QAM for Alamouti versus 4-QAM in the BICM with spatial multiplexing

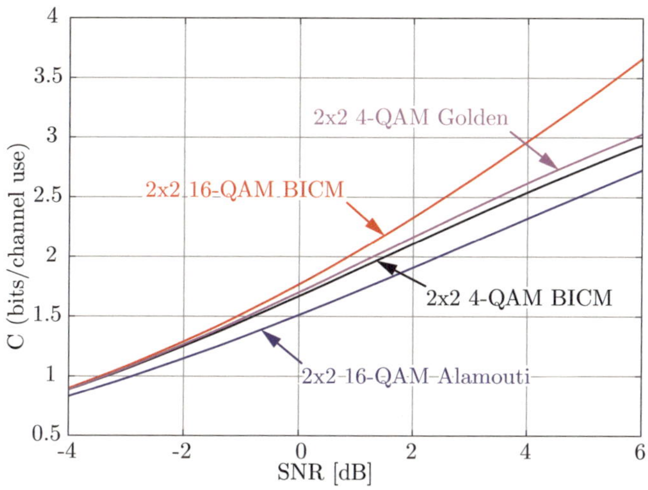

Fig. 3. Constrained capacity of 2 × 2 BICM with spatial multiplexing and in concatenation with STBCs (Alamouti and Golden) in the region of interest.

scheme). Moreover, the imposed signal structure degrades the capacity of the equivalent MIMO system: it is well known (Sandhu & Paulraj, 2000) that the unconstrained capacity (i.e. with Gaussian input symbols) of a $2 \times n_R$ MIMO system with Alamouti coding is less than or equal to that of the MIMO channel without Alamouti coding. This is also true for the case of constrained capacity, as reflected in Figure 2 (top right) where the curve corresponding to 16-QAM Alamouti presents worse performance than that of 4-QAM BICM with spatial multiplexing. For the unconstrained capacity, the equality holds only for the case $n_R = 1$ (Sandhu & Paulraj, 2000). From Figure 2 (top left), it is clear that Alamouti coding is a good choice for 2×1 MIMO systems with constrained symbols, since in this case 16-QAM Alamouti outperforms 4-QAM BICM with spatial multiplexing.

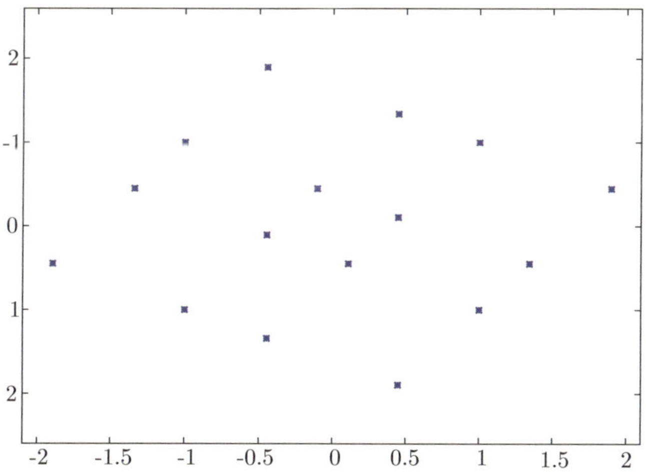

Fig. 4. Constellation at the output of the 2 × 2 Golden code when using a 4-QAM constellation at its input.

When considering 2 × 2 MIMO systems, the Golden code is an appealing alternative to the Alamouti code. The Golden code (Belfiore et al., 2005) is a non-orthogonal 2 × 2 STBC with full information rate ($R_s = n_T = 2$) that provides a capacity improvement at a little increase in complexity, as it can be seen from Fig. 3. This capacity improvement is due to the fact that the resulting constellation at the output of the Golden encoder (see Fig. 4) resembles a Gaussian distribution better than the input constellation. This effect is usually referred to as *shaping* or constellation expansion (Forney, Jr. & Wei, 1989). Although the size of the resulting constellation is larger, the Golden code does not introduce any redundancy because it employs two channel uses for the transmission of four input symbols.

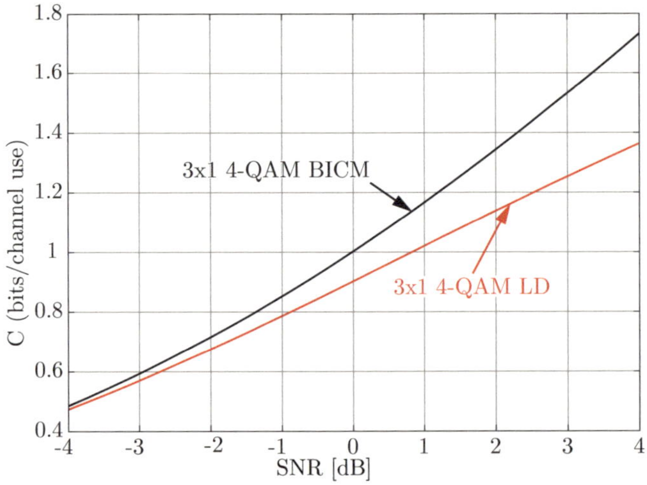

Fig. 5. Constrained capacity of 3 × 1 BICM with spatial multiplexing and in concatenation with LD STBC in the region of interest.

Finally, when $n_T > n_R$, the use of LD codes allows for the application of suboptimum detection methods without much capacity penalty. Linear Dispersion (LD) codes (Hassibi & Hochwald, 2002) are linear STBCs that transform the observation model to avoid underdetermination, at the cost of a minimum capacity loss (see Fig. 5). A stacked, real-valued, equivalent observation model can then be easily formulated (Hassibi & Hochwald, 2002). Such an equivalent observation model is not underdetermined provided that $n_R T > Q$ or, equivalently, $n_R > R_s$. Thus, concatenation with LD codes constitute a good choice when $n_T > n_R, n_T > 2$, and optimum detection is not feasible. Notice also that, similar to the Golden code, constellation expansion takes place at the output of an LD encoder, but it is controlled because symbols are produced according only to specific sequences.

3. Capacity approaching codes for MIMO transmission

In order to approach the capacity of MIMO channels, both BICM with spatial multiplexing and schemes based on concatenation with STBCs have to use an appropriate channel code. In this chapter we focus on a particular subclass of LDPC codes known as Serially-Concatenated Low-Density Generator Matrix (SCLDGM) codes (Garcia-Frias & Zhong, 2003), whose performance is similar to that of general LDPC codes but with very low encoding complexity. The convergence of any coding scheme can be predicted by tracking the mutual information of the different types of messages exchanged between the components of the receiver. This can be efficiently done by considering the EXtrinsic Information Transfer (EXIT) function (ten Brink, 1999) of each component. This procedure, termed EXIT evolution, has been successfully applied to obtain good SCLDGM codes for the Binary-Input AWGN (BIAWGN) channel (González-López et al., 2006a) and for MIMO BICM systems (Vázquez-Araújo et al., 2006; 2007).

The EXIT analysis is based on two assumptions. First, that each message can be expressed as the output LLR of a Binary-Input AWGN (BIAWGN) channel, which allows for the calculation of a bijection between the variance of the L-values and their associated mutual information, i.e., $I = J(\sigma^2)$ and $\sigma^2 = J^{-1}(I)$ (ten Brink et al., 2004). Second, that the messages passed between the components are independent and identically distributed (i.i.d.). Under these two assumptions, the EXIT functions of SCLDGM codes (and, in general, of LDPC-based codes) can be easily calculated (González-López et al., 2007).

The EXIT function of the detector also needs to be calculated. Note that in BICM the channel LLRs produced by the optimum MIMO detector (see (2)), L_{ch}, that constitute the input to the Turbo-like decoder, include the overall effect of the modulator, the channel and the detector. The EXIT function, $I_{\text{ch}}(I_A, E_b/N_0)$, depends on the channel E_b/N_0 and the mutual information of the messages from the decoder, which constitutes the input *a priori* information to the detector, I_A. The characterization of the detector is independent of any other decoding module (ten Brink et al., 2004), so it can be obtained by measuring the mutual information of the L_{ch} messages in Monte Carlo simulations. It is important to highlight that any variation in the model parameters (modulation, number of transmit and receive antennas) leads to a different equivalent channel and thus to a different detector EXIT function. Consequently, there is a different optimum code for each antenna configuration and modulation format.

For Binary-Input AWGN (BIAWGN) channels, the detector EXIT function does not depend on the information from the decoder, because each bit is transmitted in an independent channel use. Furthermore, for Single-Input Single-Output (SISO) channels with perfect CSI at reception and the usual constellations (i.e., PSK and QAM), Gray mapping results in an overall effect of modulation, channel and demodulation with an associated EXIT function equivalent to a BIAWGN channel, that is, only dependent on E_b/N_0 (i.e., it is an horizontal line for each E_b/N_0 value) (Schreckenbach et al., 2003). Thus, optimal codes for BIAWGN are also optimum for SISO channels and for any modulation, provided Gray-mapping is used. As a corollary of this result, optimal codes for OSTBC-coded systems (in particular, Alamouti-coded) with Gray mapping are the same as those optimum for the BIAWGN channel.

For the case of LD codes (for which, in terms of detection, the Golden code is indeed a particular instance), the detector produces channel LLRs according to an equivalent channel model. In this case, the EXIT function of the detector is different from the one corresponding to BICM with spatial multiplexing over the same channel model. Consequently, optimum codes for LD-coded systems differ from those obtained for BICM with spatial multiplexing systems.

Figure 6 plots the EXIT characteristic for 2×2 and 3×1 MIMO systems for either spatial multiplexing or LD coding. They correspond to SNR values at each receive antenna (equal to n_T/N_0 assuming each antenna radiates unit energy symbols) close to the convergence threshold of the best code found in each case. Let us first recall that the *area property* of an EXIT function states that the area below the curve equals the system capacity for a Binary Erasure Channel (BEC). This area property can be considered as approximate for the BIAWGN channel assumed for modelling the input and output messages of the MIMO detector. For a 2×2 system operating at SNR = 2.0 dB, it is apparent that the EXIT function corresponding to a Golden-coded system has a larger area below it than that corresponding to a BICM with

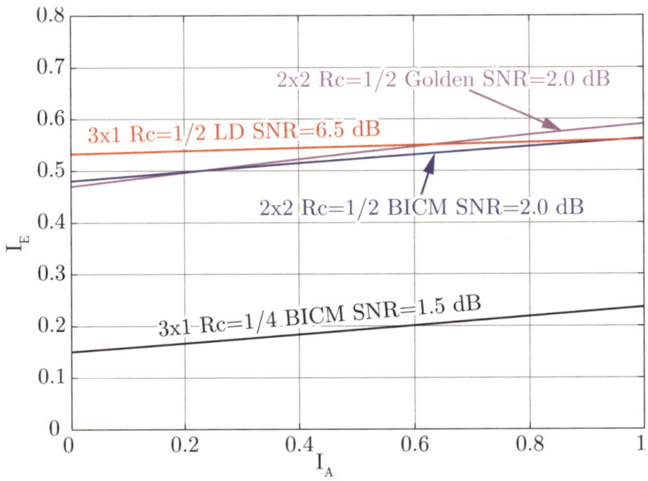

Fig. 6. EXIT characteristics corresponding to the considered MIMO schemes for a 4-QAM constellation. SNR is the signal-to-noise ratio at each receiving antenna (equal to n_T/N_0 assuming each antenna radiates unit energy symbols)

spatial multiplexing system, which is coherent with the capacity increase associated to the Golden code. Note also that the slope of the two functions is different, which leads to different optimum codes (cf. Table 1). For the 3×1 case, the EXIT function corresponding to BICM with spatial multiplexing is located far below the one corresponding to the LD-coded system, because the outer code rates are different ($R_c = 1/6$ and $1/2$, respectively) for the same overall information rate (1/6). Besides, their slopes are different. Both reasons justify that the optimum codes for these two schemes are very different from each other.

Once we have obtained the EXIT function of the detector and the decoder, system convergence can be tested by simulating the evolution of the mutual information through the iterations of both components. For a fixed E_b/N_0 we start with all information values equal to zero and, then, we iteratively compute their values. This is exactly what the decoding process does, except for using the associated information of the messages instead of their actual values. We say that the iterative receiver converges when we find a sequence of information values that finally leads to $I_o = 1$, where I_o is the mutual information associated to the *a posteriori* L-values of the uncoded bits.

Table 1 presents the best regular SCLDGM codes obtained through EXIT analysis for the different MIMO signaling schemes. The table also shows the convergence thresholds predicted by this analysis, as well as the Constrained-input Capacity Limit (CCL) corresponding to each MIMO transmission method. We have considered the antenna configurations where concatenation with STBCs is more beneficial: 2×1 for the Alamouti code, 2×2 for the Golden code, and 3×1 for the LD code. In particular, the latter case presents the characteristics of being a clearly asymmetric antenna configuration ($n_T > n_R$) as well as having a complexity low enough to appreciate the losses in i) capacity with respect to BICM with spatial multiplexing and ii) performance of suboptimum methods with respect to optimum detection in the LD-coded system. We have chosen these antenna configurations as our testbench to assess the gains provided by STBC-concatenated systems. For each antenna configuration, we optimize the SCLDGM code to maintain the overall information

Code	$n_T \times n_R$	Modul.	STBC	R_c	p	d_u^{f1}	d_u^{f2}	d_{p1}^{f2}	CCL	Thresh (dB)	Gap (dB)
#1	2×1	4-QAM	None	1/2	0.0150	3	5	38	3.30	3.87	0.57
#2	2×1	16-QAM	None	1/4	0.0200	3	8	15	2.80	3.57	0.77
#3	$2 \times n_R$	Any	Alamouti	1/2	0.0200	3	6	6	-	-	-
#4	$2 \times n_R$	Any	Alamouti	1/4	0.0400	3	9	24	-	-	-
#5	2×2	16-QAM	None	1/4	0.0275	3	9	12	1.30	1.90	0.60
#6	2×2	4-QAM	None	1/2	0.0300	3	5	32	1.56	2.02	0.46
#7	2×2	4-QAM	Golden	1/2	0.0250	3	5	38	1.42	1.87	0.45
#8	3×1	4-QAM	None	1/6	0.0350	3	10	24	0.40	1.20	0.80
#9	3×1	4-QAM	LD	1/2	0.0200	3	6	48	0.90	1.38	0.48

Table 1. Optimized SCLDGM codes for MIMO channels. "Thresh" stands for the convergence threshold and "Gap" is the gap to the constrained-input capacity limit (CCL). When no STBC is employed, a BICM with spatial multiplexing scheme is considered.

rate fixed, so we can make a fair comparison between BICM with spatial multiplexing and the concatenated schemes.

As we will see in Section 4, SCLDGM codes optimized for Single-Input Single-Output (SISO) channels also exhibit good performance when used in concatenation with the Alamouti code in $2 \times n_R$ MIMO channels (González-López et al., 2006b). This is not a surprising result since the Alamouti code actually converts a $2 \times n_R$ MIMO channel into two parallel and independent SISO channels. Recall that the detector EXIT function of Gray-mapped SISO systems with all the standard constellations is an horizontal line (i.e. it is constant for any value of I_A), resulting in the same optimum code for any constellation. This explains why we do not specify the constellation and thus we cannot provide a threshold value for the Alamouti case, since this threshold is different depending on the employed modulation and the channel model.

The code design procedure described previously assumes an ergodic channel and an infinite block length. In (Yue et al., 2008) it is shown that, when optimizing for the quasi-static channel, codes optimized for the worst-case EXIT envelope can provide a better performance. In our case, however, the slope of the worst case envelope is very similar to that of the average EXIT curve, so the codes optimized for both cases are practically identical. Thus, we will restrict ourselves to ergodic optimization. We will show in Section 4.that the resulting codes present an excellent performance when used in quasi-static channels and/or with finite block lengths. Indeed, their gaps with respect to the outage limits when applied over quasi-static channels are very similar to the gaps they present in ergodic fast-fading channels.

4. Simulation Results

Computer simulations were carried out to illustrate the actual performance of the obtained SCLDGM coded modulation MIMO systems with data blocks of finite length over ergodic and quasi-static channels. For ergodic channels, the observed thresholds are slightly worse than those predicted, since EXIT function analysis assumes infinite-length data blocks. Note that the lower the code rate (R_c) is, i) the higher the gap with respect to the Constrained-input Capacity Limit (CCL) for the best code found, and ii) the higher the gap between the theoretical threshold predicted by EXIT analysis and that observed in simulations. Similar

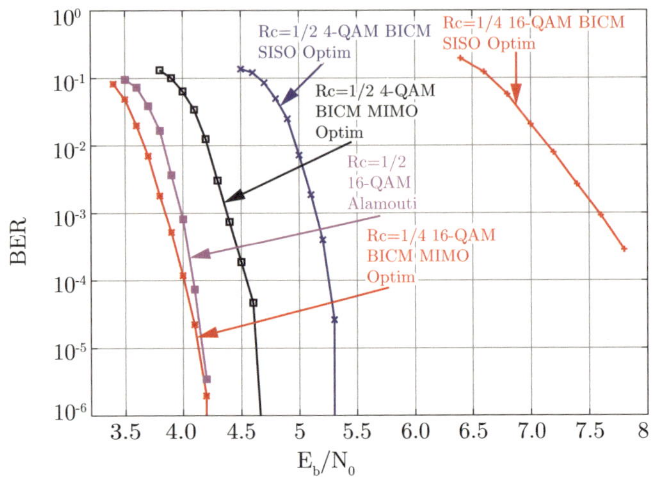

Fig. 7. Performance of a 2×1 MIMO system with a data rate of 2 information bits per channel use and i) SCLDGM + Alamouti code, ii) BICM with spatial multiplexing. Ergodic Rayleigh fading. The block length is $K = 10,000$ information bits.

conclusions hold for quasi-static channels when comparing actual performance with respect to the outage probability limit.

4.1 Ergodic channel

Figure 7 shows the performance of several SCLDGM coded modulation 2×1 MIMO systems with a data rate of two bits per channel use assuming an ergodic channel model. The block length is $K = 10,000$ information bits (we will also use this block length throughout all the simulations in ergodic channels) and 5,000 blocks are simulated for each E_b/N_0 value to obtain an adequate estimate of the Bit Error Rate (BER). The best performance is obtained when using BICM with spatial multiplexing, 16-QAM and code rate 1/4 (code #2). For a BER= 10^{-4} (which we will use as the target BER from now on) the required E_b/N_0 is 4.0 dB, which is 1.2 dB away from the CCL and 0.43 dB away from the theoretical threshold. Figure 7 also plots the performance obtained when using a concatenated Alamouti scheme with 16-QAM and code rate $R_c = 1/2$, with the code optimized for the SISO channel (code #3). The SCLDGM rate was raised up to $R_c = 1/2$ in order to maintain the data rate equal to two bits per channel use. Not surprisingly, the performance of these two schemes is very similar, because the capacity limit of a 2×1 system is the same irrespectively of whether Alamouti is used or not. Regarding receiver complexity, however, it is obvious that decoding in the concatenated scheme is considerably simpler.

The data rate of two bits per channel use can also be obtained using BICM with spatial multiplexing, 4-QAM and code rate $R_c = 1/2$. Code #1 has been specifically optimized for this particular situation and its performance is also shown in Fig. 7. Lowering the number of bits per symbol in the modulator, M_c, is interesting because it yields to a considerable reduction of the detector complexity. At the target BER, the required E_b/N_0 is 4.55 dB (1.25 dB away from the CCL for 4-QAM and 0.63 dB away from the predicted threshold). Thus, this case maintains the same gap to the CCL as the 16-QAM case, but it exhibits a 0.5 dB performance degradation due to the capacity loss resulting from changing the modulation format (see Table 1). Figure 7 also illustrates the importance of designing SCLDGM codes for each specific

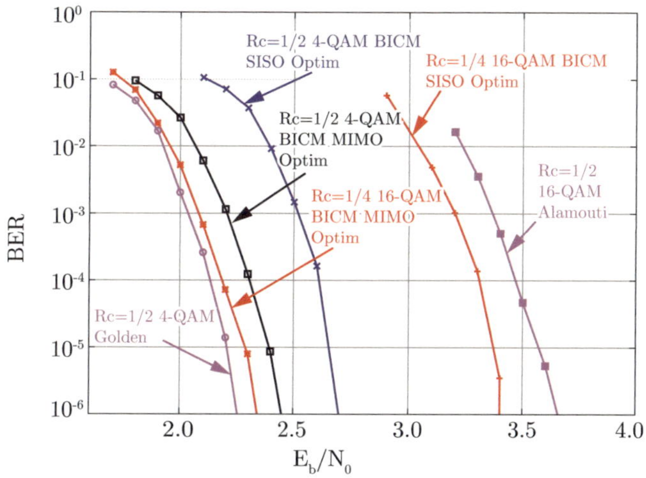

Fig. 8. Performance of a 2 × 2 MIMO system with a data rate of 2 information bits per channel use and i) SCLDGM + Golden code, ii) BICM with spatial multiplexing. Ergodic Rayleigh fading. The block length is $K = 10,000$ information bits.

coded modulation MIMO configuration for BICM with spatial multiplexing. Indeed, observe the serious degradation in performance when the SISO-optimized codes #3 and #4 are used instead of the MIMO-optimized ones. In these cases, the required E_b/N_0 is 5.25 and 8.0 dB, respectively, so that the loss in performance with respect to the MIMO-optimized codes is 0.7 dB and 4.0 dB, respectively. From these results, we can conclude that code optimization is more critical when the constellation size increases.

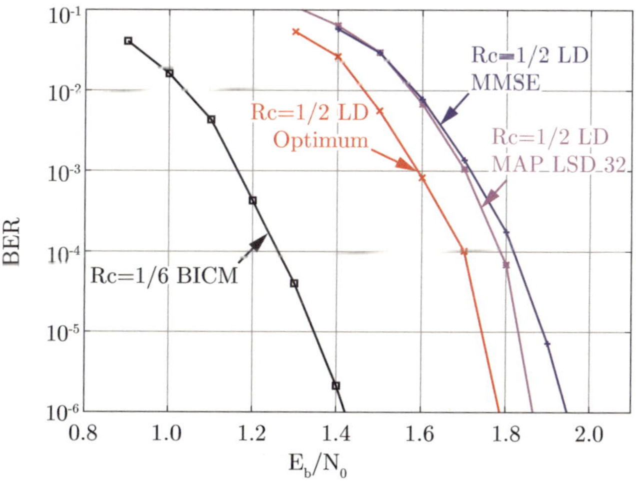

Fig. 9. Performance of a 4-QAM 3 × 1 MIMO system with a data rate of 1 information bit per channel use and i) SCLDGM + $R_s = 1/3$ STBC (LD code), ii) BICM with spatial multiplexing. Ergodic Rayleigh fading. The block length is $K = 10,000$ information bits.

The performance of a 2 × 2 MIMO system with a data rate of two bits per channel use is shown in Fig. 8 for an ergodic channel model and i) SCLDGM (code #7) + Golden code, ii) BICM with spatial multiplexing (codes #5 and #6). Although the $R_c = 1/4$ 16-QAM BICM with spatial multiplexing system has the highest capacity (its corresponding CCL is at $E_b/N_0 = 1.30$ dB), the best performance is attained by the Golden-coded system (for which its corresponding CCL is at $E_b/N_0 = 1.43$ dB). This is explained because the best $R_c = 1/4$ code found for the 16-QAM BICM with spatial multiplexing system has a threshold at $E_b/N_0 = 1.90$ dB (theoretically) and at $E_b/N_0 = 2.20$ dB (in practice), which is worse than that of the best $R_c = 1/2$ code found for the Golden-coded system (threshold at $E_b/N_0 = 1.87$ dB theoretically and at $E_b/N_0 = 2.15$ dB in practice). The $R_c = 1/2$ 4-QAM BICM with spatial multiplexing system shows worse performance, requiring $E_b/N_0 = 2.30$ dB at the target BER, which is consistent with its lower system capacity (CCL at $E_b/N_0 = 1.56$ dB). We have also included the performance obtained for these systems when the code is the optimum for a SISO model (González-López et al., 2006a). The gaps in performance with respect to the MIMO-optimized codes are significant, especially for 16-QAM (0.3 dB for 4-QAM and 1.10 dB for 16-QAM). Finally, observe the poor performance of the Alamouti coded system (1.25 dB worse than the MIMO-optimized code) even when using its optimum code (González-López et al., 2006b), which is a consequence of its system capacity loss.

Figure 9 shows the results for a 3×1 ergodic MIMO channel when using 4-QAM with a data rate of one information bit per channel use and i) SCLDGM (code #9) + LD code, ii) BICM with spatial multiplexing (code #8). The best performance is exhibited by the BICM with spatial multiplexing system employing an $R_c = 1/6$ SCLDGM code (which requires an E_b/N_0 of 1.25 dB for the target BER). As explained before, using the LD code given by (36) in (Hassibi & Hochwald, 2002) enables the application of suboptimum schemes such as LSD or SIC-MMSE at the cost of sacrificing capacity and, thus, performance. The degradation in actual performance (under optimum detection) with respect to BICM with spatial multiplexing is not severe (0.45 dB at the target BER) and is of the same order as the loss in capacity (the CCL for BICM with spatial multiplexing is at $E_b/N_0 = 0.40$ dB whereas for the LD code it is at $E_b/N_0 = 0.90$ dB). In addition, when the LD code is used, the gap of either MAP LSD or SIC-MMSE with respect to optimum detection is fairly small (around 0.1 dB). This corroborates the convenience of employing LD coding as a means of enabling suboptimum detection methods when receiver complexity is a constraint.

4.2 Quasi-static channel

We have also studied the performance of SCLDGM coded modulation MIMO schemes when transmitting over quasi-static channels. We employ the same codes as in the ergodic channel, aiming at assessing if optimization for fast fading also leads to good performance in quasi-static scenarios[1]. Figure 10 shows the performance of an SCLDGM BICM with spatial multiplexing scheme and an SCLDGM + Alamouti scheme over a 2×1 quasi-static MIMO channel. The channel block length is $B = 500$ symbol vectors, which corresponds to $K = 1000$ information bits. As it occurs in the ergodic channel, the performance of the Alamouti scheme and the BICM with spatial multiplexing and 16-QAM scheme are practically identical. The gap to the outage capacity is approximately 1.0 dB in both cases. The figure shows that the SCLDGM coded modulation scheme achieves maximum diversity, since the slope of the curve is the same as that of the outage capacity.

On the contrary, it is apparent from Fig. 10 that BICM with spatial multiplexing and 4-QAM performs worse (5 dB at a BLER of 10^{-2}) than the other two methods. Notice the lower slope

[1] Notice that optimizing for short quasi-static channels would require the development of a completely different approach, which could be skipped if the codes optimized for ergodic, fast-fading behave well in quasi-static environments.

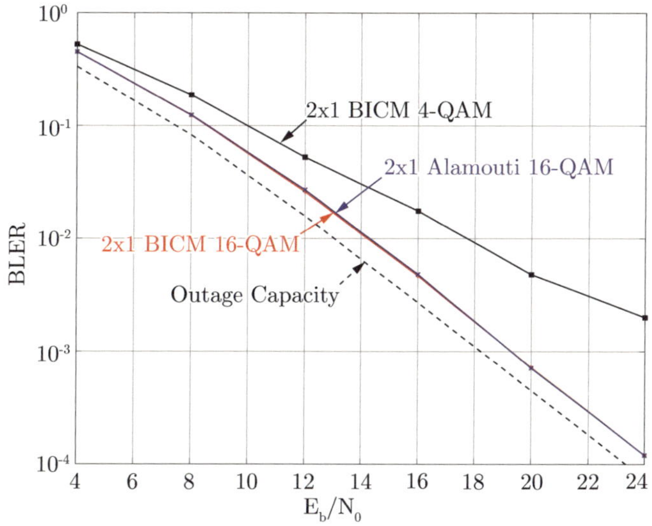

Fig. 10. Performance of a 16-QAM 2 × 1 MIMO system with a data rate of 2 information bits per channel use and i) BICM with spatial multiplexing and ii) SCLDGM + Alamouti. Quasi-static Rayleigh fading. The block length is $B = 500$ symbol vectors.

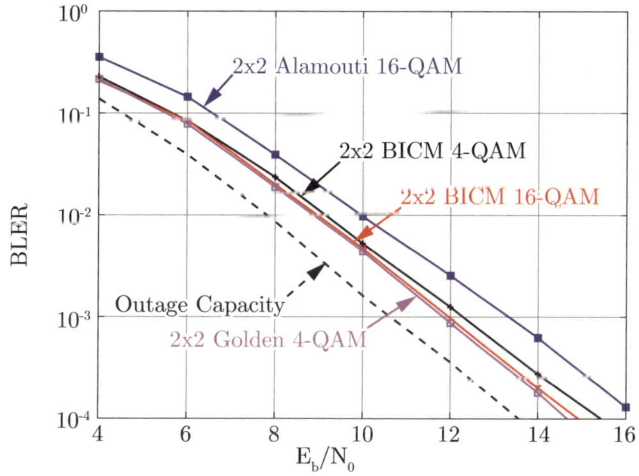

Fig. 11. Performance of a 2 × 2 MIMO system with a data rate of 2 information bits per channel use and i) BICM with spatial multiplexing, ii) SCLDGM + Golden code, iii) SCLDGM + Alamouti. Quasi-static Rayleigh fading. The block length is $B = 500$ symbol vectors.

of the BLER curve for BICM with spatial multiplexing and 4-QAM, which means that this system is not able to extract all the spatial diversity available in the channel. We conjecture two explanations for this: on the one hand, it may happen that the high rate of the channel encoder ($R_c = 1/2$) and its subsequent mapping into 4-QAM symbols does not introduce enough redundancy for the signaling scheme to obtain all the available spatial diversity; on the other hand, the degree profile of the channel encoder has been designed assuming an ergodic channel and now the channel is quasi-static.

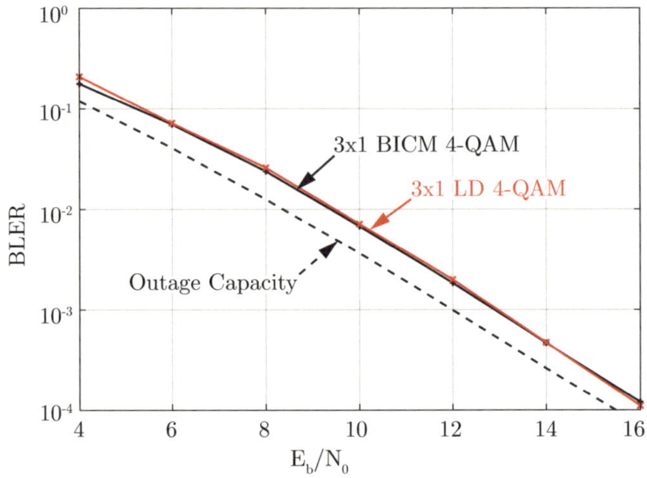

Fig. 12. Performance of a 4-QAM 3×1 MIMO system with a data rate of 1 information bit per channel use and i) SCLDGM + $R_S = 1/3$ STBC (LD code), ii) BICM with spatial multiplexing. Quasi-static Rayleigh fading. The block length is $B = 500$ symbol vectors.

Figure 11 shows the performance of an SCLDGM BICM with spatial multiplexing scheme, an SCLDGM + Golden code scheme and an SCLDGM + Alamouti scheme over a 2×2 MIMO channel. The best performance is achieved by the SCLDGM + Golden code, although the difference with the 16-QAM BICM with spatial multiplexing scheme is minimal. The gap to the outage capacity is kept at approximately 1.0 dB for the Golden and the BICM with spatial multiplexing with 16-QAM schemes and a little over 1.5 dB for the BICM with spatial multiplexing 4-QAM scheme. The distance of the Alamouti scheme to the outage capacity increases to 3.0 dB. Again, all SCLDGM coded modulation schemes achieve maximum diversity. This demonstrates that the BICM with spatial multiplexing scheme is also suitable for quasi-static channels, without having to resort to schemes that explicitly maximize the spatial diversity. Similar results can be observed in Fig. 12 for a quasi-static 3×1 MIMO channel with $B = 500$ symbol vectors ($K = 500$ information bits). The BICM with spatial multiplexing scheme is able to achieve the same diversity as the LD scheme, and both have very similar performance (approximately 1.0 dB away from the outage capacity).

5. Experimental Evaluation

For the experimental evaluation of the aforementioned schemes in a realistic indoor environment we employed a testbed developed at the University of A Coruña (García-Naya et al., 2010). A picture of the testbed is shown in Fig. 13. The testbed has been constructed using Commercial-Off-The-Shelf (COTS) modules from Sundance Multiprocessor (*Sundance Multiprocessor*, 2010) for the implementation of the baseband functionalities, and Radio Frequency (RF) font-ends from Lyrtech (*Lyrtech*, 2010). The hardware of the testbed is completed with a distributed, multilayer software architecture specifically designed to easy the interaction with the testbed hardware (Fernández-Caramés et al., 2008; García-Naya et al., 2010; García-Naya et al., 2008).

Figure 14 shows a block diagram containing the software and hardware elements utilized at the transmit side to assess the aforementioned schemes. Once the discrete-time,

Fig. 13. Picture of the testbed developed at the University of A Coruña.

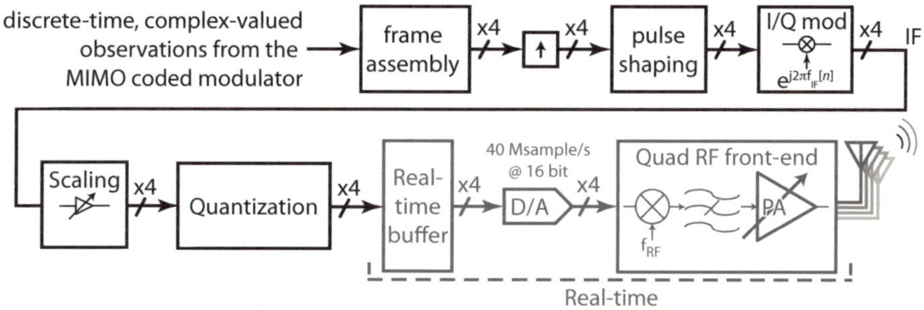

Fig. 14. Block diagram of hardware and software elements at the transmitter. Notice that depending on the scheme, two or three of the four transmit antennas are utilized.

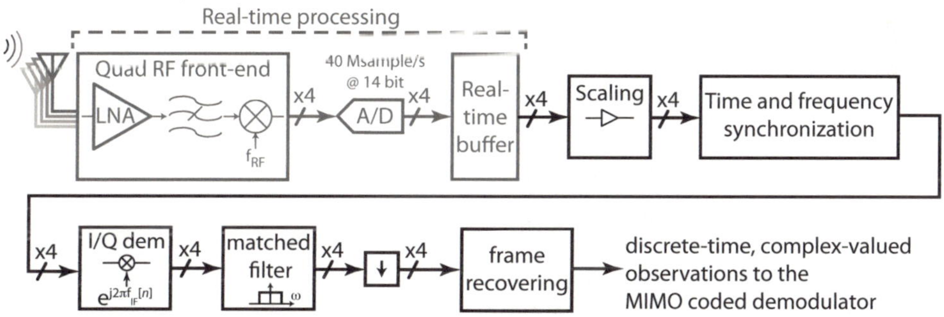

Fig. 15. Block diagram of hardware and software elements at the receiver. Notice that at the receiver, all four antennas are always utilized regarding of the scheme being acquired. Later, during the evaluation step, the corresponding signals are employed according to the scheme.

complex-valued source symbols from the encoder are generated, the following steps are carried out:

- The transmit frame is assembled. Basically, a preamble for time and frequency synchronization as well as a small silence for estimating the power spectral density of the noise are included. Then, for each transmit antenna (two or three depending on the scheme):

- Up-sampling by a factor of 15, resulting in 15 samples per symbol.
- Pulse-shape filtering using a squared root-raised cosine filter with 12% roll-off. Consequently, given that the sampling frequency of the DACs is set to 40 MHz, then the resulting signal has a bandwidth of 2.9867 MHz, which leads — according to our tests — to a frequency-flat channel response.
- The resulting signals are I/Q modulated to obtain a passband signal at a carrier frequency of 5 MHz.
- Such signals are then properly scaled in order to guarantee that the same transmit power level is achieved.
- Given the 16 bits of resolution of the DAC, the signals are properly quantized, obtaining 16-bit integer values for the samples.
- The resulting signal is then stored off-line in the buffer available at the hardware testbed.
- When the transmitter is triggered, such a buffer is read cyclically and in real-time by the DAC, which generates a signal at the intermediate frequency of 5 MHz.
- The resulting analog signal is sent to the RF front-end to be transmitted at the desired RF center frequency. In our measurements we utilized 69 different carriers in the frequencies ranging from 5 200 MHz to 5 250 MHz and from 5 480 MHz to 5 700 MHz.

At the receiver side, once the transmitter has been triggered the following steps are carried out (see Fig. 15)

- The RF front-end down-converts the signal received by the four available antennas to the 5 MHz intermediate frequency.
- The signal is then digitized by the ADCs and, in real-time, stored in the buffer. Given that the signals are being transmitted cyclically and in order to guarantee that a whole frame is acquired, twice the length of the transmit frame is acquired.
- The signals are properly scaled according to the number of resolution bits of the ADC. Notice that this factor is constant during the whole measurement campaign, thus not affecting the properties of the channel.
- In the next step, time and frequency synchronization operations are carried out.
- I/Q demodulation and filtering tasks take place. As a result, discrete-time, complex-valued observations with 15 samples per symbol are obtained. With the resulting samples (prior to the decimation stage) the instantaneous receive power as well as the instantaneous power spectral density of the noise are estimated. During the evaluation stage, using all instantaneous values estimated, the mean signal-to-noise ratio (SNR) is estimated.
- After filtering, the resulting signals are decimated, resulting in a single sample per symbol.
- Finally, the frame is properly disassembled, and the resulting observations are then sent to the MIMO coded demodulator.

5.1 Measurement Procedure

We evaluated experimentally the performance of the aforementioned MIMO schemes in a typical indoor environment (the research lab where the authors work at University of A Coruña) with a separation between the transmitter and the receiver about 9 m, employing monopole antennas both at the transmitter and at the receiver. The antenna spacing is set

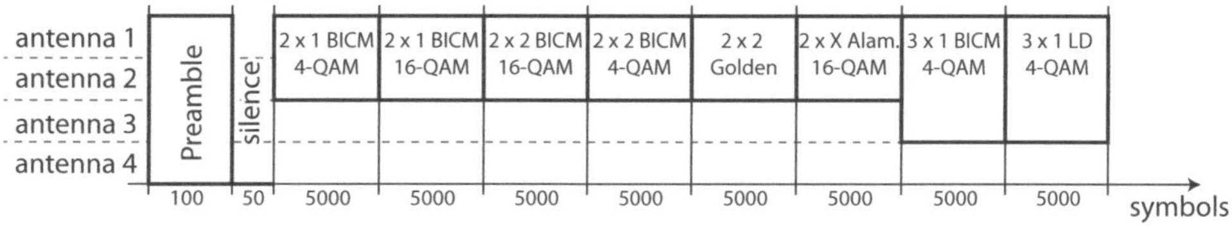

Fig. 16. Frame structure employed in the experimental evaluation.

to approximately 7 centimeters (determined by the separation of the antenna ports of the RF front-end). We experimentally observed that the channel behaved as non time-dispersive when transmitting with a bandwidth of 2.9867 MHz. This is an expected behavior since the delay spread of wireless channels in indoor environments is usually small.

We designed the frame structure shown in Fig. 16 in which, at the beginning, a pseudo random BPSK sequence (duration: 100 microseconds[2]) is used as a preamble for subsequent time and frequency synchronization at the receiver. Next, a silence is introduced (duration: 19 microseconds) with the objective of estimating the power spectral density of the noise at the receiver. Next, the eight blocks corresponding to the eight different schemes to be evaluated are transmitted. Each block occupies 5 000 symbols, resulting in a duration of 1.875 milliseconds. In total, the frame takes approximately 15 milliseconds for the transmission.

Neither the transmitter nor the receiver were moving during the transmissions. Also, experiments were carried out in a controlled scenario (at night) with no moving objects in the surroundings. This ensured the channel remained unchanged during the transmission of all the symbols corresponding to a frame containing the eight blocks.

Given that we know all transmit symbols and that we always measure in the high SNR regime (SNR $>$ 20 dB), we utilize all observations to perform a highly accurate channel estimation. Additionally, we checked if the channel actually changed from the first to the last of the eight blocks of the frame and we verified that the channel remained constant during the transmission of the whole frame. Consequently, all eight schemes experience the same channel realization, thus ensuring a fair comparison among them.

In summary, experiments were carefully designed to ensure that the channel is frequency-flat and quasi-static. This is crucial to ensure the proper performance of the decoding algorithms.

In order to simplify the evaluation stage, we always transmit at high transmit power level, avoiding non-linear effects caused by the saturation of the power amplifiers and, at the same time, guaranteeing an SNR value above 20 dB. This, on the one hand, ensures that the errors in the synchronization will not cause a significant impact on the observed results. On the other hand, the estimates of the channel will be accurate enough to be able to evaluate the results assuming perfect CSI at the receiver, thus not including the effects of the channel estimation in the results.

With the aim of obtaining different channel realizations, we make use of the following three techniques:

- Given that the Lyrtech RF front-end is frequency-agile, we measure at different RF carriers in the interval ranging from 5 200 MHz to 5 250 MHz and from 5 480 MHz to 5 700 MHz and spaced 4 MHz (greater than the bandwidth occupied by the signal), which results in 69 different frequencies, providing 69 different channel realizations.

[2] the preamble is sampled at 40 samples per symbol, resulting in 1.12 MHz of bandwidth

- For all 69 different frequencies, we measure at 7 different antenna positions. Such positions are obtaining by moving the receiver in a distance in the order of a wavelength.
- Finally, given that none of the evaluated schemes makes use of all four receive antennas, we utilize each set of receive antennas because they experience a different channel realization. For example, when a 2×1 scheme is being evaluated, we obtain four different channel realizations, one at each receive antenna. Similarly, when a 2×2 scheme is being evaluated, two different channel realizations are obtained.

After all, depending on the number of receive antennas we obtain a different number of channel realizations. In the set-ups that only use a single receive antenna, we have $69 \times 7 \times 4 = 1932$ different channel realizations. In the set-ups that make use of two receive antennas, $69 \times 7 \times 2 = 966$ different channel realizations are available.
It is interesting to examine the statistical properties of the measured channels. Figure 17 plots the histograms of the module and the angle of the estimated channel coefficients for the 2×2 measurements. It is clear from these histograms that the magnitude and phase of the measured wireless channel coefficients match quite well a Rayleigh and a uniform distribution, respectively. Additionally, it is important to look at the spatial correlation among the channel coefficients. Towards this aim we stacked in a single 4×1 vector $\mathbf{h}$ the four coefficients of the 2×2 MIMO channel measured, i.e.,

$$\mathbf{h} = \begin{bmatrix} h_{11} \\ h_{12} \\ h_{21} \\ h_{22} \end{bmatrix}, \quad (4)$$

where h_{ij} represents the complex-valued channel coefficient between the j-th transmit antenna and the i-th receive antenna. The covariance matrix of this vector is $\mathbf{C_h} = \mathrm{E}[\mathbf{h}\mathbf{h}^H]$, where $\mathrm{E}[\cdot]$ denotes the expectation operator and $(\cdot)^H$ represents the conjugate transpose operation. From the measured channel coefficients we estimated this covariance matrix and we obtained the following matrix for their absolute values

$$\mathbf{C_h} = \begin{bmatrix} 1.028395 & 0.018719 & 0.013984 & 0.028464 \\ 0.018719 & 1.156758 & 0.022994 & 0.018275 \\ 0.013984 & 0.022994 & 0.591004 & 0.029860 \\ 0.028464 & 0.018275 & 0.029860 & 0.768709 \end{bmatrix}. \quad (5)$$

Notice from (5) the low values of the non-diagonal elements which indicates that experiments were carried out in a rich scattering environment that introduced a low spatial correlation among the channel coefficients.
In summary, the statistical properties of the measured channel are very similar to those of the quasi-static channel considered in section Section 4.2

5.2 Experimental Results

Once the experiments were carried out and the received signals were recorded, experimental evaluation of the MIMO coded modulation methods was done in terms of Block Error Rate (BLER) versus E_b/N_0. Since the transmit power and the receiver noise level were fixed during each experiment, simulated AWGN was injected into the recorded received signals to change the operating E_b/N_0 value. Although the measurements were carried out at high SNR values

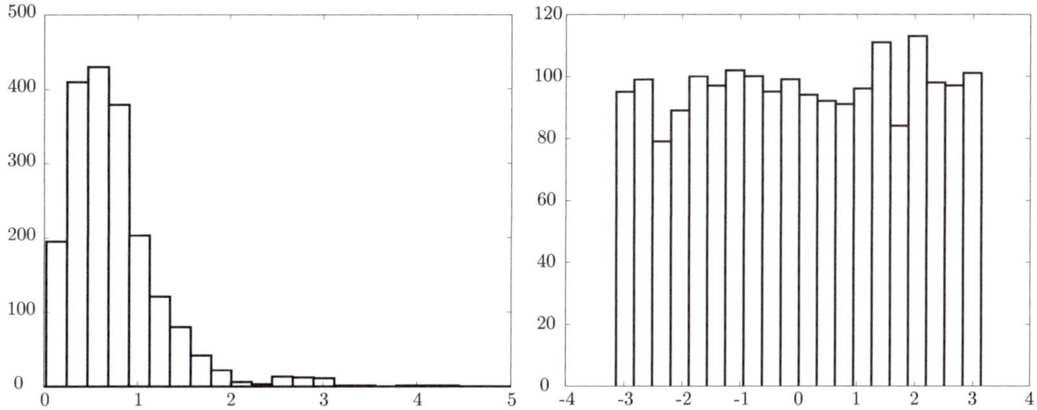

Fig. 17. Histogram of the module (left-hand) and the angle (right-hand) of the estimated channel coefficients for the 2 × 2 case.

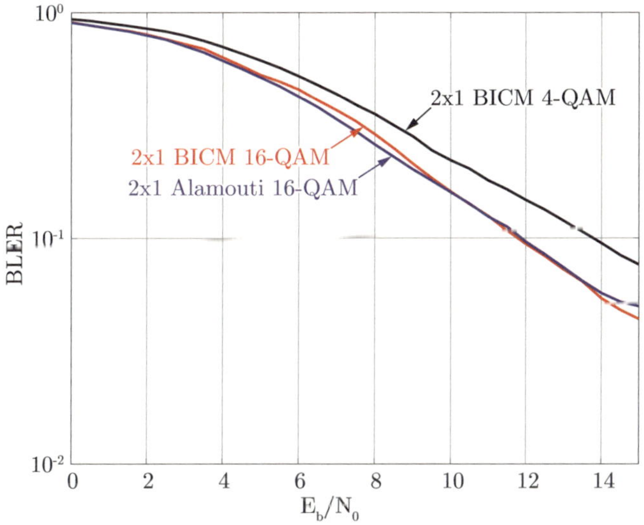

Fig. 18. Testbed results of 2 × 1 BICM with spatial multiplexing and in concatenation with STBCs.

and the target SNR is much lower than that of the measurements, we estimate the mean noise power level of the measurements as it was explained above and we take it into account when calculated the noise power level to be added in order to obtain the target mean E_b/N_0.
Figure 18 shows the experimental performance of the SCLDGM MIMO coded modulation systems for the 2 × 1 configuration and a data rate of two bits/channel use. Notice from this figure that the performance of BICM with spatial multiplexing and 16-QAM is practically the same as concatenating with the Alamouti code. This is in accordance with the constrained capacity analysis in Section 2.4and the simulation results in Section 4.2that also showed that both methods perform the same. On the contrary, it is apparent from Fig. 18 that BICM with spatial multiplexing and 4-QAM performs worse (2 dB at a BLER of 10^{-1}) than the other

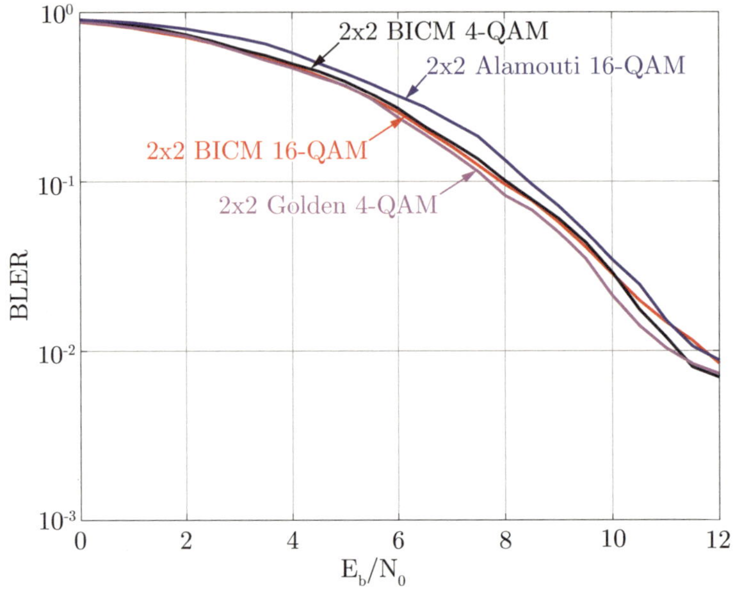

Fig. 19. Testbed results of 2 × 2 BICM with spatial multiplexing and in concatenation with STBCs.

two methods. This performance degradation was also appreciated during the simulations over a quasi-static channel (see Fig. 10). Notice the lower slope of the BLER curve for BICM with spatial multiplexing and 4-QAM, which means that this system is not able to extract all the spatial diversity available in the channel. We conjecture two explanations for this: on the one hand, it may happen that the high rate of the channel encoder ($R_c = 1/2$) and its subsequent mapping into 4-QAM symbols does not introduce enough redundancy for the signaling scheme to obtain all the available spatial diversity; on the other hand, the degree profile of the channel encoder has been designed assuming an ergodic, spatially white channel and these hypotheses are no longer true when dealing with the experimental channels.

Figure 19 shows the experimental performance of the SCLDGM MIMO coded modulation systems for the 2 × 2 configuration and a data rate of two bits/channel use. First of all, notice that the slope of all the BLER curves is twice larger than that of Fig. 18. This is because the four MIMO coded modulation methods are obtaining the full $n_T \times n_R = 4$ spatial diversity of the channel. Next, notice that from a practical point of view, the performance difference among the methods is negligible. This is in accordance with the capacity analysis in Section 2.4and the quasi-static channel simulation results in Section 4.2that also showed that the performance difference among all methods is less than 1 dB.

Finally, Fig. 20 shows the experimental performance of the SCLDGM MIMO coded modulation systems for the 3 × 1 configuration and a data rate of one bit/channel use. Again, note that the slope of the BLER curves is less than that of Fig. 19 but larger than than of Fig. 18 because the spatial diversity of a 3 × 1 MIMO channel is less than that of a 2 × 2 MIMO channel but larger than that of a 2 × 1 MIMO channel. Also note that the performance of the BICM with spatial multiplexing scheme is the same as that of the scheme concatenated with a LD code as expected from the capacity analysis in Section 2.4and the quasi-static channel simulations in Section 4.2

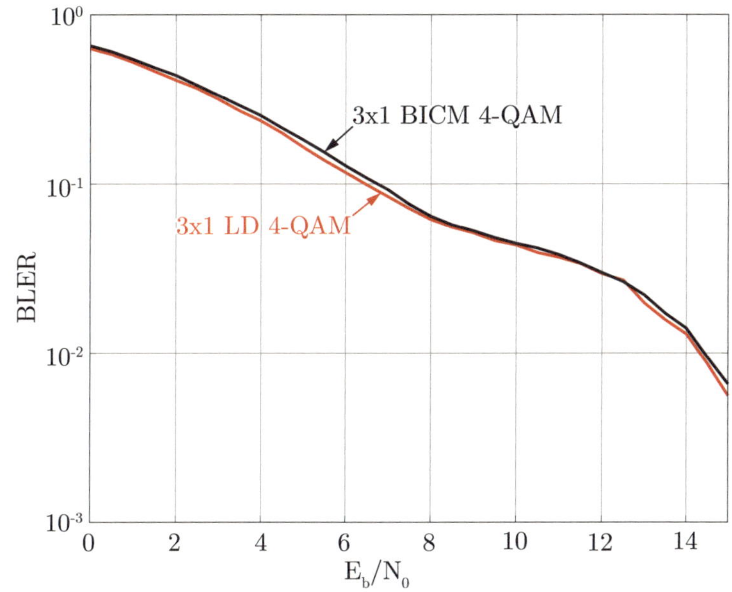

Fig. 20. Testbed results of 3×1 BICM with spatial multiplexing and in concatenation with STBCs.

6. Conclusion

We have studied MIMO coded modulation transmission schemes using either BICM with spatial multiplexing or concatenation of channel code with an STBC code, assuming turbo-like decoding at reception. Our study aims to shed light at the controversy on which of the two schemes is better to approach the capacity of MIMO channels. We have restricted ourselves to the less complex scenarios: 2×1 and two bits/channel use; 2×2 and two bits/channel use; and 3×1 and one bit per channel use. When using BICM with spatial diversity and two transmit antennas, these target data rates can be achieved with either a rate 1/2 channel encoder and 4-QAM, or with a rate 1/4 channel encoder and 16-QAM. With three transmit antennas, the BICM with spatial multiplexing uses a rate 1/6 channel encoder and 4-QAM. For the alternative coded modulation methods, the STBCs that were considered are: the Alamouti code for 2×1 and 2×2 configurations; the Golden code for 2×2; and the Linear Dispersion (LD) code for 3×1.

We have explained how to design regular SCLDGM codes for each specific MIMO coded modulation system. Code optimization has been carried out for spatially white flat-fading Rayleigh ergodic channels. By means of computer simulations, we showed the ability of the resulting MIMO coded modulation schemes to approach the ergodic channel capacity under the practical constraint of finite length codewords. We also showed by simulations that the optimized coded modulation signaling methods approach the outage capacity of quasi-static channels except for the case of 2×1, BICM with spatial multiplexing, rate 1/2 and 4-QAM.

Performance evaluation was not limited to computer simulations with synthetically generated channels. We used a MIMO hardware demonstrator developed at the University of A Coruña to evaluate the MIMO coded modulation methods in realistic environments. Different experiments were carried out at the 5 GHz Industrial, Scientific and Medical (ISM) band

considering different Tx/Rx locations and antenna positions. The results were presented in terms of Block Error Rate (BLER) versus E_b/N_0 at reception and were representative of the performance obtained over a typical indoor scenario.
Both computer simulations and experimental measurements showed that concatenation with STBCs is not necessary in order to retain maximum system capacity and spatial diversity. Simply spreading the output symbols of a BICM system among the different transmit antennas provides the same optimum performance. The utilization of STBCs is only justified to simplify the detection procedure when a large number of antennas and/or a higher modulation format is employed. This is not the case in the scenarios considered in this chapter where the detection complexity of BICM schemes is comparable to those using STBCs.

7. Acknowledgements

This work has been supported by Xunta de Galicia through grant 09TIC008105PR and by Ministerio de Ciencia e Innovación of Spain and FEDER funds of the EU under grants TEC2007-68020-C04-01 (MultiMIMO project) and CSD2008-00010 (COMONSENS project).

8. References

Alamouti, S. M. (1998). A simple transmit diversity technique for wireless communications, 16: 1451–1458.

Belfiore, J.-C., Reyaka, G. & Viterbo, E. (2005). The Golden code: A 2 × 2 full-rate space-time code with nonvanishing determinants, 51(4): 1432–1436.

Berrou, C., Glavieux, A. & Thitimajshima, P. (1993). Near Shannon limit error-correcting coding and decoding: Turbo-codes, *ICC '93*, Geneva, Switzerland, pp. 1064–1070.

Fernández-Caramés, T. M., García-Naya, J. A., González-Lopez, M. & Castedo, L. (2008). MIMO testbed middleware for transmission automation, *Proc. 50th International Symposium ELMAR*, Vol. 1, Zadar, Croatia, pp. 215–218.

Forney, Jr., G. D. & Wei, L. F. (1989). Multidimensional constellations - Part I: Introduction, figures of merit, and generalized cross constellations, 7(6): 877–892.

Gallager, R. G. (1963). *Low Density Parity Check Codes*, Research Monograph 21. MIT Press, Cambridge, Mass.

Garcia-Frias, J. & Zhong, W. (2003). Approaching Shannon performance by iterative decoding of linear codes with low-density generator matrix, 7(6): 266–268.

García-Naya, J. A., González-López, M. & Castedo, L. (2010). *Radio Communications*, In-Tech, chapter A Distributed Multilayer Software Architecture for MIMO Testbeds.

García-Naya, J. A., Perez-Iglesias, H. J., Fernández-Caramés, T. M., González-López, M. & Castedo, L. (2008). A distributed multilayer architecture enabling end-user access to MIMO testbeds, *Proc. IEEE 19th International Symposium on Personal, Indoor and Mobile Radio Communications*, Cannes, France, pp. 1–5.

González-López, M., Vázquez-Araújo, F. J., Castedo, L. & Garcia-Frias, J. (2006a). Design of Serially-Concatenated Low-Density Generator Matrix codes using EXIT charts, *Intl. Symp. on Turbo Codes & Related Topics*, Munich, Ger.

González-López, M., Vázquez-Araújo, F. J., Castedo, L. & Garcia-Frias, J. (2006b). Optimized Serially-Concatenated LDGM and Alamouti codes for approaching MIMO capacity, *PIMRC'06*, Helsinki, Finland.

González-López, M., Vázquez-Araújo, F. J., Castedo, L. & Garcia-Frias, J. (2007). Serially-Concatenated Low-Density Generator Matrix (SCLDGM) codes for transmission over AWGN and Rayleigh fading channels, 6(8): 2753–2758.

Hassibi, B. & Hochwald, B. M. (2002). High-rate codes that are linear in space and time, 48(7): 1804–1824.

Hochwald, B. M. & ten Brink, S. (2003). Achieving near-capacity on a multiple-antenna channel, 51(3): 389–399.

IEEE Standard for Local and Metropolitan Area Networks. Part 16: Air Interface for Fixed and Mobile Broadband Wireless Access Systems (2009).

IEEE Standard for Wireless LAN Medium Access Control (MAC) and PHYsical Layer (PHY) Specifications: Amendment: Medium Access Control (MAC) Enhancements for Higher Throughput (2009).

Jin, H., Khandekar, A. & McEliece, R. J. (2000). Irregular repeat-accumulate codes, *Proc. 2nd Int. Symp. Turbo Codes & Related Topics*, Brest, France, pp. 1–8.

Lyrtech (2010).
URL: *http://www.lyrtech.com*

MacKay, D. J. C. (1999). Good error-correcting codes based on very sparse matrices, 45(2): 399–431.

Sandhu, S. & Paulraj, A. (2000). Space-time block codes: A capacity perspective, 4(12): 384–386.

Schreckenbach, F., G ortz, N., Hagenauer, J. & Bauch, G. (2003). Optimized symbol mappings for bit-interleaved coded modulation with iterative decoding, *Proc. Global Telecommunications Conference (GLOBECOM 2003)*, San Francisco, CA, USA.

Sundance Multiprocessor (2010).
URL: *http://www.sundance.com*

Telatar, İ. E. (1995). Capacity of multi-antenna Gaussian channels, *Technical report*, Bell Laboratories, Lucent Technologies. Published in *European Transactions on Telecommunications*, Vol. 10, No. 6, pp. 585-595, Nov/Dec 1999.

ten Brink, S. (1999). Convergence of iterative decoding, *IEE Electronics Letters* 35(13): 1117–1118.

ten Brink, S. (2001). Convergence behavior of iteratively decoded parallel concatenated codes, 49(10): 1727–1737.

ten Brink, S. & Kramer, G. (2003). Design of Repeat-Accumulate codes for iterative detection and decoding, 51(11): 2764–2772.

ten Brink, S., Kramer, G. & Ashikhmin, A. (2004). Design of Low-Density Parity-Check codes for modulation and detection, 52(4): 670–678.

Tonello, A. M. (2000). Space-Time Bit-Interleaved Coded Modulation with an iterative decoding strategy, *Proc. IEEE VTC Fall 2000*, Boston.

Vázquez-Araújo, F. J., González-López, M., Castedo, L. & Garcia-Frias, J. (2006). Design of Serially-Concatenated LDGM coded MIMO systems, *Proc. SPAWC'06*, Cannes, France.

Vázquez-Araújo, F. J., González-López, M., Castedo, L. & Garcia-Frias, J. (2007). Serially-Concatenated LDGM codes for MIMO channels, 6(8): 2860–2871.

Vikalo, H., Hassibi, B. & Kailath, T. (2004). Iterative decoding for MIMO channels via modified sphere decoding, 3(6): 2299–2311.

Wang, X. & Poor, H. V. (1999). Iterative (turbo) soft interference cancellation and decoding for coded CDMA, 47(7): 1046–1061.

Yue, G. & Wang, X. (2005). Optimization of Irregular Repeat Accumulate codes for MIMO systems with iterative receivers, 4(6): 2843–2855.
Yue, G., Wang, X. & Madihian, M. (2008). Outage performance and IRA code design for MIMO block fading with unitary scrambling, 54(4): 2354–2363.
Zehavi, E. (1992). 8-PSK trellis codes for a Rayleigh channel, 40: 873–884.